CW00327519

Combined Chemistry

Combined Chemistry

J. Brockington, BSc, CChem, MRSC
Matthew Boulton Technical College, Birmingham

P. J. Stamper, PhD, CChem, MRSC
Matthew Boulton Technical College, Birmingham

D. R. Browning, BSc, CChem, FRSC, ARTCS
Bristol Polytechnic

A. C. Skinner, BSc, PhD, CChem, MRSC
Worcester Technical College

Longman Group UK Limited
Longman House
Burnt Mill, Harlow, Essex CM20 2JE, England
and Associated Companies throughout the World

First published 1981
Fifth impression 1986

Set in Monophoto Times

Produced by Longman Group (FE) Ltd
Printed in Hong Kong

ISBN 0-582-35183-9

Contents

Safety note

WARNING!

Chemistry can be a hazardous occupation. Practical work must be carried out under the supervision of a teacher, and the instructions carefully followed. Unauthorised experiments should never be performed.

Some chemicals are particularly dangerous. Attention has been drawn to these in Part 2 of the book by means of the following EEC hazard warning symbols.

Toxic

Toxic (poisonous) substances are dangerous by ingestion, inhalation or skin absorption. They present a serious risk of acute or chronic poisoning. Some substances cause immediate acute effects of short duration, which cease on removal from exposure and on treatment. Other substances cause chronic effects which may be cumulative and irreversible and may not be apparent until long after the original exposure.

Toxic substances require to be handled:
(i) under conditions of total enclosure;
(ii) within an enclosure under exhaust ventilation;
(iii) with appropriate protective clothing and respirators.

Oxidising agent

Oxidising substances may give rise to exothermic reactions in contact with organic matter or other easily oxidised chemicals.

Store away from organic materials and reducing agents.

Harmful/irritant

Harmful substances present a moderate risk of chronic or acute poisoning by ingestion, inhalation or skin contact.

Use under conditions to prevent contact with skin, eyes or clothing or inhalation of dust, fume or vapour.

Irritant substances are liable to cause inflammation of living tissues and irritation of the respiratory system. Prolonged contact may destroy living tissue.

Use under conditions to prevent contact with skin and eyes, and avoid breathing dust, fumes or vapour.

Explosive

Explosive substances are liable to explode when in the dry state or when subject to shock, friction or heat. They are often supplied wet and become dangerous on drying.

Corrosive

Corrosive chemicals may cause ulceration, burns or destruction of living tissue.

Protection of skin and eyes is essential.

Most corrosive chemicals will require rapid treatment to prevent serious injury or damage.

Radioactive

Radioactive substances are controlled by regulation and Codes of Practice and these should be strictly observed. Only qualified and trained personnel should be allowed to store, handle or use such products.

Flammable

Flammable substances readily ignite at temperatures above their flash point. (The *flash point* of a substance is the minimum temperature at which the vaporised substance will ignite in air when a spark is applied.)

Ensure that the correct type of fire extinguisher is readily available. Observe regulations governing the storage, handling and use of certain flammable liquids.

Highly flammable substances, with a flash point below 32 °C, are a serious fire hazard and may form explosive mixtures with air.

Eliminate potential ignition sources such as naked lights, heat, sparks (including static charges) and burning cigarettes. Store and use where there is good ventilation and spillages may be contained.

Taken from *Chemical Hazards*, wall chart: Fisons Scientific Apparatus.

Preface

Combined Chemistry is intended primarily for students on GCE 'A' level and similar courses. Combined texts of physical, inorganic and organic chemistry have always been popular at this level, partly for their convenience and partly because they offer good value for money. We trust that these traditional advantages have been retained in this new work, which has been written to meet the requirements of modern syllabuses.

In planning the book we have tried to break away from the rigid division of chemistry into three subject areas. The limitations imposed by established teaching patterns prevent the adoption of a completely integrated treatment – at least for the time being – and we have therefore divided our book into two parts. Part 1, entitled *Principles*, contains physical chemistry together with those topics, such as oxidation–reduction, strengths of acids and bases, etc, sometimes described as physical inorganic or physical organic. Part 2, *Detailed Chemistry*, consists of factual inorganic and organic chemistry, with the organic material in its logical place in group 4B of the periodic table. For ease of reference each chapter has been structured and numbered.

Nomenclature poses a particular problem to anyone studying chemistry at present, and we have devoted Chapter 1 largely to this topic. Throughout the book we have employed SI units in general use; e.g. for density, $g\ cm^{-3}$ rather than $kg\ m^{-3}$. In this we have been guided by J. G. Stark and H. G. Wallace in their excellent *Chemistry Data Book* (John Murray); both the values and the units of chemical constants in this textbook are in agreement with those in the *Chemistry Data Book*. We have quoted temperatures in degrees Celsius with the corresponding values in kelvins in parentheses; likewise pressures are given in atmospheres with the equivalents in kilopascals (1 atmosphere is approximately equal to 100 kPa). This allows for a possible change in educational practice.

At the end of each chapter will be found a selection of recent 'A' level questions, reprinted by kind permission of the examination boards concerned. One of the boards refers directly to the *Chemistry Data Book* in some of its questions, and it will be necessary to use the book when answering these questions.

When referring to chemical compounds we have used IUPAC rather than ASE names, although in the sections on nomenclature we have given both names wherever they differ. In the event of IUPAC allowing a choice between a trivial name (e.g. acetone) and a systematic name (e.g. propanone) we have selected the latter, and where both radicofunctional names (e.g. dimethyl ketone) and substitutive names (e.g. propanone) are permitted we have again taken the latter.

With a new book of this length and complexity it would be immodest if not foolish of us to claim, as did Voltaire's hero Candide, that '*Tout est pour le mieux dans le meilleur des mondes possibles*'. Nevertheless, we hope that our *magnum opus* will be found to be distinctive and stimulating. We thank all those concerned with its production, especially Mandy Keyho at Longman; and we should welcome any constructive criticisms that readers may care to offer.

John Brockington
Peter Stamper
David Browning
Aidan Skinner

May 1980

Acknowledgements

The authors and publisher would like to thank the following: Fisons Ltd. for permission to reproduce the Safety Warning information (p. xv); Dr. Michael Hudson for permission to reproduce Figs. 6.9, 6.10, 6.12, 6.13, 6.14, 6.15, 6.16, 6.17 from Crystals and Crystal Structure, Michael Hudson (Longman Group Ltd.); John Murray Ltd. for permission to use various data throughout the book, and to reproduce Fig. 2.4 from Chemistry Data Book, J. G. Stark and H. G. Wallace (John Murray Ltd.); Prentice-Hall Ltd. for permission to reproduce Figs. 3.9, 3.10, 4.17, 6.3, 8.4 from Physical Chemistry, 4th Edition, W. J. Moore (Longman Group Ltd.); Shell U.K. Limited for permission to feature Fig. 14.3 which is adapted from a Shell publication. Figs. 6.22, 6.24, 14.2 are reproduced from Introduction to Inorganic Chemistry, G. I. Brown (Longman Group Ltd.).

The authors and publisher would like to thank the following for permission to reproduce examination questions:
The Associated Examining Board (AEB) (1976 papers: Ch. 4: q. 8; Ch. 6: q. 2; Ch. 9: q. 11; 14.8: q. 6; 14.10: q. 19; Ch. 18: q. 18, 23; Misc. questions: 5, 22. 1977 papers: Ch. 2: q. 14; Ch. 5: q. 5; Ch. 6: q. 4, 12; Ch. 8: q. 10, 11, 17; Ch. 9: q. 14, 19; Ch. 10: q. 3, 6, 7, 8; Ch. 12: q. 1; Ch. 14 (organic): q. 17; (inorganic): q. 2; 14.7: q. 3; 14.10: q. 4, 6, 13; 14.11: q. 2; Ch. 15: q. 6, 9; Ch. 16: q. 1; Ch. 17: q. 1; Ch. 18: q. 11, 21.)
Joint Matriculation Board (JMB)
Oxford and Cambridge Schools Examination Board (OC)
Oxford Delegacy of Local Examinations (O)
Southern Universities' Joint Board (SUJB)
University of London University Entrance and Schools Examinations Council (L)
Welsh Joint Education Committee (WJEC)

Cover photograph by Paul Brierley

Microencapsulation: the technology of sealing measured amounts of chemicals in spherical capsules. Seen under the microscope at the stage where the outer shell dissolves in a liquid phase. Photographed using two sources (red and blue) of transmitted light. Scale: $\times 275$.

1 Principles

1 Relative masses and chemical nomenclature

1.1 Atomic and molecular masses

Although the SI unit of mass, the kilogram, is a highly practical unit for most everyday purposes, it is inconvenient for expressing the masses of atoms and other very small bodies. A hydrogen atom, for instance, has a mass of only 1.67×10^{-27} kg. Instead, it is better to select the mass of some small particle as a unit of measurement, so that the mass of an atom becomes 'x small particles', rather than a minute part of a kilogram. There is no fundamental reason why we should not employ the mass of a proton or a neutron as a unit of mass, except that neither of these standards is particularly convenient to use in the laboratory. A standard must be chosen so that the mass of any newly discovered atom can be compared with it by means of a mass spectrometer, and it is for this reason that the present standard is the mass of an atom of the isotope carbon-12 (§3.1.2).

An atom of carbon-12 is said to have a mass of twelve units. Why not one unit? Because if this were so, a hydrogen atom would have a mass of approximately $\frac{1}{12}$ unit, helium $\frac{4}{12}$ unit, lithium $\frac{7}{12}$ unit, and so on, which would be rather clumsy. By putting $^{12}C = 12$, the mass of a hydrogen atom becomes *approximately* one unit, helium approximately four units, lithium approximately seven units, etc, which is much neater. Also, and this is perhaps the most important reason, such values have been in use ever since chemistry was first put on a scientific basis in the early nineteenth century.

If a carbon-12 atom has a mass of twelve units, then *one unit is equal to one-twelfth of the mass of a carbon-*12 *atom.* The name of this unit is the *unified atomic mass unit*, symbol u; it is not definable in terms of SI units, but is approximately equal to 1.6604×10^{-27} kg.

Several standards for expressing the masses of atoms have been used in the past. The current standard, $^{12}C = 12$, has been in use only since 1960. Before this time, atoms of naturally occurring elements were used as standards. First there was $H = 1$, and then came numerous oxygen-based scales culminating in $O = 16$, but all of these were unsatisfactory because nearly all naturally occurring elements consist of a mixture of *isotopes*, i.e. atomic species with the same number of protons in the nucleus but different numbers of neutrons. If the ratio in which the isotopes occurred were constant, then a weighted average of the isotopic masses might serve as a standard, but in fact the ratio is not constant. With oxygen, for example, the ratio of oxygen-16 to oxygen-18 varies slightly, even in different samples of water from the same lake! Clearly, a single isotope is needed as a standard, and for experimental convenience carbon-12 has been chosen in preference to oxygen-16.

1.1.1 MASS NUMBER

The mass of a *nuclide*, i.e. any individual atomic species, may be expressed as a *mass number*. Because an electron has almost no mass, while a proton and a neutron each have a mass of almost exactly 1 u, the mass number of a nuclide is always a whole number, equal to the number of protons plus the number of neutrons in the nucleus of the atom.

$$A = Z + N$$

where A = mass number, Z = *atomic number*, i.e. number of protons, and N = number of neutrons.

When using symbols for elements, we write the mass number as a superscript and the atomic number as a subscript, e.g. $^{35}_{17}Cl$.

1.1.2 ATOMIC MASS, RELATIVE ATOMIC MASS AND RELATIVE MOLECULAR MASS

To every naturally occurring element can be ascribed an average mass, weighted so as to reflect the proportions in which the various isotopes are present. This is known as the *atomic mass* of the element. For example, hydrogen, consisting primarily of hydrogen-1, with traces of hydrogen-2 and hydrogen-3, has an atomic mass of 1.008 u. Chlorine, containing about 75% of chlorine-35 and 25% of chlorine-37, has an atomic mass of 35.453 u. Expressed mathematically,

$$\begin{aligned}\text{atomic mass} &= m_1 f_1 + m_2 f_2 + m_3 f_3 \dots \\ &= \Sigma mf\end{aligned}$$

where m is the mass of an isotope and f is the fraction in which it is present.

To avoid the use of a non-SI unit, it is conventional to quote the atomic mass of an element, Σmf, not in u but as the number of times it is greater than one-twelfth of the mass of a carbon-12 atom. This quantity is known as the *relative atomic mass* (A_r). The term formerly used was 'atomic weight'. By definition,

$$A_r = \frac{\text{average mass per atom of the natural nuclidic composition}}{\text{one-twelfth of the mass of an atom of the nuclide carbon-12}}$$

Relative atomic mass, like all other ratios, is a number without dimensions.

Similarly the masses of molecules, ions and all other small particles are expressed as ratios and referred to as 'relative masses'. For example, the *relative molecular mass* (M_r) of an element or compound is defined as follows:

$$M_r = \frac{\text{average mass per formula unit of the natural nuclidic composition}}{\text{one-twelfth of the mass of an atom of the nuclide carbon-12}}$$

Until recently, relative molecular mass was known as 'molecular weight'. The definition is worded in terms of 'formula unit' rather than 'molecule' so as to include electrovalent as well as covalent compounds. (For an electrovalent compound (§4.1.3) the formula unit is the lowest whole number ratio of ions in the compound; e.g. $CaCl_2$ for calcium chloride.)

The relative molecular mass of a compound is easily computed as it is the sum of the relative atomic masses of all the atoms in the formula.

1.1.3 THE MOLE CONCEPT

When working in the laboratory we cannot study individual molecules of a substance for they are far too small. Instead, we scale up to a practicable level by considering a *mole* of the substance (or some fraction or multiple of a mole), a mole being the relative molecular mass expressed in grams.

The change from a molecule to a mole is essentially a change of units:

Molecule	Mole
mass in unified atomic mass units, e.g., for water, 18 u	mass in grams, e.g., for water, 18 g

In effect, we are multiplying by a constant, called the *Avogadro constant* (*L*), which is equal to the number of molecules in a mole ($L \approx 6.023 \times 10^{23}$ mol^{-1}).

When a chemical reaction occurs, molecules react together in a simple whole number ratio which is indicated by the equation. The molar ratio in which substances react is exactly the same. For example, the equation

$$N_2\,(g) + 3H_2\,(g) \rightleftharpoons 2NH_3\,(g)$$

tells us that one mole (1×28 g) of nitrogen reacts with three moles (3×2 g) of hydrogen to form two moles (2×17 g) of ammonia if the reaction goes to completion.

The mole concept may be extended to include not just molecules but all elementary units, e.g. atoms, ions and electrons. (The terms 'gram-atom', 'gram-ion', etc, are obsolete.) The mole, which is one of the basic SI units, is defined as the amount of substance which contains as many elementary units as there are atoms in 0.012 kilograms of carbon-12.

To convert the mass of a substance, in grams, to an amount in moles we merely divide the mass by the relative molecular mass of the substance. For example, sodium hydroxide has a relative molecular mass of 40,

$$\therefore \quad 40 \text{ g of sodium hydroxide} = 1 \text{ mol},$$

$$\therefore \quad x \text{ g of sodium hydroxide} = \frac{x}{40} \text{ mol}$$

Conversely, to convert moles to grams, we multiply by the appropriate relative molecular mass.

Molarity

The concentration of a solution, expressed in moles per cubic decimetre, is known as its *molarity*. A *molar* solution, symbol M, contains one mole of the solute in one cubic decimetre of solution. A two-molar (2 M) solution contains two moles in one cubic decimetre of solution, and so on. Most laboratory reagents are 4 M (i.e. 4 mol dm^{-3}),while many of the solutions used in volumetric analysis are 0.1 M or 0.05 M.

Example 1
Calculate the mass (in grams) of anhydrous sodium carbonate needed for the preparation of 250 cm^3 of a 0.050 M solution.
Answer Sodium carbonate has a relative molecular mass of 106.0,

$\therefore$ 1 000 cm^3 of a 1.000 M solution require 106.0 g,

$\therefore$ 1 000 cm^3 of a 0.050 M solution require $106.0 \times \frac{0.050}{1.000}$ g,

$\therefore$ 250 cm^3 of a 0.050 M solution require $106.0 \times \frac{0.050}{1.000} \times \frac{250}{1\,000}$

$= \mathbf{1.325\ g}$

Example 2
1.500 g of pure sodium chloride is dissolved in water to make 250 cm^3 of solution. Calculate the molarity of the solution.
Answer Sodium chloride has a relative molecular mass of 58.4,

$\therefore$ in 250 cm^3 of solution there are $\frac{1.500}{58.4}$ mol NaCl,

$\therefore$ in 1 000 cm^3 of solution there are $\frac{1.500}{58.4} \times \frac{1\,000}{250}$
$= 0.102\,7$ mol NaCl,

i.e. the concentration of the solution is **0.1027 M**

Units in volumetric analysis

In volumetric analysis there is more than one method of performing the calculation, but it is the current practice to convert all masses to moles and all concentrations to moles per cubic decimetre (i.e. molarity). At the start of the calculation a balance reading, in grams, is expressed in moles by dividing it by the relative molecular mass of the substance weighed; conversely, at the end of the calculation, it is often necessary to revert from moles to grams by multiplying by the relative molecular mass.

The main part of the calculation is based on the principle that, at the end-point of the titration, substances are present in chemically balanced proportions as indicated by the equation. If, for example, the equation shows that x mol of substance A reacts with y mol of substance B,

$$xA + yB = \text{products}$$

then at the end-point the molar ratio of A to B is $x{:}y$,

i.e. $$\frac{\text{moles of A}}{\text{moles of B}} = \frac{x}{y}$$

The amount, in moles, of solute in a solution depends on the molarity of the solution and its volume,

i.e. moles = molarity × volume (in dm^3),

hence $$\frac{\text{molarity of A} \times \text{volume of A}}{\text{molarity of B} \times \text{volume of B}} = \frac{x}{y} \quad (1)$$

Since the unit of molarity is mol dm^{-3}, volumes should be expressed in dm^3. Nevertheless, it is permissible – indeed, normal practice – to insert the volumes in cm^3, for the resulting conversion factors in the numerator and denominator are self-cancelling.

If, however, an equation of type (2) is used, i.e.

$$\frac{\text{moles of A}}{\text{molarity of B} \times \text{volume of B}} = \frac{x}{y} \quad (2)$$

it is imperative that the volume is inserted in dm^3. In Example 3 equation (1) is used with both volumes in cm^3, while in Example 4 equation (2) is used with the volume in dm^3.

Example 3

0.869 g of potassium permanganate is dissolved in water and the solution made up to 250 cm^3. 5.880 g of an unknown iron(II) salt is dissolved in water and made up to 100 cm^3. In a titration in acidic solution 20.0 cm^3 of the iron solution requires 27.3 cm^3 of the permanganate solution. Find the percentage of iron in the salt.

Answer The relative molecular mass of potassium permanganate is 158.0,

$\therefore$ the concentration of the $KMnO_4$ solution $= 0.869 \times 4$ g dm^{-3}

$$= \frac{0.869 \times 4}{158.0} \text{ mol dm}^{-3}$$

$$= 0.0220 \text{ M}$$

The ionic equation for the reaction is as follows:

$$MnO_4^- + 8H^+ + 5Fe^{2+} = Mn^{2+} + 5Fe^{3+} + 4H_2O$$

(The construction of such equations is described in §10.1.4.)

Thus, $$\frac{\text{moles of } MnO_4^-}{\text{moles of } Fe^{2+}} = \frac{1}{5}$$

$$\therefore \quad \frac{\text{molarity of } MnO_4^- \times \text{volume of } MnO_4^-}{\text{molarity of } Fe^{2+} \times \text{volume of } Fe^{2+}} = \frac{1}{5}$$

$$\therefore \quad \frac{0.0220 \times 27.3}{\text{molarity of } Fe^{2+} \times 20.0} = \frac{1}{5}$$

$$\therefore \quad \text{molarity of } Fe^{2+} = \frac{0.0220 \times 27.3 \times 5}{20.0} = 0.1502 \text{ M}$$

The relative atomic mass of iron is 55.8,

$\therefore$ concentration of iron in solution $= 0.1502 \times 55.8 = 8.381$ g dm^{-3}

$$\therefore \quad \text{iron content of unknown salt} = \frac{\text{g dm}^{-3}\text{ of iron}}{\text{g dm}^{-3}\text{ of salt}} \times 100$$

$$= \frac{8.381}{5.880 \times 10} \times 100 = \mathbf{14.25\%}$$

Example 4
0.350 g of impure zinc dust contaminated with zinc oxide is added to an excess of iron(III) solution. The iron(II) solution produced requires 100 cm^3 of 0.0200 M potassium permanganate solution for titration. Calculate the percentage purity of the zinc.

Answer Any method for estimating the purity of a substance must be devised so that impurities do not react with the reagent. In this particular method, only metallic zinc (and not zinc oxide) reacts with the iron(III) solution:

$$Zn + 2Fe^{3+} = Zn^{2+} + 2Fe^{2+}$$

Since the iron(III) solution is in excess, one mole of zinc gives rise to two moles of iron(II).

Iron(II) is titrated against permanganate as in the previous example:

$$\frac{\text{molarity of } MnO_4^- \times \text{volume of } MnO_4^-}{\text{moles of } Fe^{2+}} = \frac{1}{5}$$

$$\therefore \quad \frac{0.0200 \times 0.100}{\text{moles of } Fe^{2+}} = \frac{1}{5}$$

$$\therefore \quad \text{moles of } Fe^{2+} = 0.0200 \times 0.100 \times 5 = 0.0100$$

From the chemical equation,

$$2 \text{ moles of } Fe^{2+} \equiv 1 \text{ mole of Zn}$$

$$\therefore \quad 0.0100 \text{ moles of } Fe^{2+} \equiv 0.0050 \text{ moles of Zn}$$

The relative atomic mass of zinc is 65.4,

$$\therefore \quad \text{mass of zinc} = 0.0050 \times 65.4 = 0.327 \text{ g},$$

$$\therefore \quad \text{purity of zinc dust} = \frac{\text{mass of zinc}}{\text{mass of sample}} \times 100$$

$$= \frac{0.327}{0.350} \times 100 = \mathbf{93.43\%}$$

1.2 Chemical nomenclature – inorganic

Chemical substances may be named in accordance with the recommendations of either the International Union of Pure and Applied Chemistry (IUPAC) or the Association for Science Education (ASE).

IUPAC nomenclature is revised from time to time. The most recent edition of *Nomenclature of Organic Chemistry* was published by Pergamon

Press in 1979, while the recommendations for inorganic chemistry last appeared in 1970 (Butterworths). The underlying philosophy is that for many simple substances *trivial names* should be retained if they are well established and unambiguous; other substances should be given systematic names based on a series of well-defined rules.

ASE published its proposals in 1972, in a booklet entitled *Chemical Nomenclature, Symbols and Terminology*. A second edition was issued in 1979. ASE recommendations, which have been accepted by examination boards in the UK, differ from those of IUPAC in that systematic names are used for all but a few substances. In addition, several of the ASE systematic names are different from those of IUPAC.

In the reference notes that follow emphasis has been placed on IUPAC nomenclature on the grounds that it is internationally recognised. Attention has been drawn, whenever necessary, to differences between IUPAC and ASE systematic naming.

1.2.1 ELEMENTS

All isotopes of an element, except hydrogen (§11.1), bear the same name.

For allotropes, IUPAC allows trivial names but also gives systematic names based on the structure of the molecule. Examples are as follows:

Allotrope	Trivial name	Systematic name
O_2	oxygen	dioxygen
O_3	ozone	trioxygen
P_4	white phosphorus	*tetrahedro*-tetraphosphorus

1.2.2 BINARY COMPOUNDS

There are two ways of naming such compounds.

(i) By means of *stoicheiometric names*, in which the proportions of the constituent elements are indicated by prefixes of Greek origin, e.g. mono, di, tri, tetra, penta and hexa. In practice, the prefix 'mono' is usually omitted. Oxides, sulphides and halides of non-metals, together with certain metal oxides, are conveniently named in this way, e.g.

N_2O	dinitrogen oxide	MnO_2	manganese dioxide
NO_2	nitrogen dioxide	Fe_3O_4	triiron tetraoxide
N_2O_4	dinitrogen tetraoxide	Pb_3O_4	trilead tetraoxide

(ii) Alternatively, *Stock's system* may be used, in which the oxidation number (§10.1.2) of an element is shown by a Roman numeral enclosed in parentheses immediately after the name. This system is becoming increasingly favoured. With only a few exceptions, all metallic oxides, sulphides and halides, together with the oxides of phosphorus, are now named in this way. Examples:

$SnCl_2$	tin(II) chloride	MnO_2	manganese(IV) oxide
$SnCl_4$	tin(IV) chloride	Fe_3O_4	iron(II) diiron(III) oxide
CrO	chromium(II) oxide	Pb_3O_4	dilead(II) lead(IV) oxide
Cr_2O_3	chromium(III) oxide	P_4O_{10}	phosphorus(V) oxide

Hydrides constitute a special case. They may be named in accordance with the above rules, e.g. sodium hydride for NaH, but many of them have well-established trivial names and both IUPAC and ASE recommend their continued use. Well-known examples are water and ammonia; others are as follows:

CH_4	methane	N_2H_4	hydrazine
SiH_4	silane	PH_3	phosphine
GeH_4	germane	AsH_3	arsine
SnH_4	stannane	SbH_3	stibine
PbH_4	plumbane	BiH_3	bismuthine

1.2.3 CATIONS

Monoatomic cations derived from elements of variable valency or oxidation state are given the name of the element with its appropriate oxidation number,

e.g. Fe^{2+} iron(II) ion Fe^{3+} iron(III) ion

In the naming of cations derived from elements of constant valency, both IUPAC and ASE recommend that the oxidation number should be omitted. Examples:

Ion	IUPAC and ASE name
Na^+	sodium ion
Ca^{2+}	calcium ion
Al^{3+}	aluminium ion

The common polyatomic cations are named as follows:

H_3O^+	oxonium ion	PH_4^+	phosphonium ion
NH_4^+	ammonium ion	NO^+	nitrosyl cation
$HONH_3^+$	hydroxylammonium ion	NO_2^+	nitryl cation

1.2.4 ANIONS

The name of a monoatomic anion is derived from that of the element, with the termination '-ide'. For example:

H^-	hydride	N^{3-}	nitride
Cl^-	chloride	P^{3-}	phosphide
O^{2-}	oxide	S^{2-}	sulphide

Polyatomic anions may have names that end in '-ide', '-ate' or '-ite'. The commonest ones that end in '-ide' are as follows:

HO^-	hydroxide	HF_2^-	hydrogendifluoride
O_2^{2-}	peroxide	NH_2^-	amide
O_2^-	hyperoxide	$NHOH^-$	hydroxylamide
HS^-	hydrogensulphide	CN^-	cyanide
I_3^-	triiodide	C_2^{2-}	acetylide (ASE: dicarbide)

Most oxoanions have names which consist of the root of the name of the characteristic element, with the termination '-ate', e.g. phosphate for PO_4^{3-}. Although it is practicable to name such anions systematically (§18.1.2) as though they were complex anions, e.g. tetraoxophosphate(V) for PO_4^{3-}, IUPAC does not advise this except for the less well-known and newly discovered ions. ASE, however, recommends the use of abbreviated systematic names for all except the commonest oxoanions, e.g. phosphate(V) for PO_4^{3-}. Other examples are as follows:

Anion	IUPAC name	ASE name
NO_3^-	nitrate	nitrate
CO_3^{2-}	carbonate	carbonate
SO_4^{2-}	sulphate	sulphate
PHO_3^{2-}	phosphonate	phosphonate
$PH_2O_2^-$	phosphinate	phosphinate
BO_3^{3-}	borate or orthoborate	borate
SiO_3^{2-}	metasilicate	silicate
ClO_3^-	chlorate	chlorate(V)
CrO_4^{2-}	chromate	chromate(VI)
$Cr_2O_7^{2-}$	dichromate	dichromate(VI)
MnO_4^-	permanganate	manganate(VII)

Certain oxoanions, containing an element in a low oxidation state, have long-established trivial names that end in '-ite'. Examples are as follows:

NO_2^-	nitrite	SO_3^{2-}	sulphite	ClO^-	hypochlorite
AsO_3^{3-}	arsenite	$S_2O_5^{2-}$	disulphite	BrO^-	hypobromite
		ClO_2^-	chlorite	IO^-	hypoiodite

Note that PHO_3^{2-}, formerly known as the phosphite ion, is now called the phosphonate ion.

ASE does not allow these names, except for nitrite and sulphite, and recommends the use of abbreviated systematic names, e.g.

AsO_3^{3-}	arsenate(III)	ClO_2^-	chlorate(III)
$S_2O_5^{2-}$	disulphate(IV)	ClO^-	chlorate(I)

1.2.5 RADICALS

For purposes of nomenclature, the term 'radical' is defined by IUPAC as a group of atoms that occurs repeatedly in several different compounds. The names of the commonest radicals are as follows. Note that the names bismuthyl and antimonyl, formerly used in naming such compounds as BiClO and SbClO, are not approved. (The naming of basic salts is described later.)

HO	hydroxyl	PO	phosphoryl
CO	carbonyl	SO	sulphinyl or thionyl
NO	nitrosyl	SO_2	sulphonyl or sulphuryl

These radicals are always regarded as forming the positive part of any compound in which they occur. They are therefore written first in the formula, and referred to first in the name, e.g.

$COCl_2$	carbonyl chloride
NOCl	nitrosyl chloride
$SOCl_2$	sulphinyl chloride or thionyl chloride

ASE recognises few radical names, and most compounds in which radicals occur are given stoicheiometric names, e.g.

	$SOCl_2$	sulphur dichloride oxide
but	$COCl_2$	carbonyl chloride

1.2.6 ACIDS

Acids are named after the anions to which they give rise. If the name of the anion ends in '-ide', the acid is named as hydrogen -ide, e.g.

HCl	hydrogen chloride
H_2S	hydrogen sulphide
HCN	hydrogen cyanide

If the IUPAC name of the anion ends in '-ite' or '-ate', that of the acid ends in '-ous acid' or '-ic acid' respectively. ASE, which seldom accepts the ending '-ite' for anions, recognises the ending '-ous acid' only for well-known compounds, and for most other oxoacids recommends the termination '-ic acid', usually with a Stock number. Examples:

Acid	IUPAC name	ASE name
HNO_2	nitrous acid	nitrous acid
HNO_3	nitric acid	nitric acid
H_2SO_3	sulphurous acid	sulphurous acid
H_2SO_4	sulphuric acid	sulphuric acid
$HClO_2$	chlorous acid	chloric(III) acid
$HClO_3$	chloric acid	chloric(V) acid

An '-ous acid' thus contains the characteristic element in a low oxidation state, while an '-ic acid' contains the element in a high oxidation state. A name of the kind 'hypo-ous acid' is used to show that the element is in a particularly low oxidation state, e.g. HClO hypochlorous acid (oxidation state of Cl = +1). It should be noted that HPH_2O_2, the compound formerly known as hypophosphorous acid, is now called phosphinic acid.

ASE does not recognise the prefix 'hypo-', but names such acids in accordance with the normal ASE system. HClO, for example, is called chloric(I) acid.

Until recently, the prefix 'per-' (not to be confused with 'peroxo-') was used for naming certain acids containing an element in a particularly high oxidation state. This prefix is now used only for $HMnO_4$, permanganic acid, $HClO_4$, perchloric acid, and the corresponding acids of bromine and iodine. ASE does not recognise the prefix at all, and recommends the names manganic(VII) acid for $HMnO_4$ and chloric(VII) acid for $HClO_4$.

Isopolyacids, i.e. oxoacids containing more than one atom of the characteristic element, are named, like simple acids, after the anions to which they give rise. For example, $H_4P_2O_7$ is called diphosphoric acid (ASE: heptaoxodiphosphoric(V) acid). The trivial name of pyrophosphoric acid is not recommended; indeed, the prefix 'pyro-' is no longer in common use. Nevertheless, the prefixes 'ortho-' and 'meta-' are permitted by IUPAC, although not by ASE, to distinguish between acids of differing water content, e.g.

Acid	IUPAC name	ASE name
H_3BO_3	orthoboric acid	boric acid
$(HBO_2)_n$	metaboric acid	polydioxoboric(III) acid
H_3PO_4	orthophosphoric acid	phosphoric(V) acid
$(HPO_3)_n$	metaphosphoric acid	polytrioxophosphoric(V) acid

The prefix 'peroxo-' is used to indicate the substitution of -O- by -O-O-, e.g.

Acid	IUPAC name	ASE name
H_2SO_5	peroxomonosulphuric acid	peroxosulphuric(VI) acid
$H_2S_2O_8$	peroxodisulphuric acid	peroxodisulphuric(VI) acid

Acids in which an atom of oxygen has been substituted by one of sulphur are known as thioacids, e.g.

Acid	IUPAC and ASE name
$H_2S_2O_3$	thiosulphuric acid
HSCN	thiocyanic acid

1.2.7 ACID DERIVATIVES

Acid chlorides, i.e. those compounds in which the HO groups of the acid have been replaced by atoms of chlorine, are named by IUPAC in accordance with the acid radicals they contain, e.g. nitrosyl chloride for NOCl. ASE, in contrast, recommends stoicheiometric names (see above, under 'radicals').

Amides, in which the HO groups of the acid have been replaced by NH_2 groups, may be named either by replacing the word 'acid' by 'amide', 'diamide', etc, or by utilising the name of the acid radical. For example, the compound $SO_2(NH_2)_2$, derived from sulphuric acid, may be called either sulphuric diamide or sulphonyl diamide. (ASE makes no recommendation.)

If not all the HO groups have been replaced by NH_2 groups, the compound may be named as either an 'amido- acid' or an '-amidic acid', e.g.

NH_2SO_3H amidosulphuric acid or sulphamidic acid

(The abbreviated name of sulphamic acid is not recommended.) ASE names the compound as aminosulphonic acid.

1.2.8 SALTS

Simple salts are named according to Stock's system for binary compounds; see above.

Acid salts, i.e. those containing acid hydrogen, are named by introducing the prefix 'hydrogen-', with di, tri, etc, if necessary, to the name of the anion, e.g.

Salt	IUPAC name	ASE name
$NaHCO_3$	sodium hydrogencarbonate	sodium hydrogencarbonate
NaH_2PO_4	sodium dihydrogenphosphate	sodium dihydrogenphosphate(V)

When naming double salts (§13.2.7), the cations (other than hydrogen) should be cited in alphabetical order, e.g.

Salt	IUPAC name	ASE name
$KNaCO_3$	potassium sodium carbonate	potassium sodium carbonate
$NaNH_4HPO_4$	ammonium sodium hydrogenphosphate	ammonium sodium hydrogenphosphate(V)

For the naming of complex salts, see §18.1.2.

For salt hydrates, two systems may be used.

(i) In the absence of structural information, the extent of hydration may be indicated by writing monohydrate, dihydrate, etc, after the name, e.g.

$AlCl_3 \cdot 6H_2O$ aluminium chloride hexahydrate

ASE does not recognise this notation.

Alternatively, the number of water molecules may be represented by an Arabic numeral – a method favoured by ASE. For example:

Salt	IUPAC and ASE name
$AlCl_3 \cdot 6H_2O$	aluminium chloride 6-water
$AlK(SO_4)_2 \cdot 12H_2O$	aluminium potassium sulphate 12-water

(ii) If the mode of hydration is known, the formula and name of the salt may be modified accordingly. For example, in aluminium chloride hexahydrate the six water molecules are coordinated to the aluminium ion. The compound is effectively a complex salt, $[Al(H_2O)_6]Cl_3$, and may be named as such, i.e. hexaaquaaluminium(III) chloride.

Basic salts, better known as oxide and hydroxide salts, are regarded as double salts containing two or more anions, one of which is O^{2-} or HO^-. When naming them, the anions should be cited in alphabetical order, e.g.

Salt	IUPAC name	ASE name
MgCl(OH)	magnesium chloride hydroxide	magnesium chloride hydroxide
BiClO	bismuth chloride oxide	bismuth(III) chloride oxide
$VO(SO_4)$	vanadium(IV) oxide sulphate	vanadium(IV) oxide sulphate

1.3 Chemical nomenclature – organic

Most organic compounds can be regarded as substituted hydrocarbons, in which one or more hydrogen atoms of the hydrocarbon have been replaced by one or more other atoms or groups of atoms, known as *characteristic groups* or *functional groups*. For example, from methane, CH_4, by the replacement of one atom of hydrogen, we have compounds such as:

$$\underset{\text{I}}{CH_3OH} \qquad \underset{\text{II}}{CH_3Cl} \qquad \underset{\text{III}}{CH_3NO_2}$$

They may be given *substitutive names*, which consist of the name of the parent hydrocarbon with a prefix or suffix to denote the characteristic group. The HO group is normally denoted by the suffix '-ol'. (The prefix 'hydroxy-' may be used if another characteristic group has priority as the suffix; see below.) Cl is denoted by the prefix 'chloro-' and NO_2 by 'nitro-'. Thus, compound I is called methanol – note the elision of the final 'e' of methane – II is chloromethane and III is nitromethane.

An alternative system of *radicofunctional names* is based on the fact that from every hydrocarbon there is derived, in theory, a hydrocarbon radical by the loss of an atom of hydrogen. For example, from methane, CH_4, there is the methyl radical, CH_3; and from ethane, C_2H_6, the ethyl radical, C_2H_5. Compound I could thus be named as methyl alcohol, and II as methyl chloride. Radicofunctional names, although often allowed by IUPAC, are now little used except for amines. (The compound CH_3NH_2 is called methylamine; a proposal to call it aminomethane has not been accepted by either IUPAC or ASE.)

If there is more than one characteristic group in the molecule, and they are of the same kind, their number is denoted by a simple multiplying prefix such as di, tri, etc. For example, the compound CH_2Cl_2 is named dichloromethane.

There may be more than one position on the parent hydrocarbon at which substitution is possible. For example, from propane, $CH_3CH_2CH_3$, two monochloropropanes are obtainable:

$$\underset{\text{IV}}{CH_2ClCH_2CH_3} \qquad \underset{\text{V}}{CH_3CHClCH_3}$$

In such an event, to ensure unambiguous naming, the carbon atoms of the chain are numbered thus:

$$\overset{1}{C}H_3\overset{2}{C}H_2\overset{3}{C}H_3$$

Compound IV is therefore called 1-chloropropane and V is called 2-chloropropane. The *locants*, i.e. numbers in the name, must be kept as low as possible; for this reason it would be incorrect to name IV as 3-chloropropane.

Difficulties may arise if an organic compound contains two or more different characteristic groups. For example, compound VI could con-

$$\underset{\text{VI}}{CH_3\overset{\overset{\displaystyle OH}{|}}{C}HNH_2} \qquad \underset{\text{VII}}{CH_3\overset{\overset{\displaystyle OH}{|}}{C}HCN}$$

ceivably be named 1-aminoethanol or 1-hydroxyethylamine. Which is right? To resolve such problems, IUPAC has published a list of characteristic groups in decreasing order of priority for citation as the principal group and hence the suffix in the name. An abridged version of the list is as follows:

Decreasing priority ↓

$-COOH$	(carboxylic acid)
$-SO_3H$	(sulphonic acid)
$-COOR$	(ester)
$-COCl$	(acid chloride)
$-CONH_2$	(amide)
$-CN$	(nitrile)
$-NC$	(isocyanide)
$-CHO$	(aldehyde)
$>CO$	(ketone)
$-OH$	(alcohol, phenol)
$-NH_2$	(amine)
$-O-$	(ether)

Compound VI contains both an HO and an NH_2 group. Because HO is above NH_2 in the list it is the principal group and is therefore cited as the suffix. The name 1-aminoethanol is correct.

Cyanohydrins, which possess both HO and CN groups, pose a similar problem; consider compound VII. Reference to the list shows that the CN group has priority over the HO group. CN must therefore be cited as suffix and HO as prefix; the name of VII is 2-hydroxypropanenitrile.

Detailed IUPAC rules, together with comments on the differences between IUPAC and ASE naming, will be found at the beginning of each of the sections 14.3–14.11.

EXAMINATION QUESTION ON CHAPTER 1

1 From Table 7 of your data book, list the *atomic number*, *relative atomic mass* and *density* of the elements aluminium, silver, platinum, lead and uranium. By reference to these numerical values and to other tables in your data book discuss the following statements.

(a) The two stable isotopes of silver are ^{107}Ag and ^{109}Ag, but its relative atomic mass is not 108.0.

(b) Samples of lead may have slightly different relative atomic masses depending on their sources.

(c) The relative atomic mass of aluminium is just over 2.0 times its atomic number but the factor for uranium is nearly 2.6.

(d) Platinum has a much higher density than lead although its relative atomic mass is lower.

The Viking I space robot which landed on the planet Mars in July 1976 used carbon-14 in tests for plant life. Given a Geiger counter, a strong light source, a sample of barium carbonate labelled with carbon-14, a nasturtium plant and the usual apparatus of a school laboratory, suggest how you would make a qualitative test for photosynthesis, the principle of which could be adapted for use in such a space probe. (OC)

2 Energy changes

2.1
Introduction

2.1.1 THE CONCEPT OF ENERGY

The popular concept of energy is valid. When you get up in the morning you feel (or ought to feel!) energetic, i.e. capable of doing work. This does not mean that you automatically will work; only that you are physically able to do so. Hence we arrive at the idea that energy is the *ability to do work* – which is the way it is defined.

Energy manifests itself in several guises, including kinetic, potential, electrical, thermal and light. In addition, compounds have the capacity to do work, i.e. they possess *chemical energy*, although this is realised only when it is converted into other energy forms such as heat. For example, if methane is burned in air the chemical energy appears in the form of heat.

The unit of energy is the joule, J. However, this is rather a small unit for many purposes to do with chemical reactions, and we adopt instead the kilojoule, kJ.

In chemistry we are concerned principally with the *changes* in chemical energy that accompany chemical reactions. It is impossible to measure the *absolute* chemical energy content of a substance, just as it is impossible to obtain the absolute potential energy of an object. All we can do is measure the change in energy when a substance undergoes a chemical reaction, or the change in energy when an object is moved.

Most chemical reactions are accompanied by the evolution or absorption of heat. Reactions in which heat is lost to the surroundings are said to be *exothermic*, while those which take in heat are *endothermic*.

2.1.2 REACTIONS AT CONSTANT PRESSURE OR CONSTANT VOLUME

The heat change accompanying a chemical reaction may depend on the prevailing conditions, such as constant pressure or constant volume. This effect is important only for those reactions in which there is a change in the number of moles of gaseous substances. Let us consider the exothermic reaction

$$CO + \tfrac{1}{2}O_2 = CO_2$$

in which the reactants and products are gaseous. If we mix one mole of carbon monoxide (volume = 22.4 dm^3 at s.t.p.) with half a mole of oxygen (volume = 11.2 dm^3 at s.t.p.) in a rigid vessel of volume 33.6 dm^3, the total pressure will be one atmosphere (100 kPa). If the reaction now goes to completion the pressure inside the vessel decreases, due to a reduction in the number of gaseous molecules. Under these conditions of constant volume the heat produced is 281.761 kJ. Let us now suppose that the same reaction is performed with the same quantities of reactants in a cylinder fitted with a freely moving piston. Before reaction occurs the total volume inside the cylinder is again 33.6 dm^3 when the external pressure is one atmosphere. After completion of the reaction the volume inside the cylinder is 22.4 dm^3, providing that there is no change in atmospheric pressure. The heat produced during the chemical reaction under these conditions is 283.000 kJ. To see why this difference arises we must consider the work which we may have to perform on the reaction mixture, or the work which we may obtain from it.

Constant volume

Because no expansion or contraction occurs when the reaction is carried out at constant volume, we can obtain no work; neither are we required to perform work on the reaction mixture. Under these conditions, the heat produced is the difference between the chemical energy of the reactant molecules and the product molecules. The production of energy as heat arises because the chemical energy of the product molecules is lower than that of the reactant molecules, i.e.

energy of products − energy of reactants = energy produced as heat (§2.2)

The heat produced under conditions of constant volume therefore represents the energy change which occurs within the molecules of the reaction mixture. Consequently, we often refer to the heat produced under conditions of constant volume as the *change in internal energy* (ΔU), pronounced 'delta U', where U represents internal energy,

i.e. $U_{\text{products}} - U_{\text{reactants}} = \Delta U$

Constant pressure

The same change in internal energy occurs under conditions of constant pressure because the same reaction is occurring. However, there is an additional factor, for work is performed on the reaction mixture in reducing the volume by one-third as the product molecules are formed. This work appears in its equivalent form of heat. Consequently, the total amount of heat produced is now

ΔU + work performed on the reaction mixture

If ΔV represents the change in volume at a constant pressure (p), then the work performed is equal to $p\Delta V$. From the ideal gas equation (§6.2.1),

$$pV = nRT$$

If p, R and T are constant, then a volume change will occur only if there is a change in the number of moles of gas (n). Doubling the number of moles doubles the volume, and halving the number of moles halves the volume. Consequently,

$$\Delta V \propto \Delta n$$

(the change in the number of moles of gas)

$$\therefore \quad p\Delta V = \Delta nRT$$

For the reaction under discussion Δn is half a mole. At the standard thermodynamic temperature of 25 °C (298 K), the work can be calculated:

$$p\Delta V = \tfrac{1}{2} \times 8.314 \times 298 = 1\,239 \text{ J}$$

The total amount of heat produced at constant pressure is therefore

$$\Delta U + p\Delta V = 281\,761 + 1\,239 = 283\,000 \text{ J} = 283.000 \text{ kJ}$$

A heat change of this sort, which accompanies a reaction performed at constant pressure, is called an *enthalpy change* and is symbolised by ΔH (delta H).

$$\therefore \quad \Delta H = \Delta U + p\Delta V$$

The enthalpy change of a chemical reaction thus represents the internal energy difference between the reactants and products, plus the work which is done on the mixture if it contracts (gain in energy) or the work which is obtained from the mixture if it expands (loss in energy). As an example of the latter, let us consider the decomposition of aqueous hydrogen peroxide:

$$H_2O_2 = H_2O + \tfrac{1}{2}O_2$$

Under constant pressure conditions there is an increase in volume due to the formation of oxygen gas, and 98.000 kJ are produced for every mole of hydrogen peroxide that is decomposed. This heat represents the change in internal energy plus the work which is performed during expansion. The work performed on expansion due to the formation of half a mole of oxygen at 25 °C (298 K) is

$$p\Delta V = \tfrac{1}{2} \times 8.314 \times 298 = 1\,239 \text{ J} = 1.239 \text{ kJ}$$

$$\therefore \quad \Delta H = \Delta U + (-1.239)$$

A negative sign is applied to $p\Delta V$ because the reaction mixture *loses* energy in doing work of expansion.

$$\therefore \quad \Delta U = 98.000 - (-1.239) = 99.239 \text{ kJ}$$

At constant volume the heat produced would thus be 99.239 kJ, because under these conditions expansion work is not performed.

2.2 Enthalpy changes

Generally, reactions are performed under conditions of constant pressure, namely that of the atmosphere. We are therefore normally concerned with enthalpy changes rather than internal energy changes, although for reactions that do not include gases, or for those in which there is no change in the number of gaseous molecules, ΔH and ΔU are equal.

The enthalpy change represents the difference in energy between the products and reactants of a reaction at constant pressure. The energy of reactants and products is commonly referred to as *enthalpy* or *heat content*, since in practice we are measuring the energy difference as heat. Thus the enthalpy change of a reaction is the difference in enthalpy, or heat content, between reactants and products. We can represent enthalpy changes with energy diagrams, Fig. 2.1.

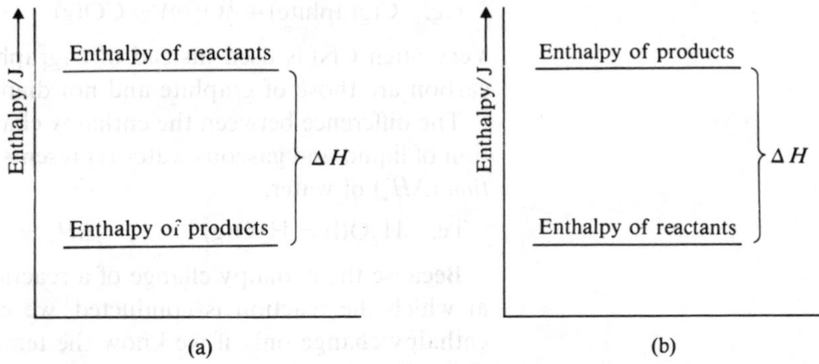

Fig. 2.1 Enthalpy diagrams for (a) an exothermic reaction, (b) an endothermic reaction.

In an exothermic reaction the enthalpy of the products is lower than that of the reactants and the enthalpy change, *taken by convention* to be

$$\Delta H = \text{enthalpy of products} - \text{enthalpy of reactants},$$

carries a negative sign. For endothermic reactions ΔH carries a positive sign.

2.2.1 REPRESENTATION OF ENTHALPY CHANGES

The enthalpy change for a chemical reaction is an *extensive property*, i.e. the enthalpy change depends on the quantities of the reactants. For example, if one mole of hydrogen combines with oxygen to form one mole of liquid water then $\Delta H = -286\ \text{kJ mol}^{-1}$, but if two moles of hydrogen react with oxygen to form two moles of water $\Delta H = 2 \times -286\ \text{kJ mol}^{-1}$. Therefore when quoting enthalpy changes we must stipulate the quantities of the reactants. This is conveniently achieved by writing the enthalpy changes beside or below the balanced chemical equation. Thus, for the formation of one or two moles of water, respectively, we write

$$H_2 + \tfrac{1}{2}O_2 = H_2O \qquad \Delta H_m = -286\ \text{kJ mol}^{-1}$$
$$\text{or} \quad 2H_2 + O_2 = 2H_2O \qquad \Delta H_m = -572\ \text{kJ mol}^{-1}$$

The ASE recommends that the units are kJ mol^{-1}, irrespective of the amounts involved. ΔH_m, termed the *molar enthalpy change*, always refers to the enthalpy change **per mole of the reaction as written**. For simplicity, however, we shall use the abbreviated symbol ΔH.

The enthalpy change for a chemical reaction depends on the physical states of the reactants and products,

$$\text{e.g.} \quad H_2 + \tfrac{1}{2}O_2 = H_2O(\text{liquid}) \qquad \Delta H = -286\ \text{kJ mol}^{-1}$$
$$H_2 + \tfrac{1}{2}O_2 = H_2O(\text{gaseous}) \qquad \Delta H = -242\ \text{kJ mol}^{-1}$$
$$C(\text{graphite}) + O_2 = CO_2 \qquad \Delta H = -394\ \text{kJ mol}^{-1}$$
$$C(\text{diamond}) + O_2 = CO_2 \qquad \Delta H = -395.9\ \text{kJ mol}^{-1}$$

To avoid any ambiguity we *always* quote the physical state of *all* reactants and products by using the following abbreviations: (s) solid, (l) liquid, (g) gas or vapour, and (aq) for substances in aqueous solution. For example,

$$H_2(g) + \tfrac{1}{2}O_2(g) = H_2O(l) \qquad \Delta H = -286 \text{ kJ mol}^{-1}$$

Different allotropes of an element or different polymorphs of a compound are indicated explicitly,

e.g. $$C(\text{graphite}) + \tfrac{1}{2}O_2(g) = CO(g) \qquad \Delta H = -111 \text{ kJ mol}^{-1}$$

Very often C(s) is used instead of C(graphite) because most reactions of carbon are those of graphite and not diamond.

The difference between the enthalpy changes accompanying the formation of liquid and gaseous water represents the *molar enthalpy of vaporisation* (ΔH_v) of water,

i.e. $$H_2O(l) = H_2O(g) \qquad \Delta H_v = +44 \text{ kJ mol}^{-1}$$

Because the enthalpy change of a reaction varies with the temperature at which the reaction is conducted, we can attach real meaning to an enthalpy change only if we know the temperature conditions. Hence we normally adopt a standard temperature of 25 °C (298 K). This implies that the initial temperature of the reactants and the final temperature of the products is 25 °C (298 K). Consider an exothermic reaction in which the initial temperature is 25 °C (298 K). During the reaction heat is produced and the temperature rises. After completion of the reaction, heat continues to be produced to the surroundings as the products cool to 25 °C (298 K). The total amount of heat produced during the heating and cooling stages is the enthalpy change for the reaction. Similar arguments apply to endothermic reactions, except that the temperature falls and heat is absorbed from the surroundings in restoring the products to 25 °C (298 K).

When an enthalpy change relates to the standard temperature it is indicated by ΔH_m(298 K),

e.g. $$Pb(s) + \tfrac{1}{2}O_2(g) = \underset{(\text{red})}{PbO(s)} \qquad \Delta H_m(298 \text{ K}) = -219 \text{ kJ mol}^{-1}$$

If the initial and final temperature in a reaction is 25 °C (298 K) and the initial and final pressure is one atmosphere (100 kPa), the enthalpy change is referred to as the *standard molar enthalpy change*, symbolised by $\Delta H_m^{\ominus}$(298 K),

e.g. $$\tfrac{1}{2}N_2(g) + \tfrac{1}{2}O_2(g) = NO(g) \qquad \Delta H_m^{\ominus}(298 \text{ K}) = +90.4 \text{ kJ mol}^{-1}$$

We shall use the common abbreviation, $\Delta H^{\ominus}$, pronounced 'delta *H* nought' or 'delta *H* standard'.

2.2.2 ENTHALPY CHANGES DURING CHEMICAL REACTIONS

Enthalpy of reaction is the heat that is produced or absorbed when a reaction occurs between the quantities of reactants specified by the coefficients in the balanced chemical equation for the reaction, e.g. $Xe(g) + 2F_2(g) = XeF_4(s)$ $\Delta H(298\ K) = -252\ kJ\,mol^{-1}$

If standard conditions prevail, we refer to the *standard* (molar) *enthalpy of reaction*,

e.g. $Li(s) + \frac{1}{2}H_2(g) = LiH(s)$ $\Delta H^{\ominus} = +128\ kJ\ mol^{-1}$

We often classify chemical reactions into various categories, e.g. combustion or neutralisation, and for some of the more important types we redefine and rename the enthalpy of reaction. *Enthalpy of combustion* (ΔH_c(298 K); under standard conditions, $\Delta H^{\ominus}_{c,m}$ or $\Delta H^{\ominus}_c$) is the heat that is produced when *one mole* of a substance is completely burned in oxygen at constant pressure. For example, the standard enthalpy of combustion of methane is $-891.1\ kJ\,mol^{-1}$, i.e.

$$CH_4(g) + 2O_2(g) = CO_2(g) + 2H_2O(l) \qquad \Delta H^{\ominus}_c = -891.1\ kJ\ mol^{-1}$$

The reaction:

$$C(s) + \tfrac{1}{2}O_2(g) = CO(g) \qquad \Delta H^{\ominus} = -111\ kJ\ mol^{-1}$$

does *not* represent the standard enthalpy of combustion of carbon because the carbon is not completely burned as the definition requires.

Enthalpy of formation (ΔH_f(298 K); under standard conditions, $\Delta H^{\ominus}_{f,m}$ or $\Delta H^{\ominus}_f$) is the heat that is produced or absorbed when *one mole* of a substance is formed from its constituent elements in their standard and most stable states, i.e. their condition at 25 °C (298 K) and one atmosphere (100 kPa) pressure. For example, the standard enthalpy of formation of ethane is $-84.7\ kJ\,mol^{-1}$ and may be represented in one of two ways:

$$2C(s) + 3H_2(g) = C_2H_6(g) \qquad \Delta H^{\ominus}_f = -84.7\ kJ\ mol^{-1}$$

or $\Delta H^{\ominus}_f[C_2H_6(g)] = -84.7\ kJ\ mol^{-1}$

The standard enthalpy of formation of carbon dioxide is equal to the standard enthalpy of combustion of carbon.

Compounds that are formed from their elements with the production of heat, e.g. ethane, are said to be *exothermic compounds*, while those formed from their elements with the absorption of heat, e.g. lithium hydride, are called *endothermic compounds*.

The standard enthalpy of formation of all elements is zero, e.g.

$$H_2(g) = H_2(g) \qquad \Delta H^{\ominus}_f = 0\,kJ\,mol^{-1}, \text{or } \Delta H^{\ominus}_f[H_2(g)] = 0\,kJ\,mol^{-1}$$

Enthalpy of neutralisation is the heat produced when an acid and a base react together in dilute aqueous solution to produce *one mole* of water. Strong acids and strong bases are almost completely dissociated in dilute solution (§9.2.1), and the only reaction that occurs on mixing the two is as follows:

$$H^+(aq) + HO^-(aq) = H_2O(l) \qquad \Delta H^{\ominus} = -57.3\ kJ\ mol^{-1}$$

e.g. $$\underbrace{H^+(aq) + NO_3^-(aq)}_{\text{solution of } HNO_3} + \underbrace{Na^+(aq) + HO^-(aq)}_{\text{solution of NaOH}}$$

$$= \underbrace{Na^+(aq) + NO_3^-(aq)}_{\text{solution of } NaNO_3} + H_2O(l)$$

Because the standard enthalpy change of this reaction is $-57.3\ \text{kJ mol}^{-1}$, the standard enthalpy of neutralisation for any strong acid by any strong base is almost constant at this value.

For weak acids and weak bases the enthalpy of neutralisation is usually lower than $-57.3\ \text{kJ mol}^{-1}$. Let us consider some examples.

Ethanoic acid and sodium hydroxide Because ethanoic acid is a weak acid (§9.2.1), it is present in solution principally in its un-ionised form and the predominant reaction is:

$$CH_3COOH(aq) + Na^+(aq) + HO^-(aq) = Na^+(aq) + CH_3COO^-(aq) + H_2O(l) \quad \Delta H^{\ominus} = -55.2\ \text{kJ mol}^{-1}$$

The enthalpy of neutralisation is lower than $-57.3\ \text{kJ mol}^{-1}$ because energy is absorbed in ionising the ethanoic acid. By contrast, for the corresponding reaction involving hydrofluoric acid, the high hydration enthalpy of the fluoride ion causes the value to be *greater* than $-57.3\ \text{kJ mol}^{-1}$:

$$HF(aq) + Na^+(aq) + HO^-(aq) = Na^+(aq) + F^-(aq) + H_2O(l) \quad \Delta H^{\ominus} = -68.6\ \text{kJ mol}^{-1}$$

Aqueous ammonia and hydrochloric acid Ammonia is a weak base and the predominant species in aqueous solution is $NH_3(aq)$ (§15.2.3). The neutralisation reaction is therefore

$$NH_3(aq) + H^+(aq) + Cl^-(aq) = NH_4^+(aq) + Cl^-(aq) \quad \Delta H^{\ominus} = -52.2\ \text{kJ mol}^{-1}$$

The essential change is thus

$$H^+(aq) + NH_3(aq) = NH_4^+(aq)$$

and the enthalpy change of this reaction is different from that of the reaction

$$H^+(aq) + HO^-(aq) = H_2O(l)$$

The standard *enthalpy of atomisation* of an element is the heat absorbed in producing *one mole* of gaseous (isolated) atoms from the element in its standard state at 25 °C (298 K) and one atmosphere (100 kPa) pressure,

e.g. $\frac{1}{2}H_2(g) = H(g) \qquad \Delta H^{\ominus} = +218\ \text{kJ mol}^{-1}$

Bond dissociation enthalpy is the heat required to break *one mole* of bonds of a specified type in a substance,

e.g. $H_2(g) = 2H(g) \qquad \Delta H^{\ominus} = +436\ \text{kJ mol}^{-1}$

Thus to break the one mole of H—H bonds in one mole of hydrogen gas 436 kJ of energy are required. For all elements that exist as diatomic molecules in the gaseous state, e.g. H_2, Cl_2 or N_2,

bond dissociation enthalpy = twice the enthalpy of atomisation.

Enthalpy of hydrogenation is the heat produced when *one mole* of an unsaturated compound is converted into the corresponding saturated compound by the addition of hydrogen,

e.g. $C_2H_4(g) + H_2(g) = C_2H_6(g)$ $\Delta H^{\ominus} = -157.3\ \text{kJ mol}^{-1}$

2.2.3 EXPERIMENTAL DETERMINATION OF ENTHALPY CHANGES

Approximate enthalpy changes for reactions in solution can be measured in expanded polystyrene beakers. Calorimeters of this sort have two great advantages, namely good thermal insulation and low specific heat capacity.

They may be used, for example, to determine the enthalpy of neutralisation of hydrochloric acid by sodium hydroxide. Dilute aqueous solutions of equal concentration (e.g. 0.1 M) are first allowed to reach laboratory temperature. Equal volumes (50 cm^3 or 100 cm^3) are then rapidly mixed together and the temperature rise noted with an accurate thermometer. The temperature rise may be determined more accurately (allowing for heat losses) by plotting a graph of temperature against time (Fig. 2.2). Readings are taken at half-minute intervals before and after mixing. The 'corrected' temperature rise can then be estimated from the graph as shown.

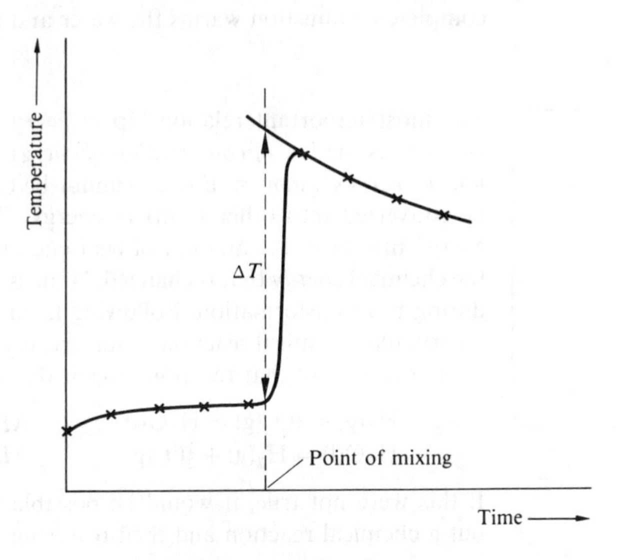

Fig. 2.2 Estimation of 'corrected' temperature rise.

When 50 cm^3 of 0.1 M HCl is mixed with 50 cm^3 of 0.1 M NaOH, the 'corrected' temperature rise is (say) 0.65 °C (0.65 K). For any system the enthalpy change, in kilojoules, is given by

$$\Delta H = mc\Delta T$$

m = total mass of reacting substances (in grams) ≈ volume of dilute aqueous solution (in cm^3).

c = specific heat capacity of dilute aqueous solutions ≈ specific heat capacity of water = $4.18 \times 10^{-3}\ \text{kJ g}^{-1}\ \text{K}^{-1}$.

ΔT = initial temperature − final temperature.

In this case, $\Delta H = 100 \times 4.18 \times 10^{-3} \times (-0.65) = -0.272\ \text{kJ}$

50 cm^3 (0.050 dm^3) of 0.1 M HCl or NaOH contains $0.050 \times 0.1 = 0.005$ mol of solute.

0.005 mol HCl reacts with 0.005 mol NaOH to give 0.005 mol H_2O.
In forming 0.005 (i.e. 1/200) mol H_2O, $\Delta H = -0.272$ kJ

$\therefore$ to form 1 mol H_2O, $\Delta H = -0.272 \times 200 = -54.4 \text{ kJ mol}^{-1}$

(The accepted value is $-57.1 \text{ kJ mol}^{-1}$ (§2.2.2).)

Accurate enthalpies of combustion are determined in a *bomb calorimeter*. This is machined out of a stainless steel block and usually has a volume of about 500 cm^3. It has a strong screw cap fitted with two terminals. A solid sample is compressed into a tablet, and into this is embedded a piece of iron wire with the ends protruding. (A liquid sample is enclosed in a plastic sachet.) The sample is accurately weighed, the ends of the wire connected to the terminals, and the sample placed in the bomb. The bomb is sealed and oxygen introduced under pressure. When the required pressure (20–30 atm) is reached, the bomb is placed inside a double walled metal container holding a known volume of water. Water at constant temperature is circulated through the double wall. The lid of the container is placed in position, followed by a thermometer (accurate to ± 0.01 °C) which passes through the lid and into the water. Current is then passed through the wire to ignite the solid. The heat produced by rapid and complete combustion warms the water and the temperature rise is noted.

2.3 The laws of energetics

The most important relationship in *energetics*, i.e. the study of energy changes, is the *law of conservation of energy*, otherwise known as the *first law of thermodynamics*: 'Energy cannot be created or destroyed, but may be converted into other forms of energy.' Thus, if we convert chemical energy into heat, the amount of heat energy obtained is exactly equal to the chemical energy that is changed. None is destroyed and none is created during the transformation. Following from this, the enthalpy change for a particular chemical reaction is numerically equal to the enthalpy change for the reverse of that reaction, except that the sign is changed,

e.g. $H_2(g) + \frac{1}{2}O_2(g) = H_2O(l) \qquad \Delta H^{\ominus} = -286 \text{ kJ mol}^{-1}$

$H_2O(l) = H_2(g) + \frac{1}{2}O_2(g) \qquad \Delta H^{\ominus} = +286 \text{ kJ mol}^{-1}$

If this were not true, it would be possible to create energy by carrying out a chemical reaction and then reversing it.

2.3.1 HESS'S LAW

Hess's law, which is an application of the law of conservation of energy to chemical changes, states: 'The enthalpy change for a chemical reaction is independent of the route by which the reaction takes place.' Let us suppose that the conversion of A into B is accompanied by an enthalpy change ΔH_1. Instead of converting A directly into B, let us now suppose that we do so through several intermediate stages, each with its own enthalpy change, e.g. $A \xrightarrow{\Delta H_2} C \xrightarrow{\Delta H_3} D \xrightarrow{\Delta H_4} B$

According to Hess's law,

$$\Delta H_1 = \Delta H_2 + \Delta H_3 + \Delta H_4$$

The enthalpy change ΔH_1 depends only on the difference in enthalpy between A and B (§2.2), so that no matter how we convert A into B the overall or total enthalpy change is always the same; we cannot destroy or create energy. We may summarise the above changes in an *enthalpy diagram* or *enthalpy cycle*:

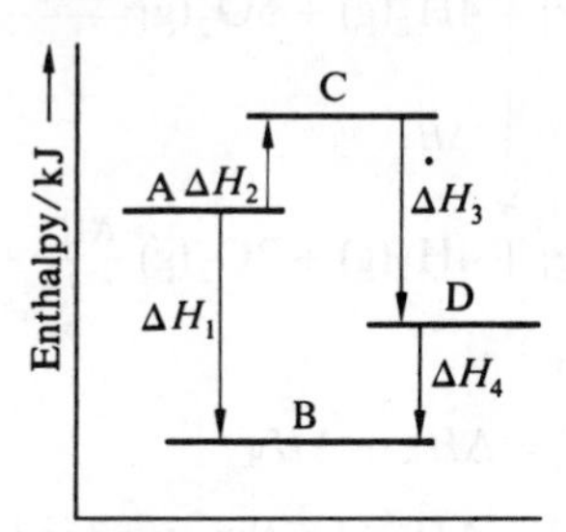

Enthalpy diagram
(Labelled horizontal lines represent relative enthalpy levels of A, B, C and D.)

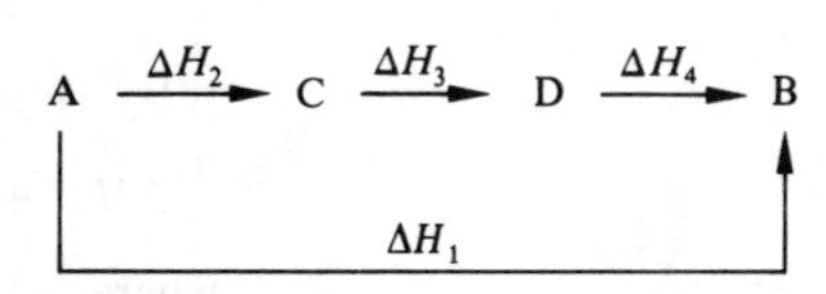

Enthalpy cycle
(These cycles may be presented in several ways. The positions of A, B, C and D do *not* represent relative enthalpy levels.)

From both the diagram and the cycle it can be seen that

$$\Delta H_1 = \Delta H_2 + \Delta H_3 + \Delta H_4$$

Changes of the type

$$W \xrightarrow{\Delta H_1} X \xrightarrow{\Delta H_2} Y \quad \text{and} \quad W \xrightarrow{\Delta H_3} Z \xrightarrow{\Delta H_4} Y$$

i.e. the conversion of W into Y via an intermediate X or Z, can be represented by the following diagram or cycle:

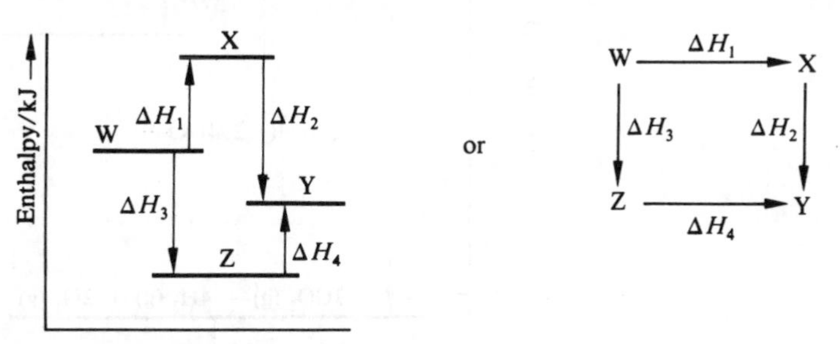

By Hess's law,

$$\Delta H_1 + \Delta H_2 = \Delta H_3 + \Delta H_4$$

Hess's law is useful for calculating enthalpy changes, especially for those reactions which are difficult to observe experimentally. We shall now perform a calculation by three methods which are in common use. The selection of a method for any problem is merely a matter of personal preference.

Example
The standard enthalpies of combustion of carbon, hydrogen and propane are −394, −286 and −2 222 kJ mol^{-1} respectively. Calculate the standard enthalpy of formation of propane.

Method 1, using an enthalpy cycle.

$$C(s) + O_2(g) = CO_2(g) \qquad \Delta H_1^\ominus = -394 \text{ kJ mol}^{-1}$$
$$H_2(g) + \tfrac{1}{2}O_2(g) = H_2O(l) \qquad \Delta H_2^\ominus = -286 \text{ kJ mol}^{-1}$$
$$C_3H_8(g) + 5O_2(g) = 3CO_2(g) + 4H_2O(l) \qquad \Delta H_3^\ominus = -2\,222 \text{ kJ mol}^{-1}$$

$$\begin{array}{ccc} 3C(s) + 4H_2(g) + 5O_2(g) & \xrightarrow{\Delta H_f^\ominus} & C_3H_8(g) + 5O_2(g) \\ \downarrow \Delta H_A & & \downarrow \Delta H_3^\ominus \\ 3CO_2(g) + 4H_2(g) + 2O_2(g) & \xrightarrow{\Delta H_B} & 3CO_2(g) + 4H_2O(l) \end{array}$$

By Hess's law,

$$\Delta H_f^\ominus + \Delta H_3^\ominus = \Delta H_A + \Delta H_B$$

where $\Delta H_A = 3 \times \Delta H_1^\ominus$, i.e. the standard enthalpy change in converting three moles of C(s) into three moles of CO_2(g),

and $\Delta H_B = 4 \times \Delta H_2^\ominus$, i.e. the standard enthalpy change for producing four moles of water.

$$\therefore \quad \Delta H_f^\ominus = 3(-394) + 4(-286) - (-2\,222)$$
$$= -104 \text{ kJ mol}^{-1}$$

This example illustrates the common practice of calculating the enthalpy of formation of a compound from the enthalpies of combustion of the compound and the elements that it contains.

Method 2, using an enthalpy diagram.

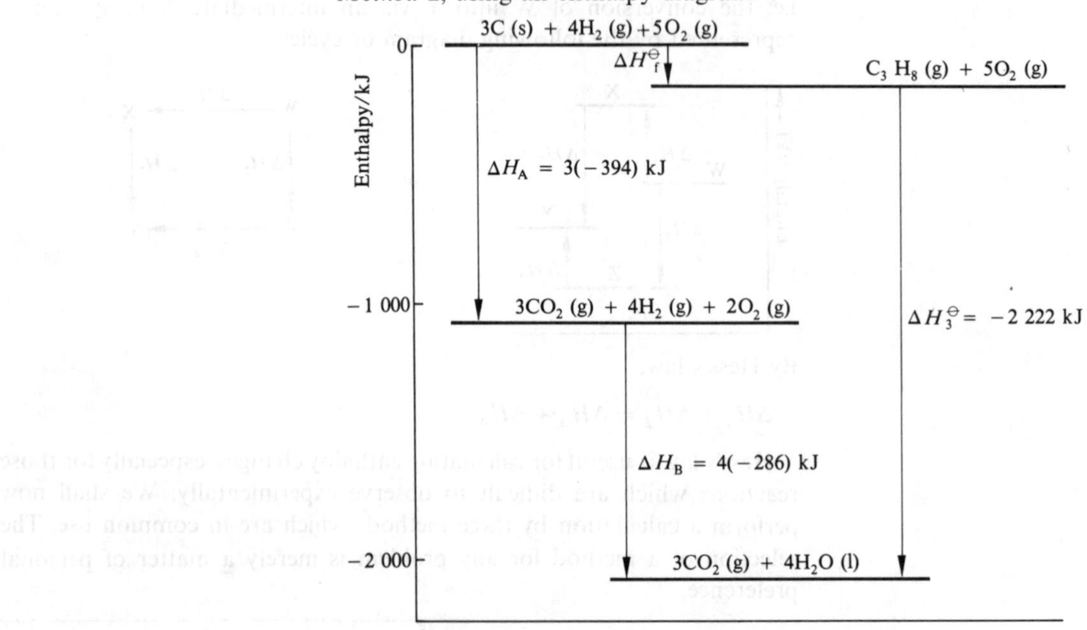

From the diagram,

$$\Delta H_f^\ominus + (-2\,222) = 3(-394) + 4(-286)$$
$$\therefore \quad \Delta H_f^\ominus = -104 \text{ kJ mol}^{-1}$$

Method 3

From the definition of ΔH as 'enthalpy of products – enthalpy of reactants', it follows that the enthalpy of a reaction is equal to the difference between the enthalpies of formation of the products and those of the reactants,

i.e. $\Delta H_{\text{reaction}} = \Sigma n \Delta H_f[\text{products}] - \Sigma n \Delta H_f[\text{reactants}]$

where $\Sigma n \Delta H_f$ is the sum of the molar enthalpies of formation of n moles of products or reactants.

In this example, the enthalpy of combustion of propane is equal to the difference between the enthalpies of formation of three moles of CO_2 and four moles of H_2O, and one mole of propane and five moles of O_2. (The latter is zero.)

i.e. $\Delta H_3^\ominus = (3 \times \Delta H_f^\ominus[CO_2(g)]) + (4 \times \Delta H_f^\ominus[H_2O(l)]) - (\Delta H_f^\ominus[C_3H_8(g)])$

$\therefore \quad -2\,222 = 3(-394) + 4(-286) - \Delta H_f^\ominus$

$\therefore \quad \Delta H_f^\ominus = -104 \text{ kJ mol}^{-1}$

2.3.2 BORN–HABER CYCLE

So far, in our construction of enthalpy cycles and diagrams, we have considered only covalent compounds, but the same principles can be applied to ionic compounds to give what are called *Born–Haber cycles*. Let us consider the formation of one mole of an ionic compound M^+X^- from a metal, M, and a non-metal, X_2,

i.e. $M(s) + \frac{1}{2}X_2(\text{g, l or s}) = M^+X^-(s) \qquad \Delta H_f^\ominus = ?$

This reaction comprises the following steps.

(i) $M(s) = M(g)$ $\quad \Delta H_1$ = enthalpy of atomisation of the metal (endothermic)

(ii) $\frac{1}{2}X_2(\text{g, l or s}) = X(g)$ $\quad \Delta H_2$ = enthalpy of atomisation of the non-metal (endothermic). (For liquid or solid non-metals this term also includes the molar enthalpies of vaporisation or sublimation respectively.)

(iii) $M(g) = M^+(g) + e^-$ $\quad \Delta H_3$ = first ionisation enthalpy of the metal (endothermic)

(iv) $e^- + X(g) = X^-(g)$ $\quad \Delta H_4$ = first electron affinity of the non-metal (exothermic)

(v) $M^+(g) + X^-(g) = M^+X^-(s)$ $\quad \Delta H_5$ = lattice enthalpy of the compound (exothermic)

Method 1

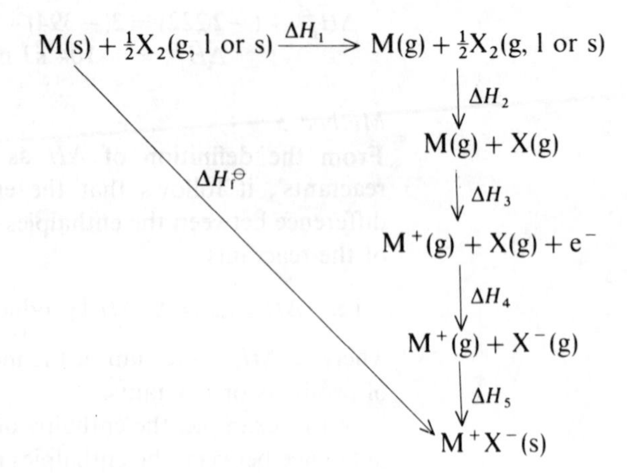

By Hess's law,

$$\Delta H_f^{\ominus} = \Delta H_1 + \Delta H_2 + \Delta H_3 + \Delta H_4 + \Delta H_5$$

Method 2

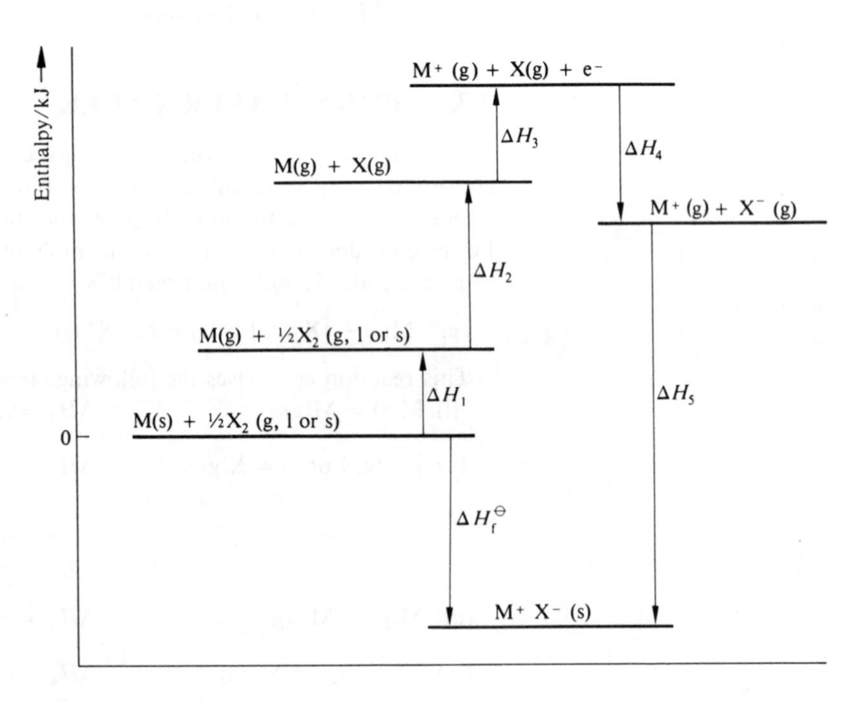

From the diagram,

$$\Delta H_f^{\ominus} = \Delta H_1 + \Delta H_2 + \Delta H_3 + \Delta H_4 + \Delta H_5$$

Lattice enthalpies, which are difficult to measure directly, can readily be calculated from Born–Haber cycles (§4.7.4) since all the other enthalpy changes can be obtained experimentally.

2.3.3 BOND DISSOCIATION ENTHALPIES

Covalent bond formation between atoms leads to a lowering of energy (§4.2.1). Thus, to break a covalent bond, energy must be absorbed and the amount that is required is equal to the energy generated during bond formation. The higher the energy required to break the bond the greater is its strength. The enthalpy required to break one mole of covalent bonds in diatomic molecules is called the *bond dissociation enthalpy* (§2.2.2). For polyatomic molecules the bond dissociation is more complex. For example, the total enthalpy change in breaking the four identical C—H bonds in one mole of methane is as follows:

$$CH_4(g) = C(g) + 4H(g) \qquad \Delta H^{\ominus} = +1\,648 \text{ kJ mol}^{-1}$$

Thus the *average bond dissociation enthalpy* for the C—H bond is

$$\frac{+1\,648}{4} = \mathbf{+412\ kJ\ mol^{-1}}\textbf{, represented as } DH^{\ominus}\mathbf{(C{-}H) = 412\ kJ\ mol^{-1}}$$

The enthalpy associated with a particular type of bond is affected by its environment within a molecule. For example, the average bond enthalpy of 412 kJ mol^{-1} for a C—H bond in methane also applies, approximately, to the C—H bonds in other alkanes, but not to the bonds in alkenes, alkynes or arenes. Nevertheless, average bond dissociation enthalpies for a series of bonds are known which do not vary too greatly from one compound to another (Table 2.1).

Table 2.1 Some average bond dissociation enthalpies in kJ mol^{-1}

$DH^{\ominus}$(C—C) = 348	$DH^{\ominus}$(C┄C) = 518 (benzene)
$DH^{\ominus}$(C═C) = 612	$DH^{\ominus}$(C—Cl) = 338
$DH^{\ominus}$(C≡C) = 837	$DH^{\ominus}$(C—H) = 412

Average bond enthalpies are useful for estimating the enthalpy change of a chemical reaction, for during the reaction bonds are broken (endothermically) in the reactants and formed (exothermically) in the products. For example, in the reaction

$$CH_4(g) + Cl_2(g) = CH_3Cl(g) + HCl(g)$$

C—H and Cl—Cl bonds are broken and C—Cl and H—Cl bonds are formed. Using the average bond dissociation enthalpies in Table 2.1 for the C—H and C—Cl bonds, and taking the bond dissociation enthalpies of Cl—Cl and H—Cl as 242 and 431 kJ mol^{-1} respectively, we see that

enthalpy change of bond breaking = $+412 + 242$ kJ mol^{-1}
and enthalpy change of bond formation = $-338 + (-431)$ kJ mol^{-1}
∴ overall enthalpy change = $412 + 242 - 338 - 431$
= -115 kJ mol^{-1}

Although this result does not agree exactly with the experimental value, -99.4 kJ mol^{-1}, it is sufficiently accurate to serve as a guide to the possibility of the reaction taking place (§2.4.2).

2.4 Elementary principles of thermodynamics

Thermodynamics is the study of the mechanical work which can be obtained from heat. In the last century thermodynamics played an important part in the development of the steam engine, but nowadays the subject has much wider applications, such as in the design of power stations. Thermodynamics can also be applied to chemical reactions in connection with the work that is obtainable from the energy changes accompanying them.

Why do some chemical reactions occur without external influence, whereas others do not? Left alone, everything in nature tends towards a state of lowest energy. A ball, for example, runs downhill so that its potential energy is reduced. The heat energy which is given out during an exothermic reaction shows that there is a lowering of chemical energy in proceeding from the reactants to the products. We may therefore be tempted to presume that a reduction in enthalpy provides the necessary impetus or driving force for a reaction to occur. However, many endothermic reactions also proceed of their own accord, so it seems that a negative enthalpy change is no reliable criterion by which to judge what reactions will take place.

There are in fact *two* factors that we must consider for any reaction, namely the enthalpy change and the *entropy change* (ΔS). The nature of entropy changes is discussed below, but for the moment let us consider the function that incorporates these two changes – the *Gibbs free energy change* (ΔG).

$$\Delta G = \Delta H - T\Delta S$$

$$\text{or } \Delta G^{\ominus} = \Delta H^{\ominus} - T\Delta S^{\ominus} \quad \text{under standard conditions,}$$

where T is the absolute temperature, $\Delta G^{\ominus}$ is the standard free energy change and $\Delta S^{\ominus}$ is the standard entropy change.

A reaction will occur of its own accord, without the provision of external energy, only if there is a decrease in free energy, i.e. *if* ΔG *is negative*, (§2.4.2).

2.4.1 THE NATURE OF ΔS, ΔH AND ΔG

Entropy is related to degree of disorder. For example, in a crystalline solid the ions or molecules are regularly arranged, so that the disorder and entropy are low. In a gas, however, the molecules are not arranged regularly, i.e. they are more randomly distributed, and the entropy is higher. If we consider a change taking place of its own accord in an isolated system, i.e. one in which energy can neither enter nor leave, then the change is *always accompanied by an increase in entropy* (or disorder). Since entropy increases, the entropy change, ΔS (or $\Delta S^{\ominus}$ under standard conditions) is positive.

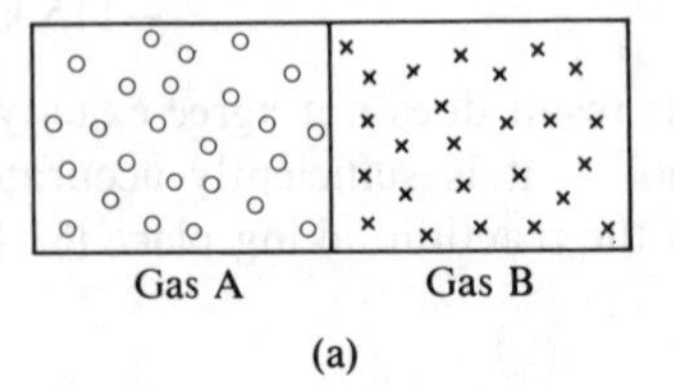

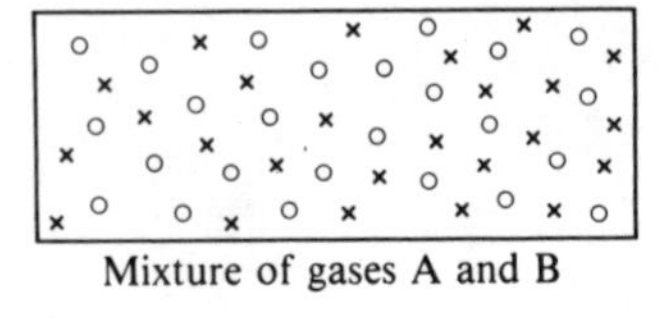

Fig. 2.3 (a) Gases A and B separated from each other, (b) Gases A and B mixed together.

Consider two gases, A and B, separated from each other by a partition in a closed container that is thermally insulated from its surroundings so that heat (energy) cannot escape or enter (Fig. 2.3(a)). This represents an isolated system. If the partition is removed the gases mix completely of their own accord. The mixture of gases (Fig. 2.3(b)) is more disordered than the unmixed gases, and hence the entropy of the mixed gases is higher than that of the unmixed gases. The mixing process thus occurs with an increase in entropy. The reverse process, i.e. the separation of the gases, does not take place of its own accord because it is accompanied by a decrease in entropy.

Consider a salt and water. The crystalline salt has a low entropy (see above). During dissolution, however, the regular arrangement of ions in the salt is broken down to produce a solution in which the ions are randomly arranged. Dissolving is therefore accompanied by an increase in disorder and hence in entropy.

Changes of state are also accompanied by entropy changes. Gases are more disordered than liquids, which in turn are more disordered than solids. Therefore the changes

solid → liquid, liquid → gas, solid → gas

are all accompanied by an increase in entropy.

A change may occur with an apparent *decrease* in entropy. For example, when a supersaturated solution of sodium thiosulphate crystallises of its own accord, the entropy appears to decrease because there is an increase in order as crystals are deposited. We should, however, consider *all* the changes occurring in the isolated system. Crystallisation is accompanied by a rise in temperature which increases the entropy of the remaining solution. This increase in entropy of the solution exceeds the decrease in entropy which occurs as a result of crystallisation. Therefore, when we consider the *complete* isolated system, there is an overall entropy increase. Similar principles apply to all changes taking place of their own accord. Thus, even if a decrease in entropy occurs in one part of a system, when we consider the complete isolated system the entropy always increases.

The enthalpy change, ΔH, represents the total amount of energy that accompanies a physical or chemical change. Not all of this energy is available for the isolated system to do work. Some is retained by the system in increasing the entropy. The term $T\Delta S$ is equal to this retained energy, and the difference, $\Delta H - T\Delta S$, expresses the amount of energy which is free or available to do work, i.e. it is the free energy change, ΔG.

2.4.2 FREE ENERGY CHANGES AND REACTION FEASIBILITY

We can now answer the fundamental question: 'Why do some chemical reactions occur of their own accord but not others?' The answer is that a change will occur only if it has the capacity to do work. This means that the change must be accompanied by the generation of *free* energy, i.e. energy which is available to do work. It is a useful generalisation to say that if $\Delta G^\ominus$ is negative a reaction will occur of its own accord, whereas if $\Delta G^\ominus$ is positive a reaction does not take place.

Free energy changes therefore enable us to predict which reactions will occur on their own and which will not. This does not mean that reactions which take place with an increase in free energy (ΔG positive) will not occur under any circumstances. Such reactions can occur if energy is provided by an external source. For example, in the manufacture of carbon disulphide, carbon is made to combine with sulphur at elevated temperatures by the absorption of heat energy:

$$C(s) + 2S(s) = CS_2(l) \qquad \Delta G^\ominus = +63.6 \text{ kJ mol}^{-1}$$

Many other industrial processes have positive $\Delta G^\ominus$ values.

Generally, the more negative the value of $\Delta G^\ominus$ for a reaction, the greater is the possibility of the reaction occurring of its own accord. In this context $\Delta G^\ominus$ is a guide to the 'driving force of a reaction'. Since $\Delta H^\ominus$ is related to $\Delta G^\ominus$ we can sometimes use $\Delta H^\ominus$ as a guide to reaction feasibility. However, great care must be exercised in using $\Delta H^\ominus$ for such purposes.

Free energy changes do not allow us to predict the *rate* of a chemical reaction. For example, the reaction

$$H_2(g) + \tfrac{1}{2}O_2(g) = H_2O(l) \qquad \Delta G^\ominus = -237 \text{ kJ mol}^{-1}$$

is extremely slow in the absence of a catalyst, despite the large free energy change.

The possibility of a reaction occurring depends on the accompanying enthalpy and entropy changes. The enthalpy change may be positive, negative or zero, while the entropy change for a reaction which is *completely isolated* from its surroundings may be positive or negative. There are therefore six possible combinations of circumstances (Table 2.2). Although in Table 2.2 and the subsequent discussion standard changes are considered, the same conclusions hold for changes under non-standard conditions.

Table 2.2 The various changes that may occur in a reaction which is completely isolated from its surroundings.

Type	$\Delta H^\ominus$	$\Delta S^\ominus$	$T\Delta S^\ominus$	$\Delta H^\ominus - T\Delta S^\ominus$	$\Delta G^\ominus$
1	0	+	+	(0) − (+)	−
2	0	−	−	(0) − (−)	+
3	−	+	+	(−) − (+)	−
4	+	−	−	(+) − (−)	+
5	+	+	+	(+) − (+)	− or +
6	−	−	−	(−) − (−)	+ or −

Type 1 $\Delta H^\ominus = 0$, $\Delta S^\ominus = +\text{ve}$

This change is exemplified by the mixing of two gases (§2.4.1). The negative value of $\Delta G^\ominus$ indicates that gases will mix of their own accord. Gases do not therefore unmix on their own because $\Delta G^\ominus$ would be positive (type 2).

Type 3 $\Delta H^\ominus = -\text{ve}$, $\Delta S^\ominus = +\text{ve}$

In reactions that produce gases from liquids or solids there is an increase in entropy,

$$\text{e.g.} \quad (NH_4)_2Cr_2O_7(s) = N_2(g) + Cr_2O_3(s) + 4H_2O(g)$$

For exothermic reactions like this, $\Delta G^{\ominus}$ is negative at all temperatures. Such reactions will always occur of their own accord, once they have started. The example quoted here is a classical one of a feasible reaction that does not occur at room temperature because heat is required to supply the necessary activation energy (§7.4.4).

Type 4 $\Delta H^{\ominus} = +\text{ve}, \quad \Delta S^{\ominus} = -\text{ve}$

The formation of liquid hydrazine from its (gaseous) elements is an example of an endothermic change accompanied by a decrease in entropy.

$$N_2(g) + 2H_2(g) = N_2H_4(l) \qquad \Delta H^{\ominus} = +50.4 \text{ kJ mol}^{-1}$$

$\Delta G^{\ominus}$ is positive, and such reactions do not occur on their own.

Type 5 $\Delta H^{\ominus} = +\text{ve}, \quad \Delta S^{\ominus} = +\text{ve}$

Let us consider the reaction

$$CaCO_3(s) = CaO(s) + CO_2(g)$$
$$\Delta H^{\ominus} = +178.0 \text{ kJ mol}^{-1}, \quad \Delta S^{\ominus} = +161.1 \text{ J K}^{-1} \text{ mol}^{-1}$$

For simplicity we will assume that the enthalpy change and entropy change are independent of temperature.

$$\text{At 25 °C (298 K)}, \; T\Delta S^{\ominus} = \frac{298 \times 161.1}{1\,000} = +48.0 \text{ kJ mol}^{-1}$$
$$\therefore \qquad \Delta G^{\ominus} = +178.0 - (+48.0) = +130.0 \text{ kJ mol}^{-1}$$

$\Delta G^{\ominus}$ has a high positive value and the reaction does not proceed.

However, at 832 °C (1105 K), $T\Delta S^{\ominus} = +178.0 \text{ kJ mol}^{-1}$

$$\therefore \qquad \Delta G^{\ominus} = 0$$

At temperatures above 832 °C (1105 K) $\Delta G^{\ominus}$ is negative and the reaction will proceed. Because $T\Delta S^{\ominus}$ increases with increasing temperature, $\Delta G^{\ominus}$ becomes decreasingly positive (or increasingly negative) as the temperature is raised, i.e. high temperatures favour these reactions.

Type 6 $\Delta H^{\ominus} = -\text{ve}, \quad \Delta S^{\ominus} = -\text{ve}$

Whenever a gas reaction is accompanied by a decrease in the number of molecules there is a decrease in entropy,

e.g. $C_2H_4(g) + H_2(g) = C_2H_6(g)$

$\Delta H^{\ominus} = -137.0 \text{ kJ mol}^{-1}, \quad \Delta S^{\ominus} = -120.0 \text{ J K}^{-1} \text{ mol}^{-1}$

At 25 °C (298 K), $T\Delta S^{\ominus} = -35.76 \text{ kJ mol}^{-1}$

$\therefore \quad \Delta G^{\ominus} = -137.0 - (-35.76) = -101.2 \text{ kJ mol}^{-1}$

$\Delta G^{\ominus}$ has a large negative value and the reaction proceeds.

However, at 868.5 °C (1141.7 K), $T\Delta S^{\ominus} = -137.0 \text{ kJ mol}^{-1}$; $\Delta G^{\ominus} = 0$. Above this temperature $\Delta G^{\ominus}$ is positive and the reverse reaction, i.e. the decomposition of ethane into ethene and hydrogen, is likely to occur.

Generally, for this type of reaction, the standard free energy change is negative only if the term $T\Delta S^{\ominus}$ is less than the enthalpy change. The standard free energy change for such reactions therefore becomes less negative (or more positive) with increasing temperature, because the $T\Delta S^{\ominus}$ term becomes larger.

2.4.3 THE THERMODYNAMICS OF METAL EXTRACTION

The change in free energy accompanying the formation of one mole of a compound from its constituent elements is called its *free energy of formation*, ΔG_f. Under standard conditions of 1 atm, we refer to the *standard free energy of formation*, $\Delta G_f^{\ominus}$.

A metal oxide, MO, may be reduced by carbon in accordance with the general equation

$$2MO + C = 2M + CO_2$$

The standard free energy change ($\Delta G^{\ominus}$) for the reaction can be obtained from the standard free energies of formation of carbon dioxide and the metal oxide in the same manner as for enthalpy changes (§2.3.1)

$$\text{i.e.} \quad \Delta G^{\ominus} = \Sigma n \Delta G_f^{\ominus}[\text{products}] - \Sigma n \Delta G_f^{\ominus}[\text{reactants}]$$

$$\therefore \quad \Delta G^{\ominus} = \Delta G_f^{\ominus}[CO_2(g)] - 2\Delta G_f^{\ominus}[MO(s)]$$

The standard free energy of formation of carbon dioxide is -395 kJ mol^{-1}, and most metal oxides have negative standard free energies of formation, i.e. $\Delta G_f^{\ominus}[MO(s)]$ is negative. For reduction to occur the standard free energy change ($\Delta G^{\ominus}$) must be negative (§2.4.2), which will happen if $2\Delta G_f^{\ominus}[MO(s)]$ is numerically smaller than $\Delta G_f^{\ominus}[CO_2(g)] = -395$ kJ mol^{-1}.

At 25 °C (298 K) carbon is unable to reduce most metal oxides because this condition is not fulfilled and $\Delta G^{\ominus}$ is positive. For some metal oxides, e.g. those of mercury and silver, $\Delta G^{\ominus}$ is negative at room temperature, but reduction does not occur because the necessary activation energy is lacking (§7.4.4). Here, though, we shall consider only the free energy changes. Many metal oxides are reduced by carbon on heating, and to see why this is so we must examine the manner in which standard free energies of formation vary with temperature. From the Gibbs free energy equation (§2.4)

$$\Delta G^{\ominus} = -\Delta S^{\ominus} \times T + \Delta H^{\ominus}$$

Now since both $\Delta S^{\ominus}$ and $\Delta H^{\ominus}$ are, to a good approximation, independent of temperature, the above equation has the form

$$y = mx + c$$

and a plot of $\Delta G^{\ominus}$ against T will be a straight line of slope $-\Delta S^{\ominus}$. The entropy change for the formation of carbon dioxide is given by

$$C(s) + O_2(g) = CO_2(g) \qquad \Delta S^{\ominus} = +3.3 \text{ J K}^{-1} \text{ mol}^{-1}$$

Therefore $-\Delta S^{\ominus} = -3.3$ J K^{-1} mol^{-1} and the standard free energy of formation of carbon dioxide ($\Delta G_f^{\ominus}[CO_2(g)]$) becomes more negative with increasing temperature (Fig. 2.4).

For metal oxides, the process is more complex, since the entropy change depends on the physical state of the metal, which alters as the temperature increases. The formation of metal oxides from their elements is always accompanied by a decrease in entropy. For example, in the reaction

$$2M(s) + O_2(g) = 2MO(s) \qquad (1)$$

the solid oxide formed is more ordered than the oxygen gas and the entropy decreases (§2.4.1). At temperatures above the melting temperature of the metal, but below that of the oxide, we must consider the reaction

$$2M(l) + O_2(g) = 2MO(s) \quad (2)$$

The liquid metal is more disordered and hence has a higher entropy than the solid metal. There is therefore a greater degree of ordering when the solid metal oxide is formed from the liquid metal than the solid metal. The entropy decrease accompanying reaction (2) is thus greater (i.e. $\Delta S^\ominus$ is more negative) than that for reaction (1). Similar considerations apply at temperatures above the boiling temperature of the metal, i.e. for the reaction

$$2M(g) + O_2(g) = 2MO(s) \quad (3)$$

In the gaseous state the metal is more disordered and hence has a higher entropy than in the liquid or solid states. The entropy decrease accompanying reaction (3) is therefore greater than that of reactions (1) or (2). This is shown schematically below.

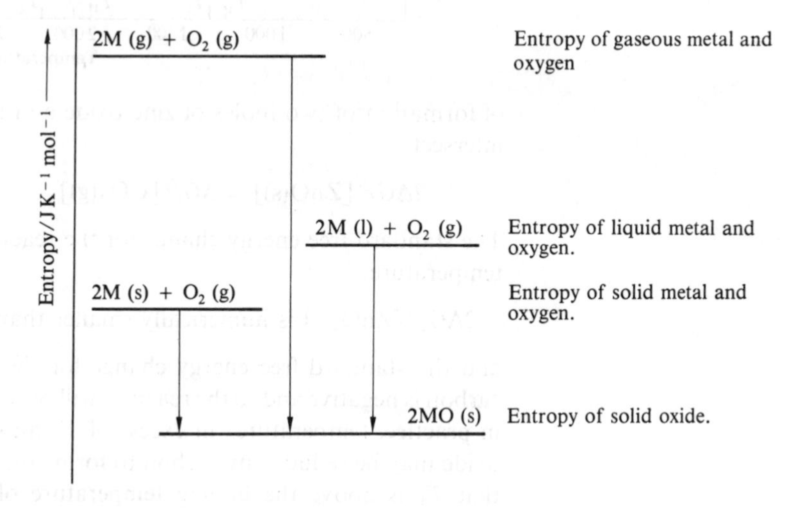

To summarise, for the conversion of metal to metal oxide, there are three constant entropy changes, one for each physical state of the metal. Therefore if we plot a graph of standard free energy of formation of a metal oxide against temperature we find that the gradient increases at points corresponding to the melting and boiling temperatures of the metal. This is shown for two metal oxides in Fig. 2.4, which is known as an *Ellingham diagram.* Notice that the graphs have positive slopes because $-\Delta S^\ominus$ is positive, and that the gradient increases as the metal melts or boils. Thus, the standard free energies of formation of metal oxides, unlike that of carbon dioxide, become decreasingly negative as the temperature rises.

Let us consider the reduction of zinc oxide,

i.e. $2ZnO + C = 2Zn + CO_2$

At temperature T_1 (Fig. 2.4) the graphs relating to standard free energies

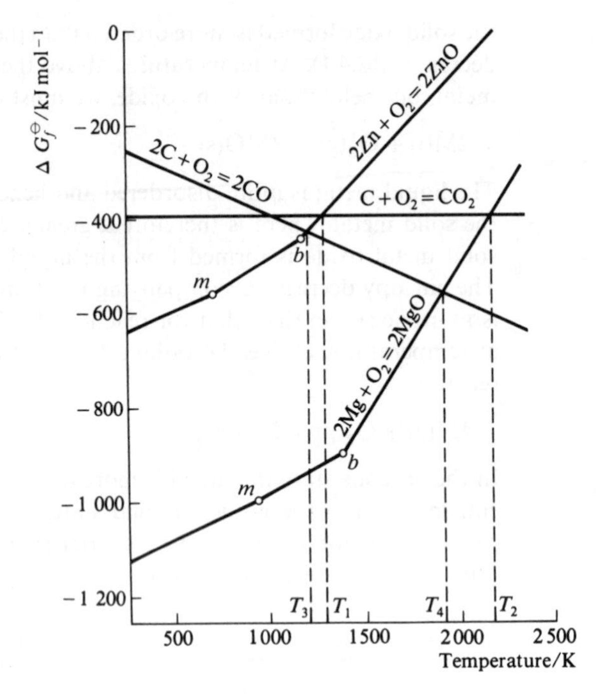

Fig. 2.4 Ellingham diagram for oxide formation. (m = melting temperature and b = boiling temperature of metal.)

of formation of two moles of zinc oxide and one mole of carbon dioxide intersect.

$$\therefore \quad 2\Delta G_f^{\ominus}[ZnO(s)] = \Delta G_f^{\ominus}[CO_2(g)]$$

The standard free energy change for the reaction is then zero. Above this temperature,

$$2\Delta G_f^{\ominus}[ZnO(s)] \text{ is numerically smaller than } \Delta G_f^{\ominus}[CO_2(g)]$$

and the standard free energy change for the reduction of zinc oxide by carbon is negative and so the reaction will proceed of its own accord. Hence, in practice, temperatures in excess of T_1 must be employed so that zinc oxide may be reduced by carbon to form zinc and carbon dioxide. Notice that T_1 is above the boiling temperature of zinc, so that the reaction occurring is

$$2ZnO(s) + C(s) = 2Zn(g) + CO_2(g)$$

Similarly,

$$2MgO(s) + C(s) = 2Mg(g) + CO_2(g)$$

Temperatures in excess of T_2 (Fig. 2.4) are required before the standard free energy change becomes negative and the reaction can proceed.

When carbon acts as a reducing agent it may be converted into the monoxide,

$$2MO + 2C = 2M + 2CO$$

Notice that for consistency we are again considering the transfer of one mole of oxygen, O_2, to the carbon from the metal oxide. The standard free energy change for this reaction is given by

$$\Delta G^{\ominus} = 2\Delta G_f^{\ominus}[CO(g)] - 2\Delta G_f^{\ominus}[MO(s)]$$

Since $\Delta G_f^{\ominus}[CO(g)]$ is -137 kJ mol^{-1}, $\Delta G^{\ominus}$ will be negative and the reduction will occur only if

$2\Delta G_f^{\ominus}[MO(s)]$ is numerically smaller than $2\Delta G_f^{\ominus}[CO(g)]$
$= -274$ kJ mol^{-1}

The standard entropy of formation of carbon monoxide ($+89.8$ J K^{-1} mol^{-1}) is greater than that of carbon dioxide ($+3.3$ J K^{-1} mol^{-1}). Therefore the line on the Ellingham diagram (Fig. 2.4) corresponding to the variation of the standard free energy of formation of carbon monoxide has a more pronounced negative slope than that for carbon dioxide. Consider the reduction of zinc oxide. At temperature T_3

$$2\Delta G_f^{\ominus}[ZnO(s)] = 2\Delta G_f^{\ominus}[CO(g)]$$

but above this temperature

$2\Delta G_f^{\ominus}[ZnO(s)]$ is numerically smaller than $2\Delta G_f^{\ominus}[CO(g)]$

and the reaction

$$2ZnO(s) + 2C(s) = 2Zn(g) + 2CO(g)$$

will proceed because $\Delta G^{\ominus}$ is negative. Thus at temperatures higher than T_3, but lower than T_1, carbon reduces zinc oxide to form zinc vapour and carbon monoxide. At temperatures in excess of T_1, however, both carbon monoxide and carbon dioxide, in addition to zinc vapour, will be formed.

Ellingham diagrams may also be constructed for the reduction of metal chlorides and metal sulphides.

2.5 Electrochemistry

2.5.1 ELECTRODE POTENTIALS

When a piece of metal is immersed in a solution of its own ions, the atoms on its surface tend to lose electrons and form hydrated cations which pass into the solution,

i.e. $M(s) = M^{n+}(aq) + ne^-$

The electrons remain on the metal, which thereby acquires a negative charge relative to the solution. The reverse process also occurs,

i.e. $M^{n+}(aq) + ne^- = M(s)$

in which case the metal acquires a positive charge with respect to the solution. After a time a state of dynamic equilibrium is established:

$$M(s) \rightleftharpoons M^{n+}(aq) + ne^-$$

Whether the piece of metal, often called an *electrode* in this context, acquires a positive or negative charge relative to the solution depends on the position of the equilibrium. If it lies to the right (ionisation more important) the metal is negatively charged, whereas if it lies to the left (deposition more important) the charge is positive. The presence of a charge on the metal means that a potential difference must exist between the electrode and the solution. This is called the *electrode potential* of the metal.

The magnitude of the electrode potential is related to the amount of charge on the metal. We shall now discuss the principal factors affecting this.

Concentration of ions in solution

For a given metal in a solution of its own ions the equilibrium is displaced to the left as the ionic concentration increases. Thus, if the metal carries a negative charge the electrode potential is decreased as some of the negative charge is neutralised by the deposition of ions. The reverse is true if the metal possesses a positive charge.

Tendency to form hydrated ions

For different metals placed in solutions of their ions at the same molar concentration, the position of the equilibrium depends on the tendency of the metal to form hydrated ions. The steps of hydrated ion formation are as follows:

$$M(s) = M(g) \qquad \text{atomisation}$$
$$M(g) = M^{n+}(g) + ne^- \qquad \text{ionisation}$$
$$M^{n+}(g) + aq = M^{n+}(aq) \qquad \text{hydration}$$

The overall enthalpy change during hydrated ion formation is thus given by

$$\Delta H_{\text{overall}} = \Delta H_{\text{atomisation}} + \Sigma\Delta H_{\text{ionisation}} + \Delta H_{\text{hydration}}$$

where $\Sigma\Delta H_{\text{ionisation}}$ is the sum of the first n ionisation enthalpies of the metal. For all metals the overall change is endothermic, but the smaller the change the greater is the tendency to form hydrated ions. Thus the alkali metals, which have low ionisation and atomisation enthalpies, form hydrated ions most readily (§4.7.2).

For metals that are negatively charged relative to the solution, the greater the tendency to form hydrated ions, the larger is the negative charge. This leads to high negative electrode potentials. For positively charged electrodes, the smaller the tendency to form hydrated ions the greater is the positive charge on the metal. This leads to high positive electrode potentials.

Temperature

The relationship between temperature and electrode potential is given by the Nernst equation (§2.5.4).

Measurement of electrode potentials

Because of their dependence on the tendency of metals to form hydrated ions, electrode potentials (together with their sign) provide a quantitative guide to the reactivities of metals in an aqueous environment. Unfortunately, we cannot measure the absolute electrode potential of a metal. If we attempt to measure the potential by means of a voltmeter connected between the electrode and the solution, we must of necessity introduce another metal into the solution. This metal will have its own electrode potential which will alter the voltage reading of the voltmeter.

In practice we choose an arbitrary second electrode against which we compare all other electrode potentials. The chosen standard is the *standard hydrogen electrode* (SHE) (Fig. 2.5). The electrode reaction occurring in the SHE is

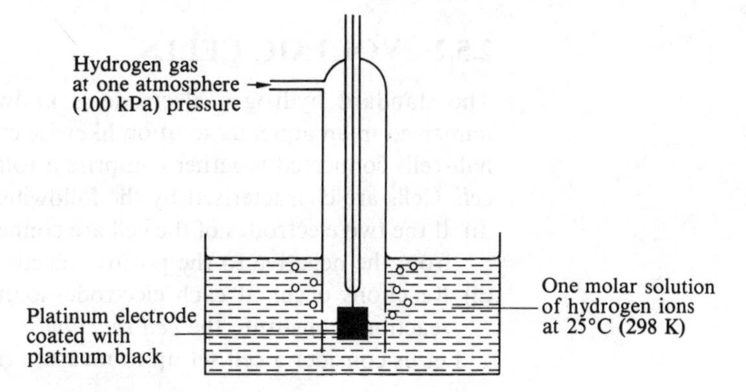

Fig. 2.5 A standard hydrogen electrode.

$$\frac{1}{2}H_2(g) \rightleftharpoons H^+(aq) + e^-$$

We arbitrarily assign a potential of 0.00 volts to the SHE. This does not mean that the *actual* potential is zero volts. By constructing a voltaic cell (details of which are discussed later) of the SHE and the electrode in its solution, we obtain the electrode potential of the metal *relative to hydrogen.*

For electrode potentials obtained in this manner to be used as a guide to the reactivity of metals, they must be measured under the same conditions of concentration and temperature. An electrode potential which relates to a metal immersed in a solution of its own ions of one molar concentration at a temperature of 25 °C (298 K) is called a *standard electrode potential*, symbolised by $E^\ominus$ (p. 43). A list of standard electrode potentials arranged in order of decreasingly negative values is known as the *electrochemical series* (Table 2.3). All the electrode processes shown in Table 2.3 are written, by convention, in the form of reduction from left to right.

Table 2.3 A selection from the electrochemical series, i.e. $E^\ominus$ values of important metals arranged in order of decreasing negative values

Electrode reaction	$E^\ominus$/V
$Li^+ + e^- = Li$	−3.04
$K^+ + e^- = K$	−2.92
$Ba^{2+} + 2e^- = Ba$	−2.90
$Ca^{2+} + 2e^- = Ca$	−2.87
$Na^+ + e^- = Na$	−2.71
$Mg^{2+} + 2e^- = Mg$	−2.38
$Al^{3+} + 3e^- = Al$	−1.66
$Zn^{2+} + 2e^- = Zn$	−0.76
$Fe^{2+} + 2e^- = Fe$	−0.44
$Sn^{2+} + 2e^- = Sn$	−0.14
$Cu^{2+} + 2e^- = Cu$	+0.34
$Ag^+ + e^- = Ag$	+0.80

When referring to a particular standard electrode potential, we often do so by writing $E^{\ominus}$ with the subscript 'oxidised form/reduced form', omitting (aq), e.g. $E^{\ominus}_{Li^+/Li}$. The addendum (aq) has also been omitted in Tables 2.3 and 2.4.

2.5.2 VOLTAIC CELLS

The standard hydrogen electrode is known as a *half-cell.* Any metal immersed in an aqueous solution likewise constitutes a half-cell. Any two half-cells connected together comprise a *voltaic cell*, often called simply a *cell.* Cells are characterised by the following features.

(i) If the two electrodes of the cell are connected by a wire, electrons flow from the negative to the positive electrode.

(ii) Reactions occur at each electrode–solution interface. The combined reactions constitute the *cell reaction.*

Let us consider a cell set up to measure the $E^{\ominus}$ value of zinc (Fig. 2.6).

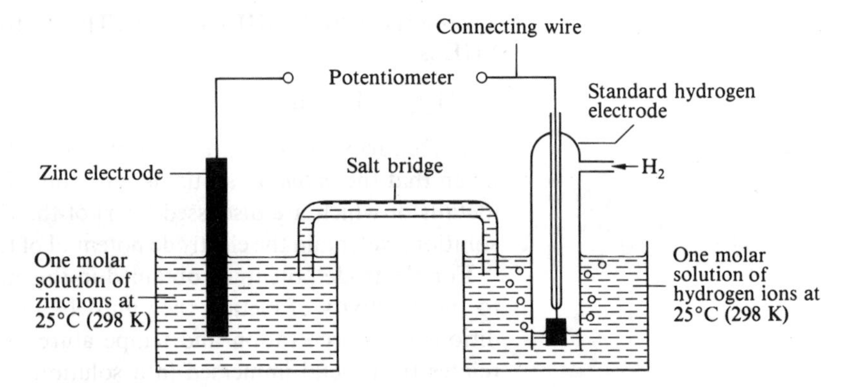

Fig. 2.6 Voltaic cell to measure the $E^{\ominus}$ value of zinc.

In order to obtain a complete circuit, the two solutions must be in electrical contact with each other. This can be achieved by placing a porous partition between the two solutions, to maintain electrical contact but prevent undue mixing. In accurate work a device called a *salt bridge* (Fig. 2.6) is used. This consists of a U-tube filled with a sol of potassium chloride or potassium nitrate in agar–agar. The device establishes an electrical contact between the solutions, but does not introduce spurious potentials at the points of contact.

The zinc electrode of the above cell is negatively charged, and *relative to this electrode* the SHE is positively charged. This is because zinc has a greater tendency than hydrogen to form hydrated ions in solution, i.e. the tendency for $Zn \rightarrow Zn^{2+}(aq) + 2e^-$ > the tendency for $\frac{1}{2}H_2 \rightarrow H^+(aq) + e^-$. Consequently, electrons flow in the external circuit from the point of negative potential (zinc) to that of positive (or lower negative) potential at the SHE.

The potential difference across the electrodes of a cell at zero current is its *electromotive force* (EMF), represented by E_{cell} or $E^{\ominus}_{cell}$ if standard conditions apply. For the above cell, $E^{\ominus}_{cell} = 0.76$ V. Since, by convention, $E^{\ominus}_{H^+/H_2} = 0.00$ V, $E^{\ominus}_{Zn^{2+}/Zn} = -0.76$ V. The magnitude of the cell EMF is thus an indication of the greater tendency for zinc to ionise than hydrogen.

If we repeat the experiment using a copper half-cell (i.e. copper immersed in a one molar solution of $Cu^{2+}(aq)$ ions) instead of zinc, we find that $E^{\ominus}_{cell}$ is 0.34 V and the copper electrode is positive. This arises because hydrogen has a greater tendency to ionise than copper; the value of $E^{\ominus}_{cell}$ again indicates the extent of the tendency.

We are now able to define the term 'standard electrode potential'. The standard electrode potential of a half-cell is defined as the cell EMF when an electrode immersed in a one molar solution of its ions at 25 °C (298 K) is coupled with a SHE. The sign of $E^{\ominus}$ is the polarity of the electrode in the cell.

Generally, the more negative the $E^{\ominus}$ value of a metal, the greater is its tendency to ionise and the higher is its reactivity. The electrochemical series thus gives a quantitative guide to the reactivity of metals when they are immersed in solutions of their own ions. For example, zinc is more reactive than copper because zinc ionises more readily than copper.

In the *Daniell cell*, which consists of a zinc half-cell and a copper half-cell, the zinc electrode is negative relative to the copper because zinc, as we have just seen, ionises more readily than copper. The conventional representation of such a cell is

$$Zn|Zn^{2+}(aq)||Cu^{2+}(aq)|Cu$$

A single vertical line represents the interface between substances in different phases. A double line represents the salt bridge connection between the two solutions. Also, by convention, the negative electrode is written on the left-hand side.

If the solutions are both one molar, the voltage between the zinc and copper electrodes, i.e. the cell EMF ($E^{\ominus}_{cell}$), is the difference between the two $E^{\ominus}$ values. We can usefully illustrate this on a 'potential' diagram.

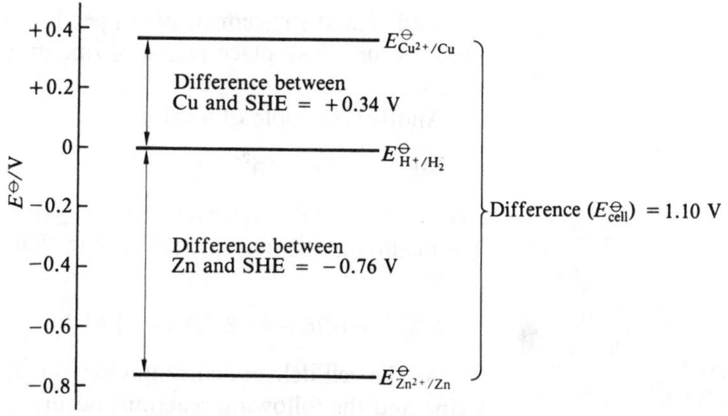

When calculating E_{cell} we again adopt a convention:

$$E_{cell} = E_{right} - E_{left}$$

where E_{right} is the electrode potential of the right-hand electrode (positive) and E_{left} is the electrode potential of the left-hand electrode (negative).

For a Daniell cell consisting of standard electrodes,

$$E^{\ominus}_{cell} = +0.34 - (-0.76) = +1.10 \text{ V}$$

If, in ignorance, we had written the cell the opposite way round, i.e. with the copper electrode on the left, then $E^{\ominus}_{cell} = -1.10$ V. The negative sign immediately indicates the mistake. E_{cell} or $E^{\ominus}_{cell}$ must always be positive.

If the two electrodes are connected by a wire, electrons flow from zinc to copper, i.e. from the point of higher negative potential. The number of electrons on the zinc is thus reduced and the equilibrium

$$Zn \rightleftharpoons Zn^{2+}(aq) + 2e^-$$

is displaced to the right and dissolution of the zinc occurs. At the copper electrode, because there is an increase in the number of electrons, the equilibrium

$$Cu \rightleftharpoons Cu^{2+}(aq) + 2e^-$$

is displaced to the left and copper is deposited. The overall cell reaction is obtained by addition:

at the left-hand electrode	$Zn = Zn^{2+}(aq) + 2e^-$
at the right-hand electrode	$2e^- + Cu^{2+}(aq) = Cu$
∴ cell reaction	$Zn + Cu^{2+}(aq) = Zn^{2+}(aq) + Cu$

The net effect is the displacement of copper from solution by zinc.

While the cell is delivering current, zinc is oxidised at the left-hand (negative) electrode and copper(II) ions are reduced at the right-hand (positive) electrode. This important relationship applies to all cells delivering current, i.e.

oxidation occurs at the negative electrode (left),
reduction occurs at the positive electrode (right).

(The two r's, i.e. *r*eduction on the *r*ight, may help in remembering this.)

Cells enable us to predict whether or not reactions occur of their own accord. The displacement of copper by zinc occurs in the Daniell cell; it also occurs if we place pieces of zinc in a solution containing copper(II) ions.

Another example of a cell is

$$Mg|Mg^{2+}(aq)||Zn^{2+}(aq)|Zn$$

Here we see that magnesium is the negative electrode because the $E^{\ominus}$ value of magnesium is more negative than that of zinc. Under standard conditions,

$$E^{\ominus}_{cell} = -0.76 - (-2.37) = +1.61 \text{ V}$$

As the cell delivers current, electrons flow from the magnesium to the zinc and the following reactions occur:

$$Mg \rightarrow Mg^{2+}(aq) + 2e^- \text{ and } Zn^{2+}(aq) + 2e^- \rightarrow Zn$$

$$\therefore \text{ cell reaction} \quad Mg + Zn^{2+}(aq) = Mg^{2+}(aq) + Zn$$

Magnesium therefore displaces zinc from solution in the cell. The same reaction also occurs if we place magnesium in a solution of a zinc salt.

The cells used for determining the $E^{\ominus}$ values of zinc and copper may be represented as

$$Zn|Zn^{2+}(aq)(1\ \text{M})||H^+(aq)(1\ \text{M})|H_2(1\ \text{atm})\ Pt$$

and $$Pt|H_2(1\ \text{atm})|H^+(aq)(1\ \text{M})||Cu^{2+}(aq)(1\ \text{M})|Cu$$

In this instance the concentrations of the various species have been included, but this is not obligatory.

The respective cell reactions are

$$Zn + 2H^+(aq) = Zn^{2+}(aq) + H_2$$

and $$Cu^{2+}(aq) + H_2 = Cu + 2H^+(aq)$$

2.5.3 REDOX ELECTRODES

As we have seen, reversible oxidation–reduction occurs with metal electrodes immersed in solutions of their ions. Reversible oxidation–reduction may also occur with non-metals in contact with their own ions,

e.g. $Cl_2 + 2e^- \rightleftharpoons 2Cl^-$

Such electrodes are often referred to as *redox electrodes*, although they do not differ in principle from the metal electrodes discussed above. Most non-metals are non-conductors of electricity and a piece of platinum is used to make electrical contact, as with hydrogen in the SHE. A chlorine electrode, for example, is similar to a SHE except that hydrogen is replaced by chlorine, and the aqueous solution contains chloride ions of concentration 1 mol dm^{-3}. The electrode is represented by $Pt|Cl_2(g)|Cl^-(aq)$.

Another type of redox electrode may involve two ionic species, one of which is the oxidised form of the other, in the same aqueous solution,

e.g. $Fe^{3+}(aq) + e^- \rightleftharpoons Fe^{2+}(aq)$

Let us consider a solution containing an oxidised and a reduced species, e.g. Fe^{3+} and Fe^{2+}, into which is placed an inert metal such as platinum. Two opposing tendencies exist.

(i) $Fe^{3+}(aq) + e^- \rightarrow Fe^{2+}(aq)$

The electron is removed from the platinum electrode, which therefore acquires a positive charge.

(ii) $Fe^{2+}(aq) \rightarrow Fe^{3+}(aq) + e^-$

The electron is transferred to the platinum, which therefore acquires a negative charge.

As with a metal electrode immersed in a solution, an equilibrium is rapidly established. If the tendency of the reduced species (e.g. Fe^{2+}) to become oxidised is stronger than that of the oxidised species (e.g. Fe^{3+}) to become reduced, then the platinum carries a negative charge. If the opposite applies the platinum is positively charged. The magnitude of the charge, and hence the potential difference between the platinum and the solution, depends on the magnitude of these two opposing tendencies. The greater the reducing power of one species relative to the oxidising power of the other, the higher is the negative charge on the platinum and the larger is the potential. The sign and magnitude of the charge on the platinum thus serves as a quantitative guide to the net oxidising or reducing power of the two species in solution. However, we cannot measure the potential difference directly. Instead, as for metal electrodes, we use a standard hydrogen electrode and define a *standard electrode potential* ($E^\ominus$), formerly 'standard redox potential'. Thus, the EMF of the cell

$$Pt|H_2(1\ atm)|H^+(aq)||Fe^{2+}(aq),\ Fe^{3+}(aq)|Pt$$

together with the sign of the charge on the platinum is called the standard electrode potential of the $Fe^{3+} + e^- \rightleftharpoons Fe^{2+}$ system, when the concentrations of the active species (Fe^{3+} and Fe^{2+}) are each one molar. We find that $E^\ominus_{cell} = 0.77$ V, and the platinum is positively charged,

$$\therefore \quad E^\ominus_{Fe^{3+}/Fe^{2+}} = +0.77\ V$$

Similarly, for the $Cr^{3+} + e^- \rightleftharpoons Cr^{2+}$ equilibrium

$$Pt|Cr^{2+}(aq),\ Cr^{3+}(aq)||H^+(aq)|H_2(1\ atm)|Pt \qquad E^\ominus_{cell} = 0.41\ V$$

$$\therefore \quad E^\ominus_{Cr^{3+}/Cr^{2+}} = -0.41\ V$$

Standard electrode potentials provide us with a sound basis for comparing the oxidising or reducing powers of different species. For example, the positive sign of the $E^\ominus$ value for the Fe^{3+}/Fe^{2+} electrode shows that hydrogen is a stronger reductant than the Fe^{2+}(aq) ion, while the negative sign of $E^\ominus$ for the Cr^{3+}/Cr^{2+} electrode (Table 2.4) indicates that the Cr^{2+}(aq) ion is a more powerful reductant than hydrogen. The Cr^{2+}(aq) ion must therefore be a stronger reducing agent than the Fe^{2+}(aq) ion. Some standard electrode potentials of various redox electrodes are given in Table 2.4. These should be considered alongside those in Table 2.3.

Table 2.4 The standard electrode potentials of some important redox electrodes

Electrode reaction	$E^\ominus$/V
$Cr^{3+} + e^- = Cr^{2+}$	−0.41
$Cu^{2+} + e^- = Cu^+$	+0.15
$Sn^{4+} + 2e^- = Sn^{2+}$	+0.15
$I_2 + 2e^- = 2I^-$	+0.54
$Fe^{3+} + e^- = Fe^{2+}$	+0.77
$Br_2 + 2e^- = 2Br^-$	+1.07
$Cr_2O_7^{2-} + 14H^+ + 6e^- = 2Cr^{3+} + 7H_2O$	+1.33
$Cl_2 + 2e^- = 2Cl^-$	+1.36
$MnO_4^- + 8H^+ + 5e^- = Mn^{2+} + 4H_2O$	+1.52
$MnO_4^- + 4H^+ + 3e^- = MnO_2 + 2H_2O$	+1.67
$S_2O_8^{2-} + 2e^- = 2SO_4^{2-}$	+2.01
$F_2 + 2e^- = 2F^-$	+2.87

(Arrows alongside the table, pointing down the list: "Increasing strength as an oxidant" and "Decreasing strength as a reductant".)

Standard electrode potentials can be used to predict the direction in which redox reactions occur. Suppose, for example, that we wish to know whether Cr^{2+} will reduce Fe^{3+}, or whether Fe^{2+} will reduce Cr^{3+}. The cell constructed from the two half-cells is

$$Pt|Cr^{2+}(aq),\ Cr^{3+}(aq)||Fe^{2+}(aq),\ Fe^{3+}(aq)|Pt$$

Because $E^\ominus_{Cr^{3+}/Cr^{2+}}$ is more negative than $E^\ominus_{Fe^{3+}/Fe^{2+}}$, the chromium half-cell forms the negative part of the cell.

$\therefore\ E^{\ominus}_{cell} = +0.77 - (-0.41) = 1.18$ V

The half-cell reactions are

$Cr^{2+}(aq) \rightarrow Cr^{3+}(aq) + e^-$ and $Fe^{3+}(aq) + e^- \rightarrow Fe^{2+}(aq)$

$\therefore$ cell reaction $Cr^{2+}(aq) + Fe^{3+}(aq) = Cr^{3+}(aq) + Fe^{2+}(aq)$

These principles may be applied to all redox reactions.

In Tables 2.3 and 2.4 all species written on the left are potential oxidants, while those on the right are potential reductants. An oxidant (on the left-hand side) will oxidise any species on the right-hand side for which the $E^{\ominus}$ value is less positive. Correspondingly, a reductant will reduce any oxidant on the left-hand side for which the $E^{\ominus}$ value is less negative. Thus, fluorine is the most powerful oxidant that there is, and fluoride ions cannot be oxidised by any other oxidant. Lithium (Table 2.3) is the strongest reductant, and lithium ions cannot be reduced by other reducing agents. Using these principles, we can also see that the permanganate ion, but not the dichromate ion, oxidises the chloride ion to chlorine.

The possibility of disproportionation (§10.1.5) occurring can also be examined by $E^{\ominus}$ values. For example,

$Cu^{2+}(aq) + e^- = Cu^+(aq)$ $\qquad E^{\ominus} = +0.15$ V

$Cu^+(aq) + e^- = Cu$ $\qquad E^{\ominus} = +0.52$ V

The cell consisting of these two half-cells is

$Pt|Cu^{2+}(aq), Cu^+(aq)||Cu^+(aq)|Cu$

$E^{\ominus}_{cell} = +0.52 - (+0.15) = 0.37$ V

The reactions occurring are

$Cu^+(aq) \rightarrow Cu^{2+}(aq) + e^-$ and $Cu^+(aq) + e^- \rightarrow Cu$

$\therefore$ cell reaction $Cu^+(aq) + Cu^+(aq) = Cu^{2+}(aq) + Cu$

i.e. $Cu^+(aq)$ ions disproportionate in solution.

The redox reactions that occur in voltaic cells also occur if the species concerned are mixed in a beaker. Standard redox potentials are therefore useful in predicting the possibility of one species oxidising or reducing another. As with thermodynamics, we can predict the possibility of reactions taking place but not the rate of reaction.

2.5.4 THE NERNST EQUATION

The variation of an electrode potential with temperature and ionic concentration is given by the *Nernst equation*,

$$E = E^{\ominus} + \frac{RT}{nF} \ln \frac{[\text{oxidised species}]}{[\text{reduced species}]}$$

where F is the Faraday constant and n is the number of electrons transferred in the redox reaction.

At 25 °C (298 K), and after converting ln to lg, the Nernst equation becomes

$$E = E^{\ominus} + \frac{0.059}{n} \lg \frac{[\text{oxidised species}]}{[\text{reduced species}]}$$

If the reduced species is a metal, then as in heterogeneous equilibria (§8.5.3) [metal] is equal to one and the Nernst equation simplifies to

$$E = E^{\ominus} + \frac{0.059}{n} \lg [M^{n+}]$$

As the ionic concentration i.e. $[M^{n+}]$ increases, the electrode potential becomes decreasingly negative for metals with negative $E^{\ominus}$ values, or increasingly positive for those with positive $E^{\ominus}$ values. Concentration changes will therefore affect the EMF of a cell. In the Daniell cell, for example, E_{cell} increases as $[Zn^{2+}]$ decreases and $[Cu^{2+}]$ increases.

The Nernst equation can be used to calculate the equilibrium constant for a cell reaction. The same equilibrium constant applies to the reaction when it occurs in a beaker. In this way we can obtain the equilibrium constant for any redox reaction and thus gain an insight into the extent to which it occurs. Let us again consider the Daniell cell reaction,

i.e. $Zn + Cu^{2+} \rightleftharpoons Zn^{2+} + Cu$

for which $K_c = \dfrac{[Zn^{2+}]}{[Cu^{2+}]}$

By applying the Nernst equation,

$$E_{Zn^{2+}/Zn} = E^{\ominus}_{Zn^{2+}/Zn} + \frac{0.059}{2} \lg [Zn^{2+}]$$

and $$E_{Cu^{2+}/Cu} = E^{\ominus}_{Cu^{2+}/Cu} + \frac{0.059}{2} \lg [Cu^{2+}]$$

At equilibrium $E_{cell} = 0$, i.e. the electrode potentials of the zinc and copper electrodes are the same. To find the equilibrium constant we must calculate $[Zn^{2+}]$ and $[Cu^{2+}]$ required to satisfy this condition. Now,

$$E_{cell} = E_{Cu^{2+}/Cu} - E_{Zn^{2+}/Zn} = 0$$

$$\therefore \quad 0 = [E^{\ominus}_{Cu^{2+}/Cu} - E^{\ominus}_{Zn^{2+}/Zn}] + \frac{0.059}{2} \lg \frac{[Cu^{2+}]}{[Zn^{2+}]}$$

$$\therefore \quad 0 = 1.10 + \frac{0.059}{2} \lg K_c$$

$$\therefore \quad K_c = 10^{37.3}$$

The very high value of K_c indicates that the reaction goes virtually to completion (§8.2.2).

2.5.5 APPLICATION OF VOLTAIC CELLS

Measurement of pH

Consider a cell comprising two hydrogen electrodes operating at the same gas pressure. The cell will develop an EMF if the concentration of hydrogen ions in one half-cell differs from that in the other. If one electrode is standard, i.e. if $[H^+(aq)]$ is one molar, then the cell EMF is directly related to the hydrogen ion concentration in the other half-cell by the Nernst equation. Thus, for the cell

$$Pt|H_2(1\ atm)|H^+(aq)(1\ M)||H^+(aq)|H_2(1\ atm)|Pt$$

$$E_{cell} = 0.059\ lg \frac{[H^+(aq)]}{[H^+(aq)(1\ M)]} = 0.059\ lg\ [H^+(aq)]$$

$$\therefore \quad E_{cell} \propto lg\ [H^+(aq)] \propto -pH \qquad (§9.2.4)$$

Hydrogen electrodes are impracticable for everyday use because of the need to supply hydrogen and the precautions necessary to prevent contamination of the platinum surface. Much more convenient for pH measurements is a cell comprising a standard calomel electrode and a glass electrode (Fig. 2.7). The calomel electrode is a *secondary reference*

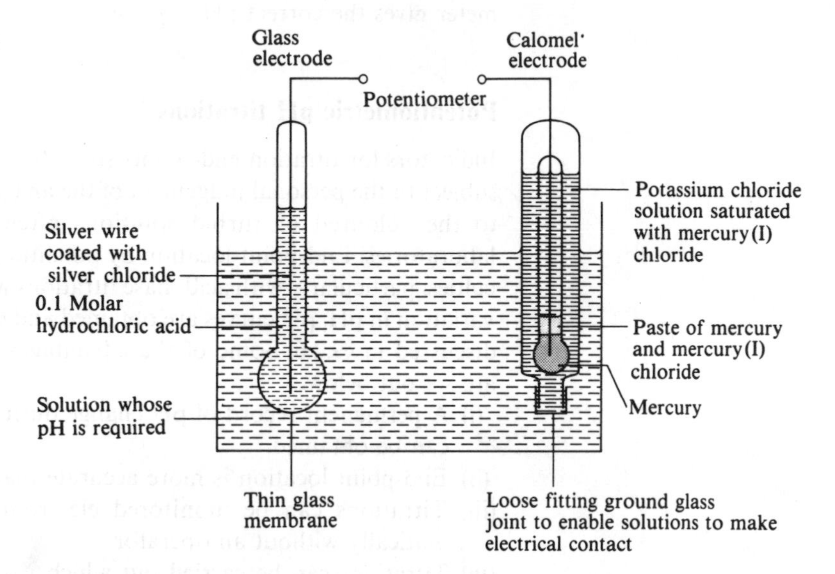

Fig. 2.7 Cell for determining the pH of a solution.

standard, i.e. its potential is accurately known at various temperatures and concentrations of potassium chloride. The potential of a calomel electrode remains constant in a range of different solutions, even when a small current is flowing through it. The reaction which occurs in the electrode is

$$Hg_2Cl_2 + 2e^- \rightleftharpoons 2Hg + 2Cl^-$$

The potential of the glass electrode is proportional to the hydrogen ion concentration of the solution into which it is placed. The mechanism of the glass electrode is complex because hydrogen ions cannot pass through the specially constructed thin glass membrane. The hydrogen ions from the solution exchange places with sodium ions present in the very thin layer of hydrated silicate which is always present when a glass surface is placed in an aqueous solution. As a result, a pH-dependent potential develops at the glass membrane, and sodium ions within the silicate lattice of the glass itself move from site to site as the means by which a tiny current is carried when the pH measurement is made. The reaction which occurs in the electrode is

$$AgCl + e^- \rightleftharpoons Ag + Cl^-$$

For pH measurement, a glass electrode and a calomel electrode are immersed in the test solution to form a cell, whose EMF is related to the hydrogen ion concentration. The EMF is measured with a *pH meter*, which is a potentiometer having a scale calibrated to read pH directly. Because of variations between one glass electrode and another, and because the response of glass electrodes changes with time, it is necessary to standardise the electrode before a pH measurement is made. For this purpose the two electrodes are immersed in a buffer solution of known pH (§9.3.3) and the pH meter scale is adjusted to the corresponding value. When the electrodes are subsequently transferred to a test solution the meter gives the correct pH.

Potentiometric pH titrations

Indicators for titration end-points suffer from the disadvantages of being subject to the personal judgement of the analyst and of being inapplicable to the coloured or turbid solutions often encountered in industrial laboratories. End-point location by potentiometers overcomes these difficulties, particularly with acid–base titrations which may be followed by the variation in pH. pH curves are obtained and end-points are determined as discussed in §9.2.6. Some of the advantages of potentiometric titrations are set out below.

(i) A permanent record of pH change during the course of the titration can be obtained.
(ii) End-point location is more accurate than with indicators.
(iii) Titrations can be monitored electronically and conducted automatically without an operator.
(iv) Titrations can be carried out which would otherwise be impossible due to lack of a suitable indicator.

Potentiometric titrations are not limited to acid–base reactions. Other ions may be estimated provided that an electrode responsive to those ions is available.

Determination of free energy changes

A voltaic cell provides one of the easiest means of measuring the free energy change (§2.4) accompanying a chemical reaction, provided that the reaction can be made to take place within a cell. The free energy change (ΔG) of the cell reaction is equal to the maximum electrical work which is theoretically obtainable. This in turn is given by

charge flowing × potential difference

The charge flowing in the external circuit is equal to the quantity of electricity carried, i.e. $n \times F$ coulombs, where F = the Faraday constant and n = the number of moles of electrons flowing in the circuit. (In the Daniell cell, for example, if one mole of zinc dissolves and one mole of copper is deposited, $n = 2$.) The sign of ΔG is negative because a cell operates with a decrease in the free energy of the cell materials.

$$\therefore \quad \Delta G = -nFE_{cell} \quad \textit{Units}\ (\text{C mol}^{-1})(\text{V}) = (\text{A s mol}^{-1})(\text{J A}^{-1}\text{ s}^{-1}) = \text{J mol}^{-1}$$

Under standard conditions,

$$\Delta G^{\ominus} = -nFE^{\ominus}_{cell}$$

Thus, the free energy change of the reaction occurring in the Daniell cell when 1 mol of zinc dissolves is

$$\Delta G^{\ominus} = -(2 \times 96\,487 \times 1.10) = -212\,271.4 \text{ J mol}^{-1}$$
$$= -212.3 \text{ kJ mol}^{-1}$$

EXAMINATION QUESTIONS ON CHAPTER 2

1 (a) What is meant by the terms (i) standard enthalpy change of formation, (ii) standard enthalpy change of combustion?

In what way is the enthalpy change of formation connected with the stability of a substance?

(b) Describe briefly how you would measure the enthalpy change of neutralisation of hydrochloric acid with sodium hydroxide solution. (Both aqueous solutions are of concentration 1 mol dm^{-3}.)

(c) If the enthalpy change of combustion of hydrogen, ethene and ethane are 286 kJ mol^{-1}, 1393 kJ mol^{-1} and 1561 kJ mol^{-1} respectively, calculate the enthalpy change of hydrogenation of ethene to ethane. (AEB)

2 State the first law of thermodynamics in terms of *heat* and *work*, defining these quantities carefully.

Calculate the work done on the gas when 3.00 mol of an ideal gas at 25 °C is compressed from 1.00 to 3.00 atm. Give an estimate (not a calculation) of what you think the difference from this value might be if the gas were (a) air, (b) carbon dioxide.

Use the bond enthalpies in Table 24 of your data book to calculate the standard enthalpy change for the reaction

$$C_2H_4(g) + HBr(g) \rightarrow C_2H_5Br(g)$$

Given that the molar enthalpy of vaporisation of bromoethane is 27.0 kJ mol^{-1}, compare your calculated value of $\Delta H^{\ominus}$ (298 K) with the one calculated from the standard enthalpies of formation and comment on your answer. (OC)

3 Explain the terms *enthalpy change* and *entropy change*, illustrating your answer with suitable examples.

Show how, under constant conditions of temperature, enthalpy change and entropy change influence the direction of chemical change.

For liquid methanol in equilibrium with methanol vapour at 338 K, the boiling point of methanol,

$$CH_3OH(l) \rightleftharpoons CH_3OH(g) \qquad \Delta H = +35.3 \text{ kJ mol}^{-1}.$$

Calculate the entropy change when methanol is evaporated. (L)

4 The *standard electrode potential*, $E^{\ominus}$, for the change between two oxidation states of the d-block element rhenium is +0.3 V.

$$ReO_4^-(aq) + 4H^+(aq) + 3e^- \rightleftharpoons ReO_2(s) + 2H_2O(l)$$

(a) Explain precisely what is meant by 'electrode potential' and 'standard electrode potential', using this change as a specific example.

(b) Describe with necessary experimental detail how this standard electrode potential could be determined.

(c) The corresponding standard electrode potential for Mn(VII)/Mn(II) in acidic solutions is +1.52 V. *Discuss* whether manganate(VII) ions, MnO_4^- would oxidise rhenium dioxide, ReO_2. (L)

5 (a) State the first law of thermodynamics.

(b) If ΔH and ΔU are the heat changes in a given process at constant pressure and constant volume respectively, write an equation which relates ΔH and ΔU.

(c) Which of the two quantities ΔH or ΔU is the more useful when studying chemical processes? Give a reason for your answer.

(d) The following are the enthalpies of hydrogenation of ethene and of benzene to ethane and to cyclohexane respectively:

$$C_2H_4(g) + H_2(g) \rightarrow C_2H_6(g) \qquad \Delta H = -132 \text{ kJ mol}^{-1}$$
$$C_6H_6(g) + 3H_2(g) \rightarrow C_6H_{12}(g) \qquad \Delta H = -208 \text{ kJ mol}^{-1}$$

(i) Use the data above to deduce the relative stabilities of ethene and benzene.

(ii) Explain your answer to (d) (i) in terms of the electronic structures of ethene and benzene (Chapter 4). (JMB)

6 (a) Define (i) *enthalpy change of neutralisation*, (ii) *enthalpy change of formation*, (iii) *endothermic compound*.

(b) State Hess's Law of constant heat summation.

(c) Comment on the statement that the enthalpy changes of neutralisation of many acids are approximately the same. Account for any exceptions.

(d) A natural gas may be assumed to be a mixture of methane and ethane only. On complete combustion of 10 litres (measured at s.t.p.) of this gas, the evolution of heat was 474.6 kJ.

Assuming ΔH combustion $[CH_4(g)] = -894 \text{ kJ mol}^{-1}$

ΔH combustion $[C_2H_6(g)] = -1560 \text{ kJ mol}^{-1}$

calculate the percentage by volume of each gas in the mixture. (AEB)

7 This question is about the equation $E = E^{\ominus} + 2.3\frac{RT}{nF}\log_{10}[X^{n+}]$,
which relates the electrode potential E, the standard electrode potential $E^{\ominus}$, and the concentration, $[X^{n+}]$, for a metal/metal ion or similar electrode. R, T and F are respectively the gas constant, the temperature and the Faraday constant.

You are given the following special materials and equipment, together with the necessary measuring instruments, etc:

source of hydrogen gas
copper and blacked platinum electrodes
saturated calomel electrode
solutions of HCl, NaOH and $Cu(NO_3)_2$ of convenient, known concentration
buffer solutions covering the range pH 2 to pH 11

Describe how you would use these materials to establish:

(a) the relationship between E and the concentration of X^{n+} when X is hydrogen, and the numerical value of $2.3\frac{RT}{nF}$,

(b) the way in which the relationship depends on $n+$, the charge on the ion. (L)

8 State Hess's Law of constant heat summation.

Define

(a) *enthalpy of formation* (*heat of formation*),

(b) *heat of neutralisation.*

Describe how you would determine the heat of neutralisation of nitric acid by sodium hydroxide solution.

Explain why approximately the same numerical result would be obtained by using hydrochloric acid, but not hydrofluoric acid, in place of nitric acid.

Using the following data collected at 25 °C and standard atmospheric pressure, in which the negative sign indicates heat evolved, calculate the enthalpy of formation of rubidium sulphate, $Rb_2SO_4(s)$

		ΔH (298 K)/kJ mol^{-1}
(i) $H^+(aq) + OH^-(aq)$	$= H_2O(l)$	-57.3
(ii) $RbOH(s)$	$= Rb^+(aq) + OH^-(aq)$	-62.8
(iii) $Rb_2SO_4(s)$	$= 2Rb^+(aq) + SO_4^{2-}(aq)$	$+24.3$
(iv) $Rb(s) + \frac{1}{2}O_2(g) + \frac{1}{2}H_2(g)$	$= RbOH(s)$	-414.0
(v) $H_2(g) + S(s) + 2O_2(g)$	$= 2H^+(aq) + SO_4^{2-}(aq)$	-907.5
(vi) $H_2(g) + \frac{1}{2}O_2(g)$	$= H_2O(l)$	-285.0

Assume that $H_2SO_4(aq)$ is fully ionised in solution. (SUJB)

9 (a) (i) Define enthalpy of formation ΔH_f of a compound.

(ii) What extra conditions must be imposed to specify the standard enthalpy of formation $\Delta H_f^{\ominus}$ of a compound?

(b) When ethanol burns in oxygen, carbon dioxide and water are formed.

(i) Write the equation which describes this reaction.

(ii) Using the data

$\Delta H_f^{\ominus}$ for ethanol(l) $= -277.0$ kJ mol^{-1}
$\Delta H_f^{\ominus}$ for carbon dioxide(g) $= -393.7$ kJ mol^{-1}

$\Delta H_f^\ominus$ for water(l) $= -285.9$ kJ mol^{-1}
calculate the value of $\Delta H_f^\ominus$ for the combustion of ethanol. (JMB)

10 (a) Describe, with the aid of a labelled diagram, an experiment to measure the standard electrode potential of silver and write an equation representing the cell reaction.

(b) Construct a cycle of the Born–Haber type for the formation of silver ions in aqueous solution from solid silver. Name the enthalpy change in each step and indicate its sign.

(c) By reference to the following data, discuss possible methods of preparation of fluorine and chlorine.

$$\tfrac{1}{2}F_2(g) + e^- \rightarrow F^-(aq) \qquad E^\ominus = +2.87\ V$$
$$\tfrac{1}{2}Cl_2(g) + e^- \rightarrow Cl^-(aq) \qquad E^\ominus = +1.36\ V$$
$$MnO_4^-(aq) + 8H^+(aq) + 5e^- \rightarrow Mn^{2+}(aq) + 4H_2O(l) \qquad E^\ominus = +1.51\ V$$
$$MnO_2(s) + 4H^+(aq) + 2e^- \rightarrow Mn^{2+}(aq) + 2H_2O(l) \qquad E^\ominus = +1.23\ V$$

(JMB)

11 (a) What is meant by the following terms: (i) enthalpy of reaction, (ii) enthalpy of formation?

(b) Use the following data to calculate the enthalpy of formation of ethane:

$$C(s) + O_2(g) = CO_2(g) \qquad \Delta H = -394.6\ kJ\ mol^{-1}$$
$$H_2(g) + \tfrac{1}{2}O_2(g) = H_2O(l) \qquad \Delta H = -285.9\ kJ\ mol^{-1}$$
$$C_2H_6(g) + 3\tfrac{1}{2}O_2(g) = 2CO_2(g) + 3H_2O(l) \qquad \Delta H = -1\,558.3\ kJ\ mol^{-1}$$

State any laws which you use in your calculations.

(c) Why does the value of the heat of reaction for a gaseous reaction depend upon whether the heat change is measured at constant pressure or at constant volume?

(d) Under given conditions, the enthalpies of combustion of diamond and of graphite are -394.9 kJ mol^{-1} and -393.0 kJ mol^{-1} respectively. To what do you attribute the difference between these two values? (ΔH is negative for exothermic reactions.) (JMB)

12 Explain what is meant by 'enthalpy change of a reaction' and 'entropy change of a reaction'.

Discuss, with examples, how these two factors, separately and together, enable the direction of a chemical change to be predicted.

What other factor or factors determine whether or not a given change will actually take place? (L)

13 (a) Define *precisely*

(i) standard enthalpy of formation (heat of formation),

(ii) enthalpy of combustion (heat of combustion).

(b) The enthalpy of formation of the carbonyl of manganese, $Mn_2(CO)_{10}$, is determined by first burning the metal carbonyl in oxygen and then dissolving the products in nitric acid:

$$Mn_2(CO)_{10} + 4HNO_3 + 6O_2 \longrightarrow 2Mn(NO_3)_2 + 10CO_2 + 2H_2O$$

Discuss the principles upon which the method depends, and outline the steps of the calculation.

(c) Using the underlying data, collected at 25 °C and standard atmospheric pressure, in which the negative sign indicates heat evolved, calculate the heat of reaction (enthalpy of reaction) under these conditions for the hydrogenation,

$C_6H_6(g) + 3H_2(g) \longrightarrow C_6H_{12}(g)$

Compare the answer to that obtained by multiplying by three the heat of hydrogenation,

$C_6H_{10}(g) + H_2(g) \longrightarrow C_6H_{12}(g)$

and comment on the difference.

Enthalpies of formation (heats of formation)

		ΔH (298 K)/kJ mol^{-1}
benzene	C_6H_6	+ 82.9
cyclohexene	C_6H_{10}	− 7.1
cyclohexane	C_6H_{12}	− 123.1

(d) Berthollet in his *Essai de Mécanique chimique* (1878) stated: 'Every chemical change accomplished without the intervention of an external energy tends towards the production of the body or system of bodies *that sets free the most heat*.'

This implies that endothermic reactions cannot occur spontaneously, and it ignores reversibility. Explain briefly the present view of these matters.

(SUJB)

14 (a) (i) Define the terms *enthalpy change of reaction*, and *enthalpy change of neutralisation*.

(ii) Explain why the quantity of heat evolved when one mole of hydrochloric acid, HCl, is neutralised by one mole of sodium hydroxide, NaOH, is the same as the quantity of heat evolved when one mole of nitric acid, HNO_3, is used instead of the hydrochloric acid.

(b) Calculate the enthalpy change of the following reaction at 298 K:

$CH_4(g) + 2O_2(g) \longrightarrow CO_2(g) + 2H_2O(l)$

given the following enthalpy changes of formation, $\Delta H_f^\ominus$, at 298 K:

Compound	$\Delta H_f^\ominus$ in kJ mol^{-1}
$CH_4(g)$	−76
$CO_2(g)$	−394
$H_2O(l)$	−286

(c) The enthalpy changes of combustion, $\Delta H_c^\ominus$, of some alkanes at 298 K are:

Alkane	$\Delta H_c^\ominus$, kJ mol^{-1}
methane CH_4	−890
ethane C_2H_6	−1 560
propane C_3H_8	−2 220
butane C_4H_{10}	−2 880
pentane C_5H_{12}	−3 510

Using graph paper, plot these values against the number of carbon atoms in the molecule of the alkane.

(i) Comment on the shape of the graph.

(ii) Estimate the enthalpy change of combustion of hexane, C_6H_{14}, at 298 K.

(iii) The graph does not pass through the origin. Determine the value of $\Delta H_c^\ominus$ when the number of carbon atoms is zero. What does this value represent?

(AEB)

3 Atomic structure and properties

3.1 Atomic structure

John Dalton (1805) imagined an atom to be a minute particle incapable of subdivision. We now know that atoms can be split, and that they consist of sub-atomic particles known as *fundamental particles.*

3.1.1 FUNDAMENTAL PARTICLES

Atoms are made up of three fundamental particles: the *proton*, the *neutron* and the *electron* (Table 3.1).

Each particle has an exceedingly small absolute mass and charge, and it is common practice to use *relative* masses and charges as shown in the bottom two lines of Table 3.1.

Table 3.1 Properties of the fundamental particles

	Proton	**Neutron**	**Electron**
symbol	p	n	e^- (or β^-)†
mass	1.67252×10^{-27} kg	1.67482×10^{-27} kg	9.1091×10^{-31} kg
charge	1.60210×10^{-19} C	0	1.60210×10^{-19} C
mass relative to electron	1836	1839	1
charge relative to proton	+1	0	−1

† β^- is used in connection with radioactive decay (§3.3.1).

In considering detailed atomic structure, account must also be taken of other particles such as the *positron* and *anti-proton.* (The positron is equal in mass to an electron but carries a positive charge, while the anti-proton has the same mass as a proton but is negatively charged.) Also, it is now believed that protons and neutrons are assemblies of much smaller particles known as *quarks.* For the most part, however, these particles are of little importance chemically, and we shall deal with atomic structure in terms only of the particles shown in Table 3.1. First we shall consider the evidence that exists for the three fundamental particles, and then discuss how the presently accepted structure of the atom was perceived.

Electrons

After the work of Michael Faraday on electrolysis it was realised that electricity consisted of small fixed amounts, and in 1891 Johnstone Stoney coined the term *electron* to recognise this fact. The smallest amount or unit of electricity, to which the name 'electron' was given, was that required to liberate one atom of a monovalent element (e.g. H, Ag or Cl) at an electrode during electrolysis. The Faraday, $9.648\,70 \times 10^4$ C (C = coulomb) is required to deposit one mole of such an element, therefore one atom requires:

$$\frac{9.648\,70 \times 10^4}{6.022\,52 \times 10^{23}} = 1.602\,10 \times 10^{-19}\ \text{C}$$

The 'electron' of Stoney was thus $1.602\,10 \times 10^{-19}$ C of electricity.

Conduction of electricity in gases

In 1897 Sir J. J. Thomson carried out a series of experiments on the conduction of electricity in gases. At ordinary pressures gases are good insulators, but at low pressures, about 1 Pa, they are quite good conductors. A suitable apparatus, called a *cathode ray tube* or a *discharge tube*, for studying the electrical conductivity of gases at these low pressures is depicted in Fig. 3.1.

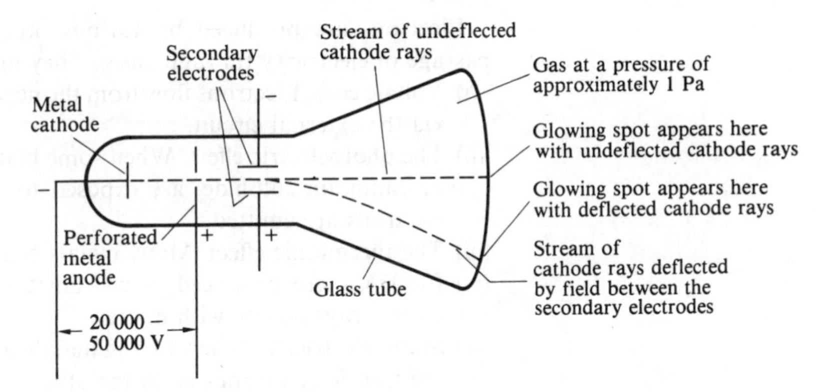

Fig. 3.1 Cathode ray tube.

When a suitably high voltage is applied across the cathode and anode the tube becomes filled with radiation called *cathode rays*. The radiation is usually invisible inside the tube, but on striking the glass of the tube behind the anode it produces a glowing spot. Cathode rays were thoroughly investigated by Thomson, with the following results.

(i) The radiation travels from the cathode to the anode, and if the latter is perforated the radiation travels straight through and impinges on the tube behind the anode.

(ii) If a metallic obstacle is placed in the radiation behind the anode a shadow is cast in the glowing spot at the end of the tube. The size of the shadow is the same as that of the metal obstacle, which indicates that the radiation is emitted at right angles to the cathode and travels in straight lines towards the anode.

(iii) A propeller or paddle wheel mounted in the tube rotates when struck by cathode rays, which indicates that the rays are particulate in nature and capable of exerting a small mechanical pressure.

(iv) If the cathode rays are passed through a secondary electrical field they are deflected away from the negative electrode and towards the positive electrode. This is observed by the movement of the glowing spot at the end of the tube, and indicates that the rays carry a negative charge.

(v) The rays are deflected by a magnetic field in a manner which indicates that they carry a negative charge.
(vi) Cathode rays are able to penetrate thin metal foil through which even the smallest atoms (e.g. helium) cannot pass.

These properties show that cathode rays consist of streams of negatively charged particles which are much smaller than atoms.

The same particles were found to be produced regardless of the gas in the tube and the nature of the electrodes. Their mass and charge were measured by several workers, including Thomson, Townsend and Millikan. The accepted values are: mass, $9.109\,1 \times 10^{-31}$ kg; charge, $1.602\,1 \times 10^{-19}$ C. Because the charge on the particles is identical with the unit of electricity proposed by Stoney, it was concluded that the particles were in fact identical with Stoney's 'electrons'. Cathode rays are now regarded as streams of electrons that travel from the cathode towards the anode with a velocity of the order of $\frac{1}{10}$ to $\frac{1}{3}$ that of light, depending on the voltage across the electrodes. The origin of cathode rays is discussed below.

Electrons are produced by various other processes, apart from the passage of electricity through gases. They include the following.

(i) Voltaic cells. Electrons flow from the negative pole to the positive pole via the external circuit.
(ii) The photoelectric effect. When some materials (e.g. caesium, selenium or cadmium sulphide) are exposed to light of suitable wavelength, electrons are emitted.
(iii) The thermionic effect. Many metals emit electrons when heated.
(iv) Electrons are produced when radiation such as X-rays, γ-rays or cosmic rays collide with matter.
(v) Static electricity. When two materials are rubbed together electrons are transferred from one to the other.

The production of electrons from such a wide variety of sources leads naturally to the conclusion that electrons are a basic constituent of all atoms.

Protons

An atom carries no overall charge, yet contains electrons. This immediately leads to the conclusion that atoms must also contain an equal amount of positive electricity.

Positively charged particles occur in a cathode ray tube, in addition to cathode rays, as Goldstein discovered in 1886. The existence of such particles can be demonstrated by means of a gas discharge tube in which the polarity of the electrodes is the reverse of that of the cathode ray tube (Fig. 3.2).

Positive rays originate from the gas and travel towards the cathode. They are very energetic, and as they traverse the rarefied gas they excite and ionise the gas and appear as a fine glowing stream. The rays also produce a dark spot on a photographic plate placed at the end of the tube, which provides a suitable means for their detection. The rays are deflected by electrical and magnetic fields in a manner which indicates

Fig. 3.2 Discharge tube for the study of positive rays. (No voltage across the secondary electrodes.)

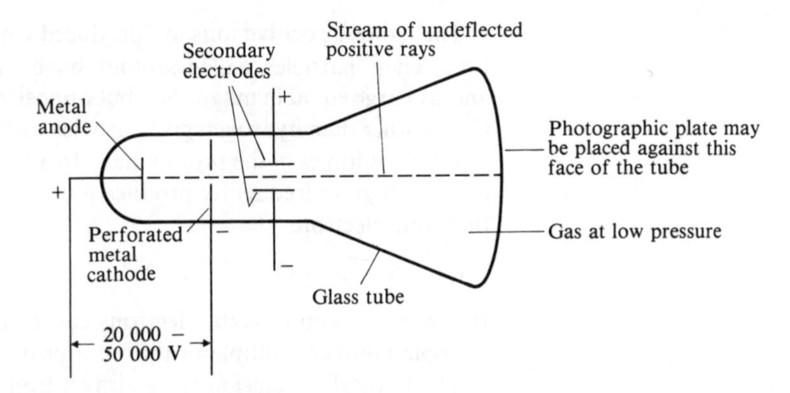

that they are positively charged. Further experiments prove that the rays are particulate in nature.

Deflection experiments can be used to measure the mass of the particles of the rays. The mass is not constant, as it is with cathode rays, but depends on the nature of the gas in the tube. For example, with hydrogen in the tube the mass is almost identical to that of a hydrogen atom. If the gas is neon *two* streams of particles are observed on deflection, containing, respectively, particles about twenty and twenty-two times heavier than a hydrogen atom.

The positively charged particles result from collisions between energetic cathode rays and gas atoms or molecules. With a diatomic molecule, e.g. hydrogen, such collisions split the molecules into atoms which are then ionised and accelerated towards the negatively charged cathode:

$$H_2 \longrightarrow 2H \longrightarrow 2H^+ + 2e^-$$

If the cathode is perforated some of the ions pass straight through and may be observed as described above.

Positive rays in monoatomic gases such as neon are streams of ions derived from gas atoms by collision with electrons travelling at a high velocity. Neon consists of two principal isotopes (§1.1) with abundances of: ^{20}Ne 90.9%, ^{22}Ne 8.8%. Ions are produced from both isotopes and in an electrical or magnetic field ^{22}Ne ions, because of their greater mass, are deflected to a lesser extent than ^{20}Ne ions, with the result that two beams of positive rays or ions are produced (Fig. 3.3).

Fig. 3.3 Positive rays derived from ^{20}Ne and ^{22}Ne on deflection in an electrical field.

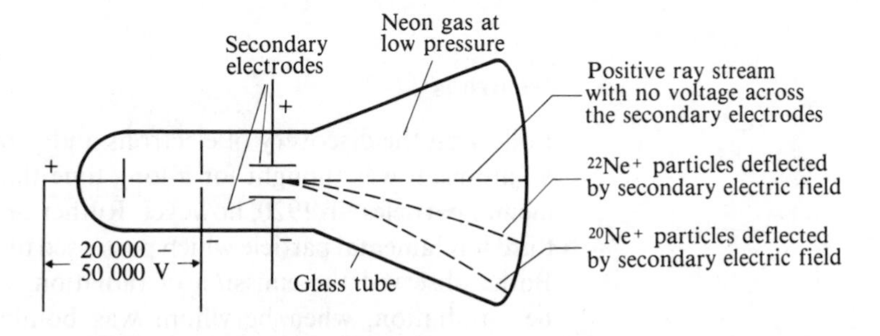

A third isotope of neon, ^{21}Ne, also exists, but because of its very low abundance (0.26%) it is usually difficult to observe in the above experiment.

The lightest positive ions are produced when hydrogen is the gas in the tube. These particles, called *protons*, have a mass of $1.672\,52 \times 10^{-27}$ kg, and a charge equal in magnitude but opposite in sign to that of an electron. Gases other than hydrogen give particles with masses that are nearly whole number multiples of the proton mass. In addition to forming monopositive ions, such gases frequently produce multiply charged ions by losing more than one electron,

e.g. $O \xrightarrow{-e^-} O^+ \xrightarrow{-e^-} O^{2+}$

In some cases up to eight electrons can be lost, but the charge is always a whole number multiple of that of a proton.

The natural conclusion to be drawn from these results is that protons are a fundamental constituent of all atoms.

The origin of cathode rays

We have seen that positive rays are formed when cathode rays collide with molecules or atoms of gas, but we have not yet discussed the origin of cathode rays. Paradoxically, they are liberated when positive rays strike the cathode. This at once begs the question as to how positive rays are formed in the first place so that an electrical discharge can start. The answer is that ionising radiation, particularly cosmic radiation, is ever-present in our environment and produces about two ion pairs (i.e. positive ion plus electron) per second in 1 cm^3 of air at atmospheric pressure. The few positive ions that are produced by such radiations are rapidly accelerated by the potential gradient in the tube towards the cathode where they produce cathode rays, which in turn produce more gas ions. In addition, some of the accelerated positive ions may ionise gas atoms or molecules thereby yielding more ions and electrons. Once this process has started the gas discharge takes place continuously, provided that a sufficiently high voltage is maintained across the electrodes.

The television tube and cathode ray oscilloscope (CRO) are familiar examples of the use to which cathode ray tubes are put. For such applications the screen behind the anode is coated with a fine powder of a material such as zinc sulphide. When struck by cathode rays it fluoresces and phosphoresces to produce a bright spot, which can be moved by altering the voltage across the deflection electrodes.

Neutrons

Following the discovery of electrons and protons as basic constituents of all atoms, it was thought for a long time that these were the only fundamental particles. In 1920, however, Rutherford predicted the existence of a third fundamental particle which possessed mass but no charge, and in 1930 Bothe observed the emission of radiation, which he mistakenly took to be γ-radiation, when beryllium was bombarded with α-particles. The radiation discovered by Bothe could expel protons at a very high velocity from certain materials such as paraffin wax, and by studying the number and range of the ejected protons J. Chadwick (1932) concluded that the radiation consisted of neutral particles, which he named *neutrons*, with a mass very close to that of a proton.

Later experiments showed that the neutron is a constituent of all atoms except protium, i.e. the hydrogen isotope, ^1_1H.

3.1.2 THE NUCLEAR THEORY OF ATOMIC STRUCTURE

In 1904 Thomson suggested that atoms consisted of electrons arranged in concentric rings, at the centre of which was a sphere of positive electricity. The positive electricity, it was thought, neutralised the negative charge of the electrons. The theory therefore accounted for the fact that atoms are electrically neutral overall, yet contain electrons. Later experiments, described below, showed that this model was basically correct, and enabled the nature of the positive electricity to be identified.

In an early experiment a stream of electrons, e.g. β-rays from a radioactive source, was directed at a thin metal foil. The stream suffered deflection by interacting with the electrons of the metal atoms of the foil, as shown in Fig. 3.4.

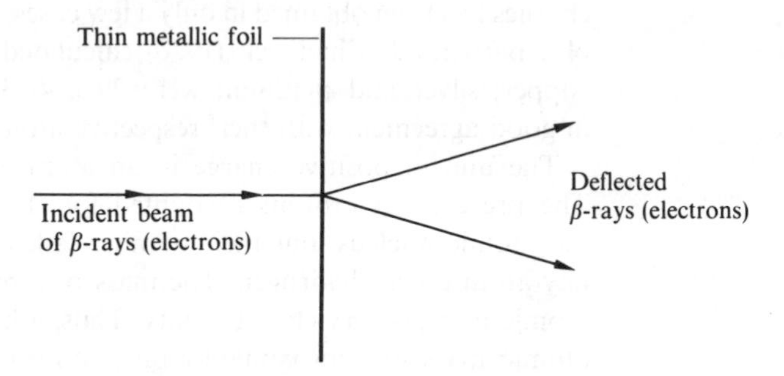

Fig. 3.4 The deflection of β-rays on passing through a thin metallic foil.

The extent of the deflection increased with the number of electrons in the metal atoms of the foil, and by measuring the deflection it was possible to estimate the number of electrons the atoms possessed. For light atoms, e.g. Mg or Al, the number of electrons was found to be approximately half the relative atomic mass.

Since an electron, relative to the total mass of an atom, has negligible mass, the question arose as to where the bulk of the mass of an atom is situated. An obvious answer was the 'positive electricity' part of the atom, and this supposition was verified by bombarding a metallic foil with α-particles, i.e. the positively charged nuclei of helium atoms. Most of the particles passed straight through the foil and suffered little or no deflection, but about one particle in twenty thousand was deflected through an angle of 90° or more. An α-particle, which has a considerable mass, is likely to suffer a deflection of this magnitude only if the particle collides with a relatively heavy mass bearing a positive charge. Calculations by Rutherford suggested that the whole of the positive charge of an atom was concentrated in the centre of the atom, at a very small point (radius about 10^{-14} cm) called the *atomic nucleus*. Because atoms have radii in the region of 10^{-8} cm, atoms are mostly empty space with the actual volume of electrons and nuclei accounting for no more than about 10^{-12} of the total.

Next came the problem of measuring the number of unit positive charges of the nucleus. In 1913 A. van den Brock suggested that the number of unit positive charges in an atomic nucleus was equal to its atomic number, which corresponded to the element's position in the periodic table.

Evidence was provided, also in 1913, by the work of H. G. J. Moseley on the X-rays which are emitted from metals when they are struck by high speed electrons. The frequency (v) of the X-rays is characteristic of the metal target, and Moseley found that v was related to the atomic number (Z) of the metal by the following equation:

$$v^{\frac{1}{2}} = a(Z - b)$$

where a and b are constants that are the same for all elements.

The change from one element to the next in the periodic table is accompanied by an increase in the frequency of X-rays, and Moseley ascribed this to an increase of one unit positive charge and one unit negative charge (electron) in the atoms concerned. He had discovered the fundamental concept upon which the periodic table is based.

Surprisingly, direct experimental proof of the number of unit positive charges has been obtained in only a few cases, but by studying the scattering of α-particles J. Chadwick (1920) calculated that the nuclear charges of copper, silver and platinum were 29.3, 46.3 and 77.4 units respectively, in good agreement with their respective atomic numbers of 29, 47 and 78.

The unit of positive charge in an atom was attributed to the proton. The presence of protons accounted for the mass being concentrated in the atomic nucleus, but their number failed to explain the total mass of any atom except hydrogen. The mass of a proton, expressed as a relative atomic mass, is very close to unity. Thus, a hydrogen atom, with a relative atomic mass of approximately one, comprises one electron of negligible mass and one proton. However, a lithium atom has three electrons, which requires the presence in the atom of three protons to preserve an overall electrical neutrality. This would give the lithium atom a relative atomic mass of three, which does not agree with the experimental value of approximately seven. Following the discovery of the neutron and the fact that it has a mass almost equal to that of the proton, the relative atomic mass of an element could be accounted for by the presence in the nucleus of both protons and neutrons. We now know that a lithium nucleus comprises *three* protons and *four* neutrons, and we generally accept that all atomic nuclei, except for protium, i.e. hydrogen-1, consist of protons and neutrons. The electrons are remote from the nucleus and occupy the space around it in cloud-like formations (§3.1.4). The evidence for this is reasonably conclusive: for example, when atoms lose or gain electrons to produce ions the nucleus is in no way affected. If ion formation affected the nucleus then such processes would cause nuclear changes. Also, the behaviour of electrons in atoms is very well described by theories which demand that the electrons exist outside and not within the nucleus (§3.1.4).

Except for protium, $^{1}_{1}H$, the existence of neutrons in the nucleus appears to be necessary for stability, since the presence of too few or too many neutrons in relation to the number of protons results in an unstable atom (§3.3.1).

The number of protons in an atomic nucleus is equal to the *atomic number* (Z) of the element, and the total number of protons and neutrons is equal to the *mass number* (A) which is a whole number very close to the relative atomic mass of the atom. The number of neutrons in the nucleus ($A - Z$) may vary without affecting the number of extranuclear electrons and hence the chemical properties of an atom, and atoms that differ from one another only in their neutron content are called *isotopes*. Thus, isotopes are atomic species that possess the same atomic number but different mass numbers; e.g. ^{20}Ne and ^{22}Ne (§3.1.1). Very few naturally occurring elements are *monoisotopic*, i.e. consist of only one isotope. Fluorine is an example. The majority, however, comprise a mixture of isotopes, and in such cases the measured relative atomic mass is the weighted mean of the isotopic masses (§1.1.2). Because their chemical properties are identical, mixtures of isotopes of the same element cannot be separated by chemical means but only by physical methods.

To distinguish between the individual isotopes of a particular element we can use the system shown in §1.1.1, or an alternative system which entails writing the name of the element followed by the mass number:

$$^{14}_{6}C \quad \text{or} \quad ^{14}C \quad \text{or} \quad \text{carbon-14}$$

$$^{12}_{6}C \quad \text{or} \quad ^{12}C \quad \text{or} \quad \text{carbon-12}$$

The model of atomic structure just described was obtained after several decades of research by many eminent scientists, and it is worthwhile philosophising for just a moment. An atom has never been seen, nor is one likely to be, so how do we know that the model is correct? The answer is that the model fits the experimental facts as they are known today. This applies almost universally throughout science: models or theories are proposed to explain or fit experimental observations.

3.1.3 THE HYDROGEN SPECTRUM

Much of our knowledge concerning the relative energy levels of atoms originates from atomic spectra, and of particular importance in this respect is the *hydrogen spectrum*. Before discussing this spectrum let us first see how spectra are obtained from radiation in general by considering a simple example.

Fig. 3.5 Origin of spectral lines.

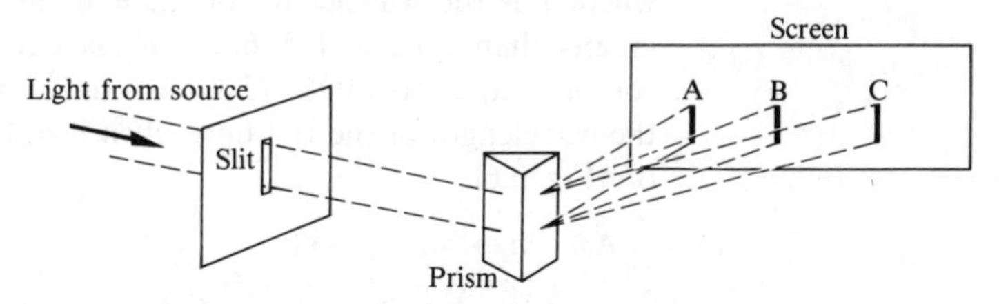

If light comprising, say, three different wavelengths is passed through a slit and then a prism we observe three different lines of light on a suitably placed screen (Fig. 3.5). The prism separates the incident light by refracting (i.e. bending) radiation of different wavelengths unequally. The shorter the wavelength of the radiation the greater is the angle through which it is refracted. The wavelengths of radiation following refraction constitute the *spectrum* of the incident radiation. (Wavelength (λ) and frequency (ν) are both used in descriptions of spectra.) In Fig. 3.5 each line on the screen represents radiation of a single frequency, with wavelengths in the order A < B < C. The lines A, B and C may be photographed so as to provide a permanent record of the spectrum. Similar results are obtained if a diffraction grating is used in place of a prism.

When an electric discharge is passed through hydrogen gas at low pressure, many of the molecules absorb energy by collision with electrons (cathode rays) and break up into atoms which then absorb more energy and emit radiation (§3.1.5). By means of a prism the emitted radiation can be separated into its component wavelengths to produce the *emission spectrum of atomic hydrogen*, commonly called the 'hydrogen spectrum'. The spectrum comprises five series of lines, each line corresponding to radiation of one particular wavelength. Each series is named after its discoverer. In the *Balmer series*, which falls in the visible and ultraviolet parts of the spectrum, the lines become progressively closer together as the wavelength decreases until eventually they converge and stop at the *series limit* (Fig. 3.6). Beyond the series limit the spectrum is continuous, i.e. it does not consist of individual lines.

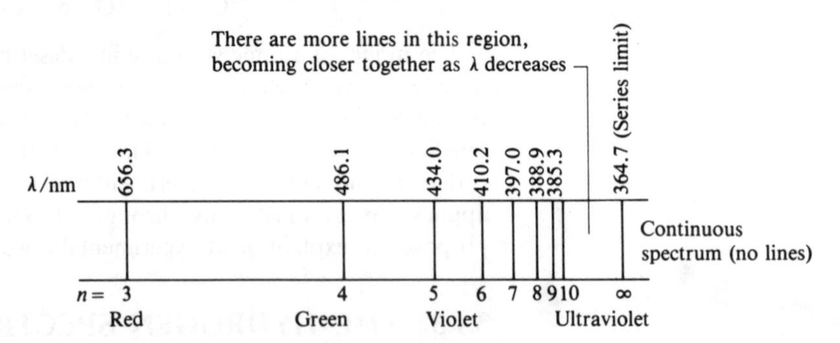

Fig. 3.6 The Balmer series. The lines labelled $n = 3, 4, 5$ and 6 are in the visible part of the spectrum. For clarity, lines in the series with an n value of greater than 10 have been omitted.

Despite the apparent complexity of this series, the wavelength of each line can be calculated from the simple equation

$$\frac{1}{\lambda} = R_H\left(\frac{1}{2^2} - \frac{1}{n^2}\right)$$

where λ is the wavelength of the emitted radiation and n is an integer greater than 2, i.e. 3, 4, 5, 6, etc. R_H is a constant, known as the *Rydberg constant*, equal to $1.0967758 \times 10^7\ m^{-1}$. When $n = 3$ the equation gives the wavelength of the red line, when $n = 4$ that of the green line, and so on (Fig. 3.6).

$$\text{As} \quad n \to \infty, \quad \frac{1}{n^2} \to 0$$

$$\therefore \quad \text{when } n = \infty, \quad \frac{1}{\lambda} = \frac{R_H}{2^2} \quad \text{and} \quad \lambda = 364.7 \text{ nm}$$

which is the wavelength of the limit in the Balmer series.

The Balmer series was discovered in 1885. Later, four other series of lines were discovered in the hydrogen spectrum, one in the ultraviolet region (the *Lyman* series) and three in the infrared (the *Paschen*, *Brackett* and *Pfund* series). Each series shows the same overall pattern as the Balmer series, i.e. lines which become closer together as the wavelength decreases until a series limit is reached (Fig. 3.7).

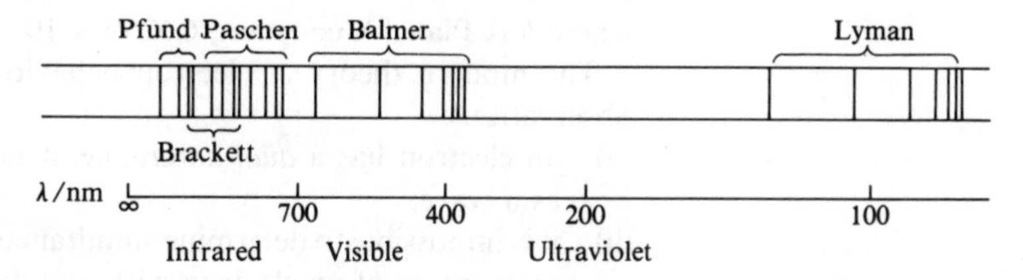

Fig. 3.7 The hydrogen spectrum. For clarity all lines have not been included, especially those towards each series limit. The Brackett series overlaps to some extent with the short wavelength end of the Pfund series and the long wavelength end of the Paschen series.

Ritz (1908) found that the lines in all five series could be represented by a general equation, similar to the one above:

$$\frac{1}{\lambda} = R_H\left(\frac{1}{n_2^2} - \frac{1}{n_1^2}\right)$$

where n_1 and n_2 are integers and $n_1 > n_2$. For each particular series n_2 remains constant throughout and n_1 varies. The Lyman series limit, where $n_1 = \infty$, is given by:

$$\frac{1}{\lambda} = R_H\left(\frac{1}{1^2} - \frac{1}{\infty^2}\right) = R_H \quad \text{or} \quad \lambda = \frac{1}{R_H}$$

The values of n for all five series in the hydrogen spectrum are as follows:

Series	n_2	n_1	Spectral region
Lyman	1	2, 3, 4, etc	ultraviolet
Balmer	2	3, 4, 5, etc	visible/ultraviolet
Paschen	3	4, 5, 6, etc	infrared
Brackett	4	5, 6, 7, etc	infrared
Pfund	5	6, 7, 8, etc	infrared

3.1.4 ELECTRONS IN ATOMS

Niels Bohr (1913) proposed a theory for the electronic structure of atoms and offered the first explanation of the atomic spectrum of hydrogen. Nowadays, although the Bohr theory has been superseded, it is noteworthy because two of Bohr's original assumptions are still applicable and form the basis of the modern theory of the origin of the hydrogen spectrum. These assumptions are:

(i) an electron in an atom cannot have a continuously varying energy, but is constrained to one of a certain number of fixed energy values, called *energy levels*;

(ii) radiant energy (radiation) is emitted only when an electron moves from a level of relatively high energy to one of lower energy.

Thus, if E_a and E_b each represent energy levels within an atom, and $E_a > E_b$, then radiant energy is emitted when an electron moves from level E_a to level E_b. The wavelength (λ) of the emitted radiation is given by:

$$E_a - E_b = \Delta E$$

and $\Delta E = \dfrac{hc}{\lambda} = h\nu$ (Planck's equation)

where h is Planck's constant, 6.6256×10^{-34} J s.

The modern theory of electron behaviour in atoms is based on two observations:

(i) an electron has a *dual nature*, i.e. it behaves both as a particle and as a wave;

(ii) it is impossible to determine simultaneously both the position and the momentum of an electron with any degree of precision.

According to the sophisticated approach of *wave mechanics*, an electron in an atom forms a *three-dimensional standing wave* about the nucleus, and its position is treated in terms of *probability* (see below).

Orbitals

An electron in an atom may be regarded as moving rapidly in space around the nucleus. In so doing the electron traces a three-dimensional volume of space called an *orbital.* Thus, we commonly refer to an electron as 'occupying an orbital' in an atom. Strictly speaking, an orbital has no existence without an electron, but for some purposes it is convenient to regard an orbital as being empty. In this sense an orbital is a volume of space which is potentially capable of being occupied by an electron.

The position of an electron in an orbital cannot be pinpointed exactly; the best we can do is to express a probability of finding it in a certain volume of space. An orbital may thus be defined as a three-dimensional volume of space surrounding the nucleus within which there is a greater than a 95% chance of finding an electron; Fig. 3.9.

Alternatively, an electron may be envisaged as a diffuse cloud with an overall charge of -1 and a density (ρ) that varies with the distance from

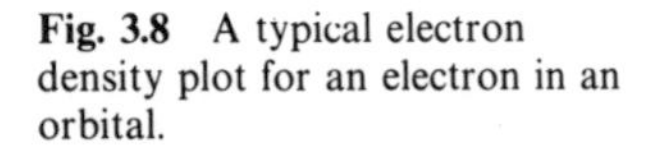
Fig. 3.8 A typical electron density plot for an electron in an orbital.

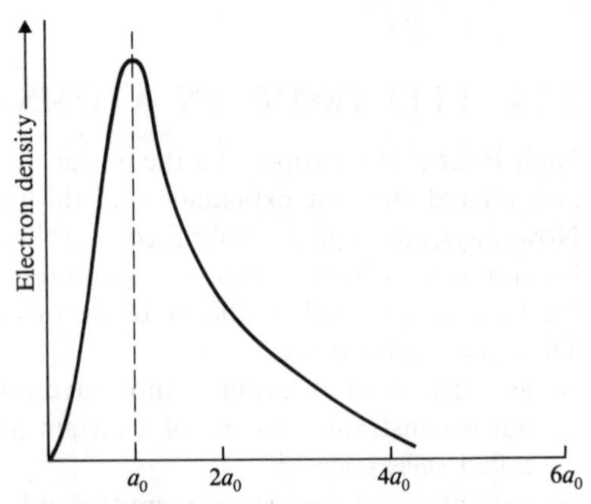

the nucleus. The density is related to the probability of finding the electron, since a high density corresponds to a region where an electron spends most of its time. A typical plot of density against distance from the nucleus for an electron in an orbital is shown in Fig. 3.8. The maximum probability of finding the electron in the orbital occurs at a distance a_0 from the nucleus, and we may use this distance to define the most probable location of the orbital.

Orbitals can be of various shapes and sizes, but provided that a certain condition is satisfied (§3.1.6) any orbital can accommodate a maximum of two electrons, i.e. only two electrons may occupy the same volume of space.

Energy levels

An electron in an orbital has associated with it a fixed amount of energy which is characteristic of that particular orbital. Orbitals are therefore often called *energy levels*, and it is common practice to refer to electron occupation of particular energy levels rather than orbitals. As a general rule, the lower the energy of an orbital, the closer to the nucleus is an electron in that orbital.

Since there is a limit to the number of orbitals and hence energy levels within an atom, an electron in an atom can have certain amounts of energy only, and not a continuously varying energy.

Labelling of orbitals

For many purposes we need to label the electrons occupying the various orbitals in an atom. Each orbital is labelled by a number, called the *principal quantum number*, and a letter.

Principal quantum numbers

The principal quantum number (n) can only have integral values 1, 2, 3 and so on up to infinity (∞). For known atoms in their *ground states*, i.e. states of lowest energy, values of $n = 7$ are the highest that are encountered. Values greater than this are important only for atoms in *excited states*, i.e. states of excess energy (§3.1.5).

The higher the principal quantum number of an orbital, the greater is its energy. Thus, an orbital with $n = 1$ represents the lowest energy orbital of an atom, an orbital with $n = 2$ the next lowest, and so on. In general, as the n value of an orbital increases so does its size, i.e. the distance from the nucleus to where the electron density is highest.

Letters

A letter is used to indicate the various types of orbital that may be encountered. Four common types exist, namely s, p, d and f. For most purposes we may regard the type as being an indication of the shape of an orbital. s Orbitals are the simplest and are spherical (Fig. 3.9).

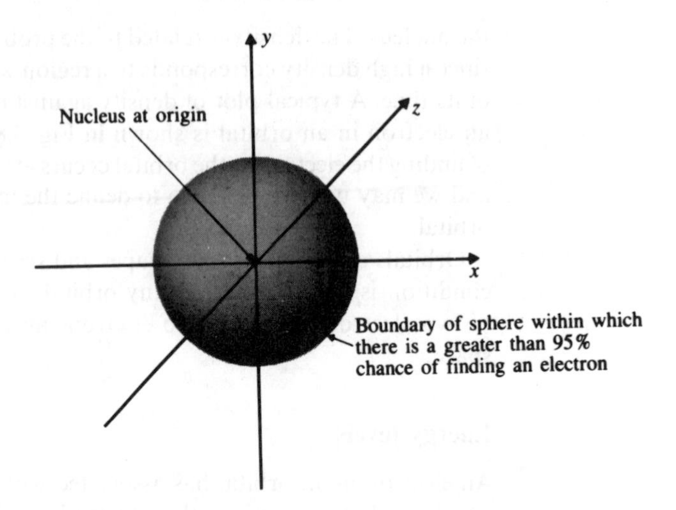

Fig. 3.9 The spherical shape of the boundary surface of a 1s orbital. The figure represents an envelope within which there is a high probability (greater than a 95% chance) of finding an electron in the orbital.

A p orbital is often described as 'dumb-bell' shaped. It consists of two pear-shaped lobes (Fig. 3.10(a)). A d orbital is more complex than a p orbital and consists of up to four lobes, while an f orbital comprises as many as eight lobes.

In labelling an orbital the principal quantum number is written first and then the letter, e.g. 1s, 3d, etc. A 1s orbital has an n value of 1 and is of the s type, while a 3d orbital is of the d type with an n value of 3.

Shells

A *shell* is defined as a complete group of orbitals possessing the same principal quantum number. A shell is numbered by the n value of its constituent orbitals, although the letters K (for $n = 1$), L (for $n = 2$), M (for $n = 3$), etc were formerly used. For example, the orbitals with a value of $n = 2$ constitute the second shell, and those with a value of $n = 3$ the third shell. The first shell is unusual in that it consists of one orbital only.

For a given shell, the value of n is equal to the number of different types of orbital that exist within the shell. The various types of orbital that exist within the first four shells are as follows:

(i) first shell – s type;
(ii) second shell – s and p types;
(iii) third shell – s, p and d types;
(iv) fourth shell – s, p, d and f types.

Composition of shells

First shell
The first shell comprises the 1s orbital (Fig. 3.9), which is the lowest energy orbital of any atom.

Second shell
In the second shell there is one 2s orbital and *three* 2p orbitals, i.e. a total

of four orbitals. A 2s orbital is spherical in shape, like a 1s orbital, but is considerably larger and higher in energy.

The three 2p orbitals in an isolated atom are *degenerate*, i.e. they possess the same amount of energy. This is often spoken of as a 'threefold degeneracy'. Since the three 2p orbitals differ only in their relative spatial orientations (Fig. 3.10(b)), it is hardly surprising that in an isolated atom, in the absence of electrical or magnetic fields, they are degenerate.

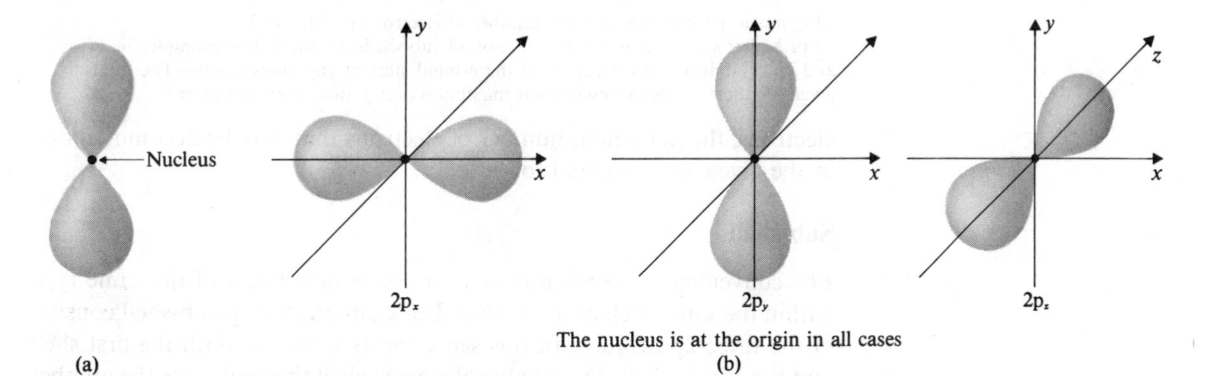

Fig. 3.10 (a) The shape of the boundary surface of a p orbital. (b) The relative orientation of the three 2p orbitals, referred to the cartesian coordinate axes, x, y and z. The distinguishing labels $2p_x$ etc, may be used, but are seldom necessary. The orbitals are mutually perpendicular to one another, with the lobes of the $2p_x$ orbital concentrated along the x axis, those of the $2p_y$ orbital along the y axis, and $2p_z$ along the z axis.

For a multi-electron atom the relative energies of orbitals of the second shell are 2s < 2p. For a hydrogen atom, energy of 2s = energy of 2p. Neither the 2s nor the 2p orbitals are occupied in the ground state of a hydrogen atom, although occupation is possible in the excited state (§3.1.5).

Third shell

There is a total of nine orbitals in the third shell.

(i) One 3s orbital, which is spherical but higher in energy and larger than the 2s orbital.

(ii) Three 3p orbitals, all equal in energy (threefold degeneracy). They are higher in energy and larger than 2p orbitals, but are similar in shape and orientation.

(iii) Five 3d orbitals, all equal in energy (fivefold degeneracy). For multi-electron atoms, their relative energies are 3s < 3p < 3d. For a hydrogen atom these orbitals are all degenerate, i.e. 3s = 3p = 3d in energy.

Fourth shell

This comprises one 4s, three degenerate 4p, five degenerate 4d and seven degenerate 4f orbitals.

The 4s, 4p and 4d orbitals are similar in shape and orientation to their counterparts in the third shell, but are higher in energy and of a larger size. For multi-electron atoms their relative energies are 4s < 4p < 4d < 4f. Complete degeneracy of all orbitals of the fourth shell arises in the hydrogen atom only.

It can be seen from Table 3.2 that up to $n = 4$ the total number of orbitals within any particular shell is given by n^2. For example, in the second shell there are $2^2 = 4$ orbitals. Since an orbital may hold a maximum of two

Table 3.2 The occupation of the first seven electron shells

shell (n)	1	2	3	4	5	6	7
sub-shells†	1s (2)	2s (2) 2p (6)	3s (2) 3p (6) 3d (10)	4s (2) 4p (6) 4d (10) 4f (14)	5s (2) 5p (6) 5d (10) 5f (14)	6s (2) 6p (6) 6d (10)	7s (2)
total number of electrons per shell	2	8	18	32	32‡	18‡	2‡

† Figures in parentheses give the number of electrons per sub-shell.
‡ For higher shells, i.e. $n = 5$, 6 or 7, not all sub-shells are filled. For example, 5g, 6f and 7p orbitals are not occupied in the ground state of any known atom. The totals given for these shells represent their maximum occupation in known atoms.

electrons, the maximum number of electrons that may be accommodated in the given shell is equal to $2n^2$.

Sub-shells

For convenience, we often refer to a group of orbitals of the same type within the same shell as a *sub-shell.* For example, the 2p sub-shell consists of the three 2p orbitals. In this sense the 1s orbital is both the first shell and the 1s sub-shell. The n value of a given shell thus indicates the number of sub-shells present.

3.1.5 THE ORIGIN OF THE HYDROGEN SPECTRUM

The hydrogen atom is unique in that all its orbitals with the same n value (i.e. all orbitals of the same shell) have the same energy. Consequently, a hydrogen atom has fewer energy levels than the atoms of other elements. They are conveniently represented by Fig. 3.11.

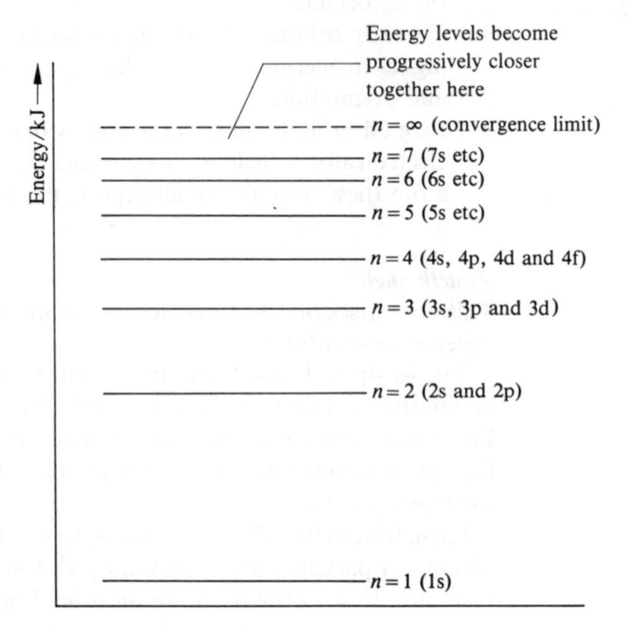

Fig. 3.11 The relative energy levels of a hydrogen atom.
Notes:
(i) A line represents the (relative) energy level of each shell according to its n value; e.g. the line labelled $n = 2$ gives the energy of orbitals of the second shell, i.e. 2s and 2p. (ii) The energy of a shell increases as the value of n increases. (iii) As n increases the energy levels become closer together until eventually they converge at $n = \infty$. For clarity, energy levels with n values greater than 7 have been omitted. (iv) An electron at the $n = \infty$ level is a free electron.

For a hydrogen atom in its *ground state*, i.e. state of lowest energy, the single electron occupies the $n = 1$ level (1s orbital). However, *electron transitions*, i.e. movements between different energy levels, can occur. By absorbing a sufficient quantity of energy, the electron can move or *be promoted* to higher energy levels ($n = 2, 3, 4$, etc) to produce an *excited* hydrogen atom. When an electron transition subsequently occurs from a higher level to the $n = 1$ level the excess energy is emitted in the form of radiation. The amount of energy emitted in such a transition (e.g. $n = 2$ to $n = 1$) is equal to the energy absorbed in promoting the electron to the higher energy level (e.g. $n = 1$ to $n = 2$). We shall now consider the relationship between some electron transitions in the hydrogen atom and the wavelength or frequency of the radiation that is emitted.

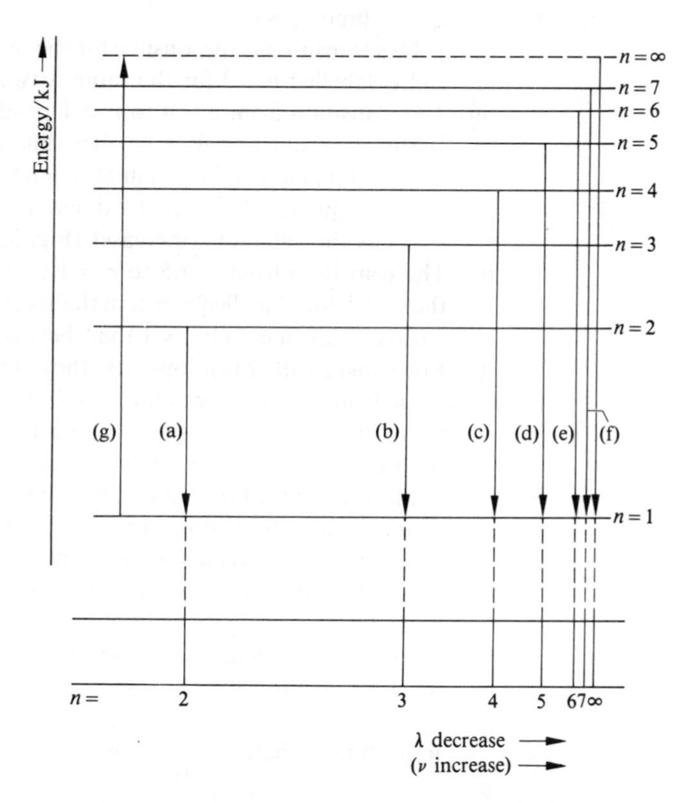

Fig. 3.12 The representation of some electron transitions in a hydrogen atom that give rise to lines in the Lyman series. Only the first six lines have been considered, i.e. those originating from the transitions represented by arrows (a) to (f). Dashed lines indicate which transitions produce which lines. Other transitions from levels higher than $n = 7$ to $n = 1$ also occur to give lines in the spectrum between the lines labelled $n = 7$ and $n = \infty$. These have been omitted for the sake of clarity.

Arrow (g) represents the transition in ionising a ground state hydrogen atom.

(i) When an electron moves from $n = 2$ to $n = 1$ (arrow (a) in Fig. 3.12) the energy change (ΔE_a) is equal to the energy difference between the two levels. The wavelength (λ) or frequency (ν) of the emitted radiation and ΔE_a are related by Planck's equation:

$$E_2 - E_1 = \Delta E_a = \frac{hc}{\lambda_a} = h\nu_a$$

where E_2 represents the energy of the $n = 2$ level and E_1 the energy of the $n = 1$ level.

Thus, the $n = 2$ to $n = 1$ transition produces radiation of a single wavelength (i.e. one line in the spectrum, §3.1.3). Since no other transition to the $n = 1$ level is accompanied by a lower energy change, the radiation corresponds to the line of longest wavelength (i.e. the first line) in the Lyman series of the hydrogen spectrum. This line is often labelled $n = 2$, to indicate the level from which the electron transition has occurred.

(ii) Similarly, the transition from $n = 3$ to $n = 1$ (arrow (b) in Fig. 3.12) causes an energy change (ΔE_b):

$$E_3 - E_1 = \Delta E_b = \frac{hc}{\lambda_b} = h\nu_b$$

where $\Delta E_b > \Delta E_a$, since the $n = 3$ level is higher in energy than the $n = 2$ level. The emitted radiation again has a single wavelength, where $\lambda_b < \lambda_a$ (and $\nu_b > \nu_a$).

This transition is responsible for the second line of the Lyman series, and is labelled $n = 3$ for the same reason as above.

(iii) The transition from $n = 4$ to $n = 1$ produces the third line. A glance at the spectrum reveals a smaller difference in wavelength between the $n = 4$ and $n = 3$ lines than between the $n = 3$ and $n = 2$ lines. This is a consequence of the smaller difference in energy between successive levels as the value of n increases (Fig. 3.11, note (iii)).

(iv) The transition from $n = 5$ to $n = 1$ (arrow (d) in Fig. 3.12) produces the fourth line, labelled $n = 5$, in the spectrum, with a smaller difference between this line and $n = 4$ than between $n = 4$ and $n = 3$.

(v) For subsequent transitions (e.g. those represented by (e) and (f), and those from levels higher than $n = 7$ to $n = 1$, not shown in Fig. 3.12), the pattern continues as successive energy levels become closer together, until eventually the lines that they produce in the spectrum converge at the series limit. The series limit is labelled $n = \infty$, since it corresponds to an electron transition from $n = \infty$ to $n = 1$. This latter transition is, of course, the reverse of ionisation, for ionisation involves promoting an electron from its ground state, $n = 1$, to $n = \infty$ (arrow (g) in Fig. 3.12). By measuring the wavelength or frequency of the Lyman series limit, we can calculate the ionisation enthalpy of a hydrogen atom from the relationship:

$$\text{ionisation enthalpy} = \frac{hc}{\lambda_{SL}} = h\nu_{SL}$$

where λ_{SL} and ν_{SL} represent the wavelength and frequency respectively of the Lyman series limit.

Thus we see that transitions from higher levels to the $n = 1$ level give rise exclusively to the lines of the Lyman series. In practice, however, when an excited hydrogen atom sheds its excess energy, the electron does not always return to the lowest level ($n = 1$). Instead, it may return to one of the other levels, $n = 2, 3, 4, 5$, etc, before eventually returning to the ground state ($n = 1$) or being excited to a higher level. Transitions occurring to the $n = 2$ level give rise to the lines of the Balmer series (Fig. 3.13). The lines in the Balmer series are labelled, as before, according to the level from which the electron transition has occurred.

The origins of the other series are as follows: the Paschen series, from

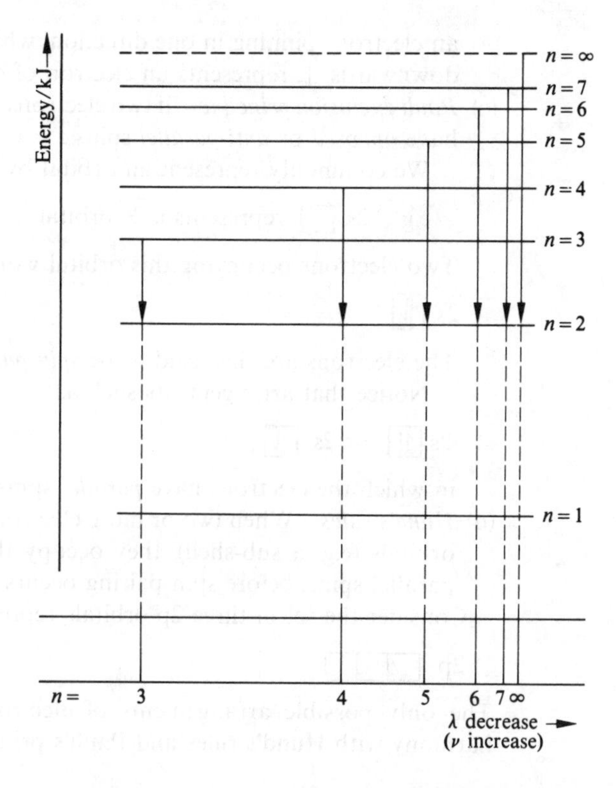

Fig. 3.13 The origin of the Balmer series. Arrows represent electron transitions, and dashed lines indicate which transitions produce which lines.

transitions to the $n = 3$ level, the Brackett series, from transitions to the $n = 4$ level and the Pfund series, from transitions to the $n = 5$ level.

We can now see the significance of the integers n_1 and n_2 in the equation established by Ritz (§3.1.3),

$$\text{i.e.} \quad \frac{1}{\lambda} = R_H\left(\frac{1}{n_2^2} - \frac{1}{n_1^2}\right)$$

n_2 is the *lower* level, to which the transition occurs, and n_1 is the *higher* level, from which the electron transition originates.

3.1.6 THE ENERGY LEVELS OF MULTI-ELECTRON ATOMS AND THE ORDER IN WHICH ATOMIC ORBITALS ARE FILLED (AUFBAU PRINCIPLE)

The *electronic configuration* of an atom, i.e. the way in which the electrons are arranged in the various orbitals in the ground state, is governed by several fundamental rules. We shall now study these rules, and afterwards see how they relate to some of the lower elements.

(a) For an atom to be in its ground state, i.e. to have the lowest possible energy, electrons must occupy the available orbitals in order of energy level. The orbital of lowest energy is always filled first, before orbitals of higher energy. This is a statement of the *aufbau principle.*

(b) Electrons behave as spinning particles, and only two types of spin are possible. An upward pointing half arrow, ↿, conveniently represents an electron spinning in one direction, while the same arrow, pointing downwards, ↓, represents an electron of opposite spin.

(c) *Pauli exclusion principle* If two electrons occupy one orbital they must have *opposed* or *anti-parallel* spins.

We commonly represent an orbital by a box,

e.g. 2s [ ] represents a 2s orbital.

Two electrons occupying this orbital would be represented as:

2s [↑↓]

The electrons are then said to be *spin paired.*

Notice that arrangements such as

2s [↑↑] or 2s [↓↓],

in which the electrons have *parallel* spins, cannot occur.

(d) *Hund's rules* When two or more electrons occupy a set of degenerate orbitals (e.g. a sub-shell), they occupy them: (i) singly, and (ii) with parallel spins, before spin pairing occurs in any one orbital.

Consider the set of three 2p orbitals represented as:

2p 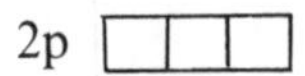

The only possible arrangements of electrons within these orbitals in harmony with Hund's rules and Pauli's principle are given in Fig. 3.14.

Fig. 3.14 Arrangements of electrons in the three degenerate 2p orbitals. Objections to the forbidden arrangements are:
* violates Hund's rule (ii): the electrons have opposite spins
** violates Hund's rule (i): the orbitals are not singly occupied
*** violates Pauli's principle: one orbital contains two electrons with parallel spins.

The only allowed arrangements	Number of electrons	Some arrangements that are forbidden, i.e. not allowed	
[↑][↑][ ]	2	[↑][↓][ ]	*
[↑][↑][↑]	3	[↑↓][↑][ ]	**
[↑↓][↑][↑]	4	[↑↑][↑][↓]	* and ***
[↑↓][↑↓][↑]	5	[↑↑][↑↓][↑]	***
[↑↓][↑↓][↑↓]	6	[↑↑][↑↓][↑↓]	***

These principles will now be applied to the electronic configuration of each of the elements up to krypton ($Z = 36$), for which purpose the orbitals occupied are 1s, 2s, 2p, 3s, 3p, 3d, 4s and 4p. Generally, their energy levels increase with increasing n value, except for the 4s orbital which is intermediate in energy between the 3p and 3d orbitals. The relative order of energy levels is thus as follows:

$$1s < 2s < 2p < 3s < 3p < 4s < 3d < 4p$$

and we shall see that the orbitals are filled in this order.

For elements beyond krypton see the periodic table (Fig. 3.15), which incorporates the complete aufbau principle.

Hydrogen (one electron) Only the lowest energy level (1s) is occupied and we write the electronic configuration as $1s^1$, to denote one electron in the 1s orbital.

Helium (two electrons) Both electrons occupy the 1s orbital, and the electronic configuration is $1s^2$.

Lithium (three electrons) Two electrons fill the 1s orbital while the other occupies the 2s orbital, which is the next lowest energy level. The configuration is written: $1s^2\ 2s^1$.

Beryllium has four electrons, with two in each of the 1s and 2s orbitals, i.e. $1s^2\ 2s^2$. After the 2s orbital, the three 2p orbitals are next lowest in energy level and these are filled for the elements from boron to neon (Table 3.3).

Table 3.3 The electronic configurations of the elements from hydrogen to neon

Element	Atomic number	Electronic configuration	or 1s	2s	2p
hydrogen	1	$1s^1$	↑		
helium	2	$1s^2$	↑↓		
lithium	3	$1s^2\ 2s^1$	↑↓	↑	
beryllium	4	$1s^2\ 2s^2$	↑↓	↑↓	
boron	5	$1s^2\ 2s^2\ 2p^1$	↑↓	↑↓	↑
carbon	6	$1s^2\ 2s^2\ 2p^2$	↑↓	↑↓	↑ ↑
nitrogen	7	$1s^2\ 2s^2\ 2p^3$	↑↓	↑↓	↑ ↑ ↑
oxygen	8	$1s^2\ 2s^2\ 2p^4$	↑↓	↑↓	↑↓ ↑ ↑
fluorine	9	$1s^2\ 2s^2\ 2p^5$	↑↓	↑↓	↑↓ ↑↓ ↑
neon	10	$1s^2\ 2s^2\ 2p^6$	↑↓	↑↓	↑↓ ↑↓ ↑↓

The second shell is full at neon and the next orbitals to be occupied are the 3s followed by 3p, which fill in a similar manner to the orbitals of the second shell to give the elements from sodium to argon.

At argon ($Z = 18$) the 3s and 3p orbitals are full and the third shell contains eight electrons, like the second shell of neon. Unlike the latter, however, the third shell is not full, for it can hold up to eighteen electrons by using 3d orbitals. Nevertheless, 3d orbitals are not the next to fill, for the lowest energy orbital after 3p is 4s. Thus, in an atom of potassium ($Z = 19$), eighteen electrons occupy the 1s, 2s, 2p, 3s and 3p orbitals, while the nineteenth enters the 4s orbital.

At calcium ($Z = 20$) the 4s orbital is full, and occupation of the 3d orbitals now commences. This gives rise to the first series of d-block elements, from scandium ($Z = 21$) to zinc ($Z = 30$). The 3d and 4s orbitals are full at zinc, leaving the 4p orbitals to be occupied for the next elements, gallium to krypton. The electronic configurations of all these elements are shown in Table 3.4.

Table 3.4 The occupation of atomic orbitals for the elements from sodium to krypton. For simplicity in representation the $1s^2$, $2s^2$ type of notation is not used

	Atomic orbitals							
Element	1s	2s	2p	3s	3p	3d	4s	4p
sodium	2	2	6	1				
magnesium	2	2	6	2				
aluminium	2	2	6	2	1			
silicon	2	2	6	2	2			
phosphorus	2	2	6	2	3			
sulphur	2	2	6	2	4			
chlorine	2	2	6	2	5			
argon	2	2	6	2	6			
potassium	2	2	6	2	6		1	
calcium	2	2	6	2	6		2	
scandium	2	2	6	2	6	1	2	
titanium	2	2	6	2	6	2	2	
vanadium	2	2	6	2	6	3	2	
chromium	2	2	6	2	6	5	1	
manganese	2	2	6	2	6	5	2	
iron	2	2	6	2	6	6	2	
cobalt	2	2	6	2	6	7	2	
nickel	2	2	6	2	6	8	2	
copper	2	2	6	2	6	10	1	
zinc	2	2	6	2	6	10	2	
gallium	2	2	6	2	6	10	2	1
germanium	2	2	6	2	6	10	2	2
arsenic	2	2	6	2	6	10	2	3
selenium	2	2	6	2	6	10	2	4
bromine	2	2	6	2	6	10	2	5
krypton	2	2	6	2	6	10	2	6

Notes

(i) The configurations of the elements scandium to zinc are discussed in §18.1.1.

(ii) In writing the electronic configurations of atoms, orbitals of the same n value are usually written together, irrespective of their relative energy levels. Thus, nickel is written as:
$1s^2\ 2s^2\ 2p^6\ 3s^2\ 3p^6\ 3d^8\ 4s^2$ and *not*: $1s^2\ 2s^2\ 2p^6\ 3s^2\ 3p^6\ 4s^2\ 3d^8$.

(iii) To save time and space, noble gas cores are often utilised in writing electronic configurations, particularly of heavy atoms. For example:
sodium [Ne] $3s^1$
iron [Ar] $3d^6\ 4s^2$
where [Ne] represents a neon core configuration, i.e. $1s^2\ 2s^2\ 2p^6$, and [Ar] stands for $1s^2\ 2s^2\ 2p^6\ 3s^2\ 3p^6$. Mercury, in similar manner, would be [Xe] $5d^{10}\ 6s^2$. Occasionally these noble gas cores are written more explicitly, for example [Ar core].

3.2 The periodic table

3.2.1 SIMILARITY OF PROPERTIES AMONG THE ELEMENTS

The chemical and physical properties of an element are determined by the electronic configuration of its atoms. In particular, it is the electronic configuration in the *outermost shell* or *outermost orbitals* that is important. The electrons in the inner, often full, shells do not participate in chemical bonding and for many purposes can be disregarded. This is not surprising, for when two atoms approach each other before chemical bond formation, it is the electrons in the outermost orbitals that are most likely to interact. For this reason we often refer to them as *bonding electrons*, or perhaps as *valency electrons*, because they are responsible for the valency of the element.

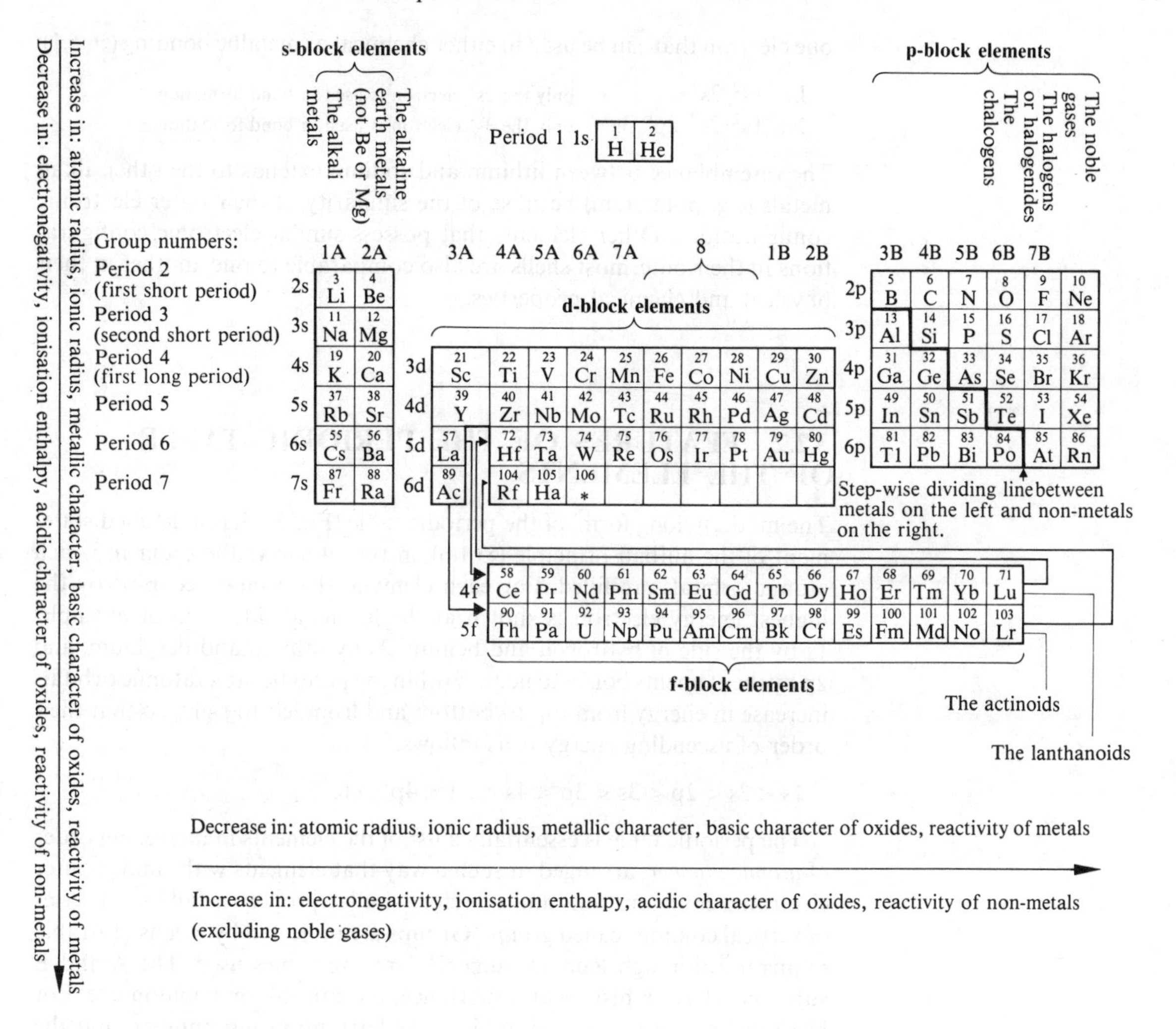

Fig. 3·15 The period table of the elements

Notes

(i) Atomic numbers are given immediately above each element.
(ii) No name has yet been allotted to element 106.
(iii) The highest energy occupied orbital of each atom is indicated by 1s, 2s, 2p, etc at the left of each block.
(iv) The table is set out in terms of increasing energy of atomic orbitals. (See aufbau principle. §3.1.6). Energy increases from top to bottom and from left to right across the table, i.e.
ls < 2s < 2p < 3s < 3p < 4s < 3d < 4p < 5s < 4d < 5p < 6s < 5d < 4f < 5d < 6p < 7s < 6d < 5f < 6d
At La ($Z = 57$) 5d < 4f with one electron in the 5d level, but at Ce ($Z = 58$) 4f < 5d and the 4f orbitals fill up to Lu ($Z = 71$), after which the 5d orbitals continue to fill. A similar pattern is observed with the 6d and 5f orbitals.
(v) Some important trends which occur in the periodic table are indicated by arrows at the bottom and left of the table. Notice that the trends which occur from left to right are the opposite of those found from top to bottom. This gives rise to the diagonal relationships discussed in §3.2.5.

We may therefore expect atoms with similar outer electronic configurations to resemble one another, both chemically and physically. For example, the strong resemblance between lithium and sodium arises because the atoms of both elements possess, in their outermost s orbitals, one electron that can be used in either chemical or metallic bonding (§6.5.1):

Li $1s^2\ 2s^1$ only the $2s^1$ electron is used in bond formation
Na $1s^2\ 2s^2\ 2p^6\ 3s^1$ only the $3s^1$ electron is used in bond formation

The resemblance between lithium and sodium extends to the other alkali metals (e.g. potassium) because of the similarity of their outer electronic configurations. Other elements that possess similar electronic configurations in their outermost shells are also comparable to one another in their physical and chemical properties.

3.2.2 FEATURES OF THE PERIODIC TABLE OF THE ELEMENTS

The modern 'long form' of the periodic table (Fig. 3.15) is a detailed statement of the aufbau principle (§3.1.6), in that it shows the order in which atomic orbitals are filled. For each element, the orbital occupied by the highest energy electron is shown at the left-hand side; see, for example, 1s by the side of hydrogen and helium, 2s by lithium and beryllium, and 2p by the elements boron to neon. Within the periodic table, atomic orbitals increase in energy from top to bottom and from left to right, so that their order of ascending energy is as follows:

$$1s < 2s < 2p < 3s < 3p < 4s < 3d < 4p \quad \text{etc.}$$

The periodic table is essentially a list of the elements in increasing order of *atomic number*, arranged in such a way that elements with similar outer electronic configurations and therefore similar properties fall into a series of vertical columns called *groups*. Groups are indicated by means of Arabic numerals, although Roman numerals are sometimes used. The A and B sub-labels have a historical importance, but are still in common use. For historical reasons, also, the noble gases have no group number, and the nine elements in the three groups headed by iron, cobalt and nickel are collectively known as group 8.

For the elements of groups 1A and 2A, and groups 1B to 7B inclusive, the group number is equal to the number of outermost electrons in the atoms concerned. Thus, the elements of group 1A have one outer electron (ns^1), those of group 2A have two such electrons (ns^2), and those of group 4B have four ($ns^2\ np^2$).

The following are exceptions to the rule that all elements in the same group have the same outer electronic configuration.

(i) Helium ($1s^2$) is placed in the same group as the noble gases, which have $ns^2\ np^6$ configurations.
(ii) The elements of group 6A show slight differences in their outer configurations, i.e. Cr $3d^5\ 4s^1$, Mo $4d^5\ 5s^1$ and W $5d^4\ 6s^2$.
(iii) Hydrogen ($1s^1$) is not placed in group 1A, even though the elements of this group all have ns^1 configurations (Table 11.3).

Despite these anomalies, the similarity in properties between the member elements of a group is so strong that inorganic chemistry is always studied on a group basis. The elements of some groups have collective names, e.g. group 1A – the alkali metals.

A horizontal row of elements is called a *period*. Each one begins with an element possessing one outermost electron (ns^1) and ends with a noble gas. The periods are numbered from the top of the periodic table, with hydrogen and helium comprising the *first period* (or period 1). The elements lithium to neon make up the *second period* and sodium to argon the *third period*.

The elements of groups 1A and 2A are collectively known as *s-block* elements, since their highest energy electrons occupy s orbitals. Correspondingly, the elements of groups 3B to 7B, which have their highest energy electrons in p orbitals, are known as *p-block* elements. The s- and p-block elements collectively are often referred to as the *representative*, *typical* or *main group* elements.

The *d-block* elements are formed by the filling of 3d, 4d and 5d orbitals. Included among these elements are the transition metals (§18.1.1). The two series of *f-block* elements, the *lanthanoids* and *actinoids*, result from outermost electrons occupying the 4f and 5f orbitals respectively. To avoid unnecessary sideways expansion of the periodic table, the f-block elements are usually placed at the bottom.

3.2.3 PERIODICITY AND TRENDS IN THE PERIODIC TABLE

When the elements are arranged in order of increasing atomic number their physical and chemical properties show a *periodic variation* or *periodicity*, which is embodied in the periodic table. Among the lighter elements (up to $Z = 20$), there is a periodic variation every eight elements, but this changes to every 18 elements up to $Z = 57$, and then to 32 for the heaviest elements. In group 1A, for example:

(i) lithium ($Z = 3$) resembles sodium ($Z = 11$) and potassium ($Z = 19$), i.e. there is a periodic variation every eight elements.

(ii) potassium ($Z = 19$) resembles rubidium ($Z = 37$) and caesium ($Z = 55$), i.e. there is a periodic variation every 18 elements.

(iii) caesium ($Z = 55$) resembles francium ($Z = 87$), i.e. there is a variation after 32 elements.

The periodic variation is a direct result of the sequence in which the orbitals are occupied.

Li $1s^2\ 2s^1$

difference = 8 (filling of 2s and 2p orbitals)

Na $1s^2\ 2s^2\ 2p^6\ 3s^1$

difference = 8 (filling of 3s and 3p orbitals)

K $1s^2\ 2s^2\ 2p^6\ 3s^2\ 3p^6\ 4s^1$

difference = 18 (filling of 3d, 4s and 4p)

Rb $1s^2\ 2s^2\ 2p^6\ 3s^2\ 3p^6\ 3d^{10}\ 4s^2\ 4p^6\ 5s^1$

difference = 18

Cs $1s^2\ 2s^2\ 2p^6\ 3s^2\ 3p^6\ 3d^{10}\ 4s^2\ 4p^6\ 4d^{10}\ 5s^2\ 5p^6\ 6s^1$

difference = 32

Fr $1s^2\ 2s^2\ 2p^6\ 3s^2\ 3p^6\ 3d^{10}\ 4s^2\ 4p^6\ 4d^{10}\ 4f^{14}\ 5s^2\ 5p^6\ 5d^{10}\ 6s^2\ 6p^6\ 7s^1$

Hence we see that the increasing number of elements in the periods results from the greater number of orbitals associated with the higher shells. Similar principles apply to the elements of other groups.

3.2.4 SOME PROPERTIES THAT SHOW A PERIODIC VARIATION

Atomic volume

Lothar Meyer in 1869 was among the first to demonstrate the periodicity of the elements. He did so by plotting atomic volume (i.e. the volume in cm^3 of one mole of the element in the solid state) against relative atomic mass. An up-dated plot, using atomic numbers, is shown in Fig. 3.16.

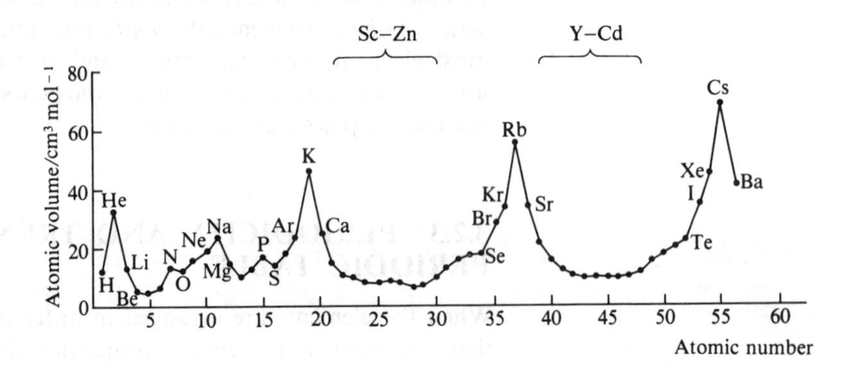

Fig. 3.16 Atomic volume curve.

The plot clearly shows a periodic variation in atomic volume, with many elements of similar properties (i.e. members of the same group) occurring at corresponding points on the curve. See, for example, the elements of groups 1A and 2A.

Atomic radius

Atomic radius can be expressed in one of three ways, as follows:

(i) *van der Waals' radius*, defined as half the shortest internuclear distance between two adjacent non-bonded atoms in a crystal of the element;

(ii) *covalent bond radius*, defined as half the internuclear distance between two identical atoms joined by a single covalent bond;

(iii) *metallic radius*, defined as half the shortest internuclear distance between two atoms in the metallic crystal.

Covalent bond radii are always shorter than van der Waals' radii, because covalent bond formation requires the overlap of orbitals in the outermost shells of adjacent atoms. Metallic radius is always larger, by up to 20%, than the corresponding covalent bond radius.

The covalent bond radius is known for most elements, and the term 'atomic radius' is generally taken to mean this quantity. A plot of atomic radius against atomic number clearly shows a periodic variation (Fig. 3.17).

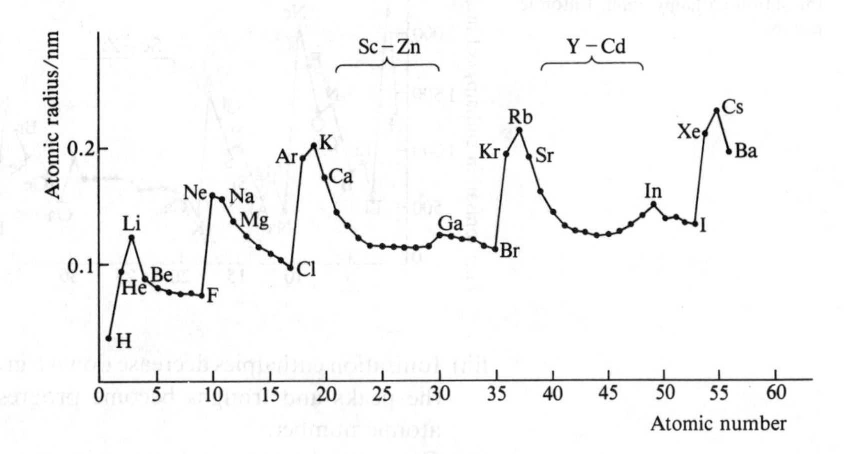

Fig. 3.17 Graph of atomic radius against atomic number. Except for the van der Waals' radii of the noble gases, the values plotted are covalent bond radii.

Two trends are apparent from Fig. 3.17.

(i) Atomic radius *increases* from top to bottom of a group, e.g.

Li 0.123, Na 0.157, K 0.203, Rb 0.216, Cs 0.235 (all in nm).

This is because, on progressing from lithium to caesium, the outermost electron occupies a successively higher shell, e.g. 2s for lithium but 3s for sodium.

(ii) Atomic radius *decreases* from left to right across a period, e.g.

Li 0.123, Be 0.089, B 0.080, C 0.077, N 0.074, O 0.074, F 0.072 (nm).

This results from an increase in the *effective nuclear charge* as atomic number rises. Let us take lithium and beryllium as examples. In the former, the outermost 2s electron is shielded from the +3 nuclear charge by the two 1s electrons. Thus, the 2s electron experiences an effective nuclear charge of one proton, i.e. $+3 - 2 = +1$. However, the two outer 2s electrons of beryllium are screened from the +4 nuclear charge by the two 1s electrons to give an effective nuclear charge of two protons, i.e. $+4 - 2 = +2$. Therefore each outer electron of beryllium experiences a greater attraction to the nucleus, and is pulled closer towards it, than the outer electron of lithium.

Ionisation enthalpy

A plot of first ionisation enthalpy against atomic number (Fig. 3.18) clearly shows a periodic variation.

Four points are worthy of comment regarding this graph.

(i) The noble gases occur at maxima and the alkali metals at minima. In alkali metal atoms, the outer electron is relatively distant from the nucleus and is well shielded from it by the inner shells (i.e. the noble gas core). Consequently, the outer electron is relatively easy to remove and the alkali metals have low ionisation enthalpies.

(ii) Within a period there is a general increase in the ionisation enthalpy with increasing atomic number (§4.7.2 and §18.1.2).

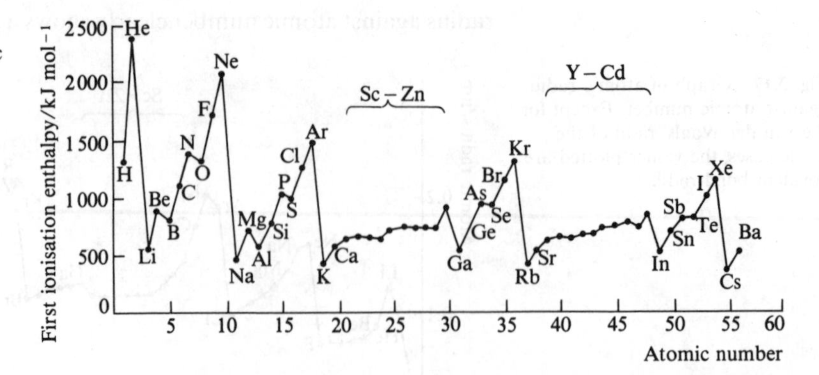

Fig. 3.18 Graph of first ionisation enthalpy against atomic number.

(iii) Ionisation enthalpies decrease down a group (§4.7.2). Thus, in Fig. 3.18, the peaks and troughs become progressively lower with increasing atomic number.

(iv) Corresponding to the decrease in ionisation enthalpy down a group the elements become more metallic, and corresponding to the increase across a period they become less metallic (§4.7.2). Thus, the most metallic elements appear on the left-hand side of the periodic table and the least metallic, or non-metallic, elements on the right-hand side. There is a step-wise dividing line across the p-block of the periodic table (Fig. 3.15) which separates the metals from the non-metals, although many of the metals adjacent to this line possess both metallic and non-metallic characteristics. Such elements, e.g. germanium and arsenic, are termed *semi-metals* (formerly 'metalloids').

Electronegativity

Electronegativity increases from left to right across a period and from bottom to top of a group (§4.6.2).

Other physical properties showing periodicity

Enthalpy of atomisation

A graph of enthalpy of atomisation against atomic number for the elements hydrogen to barium (Fig. 3.19) shows a periodic variation in the property.

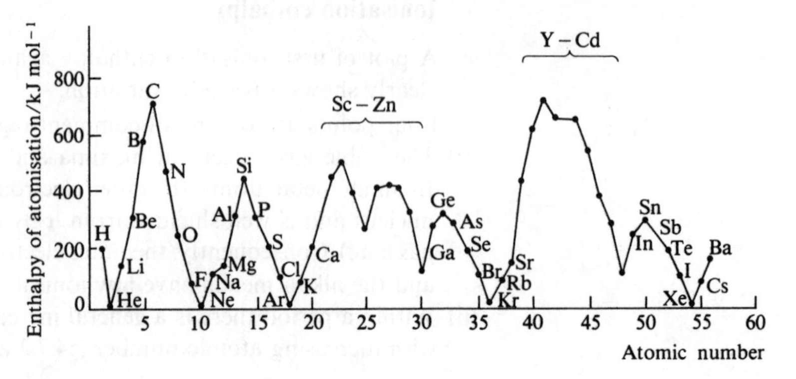

Fig. 3.19 Graph of enthalpy of atomisation against atomic number.

The reactivity of an element is very often related to its enthalpy of atomisation, since a chemical reaction can occur only after the element has been converted into free (i.e. isolated) atoms. In general, elements that possess high enthalpies of atomisation are unreactive, while those with low enthalpies of atomisation have a high reactivity. Noble gases are an exception to this rule. They exist in a monoatomic form and so have zero atomisation enthalpies, but display a lack of reactivity.

Melting temperature and boiling temperature

Neighbouring elements in a period often have totally different structures, with the result that the changes in melting temperature and boiling temperature from left to right across a period are highly erratic. Some periodicity, however, is evident within periods 2 and 3.

The noble gases and elements which form molecular crystals, e.g. oxygen and nitrogen, have low melting and boiling temperatures because of the weak van der Waals' forces that exist between the molecules. The elements carbon, silicon, boron and red phosphorus have very high melting and boiling temperatures because they form atomic crystals in which strong covalent bonds exist between the atoms. Melting or boiling can occur only by breaking these bonds, and a large amount of thermal energy is necessary for this purpose. The melting and boiling temperatures of the metallic elements vary according to the strength of the metallic bonding.

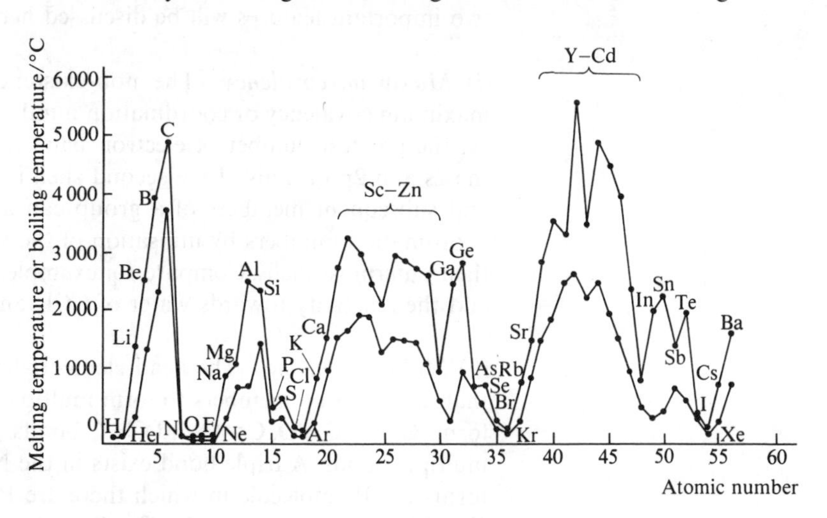

Fig. 3.20 Graphs of melting temperature (lower trace) and boiling temperature (upper trace) against atomic number.

Valency or oxidation state

The main group elements (For d-block elements see §18.1.2)

A valency or oxidation state numerically equal to the group number is displayed by all the elements of groups 1A, 2A, 3B and 4B. With the exception of the first or 'head' elements, the elements of groups 5B, 6B and 7B also achieve a valency or oxidation state equal to the group number, usually in combination with oxygen, fluorine or sometimes chlorine. With other reagents (e.g. hydrogen or sulphur) a valency of 8 minus the group number is usually observed. Some p-block elements, particularly the heavier ones of groups 3B and 4B, also show a valency equal to the group number minus two. Some examples are given in Table 3.5.

Table 3.5 Examples to illustrate the various valencies of the main group elements

	Group number						
Valency equal to	1A	2A	3B	4B	5B	6B	7B
group number	NaCl	$MgCl_2$	$AlCl_3$	CH_4	PF_5	SF_6	IF_7
8 − group number	—	—	—	—	PH_3	H_2S	HCl
group number − 2	—	—	TlCl	$PbCl_2$	—	—	—

The main group head elements

In many ways the head element of a group (i.e. the element in the second period), while resembling the second and subsequent members of the group (i.e. those elements of the third and higher periods), often shows important differences from them. In group 1A lithium and sodium are similar in most respects, but differ, for example, in that lithium forms a nitride by direct combination of the elements while sodium does not. The differences are not confined to the elements themselves but are also encountered in their compounds. Differences also occur between corresponding pairs of elements in other groups. Many of the differences are explained later, but two important features will be discussed here.

(*i*) *Maximum covalency* The non-existence of 2d orbitals limits the maximum covalency or coordination number of the head element to four, i.e. the greatest number of electron pairs that can be accommodated in the 2s and 2p orbitals of the second shell is four. In contrast, the second and subsequent members of a group can achieve higher covalencies or coordination numbers by utilisation of the vacant d orbitals that exist in their outermost shells. Compare, for example, oxygen and sulphur (§16.1.2), and the reactivity towards water of CCl_4 and $SiCl_4$ (§14.2.7).

(*ii*) *Multiple bonding* The head elements have a much greater tendency than other group members to form multiple bonds. For example, carbon forms C=O, C≡O, C=C and C≡C bonds, whereas silicon seldom forms multiple bonds. A triple bond exists in the N_2 molecule, but phosphorus forms the P_4 molecule in which there are P—P single bonds (§4.3.3 and Fig. 15.1). Oxygen forms the O=O molecule, whereas sulphur is singly bonded in S_8 molecules (§4.3.4 and §16.1.1).

Chemical behaviour

The chemical properties of the elements also show a periodic variation. From left to right across a period the reactivity of metals decreases while that of non-metals increases and reaches a maximum at group 7B. The noble gases on the extreme right of the periodic table are noted for their chemical inertness.

From top to bottom of a group we usually observe an increasing reactivity of metals and a decreasing reactivity of non-metals. These and other changes are summarised in Fig. 3.15.

Properties of compounds

Trends and periodicity in properties are not confined to the elements themselves but are shared by many of their compounds. This is well illustrated by the oxides and chlorides.

Oxides

Corresponding to the decrease in metallic character, oxides become increasingly acidic from left to right across a period,

e.g Na_2O MgO (basic) Al_2O_3 (amphoteric) SiO_2 P_4O_{10} SO_3 Cl_2O_7 (acidic)

increasingly acidic →
decreasingly basic

Oxides become increasingly basic or decreasingly acidic from top to bottom of a group as the metallic character of the elements develops,

e.g. N_2O_3 P_4O_6 (acidic) As_4O_6 Sb_4O_6 (amphoteric) Bi_2O_3 (basic)

decreasingly acidic →
increasingly basic

Halides

Many metals form ionic halides that are not hydrolysed by water. They possess high melting and boiling temperatures, and good electrical conductivity in solution or the fused state. The halides of non-metals, in contrast, are covalent and are usually hydrolysed by water. They have low melting and boiling temperatures, unless they are polymeric, and a low electrical conductivity. Some properties of period 3 chlorides are set out in Table 3.6.

Table 3.6 Some properties of the chlorides of period 3 elements

	$NaCl$	$MgCl_2$	$AlCl_3$	$SiCl_4$	PCl_3	S_2Cl_2	Cl_2
bonding	ionic	ionic	polymeric covalent	covalent	covalent	covalent	covalent
melting temperature/°C	801	714	180 (sublimes)	−70	−112	−78	−102
conductivity	good	good	nil	nil	nil	nil	nil
whether hydrolysed by water	no	no	partially (§13.2.6)	yes	yes	yes	yes

Table 3.6, viewed from right to left, shows that halides become increasingly ionic as the elements become more metallic. A similar trend

is observed from top to bottom of a group as the elements increase in metallic character,

e.g. NF_3 PF_3 AsF_3 SbF_3 BiF_3
covalent (NF₃ to SbF₃) — ionic (BiF₃)

3.2.5 DIAGONAL RELATIONSHIPS

The trends that occur from left to right across the periodic table are always accompanied by the opposite trend from top to bottom of a group (Fig. 3.15). For example, metallic character and ionic radius both *decrease* across a period, but *increase* down a group. As a result, a particular element often resembles the element which is in the period below it but one place to the right, especially in such properties as metallic character, electronegativity and atomic radius. There are thus *diagonal relationships* in the periodic table. The best and most important examples are those at the left-hand side of periods 2 and 3, comprising the following elements: Li—Mg (§12.2.8), Be—Al (§12.2.9) and B—Si (§13.2.8).

3.3 Nuclear changes

3.3.1 RADIOACTIVITY

Atoms of most nuclides (§1.1.1) are stable, in the sense that they do not change of their own accord into other groupings of protons, neutrons and electrons. There are some nuclides, however, whose atomic nuclei are so unstable that they change entirely of their own accord into the atomic nuclei of other elements. Several types of change are recognised, and the rate of change can vary enormously from one nucleus to another. During the change the unstable atomic nucleus ejects energetic particles, and often emits electromagnetic radiation of very short wavelength. Because of these radiations, unstable nuclides are said to be *radioactive*, and their spontaneous change is termed *radioactive disintegration* or *radioactive decay*.

About 40 naturally occurring nuclides and several hundred artificial ones possess this property. An example is potassium-40, which decays partly into argon-40 and partly into calcium-40. The presence of a little potassium-40 in naturally occurring potassium salts causes potassium and all its compounds to be slightly radioactive.

Ionising radiations

As a substance decays radioactively it can emit *α-rays*, *β-rays* or *γ-rays*. All are *ionising radiations*, so-called because they cause the ionisation of gases or other substances through which they pass.

α-Rays are streams of *α-particles*. An α-particle is the nucleus of a helium atom, with a relative mass of 4 and a charge of +2. It may be regarded as a dipositive helium ion and represented by the symbol He^{2+}, although it is more usually written as ${}^{4}_{2}He$.

β-Rays are streams of *β-particles*. Their behaviour in magnetic and electrical fields shows them to be negatively charged particles identical with electrons.

α- and β-Rays are said to be *corpuscular radiations*, as they are composed of small particles or *corpuscles*. γ-Rays, by contrast, are a form of *electromagnetic radiation*, comparable to X-rays but of shorter wavelength. The wavelength of γ-rays is approximately 10^{-12} m, whereas that of X-rays is approximately 10^{-10} m.

Ionising radiations may conveniently be compared under the following headings.

Velocity

α- and β-Particles are ejected from atomic nuclei with very high velocities, about one-tenth that of light for α-particles and approaching that of light for β-particles. γ-Rays, like all forms of electromagnetic radiation, travel at the velocity of light.

Energy

α-, β- and γ-Radiations are all of very high energy. The energy of α- and β-radiations is due to the high velocity of their particles, while that of γ-radiation can be attributed to the short wavelength. (Because of their shorter wavelength, γ-rays possess more energy than X-rays.)

The kinetic energy of α-particles depends on the instability of the nuclei from which they originate; the greater the instability, the greater is the energy. A particular type of nucleus may emit α-particles whose energies are identical, or it may emit groups of α-particles with energies that lie fairly close together.

The energies of β-particles vary considerably, even among β-particles from the same type of nucleus. The energies range from quite small values to a maximum which is characteristic of the emitting nucleus.

Ionising properties

When α-, β- and γ-rays travel through a gas they lose their energy in bringing about its ionisation. The mechanism of the process depends on whether the radiation is corpuscular or electromagnetic. Let us consider corpuscular radiation first. α-Particles, which are positively charged, attract the electrons of gas molecules, while β-particles, which are negatively charged, repel them. Thus, either type of particle, when in the vicinity of a gas molecule, may cause an electron to break away so that the molecule is converted into a positive ion. The electron that is displaced may remain free for a time, or it may join another gas molecule to give a negative ion. The ionisation of successive molecules has the effect of slowing down α- and β-particles so that eventually they come to rest. After α-particles have been slowed right down, they acquire electrons from their surroundings to become helium atoms. Pockets of helium originating in this way can be found in certain minerals.

For γ-radiation to ionise a gas molecule it is necessary for a photon (i.e. a particle of radiation) to collide with an electron of the gas molecule. (It is insufficient for the radiation merely to be in the vicinity of the molecule.) Such a collision displaces the electron and yields a positive ion, a process known as the *Compton effect*. However, the chances of photon–electron collisions taking place are exceedingly small, yet γ-radiation is known to be powerfully ionising. The reason is that the displaced electrons, called *recoil electrons*, have a high energy and are able to ionise many other gas molecules.

α-, β- and γ-Rays can also bring about the ionisation of liquids and solids.

Mode of travel

α-Particles, because of their high mass, travel in straight lines. β-Particles, which are light and easily deflected, have a non-linear flight path. γ-Radiation, like visible light, travels in straight lines.

Deflection

α- and β-Particles are deflected by magnetic and electrical fields, but in opposite directions. β-Particles, with their relatively small mass, are deflected more than α-particles (Fig. 3.21). γ-Rays, which do not consist of charged particles, are unaffected by magnetic and electrical fields.

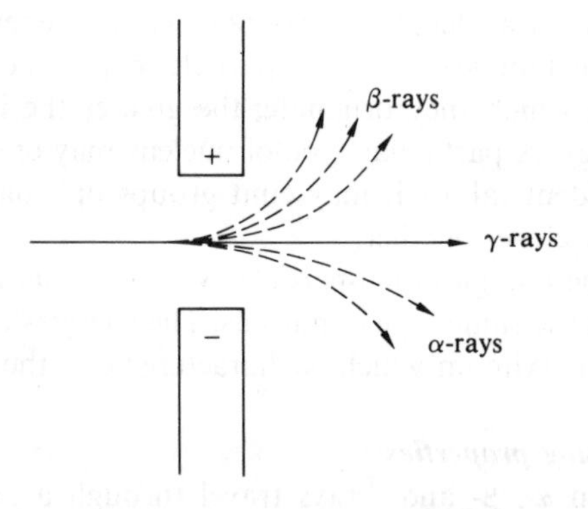

Fig. 3.21 The effect of an electrical field on α-, β- and γ-rays.

Range

α-Particles have very little penetrating power, being stopped by a sheet of paper, very thin aluminium foil or a few centimetres of air. β-Particles have a range which is much greater than that of α-particles of the same energy. Aluminium sheet of 2 mm thickness may be required to stop them.

There are two reasons for the difference. First, α-particles have a double charge, while β-particles are only singly charged. Therefore, over a certain distance, α-particles can ionise more gas (or other) molecules than can β-particles, and their energy is more rapidly dissipated. Secondly, α-particles are slower moving than β-particles and remain in the vicinity of gas molecules for longer periods of time. Consequently they are more likely to ionise the gas; once again, there is a rapid loss of energy.

γ-Radiation has a far greater range than α- and β-radiation and is able to penetrate considerable thicknesses of materials, even those of high density. The reason, as we have seen, is that energy is lost from the photons only by direct collision with electrons. Although some such collisions take place on entrance, others do not occur until the photons have travelled a considerable distance.

Rate of decay

The rate of decay of a nuclide is defined as the rate at which the number (N) of unstable nuclei decay with time (t), and is thus represented by $\frac{-\mathrm{d}N}{\mathrm{d}t}$. The rate depends only on the number of unstable nuclei,

$$\text{i.e.} \quad \frac{-\mathrm{d}N}{\mathrm{d}t} \propto N$$

$$\text{or} \quad \frac{\mathrm{d}N}{\mathrm{d}t} = -\lambda N$$

where λ is the *decay constant*, defined as the proportion of the atoms in the sample that decay in unit time.

Radioactive decay is thus a first order process (§7.4.1). A plot of the mass of an unstable nuclide against time gives a logarithmic curve, as shown by Fig. 3.22.

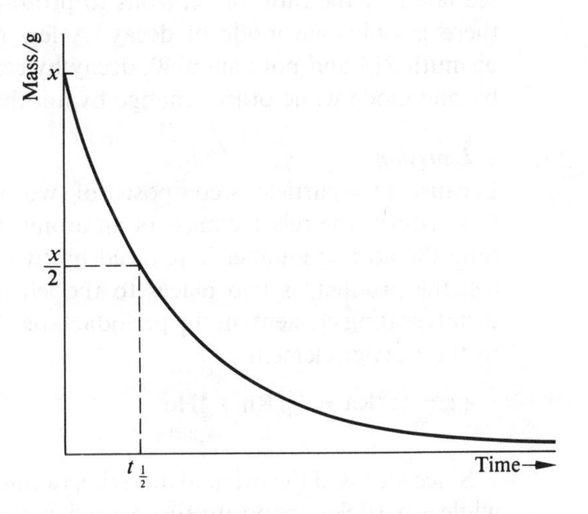

Fig. 3.22 Decay curve of an unstable nuclide of initial mass x g.

Because the decay curve is asymptotic to the x-axis, a radioactive substance never decays completely. To indicate the stability of a nuclide we therefore quote not the time taken for the entire mass to decay, but the *half-life*, $t_{\frac{1}{2}}$, which is the time required for atoms of the nuclide to decay by half. The half-lives of radioactive nuclides vary tremendously, from fractions of a second to about 10^{10} years. For example, the half-life of radium-226 is 1 580 years, whereas that of its primary decay product, radon-222, is only 3.8 days. The half-life concept is explained more fully by Fig. 3.23, which shows that the residual amount of radon-222 is reduced by half in every 3.8 days.

The rate of radioactive decay cannot be altered by any chemical or physical means. Unstable nuclides decay in the same way in compounds as in the element itself, and the rate is totally unaffected by extreme changes of temperature.

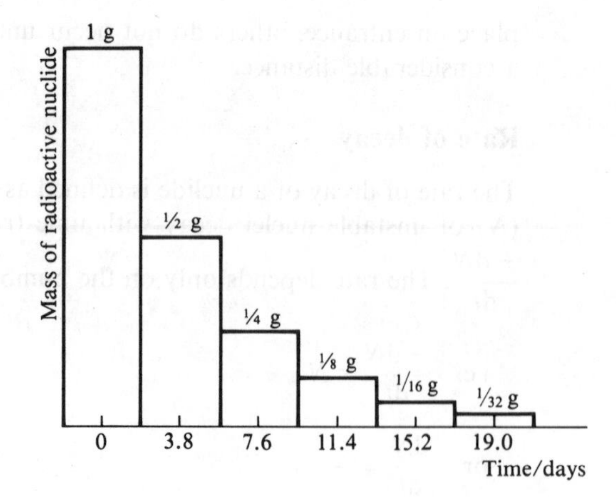

Fig. 3.23 Radioactive decay of 1 gram of radon-222 ($t_{\frac{1}{2}}$ = 3.8 days). Cf. Fig. 3.22.

Modes of decay

A nucleus may decay by α-, β- or γ-emission. There may be more than one type of emission, although a particular nucleus cannot emit both α- and β-particles simultaneously. The mode of decay is governed, as we shall see later, by the ratio of neutrons to protons in the nucleus, and usually there is only one mode of decay. A few radioactive nuclides, notably bismuth-214 and potassium-40, decay by certain of their nuclei changing by one mode while others change by another, but this is uncommon.

α-Emission

Because an α-particle is composed of two protons and two neutrons, its loss reduces the relative mass of an atomic nucleus by four. At the same time the atomic number is reduced by two, so that the *daughter nucleus* (i.e. the product) is two places to the left of the *parent nucleus* (i.e. the disintegrating element) in the periodic table. Decay of this sort is restricted to the heavier elements,

$$\text{e.g.}\quad {}^{226}_{88}\text{Ra} = {}^{222}_{86}\text{Rn} + \underset{\alpha\text{-particle}}{{}^{4}_{2}\text{He}}$$

Since atoms of the original material, radium-226, are electrically neutral, while α-particles are positively charged, it may be argued that the product, radon-222, should be formed as negative ions. In fact it is neutral, because two extranuclear electrons are lost at the same time as the α-particles. (These electrons are not regarded as β-particles since, by definition, β-particles come from the nucleus.) Similarly,

$${}^{238}_{92}\text{U} = {}^{234}_{90}\text{Th} + {}^{4}_{2}\text{He}$$

$${}^{214}_{83}\text{Bi} = {}^{210}_{81}\text{Tl} + {}^{4}_{2}\text{He}$$

In all equations for nuclear reactions the mass numbers (shown in superscript) and the nuclear charges (shown in subscript) must balance.

β-Emission (β^--Emission)

This occurs through the conversion, in the nucleus, of a neutron into a proton and an electron:

$$n = p^+ + e^-$$

The proton remains in the nucleus, thereby raising its atomic number by one. The electron, however, leaves the nucleus as a β-particle.

Thus, when a β-particle is emitted, the daughter nucleus is one place to the right of the parent nucleus in the periodic table. The mass number of the product, however, is the same as that of the original nuclide, because the sum of the protons and neutrons is unaltered. Change of this kind, in which the mass remains constant, is referred to as *isobaric change*. Examples are as follows:

$$^{234}_{90}Th = ^{234}_{91}Pa + ^{0}_{-1}e$$

$$^{214}_{82}Pb = ^{214}_{83}Bi + ^{0}_{-1}e$$

As with α-emission, the product is electrically neutral, contrary to what the equation might suggest, because the extra proton in the nucleus is balanced by the acceptance of a stray electron into the shells of the atom.

γ-*Emission*

γ-Radiation can have several causes.

(i) It is often, but not always, a consequence of α- or β-emission. The loss of an α- or β-particle may cause the remaining nucleus to be in an excited state. The excess energy, above that of the ground state of the nucleus, is subsequently emitted as γ-radiation.

(ii) Another process accompanied by γ-radiation is that of *electron capture*, i.e. the capture, by a proton, of an extranuclear electron, usually from the K shell (*K-capture*). The vacancy in the K shell ($n = 1$) is filled by an electron from a higher shell. In this way a proton is converted into a neutron. The mass number of the parent nucleus thus stays the same, but its atomic number decreases by one,

e.g. $^{85}_{38}Sr + ^{0}_{-1}e = ^{85}_{37}Rb$

γ-Emission then occurs from the excited nucleus, as before.

(iii) The γ-radiation of certain man-made nuclides may be associated with the emission of *positrons*. A positron has the same mass as an electron, but a positive charge, and may be represented as $^{0}_{1}e$ or β^+. Free positrons, like free electrons, are stable, but in the presence of any matter a positron rapidly combines with a stray electron to give γ-radiation. The process is called *pair annihilation*.
The best known example concerns phosphorus-30:

$$^{30}_{15}P = ^{30}_{14}Si + ^{0}_{1}e$$

None of the nuclides that are found in nature decay in this way.

Stability of nuclei

All nuclei containing 84 or more protons, i.e. the elements from polonium onwards, are fundamentally unstable and are radioactive. High atomic number, however, is not the sole cause of radioactivity, because certain isotopes of the lighter elements are also unstable. The factor that determines

Fig. 3.24 Nuclear stability curve.

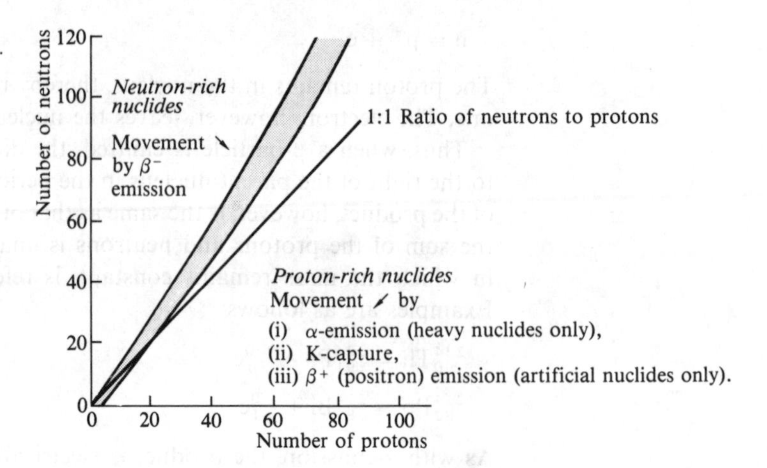

whether or not an atom is radioactive is *the ratio of neutrons to protons in the nucleus.* Stable nuclei of the lighter elements have neutron:proton ratios of about 1:1, but for the heavier elements to be stable there must be more neutrons than protons. For any atomic number, there is a restricted range of values for the neutron:proton ratio, inside which there is stability but outside which there is instability (Fig. 3.24).

The shaded region in Fig. 3.24 represents the *belt of stability.* Most nuclides whose ratios of neutrons to protons cause them to lie inside the belt are stable (potassium-40 and all isotopes of technetium and promethium are notable exceptions), but nuclides that lie outside the belt decay, by one of the processes shown in the figure, in such a way that they move into the stability range.

3.3.2 NUCLEAR REACTIONS

In a *chemical reaction* there is merely a rearrangement of extranuclear electrons. The atomic nuclei remain unaffected. In a *nuclear reaction,* by contrast, there is a change in the number of protons or neutrons in the nucleus so that a new atomic species is produced. There are two types of nuclear reaction, as follows.

(i) That in which a nucleus is unstable and undergoes a spontaneous change in order to produce a more stable nucleus. This is the basis of radioactive decay (§3.3.1).

(ii) That in which a change is brought about artificially, by the collision of two nuclei. It is this type of nuclear reaction that concerns us here.

Any atomic nucleus can be used as a *target.* The *projectile,* with which it is bombarded, may be any energetic small particle. Protons, neutrons, deuterons ($^{2}_{1}H$ nuclei), α-particles and other ions of low atomic number are all commonly used. α-Particles emitted by radioactive sources are relatively inefficient at bringing about nuclear reactions, because they are positively charged and therefore repelled by the target nucleus. All positively charged particles (protons, deuterons and α-particles), when used as projectiles, become more efficient when given a high kinetic energy by an accelerating device such as a cyclotron. This does not apply to neutrons, which are electrically neutral.

In a nuclear reaction the new atomic species that is produced is usually (but not always) radioactive. Its formation is accompanied by the emission of other particles (protons, neutrons or α-particles) or by γ-radiation,

e.g. $$^{14}_{7}N + ^{4}_{2}He = \underset{\text{(stable)}}{^{17}_{8}O} + ^{1}_{1}H$$

The mechanism of such reactions is uncertain, but it is likely that the projectile forms, with the target, a highly unstable *compound nucleus* which then decomposes:

$$^{14}_{7}N + ^{4}_{2}He \xrightarrow{\text{slow}} \underset{\text{(unstable)}}{^{18}_{9}F} \xrightarrow{\text{rapid}} ^{17}_{8}O + ^{1}_{1}H$$

Nuclear reactions are classified according to the projectile used and the particle or radiation emitted. The reaction just described is an (α,p) reaction, because α-particles are used and protons are ejected. Other examples are as follows. In each case the reaction has been classified, and the mode of decay of the product nuclide has been stated.

$$^{9}_{4}Be + ^{4}_{2}He = \underset{\text{(stable)}}{^{12}_{6}C} + ^{1}_{0}n \qquad (\alpha,n) \text{ reaction}$$

$$^{7}_{3}Li + ^{1}_{1}H = \underset{\text{(electron capture)}}{^{7}_{4}Be} + ^{1}_{0}n \qquad (p,n) \text{ reaction}$$

$$^{31}_{15}P + ^{2}_{1}H = \underset{(\beta^{-} \text{ emission})}{^{32}_{15}P} + ^{1}_{1}H \qquad (d,p) \text{ reaction}$$

$$^{14}_{7}N + ^{1}_{1}H = \underset{(\beta^{+} \text{ emission})}{^{15}_{8}O} + \gamma \qquad (p,\gamma) \text{ reaction}$$

Notice that in the last reaction no subsidiary particle is emitted; only γ-radiation.

Neutrons, because they do not have to be accelerated, are particularly useful as projectiles. A convenient source is the (α,n) reaction involving beryllium-9, listed above. The neutrons that are emitted are *fast neutrons*, and when used as projectiles they bring about comparable reactions in which other particles are emitted,

e.g. $$^{14}_{7}N + ^{1}_{0}n = ^{14}_{6}C + ^{1}_{1}H \qquad (n,p) \text{ reaction}$$

The kinetic energy of neutrons may be reduced by passage through a *moderator*, such as graphite or paraffin wax, and when the resulting *slow neutrons* are used as projectiles there may be neutron capture *without* particle emission,

e.g. $$^{107}_{47}Ag + ^{1}_{0}n = ^{108}_{47}Ag + \gamma \qquad (n,\gamma) \text{ reaction}$$

Isotopes of many other elements can be produced in this way.

3.3.3 NUCLEAR ENERGY

Einstein's equation

According to Einstein's theory of relativity, matter and energy are interconvertible. The quantitative relationship between the two is expressed by the equation:

$$E = mc^2$$

where E = energy (in joules), m = mass (in kilograms) and c = velocity of light (3×10^8 m s^{-1}). Because the constant c^2 is so high, a small amount of mass can give rise to a very large amount of energy,

e.g. $1 \text{ kg} \equiv 9 \times 10^{16}$ J

Very often, in a nuclear reaction, the mass of the products is less than that of the reactants, indicating that during the reaction some matter is converted into energy. This energy, termed *nuclear energy*, can appear as a variety of electromagnetic radiations, e.g. radiant heat, light and γ-radiation.

Mass defect and binding energy

Protons and neutrons in the nucleus of an atom are held together by very powerful binding forces. A measure of these forces is provided by the *binding energy* of a nuclide, which is defined as the energy that must be supplied to the atoms, in one mole of the nuclide, to break them down into their component sub-atomic particles.

Binding energy can readily be calculated, because it must be equal to the energy that is *released* when the atoms are formed from sub-atomic particles. Consider the case of helium-4, whose atoms are composed of two protons, two neutrons and two electrons. The relative mass of a proton is 1.007 8 and that of a neutron 1.008 7. (The mass of an electron is so small that it can be neglected.)

$\therefore$ relative atomic mass, by summation, $= 2(1.0078 + 1.0087)$
$= 4.0330$

But the *observed* relative atomic mass, 4.003 9, is less than this. The difference, 0.029 1, referred to as the *mass defect* or *mass deficit*, is the amount of matter released as binding energy when the protons and neutrons pack together to form helium nuclei. Conversion of the mass defect of 0.029 1

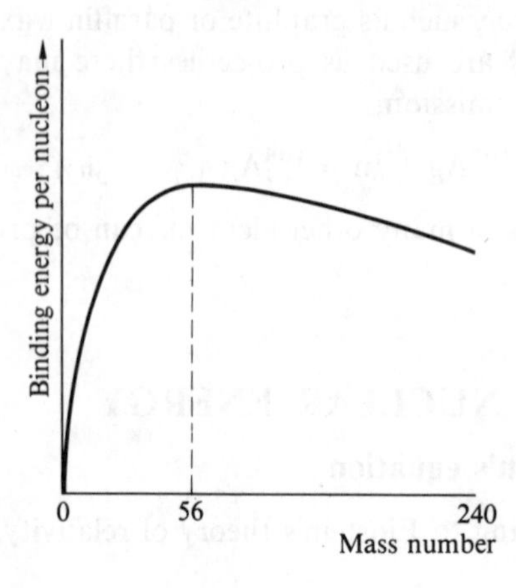

Fig. 3.25 Variation of binding energy with mass number.

to energy by Einstein's equation indicates a binding energy for the helium nucleus of approximately 2.7×10^9 kJ mol^{-1}. Division by four gives a value of 6.75×10^8 kJ mol^{-1} per nucleon (i.e. per nuclear particle). The binding energies of other atomic nuclei can be calculated in the same way and plotted against mass number (Fig. 3.25).

Figure 3.25 shows that atomic nuclei with a mass number of ~ 56 (i.e. nuclei of the isotopes of iron) have the maximum binding energy and are thus the most stable. Hence if other nuclei, to the left or right of this maximum, undergo nuclear change in such a way that they approach the maximum, energy will be released. There are, therefore, two broad approaches to the generation of nuclear energy: (i) by the *fission* or splitting of heavy nuclei; and (ii) by the *fusion* or joining together of light nuclei.

Nuclear fission

Atomic bombs and nuclear power stations are based on the fission of uranium-235 or plutonium-239.

Spontaneous fission, by which is meant the splitting of a heavy nucleus of its own accord into two smaller fragments, is a change comparable to radioactivity. A half-life can be assigned to a nucleus for its spontaneous fission, as for its radioactive decay.

However, spontaneous fission is an extremely rare event, about one million times less likely to occur than radioactive decay. Of more importance is *induced fission*, in which the change is brought about by bombarding the heavy nuclei with small particles or γ-rays. For all practical purposes, the fission of uranium-235 or plutonium-239 is induced by bombardment with slow neutrons. When a neutron joins the target nucleus an unstable compound nucleus is formed, which immediately splits into two fission products. At the same time further neutrons are released:

$$^{235}_{92}\text{U} + ^{1}_{0}\text{n} \longrightarrow \underset{\text{(unstable)}}{^{236}_{92}\text{U}} \longrightarrow \text{A} + \text{B} + 2^{1}_{0}\text{n}$$

The two fission products A and B differ from one fission to another, even for the same type of nucleus. They are generally of unequal mass, with one having a mass number of about 90 and the other 140. The following equations illustrate some likely fission products of uranium-235:

$$^{235}_{92}\text{U} + ^{1}_{0}\text{n} = ^{94}_{38}\text{Sr} + ^{140}_{54}\text{Xe} + 2^{1}_{0}\text{n} + \gamma$$

$$^{235}_{92}\text{U} + ^{1}_{0}\text{n} = ^{95}_{42}\text{Mo} + ^{139}_{57}\text{La} + 2^{1}_{0}\text{n} + 7^{\;0}_{-1}\text{e}$$

The fission of plutonium-239 follows exactly the same pattern. Most of the primary fission products are radioactive and, being neutron rich, decay by β-emission (Fig. 3.24).

Fission is always accompanied by a decrease in mass, as the following example shows:

$$\underbrace{\underset{235.0439}{^{235}_{92}\text{U}} + \underset{1.0087}{^{1}_{0}\text{n}}}_{236.0526} = \underbrace{\underset{89.9470}{^{90}_{36}\text{Kr}} + \underset{143.8810}{^{144}_{56}\text{Ba}} + \underset{2.0174}{2^{1}_{0}\text{n}}}_{235.8454}$$

relative mass

mass loss = 236.0526 − 235.8454 = 0.2072

The decrease in mass represents matter that is converted into energy. In theory, the total amount of energy that can be liberated by the fission of one kilogram of uranium-235 is equal to the energy that can be obtained by burning three million kilograms of coal; but in practice, in a nuclear power station, less than 1% of this energy is released.

A glance at the equations for the fission of uranium-235 shows that for every neutron that is absorbed two are released. Once started, therefore, this should be a *nuclear chain reaction*, i.e. a nuclear reaction that is self-propagating. In practice, however, the fission of a small piece of uranium-235 is not self-sustaining because many neutrons are lost to the environment. Nevertheless, if the uranium is above a certain *critical size*, which depends partly on mass and partly on geometry, sufficient neutrons are retained to keep the fission going.

To start the chain reaction it is unnecessary to bombard the uranium-235 with neutrons artificially, because neutrons are forever being released in the uranium by the following processes:

(i) spontaneous fission;
(ii) fission induced by γ-rays from radioactive decay;
(iii) (α,n) reactions of heavy nuclei, e.g. uranium-238:

$$^{238}_{92}U + ^{4}_{2}He = ^{241}_{94}Pu + ^{1}_{0}n$$

Naturally occurring uranium contains only 0.7% of uranium-235, the remaining 99.3% being uranium-238. (There is also a trace of uranium-234.) Before the uranium can be used in an atomic bomb or a power station it must be *enriched* in uranium-235, i.e. the ratio of $^{235}U{:}^{238}U$ must be increased. This is because uranium-238 is not fissionable in a nuclear reactor and, by absorbing neutrons, interrupts the fission of uranium-235.

Nuclear fusion

The energy of the sun and other stars is derived from various *nuclear fusion reactions*, in each of which two atomic nuclei fuse together to give a heavier one. Energy is released because fusion, like fission, is accompanied by a decrease in mass.

Nuclear fusion is a very difficult change to bring about on earth, but occurs in the *hydrogen bomb*. This is a weapon designed to liberate a huge amount of energy, in a short space of time, by the combination of deuterons and tritons (i.e. $^{3}_{1}H$ nuclei) to give helium:

$$^{2}_{1}H + ^{3}_{1}H = ^{4}_{2}He + ^{1}_{0}n$$

A small fission bomb, incorporated in the device, provides the necessary high temperature to ionise the 'heavy hydrogen' and accelerate the nuclei.

EXAMINATION QUESTIONS ON CHAPTER 3

1 (a) Describe the essential features of the atomic spectrum of hydrogen and show how it leads to the concept of the principal energy levels in an atom.

(b) Show how a comparison of the first ionisation energies of successive elements indicates the presence of energy sub-levels.

(c) Using this knowledge of the energy levels in atoms, discuss the number of elements in period 1 and in periods 2 and 3 of the periodic table. (JMB)

2 With the aid of a simple sketch, describe an experimental arrangement by means of which you could observe the emission spectrum of a gas such as hydrogen. What changes in the apparatus would be required to observe the absorption spectrum of the gas?

(a) How does the appearance of an emission spectrum differ from that of the absorption spectrum of the same gas?

(b) Discuss the relationship between the frequency of a line in the atomic (emission) spectrum and electron energy levels.

(c) How are ionisation energies of elements calculated from spectroscopic measurements?

(d) What information do ionisation energies yield about electronic energy levels? (OC)

3 Illustrate the gradation of properties across a short period of the periodic table by considering for the elements lithium to fluorine (but excluding boron) the hydrides formed by each element, their bonding and their characteristic physical and chemical properties. (O)

4 (a) What are α, β and γ emissions and how do they differ in their penetrating power and their behaviour in a magnetic field?

Explain the meaning of the two numbers before the symbol for uranium and identify P, Q, R and S in the following equations:

$$^{234}_{92}U \longrightarrow \alpha + P$$
$$^{239}_{92}U \longrightarrow \beta^- + Q$$
$$^{235}_{92}U \longrightarrow \gamma + R$$
$$^{238}_{92}U + ^{2}_{1}H \longrightarrow ^{239}_{92}U + S$$

(b) Deduce the nature of X in the following reaction:

$$^{235}_{92}U + ^{1}_{0}n \longrightarrow ^{95}_{42}Mo + ^{139}_{57}La + 2^{1}_{0}n + 7X$$

There is approximately 0.1% less mass on the right-hand side of the equation than on the left-hand side. What is the significance of this? (JMB)

5 (a) Discuss the essential features and the theoretical implications of the atomic spectrum of hydrogen. Include in your answer an explanation of why the spectrum involves many lines which can be divided into a number of groups, and why the spacing of the lines varies.

(b) Give the equations which are involved when measuring the first and second ionisation energies of magnesium.

(c) State how and why the first ionisation energies change within Group 1 (Li to Cs), and across period 3 (Na to Ar). (JMB)

6 (a) Describe the composition of the atomic nucleus and the consequences of nuclear instability.

(b) Explain why the emission of an alpha particle from ^{214}Po followed by the emission of two successive beta particles and another alpha particle produces lead.

(c) Describe and account for the nature of the atomic spectrum of hydrogen.

(d) Explain why the third row (Na–Ar) of the periodic table contains only eight elements. (JMB)

7 (a) How are the *atomic number* and *mass number* related to the structure of the nucleus of an atom?

(b) The following table gives the atomic numbers of certain elements together with the relative atomic masses of their stable isotopes.

Atomic number	Relative atomic masses
1	1, 2
5	10, 11
8	16, 18
12	24, 25, 26
17	35, 37
26	54, 57, 58
29	63, 65
40	87, 93
44	104, 107

(i) Plot a graph of atomic number against number of neutrons for each stable isotope, in order to establish a *band of stability*, with both upper and lower limits.

(ii) From the graph predict which of the following species are **unstable**

$^{24}_{10}Ne$, $^{45}_{20}Ca$, $^{50}_{23}V$, $^{64}_{29}Cu$, $^{92}_{42}Mo$,

(iii) Having determined the unstable species, explain the *number* and *types* of emission required in order to attain stable structures. What would be the atomic number and atomic mass of the stable isotopes attained?

(c) (i) What are the **two** methods by which $^{27}_{13}Al$ could lose electrons? What would be the products in each case?

(ii) If $^{27}_{13}Al$ absorbed a neutron and subsequently underwent beta-emission, what would be the resulting isotope? Give its atomic number and atomic mass. (AEB)

8 The letters **A** to **F** give the electronic configurations of six elements.

(a) To each letter append the Roman numeral of one of the statements (i) to (vi) so that the statement about the element best fits the electronic configuration.

(*You need only write the letter and the appropriate Roman numeral.*)

A	$1s^2\ 2s^22p^6\ 3s^23p^63d^5\ 4s^2$	i.e. 2, 8, 13, 2
B	$1s^2\ 2s^22p^6\ 3s^2$	2, 8, 2
C	$1s^2\ 2s^22p^6$	2, 8
D	$1s^2\ 2s^22p^6\ 3s^23p^1$	2, 8, 3
E	$1s^2\ 2s^1$	2, 1
F	$1s^2\ 2s^22p^6\ 3s^23p^4$	2, 8, 6

(i) forms no compounds with other elements
(ii) forms compounds many of which are coloured
(iii) has the highest ionisation energy in the group of the periodic table in which the element is found
(iv) forms an amphoteric oxide, X_2O_3
(v) forms a nitride on heating in nitrogen
(vi) forms only acidic oxides

(b) Give four reasons, based purely on the *properties* of the elements and their compounds (*i.e. not* theoretical considerations), why the alkali metals are placed in the same group of the periodic table.

(c) Account for the change in atomic size on
(i) descending a group of the periodic table,
(ii) crossing a period of the periodic table from left to right.

(d) Explain why tellurium is placed before iodine in the periodic table though the former has the higher relative atomic mass.

(e) Explain why vanadium, of atomic number 23, is able to form a series of ions such as V^{2+}, V^{3+}, VO^{2+} and VO_3^-. (SUJB)

9 (a) Show how the form of the periodic table is a direct consequence of the electronic structure of the elements.

(b) What is the electrochemical series? Show how it is established (experimental details are not required), and how it is related to the periodic table. (JMB)

10 The atomic spectrum of hydrogen is given by the following relationship,

$$\frac{1}{\lambda} = R_H\left(\frac{1}{n_1^2} - \frac{1}{n_2^2}\right)$$

(a) (i) What does λ represent?
(ii) What do the terms n_1 and n_2 represent?
(iii) What are the units of the constant R_H?

(b) The spectrum comprises a number of lines which may be divided into a number of series.
(i) Why does the spectrum consist of lines?
(ii) Why is there a small number of series in the spectrum?
(iii) Explain why each series converges and in what direction it converges.

(c) What method is used to generate the light source for observing the atomic spectrum of hydrogen?

(d) Name the instrument used to resolve the hydrogen spectrum. (JMB)

11 (a) Explain the terms *mass deficit* and *alpha emission.*
For the following atoms of naturally-occurring non-radioactive isotopes
^{1}H, ^{4}He, ^{7}Li, ^{9}Be, ^{11}B, ^{12}C, ^{14}N, ^{16}O, ^{19}F, ^{20}Ne, ^{23}Na, ^{24}Mg, and ^{27}Al
plot the number of protons against the number of neutrons in each nucleus. Using your graph show that ^{14}C and ^{24}Na are radioactive isotopes decaying by beta-emission. Write equations for the decay processes of these two isotopes.

(b) The mass of the nucleus of an atom of ^{127}I is $2.106\,61 \times 10^{-25}$ kg. The masses of a single proton and single neutron are $1.672\,52 \times 10^{-27}$ kg and $1.674\,82 \times 10^{-27}$ kg respectively. Calculate the mass deficit per nucleon for an iodine nucleus. (AEB)

12 (a) Outline the principles involved in determining relative atomic mass using a mass spectrometer (Chapter 5).

(b) The isotope $^{211}_{83}Bi$ undergoes radioactive decay becoming the isotope $^{207}_{83}Bi$. Trace a possible path for this conversion giving (i) the types of radiation emitted, and (ii) the relative atomic mass and atomic number of the intermediate isotopes.

(c) What is the exact nature of these radiations, and how does each of them behave when

(i) directed towards a thin aluminium sheet,
(ii) influenced by a magnetic field,
(iii) passed through a gaseous element? (AEB)

13 The following represents part of a radioactive decay series

$$\mathbf{Y} \longrightarrow \mathbf{Z} \longrightarrow {}^{207}_{82}Pb$$

(a) Identify the radioactive elements **Y** and **Z**, giving the chemical symbol, atomic number and mass number in each case.

(b) If the nucleus **Y** decays forming $^{207}_{82}Pb$ by an alternative route

$$\mathbf{Y} \longrightarrow \mathbf{Q} \longrightarrow {}^{207}_{82}Pb$$

identify the intermediate radioactive element **Q**.

(c) If the decay of **Y** to $^{207}_{82}Pb$ is allowed to take place in a closed vessel what gas would be produced?

(d) State two differences between α and β^- particles. (JMB)

4 Chemical bonding

4.1 The electronic theory of valency

According to the *electronic theory of valency*, proposed independently by W. Kossel and G. N. Lewis in 1916, a chemical bond may be formed between two atoms as a result of the electronic interactions that occur as they approach each other. The theory has three fundamental principles, which are as follows.

(i) An atom always attempts to achieve a stable electronic arrangement when it bonds with other atoms.

(ii) The tendency to reach a stable electronic arrangement may be satisfied by the transfer of one or more electrons from one atom to another. This is the principle of *electrovalency*.

(iii) The tendency may also be satisfied by the sharing of a pair of electrons between two atoms. This is the principle of *covalency*.

For the elements up to and including calcium, where electrovalent bonds are formed, the stable electronic arrangements referred to in the first principle are those of the noble gases. Where covalent bonds are formed, however, this is not always so.

4.1.1 COVALENCY

When two atoms, A and B, approach each other, the electrostatic forces of attraction between the nucleus of each atom and its electrons are supplemented by attraction between the two atoms. The nucleus of A attracts the electrons of B, and similarly the nucleus of B attracts the electrons belonging to A. The attraction is weak at first, when the atoms are far apart, but becomes stronger as the atoms move towards each other. To some extent, however, the attraction is offset by a repulsion between the electrons of A and those of B, and also between the two atomic nuclei.

Depending on the balance between these forces, the atoms may approach each other sufficiently closely for suitable orbitals to overlap, and if this happens a chemical bond is formed. Let us suppose, first, that the outermost occupied orbital of each atom holds its maximum of two electrons (Fig. 4.1).

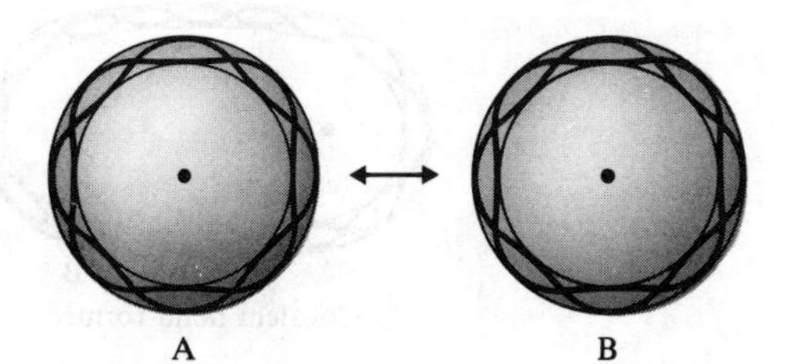

Fig. 4.1 Strong repulsion between two atoms, each with an electron pair in its outermost occupied orbital.

In such a case the repulsion between atoms A and B is so strong that it prevents these orbitals from overlapping, and there is no bond formation. (Although Figs. 4.1, 4.2 and 4.3 depict the outer envelopes of electron probability, they are not intended to offer pictorial representations of electrons in orbitals.)

If the outermost occupied orbital of one of the atoms is filled by an electron pair, and of the other by only one electron, repulsion is again too strong to permit overlap. However, if both of the outermost orbitals are singly occupied the repulsion is small enough to allow overlap of these orbitals (Fig. 4.2(a)), *provided* that the electrons have opposite spins.

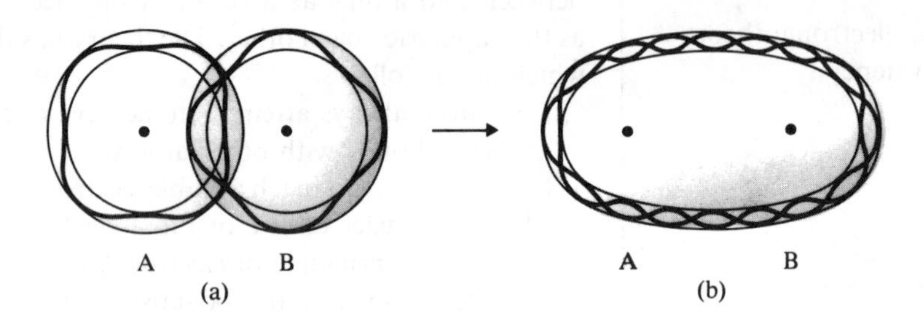

Fig. 4.2 (a) Overlap between the outermost occupied orbitals of atoms A and B, each holding an unpaired electron. (b) The resultant molecular orbital, occupied by two electrons.

The overlapping causes the two atomic orbitals to merge into a *molecular orbital*, which holds both the electrons (Fig. 4.2(b)). In a molecular orbital, the two electrons spin in opposite directions. Information on the location of bonding electrons in a molecular orbital is provided by Fig. 4.7.

Thus we see that the electron originally associated with the nucleus of A extends its wave to surround also the nucleus of B, and likewise the electron originally owned by B is shared with A. This serves to bind the atomic nuclei together, and is said to constitute a *covalent bond* between

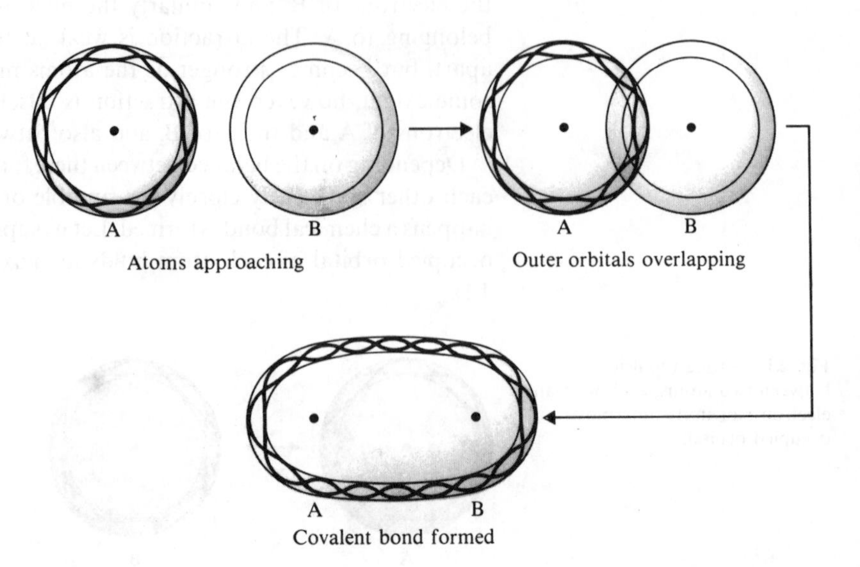

Fig. 4.3 Establishment of a coordinate bond.

A and B. Only the outermost orbital of each atom is sacrificed in forming the molecular orbital: the others remain essentially intact, although somewhat distorted, around their own nuclei.

Exactly the same result is achieved if one of the atomic orbitals is occupied by a *lone pair* of electrons (§4.5) and the other is vacant (Fig. 4.3).

In such a case, where both the electrons forming the bond originate from one atom, the bond is known as a *coordinate* or *dative covalent bond.* Once established, it is indistinguishable from an ordinary covalent bond.

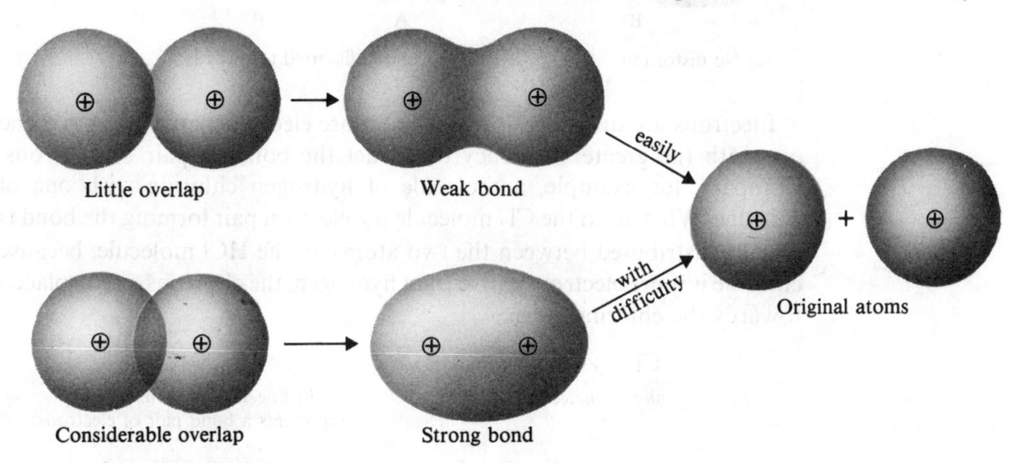

Fig. 4.4 Showing how the strength of a covalent bond depends on the extent to which atomic orbitals overlap.

The strength of a covalent bond depends on the type of atoms concerned. For instance, although carbon and silicon are chemically related elements, the C—C bond is stronger than the Si—Si bond. For this reason molecules containing long chains of carbon atoms are more stable than those containing long chains of silicon atoms, and the number of carbon compounds that exists far exceeds the number of silicon compounds. A rough indication of the strength of a bond is often given by the thermal stability of compounds that possess the bond. For example, the temperatures at which water and hydrogen sulphide start to decompose into their elements are 1 000 °C (1 273 K) and 400 °C (673 K) respectively, showing that the O—H bond in water is stronger than the S—H bond in hydrogen sulphide. A precise measure of bond strength is provided by the *bond dissociation enthalpy* (§2.3.3).

Bond strength is related to the size of the atomic orbitals that overlap and to the degree of overlap as the molecular orbital is formed (§4.2). With only a small amount of overlap the atomic orbitals are easily reproduced, but this is not so when there is substantial overlapping (Fig. 4.4).

4.1.2 POLAR COVALENCY

When two identical atoms are joined by a covalent bond, the molecular orbital is symmetrical, but whenever two different atoms are linked together the electron pair that forms the bond is attracted more strongly to one nucleus than the other, with the result that the molecular orbital becomes distorted (Fig. 4.5).

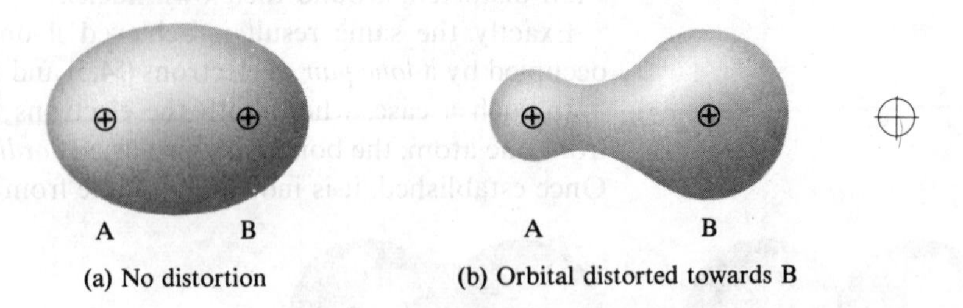

Fig. 4.5 Molecular orbitals established on linking together (a) identical atoms, (b) different atoms.

Electrons are displaced towards the more electronegative atom, i.e. the one with the greater tendency to attract the bonding pair of electrons. Compare, for example, a molecule of hydrogen chloride with one of chlorine. Whereas in the Cl_2 molecule the electron pair forming the bond is equally distributed between the two atoms, in the HCl molecule, because chlorine is more electronegative than hydrogen, the electrons are displaced towards the chlorine atom:

Cl—:—Cl *chlorine molecule* H—:—Cl *hydrogen chloride molecule*

: represents a bond pair of electrons

Consequently, the chlorine atom acquires a slight negative charge, δ^-, and the hydrogen atom a corresponding positive charge, δ^+:

$$\overset{\delta^+}{\mathrm{H}}—\overset{\delta^-}{\mathrm{Cl}}$$

δ^- is less than the charge on an electron. Because the molecule has two electrostatic poles, it is said to be a *dipole*, and hydrogen chloride is said to be a *polar covalent* substance.

4.1.3 ELECTROVALENCY

If we join an atom of a strongly electronegative element (B) to one of a weakly electronegative element (A) distortion of the molecular orbital may occur to such an extent that it becomes located almost entirely at atom B. In effect, the bonding electron of atom B stays put, while that of atom A is transferred to B. Compare the following substances:

$$\mathrm{Cl—:—Cl} \quad \overset{\delta^+}{\mathrm{H}}—:\overset{\delta^-}{\mathrm{Cl}} \quad \overset{+}{\mathrm{Na}}—:\overset{-}{\mathrm{Cl}}$$

In sodium chloride, the electron pair is located so closely to the chlorine atom that an electron is effectively transferred from sodium to chlorine. Consequently, the chlorine atom acquires a charge of −1 and the sodium atom a charge of +1. Sodium and chlorine exist no longer as neutral atoms, but as charged *ions*. They are held together in the crystal by electrostatic attraction, which is said to constitute an *ionic* or *electrovalent bond*. Electrovalency, therefore, is merely an extreme case of covalency, with the shared pair of electrons located at one atom.

4.2 The covalent bond

Two principal types of molecular orbital (MO) are recognised; in decreasing order of strength they are σ (sigma) and π (pi). σ MOs may result from the overlapping of s or p atomic orbitals, while π MOs, which are relatively weak, arise from overlapping p orbitals.

Single covalent bonds, which we shall study in this section, consist of σ MOs. Multiple covalent bonds, i.e. double and triple bonds, comprise both σ and π MOs.

σ MOs can be studied on the basis of the atomic orbitals (AOs) from which they are constructed. Numerous combinations are possible, but we shall restrict ourselves to σ MOs that result from the overlap of: (*a*) two s AOs; (*b*) two p AOs; and (*c*) one s and one p AO.

4.2.1 OVERLAP OF TWO s AOs, AS ILLUSTRATED BY THE H_2 MOLECULE

A hydrogen atom has the electronic configuration $1s^1$. As two such atoms join together to give a hydrogen molecule the 1s orbitals overlap to give a σ MO.

The 1s AO is spherically symmetrical, but when two such orbitals overlap the MO that is formed has a boundary surface that is ellipsoid in shape (Fig. 4.6). Thus, the shape of the MO is not derived just by partially superimposing the two 1s AOs.

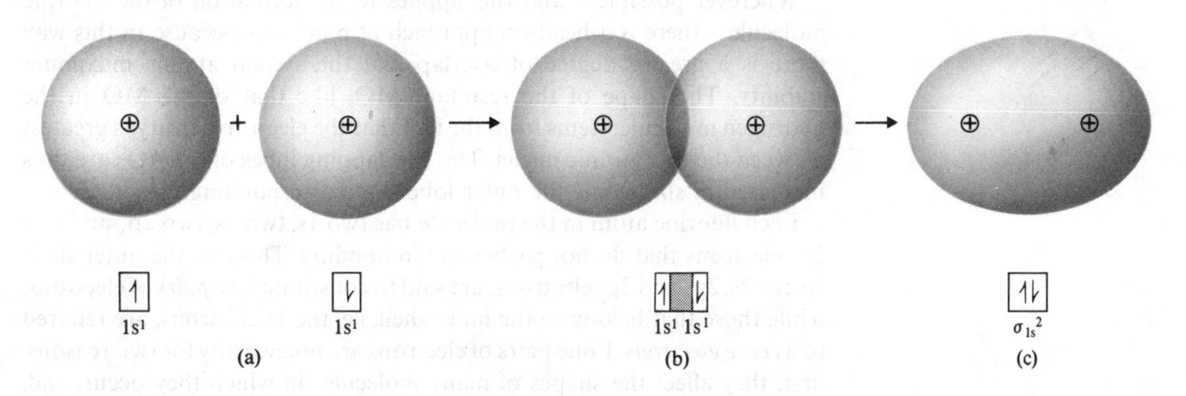

Fig. 4.6 Formation of the hydrogen molecule. (a) The approach of two hydrogen atoms, possessing electrons of opposite spins, (b) overlap of atomic orbitals, (c) the resultant molecular orbital.

The hydrogen molecule is more stable than two separate hydrogen atoms. Essentially, stability is due to the high electron density that exists between the two atomic nuclei (Fig. 4.7).

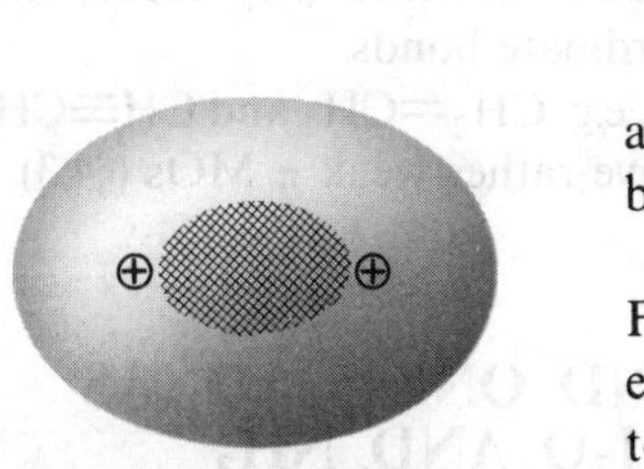

The two electrons in this region reduce repulsion between the two nuclei. Furthermore, each positive nucleus is attracted to the cloud of two electrons. Thus, the two hydrogen atoms are firmly held in close proximity to each other.

Fig. 4.7 The region of a hydrogen molecule where the probability of locating the two electrons is at a maximum.

Bond formation is accompanied by a reduction in both potential and kinetic energy. That some sort of energy is lost is clear from the great amount of heat that is produced during the combination.

4.2.2 OVERLAP OF TWO p AOs, AS ILLUSTRATED BY THE F_2 MOLECULE

Fluorine has the electronic configuration $1s^2\ 2s^2\ 2p_x^2\ 2p_y^2\ 2p_z^1$. When two fluorine atoms join together to give the fluorine molecule it is the unpaired $2p_z$ electrons of each that are affected. (Remember that $2p_x$, $2p_y$ and $2p_z$ orbitals are exactly equivalent to one another; the labelling of the axes is entirely arbitrary.) The $2p_z$ electrons are referred to as *bonding electrons* or *valency electrons*, and all the others as *non-bonding electrons*.

The overlap of p orbitals could conceivably occur in two ways (Fig. 4.8).

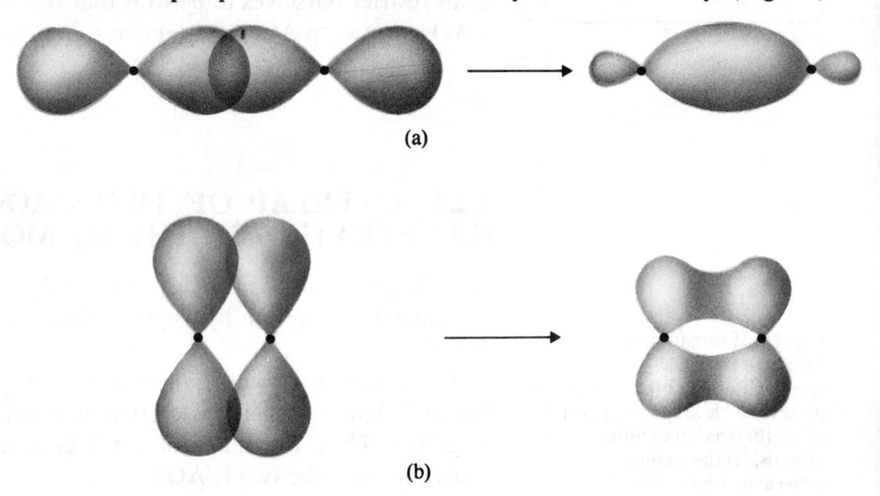

Fig. 4.8 (a) Head-on overlap of p orbitals, leading to a σ MO, and (b) sideways overlap of p orbitals, leading to a π MO.

Wherever possible – and this applies to the formation of the fluorine molecule – there is a head-on approach of p orbitals because in this way there is a greater degree of overlap and the system attains maximum stability. The shape of the resultant MO, like that of the MO in the hydrogen molecule, stems from the fact that the electron density is greatest between the two atomic nuclei. The overlapping lobes of the AOs are thus increased in size, while the outer lobes are correspondingly decreased.

Each fluorine atom in the molecule has two 1s, two 2s, two $2p_x$ and two $2p_y$ electrons that do not participate in bonding. Those in the outer shell, i.e. the 2s, $2p_x$ and $2p_y$ electrons, are said to constitute *lone pairs* of electrons, while those that belong to the inner shell, i.e. the 1s electrons, are referred to as *core electrons.* Lone pairs of electrons are noteworthy for two reasons: first, they affect the shapes of many molecules in which they occur; and, second, they may be used in forming coordinate bonds.

In certain molecules with multiple bonds, e.g. $CH_2{=}CH_2$ and $CH{\equiv}CH$, p orbitals can overlap only sideways to give rather weak π MOs (§4.3).

4.2.3 OVERLAP OF ONE s AND ONE p AO, AS ILLUSTRATED BY THE HF, H_2O AND NH_3 MOLECULES

If the AOs contributing to the MO are of different types, the MO is not symmetrical but has a shape that is derived from those of the constituent AOs. Thus, in the hydrogen fluoride molecule, where the overlapping AOs are 1s (from H) and $2p_z$ (from F), the shape is as shown in Fig. 4.9(a).

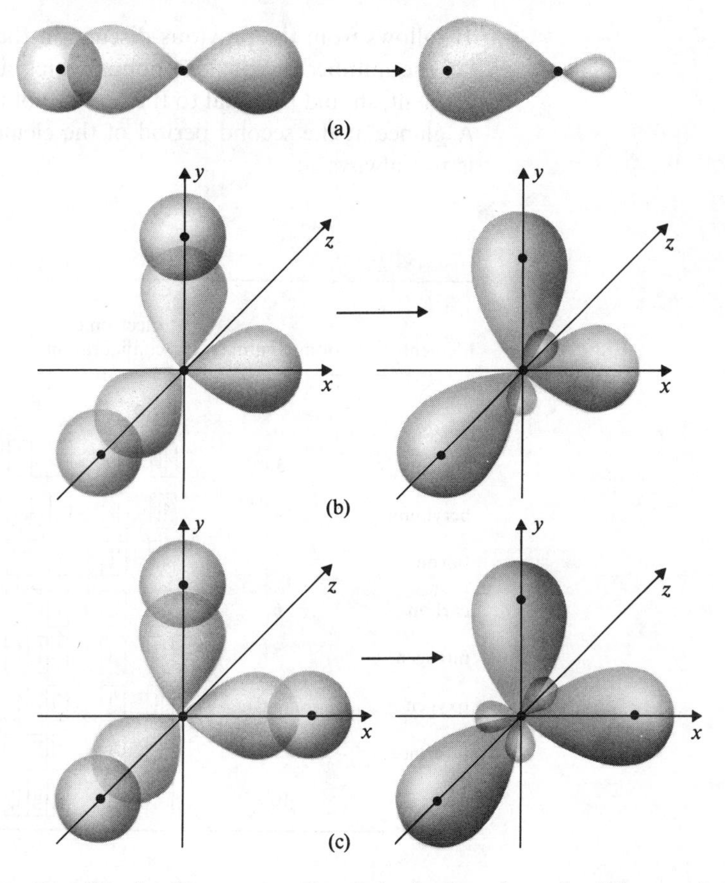

Fig. 4.9 Formation of (a) the HF molecule, (b) the H_2O molecule, and (c) the NH_3 molecule.

Bonding in the water molecule is similar, the only difference being that there are two bonds of this kind. Oxygen has the electronic configuration $1s^2\ 2s^2\ 2p_x^2\ 2p_y^1\ 2p_z^1$, and when a water molecule is formed the unpaired $2p_y$ and $2p_z$ AOs overlap the 1s AOs of two hydrogen atoms (Fig. 4.9(b)). (In Fig. 4.9(b) and (c), for the sake of simplicity, only one lobe of the $2p_x$, $2p_y$ and $2p_z$ AOs have been shown. For the same reason, the 1s and 2s AOs have been omitted.) The 2s and $2p_x$ electrons are unused in bonding, and are thus lone pairs. From this theory of bonding we might expect an H—O—H bond angle of 90°; experiment, however, indicates an angle of 104.5°. The greater bond angle stems from repulsion between the shared pairs of electrons (§4.8).

Figure 4.9(c) represents the formation of the ammonia molecule. Nitrogen has the electronic configuration $1s^2\ 2s^2\ 2p_x^1\ 2p_y^1\ 2p_z^1$, and in the formation of ammonia each of the unpaired 2p AOs overlaps the 1s AO of a hydrogen atom. The 2s electrons constitute a lone pair. Again, the bond angle (H—N—H) is greater than the expected 90° because of repulsion between electron pairs. Experiment indicates a pyramidal molecule (Fig. 4.25) with a bond angle of 107°.

4.2.4 OVERLAP OF s OR p AOs, AVAILABLE THROUGH PROMOTION, AS ILLUSTRATED BY THE $BeCl_2$ MOLECULE

It follows from the previous discussion that the *covalency* of an element, i.e. the number of covalent bonds which are formed by an atom of that element, should be equal to the number of unpaired electrons in the atom. A glance at the second period of the elements, however, shows that this is not always so.

Element	Atomic number	Electronic configuration 1s	2s	2p x y z	No. of unpaired electrons	Covalency
lithium	3	⇅	↑	[ ][ ][ ]	1	1
beryllium	4	⇅	⇅	[ ][ ][ ]	0	2
boron	5	⇅	⇅	[↑][ ][ ]	1	3
carbon	6	⇅	⇅	[↑][↑][ ]	2	4
nitrogen	7	⇅	⇅	[↑][↑][↑]	3	3
oxygen	8	⇅	⇅	[⇅][↑][↑]	2	2
fluorine	9	⇅	⇅	[⇅][⇅][↑]	1	1
neon	10	⇅	⇅	[⇅][⇅][⇅]	0	0

The explanation is that these electronic configurations relate to the ground state of the atoms, but excitation of some electrons can occur in atoms of beryllium, boron and carbon (and certain other elements, not shown here) as other atoms approach for combination, with the result that these electrons are *promoted* to other AOs at slightly higher energy levels:

	1s	2s	2p x y z	No of unpaired electrons
beryllium	⇅	↑	[↑][ ][ ]	2
boron	⇅	↑	[↑][↑][ ]	3
carbon	⇅	↑	[↑][↑][↑]	4

The number of unpaired electrons now corresponds to the number of covalent bonds that are actually formed. There is, however, a limit to the extent to which electrons can be promoted: it is not possible, for instance, for a 1s electron of lithium to be promoted to a 2p orbital because the energy requirement is greater than that available in a chemical reaction.

In any compound of beryllium, boron or carbon we may therefore expect more than one type of MO to be formed, one involving the singly filled s orbital and one involving singly filled p orbital(s). This is indeed the case as the following example illustrates.

The beryllium chloride molecule

In a $BeCl_2$ molecule the chlorine atoms are attached to the beryllium atom partly by

(i) a σ MO formed by the overlap of the 3p AOs of the chlorine atoms with the $2p_x$ AO of the beryllium atom (Fig. 4.10(a)),

(ii) a σ MO formed by the overlap of the same AOs of the chlorine atoms with the 2s AO of beryllium (Fig. 4.10(b)).

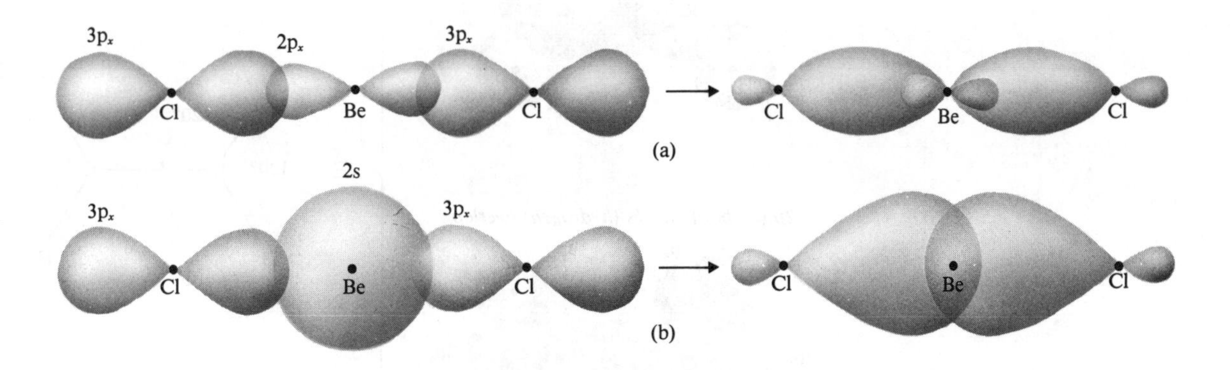

Fig. 4.10 Formation of the $BeCl_2$ molecule.

Each Be—Cl bond in the molecule consists of a combination of both the above MOs. The two MOs accommodate the four bonding electrons of the molecule – two from the beryllium atom and one from each chlorine atom. Thus, *there is no single MO that is responsible for a complete Be—Cl bond, and both bonds* (as indicated by their equal lengths) *are identical.*

The MOs in the $BeCl_2$ molecule differ from those previously discussed in that more than two AOs participate in their formation. However, the Be—Cl bonds are not multiple bonds. Since there is a total of four electrons participating in bonding, each Be—Cl bond comprises only two electrons.

4.3 Multiple covalent bonds

Multiple bonds, such as the carbon–carbon double bond in ethene and the carbon–carbon triple bond in ethyne, include both σ and π MOs. σ MOs have been discussed above; π MOs may result from sideways overlapping p AOs (Fig. 4.8).

4.3.1 THE ETHENE MOLECULE

The two carbon atoms in a molecule of ethene cannot be held together merely by a σ bond; there must be a π bond as well. This is suggested by a number of experimental facts, e.g.

(i) the high strength of the bond. The C=C bond has a bond dissociation enthalpy of 612 kJ mol^{-1}, whereas that of the C—C bond is only 348 kJ mol^{-1}.

(ii) the shortness of the bond: 0.134 nm, compared with 0.154 nm for the C—C bond.

(iii) restricted rotation about a C=C bond. Compounds such as 1,2-dichloroethene exist in two isomeric forms (§5.2.2). The carbon–carbon double bond locks the molecule in position, and prevents one isomer from changing into the other.

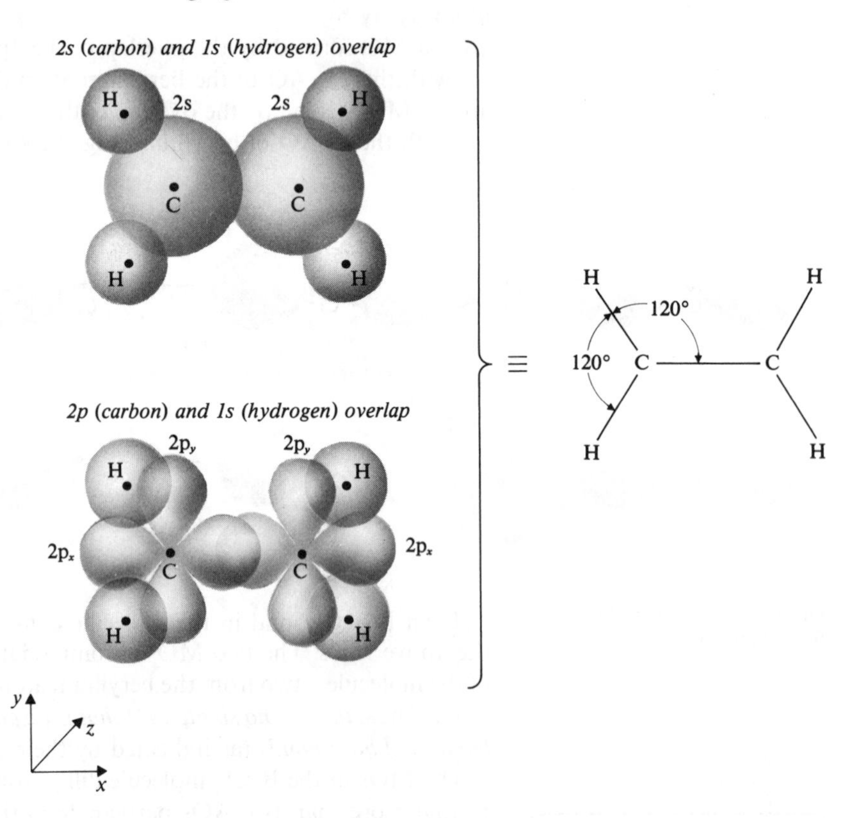

Fig. 4.11 The σ-framework of the ethene molecule.

The ethene molecule possesses a planar *σ-framework* (Fig. 4.11) in which the carbon and hydrogen atoms are held together by σ bonds, rather like the carbon and hydrogen atoms in a molecule of methane. The C—C (single) bond is due partly to overlap of 2s AOs and partly to overlap of, let us say, $2p_x$ AOs. Each C—H bond is due partly to the 2s AO of the carbon atom overlapping the 1s AO of the hydrogen atom, and partly to the $2p_x$ and $2p_y$ AOs of the carbon atom overlapping the same 1s AO of the hydrogen atom.

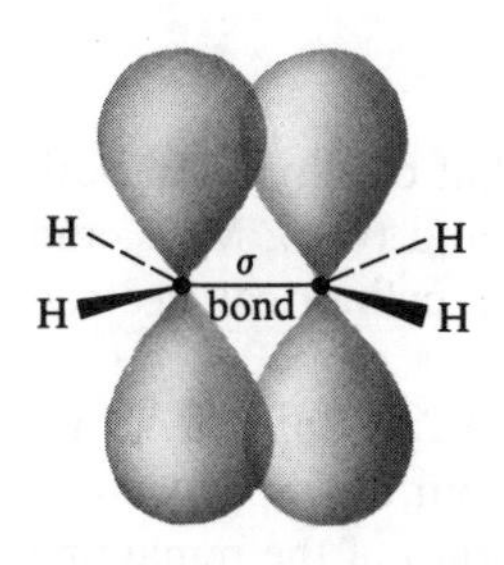

Fig. 4.12 Sideways overlap of $2p_z$ AOs in the ethene molecule.

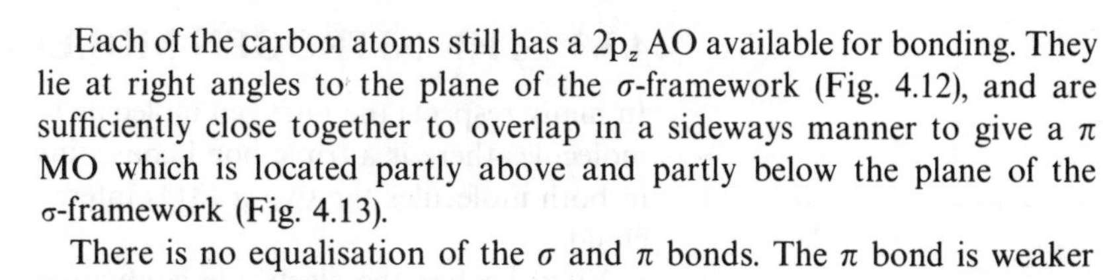

Each of the carbon atoms still has a $2p_z$ AO available for bonding. They lie at right angles to the plane of the σ-framework (Fig. 4.12), and are sufficiently close together to overlap in a sideways manner to give a π MO which is located partly above and partly below the plane of the σ-framework (Fig. 4.13).

There is no equalisation of the σ and π bonds. The π bond is weaker than the σ bond because the $2p_z$ AOs of carbon overlap to a relatively small extent. Furthermore, π electrons, unlike σ electrons, are not concentrated between the atomic nuclei. The π bond is the one that is disrupted during addition reactions (§14.4.4).

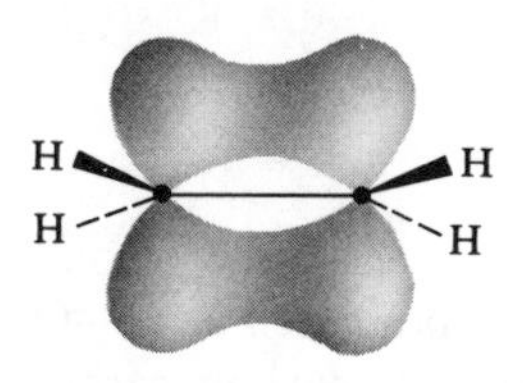

Fig. 4.13 The π MO in the ethene molecule.

4.3.2 THE ETHYNE MOLECULE

The carbon–carbon bond in ethyne is exceptionally short (0.120 nm) and exceptionally strong (bond dissociation enthalpy = 837 kJ mol^{-1}), because the two carbon atoms are held together by a triple bond, consisting of one σ and two π bonds.

The molecule has a linear σ-framework, H—C—C—H, the formation of which includes the 2s and, let us say, the $2p_x$ AOs of the carbon atoms. This leaves the $2p_y$ and $2p_z$ AOs of each carbon atom free for further bonding. They remain directed at right angles to each other and to the line of the C—C σ bond, as shown in Fig. 4.14.

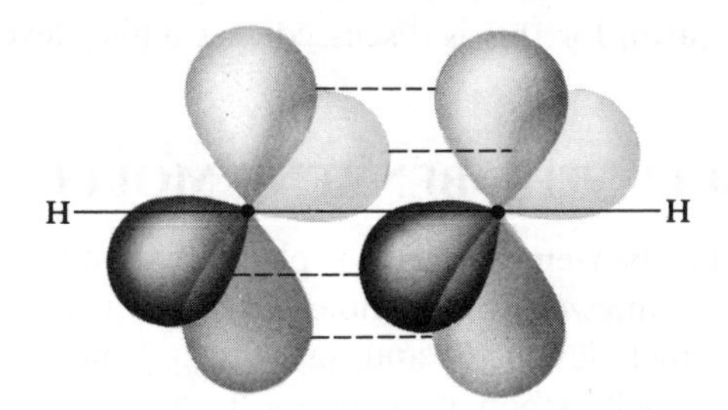

Fig. 4.14 Sideways overlap of 2p AOs in the ethyne molecule. (For the sake of clarity the distance between the carbon atoms has been exaggerated. Dotted lines indicate where overlap occurs.)

The two 2p AOs are close enough together for them both to overlap, so that two π MOs are formed. They are stronger than the π MO in ethene, and are so close to each other that they interact to give a cylindrical electron cloud around the axis of the C—C σ bond (Fig. 4.15).

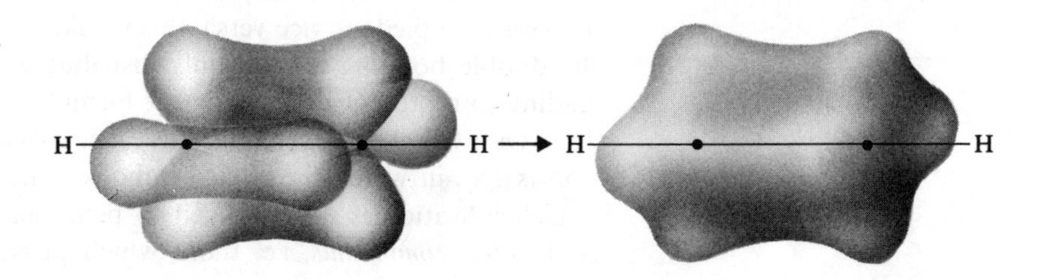

Fig. 4.15 Interaction of π MOs in the ethyne molecule.

4.3.3 THE NITROGEN MOLECULE

In many respects the nitrogen molecule resembles that of ethyne. In both molecules there is a triple bond consisting of one σ and two π MOs, and in both molecules the two π MOs interact to form a cylindrical electron cloud.

Nitrogen has the electronic configuration $1s^2\ 2s^2\ 2p_x^1\ 2p_y^1\ 2p_z^1$. If we suppose that there is head-on overlap of $2p_z$ AOs from two approaching nitrogen atoms, to give a σ MO, then sideways overlap of the remaining $2p_x$ and $2p_y$ AOs gives two π MOs, exactly as in ethyne. Interaction of the π MOs results in a cylindrical four electron cloud around the N—N axis.

4.3.4 THE OXYGEN MOLECULE

The bonding in the oxygen molecule is commonly represented as O=O, which suggests the existence of a double bond similar to the carbon–carbon double bond in ethene, and at first sight this makes sense. Oxygen has the electronic configuration $1s^2\ 2s^2\ 2p_x^2\ 2p_y^1\ 2p_z^1$. When two such atoms approach each other there could be head-on overlap of, let us say, the $2p_z$ AOs to give a σ MO. At the same time, there could be sideways overlap of the $2p_y$ AOs to give a π MO.

If this were the complete story, all electrons in the oxygen molecule would be paired and oxygen would have no unpaired electrons. Experiment shows, however, that oxygen possesses two unpaired electrons. The reason for this is discussed at a higher level.

4.4 Delocalised multiple bonds

4.4.1 THE BENZENE MOLECULE

The benzene molecule is planar, with a C—C—C bond angle of 120°. The σ-framework of the molecule (Fig. 4.16) is similar to that of ethene in that it includes the 2s and, say, the $2p_x$ and $2p_y$ AOs of the carbon atoms.

Each carbon atom has its $2p_z$ AO available for sideways overlap. Figure 4.17 shows these AOs, and makes it clear that each orbital can interact with its two immediate neighbours to an equal extent. The six $2p_z$ AOs merge together to give three *delocalised* bonding π MOs (Fig. 4.17).

The three π MOs (Fig. 4.17(b), (c) and (d)) are not at equal energy levels. The MO labelled (b) has the lowest energy, while (c) and (d) are degenerate at a higher level. Even so, all three π MOs are filled, because it is not possible for more than two electrons to enter one orbital. Because of this, the electron density is completely uniform around the ring. (If (c) but not (d) were occupied, or vice versa, this would not be so.) In simple language, the double bonding is uniformly distributed around the ring. It is misleading to represent benzene by the formula ⌬, as this suggests that the $2p_z$ AOs interact in pairs to give three *localised* π MOs. The best representation is ⏣, in which the circle symbolises the delocalised π electrons.

Delocalisation is not confined to benzene, but is also encountered in *conjugated compounds*, i.e. those which possess alternating double and

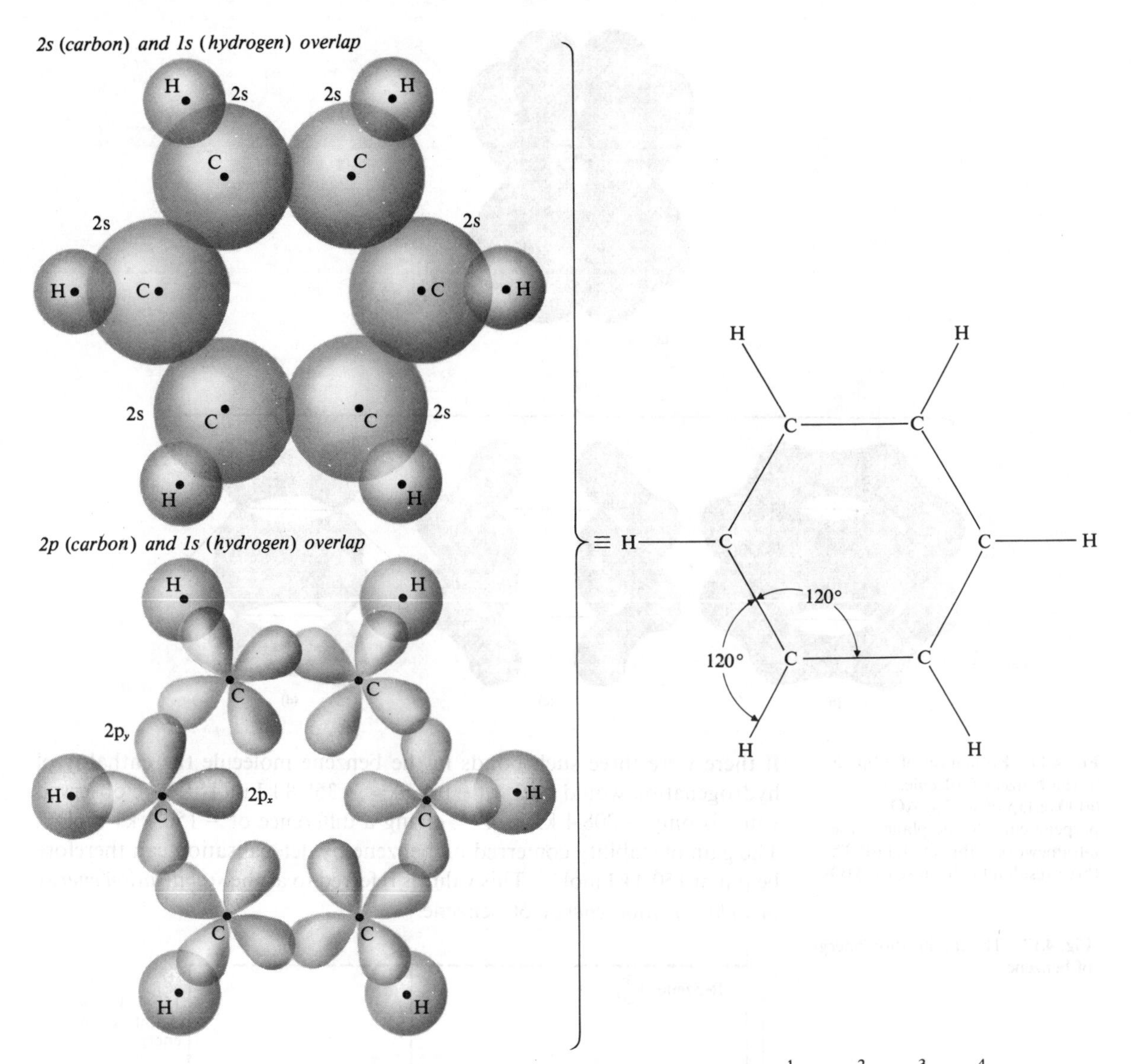

Fig. 4.16 The σ-framework of the benzene molecule.

single bonds. An example is 1,3-butadiene, $\overset{1}{C}H_2{=}\overset{2}{C}H{-}\overset{3}{C}H{=}\overset{4}{C}H_2$. While double bonding is most marked between C_1 and C_2, and between C_3 and C_4, the bond between C_2 and C_3 also has a certain amount of double bond character as the result of delocalisation.

Delocalisation occurs when it can lead to a reduction in electrical potential energy and hence to improved stability. We can determine the difference between the potential energy of, say, benzene, and the potential energy that benzene would have if the molecule possessed three localised double bonds. Consider the hydrogenation of cyclohexene:

$$\text{cyclohexene} + H_2 = \text{cyclohexane}$$

$\Delta H = -119.6\ \text{kJ mol}^{-1}$

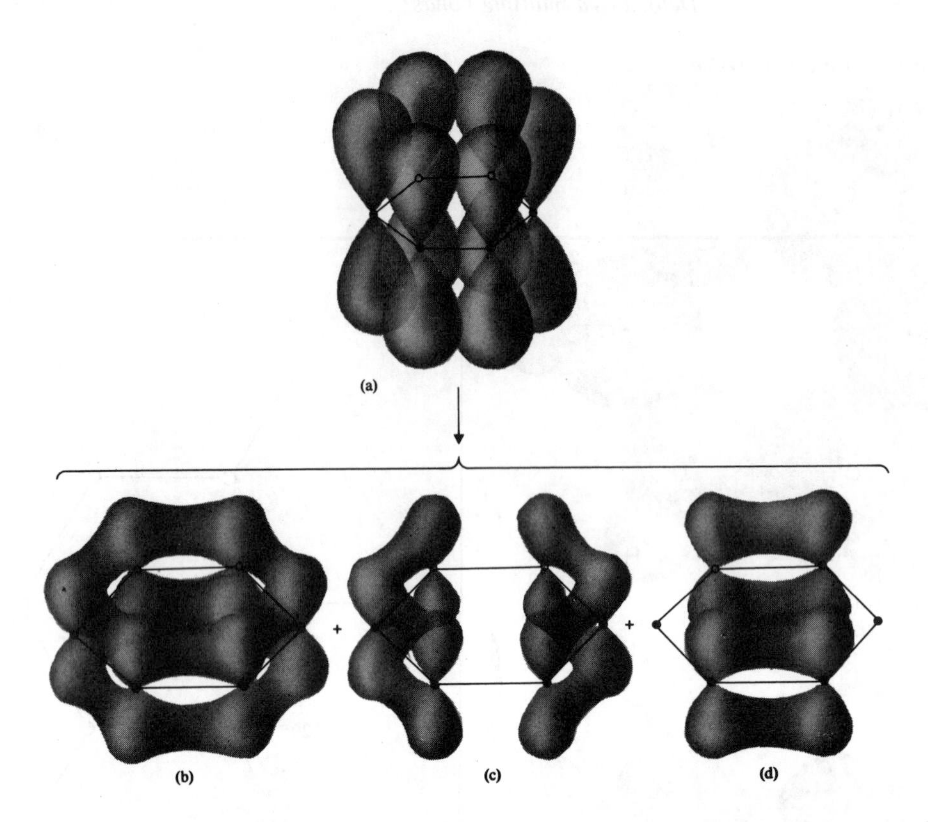

Fig. 4.17 Formation of π bonds in the benzene molecule.
(a) Overlap of six $2p_z$ AOs, perpendicular to the plane of the σ-framework. (b), (c) and (d) The three resultant delocalised π MOs.

If there were three such bonds in the benzene molecule the enthalpy of hydrogenation would be $3 \times (-119.6) = -358.8$ kJ mol^{-1}. The observed value is only -208.4 kJ mol^{-1}, giving a difference of -150.4 kJ mol^{-1}. The gain of stability conferred on benzene by delocalisation can therefore be put at 150.4 kJ mol^{-1}. This value is referred to as the *stabilisation energy* or *delocalisation energy* of benzene.

Fig. 4.18 The stabilisation energy of benzene.

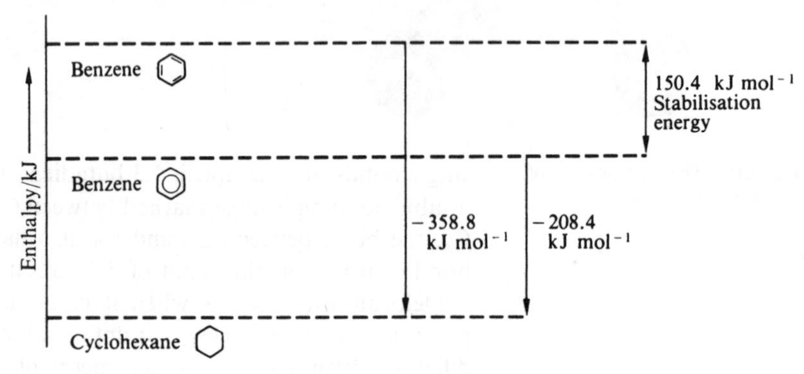

4.5 The coordinate bond

A *coordinate* or *dative covalent* bond is merely a special type of covalent bond. If two atoms A and B are joined together by a covalent bond, one electron of the bond comes from A and the other from B, but if the atoms are linked by a coordinate bond *both* the electrons originate from one atom. The atom (or ion) that provides the electrons is known as the *donor*, while the receiving atom (or ion) is called the *acceptor*. The coordinate bond is represented by an arrow from donor to acceptor, thus:

$$\underset{\text{Donor}}{\text{A:}} \longrightarrow \underset{\text{Acceptor}}{\text{B}} \qquad\qquad \text{A} \overset{\cdot}{\underset{\cdot}{—}} \text{B}$$

Coordinate bond — Covalent bond

A coordinate bond, once formed, is indistinguishable from a covalent bond. The only difference between the two lies in the origin of the electrons.

If an atom or ion is to act as a donor it must have at least one lone pair of electrons in its outer shell, and for an atom or ion to act as an acceptor there must be at least one vacant orbital in its outer shell. Coordination, like covalent bond formation, takes place because there is a lowering of energy as the bonds are formed. Within limits, which are dictated by the available space and the number of vacant orbitals, an acceptor atom forms as many coordinate bonds as it can.

As a simple example, consider the formation of an oxonium ion, H_3O^+, by the coordination of a water molecule to a proton:

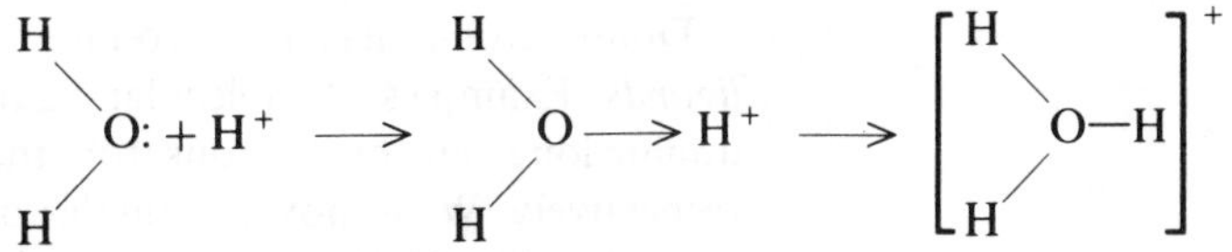

$$H_2O: + H^+ \longrightarrow H_2O \longrightarrow H^+ \longrightarrow \left[H_2O\text{—}H \right]^+$$

The oxygen atom of the water molecule has two lone pairs of electrons in its outer shell, but usually only one participates in coordination. The proton (from an acid) has a vacant 1s orbital. Coordination, as shown, enables the proton to achieve the configuration of the noble gas helium, without unduly disturbing the stable electronic arrangement of the oxygen atom. The structural formula on the right shows that no attempt is made, in the oxonium ion, to distinguish among the three O—H bonds, and that the positive charge, which belonged originally to the proton, is now distributed over the ion as a whole. Note also that coordination does not affect the oxidation number (§10.1.2) of the atoms concerned. In H_2O and H^+, and in H_3O^+, oxygen has an oxidation number of -2 and hydrogen $+1$.

In the oxonium ion oxygen has a covalency of three, although it is more usual to state that the *coordination number* of oxygen is three. The coordination number of an atom in a chemical species is defined as the number of nearest neighbours to the given atom.

The formation of H_3O^+ provides an example of *stabilisation by co-ordination*, for while a proton can exist in the form of H_3O^+ it cannot exist by itself. (Its minute radius, coupled with its full positive charge, gives it an extremely high density of charge on its surface and makes it highly reactive.) The effect is sometimes referred to as *stabilisation by complexing*, because a species that is formed by coordination is called a *complex*.

A proton may also be stabilised by the coordination of a molecule of an alcohol (§14.9.5) or ammonia. In the latter case the product is the tetrahedral ammonium ion, NH_4^+ (§15.2.3). Nitrogen, in the ammonium ion, still has an oxidation number of -3, but a coordination number of 4.

4.5.1 DONOR ATOMS

The most effective donors of lone pairs are small atoms of the non-metallic elements located at the top right-hand corner of the periodic table. (If an atom has a low atomic radius the electrons in the outer shell are close together and there is a relatively strong repulsion between them.) In the following extract from the periodic table the principal donor atoms have been printed in bold type: lesser donors appear in light type:

GROUP	1A	2A	3B	4B	5B	6B	7B
				C	N	O	F
					P	**S**	Cl
					As	Se	**Br**
							I

Donor atoms may be encountered in molecules or anions known as *ligands*. Examples of molecular ligands are H_2O, NH_3 and CO, which donate lone pairs of electrons from the oxygen, nitrogen and carbon atoms respectively. Well-known examples of anionic ligands are F^-, Cl^-, HO^- and CN^-. (The HO^- ion donates a lone pair of electrons from the oxygen atom. The CN^- ion can donate from either the carbon or the nitrogen atom, but usually does so from the carbon atom.)

4.5.2 ACCEPTOR ATOMS AND IONS

The acceptor of a lone pair of electrons is usually a metal cation. In certain other cases it may be an atom belonging to a molecule, or an atom of a transition metal.

Cations

Cations of the transition elements, such as Fe^{2+}, Fe^{3+}, Ni^{2+} and Co^{3+}, with their vacant d orbitals and relatively high *surface charge density* (§4.7.4), exert a strong attraction for ligands and enter readily into complex ion formation. For example, when anhydrous iron(II) sulphate is dissolved in water, six water molecules, acting as ligands, coordinate to the iron(II) ion to give the hydrated iron(II) ion, $[Fe(H_2O)_6]^{2+}$. The water molecule is a relatively weak ligand, and may be displaced by other ligands with a readiness that depends on their strength and concentration. Thus, the treatment of an aqueous solution of an iron(II) salt with a cyanide (e.g. NaCN) leads to the substitution of water molecules by cyanide ions. Eventually, all six water molecules are replaced,

i.e. $[Fe(H_2O)_6]^{2+} + 6CN^- \rightleftharpoons [Fe(CN)_6]^{4-} + 6H_2O$

Cations of the p-block elements, e.g. Sn^{2+} and Pb^{2+}, also enter into complex formation, but the phenomenon is seldom encountered with cations of the s-block elements. Such ions are relatively large (Fig. 4.20) and, in the case of the group 1A elements, have only a single positive charge. They therefore have a low surface charge density and exert only a feeble attraction for lone pairs of electrons.

Molecules

Many molecular substances act as electron acceptors in coordinate bond formation, because in this way more stable species are formed. Compounds of group 3B elements form a wide range of complexes, e.g. ammonia–boron trifluoride(1/1), $BF_3 \cdot NH_3$, and aluminium chloride–ethoxyethane(1/1), $AlCl_3 \cdot (C_2H_5)_2O$. Aluminium chloride, besides reacting with ethers, also dimerises in such a way that a lone pair of electrons from a chlorine atom of one $AlCl_3$ molecule is donated to the aluminium atom of another (§13.2.6).

Transition metal atoms

Metal carbonyl compounds are formed by the coordination of carbon monoxide, through a lone pair of electrons in the outer shell of carbon, to an *atom* of a transition element. Probably the best known example is tetracarbonylnickel(0), $[Ni(CO)_4]$.

4.6 The polar covalent bond

Although the bonding in some diatomic molecules (e.g. H_2) is purely covalent, pure electrovalency does not exist. Even in potassium chloride, in many respects a typical electrovalent compound, it has been estimated that the ionic character of the bond is only 52%. In some compounds, for example aluminium chloride and tin(IV) chloride, the bonding is so far removed from pure electrovalency that the compounds have low melting temperatures, low electrical conductivities in the molten state and high solubilities in organic solvents. Such substances are said to be *polar covalent*.

4.6.1 FAJANS' RULES

Kasimir Fajans in 1924 published a set of rules which stated that the electrovalent bond in a compound A^+B^- tends towards covalency when any of the following conditions apply.

The charge on A^+ or B^- is high

Although unipositive and bipositive cations are commonplace (e.g. Na^+ and Ca^{2+}), simple cations with a triple or quadruple charge are rare. The Al^{3+} ion, for instance, exists only in a few compounds (§13.1.1). The reason is that as electrons are removed from an atom the remainder are held ever more strongly by a constant positive charge on the nucleus. The removal of one or two electrons from an atom requires relatively little energy, but the loss of further electrons requires more energy than is normally available in a chemical reaction.

A similar argument applies to anions. Although uni- and binegative anions are well known, simple anions with a triple or quadruple charge (e.g. N^{3-} and C^{4-}) are uncommon. The reason is that although the introduction of an electron into a neutral atom of a non-metal is a spontaneous process in which energy is released, the corresponding process in which an electron is introduced into an anion is difficult because of repulsion between the negatively charged ion and the negatively charged electron. The greater the charge on the anion, the greater is the repulsion and the more difficult it is to accomplish the change.

A^+ is small

Cations of metals at the top of the periodic table (e.g. Li^+, Be^{2+} and Al^{3+}) have a low ionic radius (Fig. 4.20). Most salts that contain these ions have a marked degree of covalent character, for two reasons.

(i) In a small atom the electrons of the outer shell are close to the nucleus and therefore are strongly attracted to it. As a result, small atoms tend not to ionise but to use their electrons in covalent bond formation.

(ii) A small cation, especially if it possesses more than a single positive charge, has a high surface charge density (§4.7.4) and can therefore distort or *polarise* the outer orbitals of the anion B^-. Electrons are attracted from B^- to A^+, i.e. covalency is encouraged.

B^- is large

Most metal iodides are polar covalent. Even sodium iodide is highly soluble in ethanol and propanone. There are two reasons for this, both of which stem from the high ionic radius of the I^- ion.

(i) For a large atom, relatively little energy is released during the process

$$X(g) + e^- \longrightarrow X^-(g)$$

In other words, a large anion has relatively little tendency to be formed. Formation depends on the entry of one or more electrons to the outer shell, where they are retained by nuclear attraction, but if the outer shell is remote from the nucleus, and in addition shielded from it by underlying electron shells, the force of attraction is relatively weak.

(ii) A large anion, because its electrons are so far away from the nucleus, is easily polarised, especially by a small cation.

4.6.2 ELECTRONEGATIVITY

Between the extremes of pure covalency, where the molecular orbital is symmetrical, and hypothetical pure electrovalency, where the electron pair is located entirely at one atom, an infinite range of bonding is possible (Fig. 4.19).

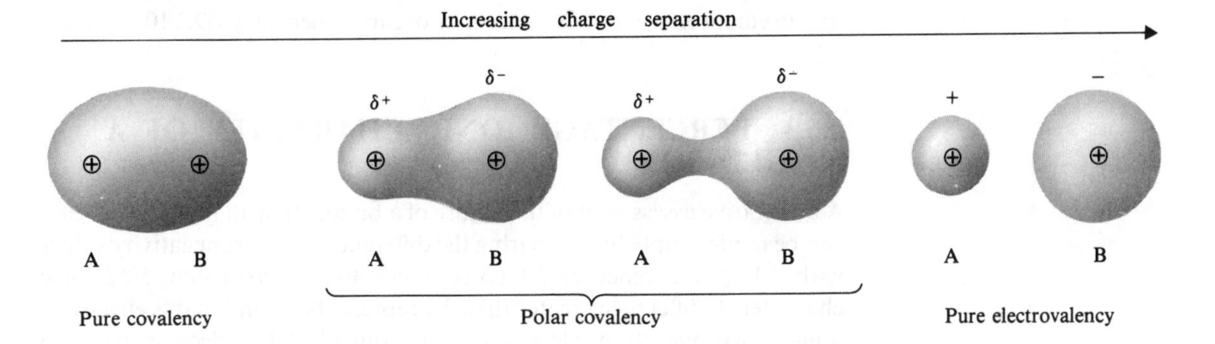

Fig. 4.19 The gradation of bonding from pure covalency to pure electrovalency.

The degree of charge separation depends on the *electronegativity* difference between the atoms A and B: the greater the difference, the greater the polarity. The electronegativity (x) of an atom is a measure of its tendency *in a molecule* to attract electrons and hence acquire a negative charge, but it is not a fundamental concept and is difficult to define precisely. Pauling arbitrarily allocated an electronegativity value of 4.0 to the most electronegative element, fluorine, and on this basis calculated values for other elements (Table 4.1).

Table 4.1 Pauling electronegativity values of important elements

1A	2A	3A	4A	5A	6A	7A	← 8 →			1B	2B	3B	4B	5B	6B	7B
H 2.1																
Li 1.0	Be 1.5											B 2.0	C 2.5	N 3.0	O 3.5	F 4.0
Na 0.9	Mg 1.2											Al 1.5	Si 1.8	P 2.1	S 2.5	Cl 3.0
K 0.8	Ca 1.0	Sc 1.3	Ti 1.5	V 1.6	Cr 1.6	Mn 1.5	Fe 1.8	Co 1.8	Ni 1.8	Cu 1.9	Zn 1.6	Ga 1.6	Ge 1.8	As 2.0	Se 2.4	Br 2.8
Rb 0.8	Sr 1.0											In 1.7	Sn 1.8	Sb 1.9	Te 2.1	I 2.5
Cs 0.7	Ba 0.9											Tl 1.8	Pb 1.8	Bi 1.9	Po 2.0	At 2.2

Electronegativity generally increases from left to right across the periodic table, and also from bottom to top. The most electronegative elements are thus concentrated in the top right-hand corner, with the least electronegative at the bottom left. The variation in electronegativity is related to variation in atomic radius. Electronegativity develops from left to right across a period because, as atomic radius decreases, there is an increase in the attractive force between the nucleus and the bonding electrons in the outer shell. Electronegativity declines from the top of a group to the bottom because the atoms become larger and the attraction between the nucleus and the peripheral electrons decreases.

The electronegativity concept is valuable throughout the whole of chemistry, and some of its principal uses are listed below.

(i) Prediction of the degree of ionic character of a covalent bond, §4.6.3.
(ii) Understanding the shapes of molecules, §4.8.
(iii) Assignment of oxidation numbers, §10.1.2.
(iv) Prediction of inductive effects in organic chemistry, §9.2.10.

4.6.3 PERCENTAGE IONIC CHARACTER OF A SINGLE BOND

A qualitative assessment of the nature of a bond, often all that is necessary, can be made simply by comparing the difference of electronegativity values with 2.1. A difference of 2.1 corresponds to approximately 50% ionic character. A difference greater than 2.1 represents a bond that is effectively ionic, since one atom effectively gains control of the electron pair. To summarise:

$A \overset{\times}{\underset{\times}{—}} B$	$\overset{\delta^-}{A} \overset{\times}{\underset{\times}{—}} \overset{\delta^+}{B}$	A^-B^+
$x_A = x_B$	$x_A - x_B < 2.1$	$x_A - x_B > 2.1$
non-polar bond	polar covalent bond	ionic bond

For example:

F—Li $\quad x_F - x_{Li} = 3.0$ ionic bond

Cl—C $\quad x_{Cl} - x_C = 0.5$ polar covalent bond

4.7 The ionic bond

Ions are formed as a result of the transfer of one or more electrons from the outer shell of an atom of a metal to the outer shell of an atom of a non-metal. They often, but not always, possess noble gas configurations. Positively charged ions are called *cations*, because they migrate to the cathode on electrolysis, while negatively charged ions, which migrate to the anode, are termed *anions*.

Compounds that are composed of ions (as opposed to molecules) are described as *ionic* or *electrovalent*. They are invariably crystalline solids of high melting temperature, in which oppositely charged ions are held together by electrostatic attraction (§6.4.2).

Here we are concerned first with the sizes of atoms and the ions to which they give rise. This subject is of paramount importance, for all trends associated with the periodic table, both down a group and across a period, can be attributed essentially to changing atomic or ionic radius. We shall continue with a study of the enthalpy changes accompanying the formation of ions, namely *ionisation enthalpy*, which is the enthalpy required to produce cations, and *electron affinity*, which is the enthalpy change accompanying the formation of anions. Afterwards we shall turn to *lattice enthalpy*, i.e. the enthalpy that is released when isolated cations and anions come together to give a crystal lattice.

4.7.1 IONIC RADII

It can be seen from Fig. 4.20 that atomic radius increases down any group of the periodic table, as more electron shells are occupied, but decreases from left to right across a period.

Ionic radius, like atomic radius, increases down a group and for the same reason, namely increased occupation of electron shells. Movement across a period is marked by a decrease in both cationic and anionic radius. We must keep these concepts separate, for whereas any cation is *smaller*

Fig. 4.20 Atomic and ionic radii in nanometres. The values quoted for metals are metallic bond radii, while those for non-metals are single covalent bond radii (§3.2.4).

1A	2A	3A	4A	5A	6A	7A	←	8	→	1B	2B	3B	4B	5B	6B	7B
																H 0.037 H^{-} 0.208
Li 0.152 Li^{+} 0.060	Be 0.112 Be^{2+} 0.031											B 0.080	C 0.077	N 0.074 N^{3-} 0.171	O 0.074 O^{2-} 0.140	F 0.072 F^{-} 0.136
Na 0.186 Na^{+} 0.095	Mg 0.160 Mg^{2+} 0.065											Al 0.143 Al^{3+} 0.050	Si 0.117	P 0.110	S 0.104 S^{2-} 0.184	Cl 0.099 Cl^{-} 0.181
K 0.231 K^{+} 0.133	Ca 0.197 Ca^{2+} 0.099	Sc 0.160 Sc^{3+} 0.081	Ti 0.146 Ti^{2+} 0.090	V 0.131 V^{3+} 0.074	Cr 0.125 Cr^{3+} 0.069	Mn 0.129 Mn^{2+} 0.080	Fe 0.126 Fe^{2+} 0.076 Fe^{3+} 0.064	Co 0.125 Co^{2+} 0.078 Co^{3+} 0.063	Ni 0.124 Ni^{2+} 0.078	Cu 0.128 Cu^{+} 0.096 Cu^{2+} 0.069	Zn 0.133 Zn^{2+} 0.074	Ga 0.141 Ga^{+} 0.108 Ga^{3+} 0.062	Ge 0.122 Ge^{2+} 0.093	As 0.121	Se 0.117 Se^{2-} 0.198	Br 0.114 Br^{-} 0.195
Rb 0.244 Rb^{+} 0.148	Sr 0.215 Sr^{2+} 0.113	Y 0.180	Zr 0.157	Nb 0.141	Mo 0.136	Tc 0.135	Ru 0.133	Rh 0.134	Pd 0.138	Ag 0.144 Ag^{+} 0.126	Cd 0.149 Cd^{2+} 0.097	In 0.166 In^{+} 0.132 In^{3+} 0.081	Sn 0.162 Sn^{2+} 0.112	Sb 0.141	Te 0.137 Te^{2-} 0.221	I 0.133 I^{-} 0.216
Cs 0.262 Cs^{+} 0.169	Ba 0.217 Ba^{2+} 0.135	La 0.188	Hf 0.157	Ta 0.143	W 0.137	Re 0.137	Os 0.134	Ir 0.135	Pt 0.138	Au 0.144 Au^{+} 0.137	Hg 0.152 Hg^{2+} 0.110	Tl 0.171 Tl^{+} 0.140 Tl^{3+} 0.095	Pb 0.175 Pb^{2+} 0.120	Bi 0.170 Bi^{3+} 0.120	Po 0.140	At 0.140

than its parent atom (compare Na^+ and Na), an anion is *larger* than its parent atom (compare Cl^- and Cl). The radius of a cation greatly depends on its charge. As successive electrons are removed from an atom, ionic radius becomes smaller and smaller, e.g. $Na > Na^+ > Na^{2+}$, etc. Thus, for a series of isoelectronic ions, i.e. ions with the same number of electrons, such as we encounter when moving from left to right across a period, radius decreases as charge increases, e.g. $Na^+ > Mg^{2+} > Al^{3+}$.

For anions the reverse is true, in that a repulsion between the electrons of a negative ion causes an increase in size, and the more electrons that are introduced the larger it is, e.g. $N^{3-} > O^{2-} > F^-$.

For a sequence of isoelectronic ions, therefore, ionic radius decreases *in order of anions followed by cations*. The ions of a complete period are not all isoelectronic, for the anions have one more occupied shell than the cations, but the same rule applies, e.g. $N^{3-} > O^{2-} > F^- > Li^+ > Be^{2+} > B^{3+}$.

4.7.2 IONISATION ENTHALPY (IONISATION ENERGY)

The *first ionisation enthalpy* (ΔH_1) of an element is defined as the enthalpy required for the removal of one mole of electrons from one mole of isolated atoms of the element in the gaseous state,

i.e. $M(g) \longrightarrow M^+(g) + e^-$

The *second ionisation enthalpy* (ΔH_2) is defined as the enthalpy needed for the removal of a second mole of electrons from one mole of isolated unipositive ions in the gaseous state,

i.e. $M^+(g) \longrightarrow M^{2+}(g) + e^-$

Higher ionisation enthalpies may be defined in a similar fashion. Thus, the *n*th ionisation enthalpy is the enthalpy required for the process:

$M^{(n-1)+}(g) \longrightarrow M^{n+}(g) + e^-$

The second ionisation enthalpy of an element is always greater than the first; the third is greater still, and so on. For example,

	ΔH_1	ΔH_2	ΔH_3	/kJ mol^{-1} at 298 K
sodium	500	4 560	6 940	
magnesium	742	1 450	7 740	
aluminium	583	1 820	2 740	

There are several reasons for this, as follows.

(i) As electrons are removed from an atom (or an ion) the remainder are more firmly held by the constant positive charge on the nucleus.

(ii) As the charge on a cation increases, its radius decreases; see above. Consequently, as we progress from M to M^+ to M^{2+}, etc, we are removing electrons which are ever closer to the nucleus (even if they belong to the same shell) and which are increasingly influenced by it. We can relate this idea to Fig. 3.12, which shows that ionisation enthalpy is the enthalpy required to remove completely an electron from an atomic orbital. As an orbital is pulled towards the nucleus its energy is lowered, and the ionisation enthalpy for the removal of an electron from it must increase, e.g. for magnesium:

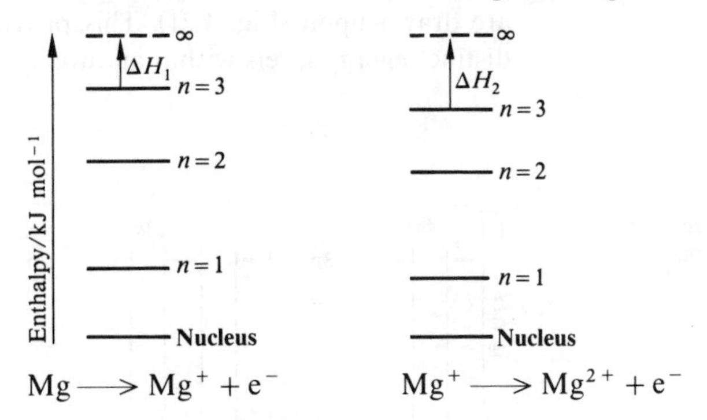

$Mg \longrightarrow Mg^+ + e^-$ $\qquad$ $Mg^+ \longrightarrow Mg^{2+} + e^-$

(iii) For the alkali metals (group 1A), the jump from ΔH_1 to ΔH_2 is particularly large because the second electron to be removed originates from a lower shell. Thus, the first ionisation enthalpy of sodium relates to the removal of an electron from the third shell, while the second ionisation enthalpy relates to an electron leaving the second shell:

∞ ΔH_1 3s 2p 2s 1s Nucleus $\qquad$ ∞ ΔH_2 3s 2p 2s 1s Nucleus

$Na \longrightarrow Na^+ + e^-$ $\qquad$ $Na^+ \longrightarrow Na^{2+} + e^-$

Similar effects are observed with elements of other groups wherever we start to draw electrons from a new shell. For example, this happens with group 2A elements as we move from the second to the third ionisation enthalpy:

∞ ΔH_1 3s 2p 2s 1s Nucleus $\qquad$ ∞ ΔH_2 3s 2p 2s 1s Nucleus $\qquad$ ∞ ΔH_3 3s 2p 2s 1s Nucleus

$Mg \rightarrow Mg^+ + e^-$ $\qquad$ $Mg^+ \rightarrow Mg^{2+} + e^-$ $\qquad$ $Mg^{2+} \rightarrow Mg^{3+} + e^-$

When successive electrons are removed from an atom a fairly regular increase in ionisation enthalpy is observed as electrons are removed from orbitals of similar energy level, but large jumps appear whenever new shells are broken into, and irregularities occur when different types of orbital are drawn upon (Fig. 4.21). This provides evidence for the existence of distinct energy levels within an atom.

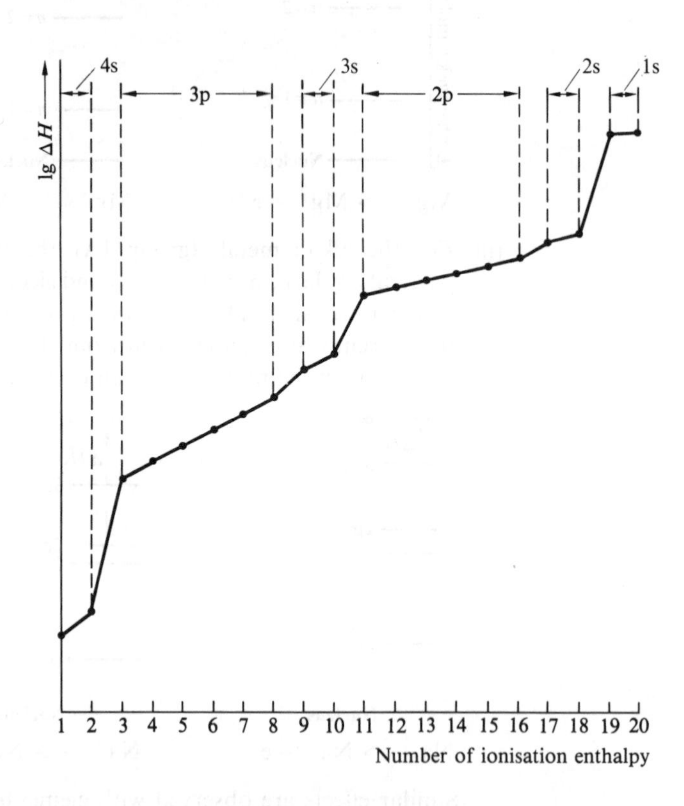

Fig. 4.21 Successive ionisation enthalpies of calcium.

All ionisation enthalpies (ΔH_1, ΔH_2, etc) decrease down a group of the periodic table. The fall in first ionisation enthalpy corresponds to an increase in atomic radius, and is well illustrated by group 1A:

	Li	Na	K	Rb	Cs
ΔH_1/kJ mol^{-1} at 298 K	525	500	424	408	382

As atoms become larger the outermost electron is further from the nucleus and therefore less attracted to it. The electron occupies a higher energy orbital so that less energy is required for its removal; compare sodium and potassium:

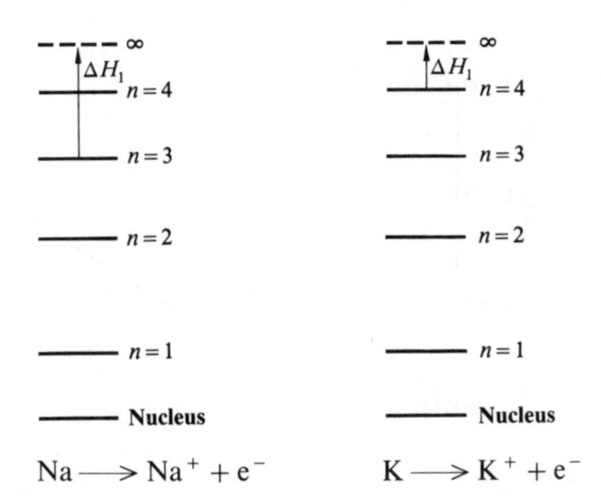

$Na \longrightarrow Na^+ + e^-$ $K \longrightarrow K^+ + e^-$

Also, as the number of electrons between the outer shell and the nucleus increases, their screening effect is greater,

e.g. Na: $\underbrace{1s^2\ 2s^2\ 2p^6}_{\text{Ten screening electrons}}\ 3s^1$ K: $\underbrace{1s^2\ 2s^2\ 2p^6\ 3s^2\ 3p^6}_{\text{18 screening electrons}}\ 4s^1$

Second and subsequent ionisation enthalpies decrease down a group for the same reasons.

Ionisation enthalpies show a general increase from left to right across any period of the periodic table, for the principal reason that atomic radius decreases. For example,

	Li	Be	B	C	N	O	F
ΔH_1/kJ mol^{-1} at 298 K	525	906	805	1 090	1 400	1 310	1 680

As before, we can argue that the attraction between the outer electron and the nucleus increases as the distance between them decreases. The energy level of the outer atomic orbital therefore falls from left to right across a period, and the ionisation enthalpy must increase. Figure 4.22 illustrates graphically the first ionisation enthalpies of the elements of the second period (Li–Ne), with a diagrammatic explanation in terms of energy levels. There are two discontinuities to be explained, i.e. the decrease from beryllium to boron, and from nitrogen to oxygen.

(i) The first ionisation enthalpy of boron is lower than that of beryllium because the electron is removed from a 2p orbital which is higher in energy than the 2s orbital of beryllium.

(ii) Whenever two electrons occupy a particular orbital they repel each other. As a result it is easier to remove one of the paired 2p electrons from an oxygen atom (Table 3.3) than it is to remove an unpaired 2p electron from a nitrogen atom.

Similar graphs are observed for the main group elements of other periods.

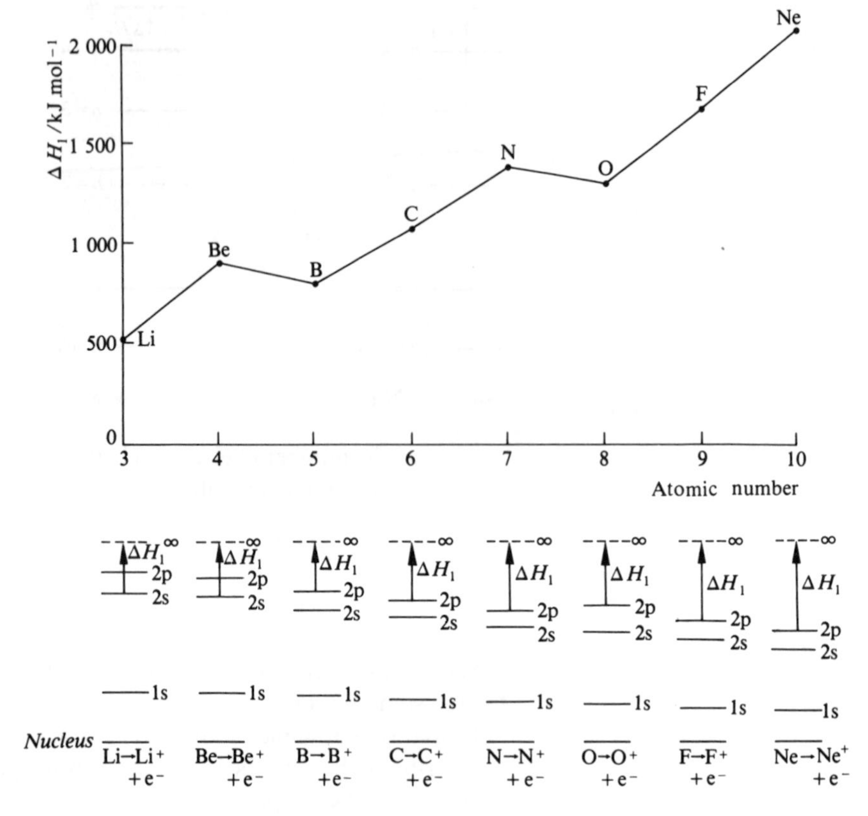

Fig. 4.22 First ionisation enthalpies of the elements of the second period.

Because it indicates quantitatively the ease of ion formation, ionisation enthalpy is a guide to the metallic character of an element. Thus, as ionisation enthalpy decreases down a group of the periodic table the elements become more metallic, and as ionisation enthalpy increases across a period the elements become less metallic. For this purpose ionisation enthalpy is to be preferred to standard electrode potential (§2.5.1). The latter comprises at least three separate enthalpy changes, only one of which (ionisation enthalpy) is directly related to metallic character.

4.7.3 ELECTRON AFFINITY

All monoatomic anions have noble gas configurations. The ions are formed from atoms of non-metals in groups 5B, 6B and 7B of the periodic table by the gain of one or more electrons to the outer shell. As an electron is accepted by an atom or anion there is an enthalpy change known as the *electron affinity*.

Electron affinity is a net enthalpy change made up of two components. One component is the enthalpy required to move an isolated electron to the negatively charged outer shell of the atom. We shall call this the 'repulsion component'. The other is the enthalpy that is released when the electron, having reached the outer shell, enters it to give an anion. We shall call this the 'attraction component'.

The *first electron affinity* of an element is defined as the enthalpy released during the addition of one mole of electrons to one mole of isolated atoms of the element in the gaseous state,

i.e. $A(g) + e^- \longrightarrow A^-(g)$

First electron affinity is an exothermic change, because the attraction component is greater than the repulsion component. The value usually decreases down a group of the periodic table as the atomic radii of the elements increase, e.g.

	F	Cl	Br	I
ΔH_1/kJ mol^{-1} at 298 K	−342	−358	−336	−308

The reason is that the smaller the atom the closer is the outer shell to the nucleus and the more strongly are electrons attracted into it.

The value for fluorine is anomalous because it includes a high repulsion component. This is a consequence of a low atomic radius and a compact outer shell.

The *second electron affinity* of an element relates to the uptake of a second electron, and is defined as the enthalpy *required* to add one mole of electrons to one mole of isolated uninegative ions,

i.e. $A^-(g) + e^- \longrightarrow A^{2-}(g)$

The change is endothermic, because the repulsion component is greater than the attraction component. Among the few values that are known are those for oxygen and sulphur:

	O	S
ΔH_1/kJ mol^{-1} at 298 K	−136	−194
ΔH_2	+850	+538
Total electron affinity	+714	+344

It will be seen that the overall changes for the formation of O^{2-} and S^{2-} ions are endothermic.

4.7.4 LATTICE ENTHALPY (LATTICE ENERGY)

The strength of an ionic bond is denoted by *lattice enthalpy*. The lattice enthalpy of an ionic compound is defined as the enthalpy required for the separation of the ions in one mole of the crystal to an infinite distance from one another,

i.e. $MX(s) \longrightarrow M^+(g) + X^-(g)$

Lattice enthalpy can also be expressed as the *stabilisation energy* that is released when isolated gaseous ions come together so as to give one mole of the crystal. There is a decrease of both kinetic and potential energy during this process. Kinetic energy decreases because ions in the solid state have very little movement. (Only a certain amount of vibration is possible, whereas gaseous ions move about rapidly.) Electrical potential energy decreases as it always does when oppositely charged bodies are brought together.

Consider the formation of sodium chloride. When one mole of sodium reacts with half a mole of chlorine to give one mole of isolated sodium and chloride ions according to the equation

$$Na(s) + \tfrac{1}{2}Cl_2(g) = Na^+(g) + Cl^-(g)$$

ΔH is the sum of:

(i) enthalpy of atomisation of sodium	$+109$ kJ mol^{-1}
(ii) $\tfrac{1}{2}$Cl—Cl bond dissociation enthalpy	$+121$ kJ mol^{-1}
(iii) first ionisation enthalpy of sodium	$+500$ kJ mol^{-1}
(iv) electron affinity of chlorine	-358 kJ mol^{-1}
Algebraic sum	$+372$ kJ mol^{-1}

The overall change is thus endothermic and unlikely to occur. (We cannot be *certain* that the change is impossible unless we consider free energy changes rather than enthalpy changes, §2.4.2.) However, when sodium ions and chloride ions come together to form solid sodium chloride sufficient lattice enthalpy is released to turn this endothermic change into an exothermic one which is much more likely to take place. To find the value of the lattice enthalpy we may subtract the enthalpy of formation of $Na^+(g)$ $Cl^-(g)$, calculated as above, from the observed enthalpy of formation of NaCl(s). By experiment, $\Delta H_f^{\ominus}[NaCl(s)] = -411$ kJ mol^{-1}; therefore the lattice enthalpy of sodium chloride $= -411 - (+372) = -783$ kJ mol^{-1}. (If we wish to quote the lattice enthalpy according to our definition, which concerns the *separation* of ions, we must write $+783$ kJ mol^{-1}.) The lattice enthalpies of other compounds can be similarly determined (§2.3.2).

The lattice enthalpy (ΔH_{LE}) of a compound depends mainly on the charge numbers (z) and radii (r) of its ions.

$$\Delta H_{LE} \propto \frac{z_c \times z_a}{d} \propto \frac{z_c \times z_a}{r_c + r_a}$$

where d represents the shortest interionic distance, and the subscripts 'c' and 'a' relate to the cation and anion respectively.

Lattice enthalpy therefore increases with ionic charge but decreases with ionic radius. Thus, for a particular series of salts, lattice enthalpy decreases from top to bottom of any group of the periodic table as r_c or r_a increases, e.g.

	LiF	NaF	KF	RbF	CsF	NaF	NaCl	NaBr	NaI
ΔH_{LE}/kJ mol^{-1} at 298 K	1035	915	814	780	729	915	783	745	696

Movement from left to right across a period is accompanied at first by an increase in lattice enthalpy as z_c increases (e.g. Na^+, Mg^{2+}, Al^{3+}), but finally by a decrease in lattice enthalpy as z_a decreases (e.g. N^{3-}, O^{2-}, F^-).

A useful concept in connection with lattice enthalpy is *surface charge density*. The surface charge density of an ion is an indication of how densely the charge is spread over its surface: it is directly related to the charge number, z, and inversely related to the surface area of the ion. Small ions with a high charge, e.g. Al^{3+}, have a high surface charge density, while large ions with a low charge, e.g. Cs^+, have a low surface charge density. For a compound in which both cation and anion have a high surface charge density, e.g. Al_2O_3, lattice enthalpy is high, while for one in which both ions have a low surface charge density, e.g. CsI, lattice enthalpy is low.

Lattice enthalpy, because it reflects the ease or difficulty of separating ions, has a major influence on the melting temperatures and solubilities of ionic compounds. High lattice enthalpy results in high melting temperature while low lattice enthalpy has the opposite effect, although low melting temperature may be associated with a partially covalent bond. Water solubility is governed by two factors, namely lattice enthalpy and hydration enthalpy, but the former is usually dominant. In general, high lattice enthalpy is responsible for low solubility while low lattice enthalpy leads to high solubility (§4.9.1).

4.8 Shapes of molecules and ions

Although the structural formula of methane is commonly written as

```
   H
   |
H—C—H
   |
   H
```

it has been known for many years that the molecule is not planar. As long ago as 1874 it was proposed that the four covalent bonds of the carbon atom were directed towards the four corners of a regular tetrahedron (Fig. 4.23). More recently, electron diffraction experiments, which are used to measure bond angles and bond lengths, have confirmed this view.

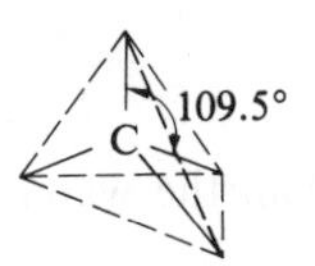

Fig. 4.23 A carbon atom located in the centre of a tetrahedron.

According to N. V. Sidgwick and H. M. Powell (*Sidgwick–Powell theory*, 1940) the shape of the methane molecule is due to electrostatic repulsion between the four pairs of electrons that constitute the C—H covalent bonds. The observed H—C—H bond angle of 109.5° is the maximum that is possible, and is due to the four bond pairs arranging themselves so as to minimise their mutual repulsions. A simple test of the theory can be conducted by tying together four balloons, each representing a bond pair of electrons. It will be found that the balloons repel one another equally, and in so doing arrange themselves so that they point towards the corners of a regular tetrahedron.

It is important to realise that *on this theory* the shape of the molecule is due primarily to repulsion between bond pairs of electrons and not to repulsion between hydrogen atoms.

The linear structure of beryllium chloride, the trigonal planar structure of boron trifluoride, and the symmetrical structures of many other simple

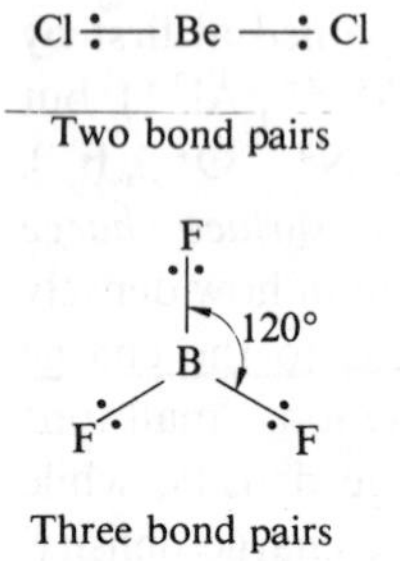

Fig. 4.24 The shapes of the beryllium chloride and boron trifluoride molecules.

molecules and ions can all be explained by the Sidgwick–Powell theory in terms of repulsion between bond pairs of electrons (Fig. 4.24).

Because ammonia and boron trifluoride have similar formulae we may expect the molecules to have similar shapes, but experiment shows that the ammonia molecule (Fig. 4.25) is not planar: the plane of the hydrogen atoms is depressed below the nitrogen atom, and the H—N—H bond angle is 106.7°.

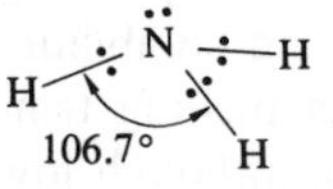

The difference between the BF_3 and NH_3 molecules stems from the fact that in the latter the central atom possesses a lone pair of electrons, which repels the three bond pairs. Consequently, the shape of the molecule is governed by repulsion between *four* pairs of electrons, similar to methane, and we might reasonably expect a bond angle of 109.5°. The fact that the observed bond angle is less than this must mean that the repulsion between lone pair and bond pair is greater than that between two bond pairs. There are two reasons for this. First, the lone pair of electrons is closer to the bond pairs than the bond pairs are to one another, and second, the lone pair is able to spread sideways, whereas each bond pair is strongly orientated along the line between the nitrogen and hydrogen atoms.

Fig. 4.25 The trigonal pyramidal shape of the ammonia molecule.

The angular shape of the water molecule (Fig. 4.26) can be explained in the same way. Again, the shape is due to repulsion between four pairs of electrons, but in this molecule they comprise two bond pairs and two lone pairs. The relatively small bond angle of 104.5° reflects the high repulsion that exists between two lone pairs of electrons.

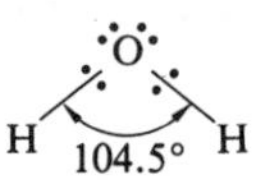

Fig. 4.26 The shape of the water molecule.

The order of repulsion between electron pairs is thus as follows:

two lone pairs > lone pair and bond pair > two bond pairs

The hydrogen sulphide molecule is structurally similar to the water molecule, and we might perhaps expect it to possess an identical shape, but experiment shows that while it is bent, like the water molecule, the bond angle is only 92.5°. Because sulphur is less electronegative than oxygen, the location of the bond pairs of electrons in the two molecules is different (Fig. 4.27). The relative remoteness of the bond pairs from the sulphur atom allows the H_2S molecule to 'close up' rather more than the water molecule.

Fig. 4.27 A comparison of the shapes of the water and hydrogen sulphide molecules.

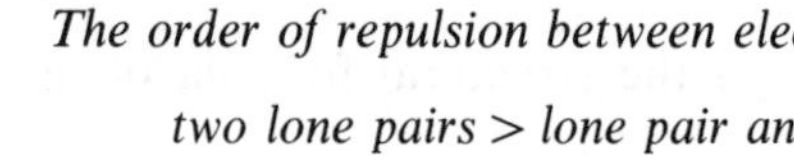

In their influence on molecular or ionic shape, multiple bonds, because they occupy only one position in space, act as single bonds. Thus, the carbon dioxide molecule is linear, and in ethene the distribution of bonds about each carbon atom is trigonal planar (Fig. 4.28).

Fig. 4.28 The shapes of the carbon dioxide and ethene molecules.

Fig. 4.29 The shape of the nitrogen dioxide molecule.

Coordinate bonds, as might be expected, behave as ordinary covalent bonds.

An unpaired electron, such as exists in NO_2, may be treated as half a lone pair. Thus, in the nitrogen dioxide molecule (Fig. 4.29) the O—N—O bond angle is 134°. If the unpaired electron were absent the bond angle would be 180° (cf. CO_2), and if it were a lone pair the bond angle would be 115.4°, as in the nitrite ion, NO_2^-.

Table 4.2 shows the shapes of many simple molecules and ions. Regular shapes are obtained only when all the electron pairs are used in bonding to identical atoms. If some of the atoms are different, departures from the

Table 4.2 The shapes of simple molecules and ions

Stoicheiometry	Electron pairs	Shape	Description	Examples, with bond angles
AB_2	2 bond pairs	B—A—B	linear	$BeCl_2$, CO_2, CH≡CH
AB_2	2 bond pairs, ½ lone pair	(A with single dot, bonded to two B)	angular (wide angle)	NO_2 (134°)
AB_2	2 bond pairs, 1 lone pair	(A with lone pair, bonded to two B)	angular, cf. trigonal planar	SO_2 (119.5°), O_3 (116.8°), NO_2^- (115.4°)
AB_2	2 bond pairs, 1½ lone pairs	·A· bonded to two B	angular	ClO_2 (117°)
AB_2	2 bond pairs, 2 lone pairs	:A: bonded to two B	angular, cf. tetrahedral	H_2O (104.5°), H_2S (92°), ClO_2^- (110°)
AB_3	3 bond pairs	A bonded to three B	trigonal planar	BF_3, SO_3, NO_3^-, CO_3^{2-} (all 120°)
AB_3	3 bond pairs, 1 lone pair	A with lone pair bonded to three B	trigonal pyramidal, cf. tetrahedral	NH_3 (106.7°), PH_3 (93°), SO_3^{2-} (∠ not known)
AB_3	3 bond pairs, 2 lone pairs	B—A—B with B below	T-shape, cf. trigonal bipyramidal	ClF_3 (87.5°)
AB_4	4 bond pairs	A bonded to four B	tetrahedral	CH_4, $[Ni(CO)_4]$, NH_4^+, SO_4^{2-}, PO_4^{3-}, ClO_4^- (all 109.5°)
AB_4	4 bond pairs, 2 lone pairs	A with two lone pairs bonded to four B	square planar, cf. octahedral	XeF_4, $[ICl_4]^-$ (all 90°)
AB_5	5 bond pairs	A bonded to B_1, B_1, B_2, B_2, B_2	trigonal bipyramidal	PF_5, PCl_5 (g) ∠B_1AB_2 90° ∠B_2AB_2 120°
AB_6	6 bond pairs	A bonded to six B	octahedral	SF_6, $[PF_6]^-$, $[AlF_6]^{3-}$ and many other complex ions (all 90°)

ideal shapes occur in accordance with the electronegativities of the elements concerned.

Because the shape of a molecule or ion is based merely on an electron count, without reference to the origin of the electrons or the orbitals to which they belong, the following simple procedure can be used to predict the shape of a species.

(i) Count the number of outer electrons on the central atom. (This is equal to the group number of the element in the periodic table.)
(ii) To this add the number of electrons contributed by the atoms bonded to the central atom.
(iii) Divide the total by two to obtain the number of electron pairs.
(iv) Determine how many are bond pairs and how many are lone pairs.
(v) Apply the shapes listed in Table 4.2.

The Sidgwick–Powell theory does not apply to transition element complexes.

4.9 General properties of ionic and covalent compounds

4.9.1 IONIC COMPOUNDS

Melting and boiling temperatures

All ionic compounds are crystalline solids with high melting temperatures (above ~500 °C (773 K)) and high boiling temperatures (above ~1 000 °C (1 273 K)). Sodium chloride is typical. (Melting temperature = 801 °C (1 074 K); boiling temperature = 1 467 °C (1 740 K)).

To understand why melting and boiling temperatures are so high we must consider the changes that occur in an ionic compound $A^+ B^-$ as it is heated from −273 °C (0 K) to its boiling temperature.

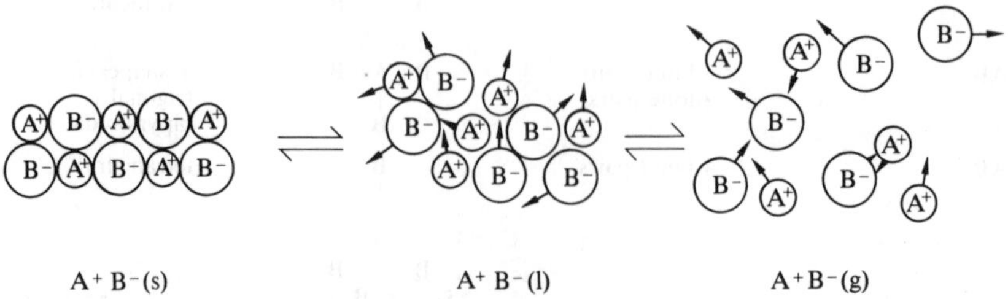

Fig. 4.30 Changes in ionic dispositions as an electrovalent compound is melted and boiled.

$A^+B^-(s) \rightleftharpoons A^+B^-(l) \rightleftharpoons A^+B^-(g)$

At −273 °C (0 K) the ions in a crystal have very little vibrational energy. On warming, the ions vibrate about a mean position, with increasing amplitude as the temperature is raised. Eventually, at the melting temperature, when sufficient energy has been supplied, the ions break away from their symmetrical arrangement in the crystal and move about at random. Even though the ions are still close together in the molten state, a considerable amount of energy is required for the melting process and the melting temperature is high.

As the temperature is raised still further, the ions gain in kinetic energy until, at the boiling temperature, they have sufficient energy to overcome the lattice enthalpy entirely. At this temperature ions can escape into the space above the liquid where they exist independently of one another or as small aggregates such as ion pairs. The boiling temperature is high because lattice enthalpy is high.

Because both melting and boiling require the overcoming of electrostatic attraction between ions, both melting and boiling temperatures are related to lattice enthalpy, which in turn is related to the charge numbers of the ions (§4.7.4). Thus, as we progress from group 1A to group 2A of the periodic table, z_c increases from $+1$ to $+2$ and the lattice enthalpy increases. Compare the melting and boiling temperatures of caesium chloride and barium chloride:

	ΔH_{LE}/kJ mol^{-1}	$\theta_{C,m}$/°C (T_m/K)	$\theta_{C,b}$/°C (T_b/K)
caesium chloride	657	645 (918)	1 287 (1 560)
barium chloride	2 042	957 (1 230)	1 557 (1 830)

Clearly, if we double z_c and double z_a lattice enthalpy increases fourfold, provided that all else remains constant. That is why magnesium oxide has the very high melting temperature of 2 800 °C (3 073 K), while sodium chloride, with an identical structure, melts at 801 °C (1 074 K).

Within a particular group of the periodic table, and for a particular series of ionic compounds, lattice enthalpy and hence melting temperature often decrease down the group. Consider the chlorides of the metals in group 1A:

	LiCl	NaCl	KCl	RbCl	CsCl
lattice enthalpy/ kJ mol^{-1}	858	783	713	688	657
$\theta_{C,m}$/°C (T_m/K)	607 (880)	801 (1 074)	767 (1 040)	715 (988)	645 (918)

The relatively low melting temperature of lithium chloride suggests that the bonding in this compound has a certain amount of covalent character.

Most electrovalent compounds which contain covalent ions decompose on heating before a melting temperature is reached. A *covalent ion* is one which is derived from a simple ion joined covalently to one or more atoms,

e.g. $O^{2-} + CO_2 \rightleftharpoons CO_3^{2-}$

$O^{2-} + SO_3 \rightleftharpoons SO_4^{2-}$

These reactions are reversible, and on heating carbonates tend to decompose into oxides and carbon dioxide, sulphates into oxides and sulphur trioxide, etc. The ease with which decomposition occurs depends on the polarising power of the associated cation. Thus, group 1A cations, with a low surface charge density, do not greatly distort these covalent ions and decomposition occurs only at a high temperature. The reverse is true where the cation has a smaller radius or a greater charge (§14.2.3).

Solubility

Ionic compounds vary considerably in their solubility in water, from extremely soluble compounds such as calcium iodide, to almost totally insoluble compounds such as barium sulphate. To understand this variation we must begin by visualising dissolution.

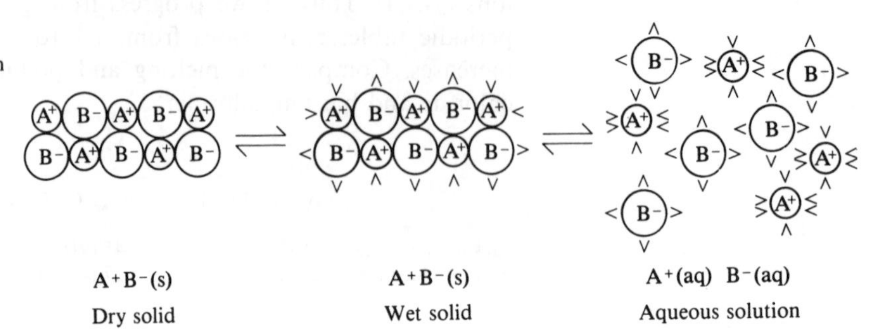

Fig. 4.31 The dissolving of an electrovalent compound $A^+ B^-$ in water (v represents a water molecule, i.e. $\overset{\delta+}{H}\cdots\overset{\delta-}{O}\cdots\overset{\delta+}{H}$).

Ions in solution are separated by a considerable distance and are able to move about randomly. To achieve this state, the crystal structure must be broken down completely. The lattice enthalpy must be overcome, and the energy that is required for this purpose originates from the hydration of ions.

When an ionic solid is placed in water, some of the water molecules, because of their charge separation, orientate themselves towards the cations and anions (Fig. 4.31). The process is entirely spontaneous, because unlike charges attract each other, and energy is released which may be sufficient to overcome the lattice enthalpy of the compound.

Every ion has its own *hydration enthalpy*, which is defined as the enthalpy released during the hydration of one mole of isolated ions in the gaseous state by an infinitely large quantity of water,

i.e. $X^{n\pm} + aq \rightarrow X^{n\pm}(aq)$

Let the sum of the hydration enthalpies of cation and anion be represented by ΔH_{hyd}. Then in general:

if $|\Delta H_{LE}| \gg |\Delta H_{hyd}|$, AB is insoluble,

if $|\Delta H_{hyd}| \gg |\Delta H_{LE}|$, AB is soluble,

and if $|\Delta H_{hyd}| \approx |\Delta H_{LE}|$, AB is likely to be soluble.

(Vertical lines denote numerical values irrespective of sign.) Although water solubility is controlled primarily by lattice enthalpy and hydration enthalpy, certain other factors, notably the entropy change (§2.4.1) also play a part. It is the increase in entropy that results in water solubility if $|\Delta H_{hyd}| \approx |\Delta H_{LE}|$.

The dissolution of AB is exothermic if $|\Delta H_{hyd}| > |\Delta H_{LE}|$, and endothermic if $|\Delta H_{LE}| > |\Delta H_{hyd}|$.

The difference between $|\Delta H_{LE}|$ and $|\Delta H_{hyd}|$ is known as the *enthalpy of solution* (ΔH_s) of the compound. This is defined as the enthalpy change when one mole of the solid is dissolved in a large quantity of water, such that the addition of more water produces no further enthalpy change.

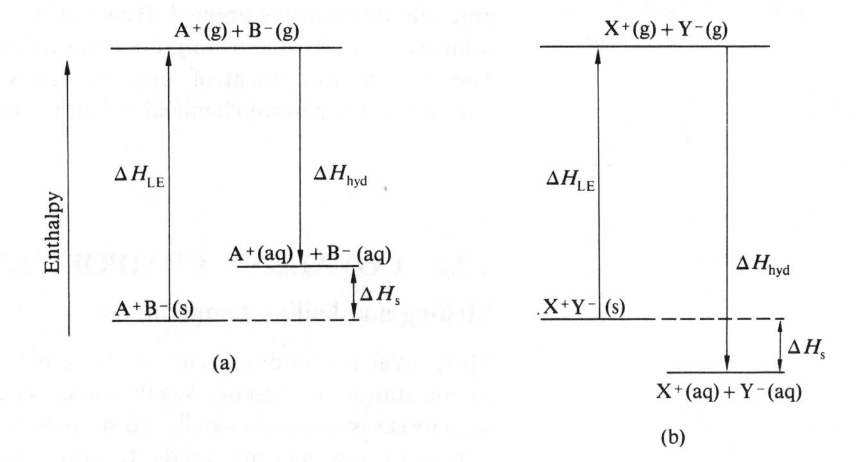

Fig. 4.32 Enthalpy changes on dissolving salts $A^+ B^-$ and $X^+ Y^-$ in water.

Figure 4.32(a) represents the dissolving of a salt for which $|\Delta H_{LE}| > |\Delta H_{hyd}|$. The process is endothermic and ΔH_s is positive. The salt XY (Fig. 4.32(b)) dissolves in water exothermically ($|\Delta H_{hyd}| > |\Delta H_{LE}|$ and ΔH_s is negative).

The hydration of ions is essentially an electrostatic process, but in the case of transition metal cations, the beryllium ion, and cations derived from metals in groups 3B, 4B and 5B of the periodic table, the ion–dipole electrostatic attraction is reinforced by coordination. Hydration is always an exothermic process because it leads to a reduction of both potential energy and kinetic energy. Potential energy is lowered because the shells of water molecules effectively shield oppositely charged ions from one another, and kinetic energy is reduced because hydrated ions are bulky and move more slowly than the free ions.

The more water molecules an ion can attract, the greater is the loss of both potential and kinetic energy and the greater is the hydration enthalpy. Hydration enthalpy is therefore directly related to the charge number of an ion and inversely related to its size. We can combine these factors and argue that the hydration enthalpy of an ion is directly related to its surface charge density. Thus, hydration enthalpy decreases down any group of the periodic table (e.g. from the small Li^+ ion to the large Cs^+ ion) as the surface charge density decreases, but increases from left to right across a period (e.g. from Na^+ to Mg^{2+} to Al^{3+}) as the surface charge density increases.

The solubility of ionic compounds in non-aqueous solvents depends on the polar nature of the solvent. Highly polar liquids, like water, are good solvents because they reduce the electrostatic attraction between ions. This applies, for example, to liquid hydrogen fluoride, hydrogen cyanide and, to a lesser extent, methanoic acid. Non-polar liquids, e.g. hydrocarbons, are poor solvents for ionic compounds.

Electrical conductivity

In the solid state, ionic compounds are non-conductors of electricity because the ions cannot move from their positions in the crystal when a potential difference is applied. However, in the fused state or in aqueous solution ions are able to migrate towards oppositely charged electrodes. The two-way movement of ions constitutes an electric current. At each electrode one or more chemical reactions occur (§10.2.1).

4.9.2 COVALENT COMPOUNDS

Melting and boiling temperatures

Most covalent compounds are made up of individual molecules, attracted to one another by van der Waals' forces. The energy needed to overcome such forces is relatively small, and so most covalent compounds are gases, liquids or low melting solids. (Melting temperature is usually below ~300 °C (573 K).)

The contrast between the low melting temperatures of covalent compounds and the high melting temperatures of ionic compounds simply reflects the fact that it is easier to separate neutral molecules from one another than charged ions. It does *not* imply that the covalent bond is much weaker than the ionic bond, for the melting of covalent substances is not accompanied by a breaking of chemical bonds, unless decomposition occurs at the same time. Substances that form extended covalent crystals, such as carbon and silica (§6.4.3), have extremely high melting temperatures because, in these few cases, melting must involve the breaking of covalent bonds.

Solubility

Like ionic compounds, covalent substances vary enormously in their solubility, but for widely different reasons.

Certain covalent compounds have a high solubility in water because, as they dissolve, they *ionise*, i.e. they react chemically with the water to form ions. Covalent compounds that behave in this way comprise acids, covalent bases and covalent salts.

Acids

All protic acids (§11.1.1) are covalent in the anhydrous state and all of them ionise in water. In general terms,

$$HA(g) \rightarrow H(g) + A(g) \rightarrow H^+(g) + A^-(g) \rightarrow H^+(aq) + A^-(aq)$$

The proton and to a lesser extent the anion exert an attraction for water molecules, which encourages ionisation and hence dissolution in water.

Bases

Ammonia and its derivatives, both inorganic and organic, dissolve in water with the formation of ammonium ions (or substituted ammonium ions) and hydroxide ions.

$$NH_3(g) + H_2O(g) \rightarrow NH_4(g) + HO(g) \rightarrow NH_4^+(g) + HO^-(g) \rightarrow NH_4^+(aq) + HO^-(aq)$$

Again, dissolving is promoted because the ions that are formed, especially HO^-, are strongly attracted to water.

Salts

Some covalent halides, e.g. aluminium chloride and iron(III) chloride, resemble acids in that they dissolve in water with ionisation. Solubility is high, and the process is exothermic. For example:

$$SnCl_4(s) \rightarrow SnCl_4(g) \rightarrow Sn(g) + 4Cl(g) \rightarrow Sn^{4+}(g) + 4Cl^-(g) \rightarrow Sn^{4+}(aq) + 4Cl^-(aq)$$

Numerous enthalpy changes occur, some exothermic and some endothermic, but the outstanding one is the high hydration enthalpy of Sn^{4+}, a small ion with a high charge. The hydration of this ion provides much of the energy that is needed for the breaking of four Sn—Cl bonds and the removal of four electrons from the tin atom.

When a covalent compound dissolves in a solvent without ionisation, solubility is governed by the polarity of the solute and solvent, the rule being that *like dissolves like.* Highly polar compounds such as glucose and ethanamide dissolve in highly polar solvents such as water and methanol, but have only a low solubility in non-polar solvents such as hydrocarbons and tetrachloromethane. Conversely, non-polar solutes, such as naphthalene, are insoluble in water but readily soluble in hydrocarbon solvents. Any pair of polar compounds will mix together because their molecules attract one another, and in the same way the molecules of two non-polar compounds will attract one another – albeit not so strongly, but sufficiently to cause mixing. However, the molecules of a polar compound will attract other molecules of the same sort rather than those of a non-polar substance; consequently there is little mixing and the solubility is low.

For the special case of covalent compounds dissolving in water without ionisation, solubility is possible only if hydrogen bonds (§6.6) can form between molecules of the solute and those of water. The lower alcohols, aldehydes, ketones, carboxylic acids and amides all dissolve in water because they satisfy this condition. The presence of a chemical grouping that can hydrogen bond with water is, in itself, no guarantee of water solubility. The higher members of any homologous series of organic compounds have a low solubility in water because of the hydrophobic effect of the long carbon chain, which outweighs any hydrophilic property of the functional group.

Electrical conductivity

Covalent compounds in the pure state, even when liquid, are non-conductors of electricity because they possess no ions to carry the current. However, in the presence of water, as we have seen, some covalent compounds can ionise. If this happens, the aqueous solutions are electrically conducting, and electrolysis occurs when a potential difference is applied.

EXAMINATION QUESTIONS ON CHAPTER 4

1 'Each of the concepts electrovalency and covalency relates to an idealised state of chemical bonding which often does not exist in real compounds'.

Discuss how far this statement is valid and give *two* examples, with suitable evidence, of cases where such 'non-ideality' in fact arises. (L)

2 (a) For each of the following shapes, give the formula of one molecule or ion having the stated shape and give the bond angle(s) found in the regular structures listed.

(i) linear
(ii) trigonal planar
(iii) tetrahedral
(iv) trigonal bipyramidal
(v) octahedral

(b) The H—N—H bond angle in ammonia is 107.3°. Explain why this angle is not one of those found in the regular structures (i)–(v). (JMB)

3 (a) Describe the essential features of the spectrum of atomic hydrogen. How is this spectrum interpreted?

(b) The first ionisation energy of the sodium atom is $496\ \text{kJ mol}^{-1}$. What does this mean? Why is the first ionisation energy of caesium (the Group I element of atomic number 55) smaller than that of sodium?

(c) What can you predict about (i) the relative sizes of the potassium and chloride ions, (ii) the relative lattice energies of lithium fluoride and potassium bromide? Give reasons for your answers. (O)

4 (a) List the three main fundamental particles which are constituent of atoms, and give their relative masses and charges.

(b) Similarly, name and *differentiate* between the radiations emitted by naturally occurring radioactive elements.

(c) Complete the following equations using your periodic table to identify the elements *X*, *Y*, *Z*, *Q*, and *R*. Add atomic and mass numbers where these are missing.

(i) ${}^{24}_{11}\text{Na} \rightarrow X + {}^{0}_{-1}\text{e}$
(ii) ${}^{14}_{7}\text{N} + {}^{1}_{0}\text{n} \rightarrow {}^{14}Y + {}_{1}Z$
(iii) $\text{Si} \rightarrow {}^{27}_{13}Q + {}^{0}_{+1}\text{e}$
(iv) $R + {}^{4}_{2}\text{He} \rightarrow {}^{13}_{7}\text{N} + {}^{1}_{0}\text{n}$

(d) Refer to c(i) and (iii) above. For each of these two processes, briefly describe **one** chemical test which could be used to confirm that a change of chemical element has occurred.

(e) If lead(II) chloride is precipitated in the presence of thorium nitrate, using an aqueous solution of lead(II) nitrate and dilute hydrochloric acid, the lead(II) chloride contains radioactive ^{212}Pb atoms. (These radioactive lead atoms are a daughter product of the thorium.)

Show how you could experimentally use this information to establish that lead(II) chloride and its saturated solution are in dynamic equilibrium. (SUJB)

5 What is meant by the terms ionic and covalent bonding? Explain why the bonds in most molecules are intermediate between these two extreme models.

Using the properties and reactions of halogen compounds as examples, illustrate the trend in bond type which occurs across the second period of the periodic table (i.e. Li to F). (JMB)

6 Explain how the shapes of simple molecules and ions can be explained in terms of electron pair repulsions. Illustrate your answer by considering examples of at least five different shapes.

Predict the shapes of the following:

(a) NH_2^-

(b) PCl_4^+

(c) ICl_4^-

Explain the following facts.

(i) The boiling point of water is higher than that of hydrogen sulphide (measured at the same pressure).

(ii) Water and ammonia react with anhydrous copper(II) sulphate but methane does not. (JMB)

7 (a) What is understood by the term *electronegativity* as applied to an element?

(b) Say how the electronegativities of the elements change

(i) as Group I (alkali metals) is descended,

(ii) as Group VII (halogens) is descended,

(iii) as Period II (lithium to fluorine) is crossed.

Comment on your answers to (i) and (ii) in the light of the electronic configuration of these elements.

(c) Why do the transitional elements, of any one series, have similar electronegativity values?

(d) In what way is the *type* of chemical bond formed between two atoms related to their electronegativities?

Having made a general statement, illustrate your answer by referring to

(i) chlorine,

(ii) magnesium chloride,

(iii) hydrogen chloride.

Draw a simple bond diagram in each case.

(e) (i) Using bond diagrams for water and hydrogen chloride, show how hydrogen chloride gas becomes ionised when dissolved in water.

(ii) $ICl + H_2O \rightarrow IOH + HCl$

$ICl + H_2O \rightarrow ClOH + HI$

Which of the two equations above is more likely to represent the hydrolysis of iodine(I) chloride?

Give reasons for your answer. (SUJB)

8 This question concerns the following elements whose electronegativities are listed

Al 1.5	C 2.5	H 2.1	N 3.0	P 2.1
B 2.0	Cl 3.0	Li 1.0	Na 0.9	S 2.5
Be 1.5	F 4.0	Mg 1.2	O 3.5	Si 1.8

(a) What is meant by the term *electronegativity*? What factors determine the electronegativity of an element?

(b) Arrange the elements according to their positions in the periodic table.

(c) What is the relationship between electronegativity and the position of elements (i) in a period, (ii) in a group?

In each case give a brief explanation.

(d) In what major respect do electronegativity and electron affinity differ?

(e) Arrange the following substances in order of increasing ionic character (i.e. putting the least ionic first).

CO_2, LiCl, MgF_2, NaCl, NH_3, S_2Cl_2.

(f) State, with reasons, how you would expect electronegativity values to vary, if at all, across the first row of transition elements. (AEB)

9 Explain the meaning of the statement that 'in its ground state the electron in a hydrogen atom is in a 1s orbital'. Further discuss the meaning of the term **covalent bond** illustrating your answer with reference to the hydrogen molecule and explaining why such a bond is stable.

Explain the principles of electron repulsions in determining the shape of the ammonia (NH_3) molecule. Give the shape and electronic structures of the carbonate ion (CO_3^{2-}) and ethene (C_2H_4) and show how the principles can further be used to account for the shapes of the species present in phosphorus pentachloride. (WJEC)

10 The following table shows the logarithms of successive ionisation energies (in kJ mol^{-1}) of the elements fluorine and magnesium.

Number of electrons removed	Log_{10} (ionisation energy)	
	Fluorine	Magnesium
1	3.32	2.87
2	3.53	3.18
3	3.78	3.89
4	3.92	4.02
5	4.04	4.13
6	4.18	4.26
7	4.25	4.34
8	4.96	4.41
9	5.03	4.50
10	—	4.55
11	—	5.20

(a) On the same graph, plot the two sets of $\log_{10}$ (ionisation energy) against the number of the electrons removed.

(b) Explain how your graphs can give information about the electronic structures of the two elements.

(c) Discuss how a knowledge of ionisation energies can help in explaining the chemical properties of elements. (L)

11 Explain, with examples, the meanings of, and any limitations in the applications of, the following concepts:
(a) diagonal relationship,
(b) electron affinity,
(c) ionic radius,
(d) electronegativity. (OC)

12 (a) Describe an experiment to determine accurately the enthalpy of combustion of a flammable substance.

(b) The enthalpy of formation of methane cannot be determined directly. Show how it may be determined indirectly from the following enthalpies of combustion: C(s): -393.5 kJ mol^{-1}; H_2(g): -285.9 kJ mol^{-1}; CH_4(g): -890.3 kJ mol^{-1}.

(c) The enthalpies of hydrogenation of cyclohexene and benzene are -120 and -208 kJ mol^{-1} respectively. Explain why the value for benzene is not three times the value for cyclohexene and estimate, with reasons, the enthalpies of hydrogenation of cyclohexa-1,4-diene and of cyclohexa-1,3-diene.

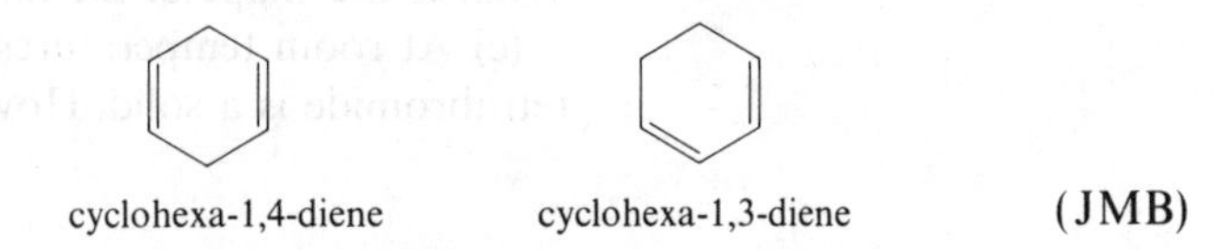

(JMB)

13 (a) Describe and suggest reasons for the trends across and down the periodic table in (i) atomic radius, (ii) atomic volume, and (iii) electronegativity.

(b) Discuss the bonding in (i) sodium chloride, and (ii) hydrogen chloride.

(c) Describe and account for the bonding you would expect to be present in (i) rubidium chloride and (ii) iodine chloride (ICl). (JMB)

14 Discuss briefly the redox properties of the halogens.

Making the approximation that if the enthalpy change of a reaction is negative it will occur, use the data below to show whether I_2 will displace Br_2 from the solid alkali metal bromide, MBr, or if the reverse reaction will be favoured.

	Br	I
Enthalpy of vaporisation (per mol X_2 molecules)	30.5	62.3
Enthalpy of dissociation (per mol X_2 molecules)	193	151
Electron affinity (per mol X atoms)	-330	-302
Lattice enthalpy of M^+Hal^- (per mol M^+X^-)	730	686

The enthalpy data are given in kJ mol^{-1}. (OC)

15 (a) For each of the following regular shapes, give the formula of one molecule which has that shape and say what bond angle(s) occur in each structure.
(i) linear,
(ii) triangular planar,
(iii) tetrahedral,

(iv) octahedral.

(b) Construct bond diagrams and so predict the bond angles in

(i) the NO_2^- ion,

(ii) the CO_3^{2-} ion.

(c) (i) Suggest, with some explanation, the geometrical shape of the PCl_5 molecule in the vapour phase.

(ii) Give reasons why molecules often have bond angles different from those angles in the regular tetrahedral shape. (SUJB)

16 Answer each of the following:

(a) What is understood by the term 'lone pair' of electrons? Show how this concept is involved in a consideration of the structure of the ammonium ion.

(b) Give the electronic structure of the chlorine molecule and of the chloride ion. Indicate the points of difference and of similarity between these two structures.

(c) Describe the bonding present in potassium chloride. What experimental evidence is there for the existence of this type of bonding in the pure material?

(d) Describe the bonding present in a molecule of carbon tetrachloride. What is the shape of the molecule?

(e) At room temperatures carbon tetrachloride is a liquid but carbon tetrabromide is a solid. How may this difference be explained? (JMB)

5 Molecular structure

5.1 The establishment of molecular structure

The structures of a great many compounds are known beyond any reasonable doubt, so that their establishment is more a matter of chemical history than current importance, but new compounds are being discovered all the time, both from natural sources and by synthesis, and their structures have to be elucidated.

The full procedure for determining the structure of a compound is lengthy and consists of the following stages:

(i) detection of the elements;
(ii) determination of percentage composition, i.e. the percentages by mass of the various elements that are present;
(iii) calculation of empirical formula;
(iv) determination of relative molecular mass;
(v) calculation of molecular formula;
(vi) deduction of structural formula.

In some cases the structure of a compound can be predicted from its synthesis, and merely confirmed by finding its percentage composition and then studying it by various techniques (§5.1.6). In this section, however, we shall consider the complete routine.

5.1.1 DETECTION OF THE ELEMENTS

For the detection of nitrogen, sulphur and halogens in organic compounds the *Lassaigne sodium fusion test* is used. To perform this test, a small sample of the unknown substance is heated in an ignition tube with a small piece of sodium. During the sodium fusion CN^-, S^{2-} and X^- (halide) ions are formed. At red heat, the tube is dropped into a boiling tube containing distilled water. The ignition tube shatters, and the soluble salts it contains are extracted into the water. The solution is then filtered, and separate portions are used in tests for nitrogen, sulphur and the halogens.

In testing for nitrogen, a portion of the extract is heated to boiling with iron(II) sulphate and then cooled. Iron(III) chloride solution is added, followed by dilute sulphuric acid until the solution is acidic. A blue precipitate or a greenish-blue coloration of Prussian blue is a positive indication of nitrogen.

$$[Fe(H_2O)_6]^{2+} + 6CN^- = [Fe(CN)_6]^{4-} + 6H_2O$$

$$Na^+ + Fe^{3+} + [Fe(CN)_6]^{4-} = \underset{\text{Prussian blue}}{Na[Fe^{II}Fe^{III}(CN)_6]}$$

A sensitive test for sulphur is provided by sodium pentacyanonitrosyl-

$Na_2[Fe(CN)_5(NO)]$

ferrate(III), commonly called 'sodium nitroprusside', a dilute solution of which turns a magenta colour in the cold in the presence of hydrogensulphide ions:

$$[Fe(CN)_5(NO)]^{2-} + HO^- + HS^- = [Fe(CN)_5(NOS)]^{4-} + H_2O$$

red-brown solution / magenta solution

The test for halogens consists of acidifying the Lassaigne extract with dilute nitric acid and adding silver nitrate solution:

$Ag^+ + Cl^- = AgCl(s)$ white precipitate, readily soluble in 4 M $NH_3(aq)$
$Ag^+ + Br^- = AgBr(s)$ pale yellow precipitate, sparingly soluble in 4 M $NH_3(aq)$
$Ag^+ + I^- = AgI(s)$ yellow precipitate, insoluble in 4 M $NH_3(aq)$

However, both CN^- and S^{2-} ions interfere in the test:

$Ag^+ + CN^- = AgCN(s)$ white precipitate
$2Ag^+ + S^{2-} = Ag_2S(s)$ black precipitate

These ions, if found in the previous tests, must therefore be removed by heating the acidified solution before adding silver nitrate:

$$H^+ + CN^- = HCN(g)$$
$$2H^+ + S^{2-} = H_2S(g)$$

If the silver nitrate test proves positive, a further test will confirm the identity of the halogen. For this purpose a fresh portion of the Lassaigne extract is acidified with dilute sulphuric acid, a few drops of tetrachloromethane added to form a separate lower phase, and dilute sodium hypochlorite solution introduced dropwise with vigorous shaking after each addition. The acidified hypochlorite oxidises bromide ions to bromine, which dissolves in the tetrachloromethane to give a reddish-brown solution. Likewise iodide ions are oxidised and give a violet colour in the tetrachloromethane, but chloride ions are not oxidised and the lower layer remains colourless.

5.1.2 DETERMINATION OF PERCENTAGE COMPOSITION

The principal techniques of quantitative analysis are as follows.

(i) Volumetric analysis (§1.1.3).
(ii) Gravimetric analysis. This is suitable for the estimation of elements that can be precipitated and weighed as highly insoluble compounds. Chlorine, for example, can be estimated gravimetrically as silver chloride, and sulphur as barium sulphate.
(iii) Spectrophotometry.

Special methods are used for the estimation of carbon, hydrogen, nitrogen and other elements in organic compounds and these will now be described.

Carbon, hydrogen and nitrogen

These elements are estimated together, by burning a sample of known mass in oxygen under carefully controlled conditions. Combined carbon is converted into carbon dioxide, hydrogen into water vapour, and nitrogen into molecular nitrogen.

Let x g of the organic compound, on complete combustion, produce y g of CO_2, z g of H_2O, and V_1 cm^3 of N_2 measured at a temperature of t °C (i.e. $(t + 273)$ K) and a pressure of p_1 atm.

$$\text{The fraction of combined carbon in carbon dioxide} = \frac{C}{CO_2} = \frac{12}{44}$$

$$\therefore \quad y \text{ g of } CO_2 \text{ contain } y \times \frac{12}{44} \text{ g carbon}$$

$$\therefore \quad \text{carbon content of the compound} = \frac{y}{x} \times \frac{12}{44} \times 100\%$$

$$\text{The fraction of combined hydrogen in water} = \frac{2H}{H_2O} = \frac{2}{18}$$

$$\therefore \quad z \text{ g of } H_2O \text{ contain } z \times \frac{2}{18} \text{ g hydrogen}$$

$$\therefore \quad \text{hydrogen content of the compound} = \frac{z}{x} \times \frac{2}{18} \times 100\%$$

Let the volume of N_2 at s.t.p. $= V_2$ cm^3.

From the gas laws,
$$V_2 = \frac{p_1 \times V_1 \times 273}{(t + 273)}$$

22 400 cm^3 (i.e. 1 mol) of N_2 at s.t.p. has a mass of 28 g

$$\therefore \quad V_2 \text{ cm}^3 \text{ of } N_2 \text{ at s.t.p. has a mass of } \frac{28}{22\,400} \times V_2 \text{ g}$$

$$\therefore \quad \text{nitrogen content of the compound} = \frac{28}{22\,400} \times \frac{V_2}{x} \times 100\%$$

For the estimation of nitrogen only, the *Kjeldahl method* is widely used. A sample of the compound is heated with concentrated sulphuric acid to convert nitrogen into ammonium sulphate. After dilution, excess alkali is added and the liberated ammonia distilled into an excess of standard acid. By back-titration with standard alkali the ammonia which is present, and hence the nitrogen in the original substance, can be determined.

Suppose, for example, that 5.378 mg of a compound gave ammonia equivalent to 4.12 cm^3 of 0.005 1 M H_2SO_4.

$$1\,000 \text{ cm}^3 \text{ of } 1 \text{ M } H_2SO_4 \equiv 2 \text{ mol } NH_3 \equiv 28 \text{ g nitrogen}$$

$$\therefore \quad 4.12 \text{ cm}^3 \text{ of } 1 \text{ M } H_2SO_4 \equiv \frac{28 \times 4.12}{1\,000} \text{ g nitrogen}$$

$$\therefore \quad 4.12 \text{ cm}^3 \text{ of } 0.005\,1 \text{ M } H_2SO_4 \equiv \frac{28 \times 4.12 \times 0.005\,1}{1\,000} \text{ g nitrogen}$$
$$= 0.000\,588\,3 \text{ g nitrogen}$$

$$\therefore \quad \text{nitrogen content of the compound} = \frac{0.000\,588\,3}{0.005\,378} \times 100 = 10.94\%$$

Oxygen

The percentage of oxygen is generally obtained by difference; i.e. the percentages of other elements are determined and their total subtracted from one hundred.

Halogens

A small sample of the compound is heated with a little fuming nitric acid and powdered silver nitrate in a sealed glass tube to about 250 °C (523 K). Halogen is converted to silver halide. After several hours the apparatus is cooled, the tube broken, and silver halide filtered off, washed, dried and weighed.

Let the mass of chlorine-containing compound = x g.
Let the mass of AgCl (relative molecular mass = 143.5) = y g.
By reasoning similar to that given above,

$$\text{chlorine content of the compound} = \frac{y \times 35.5 \times 100}{x \times 143.5}\%$$

Sulphur

A sample is heated with fuming nitric acid in a sealed tube. Sulphur is converted to the sulphate ion, which is afterwards estimated as insoluble barium sulphate by adding barium chloride solution.

Let the mass of sulphur-containing compound = x g.
Let the mass of $BaSO_4$ (relative molecular mass = 233.3) = y g.

$$\text{Sulphur content of the compound} = \frac{y \times 32 \times 100}{x \times 233.3}\%$$

5.1.3 CALCULATION OF EMPIRICAL FORMULA

The term empirical formula means literally 'a formula found by experiment'. It gives the ratio, in the smallest possible whole numbers, of the different kinds of atoms in the compound. Such a formula is of no value in itself, but merely provides a stepping-stone in arriving at the molecular and structural formulae.

The essential stage in the calculation of empirical formula is the division of the percentages of the elements by their respective relative atomic masses. This is to compensate for the fact that percentage composition depends not only on how many atoms are present, but also on their masses. An element of high relative atomic mass will automatically be present in a high proportion by mass, but a light element will account for little in terms of percentage mass although numerous atoms of it may be present. The procedure is shown in the following example.

Example

Analysis shows that a compound contains 29.11% of sodium, 40.51% of sulphur and 30.38% of oxygen. Calculate its empirical formula.

Answer

Elements:	Na	S	O
Percentages:	29.11	40.51	30.38
Divide by relative atomic masses:	$\frac{29.11}{23}$	$\frac{40.51}{32}$	$\frac{30.38}{16}$
∴ Ratio of atoms	= 1.266	1.266	1.899
To achieve a whole number ratio,			
(i) divide by the smallest, 1.266:	1	1	1.5
(ii) multiply, where necessary, by a low integer, in this case 2:	2	2	3

i.e. the empirical formula of the compound is $Na_2S_2O_3$.

With only the information given, the molecular formula, which gives the numbers of the various kinds of atoms in a molecule of the compound, cannot be deduced. The molecular formula may be the same as the empirical formula or it may be a multiple of it, but without the relative molecular mass we cannot say.

5.1.4 DETERMINATION OF RELATIVE MOLECULAR MASS

The relative molecular mass of a volatile compound can be determined by *vaporisation methods.* Several such methods are available, all designed to give the volume of vapour that is formed from a known mass of the compound in the liquid or solid state. Most substances, however, are involatile, and their relative molecular masses can be determined in solution by *colligative property methods* (§8.4.1). The most accurate method available, for compounds with relative molecular masses below 1 000, is based on *mass spectrometry.*

Vaporisation methods

At any given temperature molecules of all gases repel and attract one another to roughly the same extent, because their average kinetic energies are the same. Consequently, if N molecules of gas 1 and N molecules of gas 2 are placed in cylinders as shown by Fig. 5.1, under the same external pressure (p), they occupy approximately the same volume.

This is summarised by *Avogadro's law*, which states that: 'Equal volumes

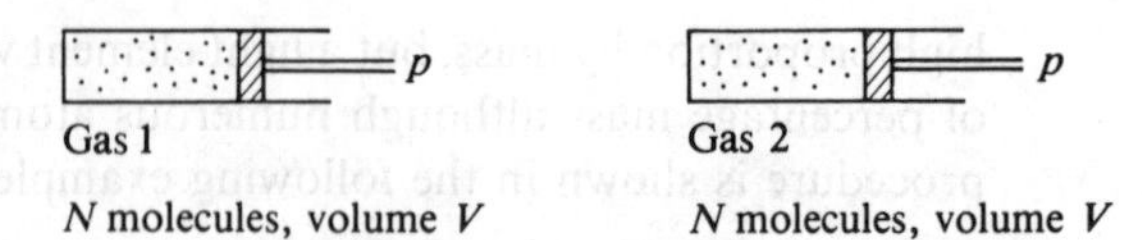

Fig. 5.1 Comparison of two gases at the same temperature, to illustrate Avogadro's law.

of all gases, under the same conditions of temperature and pressure, contain equal numbers of molecules.'

In effect, under a given set of conditions, every molecule of a gas occupies the same volume of space. Consequently, every mole of gas must occupy the same volume, called the *molar volume* (V_m), for there is a constant number of molecules in a mole. (This is the Avogadro constant, *L*.) By experiment, *one mole of any gas at s.t.p.* (273 K and 1 atm) *occupies a volume of approximately* 22.4 *dm*^{3}. Thus, if we know the volume of vapour that is formed from a certain mass of liquid or solid – and there are several types of apparatus available for the purpose – we can easily calculate the relative molecular mass of the substance.

Syringe method

The apparatus (Fig. 5.2) consists essentially of a graduated glass syringe enclosed in a glass heating jacket through which steam or other hot vapour can be passed. The end of the syringe can be covered by a septum, i.e. a self-sealing rubber cap, through which small quantities of the liquid under examination can be injected by means of a hypodermic syringe.

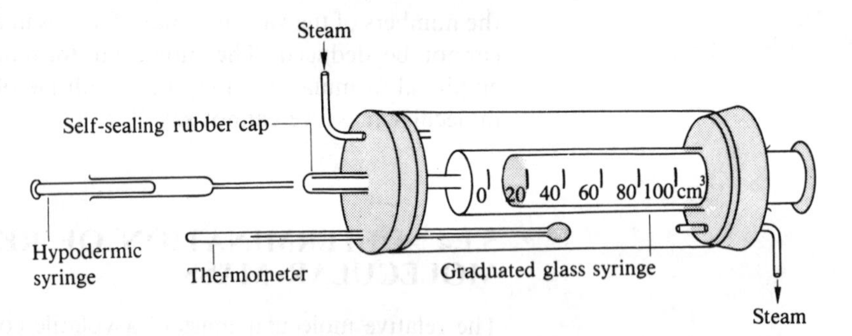

Fig. 5.2 Gas syringe with steam jacket.

About 5 cm^3 of air is drawn into the apparatus, the rubber cap is fitted, and steam is passed through the jacket until a steady state has been reached. The volume of air in the syringe is then noted, and about 0.2 cm^3 of liquid is injected. The exact mass of substance introduced is found by weighing the hypodermic syringe before and after injection.

After the liquid in the syringe has evaporated, the volume of air plus vapour is recorded. Subtraction of the original volume gives the volume of vapour, at steam temperature and atmospheric pressure, that arises from a known mass of liquid. The relative molecular mass of the substance can easily be calculated after the volume of vapour has been corrected to s.t.p.

Example

0.180 g of liquid gave rise to 60.0 cm^3 of vapour at 98.8 °C (372.0 K) and 0.987 atm. Calculate its relative molecular mass.

Answer First, we must correct the volume to s.t.p. From the gas laws,

$$\frac{p_1 \times V_1}{T_1} = \frac{p_2 \times V_2}{T_2}$$

$$\therefore \quad \frac{0.987 \times 60}{372.0} = \frac{1.000 \times V_2}{273.2}$$

from which $V_2 = 43.5 \text{ cm}^3$

If 43.5 cm³ has a mass of 0.180 g,

then 22 400 cm³ has a mass of $\frac{0.180}{43.5} \times 22\,400 = 92.7$ g,

i.e. one mole = 92.7 g; the relative molecular mass of the compound is 92.7.

Victor Meyer's method

The apparatus is in the form of a long glass vaporising tube, open at the top and with a bulb at the bottom. Most of the tube is enclosed by a heating jacket. A delivery tube leads to a smaller, graduated tube and a levelling tube (Fig. 5.3). (In an alternative form of the apparatus there is no levelling tube; instead, displaced air is collected in a graduated tube standing in a trough of water.)

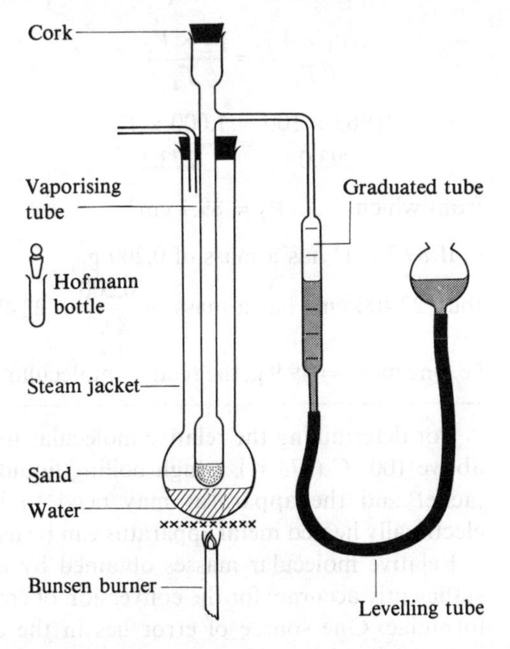

Fig. 5.3 Victor Meyer's apparatus.

With steam in the heating jacket, the apparatus may be used to determine the relative molecular mass of any liquid with a boiling temperature below 100 °C (373 K). First the cork is removed, and water is heated to produce steam for the jacket. While the apparatus is settling down, a small quantity of the liquid under examination is weighed in a *Hofmann bottle* fitted with a ground-glass stopper. When the apparatus has reached a steady state the reading on the scale is noted. The Hofmann bottle is then dropped into the apparatus and the cork is immediately inserted. The bottle drops to the bottom where its fall is broken by a layer of sand. The liquid volatilises in the bulb of the vaporising tube, and displaces an equal volume of hot air into the graduated tube. This hot air immediately cools down to room temperature so that, in effect, we measure the volume of vapour *at room temperature* that corresponds to a known mass of the substance.

After vaporisation is complete, which should not take more than a minute or two, the water levels are adjusted so that they are the same in both tubes, and the volume of air that has been expelled is noted.

Example

0.200 g of a volatile liquid displaced 100 cm^3 of air in a Victor Meyer apparatus at 19.8 °C (293.0 K) and 0.987 atm. The saturated vapour pressure of water at 19.8 °C (293.0 K) is 0.024 atm. Calculate the relative molecular mass of the liquid.

Answer Pressure of vaporised compound + water vapour pressure = atmospheric pressure

$\therefore$ pressure of vapour = 0.987 − 0.024 = 0.963 atm

Correct the volume of vapour to s.t.p.

$$\frac{p_1 \times V_1}{T_1} = \frac{p_2 \times V_2}{T_2}$$

$$\therefore \quad \frac{0.963 \times 100}{293.0} = \frac{1.000 \times V_2}{273.2}$$

from which $V_2 = 89.7\ cm^3$

If 89.7 cm^3 has a mass of 0.200 g,

then 22 400 cm^3 has a mass of $\frac{0.200}{89.7} \times 22\,400 = 49.9$ g

i.e. one mole = 49.9 g; the relative molecular mass of the compound is 49.9.

For determining the relative molecular masses of compounds that boil above 100 °C (373 K), a high boiling liquid must be used in the heating jacket and the apparatus may need to be made of silica glass. An electrically heated metal apparatus can be used to study ionic compounds.

Relative molecular masses obtained by the Victor Meyer method are sufficiently accurate for the conversion of empirical formulae to molecular formulae. One source of error lies in the diffusion of vapour from the vaporising tube to the graduated tube, where it may condense or dissolve in the water. To minimise this error the vaporising tube is made as narrow as possible; this ensures a small area of contact between vapour and air and thus reduces diffusion. The error may be further minimised by taking the final reading as soon as the water level stops falling.

Dumas' method

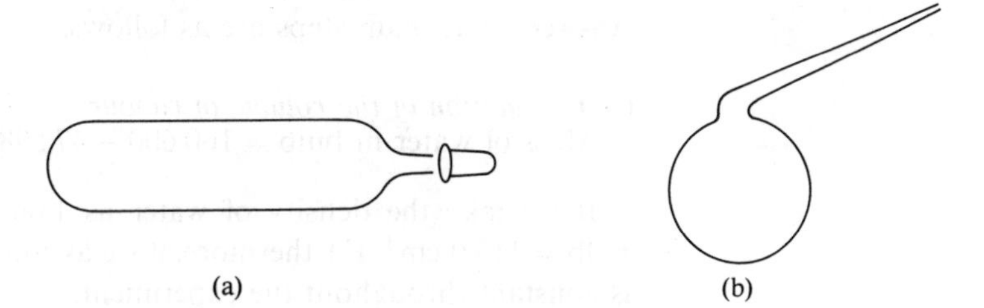

Fig. 5.4 (a) Modern Dumas bulb, (b) original Dumas bulb.

This method differs from the previous two in that the substance under investigation is weighed as a partially condensed vapour and not as a liquid.

The modern apparatus (Fig. 5.4(a)) consists of a cylindrical glass bulb, with a capacity of approximately 100–150 cm^3, fitted with a small ground-glass joint so that the bulb may be sealed by a ground-glass cap.

The bulb, with its cap, is first weighed in air. Approximately 10 cm^3 of the liquid whose relative molecular mass is to be determined is then introduced by means of a syringe, and the bulb is immersed in a beaker of boiling water (or other hot liquid) so that only the nozzle protrudes above the surface. The air in the bulb is displaced by the liquid as it vaporises. When all the liquid has vaporised, the bulb is closed by the cap, cooled down to room temperature and weighed. After this, the bulb is held with the neck under cold water, and the cap is removed. Water rushes in to fill the bulb almost completely. Finally the bulb is topped up with water from a syringe, the cap is refitted, and the bulb is again weighed.

In the calculation of relative molecular mass allowance must be made for the fact that the unknown substance is weighed as a vapour. Whenever a substance is weighed in a vessel its mass is obtained by difference, according to the equation:

mass of (vessel + substance) − mass of (vessel + air)
= mass of substance − mass of air

For solids and liquids the mass of air may be neglected, since it is very small compared with the mass of substance, but when a vapour is being weighed the mass of air in the vessel must be taken into account.

Example

The following results were obtained from a volatile compound with a Dumas bulb:

mass of bulb + air	47.598 g
mass of bulb + vapour of compound	47.913 g
mass of bulb + water	160.600 g
laboratory temperature	20.0 °C (293.2 K)
temperature of vaporisation	98.6 °C (371.8 K)
barometric pressure	0.992 atm

Given also that 1 dm^3 of dry air at s.t.p. has a mass of 1.293 g, calculate the relative molecular mass of the volatile compound.

Answer The main steps are as follows.

(*i*) *Calculation of the volume of vapour*
Mass of water in bulb = 160.600 − 47.598 = 113.0 g

If we take the density of water as 1.00 g cm^{-3}, the volume of the bulb = 113.0 cm^3. Furthermore, if we assume that the volume of the bulb is constant throughout the experiment,

volume of vapour (at 98.6 °C) = 113.0 cm^3
and volume of air (at 20.0 °C) = 113.0 cm^3.

(*ii*) *Calculation of the mass of vapour*
Mass of vapour − mass of air = mass of (bulb + vapour) − mass of (bulb + air)
= 47.913 − 47.598 = 0.315 g

∴ mass of vapour = 0.315 g + mass of air

To calculate the mass of vapour we thus need the mass of air in the bulb.

The bulb as weighed holds 113.0 cm^3 of air at 20.0 °C (293.2 K) and 0.992 atm. By applying the gas laws we can correct this volume of air to s.t.p.

$$\frac{p_1 \times V_1}{T_1} = \frac{p_2 \times V_2}{T_2}$$

$$\therefore \quad \frac{0.992 \times 113}{293.2} = \frac{1.000 \times V_2}{273.2}$$

from which $V_2 = 104.5\ cm^3$

Given that 1 000 cm^3 of air at s.t.p. has a mass of 1.293 g,

∴ 104.5 cm^3 of air at s.t.p. has a mass of $\frac{1.293}{1\,000} \times 104.5 = 0.135$ g

∴ mass of vapour = 0.315 + 0.135 = 0.450 g

(*iii*) *Calculation of relative molecular mass* The bulb when sealed holds 113.0 cm^3 of vapour at 98.6 °C (371.8 K) and 0.992 atm. By applying the gas laws we can correct the volume of vapour to s.t.p.

$$\frac{0.992 \times 113}{371.8} = \frac{1.000 \times V_2}{273.2}$$

from which $V_2 = 82.36\ cm^3$

If 82.36 cm^3 of vapour at s.t.p. has a mass of 0.450 g,

then 22 400 cm^3 of vapour at s.t.p. has a mass of $\frac{0.450}{82.36} \times 22\,400 = 122.4$ g

i.e. 1 mol = 122.4 g; the relative molecular mass of the compound is 122.4.

The Dumas method is less versatile and less accurate than the Victor Meyer method. Errors are introduced partly because of the difficulty in displacing all the air from the apparatus, and partly because, in the calculation, we make certain assumptions that are not entirely justified.

Mass spectrometry method

A very small sample of the substance, in vapour form, is bombarded with a beam of high energy electrons. Molecules of the substance lose electrons to become positively charged molecular ions:

$$M + e^- = M^+ + 2e^-$$

Such ions are unstable, and partially decompose in a variety of ways,

e.g. molecular ion (+) = smaller ion (+) + radical (neutral)
or molecular ion (+) = smaller ion (+) + molecule (neutral)

(Small amounts of negative ions are also formed.)

The various positive ions are separated according to their mass:charge ratio (*m/e*) by accelerating them electrostatically with a potential of about 2000 volts, and then passing them between the two poles of a magnet where they are deflected. Ions with a high *m/e* value are deflected less than those with a low value. By varying either the accelerating voltage or the strength of the magnetic field, each ionic species in turn is allowed to fall on a detector which provides a current when ions collide with it. The resulting current is amplified and its intensity, which is proportional to the number of ions colliding with the detector, is recorded on a mass spectrum (Fig. 5.5).

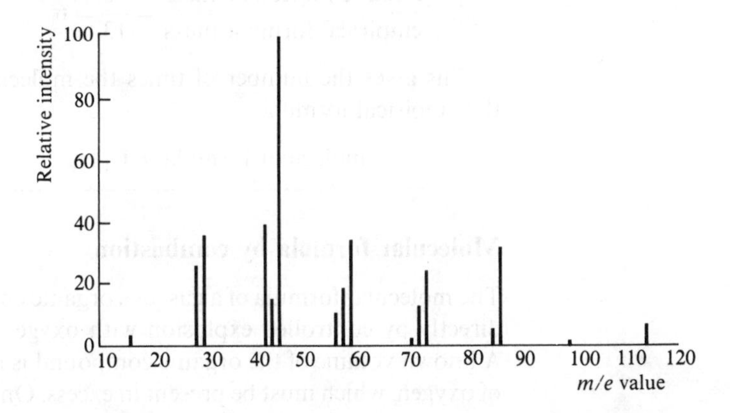

Fig. 5.5 Typical mass spectrum of an alkane.

Most ions have a single charge, i.e. $e = 1$, so that $m/e = m$. Thus the mass spectrum is in effect a chart of the masses and relative abundances of the various ions.

One of the heaviest ions will be the molecular ion, M^+, whose mass is required. However, identification of the M^+ peak is often complicated by the presence of neighbouring peaks caused by molecular ions closely similar to the main one but containing heavy or light isotopes of one or more atoms. In general, an M^+ peak can be identified because it is more intense than its neighbours, and once located it gives a highly accurate relative molecular mass.

5.1.5 CALCULATION OF MOLECULAR FORMULA

The masses of the atoms in the empirical formula are added up to give what we may call the 'empirical formula mass', which is then divided into the relative molecular mass. This gives the number of times that one is greater than the other, and if we multiply the empirical formula by this factor we get the molecular formula.

Example

Analysis of a compound gave 92.3% carbon and 7.7% hydrogen. Its relative molecular mass was found to be 78. Calculate its molecular formula.

Answer

Elements:	C	H
Percentages:	92.3	7.7
Divide by relative atomic masses:	7.7	7.7
Divide by the smallest:	1	1

i.e. empirical formula = CH

Because the relative atomic masses of carbon and hydrogen are 12 and 1 respectively,

$$\text{empirical formula mass} = 12 + 1 = 13$$

$$\therefore \quad \frac{\text{relative molecular mass}}{\text{empirical formula mass}} = \frac{78}{13} = 6$$

This gives the number of times the molecular formula is greater than the empirical formula,

$$\therefore \quad \text{molecular formula} = C_6H_6$$

Molecular formula by combustion

The molecular formula of a gaseous organic compound may be determined directly by controlled explosion with oxygen in a *Haldane gas analyser*. A known volume of the organic compound is mixed with a known volume of oxygen, which must be present in excess. On explosion, which is brought about by sparking the mixture, combined carbon is converted into carbon dioxide, and combined hydrogen into steam. The combustion products are cooled to room temperature, whereupon steam condenses to liquid water which has a negligible volume. The volume of carbon dioxide plus excess oxygen is recorded. The gases are then allowed to stand over aqueous potassium hydroxide, which absorbs the carbon dioxide to form potassium carbonate. The reduction in volume thus corresponds to the carbon dioxide content of the mixture. The residual gas is excess oxygen:

subtraction from the amount originally taken gives the volume of oxygen consumed in the combustion.

If the organic compound contains combined nitrogen, the residual gas consists of excess oxygen plus nitrogen formed on combustion. The volume of each gas is estimated by allowing the mixture to stand over an alkaline solution of 1,2,3-benzenetriol (pyrogallol), which absorbs the oxygen but leaves the nitrogen.

The calculation of molecular formula is based on Avogadro's law. Thus, if x cm^3 of a hydrocarbon gas reacts with y cm^3 of oxygen to give z cm^3 of carbon dioxide, we know that x molecules of the hydrocarbon react with y molecules of oxygen to give z molecules of carbon dioxide.

Example
10 cm^3 of a gaseous hydrocarbon were sparked with 60 cm^3 of oxygen. The 45 cm^3 of gas remaining after the explosion were reduced to 25 cm^3 on standing over aqueous alkali. Calculate the molecular formula of the hydrocarbon.

Answer Work in stages, as follows.

(i) Interpret the data.
The 45 cm^3 of gas after the explosion represent CO_2 + excess O_2.
The reduction in volume, over alkali, is 20 cm^3.
$\therefore$ the 45 cm^3 of gas mixture consist of 20 cm^3 CO_2 + 25 cm^3 excess O_2.
60 cm^3 of O_2 were originally taken, and 25 cm^3 is in excess.
$\therefore$ $60 - 25 = 35$ cm^3 of O_2 were used on combustion, i.e. 10 cm^3 of hydrocarbon react with 35 cm^3 of O_2 to give 20 cm^3 of CO_2.

(ii) Write a balanced equation for the combustion.
If we know that the hydrocarbon is an alkane we can write its formula as C_nH_{2n+2}; if it is an alkene we can represent it as C_nH_{2n}, etc. But in the absence of detailed information we must write the formula as C_xH_y.

$$C_xH_y + (x + \frac{y}{4})O_2 = xCO_2 + \frac{y}{2}H_2O$$

The corresponding equation for a nitrogen containing compound is:

$$C_xH_yN_z + (x + \frac{y}{4})O_2 = xCO_2 + \frac{y}{2}H_2O + \frac{z}{2}N_2$$

(iii) Write down, under the equation, the volumes of gases concerned.

$$1C_xH_y + (x + \frac{y}{4})O_2 = xCO_2 + \frac{y}{2}H_2O$$

10 cm^3 35 cm^3 20 cm^3

(iv) Divide the gas volumes by a suitable factor so that they may be compared with the coefficients in the equation.

As things stand, volumes and coefficients do not agree, for we have 10

cm^3 of the hydrocarbon but only one molecule appears in the equation. We therefore divide the gas volumes by 10:

$$1.0\ cm^3 \quad 3.5\ cm^3 \quad 2.0\ cm^3$$

By Avogadro's law, since 1 molecule compares with 1 cm^3,

$$x + \frac{y}{4} = 3.5 \quad \text{and} \quad x = 2.0$$

$$\text{If} \quad 2.0 + \frac{y}{4} = 3.5 \quad \text{then} \quad y = 6.0$$

i.e. the gaseous hydrocarbon has the molecular formula C_2H_6.

The determination of molecular formulae by combustion is now rarely carried out, for only the simplest organic compounds are gaseous and their molecular formulae were all established long ago. However, a Haldane gas analyser may still be useful for estimating the composition of a gas mixture, as the following example shows.

Example

10 cm^3 of a mixture of hydrogen, carbon monoxide and methane were exploded with 30 cm^3 of oxygen. The 26 cm^3 of gas remaining after the explosion were reduced to 19 cm^3 on standing over aqueous alkali. Calculate the composition by volume of the gas mixture.

Answer

(i) Interpret the data.

The 26 cm^3 of gas after explosion consist of 7 cm^3 CO_2 + 19 cm^3 excess O_2.

∴ the volume of O_2 used in the explosion = 30 − 19 = 11 cm^3, i.e. 10 cm^3 of gas mixture react with 11 cm^3 of O_2 to give 7 cm^3 of CO_2.

(ii) Write balanced equations for the combustion reactions.

$$H_2(g) + \tfrac{1}{2}O_2(g) = H_2O(l)$$

$$CO(g) + \tfrac{1}{2}O_2(g) = CO_2(g)$$

$$CH_4(g) + 2O_2(g) = CO_2(g) + 2H_2O(l)$$

(iii) Write down the volumes of the gases concerned in terms of symbols,

$$H_2(g) + \tfrac{1}{2}O_2(g) = H_2O(l)$$
$$x\ cm^3 \quad 0.5x\ cm^3$$

$$CO(g) + \tfrac{1}{2}O_2(g) = CO_2(g)$$
$$y\ cm^3 \quad 0.5y\ cm^3 \quad y\ cm^3$$

$$CH_4(g) + 2O_2(g) = CO_2(g) + 2H_2O(l)$$
$$z\ cm^3 \quad 2z\ cm^3 \quad z\ cm^3$$

(iv) Construct the same number of algebraic equations as there are unknowns.

Here, we have three unknowns, x, y and z, and therefore need three algebraic equations. We obtain them by considering:

1. the total volume of the gases

$$x + y + z = 10 \qquad (1)$$

2. the volume of oxygen consumed on combustion

$$0.5x + 0.5y + 2z = 11 \qquad (2)$$

3. the volume of carbon dioxide formed by combustion

$$y + z = 7 \qquad (3)$$

(v) Solve the simultaneous equations.
Subtract (3) from (1):

$$x = 3$$

Multiply (2) by two:

$$x + y + 4z = 22 \qquad (4)$$

Subtract (1) from (4):

$$3z = 12; \qquad \therefore z = 4$$

From (1),

$$3 + y + 4 = 10; \qquad \therefore y = 3$$

i.e. the 10 cm^3 of gas mixture consist of 3 cm^3 of hydrogen, 3 cm^3 of carbon monoxide and 4 cm^3 of methane.

5.1.6 DEDUCTION OF STRUCTURAL FORMULA

The structural formula of a compound may be deduced after an exhaustive study of its chemical and physical characteristics.

Chemical methods

Deduction from synthesis
The structure of a compound may be predicted with reasonable certainty if it is made from a starting material of known structure by a reaction which is fully understood. For example, if a compound is made from a known alcohol, ROH, by treatment with phosphorus pentachloride, then its structure will almost certainly be given by RCl.

Conversion to a compound of known structure
The hydrocarbon skeleton of an organic compound can often be identified by reducing the compound to a known hydrocarbon. For example, suppose a methyl propyl ketone on reduction yields 2-methylbutane, $CH_3CH_2CH(CH_3)_2$. Then the ketone must be methyl isopropyl

ketone (3-methyl-2-butanone), $CH_3COCH(CH_3)_2$, and not methyl propyl ketone (2-pentanone), $CH_3COCH_2CH_2CH_3$, for the latter on reduction would yield pentane, $CH_3CH_2CH_2CH_2CH_3$.

Chemical tests for specific groups
Various tests have been devised for common functional groups. The hydroxyl group (HO), for instance, may be tested for (after drying the compound) by adding sodium and looking for the evolution of hydrogen. Alternatively, phosphorus pentachloride may be used, in which case hydrogen chloride fumes are produced if the hydroxyl group is present.

Acids may be tested for with sodium hydrogencarbonate; aldehydes and ketones are detected by 2,4-dinitrophenylhydrazine; and primary, secondary and tertiary amines are distinguished by means of nitrous acid.

Not every reaction is suitable as a test. For a reaction to be suitable it must be possible for it to be performed readily in the laboratory, and it must yield an observable result within a few minutes. Tests based on colour change, precipitate formation and gas production are all widely used, but any reaction which is not obvious or which requires a high temperature or pressure or elaborate apparatus is unsuitable.

Physical methods

In recent years chemical methods of structure determination have been supplemented by *spectroscopy*, which is the study of the relationships between matter and electromagnetic radiation. Spectroscopy is also used for the qualitative and quantitative analysis of compounds.

5.2 Isomerism

Different compounds with the same molecular formula are known as *isomers*. The phenomenon, called *isomerism*, is commonly encountered among carbon compounds and complex compounds.

There are two main types of isomerism, namely *constitutional isomerism* and *stereoisomerism*. Constitutional isomers have fundamentally different molecular structures, while stereoisomers have similar constitutions but differ in their molecular geometry, i.e. in the spatial arrangements of their atoms or groups.

5.2.1 CONSTITUTIONAL ISOMERISM (STRUCTURAL ISOMERISM)

Complex compounds may exhibit ionisation isomerism and hydration isomerism (§18.1.2). Carbon compounds display four kinds of constitutional isomerism, which are as follows.

Skeletal isomerism

Otherwise known as *chain isomerism* or *nuclear isomerism*, this is concerned with the structure of the hydrocarbon skeleton. Thus, butane and 2-methylpropane are skeletal isomers:

$$CH_3CH_2CH_2CH_3 \qquad \begin{matrix} CH_3 \\ CH_3 \end{matrix}\!\!>\!CHCH_3$$

butane — 2-methylpropane

In general, skeletal isomers have similar physical properties, although a branched chain compound invariably has a lower boiling temperature than its unbranched isomer, e.g. the boiling temperatures of butane and 2-methylpropane are −0.5 °C (272.7 K) and −11.7 °C (261.5 K) respectively. Branched chain molecules are relatively easily separated from one another, partly because the molecules are compact and do not become entangled, and partly because atoms at the centres of such molecules are remote from atoms of other molecules and cannot contribute to van der Waals' forces of attraction.

Skeletal isomers differ very little in the nature of their chemical reactions, although in some cases there may be a considerable variation in reactivity. 2-Methylpropane, for example, is more reactive than butane.

Position isomerism

This arises whenever a functional group occupies different positions on the same hydrocarbon skeleton,

e.g. $CH_3CH_2CH_2OH$ (1-propanol) and $CH_3CH(OH)CH_3$ (2-propanol)

$CH_3CH(OH)CH_2CH_2CH_3$ (2-pentanol) and $CH_3CH_2CH(OH)CH_2CH_3$ (3-pentanol)

Also 1,2-(*ortho*), 1,3-(*meta*) and 1,4-(*para*) isomers,

e.g.

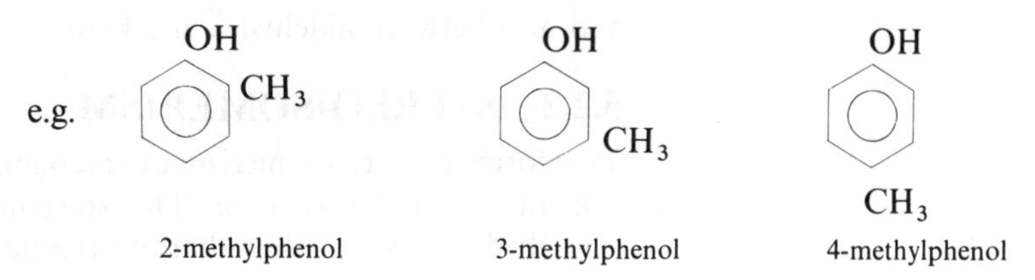

2-methylphenol — 3-methylphenol — 4-methylphenol

The position of a functional group often has little bearing on physical and chemical properties. Thus, 2-pentanol and 3-pentanol have almost identical properties. There is, however, considerably less similarity between 1-propanol and 2-propanol, for the former is a primary alcohol, while the latter is a secondary alcohol (§14.9).

Functional group isomerism

As the term implies, functional group isomers have different functional groups and therefore different physical and chemical properties.

E.g. $CH_3{-}O{-}CH_3$ (methoxymethane) and CH_3CH_2OH (ethanol)

also: CH_3CH_2COOH, (propanoic acid) $HCOOCH_2CH_3$ (ethyl methanoate) and CH_3COOCH_3 (methyl ethanoate)

Metamerism

Isomers which have different hydrocarbon radicals attached to a particular atom are called *metamers*. The atom concerned may be oxygen,

e.g. $CH_3—O—CH_2CH_2CH_3$ (1-methoxypropane) $CH_3—O—CH(CH_3)_2$ (2-methoxypropane)

$CH_3CH_2—O—CH_2CH_3$ (ethoxyethane)

or nitrogen,

e.g. $CH_3CH_2CH_2—NH_2$ $(CH_3)_2CH—NH_2$ primary amines

$CH_3CH_2—NH—CH_3$ secondary amine

$CH_3—N(CH_3)_2$ tertiary amine

or it may be the carbon atom of a carbonyl group (C=O),

e.g. $CH_3CH_2—C(=O)—H$ propanal (an aldehyde) $CH_3—C(=O)—CH_3$ propanone (a ketone)

Certain metamers, such as the ethers quoted above, are closely similar to one another, but others may differ considerably. In particular, there are important differences between primary, secondary and tertiary amines, and also between aldehydes and ketones.

5.2.2 STEREOISOMERISM

Two forms of stereoisomerism are recognised, namely *geometrical isomerism* and *optical isomerism*. The stereoisomerism of complex salts is described in §18.1.2, while that of organic compounds will be considered now. Organic stereoisomers always have the same hydrocarbon skeleton and the same functional groups.

Geometrical isomerism

Geometrical isomerism results when carbon atoms are unable to rotate freely about the bonds that join them together. It cannot occur in saturated compounds, unless they are cyclic. Thus, forms I, II and III are identical, because carbon atoms can rotate freely about a carbon–carbon single bond.

$$\underset{\text{I}}{(a)(b)(c)C—C(x)(y)(z)} \xrightarrow{\text{twist}} \underset{\text{II}}{(a)(b)(c)C—C(z)(x)(y)} \xrightarrow{\text{twist}} \underset{\text{III}}{(a)(b)(c)C—C(y)(z)(x)}$$

(It is advisable to follow this argument with the aid of molecular models.)

However, the π electrons (§4.3.1) of a double bond lock carbon atoms firmly in position. Thus, IV and V are not interconvertible unless the double bond is broken. They are said to be 'geometrical isomers' of 1,2-dichloroethene.

```
Cl      Cl         Cl      H
  \    /             \    /
   C=C                C=C
  /    \             /    \
H        H          H      Cl
    IV                 V
```

Although a carbon–carbon double bond is the commonest cause of restricted rotation, a cyclic structure may also be responsible. Thus, VI and VII, which represent 1,4-dimethylcyclohexane molecules, are also geometrical isomers.

```
CH3  CH3          CH3  H
 |    |            |   |
[cyclohexane]     [cyclohexane]
 :    :            :   :
 H    H            H   CH3
   VI                VII
```

Geometrical isomers are distinguished by the prefixes *cis* and *trans*: *cis* (Latin – 'hither') relates to the form in which the two similar atoms or groups lie on the same side of the ring or double bond, e.g. to forms IV and VI, while *trans* (Latin – 'across') denotes the isomer in which they are on opposite sides, e.g. forms V and VII. Because of this notation, geometrical isomerism is sometimes referred to as *cis–trans isomerism.*

One of the commonest examples of geometrical isomerism relates to maleic acid and fumaric acid. Both have the formula HOOCCH=CHCOOH, but maleic acid has a *cis* arrangement of groups while fumaric acid is the *trans* isomer:

```
H—C—COOH              H—C—COOH
  ||                    ||
H—C—COOH           HOOC—C—H
```

maleic acid
cis-butenedioic acid

fumaric acid
trans-butenedioic acid

Physically and chemically, they are widely different. Maleic acid has a melting temperature of 130 °C (403 K) with decomposition; fumaric acid melts at 286 °C (559 K) with decomposition. Maleic acid is highly soluble in water whereas fumaric acid is only slightly soluble. Maleic acid differs in strength from fumaric acid, with pK_1 and pK_2 values of 1.92 and 6.23 respectively as opposed to 3.02 and 4.38 for fumaric acid. The reason is that, while one proton can readily be lost, the other is held as a member of a ring by hydrogen bonding:

$$\begin{array}{l} \text{H—C—C}\overset{\displaystyle /\!\!/ \text{O}}{} \\ \quad\ \| \qquad \text{O—H} \\ \quad\ \| \qquad\ \ \vdots \\ \quad\ \| \qquad \text{O} \\ \text{H—C—C}\overset{\displaystyle /\!\!/}{} \\ \qquad\qquad \text{O—H} \end{array}$$

No such restriction applies to fumaric acid.

On heating, either alone or with ethanoic anhydride, maleic acid readily gives maleic anhydride, but fumaric acid cannot form fumaric anhydride as its carboxyl groups are too far apart. Instead, it rearranges to give maleic acid, which then dehydrates to maleic anhydride:

$$\underset{\text{fumaric acid}}{\begin{array}{c} \text{H—C—COOH} \\ \| \\ \text{HOOC—C—H} \end{array}} \xrightarrow{\text{heat}} \underset{\text{maleic acid}}{\begin{array}{c} \text{H—C—COOH} \\ \| \\ \text{H—C—COOH} \end{array}} \xrightarrow{\text{heat}} \underset{\text{maleic anhydride}}{\begin{array}{c} \text{H—C—C(=O)} \\ \| \qquad \backslash \\ \| \qquad\quad \text{O} \\ \| \qquad / \\ \text{H—C—C(=O)} \end{array}} + H_2O$$

Reactions which result in the saturation of the double bond lead to the same product from both maleic and fumaric acid. Examples are as follows:

(i) hydrogenation to succinic acid, $\begin{array}{l} CH_2COOH \\ | \\ CH_2COOH \end{array}$

(ii) hydrobromination to bromosuccinic acid, $\begin{array}{l} CHBrCOOH \\ | \\ CH_2COOH \end{array}$

The configuration of geometrical isomers is usually established by physical methods. For example, X-ray analysis shows that in one form of 1,2-dichloroethene the inter-chlorine distance is 0.35 nm, whereas in the other isomer it is 0.41 nm. The former must be *cis* and the latter *trans*.

Optical isomerism

This is due to a lack of symmetry within a molecule. The absence of symmetry in an article is a fairly uncommon occurrence, for most things possess a *plane of symmetry*, which divides the article into two symmetrical halves. For example, a person or a hat has a plane of symmetry, as represented by Fig. 5.6.

If any object has a plane of symmetry there is only one form of it, in the sense that it is identical with its mirror image. The mirror image of a hat is exactly the same as the original hat; if we were to make the 'mirror

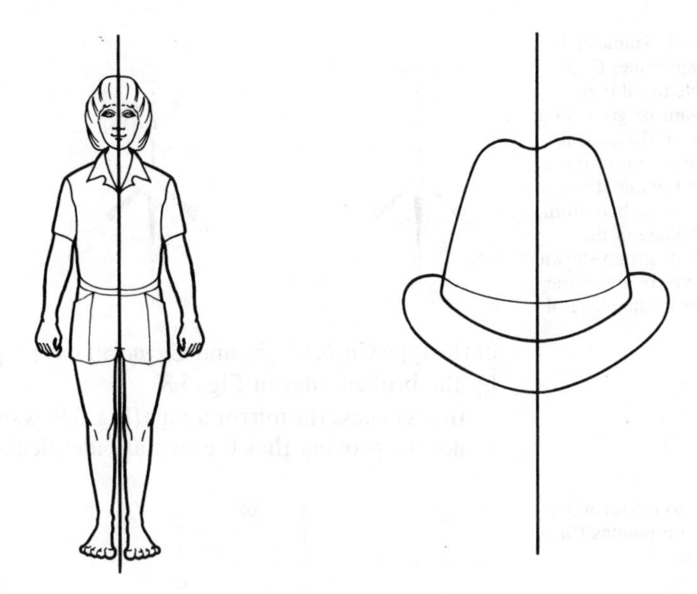

Fig. 5.6 Planes of symmetry.

image hat' it would not only look the same as the original but the two could be superimposed on each other without any difficulty. They could be neatly stacked together.

However, if an object does not have a plane of symmetry it should be possible to make a mirror image version that is quite different from the original. A left-hand glove, for instance, does not possess a plane of symmetry, and its mirror image is not another left-hand glove but a right-hand glove. The two gloves are not identical, for they are not superimposable. If we were to lay one on top of the other the fingers and thumbs would clash.

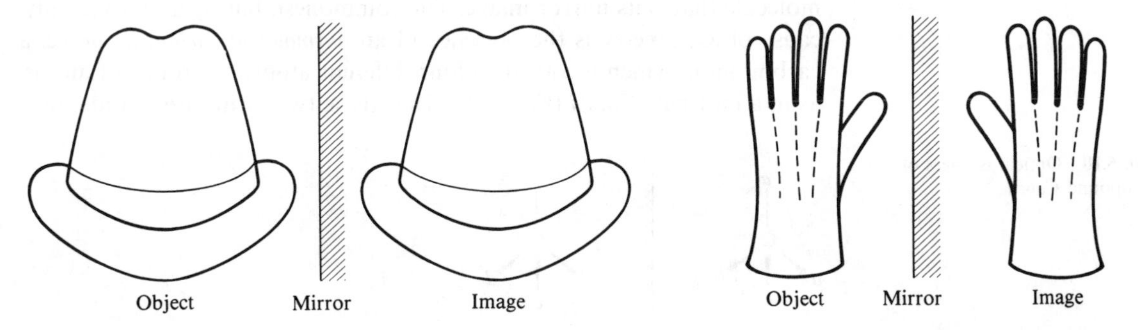

Fig. 5.7 Objects and their mirror images.

A glove or any other object lacking a plane of symmetry is said to possess *chirality*, pronounced kī-rality, which means 'handedness' (Greek: *cheir* = the hand). Any *chiral* object exists in two forms, one of which is the mirror image of the other.

Most molecules are *achiral*. They possess a plane of symmetry, so that they and their mirror image forms are identical. Tetrahedral molecules

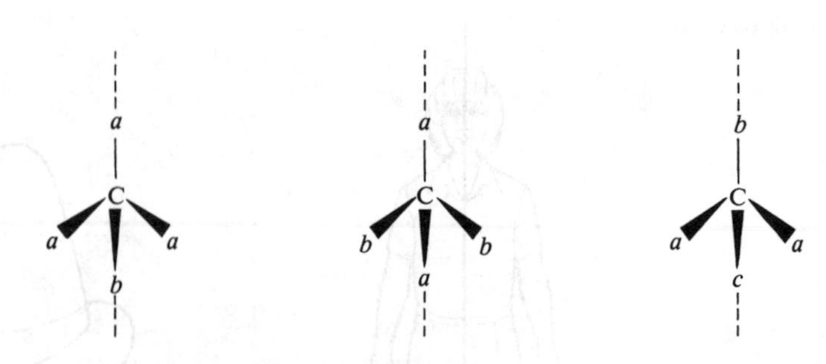

Fig. 5.8 Planes of symmetry in the tetrahedral molecules Ca_3b, Ca_2b_2 and Ca_2bc. In all these drawings the atoms or groups on the left and right of the central carbon atom extend out of the plane of the paper, while the atom or group below the carbon atom extends into the plane of the paper. The atom or group shown at the top lies directly above the carbon atom and in the plane of the paper.

of the type Ca_3b, Ca_2b_2 and Ca_2bc all have a plane of symmetry, as shown by the broken lines in Fig. 5.8.

In every case the mirror image (Fig. 5.9) is superimposable on the original molecule, proving that the two are identical. There can be no isomerism.

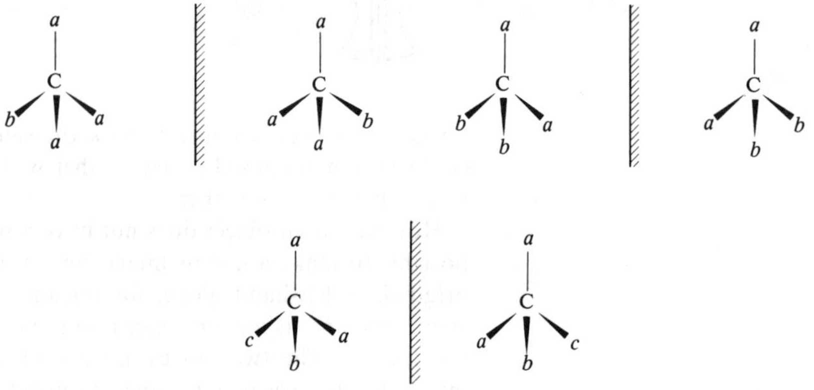

Fig. 5.9 Object to mirror image relationships of compounds Ca_3b, Ca_2b_2 and Ca_2bc.

Some molecules, however, do not have a plane of symmetry. For any such chiral molecule there can be obtained a closely related yet distinctive molecule that is its mirror image. The commonest, but by no means only, cause of asymmetry is the presence of an *asymmetric carbon atom*, i.e. a carbon atom which is joined to four different atoms or groups of atoms, as in a molecule C*abcd* (Fig. 5.10). That these two forms are not identical

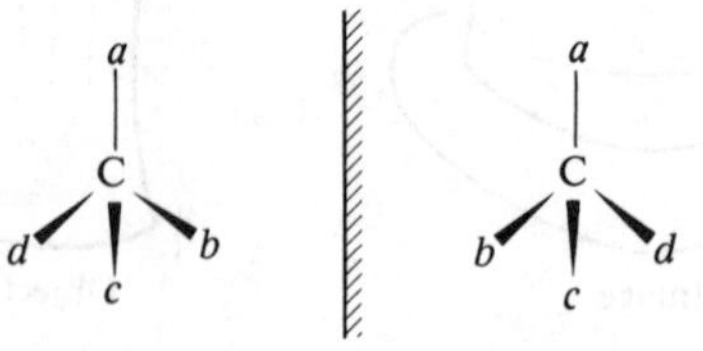

Fig. 5.10 Optical isomers of a compound C*abcd*.

is easily proved by constructing molecular models and showing that it is impossible to superimpose them without any of the groups clashing.

Thus, any compound that consists of chiral molecules must be isomeric with another compound composed of molecules of opposite chirality. Because of their structural similarity, it is hardly surprising that any two such isomers have very similar properties. These include identical melting temperatures, boiling temperatures, solubilities, refractive indices and chemical properties. The isomers differ in two respects only, namely in their physiological effects (if any) and their effect on *plane polarised light*, i.e. light that has been passed through a sheet of Polaroid so that the light vibrates in one plane only.

Their different physiological actions are shown, for example, by the widely different flavours of the two isomers of carvone. One form is 'oil of spearmint', while the isomer of opposite chirality is responsible for the flavour of caraway seeds.

The effect on polarised light is assessed with a *polarimeter*. Both isomers of a chiral substance are *optically active*, i.e. they rotate the plane of polarisation, but whereas one form rotates it to the right, the other rotates it to the left by the same amount (Fig. 5.11). The former is called a *dextrorotatory isomer*, denoted by (+) or *d*; while the latter is said to be *laevorotatory*, otherwise (−) or *l*. An equal mixture of the two isomers is *optically inactive*, i.e. it has no effect on plane polarised light, because each isomer rotates the plane of polarisation by an equal amount in opposite directions. The mixture is known as a *racemic modification*, and may be denoted by (±) or *dl*.

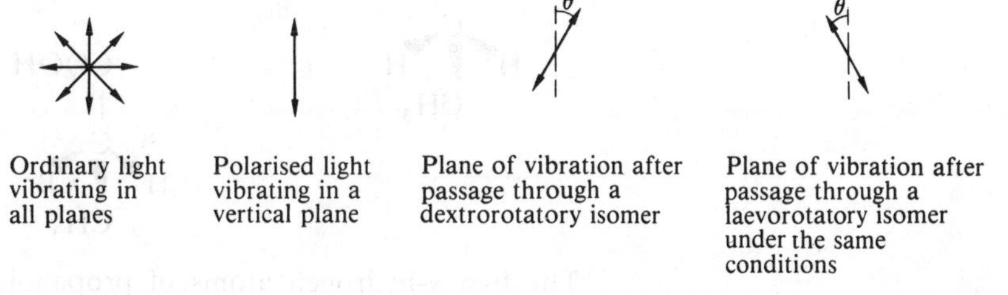

Fig. 5.11 Rotation of the plane of polarised light by optically active isomers.

Because optical activity provides the easiest way of distinguishing between them, the isomers are said to be *optical isomers*. Optical isomers which are mirror images of each other are called *enantiomers* or *enantiomorphs*.

It must be clearly understood that polarised light is in no way the *cause* of optical isomerism. Such isomerism exists in complete independence of any lighting conditions, and is purely the result of molecular asymmetry. Optical isomerism may exist in addition to any of the other types of isomerism. For example, 2-butanol, $CH_3\overset{*}{C}H(OH)CH_2CH_3$ ($\overset{*}{C}$ represents an asymmetric carbon atom), exhibits optical isomerism, but is also structurally isomeric with 1-butanol, 2-methyl-1-propanol and 2-methyl-2-propanol.

Compounds exhibiting optical isomerism

One of the best known compounds with an asymmetric carbon atom is 2-hydroxypropanoic acid, i.e. lactic acid, $CH_3\overset{*}{C}H(OH)COOH$. One enantiomer will cause dextrorotation and the other laevorotation, although which is which is hard to say. (+)-Lactic acid can be isolated from muscle tissue, and (−)-lactic acid may be obtained by the fermentation of sucrose. Both have a melting temperature of 26 °C (299 K). The (±)-form has a sharp melting temperature of 16.8 °C (290.0 K). The absence of a melting range indicates that this is a racemic compound rather than a racemic mixture (see below). It is obtained naturally from sour milk, and can also be prepared synthetically:

$$\underset{\text{ethanal}}{CH_3CHO} + HCN \rightarrow \underset{\text{2-hydroxypropanenitrile}}{CH_3CH(OH)CN} \xrightarrow{\text{hydrolysis}} CH_3CH(OH)COOH$$

$$\text{or } \underset{\text{propanoic acid}}{CH_3CH_2COOH} + Br_2 \rightarrow \underset{\text{2-bromopropanoic acid}}{CH_3CHBrCOOH} \xrightarrow{\text{hydrolysis}} CH_3CH(OH)COOH$$

It is seldom that only one enantiomer of a compound is produced by synthesis unless conditions are deliberately arranged so as to encourage this. Consider the second synthesis in more detail:

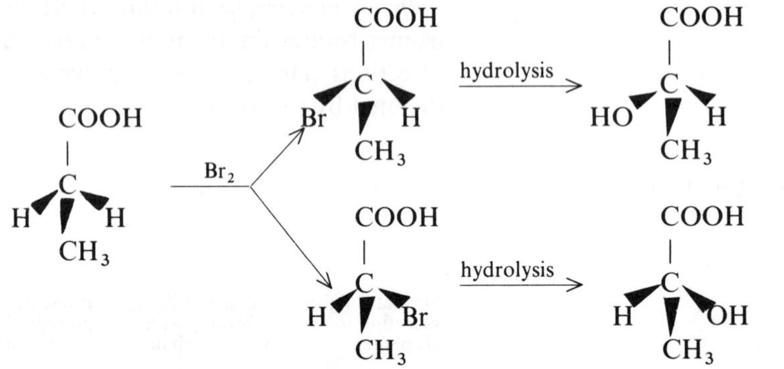

The two α-hydrogen atoms of propanoic acid stand an equal chance of becoming substituted by bromine atoms, so that both forms of 2-bromopropanoic acid are produced in equal amounts. In other words, (±)-2-bromopropanoic acid is formed, and on hydrolysis this yields (±)-lactic acid.

Racemic modifications

Racemic modifications are formed either by synthesis, as shown above, or by *racemisation*, i.e. the conversion of one or both enantiomers to an equimolecular mixture of the two. The change may be effected by heat, light, or by dissolving in a solvent. For example, if (+)-tartaric acid is heated with a little water, (±)-tartaric acid will form. The change is believed to occur via a planar configuration (Fig. 5.12).

Fig. 5.12 Racemisation of a compound *Cabcd*.

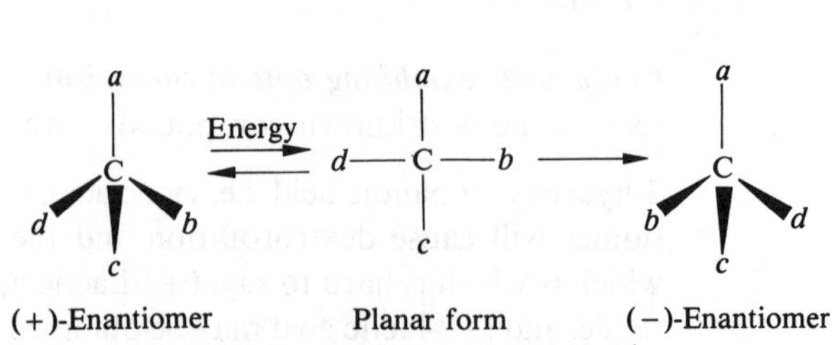

When, say, a (+)-enantiomer is heated, energy is absorbed and the molecule vibrates more and more strongly. Eventually a planar configuration is momentarily obtained. When the molecule returns from the planar state to the tetrahedral state it stands an equal chance of going to the (+)- or (−)-form, and so a racemic modification is obtained.

The separation of a racemic modification into its enantiomers is a process known as *resolution*. The technique used depends on whether the racemic modification is a *racemic mixture* of dextrorotatory and laevorotatory crystals, or a *racemic compound*, with equal amounts of (+)- and (−)-isomers in the one type of crystal.

Compounds that give racemic mixtures are rare. One such is ammonium sodium tartrate, as discovered by Louis Pasteur in 1848. At temperatures below 27 °C (300 K) the racemic modification of this salt forms two types of crystals, one the mirror image of the other. The external shape of a crystal is governed by its internal structure; thus, one set of crystals consists of the (+)-isomer, while the other is composed of the (−)-isomer. The crystals may be separated by hand. Alternatively, the two types of crystals may be grown in succession, by *inoculating* (i.e. seeding) a saturated solution first with a crystal of one enantiomer and then with a crystal of the other.

Racemic compounds are far more common than racemic mixtures. The (±)-form of lactic acid, for example, yields a single type of crystal with a structure, melting temperature and solubility that are quite different from those of the (+)- and (−)-isomers. Racemic compounds, therefore, cannot be separated by mechanical means or by the inoculation of a solution, and must be resolved by conversion into *diastereomers*. Diastereomers, unlike enantiomers, differ from one another in their physical properties, and may be separated by such techniques as crystallisation or chromatography. For instance, a racemic compound of an acid can be resolved by reaction with an optically active base, such as brucine or strychnine:

$$\underset{\text{acid - racemic}}{(\pm)\text{-A}} + \underset{\text{base - laevo}}{2(-)\text{-B}} = \underset{\text{salt - two diastereomers}}{\begin{array}{ccc} (+)\text{-A} & & (-)\text{-A} \\ | & + & | \\ (-)\text{-B} & & (-)\text{-B} \end{array}}$$

After separation, the diastereomers are hydrolysed to regenerate the free acids.

EXAMINATION QUESTIONS ON CHAPTER 5

1 (a) Outline how you would determine the relative molecular mass of a volatile liquid. In what circumstances will this method produce abnormal results for the relative molecular mass?

(b) A compound of phosphorus and fluorine contains 24.6 per cent by mass of phosphorus. 1.00 g of this compound has a volume of 194.5 cm^3 at a pressure of 1 atm and a temperature of 25 °C.

(i) Deduce the molecular formula for the compound.

(ii) What is the shape of a molecule of the compound?

(iii) 1 mole of the compound reacts with 1 mole of F^- producing a new ion. Deduce the shape of the new ion.

(iv) What type of reagent is the fluoride ion in the above reaction? (JMB)

2 (a) Give details of an experiment you would perform in order to measure the relative molar mass of a liquid with a low boiling point.

(b) A determination of the relative molar mass of nitrogen dioxide at 65 °C and pressure 101.3 kPa gave a value of 58. Calculate:

(i) the degree of dissociation of dinitrogen tetroxide at this temperature,

(ii) the volumes of dinitrogen tetroxide (N_2O_4) and nitrogen dioxide (NO_2) contained in 1 dm^3 of vapour at this temperature and under a pressure of 101.3 kPa,

(iii) the partial pressure of each component gas in the mixture.

(c) Give the names and formulae of two gaseous oxides of nitrogen other than nitrogen dioxide (NO_2) and outline how each can be prepared in the laboratory. Give **one** test in each case which would identify that particular oxide. (AEB)

3 (a) 0.90 g of an organic cyanide (RCN) was heated with excess concentrated sulphuric acid in order to convert the nitrogen content of the compound to ammonium sulphate.

The solution obtained was carefully diluted, made alkaline with excess sodium hydroxide and boiled to expel ammonia. The ammonia was then absorbed in 100 cm^3 of 0.05 M sulphuric acid. On titration, the excess sulphuric acid required 32 cm^3 of 0.10 M sodium hydroxide solution.

(i) Write an equation for the hydrolysis of RCN with sulphuric acid forming ammonium sulphate.

(ii) Write an equation for the reaction of ammonium sulphate with excess sodium hydroxide to expel ammonia.

(iii) Write an equation for the reaction of ammonia with sulphuric acid.

(iv) Calculate the percentage by weight of nitrogen in the organic cyanide and explain your working (N = 14).

(b) The organic solid, *F*, is known to contain chlorine but no other halogen.

(i) Describe how you would determine its percentage by weight of chlorine.

(ii) Explain the chemistry of your determination and show how the percentage of chlorine may be calculated from the weighings. A knowledge of all necessary relative atomic masses may be assumed. (SUJB)

4 (a) What do you understand by the terms *enantiomer* (*enantiomorph*), *racemic mixture*, *geometrical isomer*, and *structural isomer*? Use specific compounds to illustrate each term.

(b) Explain briefly what prevents the *geometrical* isomers which you have given in (a) from interconverting.

(c) Draw diagrams of each stereoisomer which may exist for each of the following structures. Indicate clearly what type(s) of isomerism is/are involved.

$C_6H_5CH{=}CHCOOH$

$C_6H_5CH_2CH(NH_2)COOH$

$C_6H_5CH{=}CHCOOCH(CH_3)C_2H_5$ (OC)

5 Write the structural formulae for:

(i) *cis*-butenedioic acid (maleic acid),

(ii) *trans*-butenedioic acid (fumaric acid).

(a) (i) What type of isomerism is shown by these two acids?

(ii) What is the cause of this type of isomerism?

(iii) Give **two** physical properties which are different for these two acids.

(iv) Give **one** chemical reaction which is different for these two acids.

(b) The addition of hydrogen bromide separately to each of these two acids gives the same products, a pair of isomers, in each case.

(i) Write structural formulae which distinguish the two isomers.

(ii) What type of isomerism is shown by these two isomers?

(iii) What is the cause of this type of isomerism? (AEB)

6 (a) Explain briefly how a knowledge of the density of a gaseous compound leads to a value for its relative molecular mass, independent of its chemical composition.

(b) Find the relative molecular mass of a liquid, 0.15 g of which on vaporisation gives 30 cm^3 of vapour measured at 373 K and a pressure of 1 atm (i.e. 10^2 kN m^{-2}). The gas constant is 0.0821 litre atm K^{-1} mol^{-1} or 8.31 J K^{-1} mol^{-1}.

(c) When the values obtained for the relative molecular masses from gas densities are compared with those obtained from the chemical formulae and accurate relative masses, it may be found that:

(i) the two values are approximately but not exactly the same, or

(ii) the gas density value is considerably less than the value found from the formula, or

(iii) the gas density value is approximately twice the value from the formula.

Discuss the reasons for these three types of discrepancy, giving examples. (L)

7 A chemist synthesised a solid organic compound which, from its method of preparation and other evidence, he knew to contain carbon, hydrogen, nitrogen, and possibly oxygen, and to have a relative molecular mass of less than about 150.

Describe a scheme by which he could determine the formula and structure of the compound. Experimental details are not required but, for quantitative experiments, you should show how results are calculated and, for chemical reactions, you should state the reagents and conditions of reaction and the conclusions that may be drawn from an examination of the products. (L)

8 (a) Describe, with the aid of a diagram, an experiment for determining the relative molecular mass of a volatile liquid. Explain how the result would be calculated from the experimental data.

(b) At a temperature of 60.2 °C, a sample of dinitrogen tetroxide had a density of 2.265 kg m^{-3} (2.265 g $litre^{-1}$) at a pressure of 101 300 N m^{-2} (1 atm). What is the percentage of NO_2 molecules in the sample under these conditions? (Take 0 °C = 273.2 K.)

(c) Write equations describing the reaction of nitrogen dioxide with (i) ice-cold water, and (ii) hot water. In each case, what is the change in the oxidation state of the nitrogen? (JMB)

9 (a) Give **two** factors which may restrict rotation about carbon–carbon bonds, and indicate the consequences which this restriction may lead to in the field of isomerism. Illustrate your answer with specific compounds.

(b) What is meant by the term *chiral molecule* and to what extent is the term confined to organic chemistry? Illustrate your answer with diagrams of specific molecules.

(c) Illustrate the importance of stereochemistry as an approach to understanding organic reaction mechanisms. (OC)

10 State what you understand by **each** of the following terms: *atomic number, isotope, atomic mass, mole.*

Draw a diagram of a mass spectrometer, labelling the parts. Explain how the instrument may be used to determine (a) molecular mass, (b) isotopic ratio.

What limitations are there in the determination of molecular mass by mass spectrometry? (WJEC)

11 (a) What do you understand by the terms (i) homologous series, (ii) isomerism, (iii) empirical formula?

Illustrate your answer to each part with an example.

(b) (i) Write the electronic structure of the carbon atom using s, p, d, etc notation.

(ii) Explain how a tetrahedral distribution of C—H bonds arises in methane using this electronic structure and the idea of orbitals.

(c) The mass spectrum of dichloromethane shows peaks (or lines) at 84, 86 and 88 a.m.u. (atomic mass units).

(i) Explain why there are three lines on the spectrum ($H = 1.0$, $C = 12.0$, $Cl = 35.5$).

(ii) The intensities of the lines at 84, 86 and 88 are in the ratio 9:6:1. Why is this? (SUJB)

6 Structure and state

6.1 States of matter

In earlier chapters we have been concerned mainly with the properties of individual atoms and molecules (microscopic behaviour), whereas in this and later chapters we shall be looking at the properties of materials in bulk (macroscopic properties).

All materials exist in one of three states of matter; solid, liquid or gas. In the gaseous state individual movements of molecules are restricted only by the walls of the vessel containing them. Liquids on the other hand, although subjected to the same restrictions as gases, are further controlled by the strength of the attractive forces within the liquid which give it cohesion and a definite volume, but not rigidity. When the volume is smaller than that of the containing vessel a surface is formed. If the cohesive forces are increased there comes a stage where the molecules are so close together that there is little actual movement, and the material assumes a definite shape as well as volume. It is then said to be in the solid state.

There are two opposing factors governing the state of substances, namely the translational movement of the molecules due to their kinetic energy (E), and the interattractive forces between the molecules. The former is increased by increasing temperature (thermal agitation). However, this is counteracted by any process which brings the molecules closer together, such as increasing pressure. Therefore high pressure and low temperature encourage the formation of solids, and the reverse is true for gases. Since the cohesive forces between molecules in the solid state can be very different in nature and strength from one solid to another, the conditions for changes of state can vary widely from one material to another.

We shall deal with gases, liquids and solids in this order as this represents that of increasing complexity. However, before looking at the behaviour of gases the basic premises of kinetic molecular theory must be stated.

(i) Small particles of any substance in bulk are characteristic of the substance.

(ii) The behaviour of large numbers of particles is statistical in nature.

The large number of molecules in one mole of a substance is statistically significant, i.e. it truly represents the average behaviour of the bulk of that material under given conditions.

6.2

The gaseous state

6.2.1 THE KINETIC THEORY OF GASES

Many of the gas laws introduced by R. Boyle, J. A. Charles, T. Graham and others were based on the concept of an *ideal gas*, which was defined as one in which there were no cohesive forces between molecules of the gas or between molecules of the gas and their container, and in which the molecules were negligibly small compared with the total volume of the gas. This is represented by the limiting case of a system in which only disruptive (or translational) forces occur. J. Clerk Maxwell and L. Boltzmann in their kinetic theory of gases expressed this ideal behaviour in mathematical form. In the development of their theory they made the following assumptions.

(i) The molecules are in constant random motion due to collisions between them.
(ii) All collisions, whether between molecules or between molecules and the walls, are perfectly elastic and therefore there is no loss of momentum.
(iii) Kinetic energy is directly proportional to the (absolute) temperature of the gas.
(iv) The molecules exert no attractive or repulsive forces on one another or on the walls of the vessel.
(v) The size of the molecules is negligibly small compared with the volume occupied by the gas.

If c is the velocity of a molecule of gas in a given direction, its value and direction are given by an equation of the form

$$c^2 = u^2 + v^2 + w^2$$

where u, v and w are the velocities along three axes at right angles to one another, namely the x-, y- and z-axes respectively.

Consider a gas contained in a cubic vessel of sides one unit of length. Let there by N molecules of gas, each of mass m, contained in the cube. It can be shown that the total pressure (p) is given by

$$p = \frac{1}{3}mN\bar{c}^2$$

where $\bar{c}^2$ is the mean square velocity of all the velocities along the three given axes.

Now $N = \frac{L}{V_m}$ where L = the Avogadro constant and V_m = molar volume, i.e. the volume (in dm^3) occupied by 1 mol (§5.1.4).

Also, $mL = M$ where M = molar mass in kg.

$$\therefore \quad p = \frac{1}{3}\frac{M}{L}\frac{L}{V_m}\bar{c}^2 = \frac{1}{3}\frac{M}{V_m}\bar{c}^2$$

$$\therefore \quad pV_m = \frac{1}{3}M\bar{c}^2$$

From this equation, which is one form of the *fundamental equation of the kinetic theory of gases*, we can deduce the gas laws.

Since the kinetic energy $= \frac{1}{2}M\bar{c}^2 = E$,

$$pV_m = \frac{2}{3}E$$

One of the basic assumptions of the kinetic theory is that $E \propto T$

$$\therefore \quad pV_m \propto T$$

This is identical with the general gas equation

$$pV_m = RT$$

where R is the gas constant.

Since $pV_m = 2/3E$,

$$E = \frac{3}{2}RT$$

$$\therefore \quad R = \frac{2E}{3T} \quad \text{for one mole}$$

E is the energy per mole,

$\therefore$ the SI units of R are $\frac{\text{J mol}^{-1}}{\text{K}}$, i.e $\text{J K}^{-1}\ \text{mol}^{-1}$

The determined value of R is 8.314 $\text{J K}^{-1}\ \text{mol}^{-1}$.

If any given volume of an ideal gas (V) contains n moles of gas,

$$V = nV_m$$

hence the fundamental equation of the kinetic theory of gases becomes

$$pV = \frac{1}{3}Mn\bar{c}^2$$

and the general gas equation becomes

$$pV = nRT$$

Although the equation $pV = RT$ will hold for one mole of all gases at sufficiently low pressures, van der Waals found that for some gases, such as carbon dioxide, at comparatively low pressures and for others at higher pressures it is no longer valid. He attributed the deviations from the gas equation, and hence from ideal behaviour, mainly to two factors:

(i) molecular interactions do occur;

(ii) gas molecules occupy a finite volume.

If $pV = nRT$, a plot of pV against p at constant temperature should give a straight line. However, for nitrogen at low temperatures (273 K), and for other gases, curves of the type shown in Fig. 6.1 are obtained.

van der Waals' equation

J. C. van der Waals modified the general gas equation to allow for these factors by correcting the pressure and volume terms.

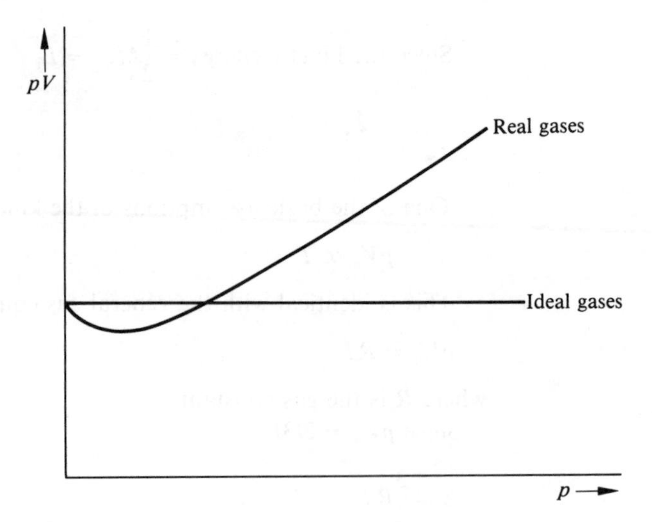

Fig. 6.1 Deviations from the general gas equation.

Pressure

When a molecule is about to strike the walls of the vessel it will only experience an attractive force inwards. Therefore the actual pressure (p) is less than the ideal one, and the ideal gas equation must be changed to allow for this.

Now, reduction in pressure on one molecule striking the surface

$$\propto \text{number of molecules } (N) \text{ in the gas}$$

The reduction in pressure on *all* the molecules striking the surface can be shown to be $\propto N^2 \propto \dfrac{1}{V^2}$

$$= \frac{a}{V^2} \quad \text{where } a \text{ is a constant}$$

The ideal pressure will be equal to the measured (actual) pressure (p) together with this pressure correction,

i.e. $\left(p + \dfrac{a}{V^2}\right)V = RT$

Volume

Since the volume of the molecules is not negligible, the free space between the molecules will be less than the total volume V. The actual molecules may be considered to be incompressible. Their total volume remains the same and can be represented by a constant b.

$\therefore \quad p(V - b) = RT \quad$ where b is a constant

Taking both factors into consideration,

$$\left(p + \frac{a}{V^2}\right)(V - b) = RT$$

This equation is known as *van der Waals' equation*. Like $pV = RT$, it is for one mole of a gas, but can be modified to allow for n moles.

Like the gas equation, Graham's law of diffusion, Gay-Lussac's law of combining volumes and Avogadro's law can all be derived from the kinetic theory of gases, and as such are true only of ideal gases and real gases at low pressure.

6.2.2 DIFFUSION

Graham's law of gaseous diffusion states that 'The rate of diffusion of a gas is inversely proportional to the square root of its relative molecular mass'. This is a direct consequence of the kinetic theory of gases since

$$pV_{\text{m}} = \frac{1}{3}M\bar{c}^2 = RT$$

$$\therefore \quad \bar{c}^2 = 3RT/M$$

$$\therefore \quad \sqrt{\bar{c}^2} = c_{\text{r.m.s.}} = \sqrt{3RT/M}$$

where $c_{\text{r.m.s.}}$, the *root mean square velocity*, replaces $\sqrt{\bar{c}^2}$ for simplicity.

Since the rate of diffusion of a gas is the rate at which the gas flows through the vessel, it is equal to the 'average' rate of movement of the gas. Therefore, at a constant temperature,

$$\text{rate of diffusion} \propto c_{\text{r.m.s.}} \propto 1/\sqrt{M}$$

Maxwell came to the same conclusion as Graham in a study of molecular velocities and their distribution. Some scatter of velocities is clearly inevitable because of the continual collision of molecules. If, in a gas containing N molecules, N_{E} is the number of molecules having a kinetic energy greater than any given value E, then

$$\frac{N_{\text{E}}}{N} = \text{e}^{-E/RT}$$

At a given temperature we obtain a distribution of molecular velocities (Fig. 6.2).

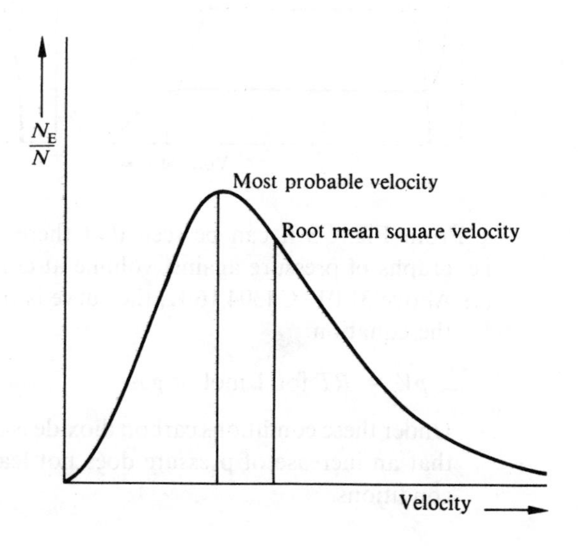

Fig. 6.2 Maxwell distribution curve.

For gas molecules,

the most probable velocity × 1.225 = the root mean square velocity

$\therefore$ most probable velocity $\propto 1/\sqrt{M}$

6.2.3 LIQUEFACTION OF GASES

We have shown earlier that two opposing factors are generally responsible for the state of a system. They are the forces of attraction between molecules and thermal agitation.

If the process of liquefaction arises merely as a result of the forces of attraction overcoming thermal agitation we should expect a gradual transition from a gas to a liquid. In fact the change usually occurs sharply and with a considerable decrease in volume, suggesting that other effects have a part to play.

One of the major contributions to our understanding of what happens on liquefaction was made by T. Andrews in his studies of carbon dioxide at high pressures. He investigated the change in volume of carbon dioxide, as pressure was increased, for a series of temperatures between 10 and 50 °C (283–323 K).

Fig. 6.3 Isotherms of carbon dioxide.

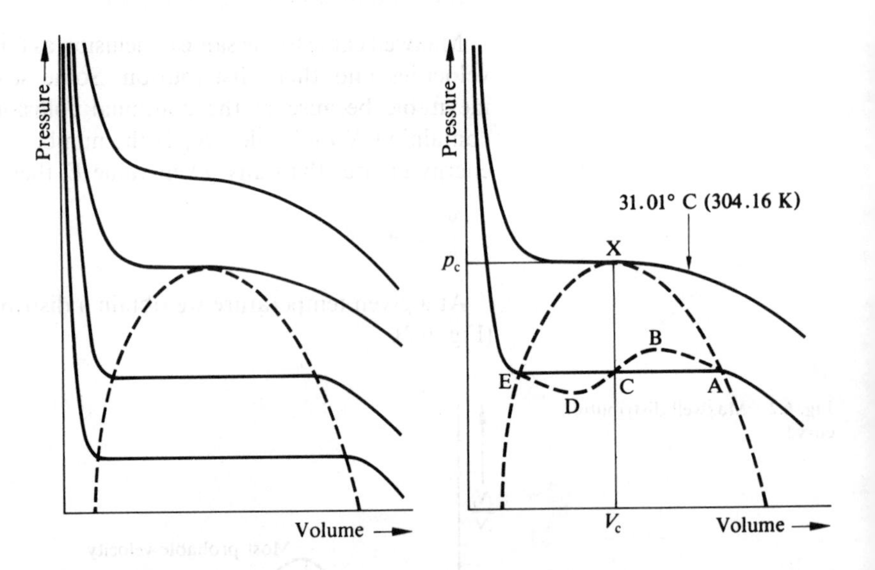

From Fig. 6.3 it can be seen that there are three types of *isotherms*, i.e. graphs of pressure against volume at constant temperature.

(a) Above 31.01 °C (304.16 K) the curve is that of a hyperbola and obeys the equation

$$pV = RT \text{ for 1 mol of gas}$$

Under these conditions carbon dioxide is always in the gaseous state so that an increase of pressure does not lead to liquefaction under any conditions.

(b) Below 31.01 °C (304.16 K) the curve has three portions.

(i) At 'low' pressures carbon dioxide is a gas which is compressed gradually by increase of pressure until at a given pressure the volume decreases suddenly with little change of pressure. At this stage liquefaction occurs.

(ii) Liquefaction along AE.
If the gas is compressed gradually and smoothly in a reaction vessel with clean, smooth inner walls it will resist liquefaction until point B when liquid will form rapidly. The reverse occurs by gradual reduction of pressure to D. The curve EDCBA represents a cubic equation in V and obeys van der Waals' equation for 1 mol

$$\left(p + \frac{a}{V^2}\right)(V - b) = RT$$

which can be rewritten as

$$V^3 - (b + RT/p)V^2 + (a/p)V - ab/p = 0$$

(iii) Further increase in pressure leads to very little change in volume since the liquid is not very compressible.

(c) At 31.01 °C (304.16 K) the curve has two coincident turning points at X. 31.01 °C (304.16 K) is known as the *critical temperature* (T_c), and the corresponding pressure and volume represented by X as the *critical pressure* (p_c) and *volume* (V_c) respectively. This suggests that the additional parameters of the van der Waals' equation over those of the general gas equation, namely finite volume of the molecules and molecular interaction, are important in studying the liquid state. These will be considered in the next section.

6.3 The liquid state

6.3.1 THE PROPERTIES OF LIQUIDS

Liquids are characterised by the following properties.

(a) Diffusion. The molecules, although in a state of random movement, are more restricted than for gases and diffusion occurs slowly.

(b) Surface tension. The pronounced unbalanced forces of attraction at the liquid boundaries give rise to surfaces.

(c) Viscosity/fluidity. Liquids have no rigidity and no elasticity.

(d) Compressibility. The molecules are much more tightly packed than in a gas. An increase in pressure therefore does not give rise to a large decrease in volume.

We shall now consider the first aspect in more detail. The statement made under (a) is not strictly true. Originally it was believed that liquids formed similar arrangements of atoms and molecules to those in gases, but because of the closer proximity of the molecules and their resultant intermolecular attraction movement was much more restricted. However, X-ray evidence does show some kind of orderly structure in liquids. This would appear to be similar to that in the solid state but less rigid. The formation of 'solid structures' is counteracted by thermal agitation, giving rise to a constant formation and breakdown of the structure and leading to restricted random motion. Diffusion should therefore occur, as in gases, but much more slowly. The picture is complicated by the fact that many liquids are almost insoluble in one another. If they do not mix, diffusion cannot occur to any extent.

6.3.2 INTERMOLECULAR FORCES

Intermolecular forces are often called *van der Waals' forces* and are normally very weak. We discussed earlier (§4.1.2) the existence of permanent dipoles in molecules such as HCl ($\overset{\delta^+}{H}$—$\overset{\delta^-}{Cl}$), but this does not explain the fact that many molecules like H_2 or even atoms like helium have no permanent dipoles and yet are capable of liquefaction. They must therefore form strong enough intermolecular attractions to overcome thermal excitation, at least at low temperatures where liquefaction occurs. An explanation was postulated by F. London. He suggested that the electrons in, say, H_2 are not always midway between the atoms but are in constant motion and can for a fraction of a microsecond be nearer one atom than the other, giving rise to a dipole with one atom slightly positive and the other slightly negative. These so-called *temporary dipoles* abound, and the positive end of one H_2 molecule will attract the negative end of another. As the molecules get closer together the forces of attraction increase until the molecules are so close that their electron clouds repel each other. Eventually equilibrium is reached, where the forces of attraction and repulsion are equal. At this stage the distance between them is known as the *collision diameter* (σ) (Fig. 6.4).

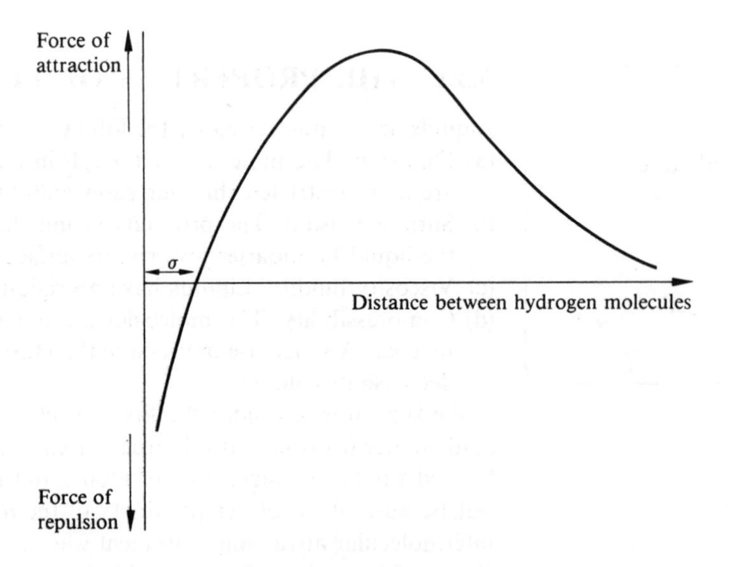

Fig. 6.4 Forces of attraction and repulsion between two hydrogen molecules at varying distances from each other.

The type of temporary dipole arising in atoms is shown in Fig. 6.5.

The forces of attraction due to the presence of temporary and permanent dipole moments are known as van der Waals' forces and are independent

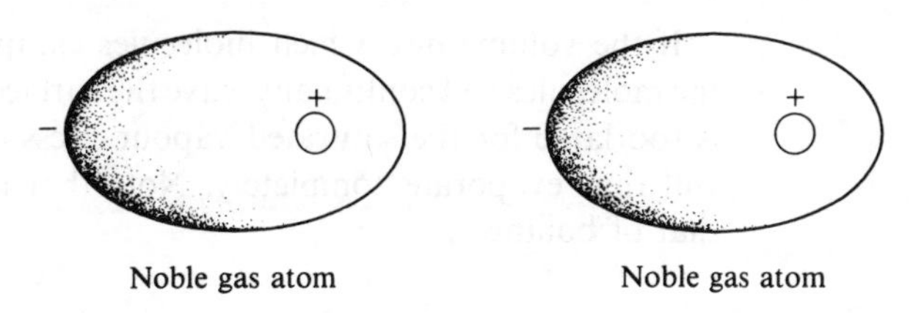

Fig. 6.5 Temporary dipoles in atoms of a noble gas.

of normal bonding forces. When these forces are strong enough to overcome thermal agitation (i.e. when the molecules are close enough together) they increase rapidly and cause liquefaction. This increase is represented by the peak in Fig. 6.4. It can be shown that a and b in the van der Waals' equation are related to the forces of attraction and repulsion, a to the forces of attraction and b to both.

The other form of bonding occurring in the liquid state, especially in water, is *hydrogen bonding* (§6.6).

6.3.3 VAPOUR PRESSURE

In all liquids there is a tendency for molecules to escape from the surface and form a vapour above it. The greater the escaping tendency the more easily the vapour forms and the higher is the resultant pressure above the liquid, i.e. the *vapour pressure.*

Since liquids obey a Maxwell distribution in the same way as gases, the molecules on the surface with the highest kinetic energy will have the greatest chance of escape. It will be shown that the parameters affecting the vapour pressure are the intermolecular forces and temperature.

The effect of temperature

The *average* kinetic energy of the molecules is proportional to the temperature, and as the temperature rises so does the tendency of molecules to escape. Therefore the vapour pressure rises. When the temperature becomes high enough for the vapour pressure of the liquid to equal that of the atmosphere, molecules can escape from all parts of the liquid. At this stage, if the atmospheric pressure remains constant, the vapour pressure and temperature also remain constant. This temperature is known as the *boiling temperature* (formerly 'boiling point') and since it is a property of the liquid it is measured at a standard pressure of 1 atm (101.325 kPa) for comparison purposes.

Below the boiling temperature molecules leaving the surface lose some of their kinetic energy and fall back again. At any given temperature, in a closed container, equilibrium (§8.1) is set up between molecules leaving and re-entering the surface. The vapour pressure of a liquid under these conditions is known as its *saturated vapour pressure* or *equilibrium vapour pressure* for the given temperature. The saturated vapour pressure varies with temperature (Fig. 8.3) according to the Clausius–Clapeyron equation (see below).

If the volume into which molecules escape is large (e.g. a laboratory) the molecules will continually leave the surface because the external volume is too large for the saturated vapour pressure to be reached. The liquid will thus evaporate completely. Note that this process is different from that of boiling.

The effect of intermolecular attraction

As the forces of cohesion increase within the liquid the escaping tendency decreases. For example, water has a much lower vapour pressure than hydrogen sulphide because the hydrogen bonds between the water molecules are much stronger than the van der Waals' forces between the H_2S molecules. The forces of attraction are increased with the size of the molecules (and hence with relative molecular mass) since the velocity decreases, leading to greater cohesion.

The above results can be related by the following equations.

(i) $$\lg p = -\frac{\Delta H_v}{2.303RT} + \text{constant}$$

where p is the vapour pressure, T is the temperature and ΔH_v is the molar enthalpy of vaporisation. This equation is known as the *Clausius–Clapeyron equation.*

(ii) At the boiling temperature (T),

$$\frac{\Delta H_v}{T} = \text{constant}$$

The constant has an approximate value of $21\ \mathrm{J\,K^{-1}\,mol^{-1}}$ for most liquids.

6.4 The solid state

The solid state is more ordered than the liquid state, most solids existing as crystalline solids in which the position of the atoms is easily found by physical methods such as X-ray diffraction. There is no translational movement and therefore no flow of solid, although vibration of atoms does occur to some extent, especially near the melting temperatures. Indeed, vibration occurs at all temperatures, but is extremely small at absolute zero and increases steadily with temperature until the melting temperature is reached.

Apart from crystalline solids there is a range of finely divided solids known as *amorphous solids* in which the atoms seem to be packed randomly as opposed to the regular structures of crystalline solids. Amorphous solids include glass, amorphous sulphur and unstretched rubber. In their disordered structures amorphous solids resemble liquids.

6.4.1 X-RAY DIFFRACTION

The structure of crystals can be investigated by X-ray diffraction. A crystalline solid may be considered to be made up of regular layers (or planes) of atoms a definite distance apart. If a beam of X-rays is allowed to fall on the crystal, the rays are diffracted by the layers, but appear to be reflected.

If the beam of X-rays of wavelength λ impinges on the crystal at an angle of incidence θ, and the distance between the layers is d, the X-rays will be 'reflected' (Fig. 6.6).

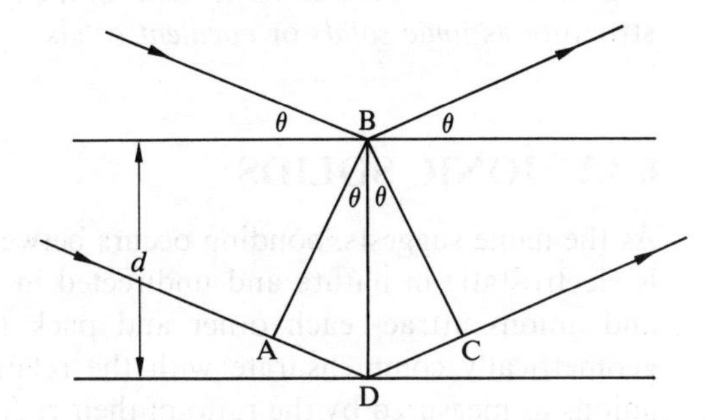

Fig. 6.6 Reflection of X-rays from the parallel planes of a crystal.

If the waves are to remain in phase and reinforce each other their path lengths must differ by a complete number of wavelengths $n\lambda$, where $n = 1, 2, 3$, etc

i.e $\quad \text{ADC} = n\lambda$

but $\quad \text{AD} = d \sin\theta \quad \text{and DC} = d \sin\theta$

$\therefore \quad \text{ADC} = 2d \sin\theta$

and $\quad 2d \sin\theta = n\lambda$

Reinforcement will occur for values of θ which are characteristic of the crystal and the value of d. They will also depend on which face of the crystal is exposed to the X-rays. A knowledge of the angles between the faces of the crystal and of its X-ray diffraction results enables us to characterise any given crystal into one of seven classes.

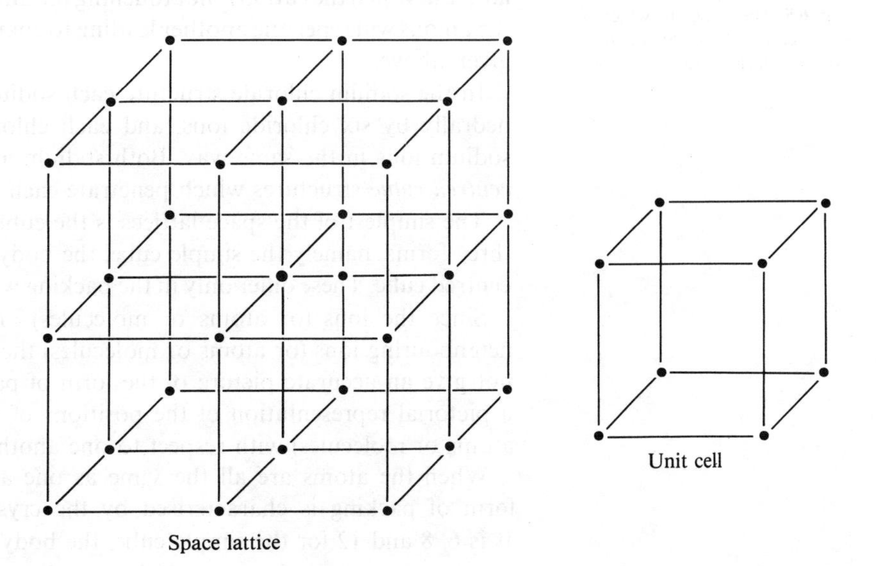

Fig. 6.7 Space lattice and unit cell.

The atoms in a crystal can be depicted by a *space lattice* which shows how the atoms are arranged in the crystal (Fig. 6.7). The imaginary lines joining the atoms are known as *lattice lines*. The space lattice is composed of a basic unit which is repeated throughout the lattice and which is known as the *unit cell* of the crystal (Fig. 6.7).

Crystals may be classified in terms of the type of bond occurring in their structure as *ionic solids* or *covalent solids*.

6.4.2 IONIC SOLIDS

As the name suggests, bonding occurs between ions in an ionic crystal; it is electrostatic in nature and undirected in space. Therefore the cations and anions attract each other and pack together in a way which is geometrically commensurate with the relative sizes of the cations and anions as measured by the ratio of their radii. The ions may be simple or complicated. When considering what happens in crystals composed of simple ions we find that there are four basic types of crystal depending on the relative number and sizes of the ions in the crystal.

Crystals in which the ratio of cations to anions is 1 : 1

The 'formula' is AB.

(i) Radius ratio $r_A/r_B > 0.732$ (A is cation, B is anion)
These compounds have a caesium chloride lattice, e.g. CsCl.

(ii) Radius ratio $0.732 > r_A/r_B > 0.414$
These compounds have a sodium chloride lattice, e.g. NaCl, KCl.

The *crystal coordination number*, i.e. the number of ions of opposite charge adjacent to a given ion, is 6 for both sodium chloride and potassium chloride. For caesium chloride the crystal coordination number is 8.

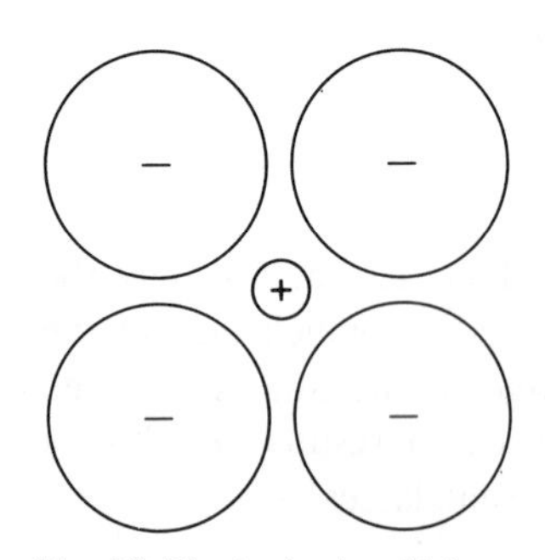

Fig. 6.8 The lattice in which a small cation does not touch adjacent large anions.

Since the cations are smaller than the anions it is possible to have a lattice in which the cation is not touching the anions (Fig. 6.8), in which case the anions will repel one another leading to changes in structure from those given above.

In the sodium chloride structure each sodium ion is surrounded octahedrally by six chloride ions, and each chloride ion is surrounded by sodium ions in the same way. Both sodium and chloride ions form *face centred cubic* structures which penetrate each other (Fig. 6.9).

The simplest of the space lattices is the cubic lattice which can exist in three forms, namely the simple cube, the body centred cube and the face centred cube. These differ only in the packing within the crystals (Fig. 6.10).

Since the ions (or atoms or molecules) are almost in contact with neighbouring ions (or atoms or molecules) the space lattice concept does not give an accurate picture of the form of packing (Fig. 6.11) but only a pictorial representation of the positions of the centres of the ions (or atoms or molecules) with respect to one another.

When the atoms are all the same as one another, as in a metal, the form of packing is characterised by the crystal coordination number. It is 6, 8 and 12 for the simple cube, the body centred cube and the face

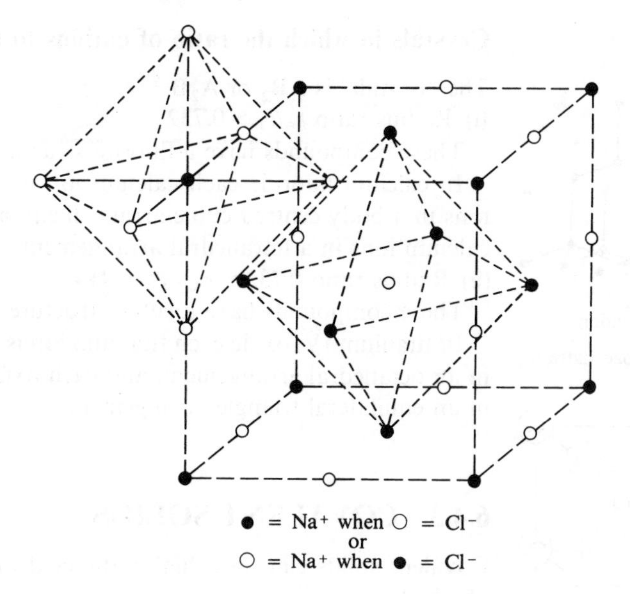

Fig. 6.9 The structure of sodium chloride.

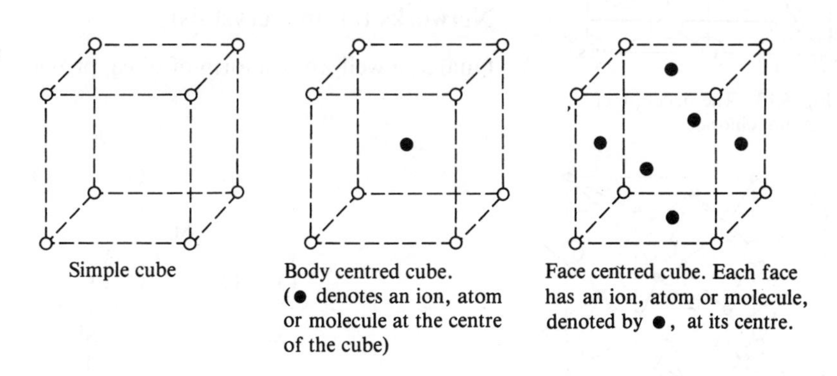

Simple cube
Body centred cube.
(● denotes an ion, atom or molecule at the centre of the cube)
Face centred cube. Each face has an ion, atom or molecule, denoted by ●, at its centre.

Fig. 6.10 Space lattices for cubic systems.

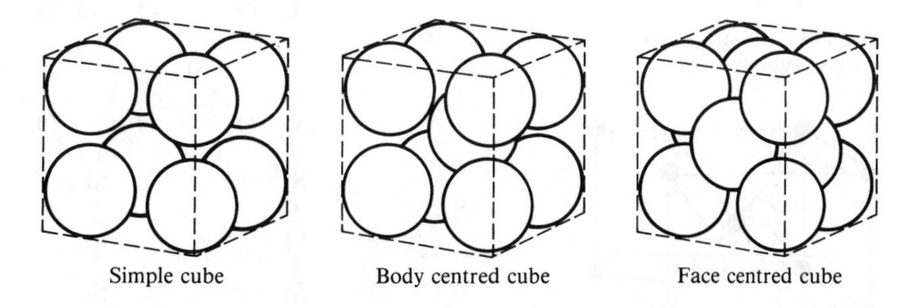

Fig. 6.11 Packing diagram for cubic lattices.

centred cube respectively. The other common coordination number is 4 and is exemplified by the tetrahedral structure of diamond (Fig. 6.12).

In caesium chloride two simple cubes penetrate each other (Fig. 6.13).

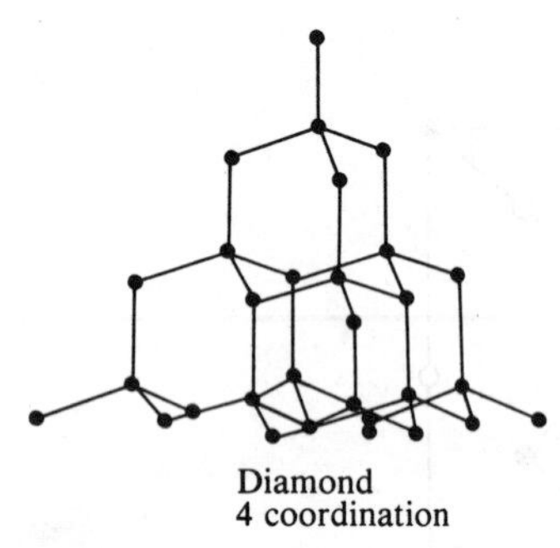

Fig. 6.12 Diamond space lattice.

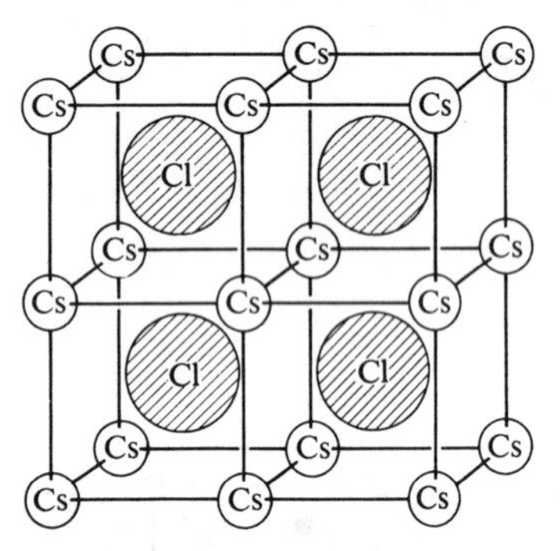

Fig. 6.13 The structure of caesium chloride.

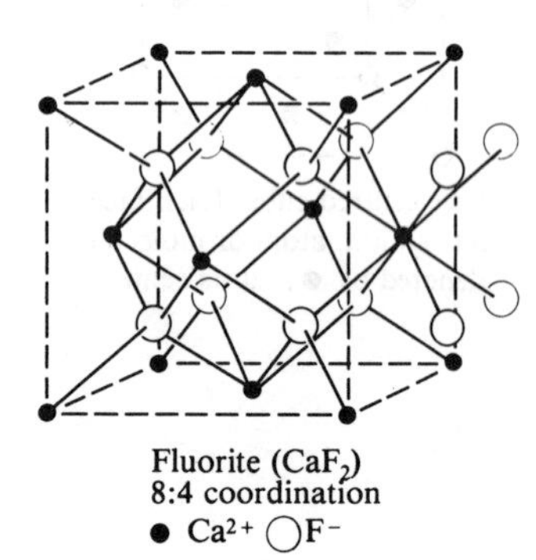

Fig. 6.14 The structure of calcium fluoride.

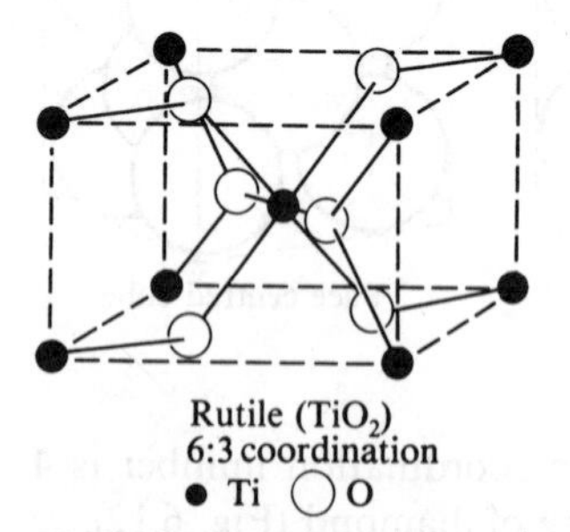

Fig. 6.15 The structure of titanium(IV) oxide.

Crystals in which the ratio of cations to anions is 1:2 or 2:1

The 'formula' is AB_2 or A_2B.
(i) Radius ratio $r_A/r_B > 0.732$

These compounds have a fluorite structure, e.g. CaF_2, BaF_2.

In calcium fluoride each calcium ion is surrounded by eight fluoride ions in a body centred cubic arrangement, and each fluoride ion by four calcium ions in a tetrahedral arrangement.
(ii) Radius ratio $0.732 > r_A/r_B > 0.414$

These compounds have a rutile structure, e.g. TiO_2, MgF_2.

In titanium(IV) oxide each titanium ion is surrounded by six oxide ions in an octahedral arrangement, and each oxide ion by three titanium ions in an equilateral triangle arrangement.

6.4.3 COVALENT SOLIDS

Covalent solids can form highly directed or non-directed crystals, both of which we shall consider.

Networks (atomic crystals)

Quartz, a well-known form of silica, provides a good example of this class.

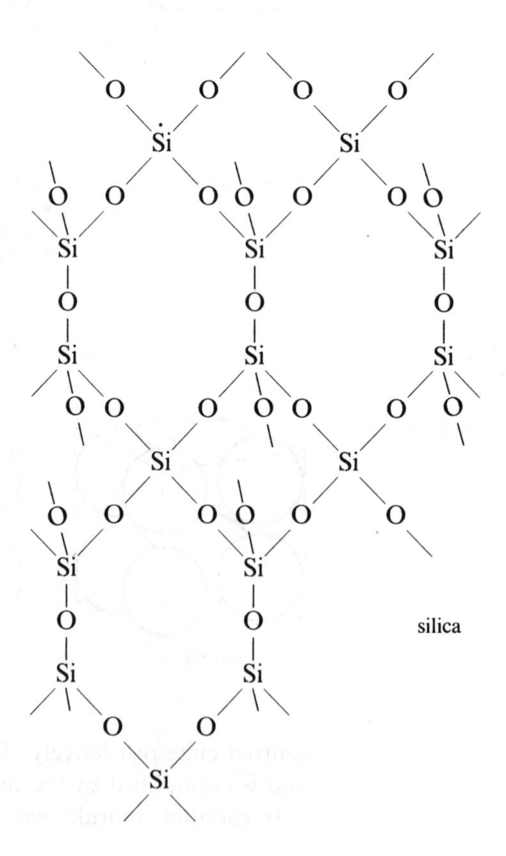

Carbon can form a two-dimensional and a three-dimensional atomic crystal (Fig. 6.16).

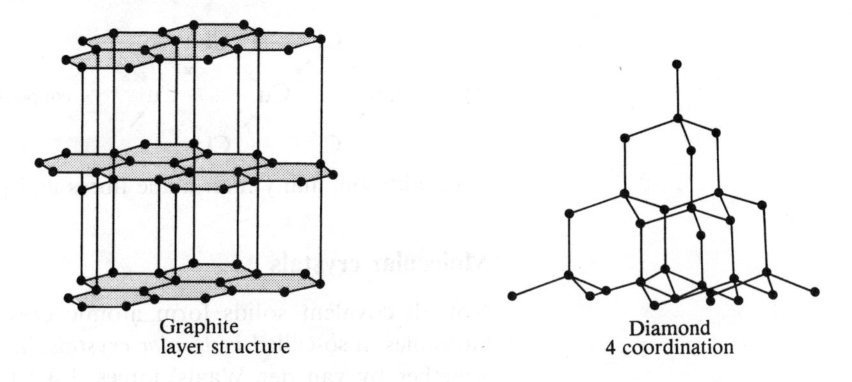

Fig. 6.16 The structures of graphite and diamond.

In diamond the carbon atoms are tetrahedrally and covalently bonded to one another in a three-dimensional framework giving a crystal of great strength. Silicon, germanium and grey tin also give crystals with the diamond structure, but form no structures comparable to that of graphite.

Graphite possesses a layered structure. Within each layer, or two-dimensional framework, the carbon atoms are covalently bonded together to form a hexagonal pattern. Each carbon atom uses three of its four outer electrons to form localised covalent bonds to its nearest neighbours. The remaining electrons, one from each carbon atom, combine to form a delocalised π orbital (§4.4.1), extending over the whole of the layer. These delocalised electrons are free to move anywhere within a given layer, but cannot move from one layer to another. They will migrate under an applied electrical potential, thus making a graphite crystal an electrical conductor in the plane of the layers, but not in a direction perpendicular to the layers.

In a crystal of graphite the layers are stacked together like a pack of cards, with only weak van der Waals' forces holding adjacent layers together. The layers can readily slide over one another, so that graphite can be used as a lubricant. Because of its open structure, which can easily be distorted, graphite, unlike diamond, is a very soft material.

A different type of layer lattice is shown by cadmium iodide. The iodine atoms lie in two parallel planes with cadmium atoms in the interstices (holes) between them. This 'sandwich' is extended two dimensionally. This is an example of a compound whose structure is intermediate between ionic and covalent.

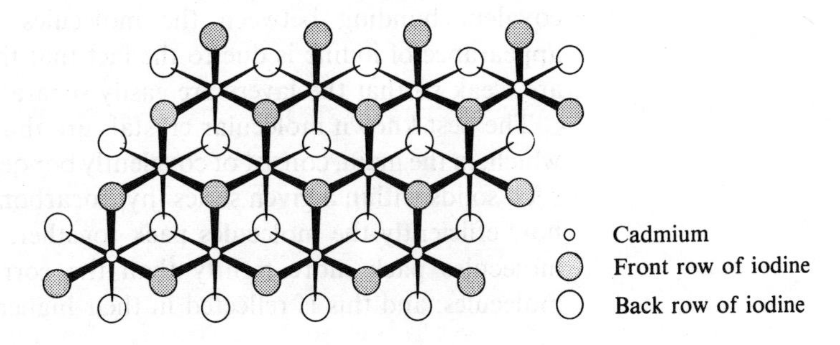

Fig. 6.17 The structure of cadmium iodide.

Apart from the two- and three-dimensional types of framework mentioned above, many covalent crystals are in the form of chains, which are held together side by side by van der Waals' forces,

```
          Cl        Cl
       ↘ /    ↘   /    ↘  /
e.g.    Cu        Cu        Cu        copper(II) chloride
       /  ↖     /  ↖     /  ↖
          Cl        Cl
```

In addition, many man-made fibres and plastics form chain solids.

Molecular crystals

Not all covalent solids form atomic crystals. Many exist as discrete molecules in so-called *molecular crystals*, in which the molecules are held together by van der Waals' forces. Like the ions in ionic crystals, the molecules pack together in a tightly packed system since the forces of attraction are non-directional. The number of molecules surrounding a given molecule is denoted by the crystal coordination number (§6.4.2). Since molecular crystals are held together by such weak forces they are soft, low melting and low boiling (§4.9.2). The 'molecules' may also be simple atoms. Both argon and iodine have face centred cubic lattices, although that of the latter is distorted along one axis.

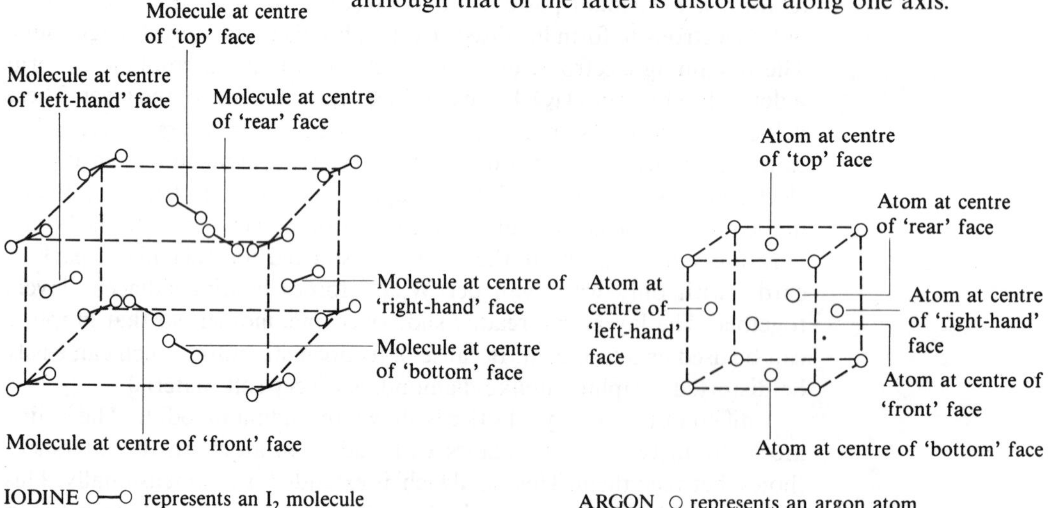

Fig. 6.18 The structures of iodine and argon.

In addition, the iodine molecules form layers with some degree of covalent bonding between the molecules within a layer. The flaky appearance of iodine is due to the fact that the bonds between the layers are weak so that the layers are easily separated.

The best known molecular crystals are those formed by organic solids which, in the main, consist of covalently bonded molecules. The strength of such solids within a given series (hydrocarbons, alcohols, etc) depends on how efficiently the molecules pack together. For example, simple linear molecules pack more tightly than the corresponding branched chain molecules, and this is reflected in their higher melting temperatures, etc.

6.5
Metals

Although there is no clear dividing line between metals and non-metals, the metals can, on the whole, be characterised by the following properties: metallic lustre, high ductility and malleability, high electrical and thermal conductivity, and high electropositive nature.

Many of these properties can be explained by the modes of packing of metal atoms and by the type of bonding which holds them together.

Most metals have a coordination number of 12 and pack together very tightly in the face centred cubic or hexagonal systems (Fig. 6.19). This type of packing is very efficient and is known as *close packing*.

A comparison of modes of packing (Fig. 6.20) shows the much greater efficiency of close packing. A slightly less efficient method of

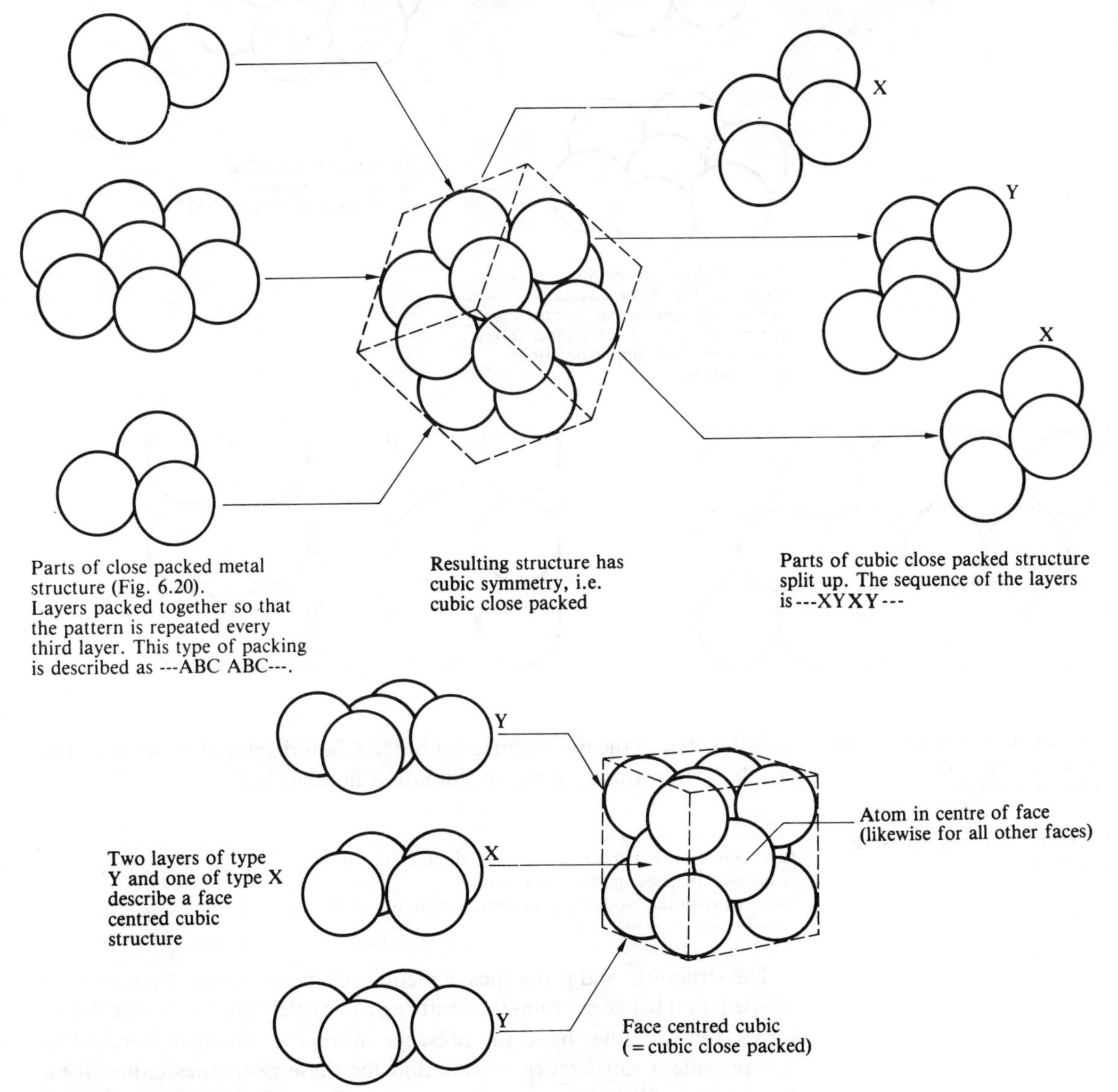
X
Y
X
Parts of close packed metal structure (Fig. 6.20).
Layers packed together so that the pattern is repeated every third layer. This type of packing is described as ---ABC ABC---.
Resulting structure has cubic symmetry, i.e. cubic close packed
Parts of cubic close packed structure split up. The sequence of the layers is ---XYXY---
Y
X
Y
Two layers of type Y and one of type X describe a face centred cubic structure
Atom in centre of face (likewise for all other faces)
Face centred cubic (= cubic close packed)

Fig. 6.19a Cubic close packing (face centred cubic).

Fig. 6.19b Hexagonal close packing.

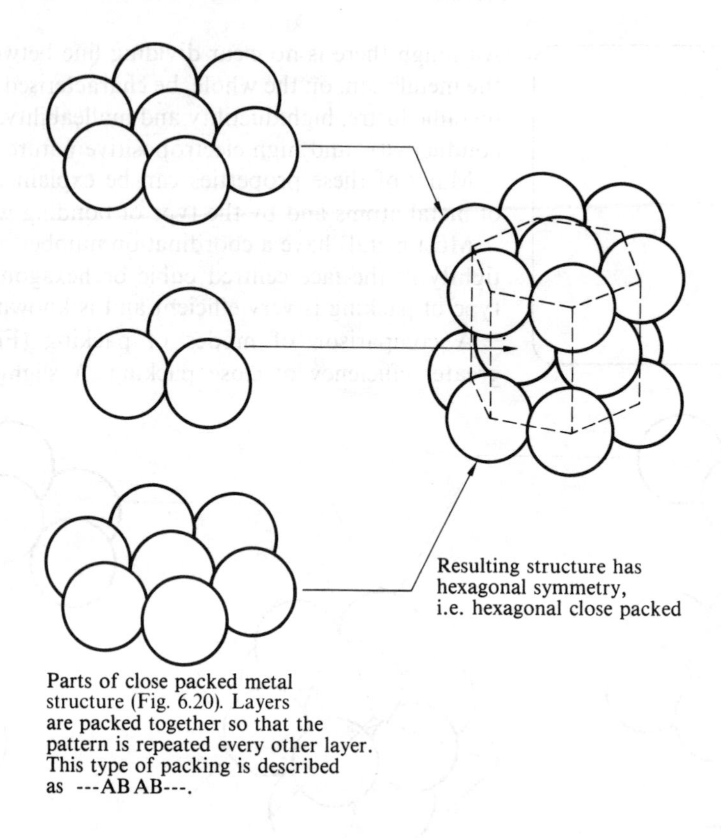
Resulting structure has hexagonal symmetry, i.e. hexagonal close packed
Parts of close packed metal structure (Fig. 6.20). Layers are packed together so that the pattern is repeated every other layer. This type of packing is described as ---AB AB---.

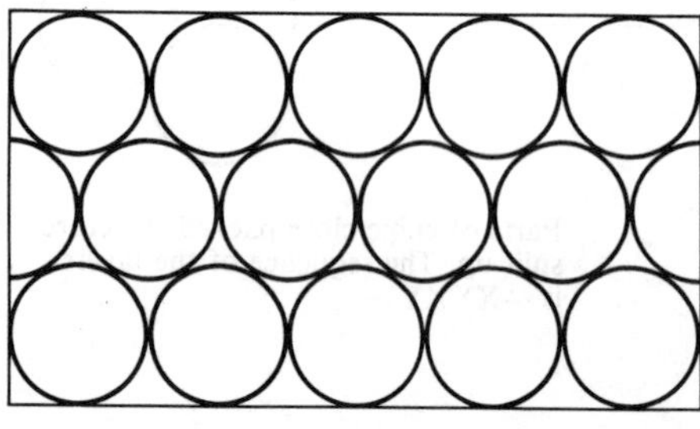

(a)

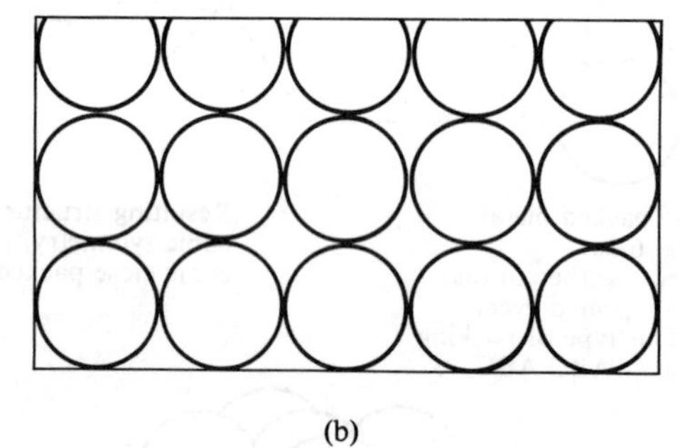

(b)

Fig. 6.20 (a) Close packing, and (b) non-close packing, of metal atoms in one layer.

packing, that of the body centred cube (§6.4.2), is displayed by some metals (Table 6.1). In this case the coordination number is 8.

Table 6.1 Examples of metallic structures

face centred cubic: copper, silver, gold, aluminium, γ-iron
hexagonal: magnesium, beryllium, zinc
body centred cubic: sodium, potassium, α-iron, tungsten

The structures and properties of metallic elements suggest that the force of attraction between atoms is too strong to be solely due to van der Waals' forces. On the other hand the presence of ionic or covalent bonds does not provide a satisfactory explanation for these properties either. Ionic bonds are unlikely because only one type of atom is present and electron transfer cannot take place. Covalent bonds are equally unlikely, because not enough electrons are present in the valence shells of the atoms to allow electron-pair formation with 8 or 12 nearest neighbours. In addition, some free movement of electrons is necessary to explain the high electrical conductivity of metals. The only conclusion to be drawn from this discussion is that we have a type of bonding different from anything we have considered before; it is called a *metallic bond.*

6.5.1 THE METALLIC BOND

The valency electrons of each metal atom become delocalised in the same way as for benzene, and form an electron cloud over the cationic network of positive metal atoms. Each atom shares $\frac{1}{12}$ or $\frac{1}{8}$ of each valency electron with each of its nearest neighbours, and receives $\frac{1}{12}$ or $\frac{1}{8}$ of an electron in return,

e.g. $Cu \rightarrow Cu^+$ share of electrons $= \frac{1}{12} + \frac{1}{12} = \frac{1}{6}$ with each of its nearest neighbours

but $Na \rightarrow Na^+$ $= \frac{1}{8} + \frac{1}{8} = \frac{1}{4}$

The metallic bond has some of the characteristics of the covalent bond but is non-directional. This is a very simple picture of the metallic bond but it is sufficient for our purposes.

6.5.2 THE PROPERTIES OF METALS

Metallic lustre

Lustre is a result of metals reflecting light strongly. This is partly because of the highly efficient close packing of atoms in metals; the resulting high density of atoms gives a correspondingly high opacity. Also, since electrons are mobile, energy transitions can occur with great ease, so that light of all frequencies can be absorbed and re-emitted shortly afterwards. Both effects lead to high reflectivity.

High ductility and malleability

The metal atoms lie in layers one on top of another. The layers are not held rigidly in position because there are no rigid bonds between the atoms. They can therefore move (or slip) over one another on the application of mild stress. Slippage over a distance of one atom restores the original structure. If the slippage is less than one interatomic distance the atoms move back to their original positions on the removal of the stress. (Metals are said to be 'elastic'.) This ease of movement of one layer of atoms over another causes the metal to be malleable. The ductility of metals is also a direct consequence of the ease of slippage.

High electrical conductivity

The delocalised electrons in the atoms of a metal occupy a series of energy levels and are paired according to the Pauli exclusion principle (§3.1.6). The energy levels are close together and the empty higher orbitals (e.g. the 3p orbitals in sodium) are available to take the electrons as they flow towards the positive pole under an applied electrical potential. As the electrons pass from the cationic lattice to the pole more electrons flow in from the negative pole to take their place. This gives rise to a high electrical conductivity through the metal. As the temperature of the metal is raised so is the violence with which the cations vibrate. This tends to decrease the mobility of the electrons and hence the conductivity of the metal.

High thermal conductivity

The atoms in a metal are relatively free to vibrate, and can transfer kinetic energy readily by their movement and by collision. This gives high thermal conductivity.

Ionic, covalent and metallic bonds are extreme types and most bonds are, in reality, intermediate in character.

6.6 Hydrogen bonding

When hydrogen is covalently bonded to strongly electronegative elements such as fluorine, oxygen or nitrogen, the bonding electron pair is attracted closer to these elements than to hydrogen, and dipoles are formed (§4.1.2).

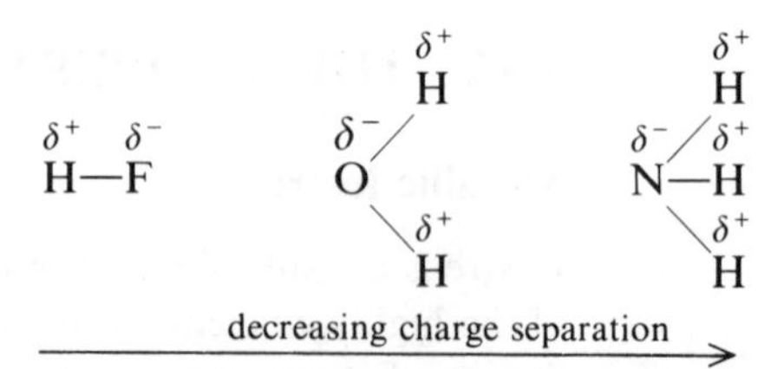

Since these are the only electrons involved with the hydrogen atom there will be very little shielding effect on the side of the hydrogen atom furthest from the electronegative element. Therefore another strongly electronegative element which has a lone pair of electrons will be attracted to the hydrogen atom. This electrostatic attraction is referred to as a *hydrogen bond.* Hydrogen bonds are strong van der Waals' forces (§6.3.2) – stronger than in compounds which do not contain these elements – and this is why a different description is used.

On this basis we might expect chlorine, which has a similar electronegativity to nitrogen, to form hydrogen bonds. However, chlorine has a larger atomic radius than nitrogen, and the lone pair occupies a larger orbital (3p compared with 2s). The chlorine lone pair therefore has a lower electron density than the nitrogen pair, and the attraction between HCl molecules is less than that between NH_3 molecules. Hydrogen chloride, in contrast to ammonia, hydrogen fluoride and water, is not hydrogen bonded. To summarise, only small, highly electronegative atoms are regarded as taking part in hydrogen bonding.

When hydrogen bonding occurs between different molecules, as in HF and H_2O (see below), it is referred to as *intermolecular hydrogen bonding*. Hydrogen bonding may also occur within a single molecule, as in the case of 2-nitrophenol:

This is known as *intramolecular hydrogen bonding*.

Although the forces of attraction giving rise to hydrogen bonds are several orders of magnitude weaker than for either ionic or covalent bonds, they are still strong enough to give compounds much higher boiling temperatures than might be expected. For example, we would expect water (relative molecular mass = 18) to have a lower boiling temperature than hydrogen sulphide (relative molecular mass = 34), but this is not so. Intermolecular hydrogen bonding between H_2O molecules is stronger than the van der Waals' forces between H_2S molecules; hence the melting temperature and boiling temperature of water are *higher* than those of hydrogen sulphide. Similar situations arise with hydrogen fluoride and ammonia, but not with methane (Fig. 6.21). These effects provide us with some of the best evidence for hydrogen bonding.

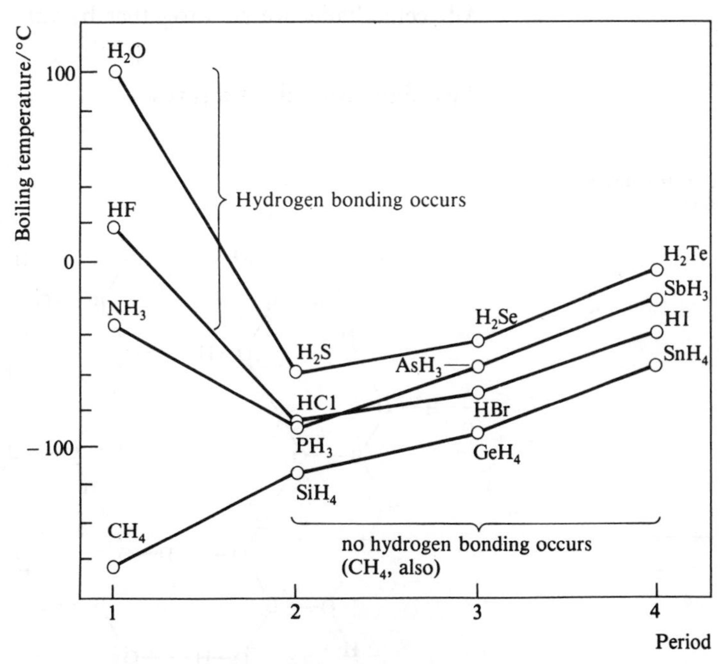

Fig. 6.21 Boiling temperatures of various hydrides.

When hydrogen bonding does not occur, as for example in H_2S, H_2Se and H_2Te, the boiling temperature increases almost linearly with relative molecular mass. Other physical properties, such as melting temperatures and enthalpies of vaporisation, show similar trends to boiling temperatures.

6.6.1 STRUCTURES INVOLVING HYDROGEN BONDING

In the solid state and to a lesser extent in the liquid state hydrogen bonding often leads to the formation of one-, two- and three-dimensional structures.

One-dimensional structures

The relative molecular mass of hydrogen fluoride in the solid or liquid phases, or in the gaseous phase just above the boiling temperature, is found to be higher than the expected value of 20. This is because the molecules are hydrogen bonded together to form chains. The chains possess a zigzag structure, which is attributed to the fact that hydrogen bonding occurs through a lone pair of p electrons on the fluorine atom, thus:

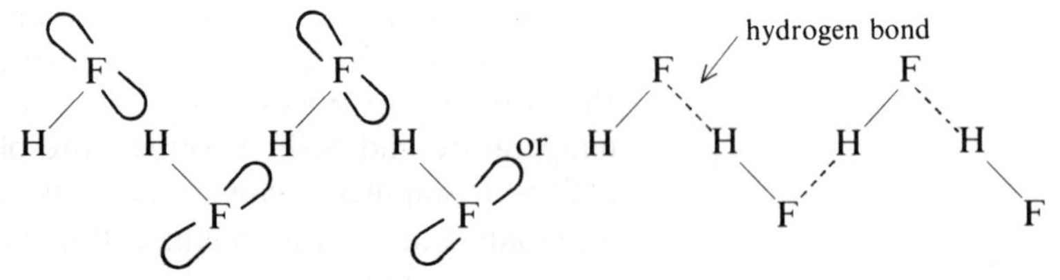

hydrogen bond

Adjacent chains are held together by van der Waals' forces.

Two-dimensional structures

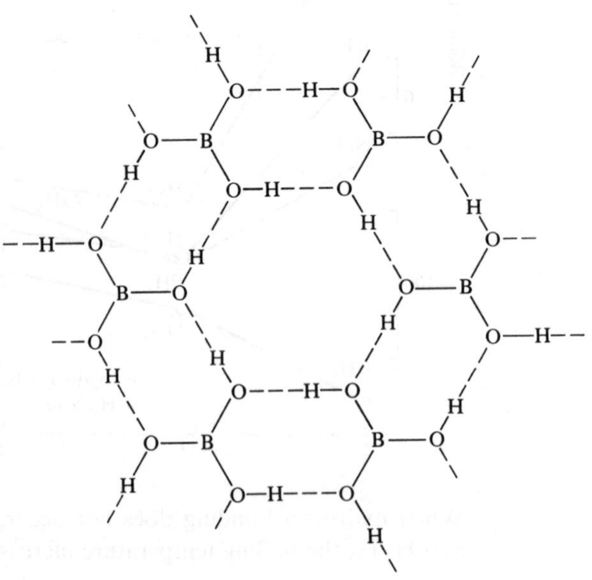

Fig. 6.22 The structure of orthoboric acid.

In, for example, orthoboric acid, H_3BO_3, the layers are held together by van der Waals' forces.

Three-dimensional structures

The classic example of this structure is ice, which has the form of the diamond lattice.

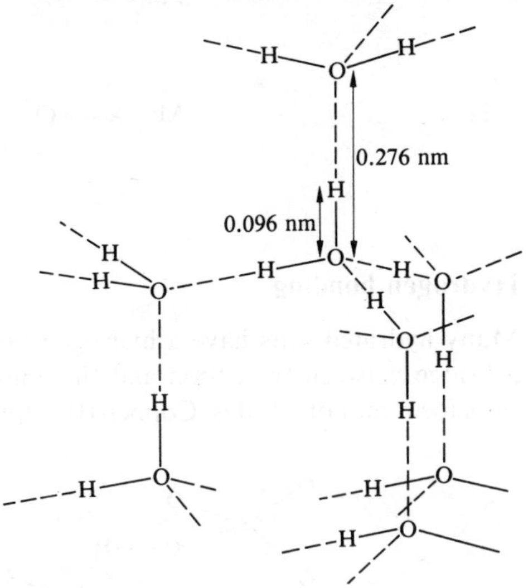

Fig. 6.23 Part of the structure of ice. The oxygen atoms are situated tetrahedrally to one another (cf. diamond). As the dimensions indicate, the hydrogen atoms are not situated midway between the oxygen atoms.

The anomalous behaviour of water provides further evidence for the existence of hydrogen bonds. At 0 °C (273 K), melting occurs with a reduction in volume instead of the normal expansion. This is caused by the open structure of ice breaking down by the disruption of hydrogen bonds. The structure partially collapses, leading to a decrease in volume. In the liquid state this is opposed by the normal force of expansion. Both are encouraged by increasing temperature. Up to 4 °C (277 K) the former predominates, and this is therefore the position of maximum density or minimum volume.

Hydrogen bonding is common in the structures of both natural and man-made fibres and is the cause of much of the strength of these materials. In cellulose, for example, carbohydrate chains are held together by hydrogen bonding involving O—H bonds, while in wool and nylon the chains are hydrogen bonded through N—H bonds.

6.6.2 WATER OF CRYSTALLISATION

Many anhydrous ionic solids, on contact with water, form hydrates in which the water appears to be chemically bound to the cation or to both the cation and the anion. Hydration may be followed by dissolution in which again there is an interaction between water and the ions. There are three types of interaction which are all essentially electrostatic in nature but decrease in strength in the order given below.

Coordinate bonds

The polarisation of water molecules (Fig. 4.31) is increased by the presence of cations. The ability of an ion to do this increases with increasing charge but decreases with increasing size. Ions possessing a high surface charge density (§4.7.4), e.g. Fe^{2+}, Be^{2+} or Al^{3+}, distort the attracted water molecules to such an extent that coordinate bonds are formed through a lone pair of electrons on the oxygen atom,

i.e. $M^{n+} \leftarrow O^{\delta-}(H^{\delta+})_2$

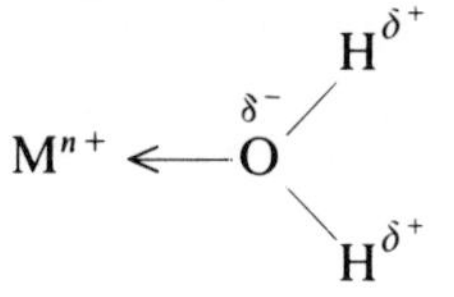

Hydrogen bonding

Many hydrated salts have a hydrogen bonded water molecule acting as a bridge between the cation and the anion. This explains why so many salts form monohydrates. Copper(II) sulphate pentahydrate is a common

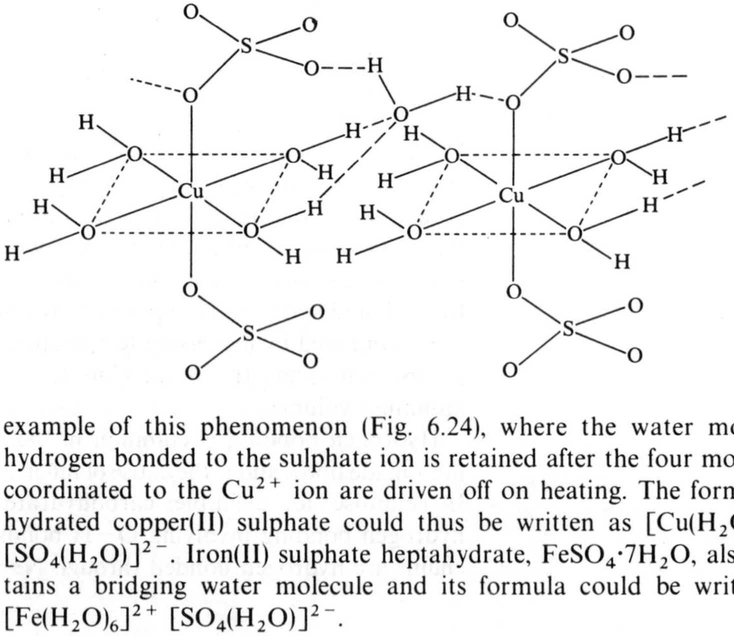

Fig. 6.24 The structure of copper(II) sulphate pentahydrate.

example of this phenomenon (Fig. 6.24), where the water molecule hydrogen bonded to the sulphate ion is retained after the four molecules coordinated to the Cu^{2+} ion are driven off on heating. The formula of hydrated copper(II) sulphate could thus be written as $[Cu(H_2O)_4]^{2+}$ $[SO_4(H_2O)]^{2-}$. Iron(II) sulphate heptahydrate, $FeSO_4{\cdot}7H_2O$, also contains a bridging water molecule and its formula could be written as $[Fe(H_2O)_6]^{2+}$ $[SO_4(H_2O)]^{2-}$.

Ion–dipole attraction

When sodium chloride is added to water the Na^+ ions and Cl^- ions attract and bind water molecules by an electrostatic mechanism (§4.9.1).

6.7
The colloidal state

6.7.1 INTRODUCTION

When a substance in one phase (or physical state) is dispersed in a substance in another phase, the product is generally a solution or a suspension. A solution is regarded as a one-phase system because the solute is molecular in size, while a suspension is a two-phase system because particles of the dispersed phase are microscopically visible. In between these extremes there is a range of materials which includes all living matter and most life-supporting materials. These materials are called *colloids* and are two-phase systems.

As with solutions, one phase is dispersed in another, the *dispersed phase* and *dispersion medium* respectively. As the particles of the dispersed phase are larger than those of the solute in a solution they cause the dispersion medium to appear cloudy. A rough approximation of particle size distribution is given below.

solute in solution	particle diameter ~ 1–1 000 nm
dispersed phase in dispersion medium	particle diameter ~ 1 000–100 000 nm
particles in suspension	particle diameter > 100 000 nm

The boundaries between the three sizes are not clear-cut. Colloids are often called 'colloidal suspensions' or 'colloidal solutions' but both names are inaccurate and misleading. The simple term 'colloid' will be used in this book.

In a colloid, both the dispersed phase and the dispersion medium can be gas, liquid or solid, but gas in gas *always* forms a true solution. Some examples are given in Table 6.2.

Table 6.2 Types of colloid systems

Dispersed phase	Dispersion medium	Type of colloid	Example
gas	gas	none possible	
	liquid	foam	whipped cream
	solid	solid foam (includes some minerals)	(meerschaum)
liquid	gas	fog	aerosols
	liquid	emulsion	mayonnaise
	solid	gel	gelatin
solid	gas	smoke	suspension of flour in air
	liquid	sol	polymer 'solutions'
	solid		alloys, some minerals

6.7.2 DIALYSIS

The original experiments in colloid science were performed by Thomas Graham and Michael Faraday in the 1850s and 1860s. Faraday was concerned with preparing sols and Graham with a technique called *dialysis.* Dialysis was used to separate colloid particles from those of salts in solution, such as sodium chloride. Graham designated the latter *crystalloids.* He achieved separation by using a membrane, such as Cellophane, whose pore size is small enough to retain colloid particles but large enough to allow the passage of salts. Nowadays, Visking tubing has proved to be an excellent dialysing medium (Fig. 6.25).

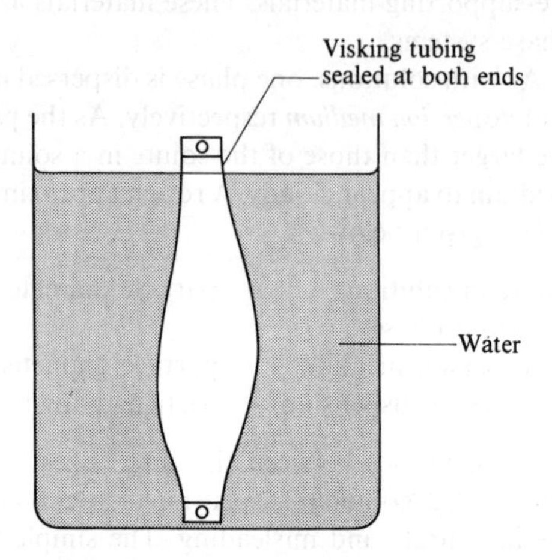

Fig. 6.25 A simple dialyser.

For example, if we put a mixture of starch and glucose in water in the tubing and place the whole in water the tubing acts as an intestine and allows the passage of glucose but not starch. In the body, glucose, but not starch unless it is digested, passes through the intestine wall.

The kidney machine is a most important form of dialyser. Tubing like Visking tubing is placed in a special salt solution. Blood is pumped through the tubing and waste water, salt and urea are dialysed out. Dialysis can be made more rapid by placing electrodes in the external solution to attract the dialysing ions (electrodialysis). The process of dialysis is essentially one of diffusion.

6.7.3 OSMOSIS

An important property of solutions, known as osmosis, also uses membranes. The membranes allow the passage of solvent, but not solute, molecules and are said to be *semi-permeable.*

When a semi-permeable membrane separates two solutions of different concentrations, solvent will flow from the one of lower concentration to that of higher concentration until an equilibrium is established, when there is no further flow of solvent. The process is *osmosis*, and leads to *osmotic equilibrium.* By using an apparatus of the type described in §8.4.2 the initial flow of solvent can be just prevented by the application of pressure on the solution or the reduction of pressure on the solvent. If the membrane separates a solution from its own solvent, the pressure is known as the *osmotic pressure* of the solution (§8.4.2).

In some cases, particularly in biological systems, the membranes are permeable to small solute molecules as well as solvent molecules. They are then known as 'leaky' membranes. In these circumstances the flow of solvent in one direction is partly reduced by that of the solute travelling in the reverse direction from the other side of the membrane. The pressure required to obtain zero volume flow is not now the true osmotic pressure.

6.7.4 PREPARATION OF COLLOIDS

Colloids are prepared by one of two general methods.

(i) *Condensation*, in which particles of molecular size are 'precipitated' out as particles of colloidal size. For example, colloidal sulphur is prepared by passing hydrogen sulphide through a dilute solution of sulphur dioxide:

$$2H_2S + SO_2 = 2H_2O + 3S$$

(ii) *Dispersion*, in which the particles are usually ground down to colloid size in a special mill known as a *colloid mill.*

6.7.5 LYOPHILIC AND LYOPHOBIC SOLS

Before considering the properties of colloids we must define the two types of colloid which exist. In order to simplify the discussion we shall consider sols only. In one case the solid has an attraction for the liquid and disperses spontaneously in it. In the other case the solid and liquid are not attracted to each other and the resultant sol resembles the pure liquid in many of its properties, e.g. viscosity and surface tension. The former are called *lyophilic* (solvent-liking) sols and the latter *lyophobic* (solvent-hating) sols. If water is the dispersion medium they become *hydrophilic* and *hydrophobic* sols respectively. Lyophobic sols may be prepared by either of the above methods, but lyophilic sols are usually prepared by a dispersion technique known as *peptisation.*

Peptisation is sometimes achieved by adding a small amount of a material (known as a peptising agent) to the dispersed phase in the dispersion medium so that, on stirring, the solid disperses evenly throughout the liquid. If for some reason the lyophilic colloid breaks down, i.e. the solid particles coalesce to form a suspension, the colloid can readily be reconstituted by stirring. This reversibility is characteristic of lyophilic colloids but does not occur in lyophobic colloids. (They are said to be irreversible.) Lyophilic colloids are usually colloids of high polymers in water.

6.7.6 PROPERTIES OF COLLOIDS

We need to consider Brownian motion and also the optical and electrical properties of colloids.

Brownian motion

Due to collisions between them, colloidal particles when viewed under a microscope can be seen to be in a constant rapid random motion, called *Brownian motion* after its discoverer Robert Brown. With larger particles the collisions of the liquid molecules on the solid cancel one another out, and the solid appears to be at rest.

Optical properties

If a thin beam of light is directed on to a true solution, it passes through and the solution appears clear because the solute and solvent particles are too small to reflect light. On the other hand, if a beam of light is passed through a colloid the liquid, when viewed at right angles to the beam, i.e. into the paper in Fig. 6.26, will appear as in the figure. This is called the *Tyndall effect*. The beam is visible because the light is scattered.

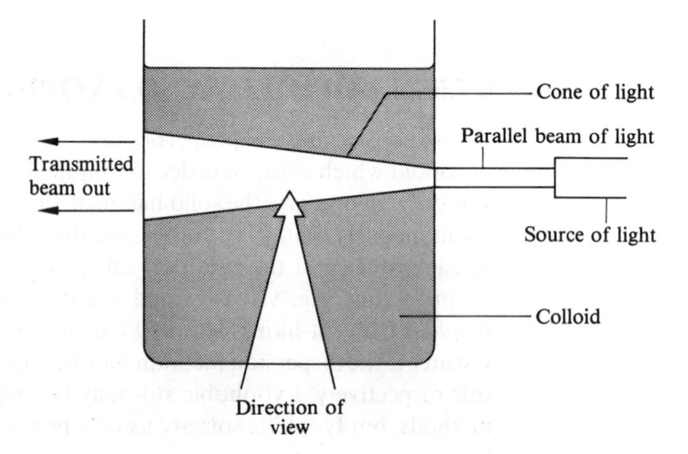

Fig. 6.26 The Tyndall effect.

A similar phenomenon occurs when light filters into a dark room through a hole in the curtains. If we look at it sideways we see a cone of light (*Tyndall cone*), caused by reflection from colloidal dust particles dispersed in the air.

Electrical properties

Particles of a sol are capable of attracting ions to form an adsorbed layer of ions round the particle (Fig. 6.27). These ions repel other particles of like charge and maintain the structure of the colloid. The sign of the charge on the colloid is dependent on a number of factors, including the presence of ions common to the colloidal particles and the dispersion medium. For example,

AgCl in a solution of chloride ions (say NaCl) attracts Cl^- ions, giving negative particles,
AgCl in a solution of silver ions (say $AgNO_3$) attracts Ag^+ ions, giving positive particles.

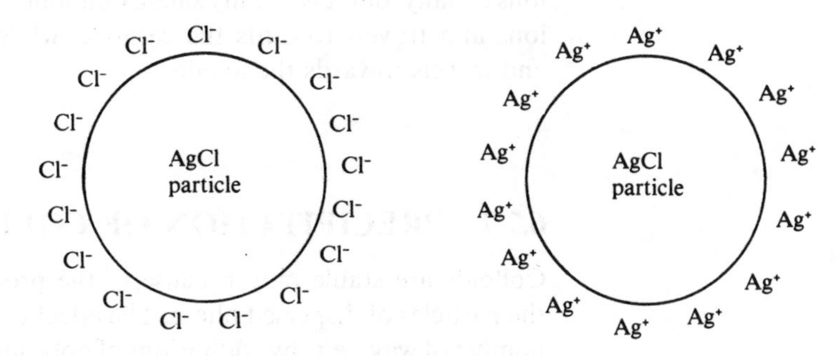

Fig. 6.27 The adsorption of ions by particles of a lyophobic sol.

Neutral sols can also occur. They are invariably lyophilic, while positive or negative particles can occur in both lyophobic and lyophilic systems.

Amino-acids and proteins can be negative, positive or neutral, depending on the pH of the colloid. In a solution containing excess H^+ ions the particles are positively charged:

$$NH_2\text{—}CHR\text{—}COOH + H^+ \rightleftharpoons \overset{\oplus}{N}H_3\text{—}CHR\text{—}COOH$$

In a solution containing excess HO^- ions the particles are negatively charged:

$$NH_2\text{—}CHR\text{—}COOH + HO^- \rightleftharpoons NH_2\text{—}CHR\text{—}COO^\ominus + H_2O$$

In a 'neutral' solution a zwitterion is formed,

$$\overset{\oplus}{N}H_3\text{—}CHR\text{—}COO^\ominus$$

The pH at which these reactions occur is dependent on the compound. When the amino-acid or protein is 'neutral' it is said to be at its *isoelectric point*. The pH of the isoelectric point varies for different amino-acids and proteins, e.g. at pH = 6.1 alanine is 'neutral', glutamic acid forms positive ions, and lysine forms negative ions.

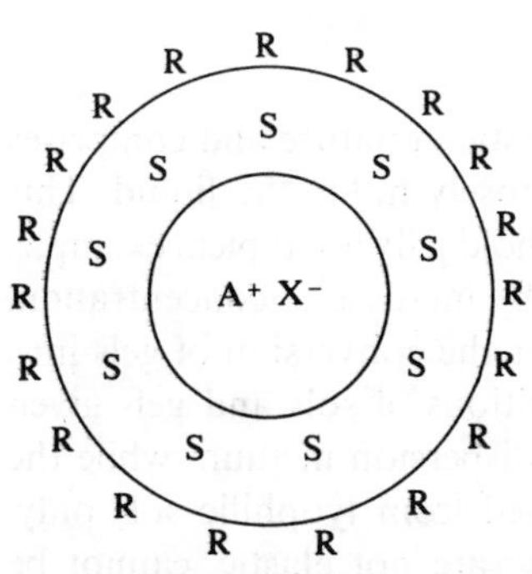

Fig. 6.28 The adsorption of ions by solvated particles of a lyophilic sol (R = adsorbed ion and S = solvent molecule).

Lyophilic sols are generally more stable than lyophobic ones, owing to the fact that lyophilic particles attract the dispersion medium. Solvated particles are formed (Fig. 6.28), and adsorption of ions can occur on top of this to give charged particles. For example, with proteins in an acidic solution with a pH less than the isoelectric point, H^+ ions surround the particles ($R = H^+$). At a pH greater than the isoelectric point HO^- ions surround the protein ($R = HO^-$).

The sign of the charge on colloid particles can be determined by *electrophoresis*. In this technique a U-tube is almost filled with the colloid, a platinum cathode is inserted in one limb and a platinum anode in the other. The particles are subjected via a d.c. power pack to a high voltage, and the charged particles travel towards the appropriate electrode. For example, in a red gold sol the particles of the dispersed phase move towards the anode, showing that they are negatively charged. In the case of alanine at pH = 6.1 there should be no migration. This is not strictly true. Migration can occur to some extent because of other factors which affect all ions equally, but it is usually small. Glutamic acid at this pH forms positive ions and travels towards the cathode, while lysine forms negative ions and travels towards the anode.

6.7.7 PRECIPITATION OF COLLOIDS

Colloids are stable only because of the protective layer of ions around the particles of dispersed phase. The effect of these can be neutralised in a number of ways, e.g. by adding ions of opposite charge or another dispersed phase of opposite charge. There is then no repulsive force and the colloids collapse. In the case of lyophilic colloids precipitation is sometimes carried out by *desolvation*. For example, a protein sol can be precipitated by adding ethanol, which has a greater attraction for the water than for the dispersed phase.

The precipitating effect of added ions increases dramatically with the charge on the ion,

e.g. $Al^{3+} > Mg^{2+} > Na^{+}$ in the ratio of 600 to 70 to 1.

This concept is expressed in the *Schulze–Hardy rule*, which states that the precipitation caused by an ion on a dispersed phase of opposite sign increases with the charge number of the ion.

In the case of lyophilic sols many more ions are required to collapse the sol than are needed for lyophobic sols because two layers often have to be removed, namely ions (R) and solvent (S) (Fig. 6.28). The solvent layer is removed as the added ions become solvated.

6.7.8 GELS

The form of gel most often encountered is elastic in nature and comprises an open network of solid particles which loosely holds the liquid. This gives some degree of rigidity. Ordinary household jelly is a typical example. Gels can be formed from sols by cooling or by increase of concentration, i.e. removal of solvent. The reverse is true for the conversion of gels into sols. These facts account for the loose definitions of sols and gels given in Table 6.2, i.e. that in gels the solid is the dispersion medium while the reverse is true of sols. Elastic gels are formed from lyophilic sols only. Lyophobic sols can also form gels but these are not elastic, cannot be used to regenerate the sol and become glassy or powdery on drying. The sols cannot be regenerated by either adding solvent or increasing temperature. A typical non-elastic gel is silica gel.

EXAMINATION QUESTIONS ON CHAPTER 6

1. The following table shows the variation of volume with pressure for 1.0 g of oxygen at 0 °C.

p/atm	V/litre	pV/litre atm
1.00	0.69982	0.6998
3.00	0.23281	0.6984
5.00	0.13942	0.6971
10.00	0.06937	0.6937

(a) (i) Plot a graph of pV against p.
(ii) On the same graph, indicate, with a dotted line, the relationship between pV and p if in the same experiment oxygen behaved as a perfect gas.
(b) Use the graph to calculate a value for the gas constant and give the appropriate units. (JMB)

2 (a) In what ways does the structure of a gas differ from the structure of a liquid?
(b) What is an *ideal gas*? Explain why real gases do not obey the equation $pV = nRT$.
(c) What is the difference between a gas and a vapour?
(d) Carbon dioxide has a *critical temperature* of 31 °C and a *critical pressure* of 7.3 MPa (73 atm).
Define the terms in italics. Sketch, and fully label, graphs of pressure/volume for carbon dioxide at 15 °C, 31 °C and 45 °C.
(e) Define the term *partial pressure*. 200 cm^3 of hydrogen at 100 kPa (1 atm) and 20 °C, and 150 cm^3 of helium at 200 kPa (2 atm) and 20 °C were mixed in a total volume of 500 cm^3. Calculate the partial pressure at 20 °C of each gas in the mixture. (AEB)

3 (a) Outline the essential assumptions underlying the simple kinetic theory of ideal gases. Explain briefly how the behaviour of real gases deviates from that postulated for ideal gases.
(b) By means of a suitable diagram, show how the particle energies are distributed in a gas. Indicate, on the same diagram, how this distribution changes with temperature.
Discuss the effect of a change in temperature on the rate of homogeneous gas-phase reactions. (L)

4 (a) (i) Explain the significance of the symbols in the equation $pV = nRT$ and state any **two** of the laws it summarises.
(ii) Calculate the value of R in appropriate units.
(iii) Why, and how, did van der Waals modify this equation?
(b) (i) State Graham's law of diffusion.
(ii) Write the molecular formulae of three hydrocarbons which diffuse at the same rate as nitrogen, nitrogen oxide (nitric oxide), dinitrogen oxide (nitrous oxide) respectively.
(iii) Calculate the ratio of the rates of diffusion of carbon monoxide and carbon dioxide under identical conditions.

[1 mol of a gas occupies 22.4×10^{-3} m^3 (22.4 l) at 101 kPa and 273 K (s.t.p.).] (AEB)

5 Discuss the assumptions on which the kinetic theory of ideal gases is based and state the equation which governs the behaviour of such gases. The van der Waals' equation for one mole of a gas,

$$(p + {}^{a}/_{V^2})(V - b) = RT \text{ (}a\text{ and }b\text{ are constants)}$$

is often used to describe the behaviour of real gases. Suggest reasons for the terms a/V^2 and b in this equation. Explain why, under the same conditions, different gases diffuse at different rates.

A wad saturated with aqueous ammonia is placed at one end of a narrow tube 1 m long and one saturated with concentrated hydrochloric acid at the other. Describe and explain what happens. (JMB)

6 Gases exert a pressure on a container, they may be compressed, they interdiffuse and they diffuse at different rates through a membrane. How may the kinetic theory of gases be used to explain these properties?

Explain, **concisely**, how real gases deviate in their behaviour from that expected on simple kinetic theory.

Under comparable conditions, 200 cm^3 of oxygen diffused through a membrane in 600 seconds and 60 cm^3 of an unknown gas diffused through the same membrane in 300 seconds. Find (a) the mean molecular mass of the gas, (b) the temperature at which the gas has the same root-mean-square velocity as oxygen at 273 K, both gases being at the same pressure. (WJEC)

7 (a) (i) What do you understand by the term *diffusion*?

(ii) Two gases can often be separated by diffusion. Explain why this is so, stating any relevant law. Why would it be easier to separate a mixture of argon and helium by diffusion than one of nitrogen and ethane?

(b) 0.254 g of iodine are introduced into an evacuated vessel of volume 100 cm^3. The vessel is closed and heated to 1473 K. It is found that the pressure of the iodine vapour in the vessel at this temperature is 1.71×10^5 N m^{-2}.

(i) Calculate the pressure expected under these conditions if it is assumed that the iodine is wholly in the form of I_2 molecules.

(ii) What qualitative inference can be drawn from the difference between your result in (i) and the observed pressure? Make any further deductions you can. (O)

8 (a) State Dalton's law of partial pressures.

If 200 cm^3 of hydrogen, 100 cm^3 of oxygen and 50 cm^3 of nitrogen each at 273 K and 101.3 kPa (1 atmosphere) were introduced into an evacuated vessel of 500 cm^3 capacity, calculate the pressure of the gas mixture (temperature remaining constant).

(b) (i) State Henry's law.

(ii) If this same gas mixture was then allowed to come into contact with a water surface, what would be the composition by volume of the dissolved gases at 273 K (given the absorption coefficients for H_2, N_2 and O_2 at s.t.p. are 0.01, 0.02, 0.04 respectively).

(iii) How would an increase in temperature affect the total volume of gas absorbed? How can the trend you suggest be explained?

(c)

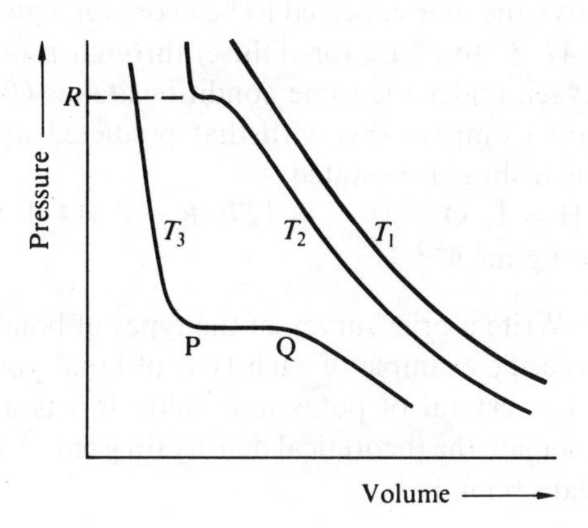

The graphs shown illustrate the behaviour of a *non-ideal* gas when the effect of pressure on volume is investigated at three different temperatures where $T_1 > T_2 > T_3$.

(i) Explain why the three curves differ.

(ii) Explain the significance of the *region* PQ and the *value R*. (AEB)

9 (a) Describe, with illustrations, the types of crystal lattice present in graphite, diamond and sodium chloride.

By reference to the type of bonding present, explain the relative hardness and electrical conductivity shown by these substances.

(b) Sodium chloride exhibits *6:6 coordination* whilst caesium chloride shows *8:8 coordination.* Explain what these italicised terms mean.

(c) When 10 g of metallic sodium reacts with excess water the energy evolved is 78.6 kJ. Similarly, 20 g of sodium oxide (Na_2O) reacts with excess water evolving 85.3 kJ. If the enthalpy change of formation of water is -284 kJ mol^{-1}, what is the enthalpy change of formation of sodium oxide? (AEB)

10 Discuss briefly, but critically, the basic postulates underlying the kinetic theory of gases and explain

(a) gas pressure,

(b) thermal expansion at constant pressure,

(c) the use of $\bar{c}^2$, the mean square velocity, in the equation

$$pV = \tfrac{1}{3}mn\bar{c}^2$$

where n molecules of ideal gas, each of mass m, occupy a volume V at a pressure p.

Calculate

(i) the kinetic energy (in joules) of the molecules in one mole of ideal gas at 47 °C,

(ii) the *root* mean square velocity of hydrogen iodide molecules, HI, at 47 °C in the gaseous phase; (in this calculation the molar mass of hydrogen iodide must be expressed in the appropriate SI unit, i.e. the *kilogram*),

(iii) the *ratio* of the *root* mean square velocities of oxygen, O_2, and hydrogen iodide at 47 °C,

(iv) the time expected to be taken for a given volume of hydrogen iodide at 47 °C to effuse (or diffuse) through a pin-hole if the same volume of oxygen under the same conditions takes 60 seconds.

(v) Compare this with that predicted by Graham's Law of diffusion, which should be stated.

H = 1, O = 16, I = 127; $R = 8.314$ J K^{-1} mol^{-1}; 0 °C = 273 K; J = kg m^2 s^{-2}. (SUJB)

11 Write a brief survey of the types of bonds which exist in solids, giving a specific example of each type of bond you mention.

The crystal of potassium chloride has a face-centred cubic structure. Calculate the theoretical density (in g cm^{-3}) of potassium chloride crystals. (Data book.) (OC)

12 (a) (i) Describe the preparation of a sol of iron(III) hydroxide from iron(III) chloride.

(ii) The iron(III) hydroxide sol is put in a cellophane bag which in turn is placed in a beaker of water. Chloride ions are soon detectable in the water but none of the red colloid appears on the water side of the cellophane. The process is known as dialysis. How do you account for the observed effects?

(b) Distinguish between the terms *lyophobic* and *lyophilic* as applied to colloids. Give **two** examples of each type.

(c) Name **two** properties which a sol of iron(III) hydroxide has in common with a gelatine sol which justify calling both of them colloids.

(d) Explain what is meant by *electrophoresis*.

Describe an experiment you would carry out with an iron(III) hydroxide sol to demonstrate *electrophoresis*. (AEB)

13 Explain the bonding present in the following solids:

(a) sodium chloride,

(b) phosphorus(V) chloride (phosphorus pentachloride),

(c) copper,

(d) ice,

(e) diamond.

In your answers relate the bonding described to the following properties of the substances concerned:

(i) melting and boiling points,

(ii) conduction of electricity,

(iii) action of water. (AEB)

14 (a) State what is understood by the term 'diffusion', and explain how Graham's law can be used to measure the relative values of the mean molecular velocities in two gases.

(b) Sketch a graph, with appropriately labelled axes, illustrating the distribution of molecular velocities in a gas. Describe how this distribution changes with an increase in temperature.

(c) Describe the differences between molecular motion in gases and liquids. Show how the distribution of molecular velocities in a liquid can be used to account for the observed variation of the vapour pressure of a liquid with temperature. (JMB)

15 (a) State Avogadro's law (hypothesis).
What is the significance of the Avogadro constant,
$L = 6.023 \times 10^{23}$ mol^{-1}?

(b) *The mole is the amount of substance of a system containing as many elementary units as there are carbon atoms in exactly 0.012 kg of carbon-12.* Amplify and explain this fundamental definition, giving examples, and mention especially the role of carbon-12.

(c) Explain how the designation $^{80}_{35}Br$ enables you to state the number of sub-atomic particles in one atom of bromine, Br, and with reference to the periodic table, its electronic configuration.

(d) Explain the importance of the Faraday constant, $F = 9.649 \times 10^4$ C mol^{-1} and calculate the charge on one of the sub-atomic particles mentioned in (c).

(e) Assuming ideal behaviour of the gases mentioned, calculate

(i) the number of molecules of hydrogen in a bulb of capacity 0.415 dm^3 evacuated to a pressure of 0.005 mmHg at 27 °C,

(ii) the kinetic energy (kJ) of the molecules in 1.50 mol of methane at 27 °C,

(iii) the ratio of the *root* mean square velocities of methane and hydrogen bromide at 27 °C.

H = 1, C = 12, Br = 80
$R = 8.314$ J mol^{-1} K^{-1}
1 mmHg = $13.60 \times 980.7 \times 10^{-2}$ N m^{-2}
J = kg m^2 s^{-2} = N m.

(f) Which of the gases hydrogen, methane and hydrogen bromide is the least ideal in behaviour under room conditions? Give *one* reason for your choice. (SUJB)

16 Explain the following in terms of the molecular kinetic theory.

(a) The space above a liquid always contains some of the vapour of that liquid.

(b) Saturated vapour pressure of a liquid.

(c) When an evaporating dish full of acetone is allowed to stand in the laboratory for a few days, the acetone will completely evaporate, even though the room temperature is far below the boiling point of acetone.

(d) It would be impossible to condense an ideal gas to a liquid, whatever temperature and pressure were employed.

10 moles of gas occupy 1 500 cm^3 at 130 atmospheres pressure and 27 °C. Calculate the pressure this gas would exert under the same conditions if it were an 'ideal gas'.

Comment on your results. (JMB)

7 Reaction Kinetics

7.1 Introduction

When we consider a chemical reaction we tend to think of how much product is obtained and what is required to make the reaction take place. However, there is another aspect of the study of chemical reactivity which is important, and that is the speed with which substances react with one another. The quantitative examination of the rates of chemical reactions is known as *reaction kinetics* or *chemical kinetics.*

To determine the rate at which a chemical reaction occurs we must be able to measure, directly or indirectly, the amount of reactant or product present in the reaction vessel at appropriate time intervals. This is best understood by looking at one or two reactions.

(i) $C_6H_5N_2^+(aq) + H_2O(l) = C_6H_5OH(aq) + N_2(g) + H^+(aq)$ (§14.11.8)

The most obvious change in this reaction is the formation of nitrogen gas which has a large volume, large enough to be readily measured by means of a gas syringe attached to the reaction vessel. This can be done at regular intervals without stopping the reaction. It is clear from the above equation that the rate of formation of nitrogen is equal to the rate of decomposition of the benzenediazonium salt.

(ii) $C_2H_5I(l) + HO^-(aq) = C_2H_5OH(aq) + I^-(aq)$ (§14.8.4)

In this case there is no apparent change in volume, but it is noticeable that the basicity of the mixture decreases because the hydroxide ions are used up in the reaction. In theory this change can be followed by using a pH meter with an electrode (§2.5.5) in the reaction vessel. Normally, however, an alternative method is adopted, in which small samples are removed at regular intervals and added in each case to a flask containing a large excess of cold water. The main reaction thus continues unchanged, since the proportion of reactants remains the same. The injection of the small sample into water causes dilution and cooling, which slows down the reaction sufficiently to allow the hydroxide ion concentration to be measured by titration with dilute hydrochloric acid. In this way the concentration of the hydroxide ion and hence of the iodoethane and the products of the reaction can be determined at chosen time intervals.

Various other properties of the materials of a chemical reaction can be used to measure reaction rates. However, many reactions are either too fast or too slow to be studied directly. For example, the reaction between equimolar amounts of aqueous sodium hydroxide and hydrochloric acid is very rapid and goes to completion. This is also true of many of the simple precipitation reactions in inorganic chemistry. Some fast reactions are violent or even explosive, although many explosive mixtures can remain inactive for long periods provided that they do not suffer sudden changes (e.g. a sudden increase in temperature) in their environment.

On the other hand, many reactions occur over years or even hundreds or thousands of years. For example, plutonium-239, which was at the centre of the Windscale controversy, has a half-life (§3.3.1) of 24 400 years. Many other radioactive elements, both natural and man-made, have very long half-lives. Many geological changes, such as weathering, are also very slow. These reactions are difficult to study kinetically. Fortunately, most reactions occur at a measurable rate, or can be made to do so by adding a catalyst (§7.5).

There are several factors that can be modified to speed up or slow down chemical reactions. The commonest are temperature, and the concentration of the reactants, but in the case of solid reactants the form and even the shape of the solid particles can be important. It is easily seen, for instance, that a large crystal tends to react more slowly than a fine powder of the same material (§16.2.2). As mentioned above, the presence of a catalyst may speed up or slow down a reaction. Natural processes can be affected by bacteria and by the presence of sunlight. For example, in the presence of light and chlorophyll, carbohydrates are formed from carbon dioxide and water, a process termed *photosynthesis*. Reactions in the body are controlled by the presence of protein catalysts called *enzymes*.

In the following sections we shall be investigating quantitative methods of studying reaction rates and the effect of the above parameters on these rates. We shall see that a knowledge of reaction kinetics provides an insight into the control of chemical reactions and information on their mechanisms.

7.2 Reaction order and molecularity

For simplicity we will consider gaseous reactions first. Suppose that gas A reacts with gas B to form a product AB:

$$A + B = AB$$

According to the kinetic theory of gases, if the concentration of A and/or B increases, the number and frequency of collisions between molecules of A and B are likely to increase proportionately. Therefore the rate of reaction (r) will be proportional to the molar concentrations of A and B ([A], [B]), and we can write a *rate expression* as follows:

$$r \propto [A][B]$$

If $A + 2B = \text{products}$

i.e. $A + B + B = \text{products}$

$$r \propto [A][B][B]$$

$$\propto [A][B]^2$$

More generally, if

$$mA + nB = \text{products}$$

$$r \propto [A]^m[B]^n$$

Let us consider two reactions:

$$H_2 + I_2 = 2HI \quad (1)$$

$$H_2 + Br_2 = 2HBr \quad (2)$$

The rates of reaction should be as follows:

$$r_1 \propto [H_2][I_2]$$

$$r_2 \propto [H_2][Br_2]$$

However, only one of these statements is true. The rate of the first reaction, as determined practically, is indeed proportional to the molar concentrations of hydrogen and iodine, but the rate of the second is *not* proportional to the concentrations of hydrogen and bromine. By experiment,

$$r_2 \propto \frac{[H_2][Br_2]^{\frac{1}{2}}}{1 + \text{constant} \times \dfrac{[HBr]}{[Br_2]}}$$

Again, in the reaction

$$4HBr + O_2 = 2H_2O + 2Br_2$$

r should be proportional to $[HBr]^4[O_2]$

but in practice $r \propto [HBr][O_2]$

It is clear that in the reactions between hydrogen and bromine and between hydrogen bromide and oxygen there are complicating features, and that these reactions are not accurately represented by the given equations. The values of m and n as 'determined' theoretically and practically are most important when considering what happens in a chemical reaction.

Let us once again consider the reaction

$$mA + nB = \text{products}$$

where, in theory, $r \propto [A]^m[B]^n$

The total number of atoms or molecules which take part in a reaction, as represented by the left-hand side of the equation, is known as the *molecularity* of the reaction. In this case,

$$\text{molecularity} = m + n$$

The sum of the powers to which the concentrations are raised in the rate expression is known as the *order of reaction*. If theory is confirmed by experiment,

$$\text{order} = m + n = \text{molecularity}$$

It may be that the result obtained practically is different from theory. Order and molecularity are thus not necessarily equal. Note that the molecularity is a theoretically contrived number, while the order is a practically determined value.

Let us once again consider the reaction between hydrogen bromide and oxygen. The molecularity of the reaction as presented is $4 + 1 = 5$, but the order of the reaction is $1 + 1 = 2$, since $r \propto [HBr][O_2]$. The reason for the discrepancy will be considered in more detail in the next section.

Molecularities of 1, 2 and 3 are termed *monomolecular*, *bimolecular* and *termolecular* respectively. The corresponding orders are first, second and third order respectively, and fractional orders are common.

Sometimes we relate orders not to the reaction itself, but to the individual reactants. For example, in the reaction between hydrogen and iodine, the reaction is said to be first order with respect to hydrogen and first order with respect to iodine, although the *overall* order of reaction is two.

7.3 Mechanisms of reaction

In the reaction

$$4HBr + O_2 = 2H_2O + 2Br_2$$

it is clear that the reaction does not occur as represented by the equation, since the molecularity and order are quite different. It has been suggested that the reaction takes place in three steps or stages, often known as *elementary processes*:

$$HBr + O_2 \rightarrow H{-}O{-}O{-}Br \qquad (1)$$

$$H{-}O{-}O{-}Br + HBr \rightarrow 2HBrO \qquad (2)$$

$$2HBrO + 2HBr \rightarrow 2Br_2 + 2H_2O \qquad (3)$$

The sum of these equations is the same as that given initially. Such a reaction is said to be a *complex reaction*.

Since the reaction clearly consists of more than one stage it would be expected that each stage would possess its own molecularity. It is important to look at this concept of molecularity more closely. The most fundamental definition of molecularity is as follows. 'In a chemical reaction the molecularity is equal to the number of particles involved in a single encounter leading to reaction.' On this basis the molecularity of

$$4HBr + O_2 = 2H_2O + 2Br_2$$

has no real meaning, but each stage has a definite molecularity, i.e. 2 in all the individual stages of the above reaction. **NB** In (3) the *collision* is represented by the simplest form of the equation,

$$HBrO + HBr \rightarrow Br_2 + H_2O$$

If we look at equation (1) we should obtain

$$r \propto [HBr][O_2]$$

which is the same as the rate expression obtained experimentally for the whole reaction. This suggests that stage (1) is the only one which affects the experimental determination. How can this be so?

Let us suppose that the formation of H—O—O—Br occurs at a rate which is measurable in the laboratory, but that the subsequent reactions are extremely fast. In other words the H—O—O—Br, as soon as it is formed, is converted to H_2O and Br_2. The rate of the whole reaction is then governed by that of the first step, which is called the *rate determining step*. The molecularity of this step is equal to the order of the whole reaction. It can be seen that the difference between the order and molecularity of this reaction has given us some insight into how the reaction occurs, i.e. into its so-called *mechanism*.

In defining molecularity (above) we have accepted that each stage of a reaction has its own molecularity. While this is true it has become the practice to define the molecularity of the reaction in terms of the rate determining step. Both definitions are acceptable, but it is important when dealing with molecularity to be sure which definition is used. We shall adopt the latter definition, but even on this basis molecularity and order can be different (see later).

The reaction between hydrogen peroxide and potassium iodide in dilute acid solution is represented by the equation

$$H_2O_2 + 2H^+ + 2I^- = 2H_2O + I_2$$

The molecularity of the reaction would appear to be 5 but the order of reaction is 2 and is given by

$$r \propto [H_2O_2][I^-]$$

Although several mechanisms have been postulated to explain this, the following is popularly accepted:

$$H_2O_2 + I^- \rightarrow H_2O + IO^- \qquad \text{rate determining step}$$
$$H^+ + IO^- \rightarrow HIO \qquad \text{fast step}$$
$$HIO + H^+ + I^- \rightarrow I_2 + H_2O \qquad \text{fast step}$$

Another interesting reaction is that between iodine and propanone in acidic solution. The overall reaction is given by:

$$CH_3COCH_3 + I_2 = CH_3COCH_2I + H^+ + I^-$$

Both the molecularity and order of reaction are 2, but

r is not proportional to $[CH_3COCH_3][I_2]$

Instead, $r \propto [CH_3COCH_3][H^+]$,

i.e. the rate is dependent on the concentration of the acid solution *in which the reaction occurs*, but not that of the iodine (§7.5.2).

Although the last two examples concern reactions in solution, the method of approach is essentially the same as for gases. When the solvent takes part in the reaction, this must be allowed for in the reaction mechanism.

In the iodination of propanone, the reaction is said to be first order with respect to propanone and with respect to the hydrogen ions. Since the iodine has no effect on the rate of the reaction, the reaction is considered to be *zero order* with respect to iodine. The reaction order must be related to particular reactants to avoid confusion. The statement that the reaction is second order is compatible with both

$$r \propto [CH_3COCH_3][I_2] \quad \text{and} \quad r \propto [CH_3COCH_3][H^+]$$

but the statement that the reaction is first order with respect to propanone and first order with respect to hydrogen ions leaves no doubt as to which substances affect the rate of reaction.

For any given *step* in the mechanism of a reaction, the rate of that step is directly proportional to the molar concentration of each reactant raised to the power of its coefficient in the equation of that step. This statement is known as the *kinetic law of mass action*. The rate of reaction as a whole is that of the rate determining step, as stated earlier.

E.g. for $H_2O_2 + 2H^+ + 2I^- = 2H_2O + I_2$

step 1 $H_2O_2 + I^- \rightarrow H_2O + IO^-$ slow step $r_1 \propto [H_2O_2][I^-]$

step 2 $H^+ + IO^- \rightarrow HIO$ fast step $r_2 \propto [H^+][IO^-]$

step 3 $HIO + H^+ + I^- \rightarrow H_2O + I_2$ fast step $r_3 \propto [HIO][H^+][I^-]$

$$r_{(\text{reaction as a whole})} \propto [H_2O_2][I^-]$$

It is not essential for the first stage to be the rate determining one. For example, in the reaction

$$5HBr + HBrO_3 = 3Br_2 + 3H_2O$$

$$r \propto [H^+]^2[Br^-][BrO_3^-]$$

which would suggest the occurrence of collisions between four ions. This is extremely unlikely and the following mechanism has been proposed. Two fast reactions occur with the formation of paired ions:

$$H^+ + Br^- \rightleftharpoons H^+Br^-$$
$$H^+ + BrO_3^- \rightleftharpoons H^+BrO_3^-$$

These are followed by a slow rate determining step:

$$H^+Br^- + H^+BrO_3^- \longrightarrow HBrO + H{-}O{-}O{-}Br$$

Two further fast reactions occur:

$$H{-}O{-}O{-}Br + HBr \longrightarrow 2HBrO$$

$$HBrO + HBr \longrightarrow Br_2 + H_2O$$

Therefore, in the rate determining step, a four ion collision is replaced by a considerably more likely collision between two paired ions.

Therefore $r \propto [H^+Br^-][H^+BrO_3^-]$
which in turn gives $r \propto [H^+]^2[Br^-][BrO_3^-]$

In some cases one of the reactants may be in considerable excess over the other reactant(s) such that any change in its concentration is negligible compared with the total amount present. It will not then appear in the rate equation. This type of reaction is common if the solvent is one of the reactants,

e.g. $RCOOR' + H_2O = RCOOH + R'OH$

$$r \propto [RCOOR']$$

In the last two cases the order of reaction and the molecularity of the rate determining step are no longer the same.

7.4

Rate equations for reactions

7.4.1 DERIVATION OF RATE EQUATIONS

The following equations are representative of the reactions as they occur, so that molecularity and order are numerically equal. The reactions are performed under isothermal conditions, i.e. conditions of constant temperature.

Let us consider the first order reaction

$$A \longrightarrow \text{products}$$

in which we have a moles of reactant to start with, x moles of which decompose after time t to give x moles of product.

The concentration of the reactant at any time t must therefore be $(a-x)$ mol dm^{-3} for a volume of 1 dm^3.

The reaction rate r can be expressed in terms of the calculus as

$$-\frac{d[A]}{dt} = \text{the rate of disappearance of a reactant}$$

$$= -\frac{d(a-x)}{dt} = \frac{dx}{dt}$$

or $$\frac{d[\text{products}]}{dt} = \text{the rate of appearance of products with time} = \frac{dx}{dt}$$

Since both values are equal to dx/dt,

then $\frac{dx}{dt} \propto [A]$ where [A] is the concentration of A at time t

$$\propto (a-x)$$

$$= k(a-x) \qquad (\text{i.e. rate} = k[A])$$

where k, the constant of proportionality, is known as the *rate constant* or sometimes as the *velocity constant*. Equations of the form *rate* $= k[A]$ are known as *rate laws*.

On integration, $$k = \frac{1}{t}\ln\frac{a}{(a-x)} = \frac{2.303}{t}\lg\frac{a}{(a-x)}$$

This is known as the *kinetic equation of a first order reaction.*

Since a and $(a-x)$ have the same units, $a/(a-x)$ has no dimensions and k has units of 1/s or s^{-1}.

For the second order reaction

$$2A \longrightarrow \text{products}$$

or $$A + B \longrightarrow \text{products}$$

(where A and B have the same initial concentration),

$$\frac{dx}{dt} = k(a-x)(a-x) = k(a-x)^2$$

where a, x, k and t have the same meaning as before.

Integration gives $$k = \frac{1}{t(a-x)} - \frac{1}{ta}$$

Both terms on the right-hand side of the equation have units of

$$\frac{1}{t \times \text{concentration}}, \text{ i.e. } \frac{1}{\text{s(mol dm}^{-3})}; \text{ therefore } k \text{ has units of mol}^{-1}\,\text{dm}^3\,\text{s}^{-1}.$$

The kinetic equations of first, second and third order reactions can be written as

first order: $\lg(a - x) = -\dfrac{k}{2.303}t + \lg a$

second order: $\dfrac{1}{(a - x)} = kt + \dfrac{1}{a}$

third order: $\dfrac{1}{(a - x)^2} = 2kt + \dfrac{1}{a^2}$ (k has units of $\text{mol}^{-2}\,\text{dm}^6\,\text{s}^{-1}$)

In each case a plot of the left-hand side of the above equations against t gives a straight line whose slope is related to k, the rate constant,

i.e. first order: $k = -2.303 \times \text{slope}$
second order: $k = \text{slope}$
third order: $k = 0.5 \times \text{slope}$

In the case of a zero order reaction the rate of reaction is independent of concentration,

i.e. $\dfrac{dx}{dt} = k = k[\text{A}]^0$ (since any value to the power 0 is 1)

A plot of x against t gives a straight line of slope k.

The above techniques are used to determine integral orders of reaction and the corresponding rate constants.

If, in any given reaction, x is plotted against t a curve is obtained. If the tangent to the curve is drawn at t (Fig. 7.1) for $t = 0$, $x = 0$, then the slope of the tangent $= dx/dt$ for concentration a. From determinations of dx/dt for various initial reactant concentrations we can obtain the corresponding order of reaction even if it is fractional.

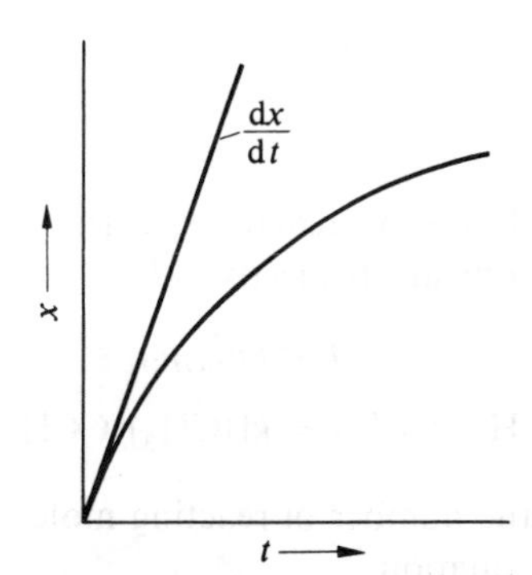

Fig. 7.1 Plot of number of moles of reactant consumed (x) against time (t).

There are many other ways of determining the order of reaction and the rate constant of a reaction, all of which demand a knowledge of a and of x at different times t. Therefore any experimental method used in studying chemical kinetics is designed for the measurement of a and x for a given reaction as is shown in example 4.

Example 1

The following information was obtained for the reaction

$3\text{A} + 2\text{B} + \text{C} = \text{D} + \text{E}$

[A]/mol dm^{-3}	[B]/mol dm^{-3}	[C]/mol dm^{-3}	Initial rate of formation of E/mol dm^{-3} s^{-1}	
0.50	0.50	0.50	2.0×10^{-5}	(1)
0.50	1.00	0.50	8.0×10^{-5}	(2)
1.00	1.00	0.50	8.0×10^{-5}	(3)
0.50	1.00	2.00	3.2×10^{-4}	(4)

What is the rate law for the reaction and what are the units of k?

Answer

rate of reaction $= k[A]^a[B]^b[C]^c$

If we take the first and second set of results we see that [A] and [C] are the same but the rate is increased four times from (1) to (2) as [B] is doubled.

$\therefore$ since $4 = 2^2$, rate $\propto [B]^2$

If we take the second and third set of results [B] and[C] are the same but [A] is changed with *no* change in the rate of reaction.

$\therefore$ rate $\propto [A]^0$, i.e. it is independent of [A]

If we take the second and fourth set of results [A] and[B] are the same but [C] is increased four times and the rate is increased four times from (2) to (4).

$\therefore$ rate $\propto$ [C]

$\therefore$ rate $= k[A]^0[B]^2[C] = k[B]^2[C]$

$\therefore$ the order of reaction is $0 + 2 + 1 = 3$

and the units of k are (mol dm^{-3})$^{-2}$ s^{-1}

Example 2

What is the molecularity and order of the following reactions, where r and k are the rate of reaction and the rate constant respectively?

(a) $2NO + 2H_2 = N_2 + 2H_2O$ $\qquad r = kp^2_{NO}p_{H_2}$

(b) $(CH_3)_3CCl + H_2O = (CH_3)_3COH + H^+ + Cl^-$ $r = k[(CH_3)_3CCl]$

Answer Since the molecularity is equal to the number of reacting molecules as given by the stoicheiometry of the equation,

(a) molecularity $= 2 + 2 = 4$

(b) molecularity $= 1 + 1 = 2$

Since the order of reaction is equal to the sum of the powers to which the concentrations of reactants are raised in the rate law,

(a) $\qquad r = kp^2_{NO}p^1_{H_2}$,

and since p can be considered as proportional to molar concentration,

order $= 2 + 1 = 3$

(b) $r = k[(CH_3)_3CCl]^1$

$\therefore$ order = 1

Example 3
Suggest a mechanism for the reaction given in 2(a).
Answer Since $r = kp_{NO}^2 p_{H_2}$ the rate determining step is theoretically

$$2NO + H_2 \longrightarrow \text{products}$$

This is most likely to occur as follows:

$$2NO + H_2 \longrightarrow N_2O + H_2O \quad \text{slow}$$

(The second molecule of H_2 then reduces N_2O to N_2 (fast).)
However, three body collisions are rare, and the following alternative mechanism is favoured:

$$2NO \longrightarrow N_2O_2 \quad \text{fast}$$
$$N_2O_2 + H_2 \longrightarrow N_2O + H_2O \quad \text{slow}$$

The rate law for the latter stage is

$$r \propto [N_2O_2][H_2]$$

but N_2O_2 is rapidly formed from two molcules of NO,

$$\therefore \quad [N_2O_2] \propto [NO]^2$$

$$\therefore \quad r = k[NO]^2[H_2]$$

NB The nature of the intermediates and final products cannot be deduced from the rate law. They are normally deduced from experiment. Indeed, a series of substances may be possible in which case the correct ones will have to be determined from other evidence.

Example 4
The rate of a chemical reaction is determined by showing how the amount of a given reactant or product varies with time. Suggest how the following reactions may be followed.

(a) $CH_3COOCH_3 + H_2O = CH_3COOH + CH_3OH$

(b) $4Fe^{2+} + O_2 + 4H^+(aq) = 4Fe^{3+} + 2H_2O$

(c) $N_2(g) + 3H_2(g) = 2NH_3(g)$

Answer
(a) In the first reaction ethanoic acid is formed and the amount increases with time. Its concentration at given times can be measured by titrating portions of the reaction mixture with sodium hydroxide.
(b) A thiocyanate (e.g. NH_4SCN) can be added to produce a deep red coloured complex with iron(III) ions. The increase in intensity of colour

and hence in the concentration of iron(III) ions can be used to follow the reaction.

(c) If the reaction is performed in a given vessel, i.e. at constant volume, there is a decrease in the number of moles from four to two with a corresponding decrease in pressure. From this decrease in pressure, measured with a manometer, the change in concentration of the reactants and products with time can be measured.

Example 5

In a given reaction the concentration of the reactant varied with time as follows:

time/s	0	600	1 200	1 800	2 400	3 000
concn/mol dm^{-3}	1.40	1.33	1.25	1.19	1.13	1.07

Show that the reaction is first order, and determine the rate constant.

Answer The equation for a first order reaction is

$$\lg (a - x) = -\frac{k}{2.303}t + \lg a$$

Since the concentration is that of the reactant, then

t	0	600	1 200	1 800	2 400	3 000
$(a-x)$	1.40	1.33	1.25	1.19	1.13	1.07
$\lg (a-x)$	0.146 1	0.123 9	0.096 9	0.075 5	0.053 1	0.029 4

Fig. 7.2 Plot of lg $(a - x)$ against t.

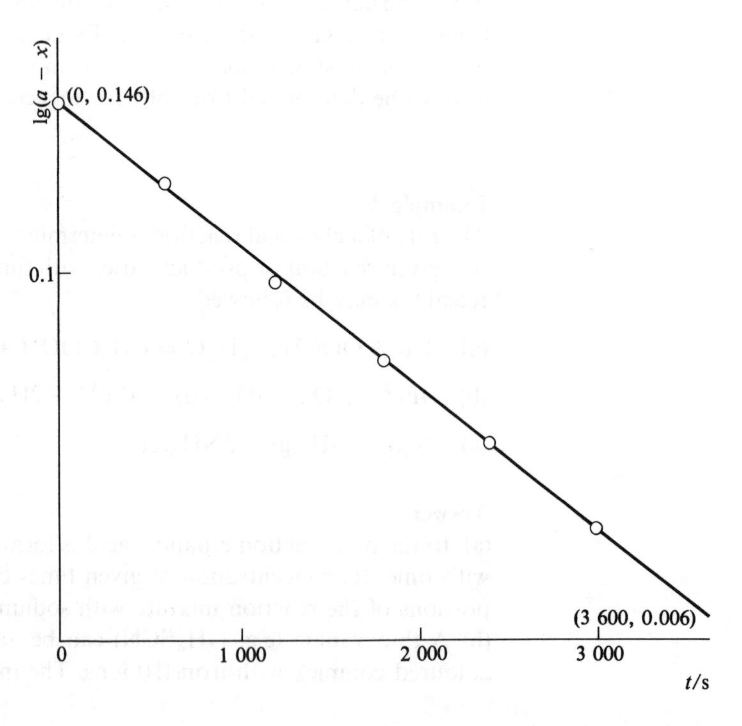

A straight line (Fig. 7.2) is obtained when lg $(a - x)$ is plotted against t; therefore the reaction is first order.

$$\text{The slope} = -\frac{k}{2.303}$$

$$= -\frac{(0.146 - 0.006)}{3\,600 - 0} = -\frac{0.140}{3\,600}$$

$$\therefore \quad k = 3.88 \times 10^{-5} \times 2.303 \text{ s}^{-1}$$

$$= 8.94 \times 10^{-5} \text{ s}^{-1}$$

7.4.2 HALF-LIFE PERIODS

In determining reaction orders, and also in studying radioactive decay (§3.3.1), the time taken for a reaction to reach the half-way stage, i.e. the time taken for the reactant concentrations to become half of their initial values, is important. This is called the *half-life* ($t_{\frac{1}{2}}$) of the reaction, and for reactions involving more than one reactant can only be obtained if the initial concentrations of all the reactants are the same.

$$t = t_{\frac{1}{2}} \quad \text{when } x = \frac{a}{2}$$

$$\therefore \text{ for a first order reaction} \qquad t_{\frac{1}{2}} = \frac{1}{k} \ln \frac{a}{\left(a - \frac{a}{2}\right)} = \frac{1}{k} \ln 2$$

NB The half-life is independent of the value of a, the initial concentration of the reactant.

Equations for half-lives for higher orders are determined in a similar manner.

7.4.3 THE ARRHENIUS EQUATION

S. A. Arrhenius found experimentally that k varies with temperature according to the equation

$$k = Ae^{-E/RT}$$

where E is an energy term and A is a constant. This equation is known as the *Arrhenius equation.*

7.4.4 ACTIVATION ENERGY

Chemical kinetics are best explained by the transition state theory. Let us consider a reaction

$$A + BC = AB + C$$

As A approaches BC along the line of centres of BC a partial bond is formed between A and B with a corresponding weakening of the bond between B and C to form a *transition state complex* (or *activated complex*):

A····B····C

Its existence is only momentary, after which it goes to form AB + C *or* reform A + BC.

As A approaches BC a given minimum amount of energy, called the *activation energy*, E_1, is necessary to overcome their mutual repulsion. This energy is afforded by the kinetic energy of the molecules. If it is equal to or greater than E_1, the kinetic energy is transformed to potential energy for the system with the corresponding formation of the activated complex (Fig. 7.3).

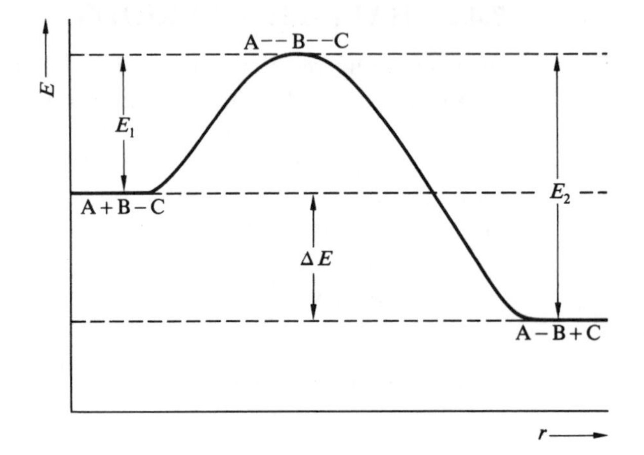

Fig. 7.3 A plot of distance between B and C along the line of centres (r) against potential energy (E).

The activated complex decomposes immediately to form either AB + C or A + BC. Since the conversion of the complex is accompanied by a greater loss of potential energy when going to products than when reverting to reactants the tendency is to form products. If E_2 is the activation energy for the reverse reaction then the energy change in the reaction is

$$\Delta E = E_2 - E_1$$

Figure 7.4 is a modified form of that given in Fig. 6.2, with energy replacing velocity. The proportion of molecules having an energy greater than a given energy E_1 is shown by the shaded portion on the graph. If E_1 represents the energy of activation the figure shows that the proportion of molecules at a temperature T_2 with energies higher than E_1 is considerably greater than that for T_1 (as shown by the shaded area). Therefore the rate of reaction is greater at T_2 than at T_1, i.e. reaction rate increases with temperature.

In the Arrhenius equation E is the activation energy of the forward reaction. A is related to the structure of the molecules and to the way in which they approach each other. It is often called the *steric factor*. If we look again at the Arrhenius equation we see that it can be put in a straight line form assuming A is a constant (i.e. independent of temperature, which, although not strictly correct, is true over reasonable ranges of temperature).

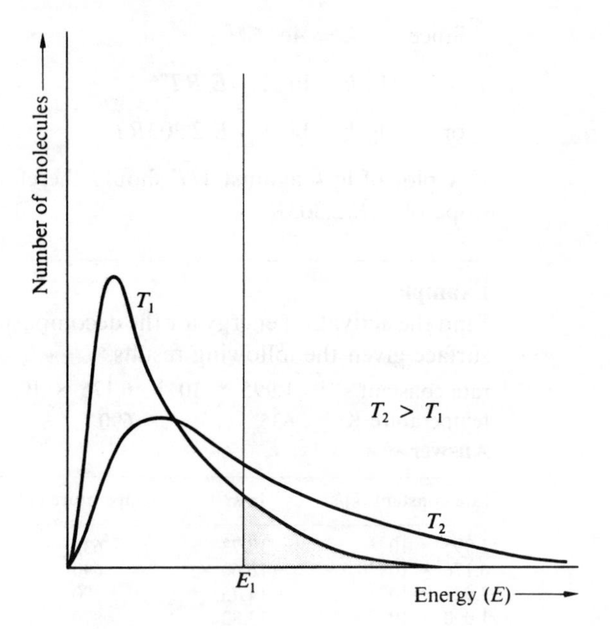

Fig. 7.4 Distribution curve of molecules with energy greater than E.

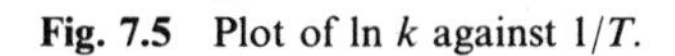
Fig. 7.5 Plot of ln k against $1/T$.

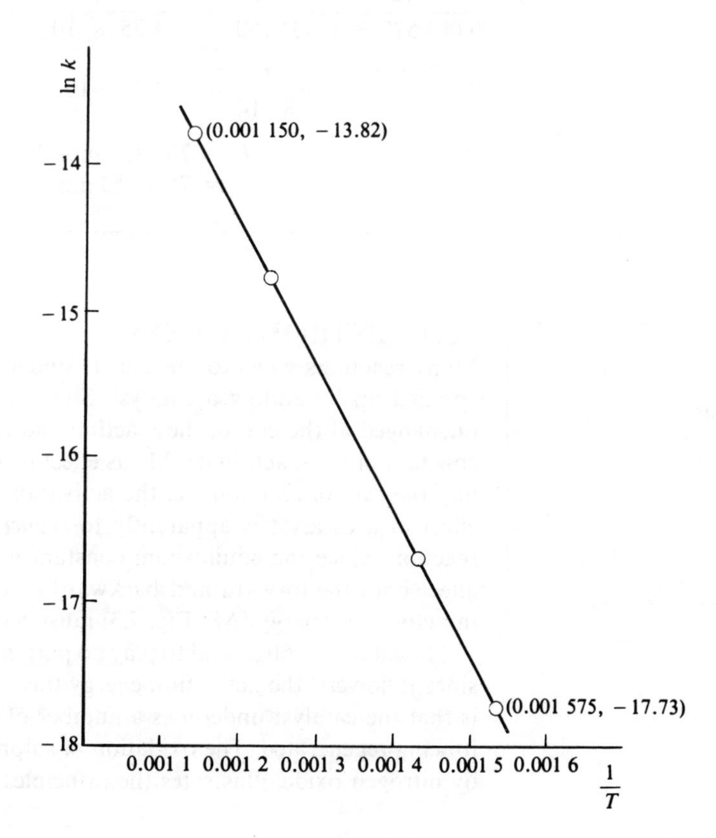

Since $k = Ae^{-E/RT}$

$$\ln k = \ln A - E/RT$$

or $$\lg k = \lg A - E/2.303RT$$

A plot of lg k against $1/T$ should therefore give a straight line with a slope of $-E/2.303R$.

Example

Find the activation energy for the decomposition of phosphine on a quartz surface given the following results.

rate constant/s^{-1}	1.995×10^{-8}	6.176×10^{-8}	3.927×10^{-7}	1.000×10^{-6}
temperature/K	635	690	800	870

Answer

rate constant (k)/s^{-1}	ln k	temperature (T)/K	$1/T/K^{-1}$
1.995×10^{-8}	−17.73	635	0.001 575
6.176×10^{-8}	−16.60	690	0.001 450
3.927×10^{-7}	−14.75	800	0.001 250
1.000×10^{-6}	−13.82	870	0.001 150

The slope of a plot of ln k against $1/T$ (Fig. 7.5) using the points (0.001 150, −13.82) and (0.001 575, −17.73) gives a value of

$$\frac{-17.73 - (-13.82)}{0.001\,575 - 0.001\,150} = -\frac{3.91}{4.25 \times 10^{-4}} = -9.2 \times 10^3$$

$$\therefore \quad -\frac{E}{8.314} = -9.2 \times 10^3$$

$$\therefore \quad E = 76\,490 \text{ J mol}^{-1} = 76.49 \text{ kJ mol}^{-1}$$

7.5 Catalysis

7.5.1 INTRODUCTION

Many reactions which occur slowly under normal circumstances can be speeded up by adding a catalyst. Since the catalyst remains chemically unchanged at the end of the reaction and does not affect the equilibrium constant of the reaction (§8.2.1) its effect is restricted to the factor controlling the rate of reaction, i.e. the activation energy. Therefore the overall effect of a catalyst is apparently to *reduce* the activation energy of the reaction. Since the equilibrium constant is not altered the catalyst must affect both the forward and backward reactions equally, and the *change* in activation energy (ΔE, Fig. 7.3) must remain unaltered.

The catalyst is often said to play no part in the overall reaction. However, since it 'lowers' the activation energy this must be untrue. What happens is that the catalyst undergoes a number of changes leading eventually to it being regenerated. The oxidation of sulphur dioxide, which is catalysed by nitrogen oxide, illustrates the principles of catalysis clearly.

The uncatalysed reaction is

$$2SO_2 + O_2 = 2SO_3$$

In the presence of nitrogen oxide two reactions occur:

(i) the oxidation of NO to NO_2 $\quad 2NO + O_2 \longrightarrow 2NO_2$
(ii) the reduction of NO_2 to NO $\quad 2NO_2 + 2SO_2 \longrightarrow 2SO_3 + 2NO$

The catalyst is chemically unchanged at the *end* of the reaction by the simple means of oxidation followed by reduction, i.e. two reactions have replaced one. In this case one reaction of high activation energy has been replaced by two of lower activation energy. Since the overall change is the same, ΔE is the same for the catalysed and the uncatalysed reaction (Fig. 7.6).

Catalysts usually need only be present in small amounts to ensure reaction, which is to be expected since the catalyst is readily and rapidly regenerated as reaction proceeds.

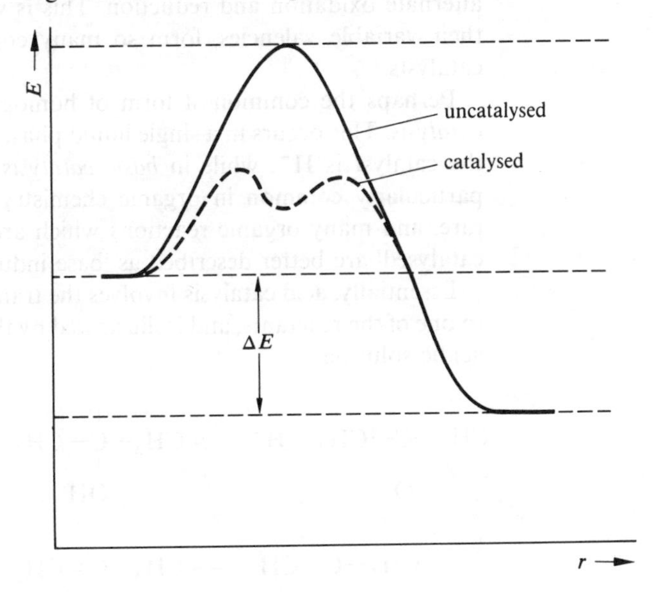

Fig. 7.6 Activation energy diagram for catalysed reactions.

It should be noted that the process of regenerating a catalyst will comprise *two or more* processes (§7.5.2).

Catalysts not only increase the rate of reaction. Some catalysts are used to slow down fast reactions deliberately, in which case they are known as *retarders.* In certain catalysed reactions additional catalysts can affect the rate adversely and are known as *poisons.* In the latter case their inclusion is not deliberate and is unwelcome. Retarders and poisons are generally known as *negative catalysts,* those which speed up reactions being termed *positive catalysts.* There are three broad classes of catalysis: homogeneous, heterogeneous and enzyme catalysis.

7.5.2 HOMOGENEOUS CATALYSIS

In *homogeneous catalysis* the reactants and the catalyst are in the same phase. In the above example they are all in the gaseous phase, whereas in the next example they are in solution. The reaction

$$S_2O_8^{2-} + 2I^- = 2SO_4^{2-} + I_2$$

is slow but may be catalysed by the presence of iron(II) ions.

(i) $S_2O_8^{2-} + 2Fe^{2+} \longrightarrow 2SO_4^{2-} + 2Fe^{3+}$
oxidation of Fe^{2+} to Fe^{3+}

(ii) $2Fe^{3+} + 2I^- \longrightarrow 2Fe^{2+} + I_2$
reduction of Fe^{3+} to Fe^{2+}

One explanation which has been proposed for the high activation energy of the uncatalysed reaction is the strong repulsion which will occur between the negative ions. This is not the case in reactions (i) and (ii).

It can be seen from the above example that the catalyst undergoes alternate oxidation and reduction. This is why transition elements, with their variable valencies, form so many compounds which are used as catalysts.

Perhaps the commonest form of homogeneous catalysis is *acid–base catalysis*. This occurs in a single liquid phase, i.e. solution. In *acid catalysis* the catalyst is H^+, while in *basic catalysis* it is HO^-. Acid catalysis is particularly common in organic chemistry. Basic catalysis is relatively rare, and many organic reactions which are commonly said to be 'base catalysed' are better described as 'base induced'.

Essentially, acid catalysis involves the transfer of a proton from the acid to one of the reactants, and is illustrated by the iodination of propanone in acidic solution.

$$CH_3{-}C(=O){-}CH_3 + H^+ \longrightarrow CH_3{-}\overset{\oplus}{C}(OH){-}CH_3 \qquad \text{slow}$$

$$CH_3{-}\overset{\oplus}{C}(OH){-}CH_3 \longrightarrow CH_3{-}C(OH){=}CH_2 + H^+ \qquad \text{fast}$$

$$CH_3{-}C(OH){=}CH_2 + I_2 \longrightarrow CH_3{-}C(I)(OH){-}CH_2I \qquad \text{fast}$$

$$CH_3{-}C(I)(OH){-}CH_2I \longrightarrow CH_3{-}C(=O){-}CH_2I + H^+ + I^- \qquad \text{fast}$$

The first step is the rate determining one, and the overall process is independent of the amount of iodine present.

In this, as in other examples of acid–base catalysis, there may be more than one intermediate stage in the reaction process, but if we consider the overall equation we see that the catalyst appears on both the left- and right-hand sides:

$$CH_3COCH_3 + H^+ + I_2 = CH_3COCH_2I + H^+ + H^+ + I^-$$

Propanone can also be iodinated under basic conditions (§14.7.5).

An example of basic catalysis is the aldol addition (§14.10.6).

7.5.3 HETEROGENEOUS CATALYSIS

Catalysis does not occur only when the catalyst is in the same phase as the reactants. When it is in a different phase from the reactants *heterogeneous catalysis* occurs. The catalyst is usually in the solid phase and the reactants in the gaseous or liquid phase.

A reaction between gaseous substances at a solid catalyst interface is prefaced by adsorption of the gases on to the solid in such a way that reaction is encouraged.

NB Substances which adhere to the surface of a solid are said to be *ad*sorbed by the solid. When these substances penetrate inside the solid they are *ab*sorbed by the solid.

There are two extreme types of adsorption, with all intermediate types capable of occurrence. In *physical adsorption* the reactants are bound to the so-called *active sites* on the surface by weak van der Waals' forces.

In *chemical adsorption* (chemisorption) the reactants are bound to the surface by bonds of the same order of strength as chemical bonds. The active sites on the surface are formed by excess attractive forces along crystal edges and grain boundaries of the solid. The more finely divided a given mass of solid, the more active sites it contains.

Reaction occurs in the following stages:

(i) diffusion of the reactants to the solid catalyst;
(ii) adsorption on the catalyst surface;
(iii) reaction between suitably arranged molecules;
(iv) desorption (reverse of adsorption) from the catalyst;
(v) diffusion of products away from the surface.

Since the positions of the active centres are determined by the crystal structure of the catalyst and are related to its chemical nature, catalysts are specific for given reactions. Finely divided nickel is used, for example, for dehydrogenation and alumina for dehydration, even of the same material.

(a)

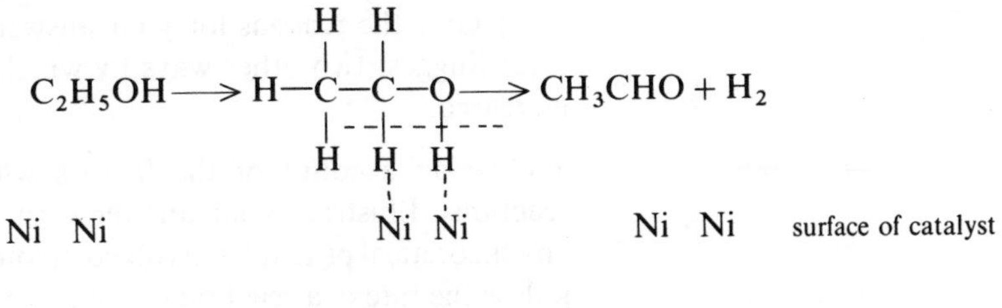

```
                      H  H
                      |  |
C2H5OH ——> H—C—C—O ——> CH3CHO + H2
                      | --|--+---
                      H  H  H
                          :  :
Ni  Ni                   Ni Ni              Ni  Ni   surface of catalyst
```

(b)

```
                H   H
                |   |
C2H5OH ——> H—C—C—H ——> CH2=CH2 + H2O
              --|---|----
                H   O—H
                :   :
Al2O3  Al2O3  Al2O3 Al2O3        Al2O3   Al2O3   surface of catalyst
```

Reaction (b) occurs above 300 °C (573 K). Just below 300 °C (573 K) $C_2H_5OC_2H_5$ is formed on Al_2O_3. In the latter case H_2O is removed from adjacent molecules by a mechanism similar to the above:

```
                                  H
                                  |
                H   H          H—C—H
                |   |             |
2C2H5OH ——> H—C—C—O   H—C—H ——> C2H5OC2H5 + H2O
                |   |   -|-------|---
                H   H   H        O—H
                        :        :
Al2O3  Al2O3            Al2O3    Al2O3       Al2O3  Al2O3
                                                surface of catalyst
```

In both cases chemisorption is involved. Many industrial processes require solid catalysts for reaction to occur at an acceptable rate.

EXAMINATION QUESTIONS ON CHAPTER 7

1 The overall reaction $2A + B \rightarrow C + D$ is found to occur by two elementary steps, via an intermediate species X.

Thus $A + B \longrightarrow X$ (1)

$X + A \longrightarrow C + D$ (2)

The rate of the overall reaction is given by

rate $= k[A]^2[B]$

(a) (i) Define molecularity.

(ii) What is the molecularity of step (1)?

(iii) What is the molecularity of step (2)?

(b) (i) Define order of reaction.

(ii) What is the overall order of the reaction?

(c) If the concentration of only one of the reactants can be doubled, which would give the greater increase in the overall rate?

(d) Give the reasons for your answer to (c).

(e) Suggest two other ways by which the rate of the reaction might be increased. (JMB)

2 Give an account of the factors which affect the rates of chemical reactions. Illustrate your answer with a range of examples, and outline any theoretical principles involved. In outline only, indicate how you would follow the rate of a reaction of your own choice and deduce its order. (L)

3 When ethanal is heated to about 800 K, it decomposes by a *second order reaction* according to the equation

$$CH_3CHO(g) \longrightarrow CH_4(g) + CO(g)$$

If, however, the ethanal is heated to the same temperature with iodine vapour, the iodine *catalyses* the reaction; this latter reaction is *first order with respect to both iodine and ethanal* and has a lower *activation energy* than the reaction with ethanal alone.

(a) Explain the meaning of the terms in italics.

(b) Suggest a set of experiments by which you could confirm that the uncatalysed reaction is second order.

(c) Indicate briefly what experiments would be necessary to show that the reaction with iodine present has a lower activation energy than the decomposition of ethanal by itself. (L)

4 The reaction between iodine and propanone (acetone) in aqueous solution,

$$CH_3COCH_3 + I_2 = CH_2ICOCH_3 + HI,$$

is catalysed by hydrogen ions.

(a) Describe briefly the experiments you would perform to determine the order of the reaction with respect to iodine, to propanone and to hydrogen ions.

(b) It is found that the rate of this reaction is proportional to the first power of the propanone concentration and to the first power of the hydrogen ion concentration, but is independent of the iodine concentration. What can you deduce from this?

(c) Explain why the usual effect of an increase of temperature on a reaction is to increase the rate at which it takes place. (O)

5 Explain what you understand by the terms *rate equation, rate constant* and *order of reaction.*

Give one possible reason why the rate equation cannot be deduced from the overall stoicheiometric equation.

For a reaction of your own choice studied during the course, describe how you determined the rate of a reaction in the laboratory and explain how the rate constant was obtained from your observations.

Using two reaction vessels of differing surface material, the reaction between A and B in the gaseous phase was investigated. The initial rate refers to the rate of removal of A (mol m^{-3} s^{-1}) in the experimental data set out below. Deduce the rate equation in each case and assuming that no errors have occurred, offer an explanation for what is found.

	$[A]$/mol m^{-3}	$[B]$/mol m^{-3}	Relative rate initially
Experiment 1	0.15	0.15	1
	0.30	0.15	4
	0.15	0.30	2
Experiment 2	0.15	0.15	1
	0.30	0.30	4
	0.60	0.30	16

(SUJB)

6 Describe **four** features which are characteristic of catalytic activity.

Comment on the following observations:

(a) There is a marked increase with time in the rate of oxidation of copper by concentrated nitric acid. However, there is no reaction at all in the presence of excess carbamide, $(NH_2)_2CO$.

(b) A mixture of hydrogen and chlorine explodes when exposed to bright sunlight but mixtures of hydrogen and bromine and of hydrogen and iodine do not. (Chapter 17.)

(c) The rate constant for the hydrolysis of methyl methanoate (methyl formate) in aqueous solution appears to increase with time, whereas it remains constant, but is larger, in the presence of an excess of hydrochloric acid. (OC)

7 The reaction $2NO(g) + O_2(g) \rightarrow 2NO_2(g)$ at normal pressures was explained by suggesting that it occurred in two fundamental steps as follows:

(i) $2NO(g) \rightleftharpoons N_2O_2(g)$

(ii) $N_2O_2(g) + O_2(g) \longrightarrow 2NO_2(g)$

The rate of the reaction is described by the relationship:

Rate $= k[NO]^2[O_2]$.

(a) What is the molecularity of the forward reaction in step (i)?

(b) Explain your answer to (a).

(c) What is the molecularity of step (ii)?

(d) What is the overall order of the reaction?

(e) Explain your answer to (d).

(f) If the pressure of only one of the reactants is to be doubled, which will cause the greater increase in rate?

(g) Suggest how the rate of the reaction may be increased without changing the pressure or the temperature.

(h) Explain briefly how the suggestion in (g) causes an increase in rate. (JMB)

8 (a) Define the term *order of a reaction.*

(b) In an experimental study of a reaction in solution between two compounds A and B, the following information was obtained for the *initial rate* of the reaction:

Initial rate in mol dm^{-3} s^{-1}	Initial concentrations of A and B in mol dm^{-3}	
	A	B
1.0×10^{-4}	1.0×10^{-1}	1.2×10^{-1}
4.0×10^{-4}	2.0×10^{-1}	1.2×10^{-1}
8.0×10^{-4}	2.0×10^{-1}	2.4×10^{-1}

What is the order of the reaction with respect to (i) the reactant A, (ii) the reactant B?

(c) The decomposition of benzene diazonium chloride in aqueous solution is a reaction of the first order which proceeds according to the equation

$C_6H_5N_2Cl = C_6H_5Cl + N_2(g)$

A certain solution of benzene diazonium chloride contains initially an amount of this compound which gives 80 cm^3 of nitrogen on complete decomposition. It is found that, at 30 °C, 40 cm^3 of nitrogen are evolved in 40 minutes. How long after the start of the decomposition will 70 cm^3 of nitrogen have been evolved? (All volumes of nitrogen refer to the same temperature and pressure.) (O)

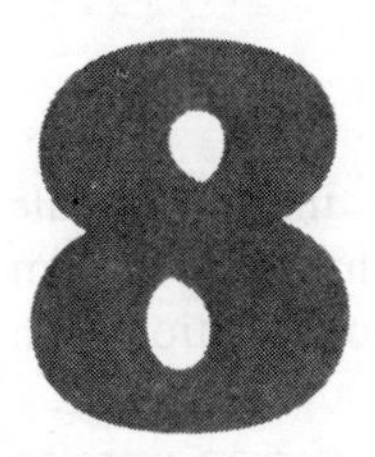

Equilibrium

8.1 Reversible reactions and dynamic equilibrium

Imagine a reaction between two reactants A and B to form two products C and D,

$$A + B \longrightarrow C + D$$ *forward reaction*

Let us now consider what would happen if the products combined to re-form A and B,

i.e. $$C + D \longrightarrow A + B$$ *reverse reaction*

There are now two opposing reactions which proceed simultaneously and the equation is normally written thus:

$$A + B \rightleftharpoons C + D$$

Such reactions, in which both forward and reverse changes occur, are called *reversible reactions*. All chemical reactions are in fact reversible. Many reactions, however, are said to 'go to completion' because the reverse reaction does not take place to any noticeable extent. When magnesium burns in oxygen, for example, the product is entirely magnesium oxide: no metallic magnesium is detectable because the magnesium oxide formed in the forward reaction does not appreciably decompose back into its elements. For reactions that go to completion we use the $=$ sign instead of the $\rightleftharpoons$ sign.

The answer to the inevitable question as to why some reactions do not go to completion lies in stability. All systems in nature try to achieve stability, i.e. the state of minimum free energy (§2.4.2). Some chemical systems (e.g. magnesium and oxygen) achieve the lowest free energy by reacting to completion, while others reach this condition by reacting to give an equilibrium mixture.

The following principles apply to all reversible reactions irrespective of the number of molecules of reactant or product. Let us reconsider the reaction

$$A + B \rightleftharpoons C + D$$

The rate of the forward reaction is related to the concentrations of A and B (§7.2). If A and B are mixed together the rate of the reaction is high at first when the concentrations of A and B are greatest. As the reaction proceeds and the concentrations of A and B decrease the rate of the forward reaction decreases. Similar arguments apply to the reverse reaction. Initially the rate of the reverse reaction is zero because the concentrations of C and D are zero, but as the forward reaction proceeds the concentrations of C and D increase and the rate of the reverse reaction increases (Fig. 8.1).

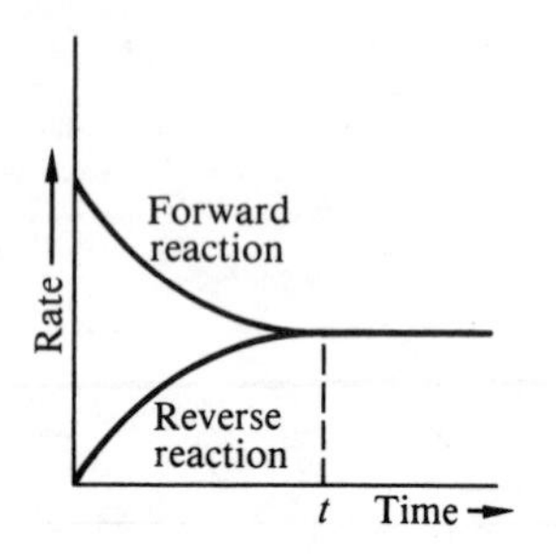

Fig. 8.1 Variation in the rates of forward and reverse reactions with time.

Thus, after time t has elapsed,

rate of forward reaction = rate of reverse reaction,

and a state of *dynamic equilibrium* is reached, where every time a molecule of A reacts with one of B, a molecule of C reacts with one of D. An *equilibrium mixture* is obtained, which does not vary in composition with time.

The situation is similar to that which would exist if a person were to walk down an escalator at the same rate that it ascends. The person would be in dynamic equilibrium with the escalator and his position would remain unaltered.

Equilibrium is seldom (if ever) reached when exactly half the reactants have been converted into products. The reaction between ethanoic acid and ethanol can be used to illustrate the above points.

$$CH_3COOH + C_2H_5OH \rightleftharpoons CH_3COOC_2H_5 + H_2O$$

When one mole of ethanoic acid, CH_3COOH, is mixed with one mole of ethanol, C_2H_5OH, reaction occurs to give an equilibrium mixture containing two-thirds of a mole of each product and one-third of a mole of each reactant. If one-third of a mole each of the acid and the alcohol are mixed together with two-thirds of a mole each of the ester (ethyl ethanoate, $CH_3COOC_2H_5$) and water, there is apparently no reaction since the composition of the mixture does not vary with time. We can show, however, that dynamic equilibrium exists by using water containing the oxygen-18 isotope; we find that some oxygen-18 atoms become incorporated into the ethanoic acid molecules. This can happen only if the reverse reaction

$$CH_3COOC_2H_5 + H_2O^{18} \longrightarrow CH_3COO^{18}H + C_2H_5OH$$

occurs. If the reverse reaction occurs then so also must the forward reaction, at an equal rate, otherwise the composition of the mixture would alter with time.

A reversible reaction will reach a state of equilibrium only in a *closed system.* This means that neither reactants nor products are added to or allowed to escape from the reaction mixture.

Equilibrium may be approached from either direction. For example, if one mole each of ethyl ethanoate and water are mixed and allowed to reach equilibrium, the equilibrium mixture will have the same composition as that produced by reacting together one mole each of ethanoic acid and ethanol. For the reaction

$$N_2 + 3H_2 \rightleftharpoons 2NH_3$$

identical equilibrium mixtures are obtained by reacting together one mole of nitrogen and three moles of hydrogen, or by heating (to the same temperature) two moles of ammonia.

8.2
Quantitative aspects of equilibrium

8.2.1 THE EQUILIBRIUM LAW

Let us again consider the ethanoic acid–ethanol equilibrium. By experiment, at 25 °C (298 K),

	CH_3COOH +	$C_2H_5OH \rightleftharpoons$	$CH_3COOC_2H_5$ +	H_2O
moles at start	1	1	0	0
moles at equilibrium	1/3	1/3	2/3	2/3
molar concentrations at equilibrium	$1/3V$	$1/3V$	$2/3V$	$2/3V$

where V represents the total volume in dm^3.

At equilibrium,

$$\frac{[CH_3COOC_2H_5][H_2O]}{[CH_3COOH][C_2H_5OH]} = \frac{2/3V \times 2/3V}{1/3V \times 1/3V} = 4.0$$

NB By convention, square brackets imply equilibrium concentrations expressed in $mol\ dm^{-3}$.

Further experiments show that irrespective of the initial concentrations of any of the participants – 'reactants' or 'products' – the equilibrium expression (as written above) at 25 °C (298 K) is always equal to 4.0.

The constant (4.0) is known as the *equilibrium constant*, K_c. The subscript c indicates that molar concentrations are used. While K_c remains constant during concentration changes and pressure changes, it is not constant when the temperature alters (§8.3.1). A value of K_c should therefore always be quoted with the temperature to which it relates.

The form of the equilibrium expression dictates the units for K_c. Here the units are:

$$\frac{(mol\ dm^{-3})(mol\ dm^{-3})}{(mol\ dm^{-3})(mol\ dm^{-3})}$$

In this particular case, because the units cancel one another, K_c is a number without dimensions.

An important question now is 'How general is this?' Do similar relationships hold for other reactions? The answer is 'Yes'. For example, consider another equilibrium, that between hydrogen, iodine and hydrogen iodide:

$$H_2 + I_2 \rightleftharpoons 2HI$$

This system has been studied in detail by A.H. Taylor and R.H. Crist, who found that, at 490.6 °C (763.8 K), the expression

$$\frac{[HI]^2}{[H_2][I_2]}$$

gave an almost constant value of 45.7 (Again, K_c is a dimensionless number.)

The value of K_c is widely different from that in the previous example. Every reversible reaction has its own value of K_c, which differs from that of other reactions.

Other expressions for the hydrogen–iodine system at equilibrium, e.g. $\dfrac{[HI]}{[H_2][I_2]}$ or $\dfrac{2[HI]}{[H_2][I_2]}$

do not give a constant value for K_c.

For *any* reversible reaction, irrespective of the number of reactants and products, if the equation is

$$aA + bB + cC + \ldots \rightleftharpoons dD + eE + fF + \ldots$$

where a, b ... f represent the smallest whole number coefficients of reactants and products in the balanced equation, then

$$\frac{[D]^d \times [E]^e \times [F]^f}{[A]^a \times [B]^b \times [C]^c} = \frac{[D]^d[E]^e[F]^f}{[A]^a[B]^b[C]^c} = K_c$$

The concentrations are those in the equilibrium mixture.

This principle is summarised by the *equilibrium law of mass action* (often called the 'equilibrium law'). 'For any reversible reaction in a state of equilibrium at a given temperature, the product of the molar concentrations of the products, each raised to the power of its coefficient in the chemical equation, divided by the product of the molar concentrations of the reactants, similarly involuted, is a constant.'

This law has been verified for a large number of reactions. Notice that by convention we place products in the numerator and reactants in the denominator. If the expression for K_c so defined is constant then its inverse must also be constant,

i.e. $$\frac{[A]^a[B]^b[C]^c}{[D]^d[E]^e[F]^f} = K_c' = \frac{1}{K_c}$$

Thus to avoid confusion when quoting equilibrium constants we should always indicate which are the reactants and which are the products. This is best done by writing down the equation for the reaction in addition to quoting the expression for K_c. Some examples should help to make the above clear:

(i) $2SO_2 + O_2 \rightleftharpoons 2SO_3$

$$K_c = \frac{[SO_3]^2}{[SO_2]^2[O_2]}$$

$$\text{Units for } K_c = \frac{(\text{mol dm}^{-3})^2}{(\text{mol dm}^{-3})^2(\text{mol dm}^{-3})} = \text{mol}^{-1}\text{ dm}^3$$

It is particularly important to quote the equation in cases like this, where there is more than one molecule of a certain substance. The reason is that alternative equations can be written, for which the values of K_c are different. For example, we can represent the oxidation of sulphur dioxide by the equation

$$SO_2 + \tfrac{1}{2}O_2 \rightleftharpoons SO_3$$

in which case the value of the equilibrium constant is the *square root* of the previous value, i.e. $$K_c = \frac{[SO_3]}{[SO_2][O_2]^{\frac{1}{2}}}$$

(ii) $N_2 + 3H_2 \rightleftharpoons 2NH_3$ (not $\frac{1}{2}N_2 + \frac{3}{2}H_2 \rightleftharpoons NH_3$)

$$K_c = \frac{[NH_3]^2}{[N_2][H_2]^3}$$

$$\text{Units for } K_c = \frac{(\text{mol dm}^{-3})^2}{(\text{mol dm}^{-3})(\text{mol dm}^{-3})^3} = \text{mol}^{-2}\ \text{dm}^6$$

Given the value of K_c for a reversible reaction, calculation can be made of the concentration of a product from given initial concentrations of reactants (or reactants plus products). For example, if 2 mol of ethanoic acid and 3 mol of ethanol are allowed to reach equilibrium at 25 °C (298 K), what is the equilibrium concentration of the ester?

Let x mol of acid react with x mol of ethanol to produce x mol of ester and x mol of water. Let the final volume be V dm^3.

	CH_3COOH	$+ C_2H_5OH$	$\rightleftharpoons CH_3COOC_2H_5$	$+ H_2O$
moles at start	2	3	0	0
moles at equilibrium	$2-x$	$3-x$	x	x
molar concentrations at equilibrium	$\frac{2-x}{V}$	$\frac{3-x}{V}$	$\frac{x}{V}$	$\frac{x}{V}$

$$K_c = \frac{[CH_3COOC_2H_5][H_2O]}{[CH_3COOH][C_2H_5OH]} = 4.0$$

$$\therefore \quad \frac{\frac{x}{V} \times \frac{x}{V}}{\frac{2-x}{V} \times \frac{3-x}{V}} = 4.0 = \frac{x^2}{(2-x)(3-x)}$$

$$\therefore \quad x^2 = 4x^2 - 20x + 24$$

Rearrangement gives

$$3x^2 - 20x + 24 = 0$$

For a quadratic equation $ax^2 + bx + c = 0$

$$x = \frac{-b \pm \sqrt{b^2 - 4ac}}{2a}$$

$$\therefore \quad x = \frac{20 \pm \sqrt{20^2 - (4 \times 3 \times 24)}}{2 \times 3}$$

from which $x = 1.57$ or 5.09

Although both roots of the quadratic equation are positive, only the former is chemically acceptable. (As the equation for the reaction makes clear, 2 mol of acid can give a maximum yield of only 2 mol of ester.) The composition of the equilibrium mixture is therefore as follows:

1.57 mol ester, 1.57 mol water, 0.43 mol acid, 1.43 mol alcohol

As confirmation, we can check that

$$\frac{1.57^2}{0.43 \times 1.43} = 4.0$$

An important consequence of the equilibrium law is that any desired concentration of product can be attained provided that the concentrations of reactants are adjusted accordingly. This can be exploited to maximise a yield of product when one of the reactants is freely available and the other is scarce.

8.2.2 THE MAGNITUDE OF K_c

The magnitude of K_c indicates the extent to which reactants have been converted into products when equilibrium has been reached. For a general reaction

$$A \rightleftharpoons B$$

if $$K_c = \frac{[B]}{[A]} = 1$$

then $[A] = [B]$

i.e. at equilibrium 50% of A has been converted into B.

For $K_c > 1$, $[B] > [A]$, and if K_c is very high then the reaction essentially goes to completion because almost total conversion of reactants into products has occurred. Equilibrium in such a case is said to lie 'towards the right', and may be symbolised by the equals sign or by writing the two half arrows thus: $\rightleftharpoons$

For $K_c < 1$, $[B] < [A]$, and if K_c is very small then the reaction may be considered 'not to occur'.

The magnitude of K_c can be very informative when deciding which of several equilibria having one or more participants in common will be the predominant one. For example, Ag^+ with Cl^-, Br^- and CN^- forms complexes of the type $[AgX_2]^-$, and with ammonia $[Ag(NH_3)_2]^+$. Suppose that to a solution of silver nitrate were added ammonia, potassium cyanide, potassium bromide and potassium chloride. How would all the participating species 'sort themselves out'? The equilibrium law requires that all the individual equilibrium concentration ratios be established. The approximate values of K_c (called *formation* or *stability constants*) for the complexes of Ag^+ with Cl^-, Br^-, CN^- and NH_3 are, respectively, 10^5, 10^0, 10^{21} and 10^7 $mol^2\ dm^{-6}$. The high value (10^{21}) for CN^- tells us that most of the silver would be complexed as $[Ag(CN)_2]^-$.

8.2.3 EQUILIBRIUM CONSTANTS FOR GASEOUS REACTIONS

So far we have expressed equilibrium constants in terms of concentrations in mol dm^{-3}, which is convenient when dealing with equilibria in solution. However, the concentration of a gas is most easily expressed by its pressure, and for gaseous reactions equilibrium constants are therefore expressed in terms of partial pressures.

The *partial pressure* of a particular gas in a gas mixture is its own individual pressure that it exerts independently of the other gases. According to *Dalton's law of partial pressures*, the total pressure of the mixture (p) is equal to the sum of the partial pressures,

i.e. $p = p_1 + p_2 + p_3 + \ldots$

Assuming ideal behaviour of gases in a mixture, we can apply the gas equation to each gas:

$$pV = nRT$$

$$\therefore \quad p = \frac{n}{V}RT$$

Thus, at a fixed temperature, the partial pressure of a gas is proportional to n/V, i.e. to the number of moles per given volume, which is of course a concentration term.

The partial pressure of a gas is governed by its mole fraction and the total pressure:

$$\text{partial pressure of a gas} = \text{total pressure} \times \text{mole fraction of the gas}$$

$$= \text{total pressure} \times \frac{\text{number of moles of the gas}}{\text{total number of moles in the gas mixture}}$$

In computing partial pressures, therefore, we do not need to know the volume of the container.

Equilibrium constants expressed in terms of partial pressures, designated K_p, are obtained in a manner similar to that for K_c. For a general case,

$$a\text{A(g)} + b\text{B(g)} + \ldots \rightleftharpoons c\text{C(g)} + d\text{D(g)} + \ldots$$

$$K_p = \frac{p_{\text{C}}^c \times p_{\text{D}}^d \times \ldots}{p_{\text{A}}^a \times p_{\text{B}}^b \times \ldots}$$

where the partial pressures are those of the equilibrium mixture.

$$\text{Units for } K_p = \frac{(\text{pressure units})^c \times (\text{pressure units})^d \times \ldots}{(\text{pressure units})^a \times (\text{pressure units})^b \times \ldots}$$

$$= (\text{pressure units})^{(c+d+\ldots)-(a+b+\ldots)}$$

For the nitrogen–hydrogen–ammonia equilibrium,

$$N_2(g) + 3H_2(g) \rightleftharpoons 2NH_3(g)$$

$$K_p = \frac{p_{NH_3}^2}{p_{N_2} \times p_{H_2}^3} (\text{pressure units})^{-2}$$

For gaseous systems of the kind $A \rightleftharpoons B$, $A \rightleftharpoons B + C$, $A \rightleftharpoons B + C + D$, etc, K_p is found from studying the thermal dissociation of the substance A. Many substances, when heated, break down into simpler substances. If the reaction goes to completion we refer to it as a *thermal decomposition.* If, on the other hand, the reaction is reversible and proceeds to equilibrium we speak of it as a *thermal dissociation.* Examples of thermal dissociation in the gaseous phase are as follows:

$N_2O_4(g) \rightleftharpoons 2NO_2(g)$

$PCl_5(g) \rightleftharpoons PCl_3(g) + Cl_2(g)$

The fraction of the original substance (e.g. N_2O_4) that is dissociated at a certain temperature is known as its *degree of thermal dissociation*, represented by α. For example, if a substance is one-quarter dissociated, $\alpha = 0.25$. When dissociation is complete, $\alpha = 1$.

A knowledge of α is required for calculation of K_p. Consider, for example, the thermal dissociation of one mole of dinitrogen tetraoxide under a pressure of p atmospheres.

$$N_2O_4(g) \rightleftharpoons 2NO_2(g)$$

moles at equilibrium $\quad 1 - \alpha \quad 2\alpha \quad$ total $= 1 + \alpha$

$$\therefore \quad \text{partial pressure of } N_2O_4,\ p_{N_2O_4},\ = p \times \frac{1-\alpha}{1+\alpha}$$

$$\text{and partial pressure of } NO_2,\ p_{NO_2},\ = p \times \frac{2\alpha}{1+\alpha}$$

$$\therefore \qquad K_p = \frac{p^2_{NO_2}}{p_{N_2O_4}} = \frac{\left(p \times \dfrac{2\alpha}{1+\alpha}\right)^2}{p \times \dfrac{1-\alpha}{1+\alpha}} \tag{1}$$

To obtain α, we argue that at constant pressure

$$\frac{\text{volume of gas at equilibrium}}{\text{volume of gas before dissociation}} = \frac{\text{number of moles at equilibrium}}{\text{number of moles before dissociation}} = \frac{1+\alpha}{1}$$

Also, since volume is inversely proportional to density at constant pressure,

$$\frac{\text{volume of gas at equilibrium}}{\text{volume of gas before dissociation}} = \frac{\text{density of gas before dissociation}}{\text{density of gas at equilibrium}} = \frac{\rho_1}{\rho_2}$$

$$\therefore \qquad 1 + \alpha = \frac{\rho_1}{\rho_2}$$

$$\therefore \qquad \alpha = \frac{\rho_1}{\rho_2} - 1 \tag{2}$$

ρ_1 and ρ_2 must refer to densities under the same conditions, e.g. at s.t.p.

The density of the original gas, ρ_1, can be calculated from the fact that one mole of any gas occupies 22.4 dm^3 at s.t.p. The density of the gas mixture at equilibrium, ρ_2, can be found by the Victor Meyer method or any other method that enables us to relate the volume of a gas to its mass (§5.1.4). A correction to s.t.p. must be made by applying the gas laws.

Example

0.022 65 g of a sample of dinitrogen tetraoxide occupied a volume of 10.0 cm^3 at a temperature of 60.2 °C (333.4 K) and a pressure of 1 atm. Calculate K_p for the dissociation of dinitrogen tetraoxide.

Answer

(i) **Density of pure N_2O_4 (ρ_1)**

Dinitrogen tetraoxide has a relative molecular mass of 92,

$$\therefore \quad 92 \text{ g of } N_2O_4 \text{ occupy } 22\,400 \text{ cm}^3 \text{ at s.t.p.}$$

$$\therefore \quad \rho_1 = \frac{92}{22\,400} = 0.004\,107 \text{ g cm}^{-3} \text{ at s.t.p.}$$

(ii) **Density of the gas mixture at equilibrium (ρ_2)**

The volume of gas at 60.2 °C (333.4 K) and 1 atm = 10.0 cm^3.

The corresponding volume at s.t.p. is obtained by applying Charles' law:

$$\frac{V_1}{T_1} = \frac{V_2}{T_2}$$

$$\therefore \quad \frac{10.0}{333.4} = \frac{V_2}{273}$$

from which $V_2 = 8.19$ cm^3,

i.e. 0.022 65 g of gas occupies 8.19 cm^3 at s.t.p.

$$\therefore \quad \rho_2 = \frac{0.022\,65}{8.19} = 0.002\,766 \text{ g cm}^{-3} \text{ at s.t.p.}$$

(iii) **Degree of thermal dissociation (α)**

From equation (2),

$$\alpha = \frac{\rho_1}{\rho_2} - 1$$

$$\therefore \quad \alpha = \frac{0.004\,107}{0.002\,766} - 1 = 0.485$$

(iv) **Equilibrium constant (K_p)**

$$p_{NO_2} = 1 \times \frac{2 \times 0.485}{1 + 0.485} = 0.653 \text{ atm}$$

$$p_{N_2O_4} = 1 \times \frac{1 - 0.485}{1 + 0.485} = 0.347 \text{ atm}$$

(Note that the sum of the partial pressures equals the total pressure.)
From equation (1),

$$K_p = \frac{p_{NO_2}^2}{p_{N_2O_4}} = \frac{0.653^2}{0.347} = 1.23 \text{ atm}$$

8.3
The factors affecting reactions at equilibrium

The composition of an equilibrium mixture remains constant only so long as conditions remain unaltered. If conditions are changed then the composition at equilibrium may alter, depending on the nature of the change.

Should a change in conditions affect the composition of the equilibrium mixture so that the concentration of products is increased while that of reactants is decreased, then the equilibrium is said to be 'disturbed to the right'. Conversely, if the concentration of reactants is increased while that of products is decreased, then the equilibrium is 'disturbed to the left'. The equilibrium is said to be 'undisturbed' when a change does not affect the composition of the equilibrium mixture.

8.3.1 THE EFFECTS OF TEMPERATURE

One effect of increasing the temperature of a reversible reaction is that equilibrium is attained more rapidly (§8.3.4). For example, the reaction between ethanoic acid and ethanol even in the presence of a catalyst (§14.10.7) may take several hours to reach equilibrium at room temperature, but at refluxing temperature equilibrium is attained in approximately twenty minutes.

A change in temperature also affects the composition of an equilibrium mixture *because it alters the value of the equilibrium constant*. The equilibrium constant is related to the temperature by the *van't Hoff equation*,

i.e. $$\lg K = -\frac{\Delta H}{2.303RT} + \text{constant}$$

where R is the gas constant and T is the temperature in kelvins. Therefore,

$$\lg K \propto -\frac{\Delta H}{T}$$

ΔH represents the enthalpy change of the *forward* reaction. Reversible reactions, like all chemical reactions, have enthalpy changes associated with them. If the forward reaction is exothermic then the reverse reaction is endothermic by exactly the same amount. Similarly, if the forward reaction is endothermic the reverse is exothermic. It is conventional to write equations for reversible reactions with the enthalpy change relating to the forward reaction,

e.g. $N_2(g) + 3H_2(g) \rightleftharpoons 2NH_3(g)$
$\Delta H^{\ominus} = -92.4 \text{ kJ mol}^{-1}$ (forward reaction is exothermic)

$$N_2O_4(g) \rightleftharpoons 2NO_2(g)$$
$$\Delta H^\ominus = +58.1\ \text{kJ mol}^{-1}\ \text{(forward reaction is endothermic)}$$

Enthalpy changes are reasonably constant over small temperature changes. Therefore, if the forward reaction is exothermic,

$$\lg K \propto +\frac{1}{T}$$

and $\lg K$ and hence K decreases as the temperature is raised. Since, essentially,

$$K = \frac{[\text{products}]}{[\text{reactants}]}$$

then if K decreases the concentration of the products decreases while that of the reactants increases, i.e. the equilibrium is disturbed to the left. If the temperature is lowered, K increases and the disturbance takes place to the right.

When the forward reaction is endothermic,

$$\lg K \propto -\frac{1}{T}$$

As the temperature is raised lg K becomes less negative and K becomes larger, resulting in an increase in the proportion of products, i.e. the equilibrium is disturbed to the right. A decrease in temperature disturbs the equilibrium to the left as lg K becomes more negative and K becomes smaller.

8.3.2 CHANGES AT CONSTANT TEMPERATURE

Under conditions of constant temperature the value of the equilibrium constant for a reaction, K_c or K_p, remains constant.

Reactions in solution

Consider the simple reaction

$$A + B \rightleftharpoons C + D$$

$$\text{where}\quad K_c = \frac{[C][D]}{[A][B]}\quad \text{at equilibrium}$$

If we add one of the products to the equilibrium mixture, then at the instant of addition the value of the concentration expression is greater than K_c and the reaction is no longer at equilibrium. The composition of the non-equilibrium mixture then alters so that the concentration of the products decreases while that of the reactants increases, i.e. equilibrium is restored by a decrease in [C][D] and an increase in [A][B] until the concentration expression again becomes equal to K_c. The equilibrium is therefore disturbed to the left. The removal of a product from the equilibrium mixture has the opposite effect in that the concentration expression now becomes smaller than K_c at the instant of removal. Adjustment takes place to produce more products and less reactants (an increase in [C][D] and a decrease in[A][B]), which is a disturbance to the right. Analogous arguments apply to the addition or removal of reactants from the equilibrium mixture. In general:

reactant	removal	displacement to the left
	addition	displacement to the right
product	removal	displacement to the right
	addition	displacement to the left

Reactions involving gases

The effect of compression or expansion

Let us consider the reaction

$$A(g) \rightleftharpoons B(g) + C(g)$$

in which there is one-third of a mole of each gas at equilibrium and an increase in the number of molecules as the forward reaction proceeds. If the equilibrium mixture is suddenly compressed to half its original volume, then at the instant of compression the total pressure will double and, because nothing has been added or removed, the partial pressures will also double. If the total pressure initially is one atmosphere, then

$$\text{before compression} \quad p_A = p_B = p_C = \tfrac{1}{3} \text{ atm}$$
$$\text{and after compression} \quad p_A = p_B = p_C = \tfrac{2}{3} \text{ atm}$$

Inserting these values into the expression for K_p, we find

$$K_p = \frac{\frac{1}{3} \times \frac{1}{3}}{\frac{1}{3}} = \tfrac{1}{3} \text{ atm} < \frac{\frac{2}{3} \times \frac{2}{3}}{\frac{2}{3}} = \tfrac{2}{3} \text{ atm}$$

before compression   *after compression*

Consequently, after compression the composition of the mixture must alter so that the expression becomes equal to the correct value of K_p, i.e. $\frac{1}{3}$ atmosphere. This is reached by more B combining with C to produce more of A, thereby decreasing the partial pressures of B and C but increasing that of A. In other words, the equilibrium is displaced to the left by an increase of pressure. A decrease in pressure has the opposite effect and displaces the equilibrium to the right.

For reactions in which there is a decrease in the number of molecules as the forward reaction occurs,

$$D(g) + E(g) \rightleftharpoons F(g)$$

we can apply similar reasoning to show that an increase in pressure displaces the equilibrium to the right, while a decrease in pressure displaces it to the left.

Reactions in which there is no change in the number of gas molecules are not affected by pressure changes. Consider the reaction

$$G(g) + H(g) \rightleftharpoons I(g) + J(g)$$

in which there is half a mole of each product and a quarter of a mole of each reactant at a total pressure of one atmosphere at equilibrium.

Before compression $p_G = p_H = \frac{1}{6}$ atm; $p_I = p_J = \frac{1}{3}$ atm
After doubling the pressure by compression $p_G = p_H = \frac{1}{3}$ atm; $p_I = p_J = \frac{2}{3}$ atm

Inserting these values into the expression for K_p, we find

$$\frac{\frac{2}{3} \times \frac{2}{3}}{\frac{1}{3} \times \frac{1}{3}} = 4 = K_p = \frac{\frac{1}{3} \times \frac{1}{3}}{\frac{1}{6} \times \frac{1}{6}}$$

after compression *before compression*

The composition of the equilibrium mixture, therefore, will not alter, because the numerical value of the equilibrium expression is unaffected by pressure.

The addition or removal of a reactant or product at constant pressure

Let us consider the reaction

$$A(g) \rightleftharpoons B(g) + C(g)$$

in which there is one-third of a mole of each gas in equilibrium at a total pressure of one atmosphere. If we now add one-third of a mole of A, the partial pressures of the gases are as follows:

$$p_A = 1 \times \frac{\frac{2}{3}}{\frac{1}{3} + \frac{1}{3} + \frac{2}{3}} = \frac{1}{2} \text{ atm}$$

$$p_B = p_C = 1 \times \frac{\frac{1}{3}}{\frac{4}{3}} = \frac{1}{4} \text{ atm}$$

Inserting these into the expression for K_p, we find

$$\frac{\frac{1}{4} \times \frac{1}{4}}{\frac{1}{2}} = \frac{1}{8} \text{ atm} < K_p = \frac{1}{3} \text{ atm}$$

The composition of the mixture will alter so that the expression becomes equal to the correct value of K_p (i.e. $\frac{1}{3}$ atm). This is reached by a displacement of the equilibrium to the right, thereby increasing the partial pressures of B and C and decreasing that of A. The removal of A from the equilibrium mixture has the opposite effect and displaces the equilibrium to the left. Similar reasoning can be applied to the addition or removal of products. Addition displaces equilibrium to the left and removal displaces it to the right.

We can also apply these arguments to reactions in which there are fewer product molecules than reactants, and in which the numbers are equal. The arguments are analogous to those applied to reactions in solution. The similarity is hardly surprising since partial pressure is proportional to concentration.

8.3.3 LE CHATELIER'S PRINCIPLE

Le Chatelier's principle states that if a change of concentration, pressure or temperature occurs to a system at equilibrium, the system will tend to adjust itself so as to counteract, as far as possible, the effect of that change.

The principle is attractive because, if applied correctly, it enables us to predict qualitatively the outcome of altering the conditions under which an equilibrium has been established. For example, in the reaction

$$A + B \rightleftharpoons C + D$$

if we add more of product D, the composition of the mixture alters so as to reduce the concentration of added D. There is therefore a disturbance of equilibrium to the left. Similarly, if we remove D a disturbance to the right occurs as once again the system attempts to cancel out the change.

For the adjustment of gaseous reactions to pressure changes, consider

$$A(g) \rightleftharpoons B(g) + C(g)$$

Clearly, the decomposition of A into B and C leads to an increase in the number of molecules and therefore an increase in pressure, while the combination of B and C results in a decrease in the number of molecules and a decrease in pressure. Now suppose that the pressure on the equilibrium mixture is increased. According to Le Chatelier's principle, the system will adjust itself so as to relieve the pressure. This can be caused only by a reduction in the number of molecules; therefore there is a disturbance to the left, i.e. more of A is formed. A decrease in pressure has the opposite effect in that it induces a pressure increase, leading to the formation of more B and C.

Analogous arguments may be applied to reactions in which there is a reduction in the number of gaseous molecules. Reactions in which the number of product molecules is equal to the number of reactant molecules,

e.g. $H_2(g) + I_2(g) \rightleftharpoons 2HI(g)$

are unaffected by applied pressure changes. The system cannot adjust itself to relieve the change because there is no overall change in the number of molecules it contains.

An increase in temperature alters the composition at equilibrium by preferentially favouring the reaction which absorbs heat. In other words, the *endothermic* change is promoted to a greater extent than the exothermic one, for in this way the additional heat is partly removed. If, say, the forward reaction is endothermic, then a disturbance to the right occurs. By similar reasoning, a decrease in temperature favours the exothermic reaction.

A few examples of the effects of changing conditions on some reversible reactions will now be considered. The long arrows indicate the changes which are necessary to disturb the equilibria to the right.

$$N_2(g) + 3H_2(g) \rightleftharpoons 2NH_3(g) \qquad \Delta H^\ominus = -92.4 \text{ kJ mol}^{-1}$$

$$\xrightarrow[\text{removal of } NH_3\text{, increase of } p]{\text{lowering of } T\text{, addition of } N_2 \text{ or } H_2\text{,}}$$

$$PCl_5(g) \rightleftharpoons PCl_3(g) + Cl_2(g) \qquad \Delta H^\ominus = +93\ \text{kJ mol}^{-1}$$

$$\xrightarrow[\text{removal of } PCl_3 \text{ or } Cl_2,\ \text{lowering of } p]{\text{increase of } T,\ \text{addition of } PCl_5,}$$

$$H_2(g) + I_2(g) \rightleftharpoons 2HI(g) \qquad \Delta H^\ominus = +51.8\ \text{kJ mol}^{-1}$$

$$\xrightarrow[\text{removal of HI (change of } p \text{ no effect)}]{\text{increase of } T,\ \text{addition of } H_2 \text{ or } I_2,}$$

8.3.4 THE TIME REQUIRED FOR EQUILIBRIUM TO BE ESTABLISHED

A catalyst that speeds up a forward reaction also speeds up the reverse reaction by exactly the same amount. The value of the equilibrium constant for the reaction is not altered, and the composition of the equilibrium mixture is unchanged. The only observable effect is that in the presence of a catalyst the time taken for equilibrium to be established is reduced (Fig. 8.2).

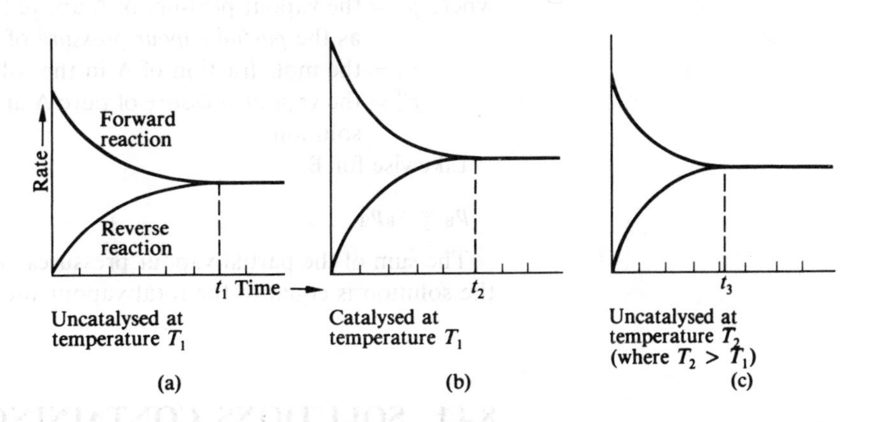

Fig. 8.2 The effect of a catalyst and an increase in temperature on the time required for equilibrium to be established.

In the absence of a catalyst equilibrium is attained after time t_1. In the presence of a catalyst, where the rate of decrease of reactant concentration and the rate of increase of product concentration are both greater than in the uncatalysed reaction, equilibrium is reached after time t_2. As the graph (b) indicates, time t_2 is shorter than time t_1. Similar arguments apply to an increase in temperature, where again the rates of both forward and reverse reactions are increased and equilibrium is reached after time t_3, where $t_3 < t_1$ (t_3 is not necessarily equal to t_2).

To summarise, the time required for equilibrium to be established depends on the temperature and the use of catalysts. While a change of temperature affects the composition of the equilibrium mixture, the introduction of a catalyst does not.

8.4 Raoult's law

A *solution* is a uniform mixture of two or more substances. Several types of solution exist, but in this section we are concerned with two-component solutions, i.e. solutions comprising two chemically independent substances. In particular, we shall study solutions of solids in liquids and of liquids in liquids.

If A and B represent two substances that are chemically similar to each other (e.g. heptane and octane, or benzene and methylbenzene), the forces of attraction and repulsion between molecules of A and B in the solution are similar to those between molecules of A and A and between molecules of B and B. Such a solution is said to be 'ideal'. If the forces are dissimilar the solution is said to be 'non-ideal' or 'real'.

A liquid in a solution has a lower vapour pressure (§6.3.3) than the pure liquid at the same temperature. F. M. Raoult (1887) investigated the relationship between the vapour pressure and composition of solutions and formulated the following law which bears his name. *The vapour pressure of a component in an ideal solution is equal to the product of the mole fraction of the component in the solution and the vapour pressure of the pure component at the same temperature.* Ideal solutions are rare, but dilute non-ideal solutions approximate very closely to ideal solutions. Stated mathematically, for component A we obtain:

$$p_A = x_A p_A^0$$

where p_A = the vapour pressure of A above the solution (often referred to as the *partial vapour pressure* of A)
x_A = the mole fraction of A in the solution
p_A^0 = the vapour pressure of pure A at the same temperature as the solution.

Likewise for B,

$$p_B = x_B p_B^0$$

The sum of the partial vapour pressures of A and B, i.e. $p_A + p_B$, above the solution is equal to the total vapour pressure of the solution.

8.4.1 SOLUTIONS CONTAINING AN INVOLATILE COMPONENT

Lowering of vapour pressure

If one component of a solution is involatile it is regarded as the *solute*. It may be either a solid or a liquid. The volatile component is referred to as the *solvent*. Since the solute is involatile (p_{solute}^0 is very low) it does not make a major contribution to the vapour pressure above the solution. Consequently, for these solutions Raoult's law may be more simply stated as follows. *The vapour pressure of a solution* (i.e. the partial vapour pressure of the solvent above the solution) *is equal to the product of the mole fraction of the solvent in the solution and the vapour pressure of the pure solvent at the same temperature.* Again, the law strictly applies only to ideal solutions, but holds well for dilute non-ideal solutions.

In the discussion that follows, the subscript $_T$ represents the solvent and $_E$ the solute. If n_T represents the number of moles of solvent and n_E the number of moles of solute in the solution, then

$$\text{mole fraction of solvent } (x_T) = \frac{n_T}{n_T + n_E} \tag{1}$$

$$\text{mole fraction of solute } (x_E) = \frac{n_E}{n_T + n_E} \tag{2}$$

$$\therefore \quad x_T + x_E = \frac{n_T + n_E}{n_T + n_E} = 1 \tag{3}$$

From Raoult's law,

$$p_T = x_T \times p_T^0 = \text{vapour pressure of the solution} \tag{4}$$

Because the mole fraction of the solvent in the solution is less than one, the vapour pressure of the solution is lower than that of the pure solvent, i.e. $x_T \times p_T^0 < p_T^0$. Rearranging equation (4) and substituting in equation (3) we obtain

$$\frac{p_T}{p_T^0} = x_T = 1 - x_E$$

$$\therefore \quad 1 - \frac{p_T}{p_T^0} = x_E = \frac{p_T^0 - p_T}{p_T^0} \tag{5}$$

The right-hand expression in equation (5) is known as the *relative lowering of vapour pressure*, and so an alternative statement of Raoult's law is as follows. *The relative lowering of vapour pressure is equal to the mole fraction of the solute in the solution.*

If the solution is dilute, which in practice is very often the case, especially for non-ideal solutions, then n_E is very small compared with n_T, and $n_T + n_E$ is approximately equal to n_T. From equations (2) and (5) we obtain

$$x_E \approx \frac{n_E}{n_T} \approx \frac{p_T^0 - p_T}{p_T^0} \tag{6}$$

For a particular solvent at a given temperature p_T^0 is a constant.

$$\therefore \quad p_T^0 - p_T \propto x_E \quad \text{and} \quad p_T^0 - p_T \propto \frac{n_E}{n_T}$$

Thus, the lowering of vapour pressure ($p_T^0 - p_T$) is proportional to the mole fraction of the solute in the solution, which is yet another statement of Raoult's law.

For a fixed quantity of a given solvent, n_T is constant,

$$\therefore \quad p_T^0 - p_T \propto n_E \tag{7}$$

i.e. the lowering of vapour pressure is proportional to the number of moles of solute in the solution. Therefore, at a constant temperature, equimolar amounts of any solute dissolved in the same amount of the same solvent produce the same lowering of vapour pressure. For example, one mole of any solute, comprising 6.023×10^{23} molecules, dissolved in 1 000 g of ethanol always produces the same lowering of vapour pressure. Such a property, which depends only on the number of solute particles irrespective of their size, shape or charge in a given quantity of the solvent, is referred to as a *colligative property*.

The relative molecular mass of the solute in a solution can be determined from the lowering of vapour pressure. Let the masses of solvent and solute in the solution be represented by m_T and m_E grams respectively, and let M_E and M_T represent the molar masses (i.e. relative molecular masses expressed in grams) of solute and solvent respectively. Since

$$n_T = \frac{m_T}{M_T} \quad \text{and} \quad n_E = \frac{m_E}{M_E}$$

equations (5) and (6) become, respectively,

$$\frac{\dfrac{m_E}{M_E}}{\dfrac{m_E}{M_E} + \dfrac{m_T}{M_T}} = \frac{p_T^0 - p_T}{p_T^0} \tag{8}$$

and

$$\frac{\dfrac{m_E}{M_E}}{\dfrac{m_T}{M_T}} = \frac{m_E \times M_T}{m_T \times M_E} \approx \frac{p_T^0 - p_T}{p_T^0} \tag{9}$$

Hence, if the vapour pressure of a solution, its concentration, the vapour pressure of the pure solvent at the same temperature and the molar mass of the solvent are all known, the molar mass of the solute can be calculated. For accurate work equation (8) rather than the approximate equation (9) should be used. Both equations are valid only so long as the solute does not dissociate or associate in solution. A modern instrument for determining molar mass by the lowering of vapour pressure is described later.

Elevation of boiling temperature

At a fixed pressure the boiling temperature of a solution is higher than that of the pure solvent, a phenomenon known as the *elevation of boiling temperature*. Figure 8.3 shows the vapour pressure curve of a solvent, which boils at temperature T_0 when its vapour pressure is equal to the external pressure over the liquid (§6.3.3). Solution X, which contains an involatile solute, has a lower vapour pressure than the pure solvent at any given temperature and therefore does not boil until temperature T_X is reached. Solution Y contains the same solute as solution X but is more concentrated. Because the vapour pressure of solution Y is lower than that of solution X or the pure solvent at a given temperature, solution Y does not boil until temperature T_Y is reached.

Figure 8.3 clearly indicates that at a fixed external pressure the boiling temperature of a solution is higher than that of the pure solvent and that

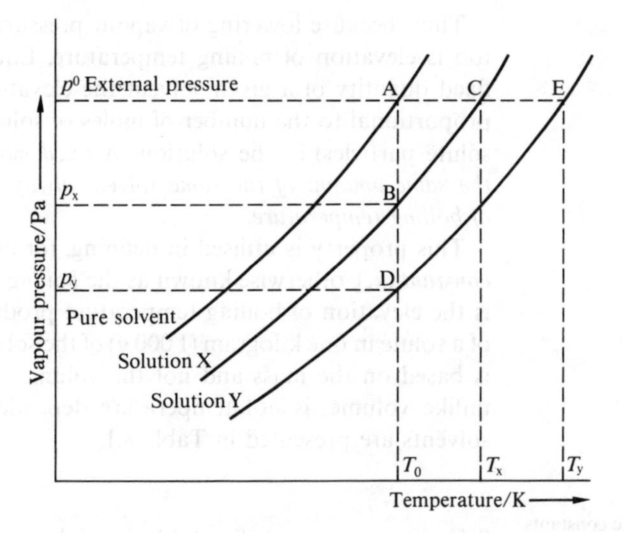

Fig. 8.3 The vapour pressure curves of: (i) a pure liquid (solvent); (ii) a dilute solution X; (iii) a more concentrated solution Y.

the elevation of boiling temperature increases with concentration.

The relationship between elevation of boiling temperature and concentration of the solution can be approximately derived from Fig. 8.3. The areas bounded by ABC and ADE closely approximate to similar triangles, hence

$$\frac{\mathrm{AB}}{\mathrm{AD}} = \frac{\mathrm{AC}}{\mathrm{AE}} \tag{10}$$

AB $= p^0 - p_X$ = the lowering of vapour pressure of solution X
AD $= p^0 - p_Y$ = the lowering of vapour pressure of solution Y
AC $= T_X - T_0$ = the elevation of boiling temperature of solution X
AE $= T_Y - T_0$ = the elevation of boiling temperature of solution Y

Therefore, from equation (10),

$$\frac{p^0 - p_X}{p^0 - p_Y} = \frac{T_X - T_0}{T_Y - T_0} \tag{11}$$

Hence the lowering of vapour pressure is proportional to the elevation of the boiling temperature,

i.e. $p^0 - p_X \propto T_X - T_0$

or $p^0 - p_X = (T_X - T_0) \times c$ (12)

and $p^0 - p_Y \propto T_Y - T_0$

or $p^0 - p_Y = (T_Y - T_0) \times c$ (13)

where c is a constant.

Equation (12) divided by equation (13) yields equation (11).

Using ΔT_b to represent elevation of boiling temperature, we can write

$$\Delta T_b \propto p_T^0 - p_T \tag{14}$$

and, from equation (7),

$$\Delta T_b \propto n_E \tag{15}$$

Thus, because lowering of vapour pressure is a colligative property, so too is elevation of boiling temperature. Equation (15) shows that for a fixed quantity of a given solvent the elevation of boiling temperature is proportional to the number of moles of solute (and hence the number of solute particles) in the solution. *A fixed molar quantity of* any *solute in the same amount of the same solvent always produces the same elevation of boiling temperature.*

This property is utilised in defining, for every solvent, its *ebullioscopic constant* (k_b), otherwise known as the *boiling point elevation constant.* This is the elevation of boiling temperature produced by dissolving one mole of a solute in one kilogram (1 000 g) of the solvent. Notice that the constant is based on the mass and not the volume of the solvent, because mass, unlike volume, is not temperature dependent. The constants for some solvents are presented in Table 8.1.

Table 8.1 Ebullioscopic constants (k_b) and cryoscopic constants (k_f) for some common solvents.

Solvent	k_b/K mol^{-1} kg	Solvent	k_f/K mol^{-1} kg
water	0.52	water	1.86
benzene	2.53	benzene	5.12
propanone	1.71	camphor	40.0
tetrachloromethane	5.02	tetrachloromethane	30.0

It is important to realise that these constants are properties of the *solvents* and are independent of the nature of the solute. In general,

for 1 mol of solute dissolved in 1 000 g of solvent $\Delta T_b = k_b$

i.e. for M g of solute dissolved in 1 000 g of solvent $\Delta T_b = k_b$

$\therefore$ for m_E g of solute dissolved in 1 000 g of solvent $\Delta T_b = k_b \times \dfrac{m_E}{M}$

and for m_E g of solute dissolved in m_T g of solvent $\Delta T_b = k_b \times \dfrac{m_E}{M} \times \dfrac{1\,000}{m_T}$ (16)

where M is the molar mass of the *solute* (units, g mol^{-1})
m_E is the mass of *solute* in the solution (units, g)
m_T is the mass of *solvent* in the solution (units, g)

Equation (16) enables us to verify the units of k_b (Table 8.1). ΔT_b is measured in kelvins, K,

$$\therefore \quad k_b = \text{K} \times \frac{\text{g mol}^{-1}}{\text{g}} \times \frac{\text{g}}{1\,000} = \text{K mol}^{-1}\text{ kg}$$

Provided that the ebullioscopic constant of the solvent is known, the molar mass of a solute can be determined by measuring the elevation of boiling temperature of a solution of known concentration. Correct values are obtained only if:

(i) the solute does not dissociate or associate in the solvent;
(ii) the solution is dilute and obeys Raoult's law;

(iii) the solute is involatile;
(iv) the solute and solvent do not chemically react together.

Depression of freezing temperature

The freezing temperature of a solvent is the temperature at which its solid and liquid forms exist in equilibrium at a given external pressure. The freezing temperature of a *solution* at a given pressure is the temperature at which the solution and solid solvent exist in equilibrium. For example, if a solution is cooled down then the freezing temperature is the temperature at which the solid solvent just begins to separate. At this temperature the solution and the solid solvent are in equilibrium with each other.

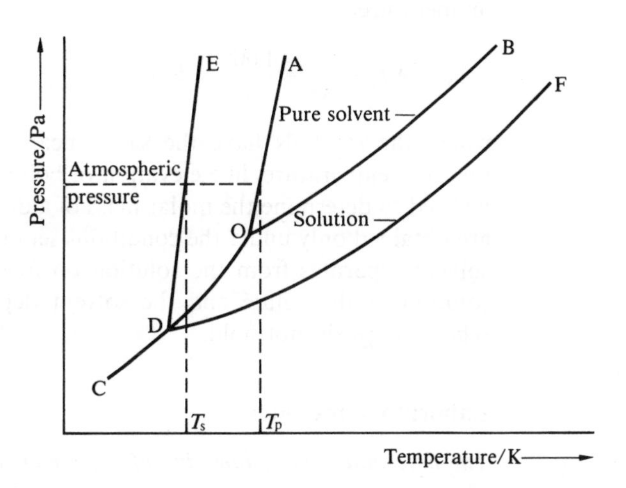

Fig. 8.4 Vapour pressure curves for a solvent and a solution.

Freezing temperature, like boiling temperature, varies with the external pressure. In Fig. 8.4, OA represents the variation with pressure of the freezing temperature of a solvent. The line OA for water is unusual in that it has a negative slope unlike the usual positive slope shown. OB and OC represent the variation with temperature of the vapour pressure of the liquid and solid solvent respectively. DF represents the variation of the vapour pressure of a solution. DE represents the temperature and pressure conditions under which a solution and the solid solvent exist in equilibrium. Therefore, under atmospheric pressure, the freezing temperature of the solvent is T_p, while the solution freezes at temperature T_s. The freezing temperature of a solution is thus lower than that of the pure solvent, and the difference $T_p - T_s$ is referred to as the *depression of freezing temperature* and is symbolised by ΔT_f. Depression of freezing temperature arises therefore from the lowering of vapour pressure that occurs when a solute is dissolved in a solvent. Since lowering of vapour pressure is a colligative property we can argue that depression of freezing temperature is also a colligative property.

Depression of freezing temperature is proportional to the lowering of vapour pressure (cf. elevation of boiling temperature),

$$\therefore \quad \Delta T_f \propto p_T^0 - p_T$$

and, for a fixed quantity of solvent,

$$\Delta T_f \propto n_E$$

A fixed molar quantity of *any* solute dissolved in the same quantity of the same solvent therefore produces the same depression of freezing temperature. In the same way that we can define for every solvent an ebullioscopic constant, so we can define a *cryoscopic constant* or *freezing point depression constant* (k_f). This is the depression of freezing temperature produced by dissolving one mole of a solute in one kilogram of that solvent. Values for some common solvents are set out in Table 8.1. Notice that k_f has the same units as k_b, because the equation relating to depression of freezing temperature has the same form as that for elevation of boiling temperature,

$$\text{i.e.} \quad \Delta T_f = \frac{m_E}{M} \times \frac{1\,000}{m_T} \times k_f$$

where the symbols have the same meanings as before. Depression of freezing temperature, like elevation of boiling temperature, can therefore be used to determine the molar mass of a dissolved solute. Correct values are obtained only under the conditions set out above, and only if the solid solvent separates from the solution on freezing. In cases where a solid solution of the solute and the solvent deposits on freezing the above relationships do not hold.

Laboratory methods

The determination of lowering of vapour pressure

The direct measurement of vapour pressure is a tedious and often difficult task. Nowadays, however, instruments called *vapour pressure osmometers* are available. The apparatus (Fig. 8.5) consists of a chamber in which there

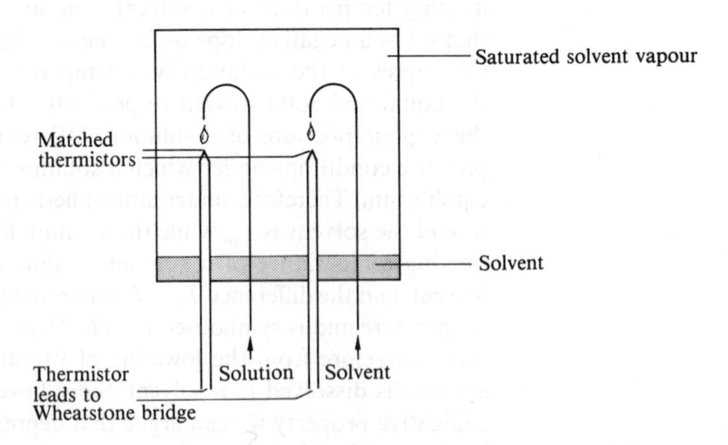

Fig. 8.5 Vapour pressure osmometer.

are two identical *thermistors*, i.e. semi-conducting devices whose resistances decrease with increasing temperature. A drop of solution of known concentration of the solute whose relative molecular mass is required is placed on one thermistor and a drop of the solvent is placed on the other. A quantity of the same solvent on the floor of the chamber ensures that the atmosphere is saturated with solvent vapour. Because the vapour pressure of the solution is lower than that of the pure solvent, solvent vapour condenses on the thermistor carrying the solution. Condensation is accompanied by the production of the latent enthalpy of vaporisation of the solvent, which increases the temperature and so reduces the resistance of the thermistor. The other thermistor is unaffected. The lower the vapour pressure of the solution the greater is the amount of vapour condensing and the greater is the increase in temperature,

i.e. lowering of vapour pressure $\propto$ quantity of solvent condensing
$\propto$ increase in temperature $\propto$ decrease in resistance

The change in resistance between the two thermistors is obtained by connecting them to the two arms of a d.c. Wheatstone bridge.

The determination of elevation of boiling temperature

Because the boiling temperature of a liquid varies with pressure, it is advisable to record the boiling temperature of the pure solvent as well as that of the solution.

An accurate determination of boiling temperature cannot be carried out merely by placing a thermometer in the boiling liquid, for temperature fluctuations occur as the result of *superheating*, i.e. the heating of the liquid above its boiling temperature without boiling taking place. Superheating leads to an unstable condition and boiling eventually occurs, often with

Fig. 8.6 The Landsberger–Walker apparatus.

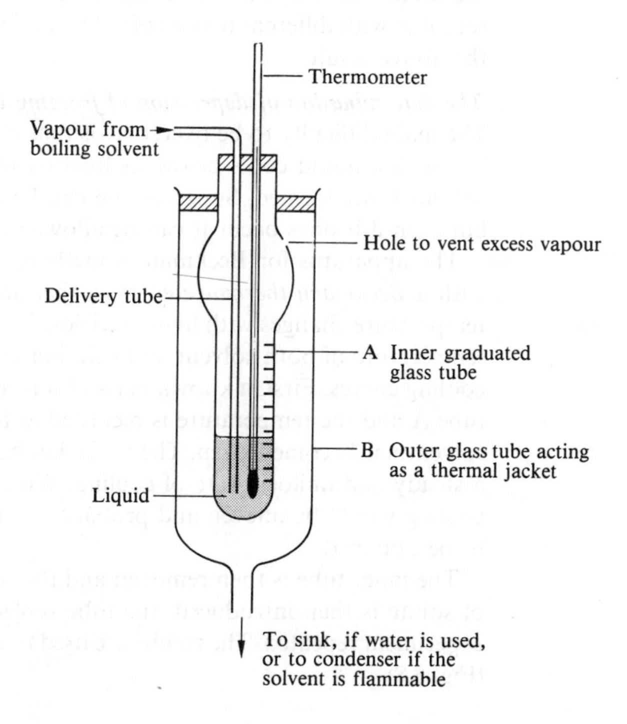

bumping; at the same time the liquid returns to its normal boiling temperature.

Superheating is promoted by direct heating and lack of agitation, and can be eliminated if these causes are removed. In the *Landsberger–Walker* apparatus (Fig. 8.6) the vapour from the boiling solvent contained in a separate vessel is used as a source of heat. As the boiling solvent vapour passes into tube A it heats and agitates the liquid. The heat is derived from the latent enthalpy of vaporisation which is produced as the vapour condenses.

A small quantity of solvent is first placed in tube A and the boiling solvent vapour passed in until the thermometer reading is steady. This is the boiling temperature of the solvent. A known mass of solute is then dissolved in the solvent in tube A and vapour passed in again until a steady temperature is attained. This is the boiling temperature of the solution. The passage of vapour is then stopped, the delivery tube and thermometer removed, and the volume of the solution noted. Now,

mass of solution = density of boiling solution × volume

If the solution is dilute we can assume that:

(i) the mass of the solution is equal to the mass of the solvent in the solution;

(ii) the density of the boiling solution is equal to that of the boiling solvent, which can be obtained from tables.

∴ mass of solvent (m_T) = density of boiling solvent × volume

Since the mass of solute, mass of solvent, elevation of boiling temperature and ebullioscopic constant, k_b (obtainable from tables) are known, the molar mass of the solute can be calculated. The experiment can be repeated with different masses of solute and volumes of solvent to confirm the above result.

The determination of depression of freezing temperature

The main difficulty to be overcome is that of *supercooling*, a phenomenon in which a liquid cools below its freezing temperature without the solid solvent being formed. Supercooling can be avoided by vigorous stirring, but even if it does occur it can be allowed for.

The apparatus for Beckmann's method, which ideally should be used with a *Beckmann thermometer*, i.e. a thermometer able to measure small temperature changes with high precision, is shown in Fig. 8.7. The freezing temperature of both solvent and solution are obtained by the plotting of cooling curves. First, a known mass of solvent is introduced into the inner tube A and the temperature is recorded at half minute intervals until the solvent has become frozen. The air jacket between tubes A and B ensures a steady and uniform rate of cooling. Without such an arrangement the cooling would be uneven and probably far too rapid for accurate results to be obtained.

The inner tube is then removed and the solvent melted. A known mass of solute is then introduced, the tube replaced in the air jacket and the experiment repeated. The results are used to obtain temperature/time plots (Fig. 8.8).

Fig. 8.7 Beckmann's apparatus.

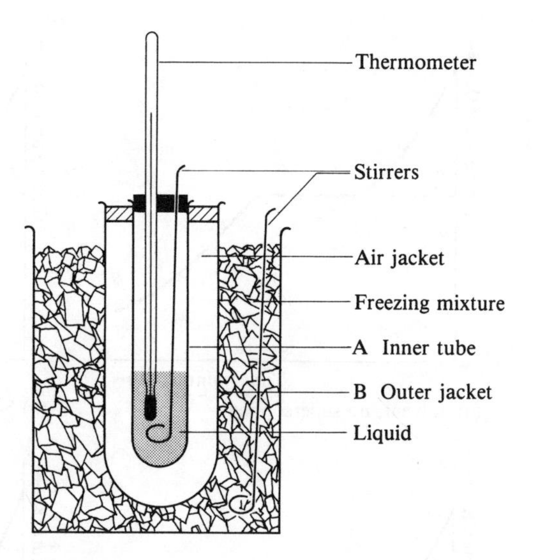

Figure 8.8(a) shows a typical cooling curve for a solvent when no supercooling occurs. When the solid and liquid forms exist in equilibrium the temperature remains constant until complete solidification has occurred. Temperature T_f is therefore the freezing temperature of the solvent. If supercooling occurs then a plot of the kind shown in Fig. 8.8(b) is obtained. A supercooled liquid is in an unstable state and eventually the solid solvent separates out. When this happens the temperature rises to the freezing temperature, T_g, as latent enthalpy of fusion is produced.

Solutions in which no supercooling occurs produce a typical cooling curve shown in Fig. 8.8(c). As freezing begins the temperature often remains steady for a short time, in which case T_h represents the freezing temperature of the solution. The separation of solid solvent has the effect of increasing the concentration and thus lowering the freezing temperature of the residual solution. Consequently, as cooling continues the residue becomes more concentrated and the freezing temperature decreases. When supercooling occurs the maximum temperature reached after supercooling is the freezing temperature of the solution; T_j in Fig. 8.8(d).

Rast's method determines the depression of freezing temperature of camphor brought about by a known amount of solute. The method is of particular importance for organic compounds, because it offers three great advantages.

(i) Camphor has a remarkably high cryoscopic constant; 40.0 K mol^{-1} kg compared with only 1.86 K mol^{-1} kg for water, so that it is unnecessary to use a Beckmann thermometer.
(ii) Camphor is a good solvent for a wide variety of organic compounds.
(iii) Because of the high value of k_f the determination requires only a small amount of the solute.

At ordinary temperatures camphor is a solid, and in the Rast method melting temperatures rather than freezing temperatures are determined. A melting point apparatus is used to find first the melting temperature

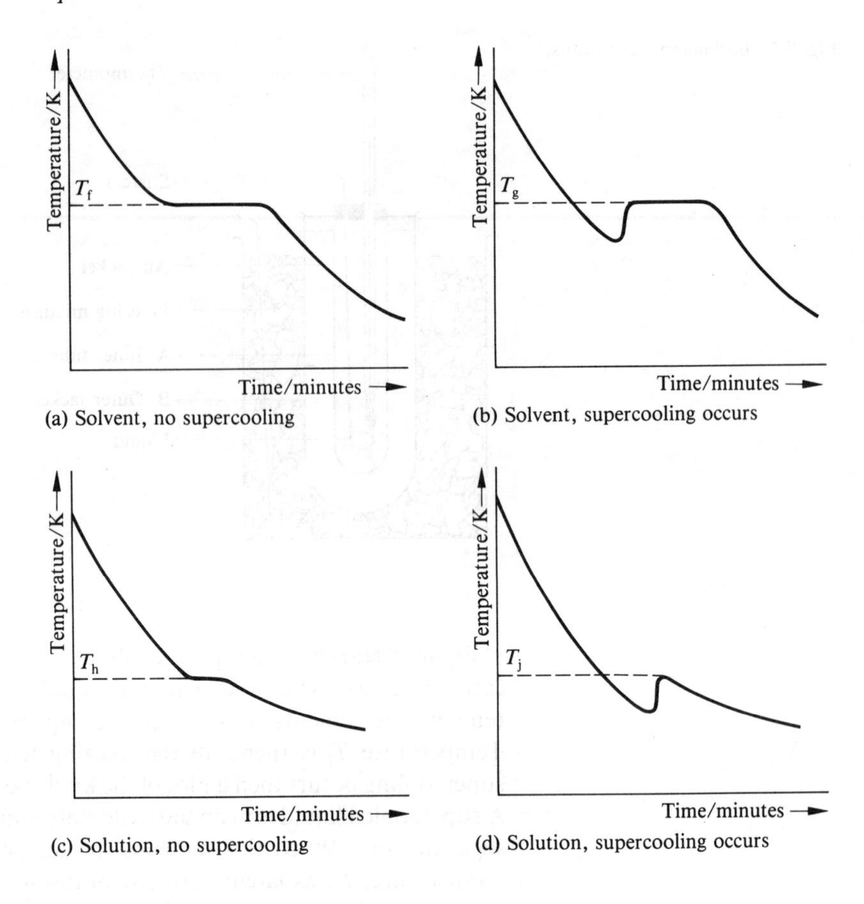

Fig. 8.8 Idealised cooling curves for a solvent and a solution. T_f, T_g, T_h and T_j represent the freezing temperatures.

of pure camphor and then the melting range of a camphor–solute mixture. The mixture, which must be intimate, is made by melting the two substances together in a sealed tube **(CARE!)** so as to avoid losses due to volatilisation. After fusion and cooling, the contents of the tube are transferred to a mortar and ground to a fine powder. The melting range is determined, and the *upper* point of the range, i.e. the temperature at which the sample becomes completely molten, is taken as the freezing temperature of the solution.

8.4.2 OSMOTIC PRESSURE

A close parallel exists between the osmotic pressure (§6.7.3) of an ideal solution, or a dilute non-ideal solution, and the behaviour of an ideal gas.

(i) The osmotic pressure of a solution (Π) is proportional to the concentration of the solution, or inversely proportional to its volume (V). (Concentration and volume are inversely related.)

$$\therefore \quad \Pi \propto \frac{1}{V}$$

which compares with Boyle's law, $p \propto \frac{1}{V}$.

(ii) The osmotic pressure of a solution is proportional to its temperature (T) in kelvins,

i.e. $\Pi \propto T$

which corresponds to the law of pressures, $p \propto T$.

(iii) A solution containing one mole of solute in 22.4 dm^3 of solution exerts an osmotic pressure of one atmosphere ($\sim 10^2$ kPa or, more accurately, 101.325 kPa) at 0 °C (273 K). This compares with the molar volume of an ideal gas (§5.1.4).

These relationships hold for solutes that do not dissociate or associate in solution. Therefore,

for 1 mol of solute in 22.4 dm^3 of solution at 273 K, $\Pi = 1$ atm
for M g of solute in 22.4 dm^3 of solution at 273 K, $\Pi = 1$ atm
for m_E g of solute in 22.4 dm^3 of solution at 273 K, $\Pi = \frac{m_E}{M}$ atm
for m_E g of solute in V dm^3 of solution at 273 K, $\Pi = \frac{m_E}{M} \times \frac{22.4}{V}$ atm
for m_E g of solute in V dm^3 of solution at T K, $\Pi = \frac{m_E}{M} \times \frac{22.4}{V} \times \frac{T}{273}$ atm

where the symbols are those previously defined.

If pressure is quoted in kPa,

$$\Pi = \frac{m_E}{M} \times \frac{22.4}{V} \times \frac{T}{273} \times 101.325 \text{ kPa}$$

At a constant temperature and for a fixed amount of solvent we may write

$$\Pi = \frac{m_E}{M} \times \text{constant} = \text{moles of solute} \times \text{constant}$$

$$\therefore \quad \Pi \propto \text{moles of solute } (n_E)$$

Osmotic pressure is therefore a colligative property (§8.4.1), and can be used in determining the molar mass of a dissolved solute, if the concentration and temperature are known. The method is especially useful for solutes of high molar mass, e.g. polymers, proteins or carbohydrates. These solutes, because of their high molar mass, produce a small elevation of boiling temperature or depression of freezing temperature $\left(\Delta T \propto \frac{1}{M}\right)$, and proper measurement is difficult. Osmotic pressure, while relatively low, is still sufficient to be measured accurately, especially with modern instruments.

The determination of osmotic pressure

A typical modern instrument for determining osmotic pressure consists of a stainless steel cell which is divided into two compartments by a rigidly fixed semi-permeable membrane (§6.7.3); Fig. 8.9.

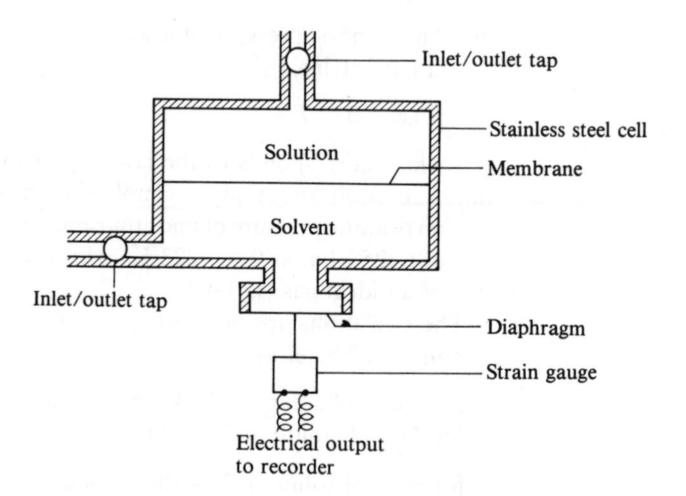

Fig. 8.9 Stainless steel cell of a modern osmometer.

A flexible diaphragm is contained in the wall of the solvent compartment. When both compartments are filled and sealed by the taps osmosis occurs and solvent flows into the solution (§6.7.3). This reduces the pressure in the solvent compartment, causing the diaphragm to distort. The pressure eventually becomes low enough in the solvent compartment to prevent osmosis occurring. The degree of diaphragm distortion is related to the reduction in pressure required to stop osmosis occurring, and this in turn is related to the osmotic pressure of the solution. Diaphragm distortion is measured by a device called a 'strain gauge', which provides an electric current that is proportional to the amount of distortion. The gauge is calibrated directly in terms of osmotic pressure.

The limitations of colligative property methods

At high concentrations a solution of a solute in a solvent exhibits non-ideal behaviour. As a result Raoult's law ceases to be obeyed; also, osmotic pressure ceases to be proportional to concentration. The reason for these deviations is not clear, but appears to be associated with solvation of the solute. This has the effect of removing solvent molecules from the solution, thus giving the impression of increased concentration. The lowering of vapour pressure, elevation of boiling temperature, depression of freezing temperature and osmotic pressure are therefore greater than expected, since the extent or magnitude of each of these properties is proportional to concentration.

However, at low concentrations the quantity of solvent removed by solvation is small in relation to the amount of solvent present, and deviations are small enough to be negligible. Dilute non-ideal solutions therefore closely approach ideal behaviour (Fig. 8.10).

The number of solute particles is altered if the solute ionises, dissociates or associates in solution. Ionisation and dissociation (§9.1.1) increase the number of particles, while association (§8.5.5) leads to a decrease. Provided that allowance is made for these changes, the relationships discussed previously can still be applied, and enable the degree of dissociation or

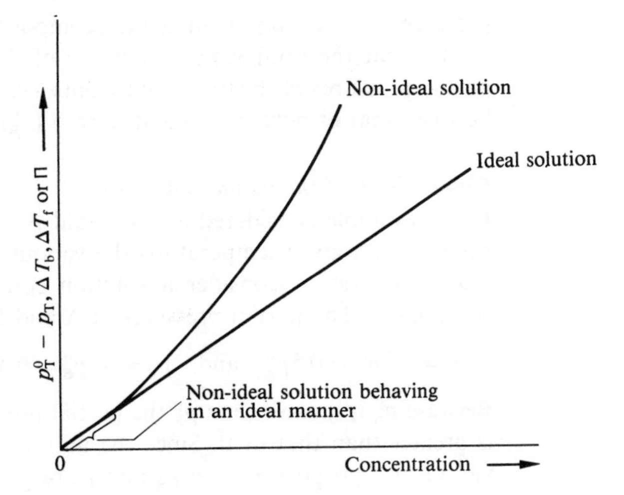

Fig. 8.10 The variation of lowering of vapour pressure ($p_T^0 - p_T$), elevation of boiling temperature (ΔT_b), depression of freezing temperature (ΔT_f) and osmotic pressure (Π) with concentration for ideal and non-ideal solutions.

association of the solute to be calculated (§9.1.2). Most cases of association lead to the almost complete formation of double molecules, so that the observed molar mass is twice that expected.

8.4.3 SOLUTIONS CONTAINING TWO VOLATILE COMPONENTS

Ideal solutions

According to Raoult's law (§8.4), the partial vapour pressure exerted by each of two liquids A and B in an ideal solution is proportional to its mole fraction in the solution. Figure 8.11 is a graph of the partial vapour

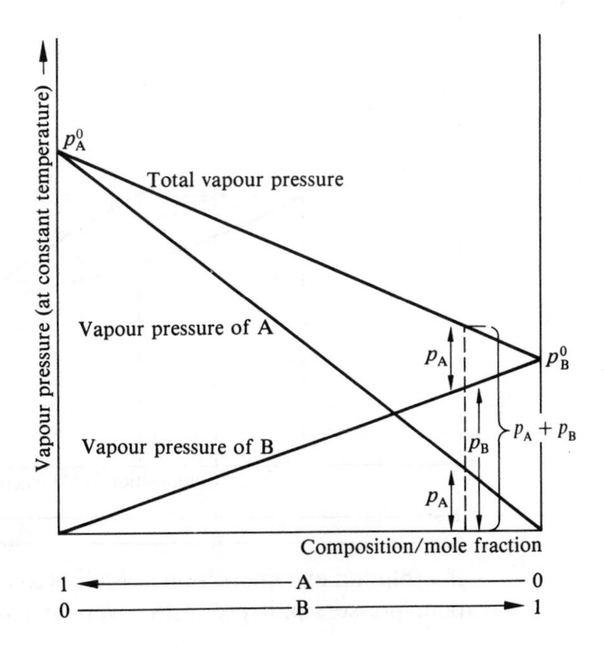

Fig. 8.11 Graph of vapour pressure against composition for ideal solutions of two liquids A and B.

pressure of each component against composition at constant temperature. Notice that the total vapour pressure of the solution is the sum of the partial pressures of the two components, i.e. $p_A + p_B$ (§8.4), and always lies between that of pure A (p_A^0) and pure B (p_B^0).

Composition of the liquid and vapour

In the example considered above liquid A is more volatile than liquid B because, at a given temperature, the vapour pressure of A is greater than that of B. Let us consider a solution containing half a mole of each component. The partial pressures of A and B are as follows:

$$p_A = x_A p_A^0 = 0.5 p_A^0 \quad \text{and} \quad p_B = x_B p_B^0 = 0.5 p_B^0$$

Because p_A^0 is greater than p_B^0 the partial pressure of A above the mixture is greater than that of B. Since the number of moles of A and B in the vapour are proportional to p_A and p_B respectively, the mole fractions of A (x_A^1) and B (x_B^1) in the *vapour* are given by:

$$x_A^1 = \frac{p_A}{\text{total vapour pressure}} = \frac{p_A}{p_A + p_B}$$

$$\text{and} \quad x_B^1 = \frac{p_B}{p_A + p_B}$$

Since p_A is greater than p_B, the vapour at equilibrium contains more A than B. *It is usually observed that the vapour above a solution is always richer in the more volatile component.* Conversely, the residual liquid at equilibrium is richer in B than the original solution. A graph of liquid and vapour compositions against total vapour pressure at constant temperature is shown in Fig. 8.12.

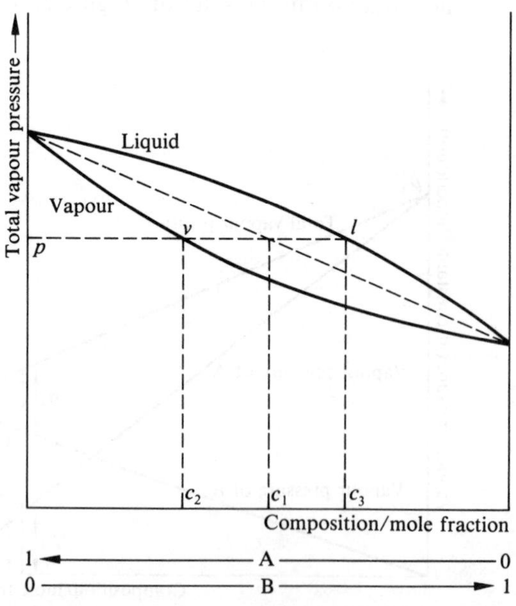

Fig. 8.12 Graph of total vapour pressure against composition of vapour and liquid at constant temperature.

If a solution of composition c_1 is allowed to reach equilibrium at a total vapour pressure p, it produces a vapour v of composition c_2 and leaves a liquid l of composition c_3. This property can be used to separate A from B by the technique of *distillation*. For example, if the vapour is removed and condensed, a liquid of composition c_2 is obtained. The vapour to which this liquid gives rise is even richer in A. By repeating the process several times pure A can ultimately be obtained.

Distillation equilibria

The boiling temperature of a liquid is inversely related to its vapour pressure, i.e. the lower the vapour pressure of a liquid at a given temperature the higher is its boiling temperature. Since the total vapour pressure of an ideal solution of A and B (see above) lies between that of pure A and pure B, it follows that the boiling temperature of the liquid mixture lies between the boiling temperatures of A and B. A plot of liquid composition against boiling temperature at a constant pressure of, say, one atmosphere (10^2 kPa) is depicted by the curve marked *liquidus* in Fig. 8.13.

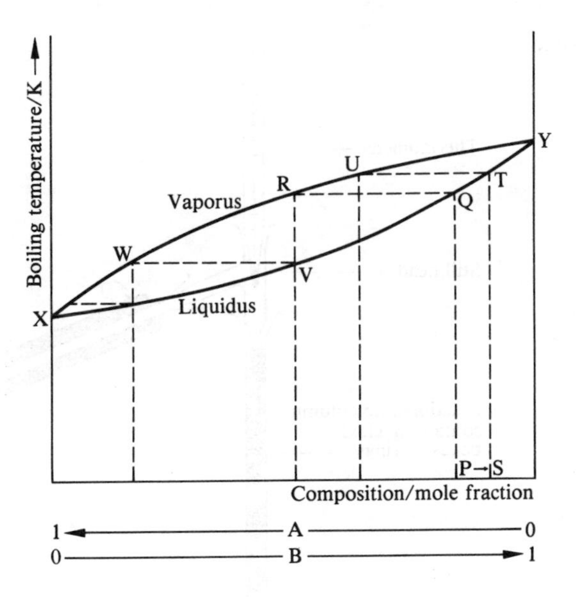

Fig. 8.13 Graph of boiling temperature against composition for a mixture of two liquids, A and B. (Points X and Y represent the boiling temperatures of pure A and pure B respectively.)

The curve marked *vaporus* gives the composition of the vapour in equilibrium with any boiling liquid mixture. Notice that Figs. 8.12 and 8.13 are inversely related to each other.

Suppose a mixture P, rich in B, were to be distilled. At a temperature corresponding to Q on the liquidus the mixture would start to boil, giving a vapour richer in A. The exact composition of this vapour corresponds to R on the vaporus: this is the composition of the initial distillate, and it will be seen that a complete separation is not achieved; merely an enrichment in A.

Consider what happens as distillation continues. If the vapour coming off is richer in A, it follows that the residue in the distillation flask becomes richer in B, i.e. it moves towards the right of the graph, and after a few minutes P may have moved to S. This boils at T giving a vapour corresponding to U, and it will be seen that while this still represents an enrichment in A the distillate contains more B than the initial distillate.

Hence, simple distillation not only fails to give a complete separation, but also becomes increasingly valueless the longer it proceeds until, finally, complete distillation yields the original mixture again.

Repeated simple distillations, however, are capable of giving a small amount of pure A. Suppose once more that we start with a mixture P; we have seen that a single distillation gives enrichment to R. On redistilling the distillate, R boils at V to give a distillate W, corresponding to about 0.8 mol A, and a third distillation should yield a small amount of almost pure A.

This technique is inconvenient as well as wasteful, and it is therefore the usual practice to perform in effect a large number of simple distillations simultaneously in the same apparatus, using a *fractionating column.* The process is called *fractional distillation*, and entails introducing the fractionating column between the distillation flask and the condenser (Fig. 8.14).

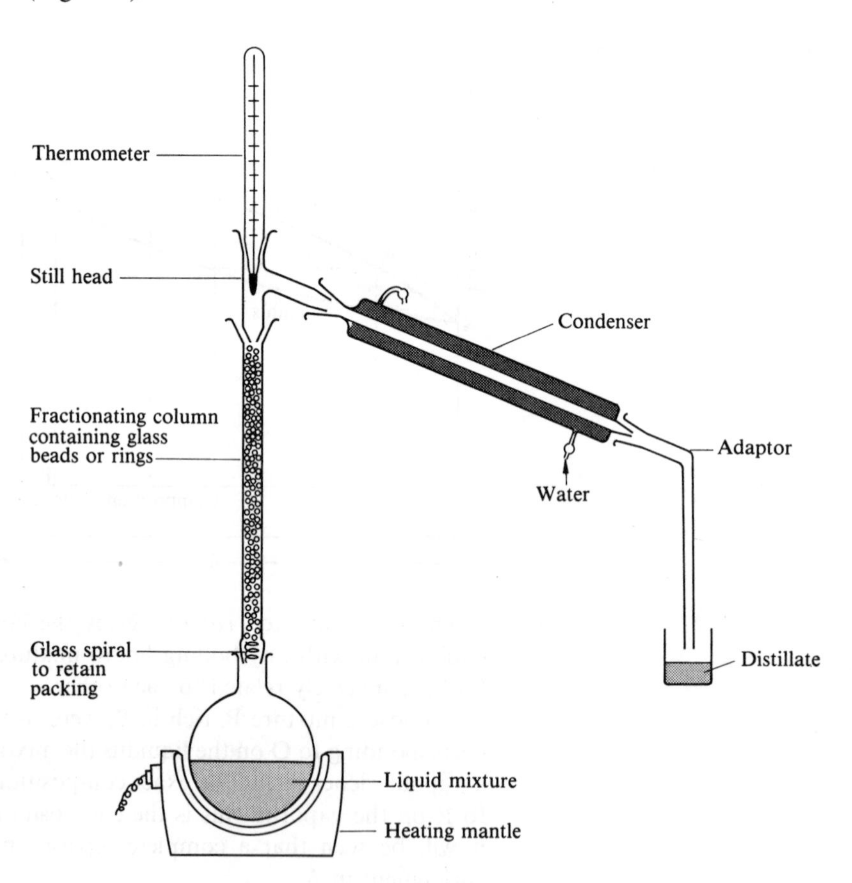

Fig. 8.14 Laboratory apparatus for fractional distillation.

In its most effective form, the fractionating column consists of a long glass tube packed with small glass beads or rings. On heating, the mixture in the distillation flask boils to give a vapour that condenses at the base of the fractionating column. There it remains for a short time until it is reboiled, partly by radiant heat from the flask below and partly by latent enthalpy of vaporisation given out by further vapour as it condenses. It passes a little further up the column and again condenses. Thus, there is repeated boiling and condensing all the way up the column. In effect a great many simple distillations are performed and the vapour emerging and condensing from the top of the column should be almost pure A. (In practice, complete separation is never achieved.) A suitably placed thermometer records a constant temperature, namely the boiling temperature of A.

The residue in the flask becomes gradually richer in B until eventually it is just B. At this point an abrupt change is observed. Distillation temporarily ceases; then the thermometer reading increases rapidly to the boiling temperature of B as pure B starts to distil. Obviously, the receiving flask should be changed as this happens.

Non-ideal solutions

Positive deviations from Raoult's law

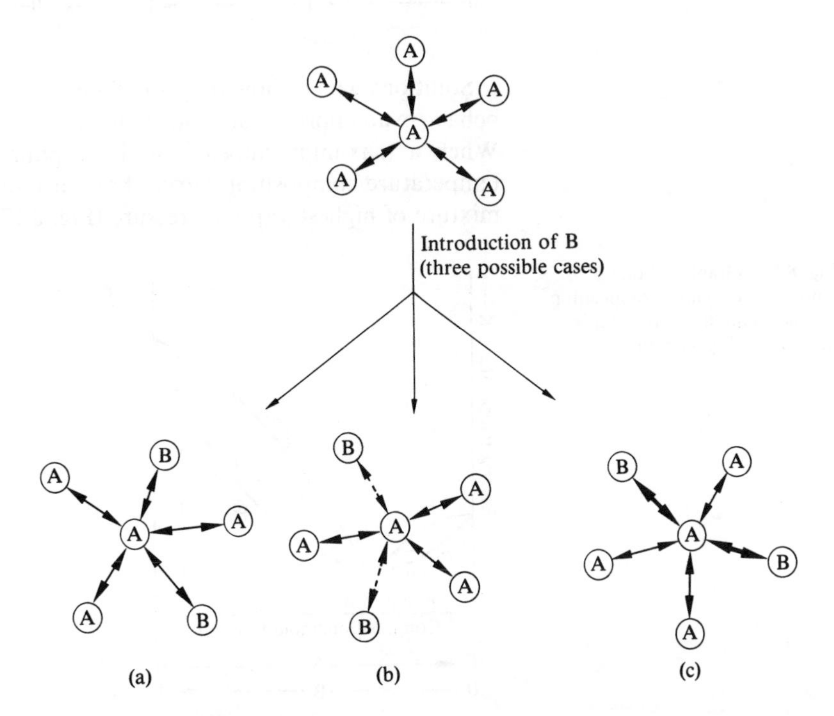

Fig. 8.15 Mixtures of two liquids, A and B. (a) An ideal solution. (b) A non-ideal solution in which the forces of attraction between a molecule of A and a molecule of B are relatively weak. (c) A non-ideal solution in which the forces of attraction between a molecule of A and a molecule of B are relatively strong.

A mixture of two liquids A and B shows ideal behaviour, i.e. obeys Raoult's law, only if the forces of attraction and repulsion between a molecule of A and a molecule of B are similar to those between two molecules of A or between two molecules of B (Fig. 8.15(a)). If the forces of attraction between a molecule of A and a molecule of B are relatively weak, the overall cohesive forces on a molecule of A are less than in the ideal case (Fig. 8.15(b)). It therefore becomes easier for a molecule of A to escape from the liquid to the vapour phase, i.e. the partial vapour pressure of A is increased. By the same argument, the partial vapour pressure of B is also raised. The total vapour pressure of the solution is therefore greater than that calculated from Raoult's law (Fig. 8.16(a)). This is known as a *positive deviation from Raoult's law.* If the extent of deviation is large then the total vapour pressure curve may show a maximum point (Fig. 8.16(b)).

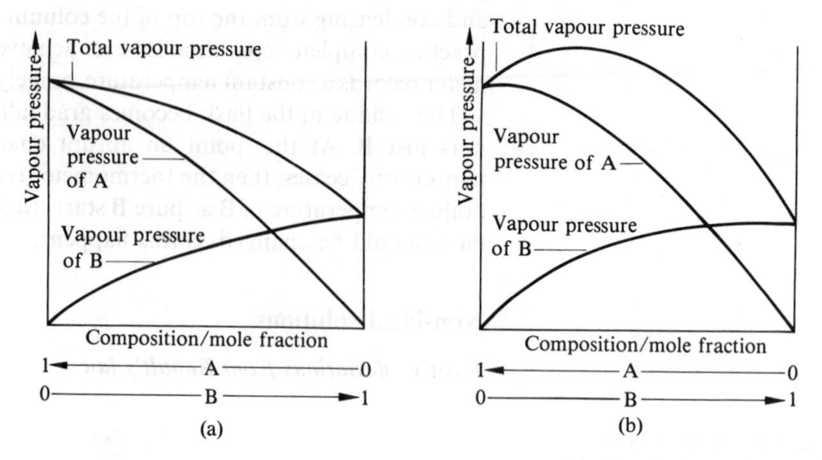

Fig. 8.16 Positive deviations from Raoult's law.

Solutions which show only small deviations, as typified by Fig. 8.16(a), behave in an approximately ideal manner and may be treated accordingly. When a maximum appears in the vapour pressure curve the boiling temperature/composition curve shows a minimum corresponding to the mixture of highest vapour pressure (Fig. 8.17).

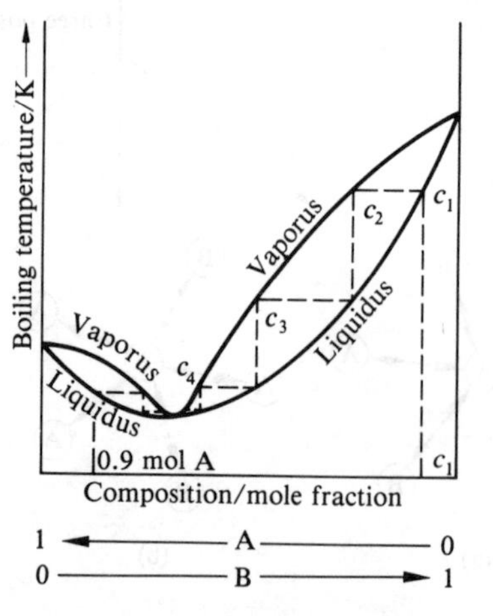

Fig. 8.17 Graph of boiling temperature against composition for the solution of two liquids depicted in Fig. 8.16(b).

Consider what happens on distilling: (i) a mixture corresponding to the minimum; (ii) a mixture to the left of the minimum; (iii) a mixture to the right of the minimum.

(i) If a mixture corresponding to the minimum point is distilled, it passes over unchanged at a constant temperature. Such a mixture is called an *azeotrope* or *constant boiling mixture.* Because its boiling temperature is below that of either constituent, it can also be called a *minimum boiling temperature mixture.*

(ii) When a mixture to the left of the minimum point (e.g. 0.9 mol A and 0.1 mol B) is distilled, the graph shows that the vapour coming off is richer in B which, in this instance, is the *less volatile* component.

On fractional distillation the azeotrope will distil over. As the composition of the distillate moves to the right, so that of the residue moves to the left. Eventually only A is left in the flask; when this happens there is an abrupt rise in boiling temperature as pure A distils.

(iii) Distillation of a mixture to the right of the minimum point, e.g. of concentration c_1, brings about an enrichment in A to concentration c_2. On fractional distillation with an efficient column the concentration rises to c_3, then to c_4, and finally to that of the azeotrope. The concentration of A in the distillate cannot rise beyond this, and the azeotrope comes over until no more A is left in the flask, just pure B. This then distils, accompanied by a sharp rise in the boiling temperature.

Ethanol and water form a system of this sort, ethanol boiling at 78.3 °C (351.5 K), water at 100.0 °C (373.2 K) and the azeotrope, which contains 95.6% (by mass) of ethanol and 4.4% of water, at 78.2 °C (351.4 K). This explains why pure ethanol cannot be obtained by distilling the dilute solution prepared by fermentation. The most that can be expected is the azeotrope, from which pure ethanol can be recovered by treatment with calcium.

Negative deviations from Raoult's law

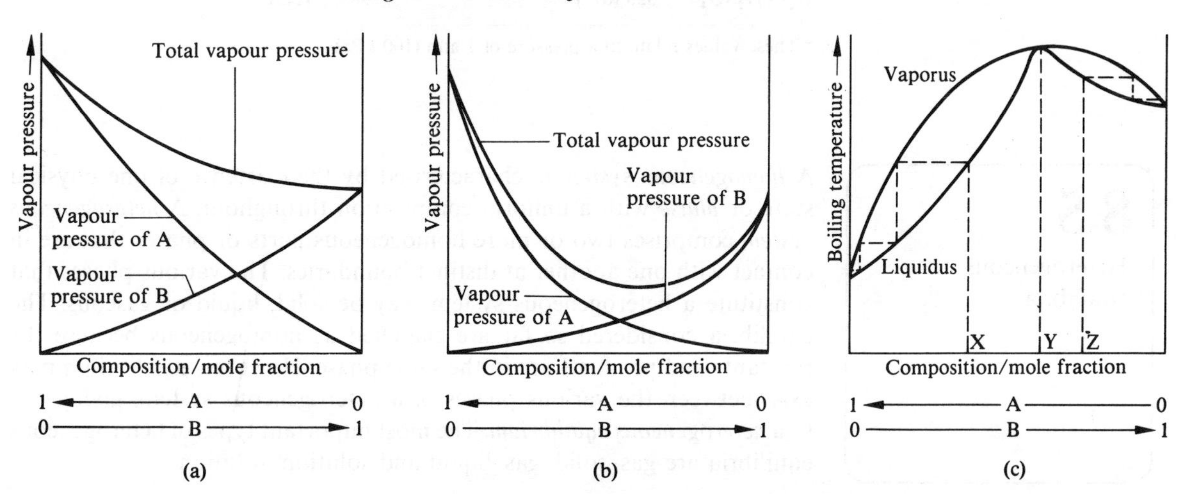

Fig. 8.18 Negative deviations from Raoult's law.

If the forces of attraction between a molecule of A and a molecule of B are strong in relation to the ideal case, the overall cohesive forces on a molecule of A are increased (Fig. 8.15(c)), as are those on a molecule of B. Thus, the tendency for molecules of A and B to escape from the solution will be lower than in the ideal case. The partial vapour pressure of each component in the mixture, and hence the total vapour pressure of the solution, is therefore lower than that calculated from Raoult's law. This is referred to as a *negative deviation from Raoult's law*. Small deviations produce vapour pressure curves typified by Fig. 8.18(a). Such solutions behave in an approximately ideal manner and may be treated accordingly. When large deviations occur the vapour pressure curve shows a minimum point (Fig. 8.18(b)) and the boiling temperature/composition curve shows a maximum, corresponding to the mixture of minimum vapour pressure (Fig. 8.18(c)).

Let us consider the behaviour of mixtures X, Y and Z on distillation.

(i) Mixture Y, the azeotrope or constant boiling mixture, distils unchanged. In contrast to the previous azeotrope, this is a *maximum* boiling temperature mixture.

(ii) When X is distilled the vapour is richer in A, and on fractional distillation pure A is obtained – for a while. But the composition of the residue moves to the right, and eventually the azeotrope distils over.

(iii) Fractional distillation of Z yields pure B at first. The composition of the residue moves to the left, and ultimately the azeotrope is obtained. The best known systems of this kind are given by water and inorganic acids (Table 8.2).

Table 8.2 Some common azeotropes with a maximum boiling temperature

System	Azeotrope	
	Boiling temperature/°C (K)†	Composition
H_2O/HNO_3	120.5 (393.7)	68.0% HNO_3
H_2O/HCl	108.6 (381.8)	20.2% HCl
H_2O/H_2SO_4	338 (611)	98.7% H_2SO_4

† These values relate to a pressure of 1 atm (100 kPa).

8.5 Heterogeneous equilibria

A *homogeneous system* is characterised by the existence of one physical state or *phase* with a uniform composition throughout. A *heterogeneous system* comprises two or more homogeneous parts or phases that are in contact with one another at distinct boundaries. The various phases that constitute a heterogeneous system may be solid, liquid or gaseous. The equilibria considered so far are classified as homogeneous because the reactants and products are in the same phase. However, equilibrium may exist between the various phases of a heterogeneous system, giving rise to a *heterogeneous equilibrium.* The most important types of heterogeneous equilibria are gas–solid, gas–liquid and solution–solution.

8.5.1 GAS–SOLID EQUILIBRIA

If we consider the heterogeneous equilibrium

$$CaCO_3(s) \rightleftharpoons CaO(s) + CO_2(g)$$

and apply the equilibrium law we expect:

$$\frac{[CaO][CO_2]}{[CaCO_3]} = \text{a constant}$$

If we add more CaO to the equilibrium mixture we should expect to obtain a higher concentration of $CaCO_3$, and similarly if we add more $CaCO_3$ we should expect to obtain a higher concentration of CaO and CO_2 (§8.3.2). In practice, however, we find that this is not the case. The implication of this is that the effective concentrations of $CaCO_3$ and CaO are constant at a given temperature, irrespective of how much of these materials are present. Therefore, at a given temperature, the concentration of CO_2 is equal to a constant, which is called the equilibrium constant for the reaction,

i.e. $K_c = [CO_2]$

or $K_p = p_{CO_2}$

Thus at a given temperature the partial pressure of CO_2 in equilibrium with $CaCO_3$ and CaO, often called the *dissociation pressure* of $CaCO_3$, has a constant value.

The above principles apply to all gas–solid equilibria, in that the concentrations or partial pressures of all solids are effectively constant and *by convention are equated to unity*. This considerably simplifies the equilibrium constant expressions as the following examples illustrate.

$$3Fe(s) + 4H_2O(g) \rightleftharpoons Fe_3O_4(s) + 4H_2(g) \qquad K_p = p^4_{H_2}/p^4_{H_2O}$$

$$NH_2COONH_4(s) \rightleftharpoons 2NH_3(g) + CO_2(g) \qquad K_p = p^2_{NH_3} \times p_{CO_2}$$

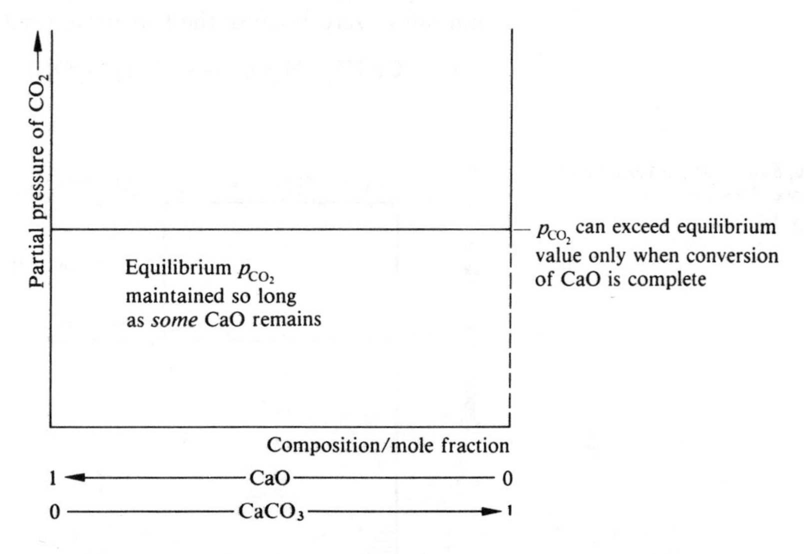

Fig. 8.19 Graph of partial pressure of carbon dioxide against composition for the $CaCO_3/CaO/CO_2$ system.

Returning to the $CaCO_3/CaO/CO_2$ system, let us consider the effect of adding CO_2 to the equilibrium mixture contained in a closed container at a fixed temperature. At the instant of addition the partial pressure of CO_2 is greater than the dissociation pressure of $CaCO_3$, which is equal to K_p, and so the system is no longer at equilibrium. The excess CO_2 is removed by reaction with CaO to form $CaCO_3$ until the equilibrium partial pressure is restored. Continuous addition of CO_2 will result in more and more $CaCO_3$ being formed; therefore, so long as CaO is present, the partial pressure of CO_2 at equilibrium will not exceed the dissociation pressure of $CaCO_3$. Only when all of the CaO has been converted into $CaCO_3$ is it possible for the partial pressure of CO_2 to exceed the equilibrium value, for then there is nothing to remove the excess gas.

If CO_2 is removed from the equilibrium mixture then more $CaCO_3$ decomposes to restore the partial pressure of CO_2 to its equilibrium value. Only when all the $CaCO_3$ has decomposed is it possible for the partial pressure of CO_2 to fall below the equilibrium value. These conclusions are represented graphically in Fig. 8.19.

Salt hydrates

Salt hydrates behave in a manner similar to $CaCO_3$. For example, if copper(II) sulphate pentahydrate, $CuSO_4{\cdot}5H_2O$, is placed inside an evacuated container it exerts a small pressure due to the presence of water vapour arising from the equilibrium:

$$CuSO_4{\cdot}5H_2O(s) \rightleftharpoons CuSO_4{\cdot}3H_2O(s) + 2H_2O(g)$$

The pressure of the vapour, called the dissociation pressure of $CuSO_4{\cdot}5H_2O$, is constant at a given temperature. If the equilibrium water vapour is removed then more $CuSO_4{\cdot}5H_2O$ decomposes to restore the dissociation pressure. Continuous removal of vapour would result in the pentahydrate decomposing completely into the trihydrate; only then would the vapour pressure fall below the equilibrium value. The pressure does not fall to zero because the trihydrate itself decomposes,

i.e. $CuSO_4{\cdot}3H_2O(s) \rightleftharpoons CuSO_4{\cdot}H_2O(s) + 2H_2O(g)$

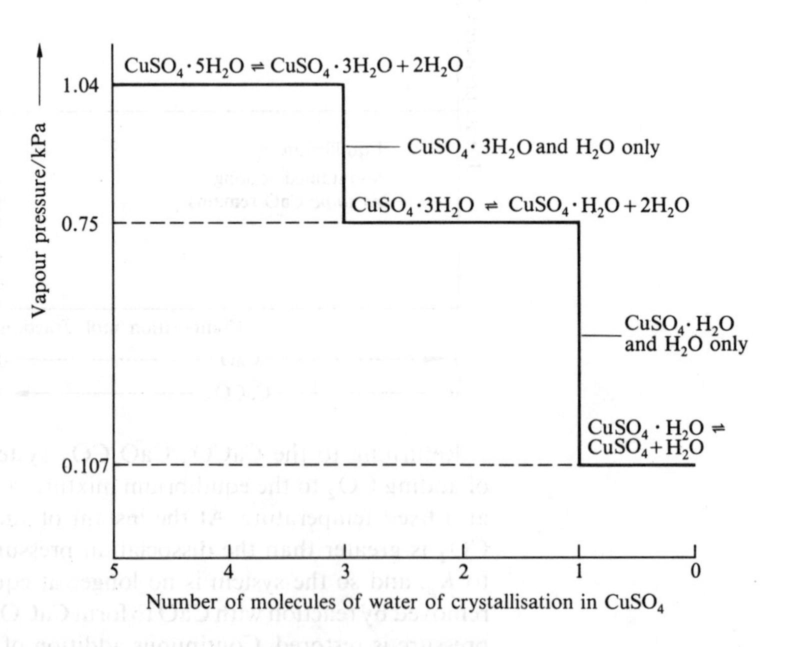

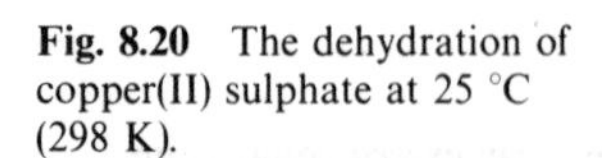
Fig. 8.20 The dehydration of copper(II) sulphate at 25 °C (298 K).

The trihydrate therefore has its own dissociation pressure, which is lower than that of the pentahydrate. If the equilibrium water vapour is removed the dissociation pressure is maintained by decomposition of the trihydrate until it has all decomposed into the monohydrate. Only after this can the vapour pressure fall below the equilibrium value. Again the pressure does not fall to zero because the monohydrate decomposes and thus has its own dissociation pressure:

$$CuSO_4{\cdot}H_2O(s) \rightleftharpoons CuSO_4(s) + H_2O(g)$$

Further continuous removal of water vapour causes the monohydrate to decompose until eventually the anhydrous salt remains. These findings are presented graphically in Fig. 8.20.

Opposite arguments apply to the addition of water vapour to copper(II) sulphate. The anhydrous salt will not absorb water until the vapour pressure exceeds the dissociation pressure of the monohydrate. In taking up water the anhydrous salt removes water vapour to reduce the vapour pressure to the equilibrium value for the monohydrate. Likewise, the monohydrate does not absorb water until the vapour pressure exceeds that of the trihydrate, and the trihydrate absorbs water only if the vapour pressure exceeds the dissociation pressure of $CuSO_4{\cdot}5H_2O$.

Suppose that at 25 °C (298 K) the vapour pressure is continuously maintained at between 1.04 kPa and 0.75 kPa, i.e. the dissociation pressures of the pentahydrate and trihydrate respectively. What would be present? The pentahydrate would not, for the reasons just discussed. If the monohydrate were present then it would take up water, to form the trihydrate, in an attempt to lower the vapour pressure to the equilibrium value. Therefore all that exists is the trihydrate and water vapour (Fig. 8.20). Similarly, at pressures between 0.75 kPa and 0.107 kPa the monohydrate and water vapour exist, while between 0.107 kPa and 0 kPa the anhydrous salt and water vapour are present.

Efflorescence and deliquescence

The principles governing *deliquescence* (the absorption by salts of sufficient atmospheric water to dissolve) and *efflorescence* (loss of water from a hydrated salt to the atmosphere) are outlined above. Whether a particular salt deliquesces or effloresces depends on the water vapour pressure of the atmosphere. At 20 °C (293 K) this is approximately 15 mmHg ($\sim$2 kPa).

Copper(II) sulphate pentahydrate does not deliquesce because the saturated solution tends to lose water to the atmosphere. The saturated solution exerts a vapour pressure because of the equilibrium:

$$CuSO_4(aq) \rightleftharpoons CuSO_4{\cdot}5H_2O(s) + aq(g)$$

The equilibrium vapour pressure at 20 °C (298 K) is 2.13 kPa. If we reduce the vapour pressure below this level (i.e. if we remove water), more water vapour is produced by displacing the equilibrium to the right. In this way the equilibrium vapour pressure is restored. Therefore, if the saturated solution is exposed to the atmosphere (vapour pressure = 2 kPa) it will lose water until the dry pentahydrate is formed.

Let us now consider what would happen if we increased the vapour pressure to a value greater than 2.13 kPa. The salt would now absorb water vapour to reduce the vapour pressure to the equilibrium value. In so doing a saturated solution would be formed, i.e. the equilibrium would be displaced to the left. Generally, if the water vapour pressure of the atmosphere ($\sim$2 kPa) is greater than the vapour pressure of a saturated solution of a particular salt, then that salt deliquesces. Under ordinary conditions $CuSO_4 \cdot 5H_2O$ is not deliquescent but $CaCl_2 \cdot 6H_2O$ is because the vapour pressure of its saturated solution is only 1.00 kPa.

Let us now consider sodium carbonate decahydrate, which exerts a vapour pressure due to the equilibrium:

$$Na_2CO_3 \cdot 10H_2O(s) \rightleftharpoons Na_2CO_3 \cdot H_2O(s) + 9H_2O(g)$$

If we reduce the vapour pressure below the equilibrium value (dissociation pressure) of 2.13 kPa, more of the decahydrate decomposes to restore the equilibrium. Therefore, in the atmosphere, the decahydrate continually loses water until complete conversion into the monohydrate has occurred. Generally, a hydrated salt will effloresce if its equilibrium vapour pressure is greater than the vapour pressure of the atmosphere. Sodium sulphate decahydrate, $Na_2SO_4 \cdot 10H_2O$, also effloresces because its dissociation pressure is 3.23 kPa, but $CuSO_4 \cdot 5H_2O$, with a dissociation pressure of 1.04 kPa, does not.

8.5.2 GAS–LIQUID EQUILIBRIA

Let us consider the equilibrium that exists between a gas and its solution in water:

e.g. $N_2(g) \rightleftharpoons N_2(aq)$

By applying the equilibrium law,

$$\frac{[N_2(aq)]}{[N_2(g)]} = \text{a constant}$$

If the concentration of nitrogen gas is expressed as a pressure,

$$[N_2(aq)] = \text{a constant} \times p_{N_2}$$

Generally:

$$\text{mass of gas dissolved in 1 dm}^3 = \text{a constant} \times \text{pressure of gas}$$

This is a mathematical expression of *Henry's law* (1803), which states that at a fixed temperature the mass of gas dissolved in a given volume of solvent is proportional to the pressure of gas in equilibrium with the solution.

Volume is more convenient than mass when dealing with gases and since, at a fixed temperature and pressure,

$$\text{mass of gas} \propto \text{volume of gas}$$

$$\therefore \quad \text{volume of gas (reduced to s.t.p.) dissolved in 1 dm}^3 = \text{a constant} \times \text{pressure}$$

or: volume of gas dissolved in unit volume = a constant × pressure

The proportionality constant in the last equation is called the *absorption coefficient* (α) of the gas. It is the volume of gas, measured at s.t.p., that dissolves in unit volume of the solvent at a stated temperature under a pressure of one atmosphere (10^2 kPa). For example, the absorption coefficient of argon in water at 20 °C (293 K) is 0.056. This means that 0.056 dm^3 of argon, measured at s.t.p., will dissolve in 1 dm^3 of water at 20 °C (293 K).

Henry's law applies not only to single gases but also to mixtures of gases. In the latter case the solubility of each component is proportional to its own partial pressure, and not the total pressure of the gas.

8.5.3 SOLUTION–SOLID EQUILIBRIA

Similar principles to those of gas–solid equilibria apply. The concentration of all solids at equilibrium is constant and by convention equated to unity,

e.g. $Fe^{2+}(aq) + Ag^{+}(aq) \rightleftharpoons Fe^{3+}(aq) + Ag(s)$

$$K_c = \frac{[Fe^{3+}(aq)]}{[Fe^{2+}(aq)][Ag^{+}(aq)]}$$

This type of equilibrium is most usefully applied to saturated solutions of sparingly soluble salts,

e.g. $PbI_2(s) \rightleftharpoons \underbrace{Pb^{2+}(aq) + 2I^{-}(aq)}_{\text{ions in solution}}$

hence $K_c = [Pb^{2+}(aq)][I^{-}(aq)]^2$

That a solid salt exists in dynamic equilibrium with its ions in solution can be demonstrated by the use of radioactive tracers. For example, if lead(II) iodide labelled with the radioactive isotope iodine-131 is added to a saturated solution of lead(II) iodide containing non-radioactive iodine, then after a time some radioactivity is imparted to the solution. This can occur only if the above equilibrium is dynamic. If it were not dynamic then the labelled lead(II) iodide would not dissociate in the saturated solution.

Generally, for a sparingly soluble salt of formula X_nY_m,

$X_nY_m(s) \rightleftharpoons nX^{m+}(aq) + mY^{n-}(aq)$

The product $[X^{m+}]^n[Y^{n-}]^m$ is a constant known as the *solubility product*, K_s, of the salt (§9.3.1).

8.5.4 LIQUID–LIQUID EQUILIBRIA

Certain pairs of liquids, e.g. ethanol and water, dissolve in each other to form a solution. Molecules of ethanol strongly attract those of water (§4.9.2) and the two liquids are *miscible* in all proportions.

Other pairs of liquids, e.g. phenylamine (aniline) and water, do not give a homogeneous mixture. The attraction between water molecules themselves is much greater than between water molecules and phenylamine molecules. Consequently, when phenylamine and water are shaken together, the mixture separates into two liquid phases (or layers) with a clearly defined phase boundary between them. The lower, denser phase is apparently phenylamine, and the upper phase water. The two liquids are said to be *immiscible* with each other.

However, analysis shows that some dissolving takes place. The lower phase is actually a dilute (but saturated) solution of water in phenylamine, while the upper one is a dilute (but saturated) solution of phenylamine in water. A dynamic equilibrium exists between the two phases. In view of this, phenylamine and water are more accurately described as being *partially miscible*. No two liquids are totally immiscible in each other. In some cases, however, the solubility of one liquid in another is so low that for all practical purposes they can be regarded as immiscible. Such is the case with mercury and water or hydrocarbon oils and water.

Solubility of partially miscible liquids

The mutual solubility of two partially miscible liquids usually increases with temperature, to such an extent that the liquids may become miscible in all proportions. This is easily demonstrated by mixing together phenol and water in roughly equal proportions. At room temperature the mixture separates into two liquid phases, the lower one rich in phenol and the upper one rich in water. As the temperature rises little appears to happen, but analysis shows that water passes into the phenol-rich layer and phenol into the watery layer. At a certain temperature the composition of the two layers becomes identical, and when this happens the phase boundary suddenly disappears. On cooling down, the liquid becomes milky at this temperature (although there may be some delay) and separates once more into two phases.

Distillation of immiscible liquids

If a mixture of two immiscible liquids is agitated, each liquid can reach the surface and exert its normal equilibrium vapour pressure, as though it were present by itself,

$$\therefore \quad \text{total pressure} = p_A^0 + p_B^0$$

If the mixture is heated the total pressure becomes equal to the external pressure, e.g. that of the atmosphere, and the mixture boils. For example, at a temperature of 99 °C (372 K) phenylamine has a vapour pressure of 0.036 atm and water 0.964 atm. Therefore neither liquid, by itself, will boil at this temperature. But the combined vapour pressure is one atmosphere, which enables the mixture to boil. Generally, *the boiling temperature of the mixture is below that of either component*. Furthermore, the boiling temperature remains constant for as long as both liquids are present.

This is the principle behind *steam distillation*, in which a current of steam is passed through a liquid which is immiscible with water to purify it.

The method is particularly useful for high boiling compounds or those which decompose below their boiling temperatures. It is a method of preliminary purification only, for the liquid must subsequently be separated from water, dried and distilled.

The relative amounts of two liquids A and B that distil can be calculated from their molar masses and their vapour pressures at the temperature of distillation. The number of moles of A that distils (n_A) is proportional to its vapour pressure (p_A^0). Likewise, the number of moles of B in the vapour (n_B) is proportional to the vapour pressure of B (p_B^0).

Fig. 8.21 Apparatus for steam distillation.

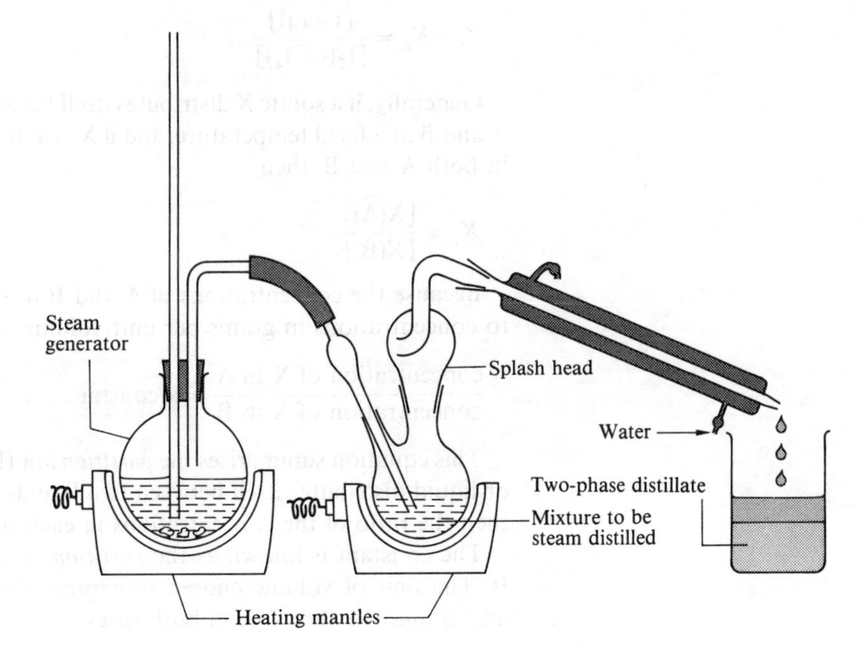

$$\therefore \quad \frac{n_A}{n_B} = \frac{p_A^0}{p_B^0}$$

$$\therefore \quad \frac{n_A \times M_A}{n_B \times M_B} = \frac{p_A^0 \times M_A}{p_B^0 \times M_B}$$

where M_A and M_B are the molar masses of A and B respectively.

Now since $n_A M_A$ is equal to the mass of A in the distillate, and $n_B M_B$ similarly is equal to the mass of B, then

$$\frac{\text{mass of A}}{\text{mass of B}} = \frac{p_A^0 \times M_A}{p_B^0 \times M_B}$$

If the molar mass of one liquid only is known, that of the other can be calculated from the masses which are steam distilled. However, this is not an accurate method of determining molar mass.

8.5.5 SOLUTION–SOLUTION EQUILIBRIA

Partition law

In equilibria involving a dissolved solid distributed between two mutually immiscible liquids we can again apply the equilibrium law. For example, consider a solution of iodine in tetrachloromethane, represented as $I_2(CCl_4)$, which is shaken with water. An equilibrium is established in which the iodine is distributed between the two liquids,

i.e. $I_2(CCl_4) \rightleftharpoons I_2(aq)$

$$\therefore \quad K_c = \frac{[I_2(aq)]}{[I_2(CCl_4)]}$$

Generally, if a solute X distributes itself between two immiscible solvents A and B at a fixed temperature, and if X is in the same molecular condition in both A and B, then

$$K_c = \frac{[X(A)]}{[X(B)]}$$

Because the concentrations of A and B in mol dm^{-3} are proportional to concentrations in grams per unit volume, we may write

$$\frac{\text{concentration of X in A}}{\text{concentration of X in B}} = \text{a constant}$$

This equation summarises the *partition law* (Nernst, 1891): 'When a solid or liquid distributes itself between two liquids in contact with each other, then the ratio of the concentrations in each liquid phase is constant'.

The constant is known as the *partition coefficient* of X between A and B. The unit of volume chosen to express the concentrations of X in A and B **must** be the same in both cases.

Ether extraction

An organic compound is often prepared in aqueous solution together with various inorganic compounds. It may be separated from these by shaking the solution with ether (ethoxyethane) in a separating funnel, a technique known as *ether extraction.* Ionic (inorganic) compounds are almost insoluble in ether and remain in the aqueous phase, while covalent (organic) compounds distribute themselves between the ether and the water in accordance with their partition coefficients.

It is always preferable to conduct repeated extractions with small volumes of ether, rather than just one extraction with a large volume of ether. Suppose that 20 g of organic solid are dissolved in 500 cm^3 of water and suppose, first, that extraction is with 300 cm^3 of ether. Let the partition coefficient be 4:1 in favour of the ether.

Let the concentration in ether be x g dm^{-3}. The 300 cm^3 of ether will extract $\frac{300}{1\,000} \times x = 0.3x$ g of solid, and the amount left in the water will be $20 - 0.3x$ g. This is in 500 cm^3; therefore in 1 dm^3 there will be $40 - 0.6x$ g.

$$\therefore \quad \frac{\text{concentration in ether}}{\text{concentration in water}} = \frac{x}{40 - 0.6x} = \frac{4}{1}$$

from which $\quad x = 47.1 \text{ g dm}^{-3}$

If there are 47.1 g in 1 dm^3, then 300 cm^3 of ether extract 47.1 × 0.3 = **14.13 g** of solid.

Now suppose that the 300 cm^3 of ether were to be used in three 100 cm^3 portions. By the same reasoning, for the first extraction,

$$\frac{\text{concentration in ether}}{\text{concentration in water}} = \frac{x}{40 - 0.2x} = \frac{4}{1}$$

from which $\quad x = 88.9 \text{ g dm}^{-3}$

100 cm^3 of ether therefore extract 8.89 g of solid, and 11.11 g remain in the water.

For the second extraction,

$$\frac{\text{concentration in ether}}{\text{concentration in water}} = \frac{x}{22.22 - 0.2x} = \frac{4}{1}$$

from which $\quad x = 49.4 \text{ g dm}^{-3}$

100 cm^3 of ether extract 4.94 g, and 6.17 g remain in the water.

For the third extraction,

$$\frac{\text{concentration in ether}}{\text{concentration in water}} = \frac{x}{12.34 - 0.2x} = \frac{4}{1}$$

from which $\quad x = 27.4 \text{ g}$

100 cm^3 of ether therefore extract 2.74 g.

Total extraction is therefore 8.89 + 4.94 + 2.74 = **16.57 g**, which compares favourably with the 14.13 g obtained by the single extraction.

Modification of the partition law to allow for association

Some solutes associate in one of the solvents but not in the other. This is the case with carboxylic acids distributed between water and benzene. In the latter solvent, association takes place by hydrogen bonding to give dimeric molecules, i.e.

```
          O···H—O
         //        \
    R—C              C—R
         \        //
          O—H···O
```

At equilibrium we have:

$$2RCOOH(aq) \rightleftharpoons (RCOOH)_2(C_6H_6)$$

$$\therefore \quad K_c = \frac{[(RCOOH)_2(C_6H_6)]}{[RCOOH(aq)]^2}$$

If we were to separate the two solutions and analyse them we should find:

$$\frac{\text{concentration of carboxylic acid in } C_6H_6}{(\text{concentration of carboxylic acid in water})^2} = \text{a constant}$$

or

$$\frac{(\text{concentration of carboxylic acid in } C_6H_6)^{\frac{1}{2}}}{\text{concentration of carboxylic acid in } H_2O} = \text{a constant}$$

(partition coefficient)

where the concentrations are measured in grams per unit volume.

For the general case in which a solute Y does not associate in solvent B, but associates in solvent A to form molecules of the type Y_n,

i.e. $nY(B) \rightleftharpoons Y_n(A)$

$$\frac{(\text{concentration of Y in A})^{\frac{1}{n}}}{\text{concentration of Y in B}} = \text{a constant}$$

The value of n therefore reflects the number of molecules linked together. The same relationship holds for dissociation of the solute, but in this case n is less than one.

The investigation of complexes

If two substances X and Y react together in aqueous solution to give an unstable substance XY_n,

i.e. $X + nY \rightleftharpoons XY_n$ (1)

it may be difficult to investigate the nature of XY_n without destroying it. However, this can be done by adding to an aqueous solution of X of known concentration an excess of Y, followed by a solvent that is immiscible with water and in which only Y dissolves. The following equilibrium, for free (i.e. uncombined) Y, is established after shaking:

$Y(aq) \rightleftharpoons Y(\text{solvent})$

After separation, the two solutions are analysed to determine the following:

(i) the molar concentration of free Y in the solvent $= c_s$;
(ii) the total molar concentration of Y in the aqueous solution $= c_t$. (The equilibrium (1) is disturbed to the left during analysis.)

These two concentrations relate to the total system at equilibrium, i.e. the aqueous solution plus solvent.

Knowing the partition coefficient of Y between water and the solvent, we can calculate the concentration of free Y in the aqueous layer, i.e.

concn. of free Y in $H_2O = c_f =$ partition coefficient × concn. of Y in solvent

$\therefore \quad c_t - c_f =$ molar concentration of Y in XY_n

If all of X is combined at equilibrium to form XY_n, then the molar ratio X:Y and hence the value of n can be calculated.

For example, the constitution of the complex ion formed by adding ammonia to aqueous copper(II) sulphate can be investigated by the addition of trichloromethane, in which ammonia (but not $CuSO_4$) is soluble. Free ammonia distributes itself between the two solvents in accordance with its partition coefficient, while combined ammonia remains with the copper in the aqueous phase. The concentration of ammonia in each phase is determined by titration with standard acid.

Let us assume that at laboratory temperature the partition coefficient for ammonia between water and trichloromethane is 25.0. Suppose that after adding 25.0 cm³ NH_3(aq) to 25.0 cm³ 0.05 M $CuSO_4$(aq), followed by a quantity of trichloromethane, analysis shows that:

$$\text{concn. of } NH_3\text{(free) in } CHCl_3 = 0.0128 \text{ M}$$
$$\text{and concn. of } NH_3\text{(total) in } H_2O = 0.416 \text{ M}$$

From the partition law,

$$\frac{\text{concn. of } NH_3\text{(free) in } H_2O}{\text{concn. of } NH_3\text{(free) in } CHCl_3} = 25.0$$

$$\therefore \quad \text{concn. of } NH_3\text{(free) in } H_2O = 25.0 \times 0.0128 = 0.320 \text{ M}$$

Now, in the aqueous phase,

$$\text{concn. of } NH_3\text{(total)} = \text{concn. of } NH_3\text{(combined)} + \text{concn. of } NH_3\text{(free)}$$

$$\therefore \quad \text{concn. of } NH_3\text{(combined)} = 0.416 - 0.320 = 0.096 \text{ M}$$

$$\text{After addition of } NH_3\text{(aq), concn. of } CuSO_4 = 0.025 \text{ M}$$

$$\therefore \quad [Cu^{2+}] = 0.025 \text{ M}$$

$$\therefore \quad \frac{\text{concn. of } NH_3\text{(combined)}}{\text{concn. of } Cu^{2+}} = \frac{0.096}{0.025} = 3.84$$

i.e. the complex ion has the formula $[Cu(NH_3)_4]^{2+}$.

The generalisations made above concerning the formation of XY_n type compounds are valid only if Y does not associate in the solvent. If association does occur then the correct expression for the partition coefficient must be applied.

EXAMINATION QUESTIONS ON CHAPTER 8

1 (a) Explain what you understand by the term *equilibrium constant*.

(b) Write a brief account of the influence of change of pressure and temperature on the position of equilibrium of a balanced homogeneous gaseous reaction, the rate at which equilibrium is attained and the value of the equilibrium constant.

(c) Carbon monoxide will react with steam under appropriate conditions according to the following reversible reaction

$CO(g) + H_2O(g) \rightleftharpoons CO_2(g) + H_2(g)$; $\Delta H = -40 \text{ kJ mol}^{-1}$

(i) Calculate the number of moles of hydrogen in the equilibrium mixture when three moles of carbon monoxide and three moles of steam are placed in a reaction vessel of constant volume and maintained at a temperature at which the equilibrium constant has the numerical value of 4.00.

(ii) Calculate the mole fractions of reactants and products.

(iii) Sketch a graph to show how the amount of carbon monoxide changes during the course of the reaction.

(iv) Sketch similar graphs on the same scale to illustrate what happens when the reaction is repeated exactly as before except that in the first the temperature is raised, in the second the pressure is lowered and in the third an industrial catalyst is introduced, the other factors remaining constant. Add a brief note to clarify what you have shown. (SUJB)

2 What do you understand by the terms *partial pressure* and *concentration* as applied to gases? Deduce the relationship between these two quantities for an ideal gas.

Write down expressions for the equilibrium constants K_p and K_c for the reaction

$SO_2Cl_2(g) \rightleftharpoons SO_2(g) + Cl_2(g)$

What is the effect (at constant temperature) on the position of this equilibrium and on K_p of:

(a) adding a catalyst,

(b) compressing the system?

At a temperature of 375 °C and an overall pressure of 101 325 N m^{-2} a sample of SO_2Cl_2 in the gas phase was found to be 84% dissociated. What is the value of K_p for the above reaction under these conditions? (OC)

3 Write a *short account* of the factors affecting the position of equilibrium of a balanced reaction, the rate at which equilibrium is attained and the value of the equilibrium constant.

(a) Using partial pressures, show that for gaseous reactions of the type

$XY(g) \rightleftharpoons X(g) + Y(g)$

at a given temperature, the pressure at which XY is exactly *one-third* dissociated is numerically equal to *eight* times the equilibrium constant at that temperature.

(b) When one mole of ethanoic acid (acetic acid) is maintained at 25 °C with 1 mole of ethanol, one-third of the ethanoic acid remains when equilibrium is attained. How much would have remained if one-half of a mole of *ethanol* had been used instead of one mole at the same temperature? (SUJB)

4 Ammonia, when in contact with a heated tungsten wire, decomposes into its constituent elements. The rate of this decomposition reaction may be studied by measuring the pressure of gas inside a closed reaction vessel at suitable time intervals. In an experiment carried out in a vessel kept at constant temperature, the following results were obtained:

Time (seconds)	0	100	200	400	600	800	1 000
Pressure (mmHg)	200.0	211.0	222.1	244.0	266.3	287.9	310.0

(a) Write down a balanced equation for the decomposition of ammonia.

(b) Plot a graph showing the increase in pressure with time, for the decomposition of ammonia.

(c) Calculate the time, from the beginning of the experiment, when three-quarters of the original quantity of ammonia would have decomposed. Briefly state the reason for your answer.

(d) Draw a labelled diagram of a suitable apparatus in which you could carry out the decomposition experiment. Assume that the temperature of the experiment is 100 °C although the temperature of the wire may be higher than this. (L)

5 Discuss the origin of the various possible types of curve arising when the vapour pressure of a mixture of two miscible volatile solvents is plotted against the composition expressed in mole fraction. You should refer to the ideal case and also to two non-ideal cases. For each case mentioned above, draw also the boiling point/composition curve. In the ideal case, show how the curve may be used to explain the process of fractional distillation; and for one non-ideal case, show how the curve may be used to explain the existence of an azeotropic mixture. (JMB)

6 Bromobenzene and water are immiscible liquids.

(a) Sketch curves to show how the vapour pressure of the system varies with (i) composition, (ii) temperature, showing in each case the contributions made by separate components, and explain briefly what is illustrated. Derive an expression relating the ratio of the masses of the components obtained on distillation to their molar masses (relative molecular masses) and vapour pressures.

Explain why steam distillation is important in the laboratory.

(b) Describe briefly how the behaviour of the system would differ if the two liquids were miscible.

(c) An aromatic compound distils in steam at 98.2 °C under 731.9 mmHg pressure. The ratio of the mass of compound to mass of water in the distillate is 0.188. Calculate its molar mass (relative molecular mass) given that the saturated vapour pressure of water at 98.2 °C is 712.4 mmHg, and state the units in which it is expressed ($H_2O = 18$).

Note: the non-SI unit of pressure, mmHg, is used in this calculation. (SUJB)

7 At standard atmospheric pressure nitric acid (boiling point 87 °C) and water form a constant boiling mixture, having a boiling point of 122 °C and composition 65% by mass of nitric acid.

(a) Define the term *constant boiling mixture.*

(b) Sketch and label fully the boiling point/composition diagram for nitric acid and water.

(c) State Raoult's law. Explain what is meant by the statement that a *nitric acid/water mixture shows negative deviation from the law.*

(d) What changes take place when nitric acid is added to water?

(e) State qualitatively what happens to the temperature and composition of the residual mixture during the distillation of a nitric acid/water mixture containing initially 20% by mass of nitric acid.

(f) Name a pair of liquids which give a positive deviation from Raoult's law and another pair which obey Raoult's law very closely. (AEB)

8 Write a brief account of the influence of change of pressure and temperature on the position of equilibrium of a balanced reaction, the rate at which equilibrium is attained and the value of the equilibrium constant.

Write down equilibrium constants, K_p or K_c as appropriate, indicating their units (if any), for the following equilibrium

$P + Q \rightleftharpoons 2R + S$

where, under the conditions of the experiments,

(i) P, Q, R and S are all gases,

(ii) P, Q, R and S are all liquids,

(iii) P and S are gases, Q and R, solids.

The vapour of dinitrogen tetroxide is partially dissociated

$N_2O_4 \rightleftharpoons 2NO_2$

4.80 g of dinitrogen tetroxide occupies a volume of 1.50 dm^3 at normal atmospheric pressure, 1.00×10^5 Pa (1 atmosphere) and 27 °C.

Calculate:

(i) the degree of dissociation,

(ii) the equilibrium constant, K_p, at this temperature.

N = 14, O = 16

$R = 8.314 \text{ J mol}^{-1} \text{ K}^{-1}$

$\text{J} = \text{kg m}^2 \text{ s}^{-2} = \text{N m}$

Pa (pascal, N m^{-2}) is the SI unit of pressure. (SUJB)

9 (a) The industrial manufacture of ammonia from its elements may be summarised by the equation

$N_2(g) + 3H_2(g) \rightleftharpoons 2NH_3(g) \qquad \Delta H^{\ominus} = -92.5 \text{ kJ}$

The conditions used for this conversion are:

(i) a high pressure of about 2.5×10^7 Pa (250 atmospheres),

(ii) a temperature of 400 to 500 °C,

(iii) a finely-divided catalyst of iron.

Discuss these operating conditions in the light of the chemical principles involved.

(b) When a mixture of hydrogen and nitrogen in the ratio 3 mol of hydrogen to 1 mol of nitrogen is allowed to attain equilibrium at 1.00×10^7 Pa (100 atmospheres) pressure and 400 °C, the equilibrium mixture contains 25% of ammonia by volume.

Calculate the value of K_p in Pa^{-2} for the reaction

$N_2(g) + 3H_2(g) \rightleftharpoons 2NH_3(g)$ at 400 °C.

(c) Explain the following reactions in terms of acid-base theory:

(i) $NaNH_2 + NH_4Cl \rightarrow NaCl + 2NH_3$ in liquid ammonia,

(ii) $NH_3(g) + BCl_3(g) \rightarrow NH_3 \cdot BCl_3(l)$. (AEB)

10 (a) In the hydrolysis of ethyl ethanoate (ethyl acetate) state how the *position of equilibrium*, and the *rate of attainment of equilibrium* are affected by (i) temperature, (ii) the use, as reagents for the hydrolysis, of (1) water and (2) dilute sulphuric acid.

(b) For the equilibrium between propanoic acid and ethanol

$C_2H_5OH(l) + C_2H_5COOH(l) \rightleftharpoons C_2H_5COOC_2H_5(l) + H_2O(l)$

the equilibrium constant in terms of concentration, K_c, is 7.5 at 50 °C.

(i) Write an expression for K_c for this reaction.
(ii) What mass of ethanol must be mixed with 74 g of propanoic acid at 50 °C in order to obtain 80 g of ethyl propanoate in the equilibrium mixture? (AEB)

11 (a) State the law defining the equilibrium distribution of a solute between two immiscible solvents.
(b) (i) **Outline** the method you would use to determine the distribution coefficient of iodine between water and tetrachloromethane (carbon tetrachloride) at room temperature.
(ii) State how and why the distribution of iodine would be affected by the addition of potassium iodide to the aqueous layer in (i).
(c) Industrially, silver is extracted from molten lead using molten zinc which is insoluble in lead. The solubility of silver is 300 times greater in zinc than it is in an equal volume of lead. If 0.005 litre of molten zinc is added to a liquid solution of 2 g of silver in 0.1 litre of lead, calculate the percentage of silver extracted by the zinc when the system has attained equilibrium. (AEB)

12 What do you understand by the term *dynamic equilibrium*? Describe a simple experiment which would demonstrate whether a state of dynamic equilibrium exists in a selected system.
The reaction
$ZnO(s) + CO(g) \longrightarrow Zn(g) + CO_2(g)$
has an equilibrium constant of 1.00 atm at 1 500 K. Calculate the equilibrium partial pressure of zinc vapour in a reaction vessel if an equimolar mixture of CO and CO_2 is brought into contact with solid zinc oxide at 1 500 K and at a total pressure of 1.0 atm.
This reaction is highly endothermic. Zinc is manufactured by bubbling the effluent gas from such a reaction vessel through molten lead at 700 K. Suggest reasons why this is done. (OC)

13 In working out each of the following calculations state clearly the physico-chemical principles upon which it depends.
(a) An organic compound is four times as soluble in tetrachloromethane (carbon tetrachloride) as in water. By calculation show that two successive extractions with 50 cm^3 of tetrachloromethane are together more productive than one extraction with 100 cm^3 of tetrachloromethane, in recovering the compound from its solution in 200 cm^3 of water. State any assumptions relating to the underlying principle clearly.
(b) The molar masses of nitrobenzene and of water are 123 and 18 respectively. During steam distillation of nitrobenzene the mixture boiled at 99.25 °C under atmospheric pressure, 1.00×10^5 Pa (760 mmHg), the partial vapour pressure of water at this temperature being 9.75×10^4 Pa (741 mmHg). Calculate the percentage by weight of nitrobenzene in the steam-distillate.
Pa (pascal, $N\ m^{-2}$) is the SI unit of pressure.
1 mmHg = 13.60×9.807 Pa.
(c) A complex plant sugar, a polysaccharide, develops an osmotic pressure of 7.1×10^2 Pa (5.40 mmHg) at 27 °C when a solution of concentration 5.0 g dm^{-3} is investigated by the Berkeley and Hartley method.

Calculate the number of simple sucrose sugar units, $C_{12}H_{22}O_{11}$, in each molecule of the polysaccharide, using information from *the following data only*.

$C_{12}H_{22}O_{11} = 342$; $H_2O = 18$

$R = 8.314\ \text{J mol}^{-1}\ \text{K}^{-1}$ ($= 0.0821\ \text{dm}^3\ \text{atm mol}^{-1}\ \text{K}^{-1}$)

$\text{J} = \text{kg m}^2\ \text{s}^{-2} = \text{N m}$

Pa (pascal, N m^{-2}) is the SI unit of pressure.

Standard atmospheric pressure = 760 mmHg. (SUJB)

14 (a) Assuming that the vapour pressure of water at any constant temperature is reduced on adding an involatile solute such as sugar by an amount proportional to the concentration of the solute, show with the aid of a diagram that the elevation of the boiling point of such a solution is proportional to the concentration of the solute.

(b) Give a labelled sketch of the apparatus you would use to determine the depression of the freezing point of a solution of sugar.

(c) The freezing point of pure benzene is 5.533 °C. The freezing point of a solution of 6.40 g of naphthalene ($C_{10}H_8$) in 1 000 g of benzene is 5.277 °C, while that of a solution of 15.25 g of benzoic acid, $C_7H_6O_2$, in 1 000 g of benzene is 5.175 °C. What conclusions can you draw from this information? (O)

15 Describe, with the aid of a diagram, how you would determine the relative molecular mass of a substance by either the elevation of boiling point or the depression of freezing point. Under what three conditions is the method unsatisfactory?

18.0 g of carbamide (*urea*), $CO(NH_2)_2$, dissolved in 200 g of water raise the boiling point of water by 0.78 °C.

45.0 g of unknown substance X dissolved in 250 g of water cause an elevation of boiling point of 0.52 °C.

If X contains 40.0% carbon, 6.7% hydrogen and 53.3% oxygen by mass, what is its molecular formula? (JMB)

16 State how the solubility of a gas in a liquid varies with pressure (Henry's law). What types of system would you expect to deviate substantially from this law? How does the solubility of (a) nitrogen, and (b) hydrogen chloride, in water vary as the temperature is raised?

Using information from your data book, calculate the molar ratio of nitrogen and oxygen which will dissolve in water in contact with air at 20 °C.

Show that the data for the solubility of oxygen in water at 0 °C, given in the data book tables 35 (a) and (b), are compatible with one another. The density of water at 0 °C is $0.9999\ \text{kg dm}^{-3}$. (OC)

17 (a) State:

(i) Henry's law,

(ii) Dalton's law of partial pressures,

(iii) Raoult's law.

(b) Give explanations of the following statements:

(i) When oxygen is dissolved in water Henry's law is obeyed whereas it is not obeyed when ammonia is dissolved in water under the same conditions,

(ii) addition of glucose to water lowers the freezing point of water,

(iii) a system containing bromobenzene and water boils at 98 °C at normal atmospheric pressure, 101 kPa (760 mmHg). (The vapour pressures of bromobenzene and water at 100 °C are 20 kPa (150 mmHg) and 100 kPa (752 mmHg) respectively.)

(c) The saturation vapour pressures of a liquid X and a liquid Y at the same temperature are respectively 66 kPa (497 mmHg) and 80 kPa (602 mmHg). A mixture of X and Y in which the mole fraction of X is 0.4, at the same temperature, has a vapour pressure of 92 kPa (692 mmHg). Does this mixture obey Raoult's law? Explain how you reach your conclusion. (AEB)

18 State carefully Raoult's law of vapour pressure lowering.

(a) *Vapour pressure lowering, boiling point elevation, freezing point depression and osmotic pressure are all closely related properties of solutions.* Comment and indicate why different methods for determining molar masses (relative molecular masses), which are based on them, are preferred according to the circumstances.

(b) Describe how you would determine the molar mass of a compound in the laboratory by measuring

either the elevation of the boiling point

or the depression of the freezing point

of a suitable solvent on forming a solution of the compound. Include the relevant vapour pressure/temperature graph in your account and link it briefly to Raoult's law.

(c) A 4.0% solution of ribitol in water (i.e. 4 g in 100 g of solution) has the same boiling point as a 4.5% solution of glucose ($C_6H_{12}O_6 = 180$) in water. Calculate the molar mass of ribitol.

(d) If the apparent degree of ionisation of potassium chloride ($KCl = 74.5$) in water at 290 K is 0.86, calculate the mass of potassium chloride which must be made up to 1 dm^3 (1 l) of aqueous solution to have the same osmotic pressure as the solution of glucose at that temperature (Chapter 9). (SUJB)

19 Describe how you would measure the relative molecular mass of a compound either by the method of boiling point elevation or by the method of freezing point depression.

These two properties, among others, are described as colligative properties. Explain the meaning of the term *colligative property* and give **two** other examples of colligative properties.

A solution of mercury(II) nitrate containing 3.270 g in 600 g of water has a freezing point of −0.093 °C and a solution of mercury(II) chloride containing 8.131 g in 750 g of water has a freezing point of −0.075 °C.

Calculate the apparent relative molecular mass of each of the salts and discuss the significance of your results.

(Freezing point depression constant = 1.86 K mol^{-1} (kg of water).) (JMB)

20 (a) What is meant by the terms *colligative property* and *osmosis*? State how the osmotic pressure of a system is related to (i) the concentration of the solute, (ii) temperature, and (iii) the relative molecular mass of the solute.

(b) Describe a method used to give a fairly accurate determination of the relative molecular mass of a solute using osmosis. What kind of solutes are best suited for such a determination?

(c) An aqueous solution of a substance X at 27 °C exhibited an osmotic pressure of 779 kPa (7.69 atmospheres). What would be the freezing point of the same solution at standard atmospheric pressure? The molecular depression constant for water is 1.86 K kg/mol. (X is known not to associate or dissociate in aqueous solution.) (AEB)

21 (a) (i) Describe with the aid of a sketch an apparatus for the determination of the elevation of the boiling point of a solvent by a solute at a known concentration.

(ii) Explain how such measurements can be used to estimate the relative molecular mass of the solute.

(iii) State briefly what factors or circumstances could give rise to a misleading answer in such a determination of a relative molecular mass.

(b) 0.600 g of iron(III) chloride are introduced into a vessel of internal volume 200 cm^3. The vessel is evacuated and closed, and then heated to 600 K. The iron(III) chloride completely vaporises, and the pressure in the vessel is found to be 4.60×10^4 Pa. Calculate the relative molecular mass of iron(III) chloride in the gaseous state at this temperature, and comment briefly on your result.

($1 \text{ Pa} = 1 \text{ N m}^{-2}$.) (O)

22 (a) Define the term *osmotic pressure* and explain how the phenomenon of *osmosis* arises.

(b) Describe in outline an accurate method for measuring the osmotic pressure of a solution of cane sugar at 25 °C.

(c) Explain why osmotic pressure (Π) is the only colligative property which offers a practical method for the determination of relative molar masses (molecular weights) above 10 000 g mol^{-1} in value and why, during the calculation, a plot of Π/c against concentration c, at a known temperature is extrapolated to zero concentration.

(d) The osmotic pressures (Π) of a series of solutions of different concentration (c) of a sample of polystyrene in butanone are measured at 27 °C. The height of butanone records the pressure (cm) for each concentration (g cm^{-3}), the density (d) of butanone at 27 °C being 0.80 g cm^{-3}. If the intercept at $c = 0$ of a plot of Π/c against c is 110 cm^4 g^{-1}, calculate the average molar mass of the sample of polystyrene.

$R = 8.31 \text{ J mol}^{-1} \text{ K}^{-1}$

$\text{J} = \text{kg m}^2 \text{ s}^{-2}$

$g = 981 \text{ cm s}^{-2}$ (SUJB)

9 Ions in solution

9.1 The origin of ions in solution

9.1.1 IONISATION AND DISSOCIATION

Some covalent compounds *ionise* on dissolving in an ionising solvent such as water,

e.g. $CH_3COOH \rightleftharpoons H^+(aq) + CH_3COO^-(aq)$
two ions produced

$HCl \rightleftharpoons H^+(aq) + Cl^-(aq)$
two ions produced

$H_2SO_4 \rightleftharpoons 2H^+(aq) + SO_4^{2-}(aq)$
three ions produced

Ionisation is seldom complete, and a dynamic equilibrium is rapidly established in which the concentrations of the various species are constant. The term *degree of ionisation* (α) denotes the fraction of molecules that are ionised in solution at equilibrium. For example, if 0.1 mol of a solute is added to water and half of the molecules (i.e. 0.05 mol) are ionised at equilibrium, then $\alpha = 0.5$. Values for α are often multiplied by 100 and expressed as percentages. Thus, in the above example, the *percentage ionisation* of the solute is 50%.

Electrovalent or ionic compounds *dissociate* when added to water. Such compounds in the solid state already consist of ions, and on dissolving in water all that happens is that the ions become separated. Ions in solution attract one another, giving the impression, in colligative properties and conductance experiments, of incomplete dissociation. In dilute solutions, where the ions are relatively remote from one another, the attraction is negligible, but in more concentrated solutions where the ions are closer together the attraction is important. A measure of the attraction is provided by α, referred to in this context as the *degree of dissociation* or *apparent degree of dissociation.* The overall effect in solution is that the ionic compound appears to be incompletely dissociated, and this allows us to treat the phenomena arising from dissociation in precisely the same way as those due to ionisation.

9.1.2 DETERMINATION OF α

Colligative properties (§8.4.1) depend on the concentration of particles in solution. Ionisation increases the number of solute particles, all of which – ions as well as molecules – are equally effective in so far as colligative properties are concerned. For example, each of the two ions formed when a molecule of ethanoic acid ionises is as effective as the molecule itself. The lowering of vapour pressure, depression of freezing temperature, elevation of boiling temperature or osmotic pressure is therefore greater than would be expected on the basis of the solute not ionising.

Suppose that one mole of a solute AB producing two ions per molecule is added to water, and that α is the degree of ionisation. At equilibrium:

$$AB \rightleftharpoons A^+ + B^-$$

moles at equilibrium $1-\alpha \quad \alpha \quad \alpha$

Total number of moles at equilibrium $= (1-\alpha) + \alpha + \alpha = 1 + \alpha$

To assist in the determination of α, a factor known as the *van't Hoff factor* (i) is defined as follows:

$$i = \frac{\text{observed effect}}{\text{calculated effect}}$$

The 'observed effect' is an *experimental* value of a colligative property, whereas the 'calculated effect' is the value of that colligative property which the solution would possess if the solute did not ionise. The observed effect is proportional to $(1+\alpha)$ mol of solute particles, while the calculated effect is proportional to 1 mol of solute particles.

$$\therefore \quad \frac{\text{observed effect}}{\text{calculated effect}} = \frac{(1+\alpha) \times c}{1 \times c} = \frac{1+\alpha}{1}$$

where c is a constant.

$$\therefore \quad \alpha = i - 1$$

For a solute of the type A_2B producing three ions per molecule, the corresponding ionic equilibrium is

$$A_2B \rightleftharpoons 2A^+ + B^{2-}$$

moles at equilibrium $1-\alpha \quad 2\alpha \quad \alpha$

Total number of moles at equilibrium $= (1-\alpha) + 2\alpha + \alpha = 1 + 2\alpha$

$$\text{The van't Hoff factor} \quad (i) = \frac{1+2\alpha}{1}$$

$$\therefore \quad \alpha = \frac{i-1}{2}$$

Corresponding expressions for α can be derived for solutes providing four or more ions in solution.

Colligative properties together with the relevant van't Hoff factor are useful for obtaining α. For example, a solution containing 47.25 g of chloroethanoic acid in 1 000 g of water depressed the freezing temperature by 0.97 °C. Given that the cryoscopic constant for water is 1.86 K mol^{-1} kg, what is α?

Chloroethanoic acid produces two ions in solution:

$$CH_2ClCOOH \rightleftharpoons H^+(aq) + CH_2ClCOO^-(aq)$$

If the acid (relative molecular mass = 94.5) did not ionise, the depression of freezing temperature (calculated effect) would be:

$$\Delta T = \frac{47.25}{94.5} \times \frac{1\,000}{1\,000} \times 1.86 = 0.93\ ^\circ\text{C} \qquad (\S 8.4.1)$$

$$\therefore \quad i = \frac{1+\alpha}{1} = \frac{0.97}{0.93} = 1.043$$

$$\therefore \quad 1+\alpha = 1.043$$

$$\therefore \quad \alpha = 0.043 \text{ (or } 4.3\%)$$

An alternative method for determining α is based on conductance measurements. Values of α are needed for the determination of pK and pH values (§9.2.1 and §9.2.4).

9.2 Acids and bases

One of the first definitions of acids and bases was put forward by S. Arrhenius in 1887. He defined an *acid* as a compound which, in aqueous solution, produces hydrogen ions, $H^+(aq)$ (§11.1.1). A *base* was defined as a substance that gives rise to hydroxide ions, HO^-. Neutralisation, on this theory, amounts to:

$$H^+(aq) + HO^-(aq) = H_2O(l)$$

An acid that produces one hydrogen ion from each molecule of acid that ionises is said to be *monoprotic*. If two or more hydrogen ions are produced per molecule of acid the terms *diprotic*, *triprotic* and so forth are used. If one molecule of a monoprotic acid is neutralised by one molecule of a base then the base is said to be *monoacidic*. If two or more molecules of acid are required for neutralisation the base is said to be *diacidic*, *triacidic* and so forth.

Although somewhat restricted in scope, the Arrhenius definition enables much of the aqueous chemistry of acids and bases to be explained. Wider definitions of the terms 'acid' and 'base' were later proposed by Brønsted and Lowry (§9.2.7) and G. N. Lewis (§9.2.9).

9.2.1 DISSOCIATION CONSTANTS AND pK VALUES

Acids

Acids such as nitric acid and hydrochloric acid give a high concentration of hydrogen ions in aqueous solution (relative to the concentration of the un-ionised acid) and are called *strong acids*. Those such as ethanoic acid and hydrogen cyanide give only a low concentration of hydrogen ions (relative to that of the acid) and are called *weak acids*. The differences arise from the extent to which acid molecules ionise in aqueous solution. In a dilute solution of hydrochloric acid ionisation is virtually complete, whereas in a dilute solution of ethanoic acid only a few per cent of ethanoic acid molecules form ions.

The terms 'weak' and 'strong' are relative and the range of acid strengths is wide. A quantitative expression of acid strength can be obtained in terms of the *dissociation constant* of the acid.

In an aqueous solution of a weak monoprotic acid HA a dynamic equilibrium is reached:

$$HA + H_2O \rightleftharpoons H^+(aq) + A^-(aq)$$

Application of the equilibrium law gives:

$$\frac{[H^+(aq)][A^-(aq)]}{[HA][H_2O]} = \text{constant at a given temperature}$$

The concentration of water in dilute solutions differs little from that in pure water and may be taken as a constant. For example, the concentration of H_2O in pure water is 55.54 mol dm^{-3}, whereas in hydrochloric acid containing 0.1 mol of HCl per dm^3 the concentration of H_2O is 55.37 mol dm^{-3}. The above equations can therefore be simplified as follows:

$$HA \rightleftharpoons H^+ + A^-$$

$$\therefore \quad \frac{[H^+][A^-]}{[HA]} = \text{constant } (K_a),$$

where K_a is the *dissociation constant* or *ionisation constant* of the acid HA. Values of K_a provide a *direct* indication of acid strength, i.e. the greater the dissociation constant the stronger is the acid. Some typical values for relatively weak acids at 25 °C (298 K) are quoted in Table 9.1.

Table 9.1 K_a and pK_a values of some common acids at 25 °C (298 K)

Acid	K_a/mol dm^{-3}	pK_a
CCl_3COOH	2.24×10^{-1}	0.65
$CHCl_2COOH$	5.13×10^{-2}	1.29
H_3PO_4	7.08×10^{-3}	2.15 (pK_1)
$CH_2ClCOOH$	1.38×10^{-3}	2.86
HNO_2	4.57×10^{-4}	3.34
$HCOOH$	1.78×10^{-4}	3.75
CH_2ClCH_2COOH	7.94×10^{-5}	4.10
C_6H_5COOH	6.31×10^{-5}	4.20
CH_3COOH	1.74×10^{-5}	4.76
HClO	3.72×10^{-8}	7.43
HBrO	2.00×10^{-9}	8.70
HCN	3.98×10^{-10}	9.40
C_6H_5OH	1.00×10^{-10}	10.0
HIO	3.02×10^{-11}	10.52
H_2O	1.81×10^{-16}	15.74
C_2H_5OH	$\sim 10^{-16}$	~ 16

Decreasing acid strength ↓

K_a as defined above does not apply to strong acids.

Discussions of acid strengths in terms of negative indices can be avoided if pK_a values are used. By definition,

$$pK_a = -\lg K_a$$

Thus, for ethanoic acid,

$$K_a = 1.74 \times 10^{-5} \text{ mol dm}^{-3}$$

$$\text{and } pK_a = -\lg(1.74 \times 10^{-5})$$

$$= -(0.24 - 5) = 4.76$$

pK_a values for all weak acids are positive. Because the sign is changed on going from K_a to pK_a, the latter is *inversely* related to acid strength, i.e. the smaller a pK_a value the stronger is the acid (Table 9.1).

For a diprotic acid, the degree of dissociation for the second stage of ionisation is less than that for the first, owing to the difficulty of removing a proton from a negatively charged ion. For a triprotic acid, the degree of dissociation for the third stage of ionisation is invariably small. Thus, if an acid has a basicity greater than one, each stage of the ionisation has its own dissociation constant and its own pK value. For such acids we no longer use the notation pK_a, but pK_1, pK_2, etc (§15.2.10).

Bases

Substances which give rise to a high concentration of hydroxide ions in aqueous solution (relative to the concentration of the substance itself) are referred to as *strong bases*. They comprise ionic compounds, e.g. sodium hydroxide, which are completely dissociated in solution. By contrast, *weak bases* give only a low concentration of hydroxide ions in solution. They are covalent compounds that ionise to a limited extent,

e.g. for $NH_3 + H_2O \rightleftharpoons NH_4^+(aq) + HO^-(aq)$

$\alpha = 0.013$ for a 0.1 M solution at 25 °C (298 K).

By application of the equilibrium law,

$$\frac{[NH_4^+(aq)][HO^-(aq)]}{[NH_3][H_2O]} = \text{constant at a given temperature}$$

For the reason given above this can be simplified to:

$$\frac{[NH_4^+][HO^-]}{[NH_3]} = \text{constant } (K_b)$$

where K_b stands for the *dissociation constant* or *ionisation constant* of the base. pK_b is defined analogously to pK_a. Some values are given in Table 9.2.

Table 9.2 K_b and pK_b values of some common bases at 25 °C (298 K)

Base	K_b/mol dm^{-3}	pK_b
$(CH_3)_2NH$	5.25×10^{-4}	3.28
CH_3NH_2	4.37×10^{-4}	3.36
$(CH_3)_3N$	6.31×10^{-5}	4.20
$C_6H_5CH_2NH_2$	2.34×10^{-5}	4.63
NH_3	1.78×10^{-5}	4.75
NH_2NH_2	8.51×10^{-7}	6.07
NH_2OH	6.61×10^{-9}	8.18
$C_6H_5NH_2$	4.17×10^{-10}	9.38
CH_3CONH_2	7.9×10^{-16}	15.1

Decreasing base strength ↓

Instead of pK_b values for bases, it is common practice to quote pK_a values for the corresponding conjugate acids, e.g. pK_a for NH_4^+ (§9.2.7).

9.2.2 VARIATION OF α WITH CONCENTRATION (Ostwald's dilution law)

Ionisation requires water (or other ionising solvent). In the absence of water most acids are un-ionised, as shown by the fact that they are non-conductors of electricity. The more water that is present, the greater is the degree of ionisation, as may be shown by applying the equilibrium law to the ionisation. Consider a solution of a weak monoprotic acid HA, having a concentration of c mol dm^{-3} and a degree of ionisation α.

	HA	$\rightleftharpoons$	H^+	+ A^-
molar concentrations at start	c		0	0
molar concentrations at equilibrium	$c(1-\alpha)$		$c\alpha$	$c\alpha$

$$\therefore \quad K_a = \frac{[H^+][A^-]}{[HA]} = \frac{c^2\alpha^2}{c(1-\alpha)} = \frac{c\alpha^2}{1-\alpha}$$

This relationship, known as *Ostwald's dilution law*, is applicable to weak acids and, by analogy, to weak bases. Since K_a is constant at a given temperature, the dilution law shows that any decrease in concentration must be accompanied by an increase in the degree of ionisation.

$$\text{From} \quad K_a = \frac{c\alpha^2}{1-\alpha}, \qquad c\alpha^2 = K_a - K_a\alpha$$

Rearrangement yields a quadratic equation,

$$c\alpha^2 + K_a\alpha - K_a = 0$$

which can be solved for α by using the standard formula.

For very weak acids such as ethanoic acid, provided that the solution is not too dilute, α is very small compared with one and

$$1 - \alpha \approx 1$$

$$\therefore \quad K_a \approx c\alpha^2$$

$$\text{and} \quad \alpha \approx \sqrt{K_a/c}$$

This is referred to as the 'approximate form of Ostwald's dilution law'. Although extensively used for very weak acids and bases, it must not be used if the solution is very dilute, since α is then comparable with one and the approximation cannot be made.

The dilution law shows that any increase in hydrogen ion concentration to be expected from increasing the concentration of a weak acid is partially counteracted by a decrease in the degree of ionisation. Consider, for example, the increase in hydrogen ion concentration of, first, hydrochloric acid and, second, ethanoic acid as the acid concentration is raised from 0.1 mol dm^{-3} to 1.0 mol dm^{-3}.

Hydrochloric acid

If we assume that the acid is completely ionised in solution, each mole of HCl produces one mole of hydrogen ions,

$\therefore \quad [H^+] = c$

$\therefore$ for 0.1 M HCl, $[H^+] = 0.1\ \text{mol dm}^{-3}$

and for 1.0 M HCl, $[H^+] = 1.0\ \text{mol dm}^{-3}$

i.e. there is a tenfold increase in hydrogen ion concentration.

Ethanoic acid

(i) At a concentration of 0.1 M

$$\alpha = \sqrt{K_a/c} = \sqrt{1.74 \times 10^{-5}/0.1}$$
$$= 1.3 \times 10^{-2} \qquad \text{(i.e. ionisation = 1.3\%)}$$

For weak acids, $[H^+] = c\alpha$

$$\therefore \quad [H^+] = 1.3 \times 10^{-3}\ \text{mol dm}^{-3}$$

(ii) At a concentration of 1.0 M

By similar reasoning, $\alpha = 4.2 \times 10^{-3}$ (i.e. ionisation = 0.42%)

and $[H^+] = c\alpha = 4.2 \times 10^{-3}\ \text{mol dm}^{-3}$

Thus, for ethanoic acid, there is little more than a threefold increase in the hydrogen ion concentration.

9.2.3 IONIC PRODUCT OF WATER

The purest water obtainable conducts electricity to a small extent because of self-ionisation:

$$H_2O(l) \rightleftharpoons H^+(aq) + HO^-(aq)$$

Application of the equilibrium law gives, at a fixed temperature,

$$\frac{[H^+(aq)][HO^-(aq)]}{[H_2O]} = \text{constant } (K_c)$$

Because α is extremely small the concentration of H_2O is far greater than that of the ions present and remains to all intents and purposes constant. The above equations can therefore be rewritten as follows:

$$H_2O \rightleftharpoons H^+ + HO^-$$

$$\therefore \quad [H^+][HO^-] = K_c \times [H_2O] = \text{constant } (K_w)$$

The constant K_w is known as the *ionic product* of water. Its value ranges from $1.14 \times 10^{-15}\ \text{mol}^2\ \text{dm}^{-6}$ at 0 °C (273 K) to $5.13 \times 10^{-13}\ \text{mol}^2\ \text{dm}^{-6}$ at 100 °C (373 K). *At 25 °C (298 K) the value of* K_w *is* $1.008 \times 10^{-14} \approx 1 \times 10^{-14}\ \text{mol}^2\ \text{dm}^{-6}$.

In pure water, because ionisation produces equimolar concentrations of hydrogen ions and hydroxide ions,

$$[H^+] = [HO^-] = \sqrt{10^{-14}} = 10^{-7}\ \text{mol dm}^{-3}$$

The constant value of K_w at a given temperature controls the hydrogen ion concentration and hydroxide ion concentration for *all* aqueous solutions. If an acid or a soluble base (i.e. an alkali) is added to water, $[H^+]$ and $[HO^-]$ will no longer be equal to each other but, at 25 °C (298 K), *the product of* $[H^+]$ *and* $[HO^-]$ *will always be* 10^{-14} $mol^2\ dm^{-6}$. No matter how concentrated a solution of an acid or a base may be, the former will always contain some HO^- ions and the latter some H^+ ions, and if the concentration of either species is known then that of the other can be calculated. Thus, when $[H^+] = 10^{-5}$ mol dm^{-3}, the concentration of HO^- ions, as required by K_w, is 10^{-9} mol dm^{-3}.

Solutions containing greater than 10^{-7} mol dm^{-3} of HO^- ions are said to be *basic* or *alkaline* because the properties of such solutions are governed by the relatively large concentration of HO^- ions present. Similarly, solutions containing more than 10^{-7} mol dm^{-3} of H^+ ions are said to be *acidic* because the excess of H^+ ions governs the properties of such solutions. *Neutral* solutions are those in which $[H^+] = [HO^-] = 10^{-7}$ mol dm^{-3} at 25 °C (298 K).

9.2.4 THE pH SCALE

Most hydrogen ion concentrations fall between values of 10^0 (i.e. 1) and 10^{-14} mol dm^{-3}. The Swedish physical chemist S. P. Sørensen (1909) realised that this wide range of hydrogen ion concentrations could conveniently be expressed by the numbers 0 to 14 by referring to the negative logarithm to the base 10 of the hydrogen ion concentration, i.e. to the *pH* of the solution.

$$\text{pH} = -\lg[H^+] \qquad \text{(cf. definitions of p}K_a \text{ and p}K_b)$$

The set of positive numbers from 0 to 14 constitutes the *pH scale.*

For pure water at 25 °C (298 K) in which $[H^+] = 10^{-7}$ mol dm^{-3},

$$\text{pH} = -\lg[H^+] = -\bar{7} = 7$$

Acidic solutions, in which $[H^+] > 10^{-7}$ mol dm^{-3}, have pH values below 7. The more acidic the solution, the lower is the pH. For basic solutions, in which $[H^+] < 10^{-7}$ mol dm^{-3}, the pH is above 7. The more basic the solution, the higher is the pH.

$[H^+]$/mol dm^{-3}	10^0	10^{-7}	10^{-14}
pH	0	7	14
	ACIDIC SOLUTIONS	NEUTRAL SOLUTION	BASIC SOLUTIONS
	← increasing acidity		increasing basicity →

pH of acidic solutions

For strong monoprotic acids, as we have already seen, the hydrogen ion concentration is equal to the molar concentration of the acid. For example, in a 0.001 M solution of hydrochloric acid,

$[H^+] = c = 10^{-3}$ mol dm^{-3} $\therefore$ pH = 3

Because the pH scale is logarithmic, a change of one pH unit corresponds to a tenfold change in hydrogen ion concentration. Thus, a tenfold dilution of 0.001 M hydrochloric acid changes the hydrogen ion concentration from 10^{-3} to 10^{-4} mol dm^{-3}, and the pH from 3 to 4. A further tenfold dilution results in a pH of 5. The reasoning behind this relationship between dilution and pH must not be taken to extremes, however. As such a solution becomes more and more diluted, its composition approaches nearer and nearer to that of pure water. Extreme dilution therefore takes the pH nearer and nearer to, but never beyond, the value for pure water, i.e. pH 7 (Figure 9.1).

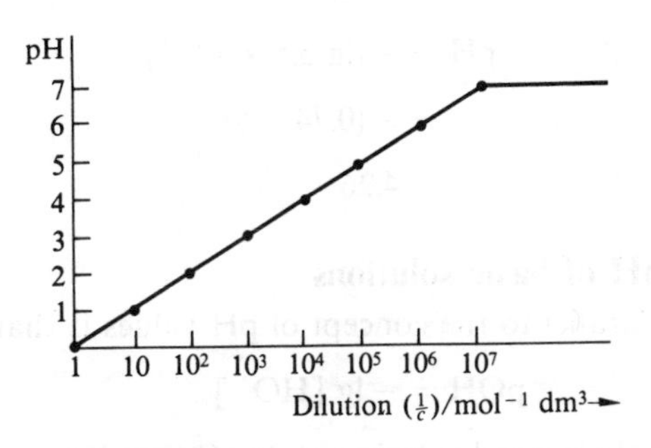

Fig. 9.1 Variation in the pH of a solution of a strong acid with increasing dilution.

pH values are not necessarily whole numbers. For example, in a solution of hydrochloric acid of concentration 4.3×10^{-4} mol dm^{-3},

$$[H^+] = c = 4.3 \times 10^{-4} \text{ mol dm}^{-3}$$

$$\begin{aligned} \therefore \quad pH &= -\lg [H^+] = -(\lg 4.3 \times 10^{-4}) \\ &= -(\lg 4.3 + \lg 10^{-4}) \\ &= -(0.63 - 4) \\ &= 3.37 \end{aligned}$$

For solutions of weak monoprotic acids the hydrogen ion concentration does *not* equal the molar concentration of the acid. For example, in a 0.1 M solution of ethanoic acid in which the degree of ionisation is 0.013, the hydrogen ion concentration is not 0.1 mol dm^{-3} since only 1.3% of the acid becomes ionised.

$$\therefore \quad [H^+] = c\alpha = 0.1 \times 0.013 = 1.3 \times 10^{-3} \text{ mol dm}^{-3}$$

$$\begin{aligned} \therefore \quad pH &= -(\lg 1.3 \times 10^{-3}) \\ &= -(\lg 1.3 + \lg 10^{-3}) \\ &= -(0.11 - 3) \\ &= 2.89 \end{aligned}$$

This value contrasts sharply with a pH of 1 for a 0.1 M solution of hydrochloric acid or nitric acid.

Because of the logarithmic nature of the pH scale, pH values for a mixture of acidic solutions cannot be obtained by simple addition of the pH of each component of the mixture. 1 dm^3 of solution of pH 4 added to 1 dm^3 of solution of pH 5 will not result in 2 dm^3 of solution of pH 9; neither will the pH be 4.5, although it must lie between 4 and 5. The pH of the mixture can be calculated as follows:

concentration of H^+ ions in solution of pH 4 = $10^{-4} = 0.0001$ mol dm^{-3}
concentration of H^+ ions in solution of pH 5 = $10^{-5} = 0.00001$ mol dm^{-3}

The result is 2 dm^3 of solution containing 0.000 11 mol of H^+ ions.

$$\therefore \quad [H^+] = \frac{0.000\,11}{2} = 0.000\,055 = 5.5 \times 10^{-5} \text{ mol dm}^{-3}$$

$$\therefore \quad pH = -(\lg 5.5 \times 10^{-5})$$
$$= -(0.74 - 5)$$
$$= 4.26$$

pH of basic solutions

Parallel to the concept of pH values is that of *pOH* values, for which

$$pOH = -\lg [HO^-]$$

In neutral solution at 25 °C (298 K),

$$[HO^-] = [H^+] = 10^{-7} \text{ mol dm}^{-3}$$

hence pOH = pH = 7

The relationship between pOH and pH values follows from the fact that for all aqueous solutions at 25 °C (298 K)

$$[H^+][HO^-] = K_w = 10^{-14} \text{ mol}^2 \text{ dm}^{-6}$$

Taking negative logarithms to the base 10 gives:

$$-\lg [H^+] + (-\lg [HO^-]) = -\lg 10^{-14}$$

i.e. $\quad pH + pOH = 14$

pOH values are useful in calculating the pH of basic solutions. For example, for a solution of sodium hydroxide of concentration 0.05 mol dm^{-3}, assuming complete dissociation,

$$[HO^-] = c = 0.05 \text{ mol dm}^{-3}$$

$$\therefore \quad pOH = -(\lg 5 \times 10^{-2})$$
$$= -(0.7 - 2)$$
$$= 1.3$$

$$\therefore \quad pH = 14 - pOH = 14 - 1.3 = 12.7$$

pH of mixed solutions of acid and base

If the overall mixture is acidic we calculate its hydrogen ion concentration and hence pH. Should the resulting solution be basic it is best to calculate pH via hydroxide ion concentration and pOH.

Example
What is the pH of a solution obtained by mixing 60 cm^3 of sodium hydroxide solution containing 0.1 mol dm^{-3} with 40 cm^3 of hydrochloric acid containing 0.05 mol dm^{-3}?

Answer Base will be present in excess in the final solution because:

$$\text{amount of } HO^- \text{ added} = \frac{60}{1\,000} \times 0.1 = 0.006 \text{ mol}$$

$$\text{amount of } H^+ \text{ added} = \frac{40}{1\,000} \times 0.05 = 0.002 \text{ mol}$$

$$\text{Concentration of } HO^- \text{ in the final solution} = 0.004 \text{ mol in } 100 \text{ cm}^3$$

$$= 4 \times 10^{-2} \text{ mol dm}^{-3}$$

$$\text{pOH} = -\lg [HO^-] = -(\lg 4 \times 10^{-2})$$

$$= -(0.6 - 2)$$

$$= 1.4$$

$$\therefore \quad \text{pH} = 14 - \text{pOH} = \mathbf{12.6}$$

9.2.5 ACID–BASE INDICATORS

Acid–base indicators are substances which are capable of showing visually the sharp pH change that occurs at the 'end' of a titration (see below). Such indicators are organic acids and they ionise in water:

$$\text{HInd} \rightleftharpoons H^+ + \text{Ind}^-$$

HInd and Ind^- represent, respectively, the un-ionised and ionised forms of the indicator. It is an essential requirement of an indicator that HInd and Ind^- must be of different colours, and that the colours must be intense enough, even when the indicator is present in very dilute solution, to be seen clearly.

In acidic solution the equilibrium is displaced to the left and the resultant colour of the indicator is that of the un-ionised acid, i.e. HInd. In basic solution H^+ ions are removed by reaction with HO^- ions. As a result the equilibrium is displaced to the right and the colour observed is that of the anion, Ind^-.

Range

Most indicators change colour over a range of approximately 2 pH units, called the *range of the indicator* (Table 9.3).

Table 9.3 pH ranges of some acid–base indicators

Indicator	Range	Colour change (acid–base)
methyl orange	3.1–4.4	red–orange
methyl red	4.4–6.0	red–yellow
litmus	4.5–8.3	red–blue
bromothymol blue	6.0–7.6	yellow–blue
phenolphthalein	8.3–9.8	colourless–pink

To be of any use for acid–base titrations indicators should have as narrow a range as possible. Litmus, with an exceptionally wide range of 3.8 pH units, does not give an accurate end-point and is seldom used in titrations. Its colour does not change sharply from red to blue but goes through various shades of purple.

Choice of indicator

Different titrations have different pHs at the *equivalence point*, i.e. the theoretical end-point, at which acid and base are present in equivalent amounts as required by the chemical equation. When selecting an indicator the pH at the equivalence point must be known, for it is essential that this pH coincides with the pH range of the indicator. For example, in titrations of strong acids against weak bases the equivalence point lies below pH 7 and methyl orange (Table 9.3) is a suitable indicator to use. If phenolphthalein were inadvertently selected for this titration the result would be inaccurate because the colour change would not be observed until an excess of base had been added. An *indicator error* would be introduced, this being the difference between the visual end-point and the theoretical equivalence point.

Screened indicators

The sharpness of the colour change can often be improved by the addition to an indicator of a suitable dye, called a *screening dye*. The mixture is known as a *screened indicator*. The screening dye is not itself an indicator, but merely gives a colour combination which is more easily observed than that of the indicator alone. For example, methyl orange can be screened with a blue dye called xylene cyanol FF.

In basic solution the colour is blue + orange = green.

In acidic solution the colour is blue + red = purple.

At the end-point the colour is blue + orange-red = grey.

The end-point is detected easily, because the colour contrast between grey and green or purple is greater than that between orange-red and orange or red.

Universal indicators

By mixing together carefully selected indicators, e.g. methyl orange, methyl red, bromothymol blue and phenolphthalein (Table 9.3) a *universal indicator* can be obtained, which displays different colours at different pHs.

Universal indicators are not used in titrations as they are not highly accurate, but they are invaluable (e.g. in industrial process control) wherever the approximate pH of a solution is required rapidly. All that is necessary is to add a few drops of the indicator to the test solution and then compare its colour with a reference chart. Both wide range and narrow range universal indicators are available. *pH papers*, which are absorbent strips of paper impregnated with universal indicator, offer an alternative to the indicator solution.

9.2.6 pH CURVES

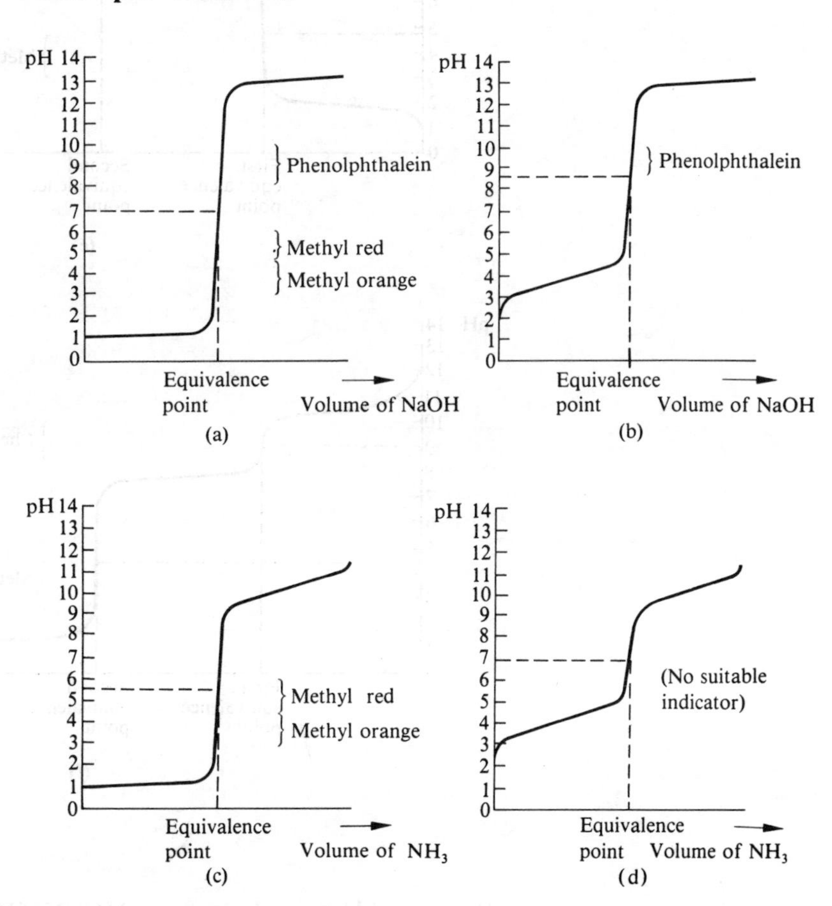

Fig. 9.2 Titration curves for (a) HCl/NaOH, (b) CH_3COOH/NaOH, (c) HCl/NH_3, (d) CH_3COOH/NH_3,

A plot of pH against volume of titrant for an acid/base titration is known as a *pH curve* or *titration curve*. The curves can be established either by using a pH meter (§2.5.5) to record pH changes during titration, or by calculation. For strong acid/strong base titrations the calculations are similar to that in §9.2.4; for titrations of weak acids or weak bases the calculations require a knowledge of K_a or K_b.

All the pH curves in Fig. 9.2 except (e) and (f) refer to the addition of 0.1 M monoacidic base to 0.1 M monoprotic acid. The equivalence point, reached when acid and base are present in equimolar proportions, is the same in all cases. The curves have been extended beyond the equivalence points to a titre at which the molar ratio of acid to base is 1:2.

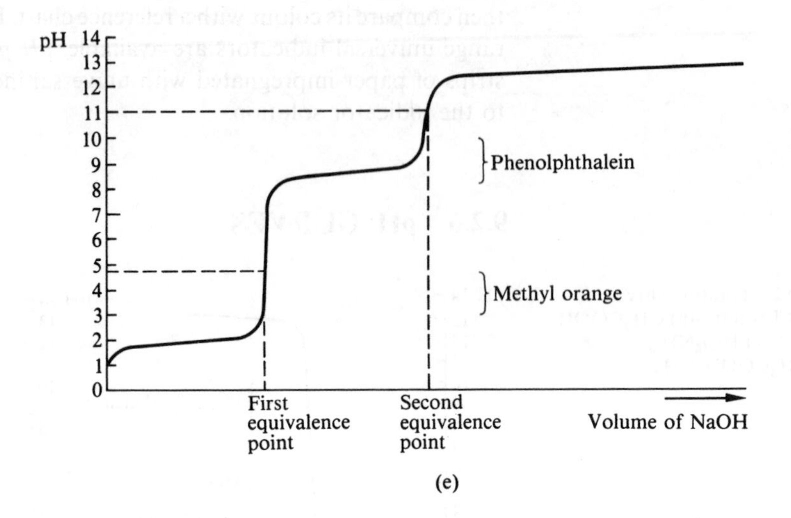

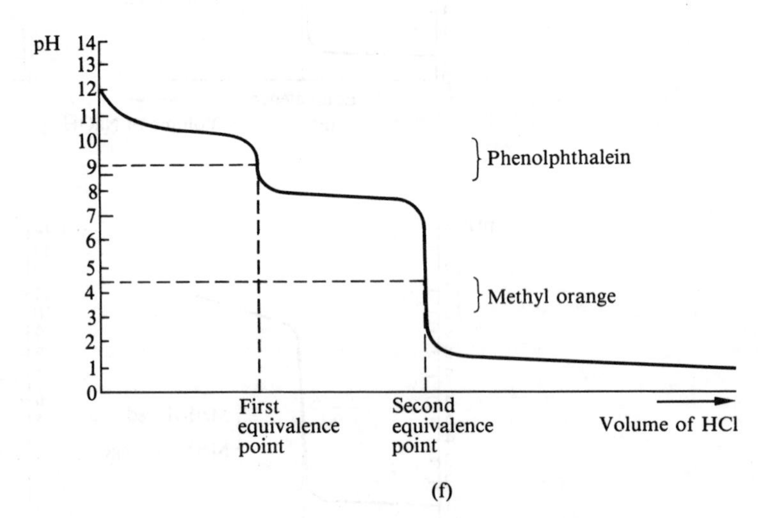

Fig. 9.2
(e) $(COOH)_2$/NaOH and
(f) Na_2CO_3/HCl.

Strong acid/strong base, e.g. HCl/NaOH (Fig. 9.2(a))

In this type of titration the addition of base has almost no effect on pH until most of the acid has been neutralised. In the region of the equivalence point, however, a small addition of titrant produces a relatively large increase in pH. Beyond the equivalence point the curve flattens off and finally reaches the pH of the strong base solution. The equivalence point is midway between the two 'flat' portions of the curve, and the whole curve is symmetrical about this point.

Figure 9.2(a) relates to 0.1 M HCl (which has a pH of 1) and 0.1 M

NaOH (pH 13). If the acid and alkali are more concentrated than this the 'vertical' portion is more extensive, and if the solutions are more dilute it is shorter. If the titration is performed so that acid is added to base the pH curve is the mirror image of that shown.

The shape of the curve is due essentially to the logarithmic nature of the pH scale. For example, when 0.1 M HCl is half neutralised its $[H^+]$ falls from 0.1 to 0.05 mol dm^{-3}, but the pH rises from 1.0 to only 1.3. When 99% of the acid is neutralised the $[H^+]$ is 0.001 mol dm^{-3} and the pH is 3.0. The remaining jump of 4 pH units to the equivalence point corresponds to the neutralisation of the last 1% of hydrogen ions.

Ideally, for strong acid/strong base titrations, an indicator is needed that changes colour at about pH 7. Apart from litmus, whose use is precluded on the grounds of wide range (see above), there is no common indicator that changes colour at this point. However, the 'vertical' portion of the pH curve is so extensive that the use of indicators such as phenolphthalein, methyl red or methyl orange, which change colour at pHs remote from 7, leads to negligible indicator errors. Unless the solutions are extremely dilute, any common indicator can be used for the titration of strong acid against strong base.

Weak acid/strong base, e.g. $CH_3COOH/NaOH$ (Fig. 9.2(b))

There are a number of differences between this and the previous curve, as follows.

(i) The initial pH is higher, owing to the relatively low hydrogen ion concentration from the weak acid. (The pH of 0.1 M ethanoic acid is 2.89.) The weaker the acid, the higher is the initial pH.

(ii) As base is added, there is an initial rapid rise in pH followed by a gradual rise that is steeper than in the previous case.

(iii) Because the second part of the curve (i.e. beyond the equivalence point) is the same as in (a), the curve as a whole is unsymmetrical.

(iv) The pH at the equivalence point is greater than 7. The reason is that at the equivalence point the solution contains, in effect, a salt of a weak acid and a strong base. Such salts are hydrolysed to give an alkaline solution (§9.2.8).

(v) The 'vertical' portion of the curve is less extensive; the weaker the acid used in the titration the smaller is the extent of the pH change around the equivalence point. Consequently, the indicator for such titrations must be carefully selected. Phenolphthalein is the obvious choice because its range encompasses the equivalence point.

Strong acid/weak base, e.g. HCl/NH_3 (Fig. 9.2(c))

The pH curve in this case is related to that in (b) and analogous arguments apply. It should be noted particularly that the pH at the equivalence point is below 7 (§9.2.8), so that methyl red or methyl orange is a suitable indicator.

Weak acid/weak base, e.g. CH_3COOH/NH_3 (Fig. 9.2(d))

When both the acid and the base are weak, the pH change marking the equivalence point is small and gradual. Under such conditions nearly all indicators change colour only gradually, and such titrations are impracticable unless they are monitored by conductance measurements.

Diprotic acid/monoacidic base, e.g. $(COOH)_2/NaOH$ (Fig. 9.2(e))

The pH curve displays two sharp rises in pH. The first corresponds to the formation of the hydrogenanion by the addition of one mole of HO^- ions for every mole of acid:

$$\begin{matrix} COOH \\ | \\ COOH \end{matrix} + HO^- \rightleftharpoons \begin{matrix} COO^- \\ | \\ COOH \end{matrix} + H_2O$$

The hydrogenanion, although still acidic, is weaker than the dibasic acid. This is natural, since removal of a proton from an anion is more difficult than removal of a proton from a molecule of the original acid.

Further titration, i.e. the addition of a second mole of HO^- ions, leads to complete neutralisation and a second sharp increase in pH marking the equivalence point:

$$\begin{matrix} COO^- \\ | \\ COOH \end{matrix} + HO^- \rightleftharpoons \begin{matrix} COO^- \\ | \\ COO^- \end{matrix} + H_2O$$

Thereafter, the pH remains steady at a value for the solution of the strong base.

The first equivalence point may be detected with methyl orange and the second by phenolphthalein. For consistency with the previous curves, the construction of the entire pH curve (Fig. 9.2(e)) requires the addition of *four* moles of base to one of acid, i.e. two moles of base to neutralise the acid and a further two moles of base beyond the second equivalence point.

Diacidic base/monoprotic acid, e.g. Na_2CO_3/HCl (Fig. 9.2(f))

Figure 9.2(f) relates to the titration of sodium carbonate solution by hydrochloric acid. In the presence of phenolphthalein a first equivalence point can be detected, corresponding to the formation of sodium hydrogencarbonate:

$$Na_2CO_3 + HCl = NaHCO_3 + NaCl \quad (1)$$

By adding methyl orange and continuing the titration a second equivalence point can be reached:

$$NaHCO_3 + HCl = NaCl + CO_2 + H_2O \quad (2)$$

The fact that sodium carbonate can be determined by titration to the hydrogencarbonate allows this compound to be estimated in mixtures with, for example, sodium hydroxide or sodium hydrogencarbonate.

Example
25.0 cm^3 of a solution containing sodium carbonate and sodium hydroxide required 22.5 cm^3 of 0.1 M hydrochloric acid for titration with phenolphthalein as indicator. When titration was continued after adding methyl orange a further 6.1 cm^3 of 0.1 M acid was required. Calculate the molar concentration of each solute.

Answer In the first titration the sodium hydroxide reacts completely:

$$NaOH + HCl = NaCl + H_2O$$

Let the volume of acid required = x cm^3. At the same time the sodium carbonate half reacts; equation (1). Let the volume of acid = y cm^3; $x + y = 22.5$.

In the second part of the titration the sodium carbonate completes its reaction; equation (2). Volume of acid = y = 6.1 cm^3.

Therefore, for the total reaction of sodium carbonate, i.e. for

$$Na_2CO_3 + 2HCl = 2NaCl + CO_2 + H_2O,$$

volume of acid = $2y$ = 2 × 6.1 = 12.2 cm^3.

$$\therefore \quad \frac{\text{molarity}(Na_2CO_3) \times \text{volume}(Na_2CO_3)}{\text{molarity}(HCl) \times \text{volume}(HCl)} = \frac{\text{molarity}(Na_2CO_3) \times 25.0}{0.1 \times 12.2} = \frac{1}{2}$$

from which molarity(Na_2CO_3) = 0.024 4

The volume of acid required for reaction with sodium hydroxide = $(x + y) - y = 22.5 - 6.1 = 16.4$ cm^3.

$$\therefore \quad \frac{\text{molarity}(NaOH) \times \text{volume}(NaOH)}{\text{molarity}(HCl) \times \text{volume}(HCl)} = \frac{\text{molarity}(NaOH) \times 25.0}{0.1 \times 16.4} = \frac{1}{1}$$

from which molarity (NaOH) = 0.065 6

9.2.7 THE BRØNSTED–LOWRY THEORY

J. N. Brønsted and T. M. Lowry (1922–23) defined an *acid* as any species capable of donating a proton, and a *base* as any species capable of accepting a proton. Neutralisation, therefore, comprises the transfer of a proton from an acid to a base,

$$\text{e.g.} \quad \underset{\text{acid}}{H_3O^+} + \underset{\text{base}}{HO^-} = 2H_2O$$

The Brønsted–Lowry theory enables the importance of the solvent in solutions of acids and bases to be interpreted. We shall examine this aspect by considering aqueous solutions of an acid (HCl) and a base (NH_3).

When hydrogen chloride is dissolved in water the following equilibrium is established:

$$HCl + H_2O \rightleftharpoons H_3O^+ + Cl^-$$

In the forward reaction the HCl molecule acts as an acid and donates a proton to a molecule of water which thus acts as a base. The oxonium ion, H_3O^+, is further hydrated to form the hydrogen ion, $H^+(aq)$, (§11.1.1) but when considering the Brønsted–Lowry theory it is convenient to write H_3O^+. For the reverse change the H_3O^+ ion functions as an acid and the Cl^- ion as a base. Thus, from the base H_2O we obtain the acid H_3O^+, and from the acid HCl we obtain the base Cl^-. The oxonium ion is referred to as the *conjugate acid* of the base H_2O, and the Cl^- ion is known as the *conjugate base* of the acid HCl. The general pattern is:

$$\text{acid A} + \text{base B} \rightleftharpoons \text{acid B} + \text{base A}$$

where base A is the conjugate base of acid A, and acid B is the conjugate acid of base B. Applying this to aqueous ammonia,

$$H_2O + NH_3 \rightleftharpoons NH_4^+ + HO^-$$

we see that the ammonium ion is the conjugate acid of the base NH_3, and the hydroxide ion is the conjugate base of the acid H_2O.

Water can therefore function as either an acid or a base. When a stronger acid, e.g. HCl, is present water acts as a base, while in the presence of a stronger base, e.g. NH_3, water functions as an acid. Although an acid like ethanoic acid is regarded as a weak acid, it is nevertheless a stronger acid than water. Proton transfer to water molecules therefore occurs and a solution of ethanoic acid is acidic. In the case of strong acids, e.g. HCl, HNO_3 and $HClO_4$, proton transfer to water molecules is virtually complete and such acids are of approximately equal strength in aqueous solution.

Let us consider the general case of the ionisation in water of an acid, HA:

$$HA + H_2O \rightleftharpoons H_3O^+ + A^-$$

If the acid, HA, is strong the equilibrium lies considerably to the right and the reverse reaction occurs to a limited extent only. Therefore, the conjugate base, A^-, is a weak base because it does not readily accept a proton from the H_3O^+ ion. By contrast, if the acid, HA, is weak the equilibrium lies to the left, which means that A^- is a strong base and readily accepts a proton from the H_3O^+ ion. Thus, *the conjugate base of a strong acid is a weak base, and the conjugate base of a weak acid is a strong base.* For example, the chloride ion is a weak base, while the ethanoate ion is a strong base. Generally, the base strength of anions increases with decreasing strength, and hence increasing pK_a values, of acids.

A similar argument applies to bases.

$$B + H_2O \rightleftharpoons BH^+ + HO^-$$

If B is a weak base the conjugate acid, BH^+, is a strong acid. Where B is a strong base, BH^+ is a weak acid. Generally, the strength of conjugate acids decreases with increasing strength, and hence decreasing pK_b values, of bases.

The Brønsted–Lowry theory allows us to quote the strength of a base by using, instead of pK_b, the pK_a value of its associated acid. For example, the position of the equilibrium

$$NH_4^+ + H_2O \rightleftharpoons NH_3 + H_3O^+$$

depends on the strength of the ammonium ion as an acid as well as the base strength of its conjugate base, ammonia. The respective strengths of the acid and base are expressed by their K_a and K_b values,

$$\text{i.e.} \quad K_a = \frac{[NH_3][H_3O^+]}{[NH_4^+]} \quad \text{and} \quad K_b = \frac{[NH_4^+][HO^-]}{[NH_3]}$$

$[H_3O^+] = [H^+]$ (see above).

Therefore:

$$K_a \times K_b = \frac{[NH_3][H_3O^+][NH_4^+][HO^-]}{[NH_4^+][NH_3]} = [H_3O^+][HO^-]$$

$$= K_w$$

At 25 °C (298 K),

$$\therefore \quad K_a \times K_b = 10^{-14}\ \text{mol}^2\ \text{dm}^{-6}$$

$$\therefore \quad pK_a + pK_b = 14$$

The pK_b of ammonia is 4.75; therefore, the pK_a of the ammonium ion is 9.25.

The Brønsted–Lowry theory also enables us to interpret acid–base behaviour in solvents other than water. For example, in liquid ammonia the ammonium ion behaves as an acid, and the amide ion, NH_2^-, as a base (§15.2.3).

9.2.8 SALT HYDROLYSIS

In aqueous solution some salts react with water molecules, an effect referred to as *salt hydrolysis*. The resulting solution may be neutral, basic or acidic. A salt is hydrolysed whenever its anion is a strong conjugate base, or its cation is a strong conjugate acid.

The anion is a strong conjugate base

The salts of strong bases, e.g. sodium hydroxide, and weak acids, e.g. ethanoic acid or carbonic acid, typify this class. Because the acid is weak, the anion, A^-, is a strong conjugate base and removes protons from water molecules. The equilibrium:

$$A^- + H_2O \rightleftharpoons HA + HO^-$$

therefore lies to the right and the solution is basic.

A solution of sodium ethanoate is less basic than one of sodium carbonate, because the carbonate ion, CO_3^{2-}, with its double negative charge, is a stronger base than the singly charged ethanoate ion, i.e. CO_3^{2-} has a greater attraction for protons than has CH_3COO^-. The cations of these salts, e.g. Na^+, show virtually no tendency to react with water molecules.

The cation is a strong conjugate acid

The salts of weak bases, e.g. ammonia or amines, and strong acids typify this category. Because the base is weak, its conjugate acid is strong and thus readily donates protons to water molecules,

e.g. $NH_4^+ + H_2O \rightleftharpoons NH_3 + H_3O^+$

The resulting solution is therefore acidic. Ammonium chloride serves as an example. Its anion, Cl^-, is the weak conjugate base of a strong acid and has little tendency to attract protons from water molecules.

Hydrated metal ions, such as $[Al(H_2O)_6]^{3+}$ (§13.2.4), also fall into this category. For example, aluminium chloride may be regarded as the salt formed from the weak base, aluminium hydroxide, and the strong acid, hydrochloric acid.

The anion is a strong conjugate base and the cation is a strong conjugate acid

The salts of weak acids and weak bases fall within this category. Here we must consider the acid strength of the cation relative to the base strength of the anion.

The anion and cation are of equal strength

Consider ammonium ethanoate. The pK_a value of ethanoic acid (4.76) is almost identical with the pK_b value of ammonia (4.75). Consequently, the base strength of the ethanoate ion is equal to the acid strength of the ammonium ion. The tendency of ethanoate ions to accept protons from water molecules is therefore equal to the tendency of ammonium ions to donate protons to water molecules. Thus, in a solution of ammonium ethanoate, the increase in concentration of hydroxide ions due to the equilibrium:

$$CH_3COO^- + H_2O \rightleftharpoons CH_3COOH + HO^-$$

is equal to the increase in concentration of oxonium ions arising from the equilibrium:

$$NH_4^+ + H_2O \rightleftharpoons NH_3 + H_3O^+$$

Because the concentrations of oxonium ions and hydroxide ions remain the same the solution is neutral. Because hydroxide and oxonium ions are produced during hydrolysis, the extent to which water self-ionises must decrease so that the ionic product is always equal to 10^{-14} mol^2 dm^{-6}.

To summarise, the salts of weak acids and weak bases of comparable strengths (i.e. with the same numerical values of pK_a and pK_b respectively) are neutral in aqueous solution.

The base strength of the anion is greater than the acid strength of the cation

Consider ammonium carbonate. The base strength of the carbonate ion is greater than the acid strength of the ammonium ion. The increase in concentration of hydroxide ions arising from the equilibrium:

$$CO_3^{2-} + H_2O \rightleftharpoons HCO_3^- + HO^-$$

is therefore greater than the increase in concentration of oxonium ions produced by the hydrolysis of ammonium ions. Consequently, solutions of ammonium carbonate are basic.

The base strength of the anion is less than the acid strength of the cation

Here the opposite argument applies. For example, the base strength of the methanoate ion is lower than the acid strength of the ammonium ion, as indicated by their relative pK values. A solution of ammonium methanoate is therefore acidic because the increase in concentration of oxonium ions resulting from hydrolysis of the ammonium ion is greater than the increase in hydroxide ion concentration arising from hydrolysis of the anion.

Acid salts

The situation regarding salts such as $NaHCO_3$ and $NaHSO_3$ is complicated by the fact that ionisation and hydrolysis occur simultaneously,

e.g. $HCO_3^- + H_2O \rightleftharpoons H_3O^+ + CO_3^{2-}$ ionisation

$HCO_3^- + H_2O \rightleftharpoons H_2CO_3 + HO^-$ hydrolysis

If the acid is very weak, as in the case of carbonic acid, then hydrolysis occurs to a greater extent than ionisation and the solution is basic. A solution of sodium hydrogencarbonate, for example, has a pH of approximately 8.5. With stronger acids, ionisation is more extensive than hydrolysis and the resulting solution is acidic. For example, solutions of sodium hydrogensulphate or sodium hydrogensulphite have pHs less than 7. The latter is less acid than the hydrogensulphate because 'sulphurous acid' (§16.2.6) is a weaker acid than sulphuric acid.

Cation and anion possess no marked acidic or basic strength

Aqueous solutions of salts derived from strong acids and strong bases are neutral. For example, both the sodium ion and the chloride ion have virtually no tendency to react with water molecules, with the result that sodium chloride does not undergo hydrolysis and gives solutions with a pH of 7.

9.2.9 THE G. N. LEWIS THEORY

In 1923 G. N. Lewis proposed the following all-embracing definitions of acids and bases. An *acid* is a species capable of accepting a lone pair of electrons to form a coordinate (and hence a covalent) bond. A *base* is a species capable of donating a lone pair of electrons to form a coordinate (and hence a covalent) bond. For example, the following reactions involve a Lewis acid and a Lewis base:

$$H_2O: + H^+ = \left[\begin{array}{c} H{-}O{\rightarrow}H \\ | \\ H \end{array}\right]^+ = \left[\begin{array}{c} H{-}O{-}H \\ | \\ H \end{array}\right]^+$$

$$6\,H_2O: + Al^{3+} = [Al(H_2O)_6]^{3+} \qquad (§13.2.4)$$

Lewis base   Lewis acid

The Lewis theory is broadly based and does not specify the nature of the acceptor or the donor; nor does it require the participation of a solvent. Thus, the reaction between the oxide ion and the sulphur trioxide molecule,

i.e. $$\underset{\text{Lewis base}}{O^{2-}} + \underset{\text{Lewis acid}}{SO_3} = SO_4^{2-}$$

occurs even in the absence of a solvent.

The Lewis theory helps to explain the similar catalytic effect of otherwise unrelated substances. For example, both concentrated sulphuric acid and aluminium oxide are able to catalyse the dehydration of an alcohol to an alkene (§14.9.5). The reason is that a molecule of alcohol can coordinate either to a proton (from the H_2SO_4) or to aluminium oxide; in other words, both H^+ and Al_2O_3 are Lewis acids.

Electrophiles and nucleophiles

In organic chemistry the Lewis acids that primarily concern us are those that can accept a lone pair of electrons from a negatively charged carbon atom – or, more often, from a carbon–carbon double bond or triple bond. Such Lewis acids are known as *electrophilic reagents*, or *electrophiles.* Similarly, the Lewis bases that most concern us are those that are capable of donating a lone pair of electrons to a positively charged carbon atom; they are called *nucleophilic reagents*, or *nucleophiles.*

An electrophilic reagent derives its name from the fact that it is electron seeking, i.e. it always seeks out an electron-rich centre for attack. The essential requirements of an electrophile are as follows.

(i) It must possess a positive charge; either a full positive charge (+) or a partial positive charge (δ^+).
(ii) It must be able to accept a lone pair of electrons.
(iii) It must be able to form a strong covalent bond with a carbon atom.

An example of an electrophile is the nitryl cation, NO_2^+. Most common cations, e.g. Na^+, Ca^{2+}, Al^{3+} and NH_4^+, are not electrophiles because they do not satisfy conditions (ii) or (iii).

A nucleophilic reagent is so called because it seeks out, for its attack, a centre that is positively charged, cf. the atomic nucleus. The essential requirements of a nucleophile are as follows.

(i) It must possess a negative charge, $-$ or δ^-.
(ii) It must be able to donate a lone pair of electrons.
(iii) It must be able to form a strong covalent bond with a carbon atom.

Examples are HO^-, $C_2H_5O^-$, CH_3COO^-, CN^- and NH_3. In addition, we may include certain substances which are not in themselves nucleophiles, but which act as such in the presence of a catalyst. Examples are HCN, CH≡CH, and most aldehydes, ketones and esters.

Nucleophiles (and to a lesser extent electrophiles) vary considerably in their reactivity. The most reactive, which are referred to as 'good entering groups', have the following characteristics:

(i) a complete negative charge, as opposed to a partial one;
(ii) the ability to be readily polarised by the positive charge on the substrate, i.e. the species that is being attacked. *Polarisation*, in this context, relates to the distortion of electron shells. For example, Cl^- and I^- ions have closely related electronic structures ($Cl^- = 2, 8, 8$; $I^- = 2, 8, 18, 18, 8$) and we might expect them to have similar nucleophilic properties. In fact I^- is much the better nucleophile, because the ready distortion of its large electron cloud towards the positive centre greatly facilitates its coordination.

9.2.10 FACTORS AFFECTING THE STRENGTHS OF OXOACIDS IN AQUEOUS SOLUTION

Many compounds containing at least one hydroxyl group are acidic and are often called *oxoacids*: The atom to which the hydroxyl groups are bonded, referred to as the *central atom*, is usually that of a non-metal, and is often bound to one or more oxygen atoms. Sometimes, however, the hydroxyl groups are bonded to a metal atom in a high valency state. Some examples of oxoacids are as follows:

```
             O    O—H    O      O              O
              \\  /       \\   //              //
Cl—O—H          S          Mn        CH3—C
              //  \       //   \              \
             O    O—H    O      O—H            O—H

             H—O     O—H
                \   /
                  P
                //  \
               O     O—H
```

In aqueous solution oxoacids ionise,

i.e. $$X{-}O{-}H \rightleftharpoons XO^-(aq) + H^+(aq)$$

where X represents the atom or group of atoms to which the hydroxyl group is bonded.

There are two principal factors governing the strength of oxoacids, as follows.

(i) The stability of the anion, $XO^-(aq)$, relative to that of the un-ionised acid, XOH. If the anion is more stable than the acid, acidity is encouraged because the anion will be reluctant to recombine with a proton. The greater the stability of the anion, in comparison with the molecule, XOH, the stronger is the acid.

(ii) The electron withdrawing or releasing effect of the atom or group X. If X draws electrons towards itself, the electrons in the O—H bond are in turn displaced towards the oxygen atom:

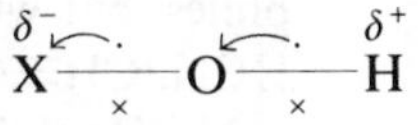

$$\overset{\delta^-}{X} \leftarrow O \leftarrow \overset{\delta^+}{H}$$

This results in a weakening of the O—H bond, which facilitates the loss of a proton. The greater the electron withdrawing effect of X, the easier proton loss can occur and the stronger is the acid.

By contrast, if X has an electron *releasing* effect, polarisation of the O—H bond is reduced. The bond is strengthened, proton loss can occur only with difficulty, and the acid becomes weaker.

Inductive effect

There are two common electron withdrawing or releasing effects, namely the inductive effect and the mesomeric effect.

An alkyl group (e.g. CH_3 or C_2H_5) is said to have a *positive inductive effect* (+I) because, in comparison with a hydrogen atom, it releases electrons towards a carbon atom to which it is attached. A halogen atom, by contrast, has a *negative inductive effect* (−I) because, compared with a hydrogen atom, it draws electrons away from an adjoining carbon atom. Both +I and −I effects are symbolised by drawing arrows on the appropriate covalent bonds:

$$Cl \leftarrow C \qquad CH_3 \rightarrow C$$

The inductive effect is due essentially to electronegativity differences. Because chlorine is more electronegative than carbon, the C—Cl bond is polar covalent in character, i.e. the electrons of the bond are located relatively near the chlorine atom (§4.1.2). Similarly, the electron releasing effect of a methyl group is believed to stem from the electronegativity difference between carbon and hydrogen. Because of this difference the electron density on the carbon atom is increased, and the electrons of a CH_3—C bond are repelled towards the adjoining carbon atom.

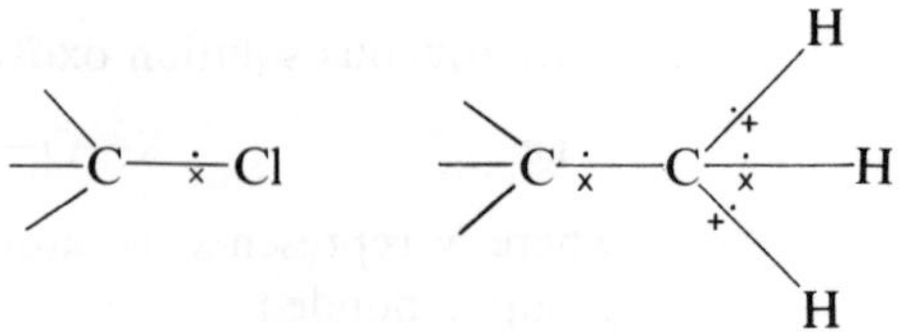

$$-C \cdot\!\times Cl \qquad -C \cdot\!\times C \cdot\!\times H \quad (\text{with two further H atoms on the right-hand C})$$

The electronegativity difference between carbon and a halogen is much greater than that between carbon and hydrogen. Consequently, the −I effect of a halogen is much greater than the +I effect of an alkyl group. Both +I and −I effects are transmitted along a hydrocarbon chain, but become progressively weaker with increasing distance from the cause of the effect.

The mesomeric effect is discussed elsewhere (§14.8.4 and 14.10.5).

We shall now apply these principles in a qualitative discussion of the relative acid strengths of some well known compounds.

Alcohols and phenols

Alcohols are less acidic than water because alkyl groups release electrons to the oxygen atom (+I effect).

R→O—H

Phenols are more acidic than water because the benzene ring effectively withdraws a lone pair of electrons from the oxygen atom (mesomeric effect).

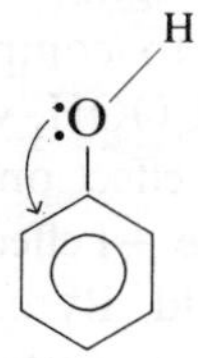

phenol molecule

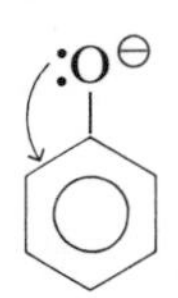

phenoxide ion

The effect is present in the phenol molecule and the phenoxide ion and in both cases promotes an increase in acidity. In the molecule, the effect results in a weakening of the O—H bond, and in the ion it allows the negative charge to become delocalised over the benzene ring. (No such delocalisation can occur in an alkoxide ion.) The latter is particularly important, because delocalisation of charge always results in a considerable increase in stability.

Carboxylic acids

Carboxylic acids are more acidic than water. This is because they possess the acyl group, RCO, which contains a highly electronegative oxygen atom. Consequently, in comparison with a hydrogen atom, the acyl group draws electrons away from the hydroxyl oxygen atom.

R—C(=O)←O—H

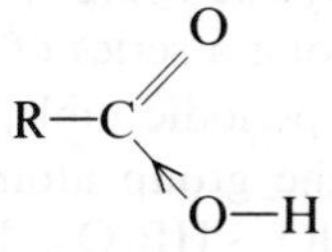

When discussing the relative strengths of different carboxylic acids we can ignore this effect because it is common to all of them. All we need to consider is whether the group R withdraws or repels electrons from the carboxyl group. Such effects are transmitted through the molecule to the hydroxyl oxygen atom, i.e. R—<—C(=O)—O—H R withdraws electrons   R—>—C(=O)—O—H R repels electrons

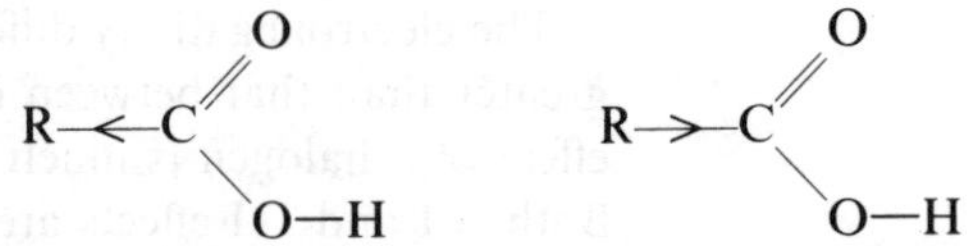

If R withdraws electrons it increases acid strength, because it weakens the O—H bond and also stabilises the carboxylate ion by removing some of the negative charge on the oxygen atoms of the ion:

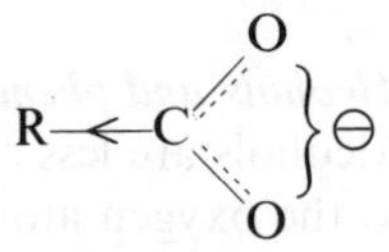

R—<—C(O)(O) $\ominus$

carboxylate ion

Generally, the greater the electron withdrawal effect of R, the stronger is the acid. Conversely, if R repels electrons, acid strength decreases. (It has been suggested that other factors, in particular entropy effects, have a bearing on acid strength.)

It is interesting to compare the strengths of carboxylic acids with methanoic acid, HCOOH, where the hydrogen atom of the C—H bond has no appreciable effect on the electron density of the hydroxyl oxygen atom. Because of the +I effect of the methyl group, ethanoic acid is weaker than methanoic acid. By contrast, chloroethanoic acid, $CH_2ClCOOH$, is a stronger acid than methanoic acid, due to the −I effect of the chlorine atom. Dichloroethanoic acid, $CHCl_2COOH$, is stronger than chloroethanoic acid because of the greater electron withdrawing capacity of its two chlorine atoms. By similar reasoning trichloroethanoic acid, CCl_3COOH, is stronger still. 3-Chloropropanoic acid, CH_2ClCH_2COOH, is weaker than chloroethanoic acid because the chlorine atom is relatively remote from the hydroxyl group. Bromine-substituted acids are weaker than chloro-acids, and iodo-acids are even weaker (although still stronger than unsubstituted acids). This is in accordance with the electronegativity order of the halogens: $Cl > Br > I$.

Benzoic acid, C_6H_5COOH, is a weaker acid than methanoic acid because, in contrast to phenol, the benzene ring *donates* electrons to the carboxyl group. Nitrobenzoic acids are stronger than benzoic acid because the nitro group withdraws electrons from the benzene ring.

Inorganic hydroxy compounds

Among a series of oxoacids formed by the members of a given group of the periodic table, the acid strength diminishes as the electronegativity of the group atom decreases. For example, in terms of acid strength, $HClO > HBrO > HIO$, and $H_2SO_4 > H_2SeO_4$. The decreasing electronegativity of the group atom increases the electron density on the hydroxyl oxygen atom and strengthens the O—H bond.

The strength of oxoacids increases with the number of oxygen atoms bonded to the central atom. For example, in terms of acid strength, $HClO < HClO_2 < HClO_3 < HClO_4$, and $HNO_2 < HNO_3$. The more extensive delocalisation of charge on the higher anions and their greater hydration enthalpy account for these trends.

9.2.11 FACTORS AFFECTING THE STRENGTHS OF BASES IN AQUEOUS SOLUTION

Here we shall consider ammonia and its derivatives, which ionise in aqueous solution to produce hydroxide ions:

$$R_3N + H_2O \rightleftharpoons R_3NH^+ + HO^-$$

The degree of ionisation, and hence base strength, is related to the readiness with which the nitrogen atom can donate its electron pair to a proton,

$$\text{i.e.}\quad R_3N\!: + H^+ = R_3NH^+$$

If R releases electrons it increases the ability of the nitrogen atom to donate electrons and hence increases base strength. If R withdraws electrons the opposite is true.

Aliphatic amines

The $+$I effect of the methyl group results in methylamine, CH_3NH_2, being a stronger base than ammonia. Dimethylamine, $(CH_3)_2NH$, possesses two electron releasing methyl groups and is a stronger base than methylamine. Generally, base strength varies in the order

$$NH_3 < RNH_2 \text{ (primary amine)} < R_2NH \text{ (secondary amine)},$$

provided that both amines contain the same alkyl group.

Extending this argument to tertiary amines, we might expect trimethylamine, $(CH_3)_3N$, to be an even stronger base because there are three methyl groups. Trimethylamine, however, is a *weaker* base than dimethylamine. The reason is that the base strength of an amine depends not only on the electron releasing properties of R, but also on the extent to which its cation can become hydrated. Hydration stabilises cations and therefore increases the base strength of amines. The cation derived from the tertiary amine, $(CH_3)_3NH^+$, is less extensively hydrated and therefore less stabilised than that from the secondary amine, $(CH_3)_2NH_2^+$. This is because there are fewer hydrogen atoms to which water molecules can become hydrogen bonded:

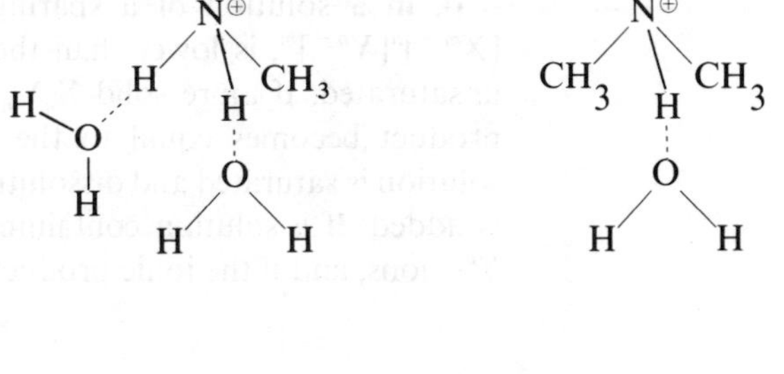

CH$_3$ N$^\oplus$ H CH$_3$ H H O H H O H H

CH$_3$ N$^\oplus$ CH$_3$ CH$_3$ H O H H

This effect is sufficient to account for the lower base strength of tertiary amines compared with secondary amines.

To summarise, as successive alkyl groups are introduced into ammonia, the base strength of the amines depends on the balance between the inductive effect and the extent to which the cations are stabilised by hydration:

NH_3 RNH_2 R_2NH R_3N

+I effect of R increases base strength →

decreasing stabilisation of cation by hydration reduces base strength

Aromatic amines

Phenylamine, $C_6H_5NH_2$, is a much weaker base than ammonia because the benzene ring effectively withdraws electrons and hence reduces the electron density on the nitrogen atom (mesomeric effect). Diphenylamine, $(C_6H_5)_2NH$, is even weaker than phenylamine, because there are two benzene rings reducing the electron density, and triphenylamine, $(C_6H_5)_3N$, is an exceedingly weak base.

When the nitrogen atom is situated in a side chain, as in the case of (phenylmethyl)amine, $C_6H_5CH_2NH_2$, the electron withdrawing effect of the ring is largely isolated from the nitrogen atom by the intervening saturated carbon atom. Consequently, the base strength of (phenylmethyl)-amine is roughly the same as that of ammonia and much greater than that of phenylamine.

9.3 Further considerations of ionic equilibria

9.3.1 SOLUBILITY PRODUCT

The solubility product (K_s) of a sparingly soluble salt, X_nY_m, is the ionic product in saturated solution,

i.e. $$K_s = [X^{m+}]^n[Y^{n-}]^m \quad (\S 8.5.3)$$

where $[X^{m+}]$ and $[Y^{n-}]$ represent molar concentrations in the saturated solution. The solubility product of a given salt is constant at a fixed temperature,

e.g. at 25 °C for AgCl, $K_s = [Ag^+][Cl^-]$

$= 1.8 \times 10^{-10}\ mol^2\ dm^{-6}$

for Ag_2CrO_4, $K_s = [Ag^+]^2[CrO_4^{2-}]$

$= 1.3 \times 10^{-12}\ mol^3\ dm^{-9}$

If, in a solution of a sparingly soluble salt, the ionic product, i.e. $[X^{m+}]^n[Y^{n-}]^m$, is lower than the solubility product then the solution is unsaturated. If more solid X_nY_m is added it will dissolve until the ionic product becomes equal to the solubility product, at which point the solution is saturated and dissolution of the salt ceases no matter how much is added. If a solution containing X^{m+} ions is added to one containing Y^{n-} ions, and if the ionic product exceeds the solubility product, then the salt X_nY_m will be precipitated until the ionic product becomes equal to the solubility product. (The precipitation of X_nY_m reduces the concentrations of X^{m+} and Y^{n-} ions in solution.)

Tabulated values of solubility products appear in data books and can be used to calculate the solubilities of salts. For example, in a very dilute solution of silver chloride dissociation is complete, and one mole each of silver ions and chloride ions originate from one mole of dissolved silver chloride,

i.e. concentration of dissolved $AgCl = [Ag^+] = [Cl^-]$

In a saturated solution,

$$K_s = [Ag^+][Cl^-] = [Ag^+]^2 = [Cl^-]^2 = [AgCl]^2$$
$$= 1.8 \times 10^{-10}\ \text{mol}^2\ \text{dm}^{-6}$$

$\therefore$ the solubility of $AgCl = \sqrt{1.8 \times 10^{-10}} = 1.34 \times 10^{-5}$ mol dm $^{-3}$

Similarly for silver iodide, where $K_s = 8.3 \times 10^{-17}$ mol^2 dm^{-6}, the solubility is 9.11×10^{-9} mol dm^{-3}.

In a solution of silver chromate,

$$Ag_2CrO_4 = 2Ag^+ + CrO_4^{2-}$$

Therefore, in solution, $[Ag^+] = 2[CrO_4^{2-}]$

Each mole of dissolved silver chromate yields two moles of silver ions and one mole of chromate ions,

$\therefore$ concentration of dissolved $Ag_2CrO_4 = \frac{1}{2}[Ag^+] = [CrO_4^{2-}]$

In a saturated solution,

$$K_s = [Ag^+]^2[CrO_4^{2-}] = [Ag^+]^2 \times \tfrac{1}{2}[Ag^+] = \tfrac{1}{2}[Ag^+]^3$$
$$= (2[CrO_4^{2-}])^2 \times [CrO_4^{2-}] = 4[CrO_4^{2-}]^3$$
$$= 1.3 \times 10^{-12}\ \text{mol}^3\ \text{dm}^{-9}$$

If $4[CrO_4^{2-}]^3 = 1.3 \times 10^{-12}$ mol^3 dm^{-9},

then $[CrO_4^{2-}] = \sqrt[3]{\frac{1}{4} \times 1.3 \times 10^{-12}} = 0.69 \times 10^{-4}$ mol dm^{-3}

$\therefore$ the solubility of $Ag_2CrO_4 = 0.69 \times 10^{-4}$ mol dm^{-3}

(As a check, $\frac{1}{2}[Ag^+]^3 = 1.3 \times 10^{-12}$ mol^3 dm^{-9}

$\therefore$ $[Ag^+] = \sqrt[3]{2 \times 1.3 \times 10^{-12}} = 1.38 \times 10^{-4}$ mol dm^{-3}

which is exactly twice the chromate ion concentration.)

These calculations illustrate two important points.

(i) The solubility of silver chloride is higher than that of silver iodide, as indicated by the lower solubility product of the latter.

(ii) The solubility of silver chromate is higher than that of silver chloride, despite the fact that the chromate has a lower solubility product. Solubility products therefore give a direct comparison of the solubility of two salts only if the ratio of cations to anions produced in solution is the same in both cases, e.g. AgCl and AgI. If the ratios are different, as in the case of AgCl and Ag_2CrO_4, then the solubility products do not give a comparison of solubilities that is immediately obvious, and a calculation must be performed.

Consider an ion X that reacts with ions Y or Z to produce sparingly soluble salts XY and XZ respectively. If, to a solution containing Y and Z ions, we add X ions, the less soluble salt is precipitated completely before the other begins to precipitate. Thus, if XY is less soluble than XZ, XY is precipitated before XZ. This principle is used to estimate ionic chlorides by titration with silver nitrate, using potassium chromate as an indicator. Not until all the chloride ions have been precipitated as silver chloride is there a brick-red precipitate of silver chromate to indicate the equivalence point of the titration.

9.3.2 THE COMMON ION EFFECT

Let us consider the following equilibrium in solution:

$$AB \rightleftharpoons A^+ + B^-$$

where AB represents either a sparingly soluble salt or a covalent substance which ionises in water. If, to this solution, we add another solution containing either A^+ or B^- ions the equilibrium is displaced to the left. This is known as the *common ion effect.*

When AB is a weak acid or weak base the addition of a common ion decreases the degree of ionisation (α). By using K_a or K_b expressions we can calculate the reduction in α. For example, let us calculate the degree of ionisation of a 0.1 M solution of ethanoic acid after the addition of 0.1 mol of sodium ethanoate. In the absence of the salt the degree of ionisation calculated from Ostwald's dilution law (§9.2.2) is 1.3×10^{-2}.

Because ethanoic acid is a weak acid, the concentration of ethanoate ions to which it gives rise is negligible in comparison with the concentration of ethanoate ions from the almost completely dissociated sodium ethanoate. Therefore, in the expression

$$K_a = \frac{[H^+][CH_3COO^-]}{[CH_3COOH]}$$

we may, with acceptable accuracy, equate $[CH_3COO^-]$ to 0.1 mol dm^{-3}.

Since the acid is weak, and its ionisation is suppressed by the presence of ethanoate ions from the salt, the concentration of *un-ionised* acid at equilibrium is almost equal to the molar concentration of acid in the solution. Therefore, $[CH_3COOH] = 0.1$ mol dm^{-3}, and

$$[H^+] = \frac{K_a \times 0.1}{0.1} = 1.74 \times 10^{-5} \text{ mol dm}^{-3}$$

Since $[H^+] = \alpha \times \text{molarity of acid}$ (§9.2.2)

$$\alpha = \frac{1.74 \times 10^{-5}}{0.1} = 1.74 \times 10^{-4}$$

Thus, the degree of ionisation is reduced considerably by the addition of a common ion.

Adding an ion common to a sparingly soluble salt precipitates that salt. For example, the addition of sodium chloride to a saturated solution of silver chloride produces a cloudiness as solid silver chloride is formed. The introduction of a common ion therefore decreases the solubility of the salt, to an extent which is readily calculated. For example, let us calculate the solubility of silver chloride in a solution that contains 0.01 mol dm^{-3} of sodium chloride. In the mixed solution, the concentration of chloride ions arising from the silver chloride is negligible in comparison with that arising from the almost completely dissociated sodium chloride.

$$\therefore \quad [Cl^-] = 0.01 \text{ mol dm}^{-3}$$

$$\therefore \quad K_s = [Ag^+] \times 0.01 = 1.8 \times 10^{-10} \text{ mol}^2 \text{ dm}^{-6}$$

$$\therefore \quad \text{the solubility of silver chloride} = [Ag^+] = 1.8 \times 10^{-8} \text{ mol dm}^{-3}$$

This compares with a value of 1.34×10^{-5} mol dm^{-3} in water. The solubility of silver chloride is thus considerably reduced by the presence of the common chloride ion.

The common ion effect also works in reverse, in the sense that if the concentration of either A^+ or B^- ions is *reduced*, the equilibrium $AB \rightleftharpoons A^+ + B^-$ is displaced to the right. Consequently, the degree of ionisation of a covalent solute or the solubility of a sparingly soluble salt is increased.

The above principles are utilised in the selective precipitation of metal hydroxides by means of ammonia. For example, if aqueous ammonia is added to a solution of zinc ions a precipitate of zinc hydroxide is obtained, but in the presence of ammonium chloride there is no precipitate. Being a salt, ammonium chloride is almost completely dissociated, and the presence of a common ion (NH_4^+) suppresses the ionisation of ammonia:

$$NH_3 + H_2O \rightleftharpoons NH_4^+ + HO^-$$

The concentration of hydroxide ions is therefore reduced and the solubility of zinc hydroxide is increased. Zinc hydroxide will not precipitate if the hydroxide ion concentration is reduced to such an extent that the ionic product, i.e. $[Zn^{2+}][HO^-]^2$, is lower than the solubility product. (In this event the solution is unsaturated; §9.3.1.)

By contrast, aluminium hydroxide, which has a much lower solubility product than zinc hydroxide, is precipitated by aqueous ammonia in the presence of ammonium chloride. This is because the concentration of hydroxide ions is still high enough to exceed the solubility product of aluminium hydroxide. Hence zinc ions and aluminium ions in solution can be separated from each other.

9.3.3 BUFFER SOLUTIONS

Solutions containing weak acids or weak bases and their respective salts are called *buffer solutions*. They maintain their pH virtually unchanged even when moderate amounts of strong acid or strong base are introduced into the solution, i.e. they act as a 'buffer' towards pH changes. For example, the addition of 10^{-2} mol of hydrochloric acid to 1 dm^3 of a solution containing 0.1 mol dm^{-3} of both ethanoic acid and sodium ethanoate changes the pH from 4.76 to 4.67. The same addition to 1 dm^3 of pure water changes the pH from 7.00 to 2.00.

The action of a buffer solution comprising a weak acid, HA, and one of its salts can be understood in terms of K_a.

$$K_a = \frac{[H^+][A^-]}{[HA]}$$

To a good approximation $[A^-]$ represents the concentration of anion from the salt, and [HA] is equal to the molar concentration of the acid (§9.3.2). If we attempt to increase the hydrogen ion concentration by adding a strong acid the above expression momentarily becomes greater than K_a. But this situation cannot last because at equilibrium the expression must always be numerically equal to the dissociation constant, K_a. What happens is that virtually all of the added hydrogen ions combine with anions, A^-, to form molecules of the weak acid so that the expression again becomes equal to K_a. The effect of added hydrogen ions on the pH of the solution is therefore negligible.

Let us reconsider the example quoted above. The initial hydrogen ion concentration of a buffer solution containing 0.1 mol dm^{-3} each of ethanoic acid and sodium ethanoate is 1.74×10^{-5} mol dm^{-3} (§9.3.2), which corresponds to a pH of 4.76. The 10^{-2} mol of added hydrochloric acid will combine with 10^{-2} mol of ethanoate ions to produce a further 10^{-2} mol of ethanoic acid.

$$\therefore \quad K_a = \frac{[H^+][CH_3COO^-]}{[CH_3COOH]} = \frac{[H^+] \times (0.1 - 0.01)}{(0.1 + 0.01)}$$

$$= 1.74 \times 10^{-5} \text{ mol dm}^{-3}$$

from which $[H^+] = 2.13 \times 10^{-5}$ mol dm^{-3} and pH = 4.67.

If a small amount of a strong base is added to the buffer solution, the hydroxide ions are neutralised by the weak acid and so once again the pH is virtually unaffected.

A buffer comprising a weak base and one of its salts, e.g. aqueous ammonia and ammonium chloride, behaves in a similar manner. Thus, if a strong acid is added it is neutralised by the ammonia, and if a strong base is introduced virtually all the additional hydroxide ions react with ammonium ions to form molecules of ammonia and water so as to keep K_b constant.

Buffer solutions are used extensively where constancy of pH is essential despite the addition or formation *in situ* of acid or base. They are used in kinetics where the rate of a reaction to be measured is pH dependent, and in equilibrium experiments where hydrogen ions are a participating species. An especially important application is the standardisation of pH meters (§2.5.5). Ordinary distilled water cannot be relied upon as a standard for pH 7 because it absorbs aerial carbon dioxide and usually has a pH of about 5.5.

Calculation of the pH of buffer solutions

A buffer solution comprising a weak acid and one of its salts is acidic and sometimes referred to as an *acid buffer*. Its pH can be calculated using the approximations mentioned in §9.3.2.

$$K_a = \frac{[H^+][A^-]}{[HA]} = \frac{[H^+][salt]}{[acid]}$$

Rearranging and taking negative logarithms,

$$-\lg [H^+] = -\lg K_a - \lg \frac{[acid]}{[salt]}$$

$$\therefore \qquad pH = pK_a - \lg \frac{[acid]}{[salt]} = pK_a + \lg \frac{[salt]}{[acid]}$$

This is known as the *Henderson–Hasselbalch equation*. If the molar concentrations of salt and acid are the same, then:

$$pH = pK_a, \qquad \text{since} \quad \lg 1 = 0$$

This provides a convenient way of obtaining pK_a and hence K_a of a weak acid.

A similar relationship applies to a buffer solution comprising a weak base and one of its salts, sometimes known as an *alkaline buffer*.

$$K_b = \frac{[BH^+][HO^-]}{[B]} = \frac{[salt][HO^-]}{[base]}$$

$$\therefore \qquad pOH = pK_b - \lg \frac{[base]}{[salt]} = pK_b + \lg \frac{[salt]}{[base]}$$

$$\text{or} \qquad pH = 14 - \left(pK_b + \lg \frac{[salt]}{[base]}\right)$$

9.4 The electrical conductivity of solutions

Solutions containing ions conduct electricity whenever a potential difference is applied across electrodes immersed in the solution. Various reactions occur at the electrodes during the passage of current (§10.2.1), but in this section we shall discuss only the behaviour of the ions as carriers of the electric current.

9.4.1 QUANTITIES AND THEIR UNITS

Like most electrical conductors, solutions obey Ohm's law:

$V = I \times R$

where V = electrical potential across the electrodes (volts)
I = current (amperes)
R = resistance (ohms).

As for any other conductor,

$$R \propto \text{length of conducting solution } (l)$$

Also: $R \propto \dfrac{1}{\text{cross-sectional area of conducting solution } (a)}$

$$\therefore \quad R \propto \frac{l}{a}$$

$$\therefore \quad R = \rho \times \frac{l}{a}$$

The proportionality constant (ρ) is called the *resistivity*. Physically, resistivity is the resistance of one metre cube of the conductor since, when $l = 1$ m and $a = 1$ m^2, $R = \rho$.

Conductance (G), defined as R^{-1}, i.e. $\frac{1}{R}$, has the units of Ω^{-1}, although the unit siemens, S, is sometimes used. $\frac{1}{\rho}$ defines the *electrolytic conductivity* of a solution ($\varkappa$, pronounced 'kappa').

$$\therefore \quad \frac{1}{R} = \frac{1}{\rho} \times \frac{a}{l}$$

$$\therefore \quad G = \varkappa \times \frac{a}{l}$$

$$\therefore \quad \varkappa = G \times \frac{l}{a}$$

We can find the physical significance of $\varkappa$ by putting $l = 1$ m and $a = 1$ m^2, in which case $\varkappa = G$ = the conductance of one metre cube of the solution. Because $\varkappa$ relates to a definite length and cross-sectional area, it forms the basis for comparing the conducting powers of different solutions. The units of $\varkappa$ are readily derived:

$$\varkappa = \Omega^{-1} \frac{\text{m}}{\text{m}^2} = \Omega^{-1}\,\text{m}^{-1}$$

Electrolytic conductivity is used principally as a guide to the degree of ionisation or dissociation (α) of an electrolyte. Arrhenius argued that the electrolytic conductivity of a solution depends on: (i) the nature of the ions, (ii) the temperature and (iii) the number of ions in unit volume of solution.

If (i) and (ii) are fixed, then $\varkappa$ depends only on (iii). $\varkappa$ is therefore influenced by α and also by the concentration of the solution, and to compensate for the latter we must consider volumes of solution that contain equal amounts of solute, say one mole. For example, we should compare 1 dm^3 of a 1 M solution with $\frac{1}{0.1}$ dm^3 (i.e. 10 dm^3) of a 0.1 M solution. Generally, for a concentration c, we should consider a volume of $\frac{1}{c}$.

Thus, if we obtain $\varkappa$ for a solution, and we wish to correct for the effect of concentration, we divide $\varkappa$ by the concentration (c). Because, in the definition of $\varkappa$, the volume of conducting solution is 1 m^3, we quote the concentration (c) in mol m^{-3}.

$$\frac{\varkappa}{c} = \Lambda$$

where Λ (lambda), the *molar conductivity*, is the electrolytic conductivity of a solution containing one mole of electrolyte. The units of Λ are derived as follows:

$$\Lambda = \Omega^{-1}\ m^{-1}/\text{mol}\ m^{-3} = \Omega^{-1}\ m^2\ \text{mol}^{-1}$$

9.4.2 THE MEASUREMENT OF CONDUCTANCE

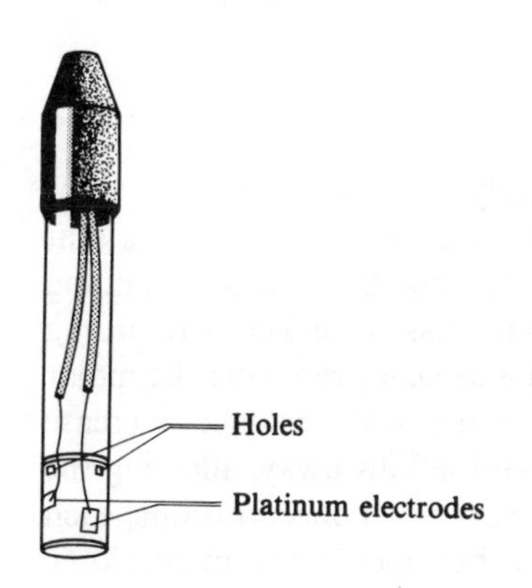

Fig. 9.3 A dip-in conductance cell.

Conductance readings are conveniently taken with a 'dip-in' type of cell (Fig. 9.3) connected to a *conductance bridge*. This is basically a Wheatstone bridge for measuring resistance, but is calibrated to give conductance readings directly. The bridge differs from the usual type in using alternating current instead of direct current; the latter would cause electrolysis and so alter the concentration of the solution. The conductance recorded is that of the solution contained between the two electrodes of the cell.

Should it be necessary to obtain the molar conductivity (Λ), the electrolytic conductivity ($\varkappa$) must first be calculated from:

$$\varkappa = G \times \frac{l}{a}$$

where l is the distance between the electrodes and a is the cross-sectional area of the conducting part of the solution. In practice l and a are difficult to measure directly. Instead, the ratio l/a, called the *cell constant*, is obtained by means of standard solutions of potassium chloride, whose electrolytic conductivities are accurately known and available from tables. For example, for 0.01 M KCl, $\varkappa = 0.141\ \Omega^{-1}m^{-1}$ at 25 °C (298 K). If the measured conductance of this solution in a certain cell is $9.53 \times 10^{-4}\ \Omega^{-1}$,

$$\text{cell constant} = \frac{0.141}{9.53 \times 10^{-4}} = 148\ m^{-1}$$

Cell constants are usually marked on cells.

9.4.3 THE VARIATION OF MOLAR CONDUCTIVITY WITH CONCENTRATION

Electrolytes, i.e. substances which in aqueous solution conduct electricity, broadly fall into two categories, namely 'strong' and 'weak'. Strong electrolytes have much higher electrolytic conductivities than weak electrolytes at the same concentrations. Strong electrolytes include strong acids and most ionic compounds where α is close to one. Weak acids and bases are only feebly ionised in solution (α is less than 0.1) and are therefore weak electrolytes.

The molar conductivities of both types of electrolyte vary with concentration (Fig. 9.4).

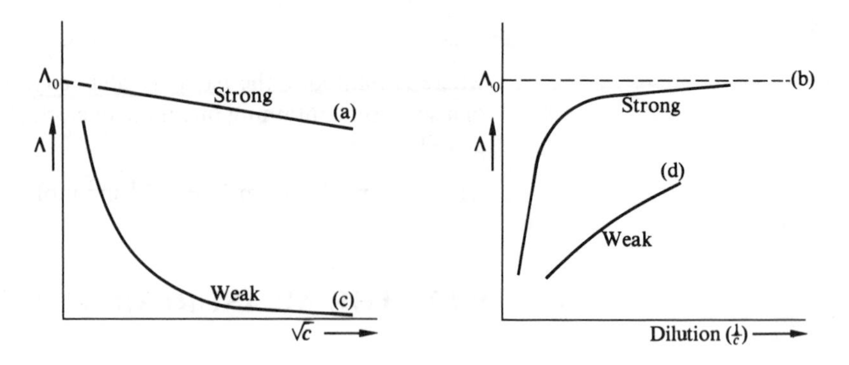

Fig. 9.4 The variation of molar conductivity with concentration.

Strong electrolytes

Strong electrolytes in solution are completely dissociated at all concentrations. Why then does Λ vary with concentration? The answer is that the ions influence one another's behaviour by attracting and surrounding themselves with ions of opposite charge. This has the effect of retarding the movement of ions through the solution and hence reducing the molar conductivity. With decreasing concentration interionic distances increase and the ionic 'atmosphere' surrounding each ion falls away, allowing the ions to move more rapidly. Eventually, at very low concentrations, each ion loses its cloud of oppositely charged ions, becomes free from interionic effects, and reaches its maximum velocity. The molar conductivity of a solution therefore increases with decreasing concentration (or increasing dilution); Fig. 9.4(a) and (b). By extrapolation we can obtain the molar conductivity in extremely dilute solution in which interionic effects are negligible. This value is known as the *molar conductivity at infinite dilution* (Λ_0). The form of plot (a) in Fig. 9.4 shows that, for strong electrolytes,

$$\Lambda = \Lambda_0 - b\sqrt{c}$$

where b is a constant.

For strong electrolytes α is not, in fact, related to the degree of ionisation or dissociation but is a measure of the extent of the influence that ions have on the behaviour of one another. As concentration decreases, α approaches one and interference decreases, until eventually at infinite dilution α becomes equal to one, indicating no interference.

Weak electrolytes

Because weak electrolytes, such as ethanoic acid, produce relatively few ions in solution, we may regard them as being completely free from the interionic effects discussed above. Consequently, molar conductivity is dependent only on the extent to which the solute is split into ions, i.e. on its degree of ionisation (α). As α increases with dilution $\left(\frac{1}{c}\right)$ so does Λ.

The variation of α and hence Λ with $\frac{1}{c}$ or $\sqrt{c}$ is not linear (Fig. 9.4(d) and (c)) because of the α^2 term in Ostwald's dilution law,

$$\text{i.e.}\quad \frac{1}{c} \propto \frac{\alpha^2}{(1-\alpha)} \quad \text{or} \quad \sqrt{c} \propto \sqrt{\frac{1-\alpha}{\alpha^2}}$$

At infinite dilution, i.e. as c tends to zero, α approaches one and Λ reaches its maximum value (Λ_0).

9.4.4 DEGREE OF IONISATION FROM MOLAR CONDUCTIVITIES

For a weak electrolyte, at a dilution V, the molar conductivity (Λ_V) is related to the degree of ionisation,

$$\text{i.e.}\quad \Lambda_V \propto \alpha$$

$$\therefore \quad \Lambda_V = k\alpha, \quad \text{where } k \text{ is a constant.}$$

At infinite dilution, where $\alpha = 1$,

$$\Lambda_0 = k$$

$$\therefore \quad \alpha = \frac{\Lambda_V}{\Lambda_0}$$

This approach can also be used for strong electrolytes to find the (apparent) degree of dissociation (e.g. for NaCl) or the degree of ionisation (e.g. for HCl). At low concentrations, α values for strong electrolytes determined in this way agree reasonably well with the values obtained from colligative property measurements (§9.1.2). However, discrepancies arise at high concentrations. This is because α as determined by conductivity is related to the freedom of movement of *ions* in the solution, whereas colligative properties relate to the freedom of movement of *solvent molecules*. For weak electrolytes, where ionic interference is negligible, conductivity and colligative property values of α agree closely at all concentrations.

EXAMINATION QUESTIONS ON CHAPTER 9

1 The following statements have been made at various times about acids:

(1) An acid is a substance containing hydrogen which can be replaced by a metal.

(2) An acid is a compound containing hydrogen which dissolves in water to give an excess of hydrogen ions over hydroxide ions.

(3) An acid is a substance capable of donating a proton to a base.

Briefly discuss to what extent you consider each of these statements to provide a satisfactory definition of the term 'acid'.

Consider the application of the statements to the following species in aqueous solution: HCl, OH^-, NH_4^+ and HCO_3^-, and to the following species in concentrated sulphuric acid: H_2O and HNO_3. (L)

2 (a) Define the terms *Brønsted acid* and *Brønsted base*.

(b) Explain whether **each** of the following species normally acts as an acid or a base or both in (i) aqueous solution, and (ii) liquid ammonia: NH_4^+, HSO_4^-, NH_2^-, NH_3.

(c) Discuss the factors which have to be taken into account in determining the equivalence point (end point) in the titration of an aqueous solution of ethanoic acid (acetic acid) with sodium hydroxide solution. (WJEC)

3 (a) Explain what is meant by the terms 'strong acid' and 'weak acid'. Arrange the following in order of increasing acid strength (in water): C_2H_5OH, C_6H_5COOH, $C_6H_5CH_2OH$ and $C_6H_5SO_3H$.

(b) Explain briefly what is meant by a 'buffer solution'. How may such a solution be prepared?

(c) Calculate the pH of a solution containing 6 g of ethanoic acid in 1 litre of solution (K_a for ethanoic acid = 1.8×10^{-5} mol l^{-1} at 25 °C).

(d) By consideration of salt hydrolysis show whether a solution of sodium ethanoate in water is acidic, neutral or alkaline. (JMB)

4 The pK_a for the ionisation of ethanoic (acetic) acid in aqueous solution at 25 °C is 4.8. Write an equation for the ionisation of ethanoic (acetic) acid in aqueous solution and hence derive a mathematical expression relating pK_a to the concentrations of the components of this system. Why is it necessary to state the solvent and temperature when giving a value of pK_a?

From the expression derived show that when the pH of a solution of ethanoic (acetic) acid is 4.8 the concentration of undissociated acid is equal to the concentration of anions.

Describe, giving brief experimental details, how the pK_a for ethanoic (acetic) acid may be determined experimentally. Given that the pK_a for methanoic (formic) acid and ethanoic (acetic) acid are 3.8 and 4.8 respectively, explain qualitatively the pK_a values of the following organic acids:

(i) 2-chloroethanoic (chloroacetic) acid 2.9
(ii) benzoic acid 4.2
(iii) phenol 10.0 (JMB)

5 (a) Explain concisely why an aqueous solution of sodium ethanoate has a pH greater than 7.

(b) If the solubility of lead chloride is determined in solutions containing an increasing concentration of sodium chloride, it is found that as the concentration of the sodium chloride increases from zero, the solubility of the lead chloride first decreases, passes through a minimum, and then increases. Suggest an explanation for these observations.

(c) An aqueous solution is 1.00 molar in ammonium chloride, 0.0100 molar in manganese(II) sulphate, and also contains some ammonia. Calculate the minimum concentration of ammonia which will just bring about the precipitation of manganese(II) hydroxide, the solubility product of which is 4.00×10^{-14} mol^3 dm^{-9}. The dissociation constant K_b of ammonia is 1.80×10^{-5} mol dm^{-3}, where

$$K_b = \frac{[NH_4^+][OH^-]}{[NH_3]}$$

(Assume that the ammonium ions in the solution derive solely from the ammonium chloride.) (O)

6 (a) You are given a 0.1 M solution of an unknown monobasic acid in water. Describe an experiment you would carry out in order to decide if it is a strong or a weak acid, and show how the results of this experiment would be interpreted.

(b) What is a buffer solution? Explain the action of a buffer solution using as the example mixtures of ethanoic (*acetic*) acid and sodium ethanoate (*acetate*).

(c) Explain why aqueous solutions of sodium ethanoate are alkaline whilst those of ammonium ethanoate are approximately neutral. (JMB)

7 What are the Brønsted–Lowry definitions of an *acid* and a *base*?

Indicate, with reasons, whether each of the following compounds will give an acidic or an alkaline solution when dissolved in water:

(a) sodium chlorate(VII) (perchlorate),

(b) potassium hydrogen benzene-1,2-dicarboxylate (hydrogen-phthalate),

(c) aluminium chloride,

(d) sodium ethanedioate (oxalate),

(e) carbon dioxide.

Why does the ionic product of water (data book) increase appreciably as the temperature is increased? What is the concentration of hydrogen ions and the pH of pure water at 100 °C? (OC)

8 (a) Sodium hydroxide solution of concentration 0.1 mol dm^{-3} is added to

(i) 25.0 cm^3 of hydrochloric acid of concentration 0.1 mol dm^{-3},

(ii) 25.0 cm^3 of ethanoic (acetic) acid solution of concentration 0.1 mol dm^{-3},

(iii) 25.0 cm^3 of phosphoric(V) acid solution of concentration 0.1 mol dm^{-3}.

For each case draw curves to show approximately how the pH of the solution varies with the volume of sodium hydroxide solution added.

Explain how the shapes of the curves influence the choice of indicators used when the above acids are titrated with sodium hydroxide solution. Your answer should include the names of suitable indicators.

(b) Define

(i) the pH of a solution,

(ii) the pK_a of an acid,

(iii) the ionic product, K_w, of water.

(c) Calculate the pH of a solution of ethanoic acid of concentration 0.1 mol dm^{-3} at 298 K, given that the pK_a of the acid at this temperature is 4.76.

Explain any assumptions you make. (AEB)

9 (a) What is meant by the terms *electrolyte conductivity* and *molar conductivity*? Describe how you would determine the conductivity of a strong electrolyte of known concentration.

Show graphically how the molar conductivity will vary with dilution for:

(i) strong electrolytes,
(ii) weak electrolytes.

(b) The molar conductivity of a solution of sodium chloride containing 12.5 g solid in 1 kg of water was found to be 96.2×10^4 S m^2 mol^{-1} (96.2 Ω^{-1} cm^2 mol^{-1}) at 298 K, whereas the corresponding value for infinite dilution was 12.6×10^5 S m^2 mol^{-1} (126 Ω^{-1} cm^2 mol^{-1}).

Calculate the freezing point of this solution. (The molar mass of sodium chloride is 58.5. The molar freezing point depression constant for water is 1.86 K kg mol^{-1}.) (AEB)

10 (a) What do you understand by the term *molar conductivity* as applied to an electrolyte?

(b) Show graphically how the molar conductivity changes with increasing dilution for (i) a strong electrolyte, (ii) a weak electrolyte. Give one example of each type of electrolyte.

Account briefly for the changes you have suggested.

(c) The molar conductivity of aqueous ethanoic (acetic) acid of concentration 0.1 mol/l was 4.6 cm^2 S/mol (cm^2/Ω mol), and at infinite dilution 352 cm^2 S/mol (cm^2/Ω mol).

Calculate (i) the degree of dissociation of the acid at this concentration, (ii) the pH of the acid solution.

(d) What substance would you add to a solution of ethanoic acid in order to make a solution the pH of which is almost unaffected by the addition of small quantities of acid or alkali?

What is the name given to such a mixture? Explain briefly how it works. (AEB)

11 The *molar conductivity* of ethanoic acid has numerical values of 387 at *infinite dilution* and 68 at a *dilution* of 2 024 litres.

(a) Define the terms in italics.
(b) State, with reasons, the units of molar conductivity.
(c) Calculate the degree of dissociation of ethanoic acid at the given dilution.
(d) Calculate the pH of ethanoic acid at the given dilution.
(e) Derive an expression for Ostwald's dilution law. Calculate the dissociation constant of ethanoic acid, stating the units.
(f) State briefly why strong electrolytes do not conform to Ostwald's law. (AEB)

12 (a) Explain the term *solubility product* and state why the concept does not apply to (i) sodium chloride, (ii) benzoic acid separately.

(b) Derive an expression for the solubility product of aluminium hydroxide, stating the units.

(c) You are given the following *numerical values* only for the solubility products of various salts at 25 °C.

K_{sp} silver chloride $= 2 \times 10^{-10}$
K_{sp} lead(II) bromide $= 3.9 \times 10^{-5}$
K_{sp} silver bromate(V) $= 6.0 \times 10^{-5}$
K_{sp} magnesium hydroxide $= 2.0 \times 10^{-11}$

(i) State the units for each of the above solubility products.

(ii) Which of the following pairs of solutions (all of concentration 1×10^{-3} mol dm^{-3}) will form a precipitate when equal volumes are mixed at 25 °C. Give reasons for your answers.

silver nitrate and sodium chloride
lead(II) nitrate and sodium bromide
silver nitrate and potassium bromate(V)
magnesium sulphate and sodium hydroxide.

(d) Choose **one** of the precipitates above and explain the steps you would take in the recovery and treatment of the precipitate, if a quantitative assessment of the precipitate is necessary. (AEB)

13 (a) Explain what is meant by the term *solubility product*, using calcium sulphate as your example.

(b) Given that the solubility product of calcium sulphate, K_{sp}, is 2.0×10^{-5} mol^2 l^{-2} at 298 K, calculate the solubility, in mol l^{-1} at that temperature, of calcium sulphate in:

(i) water,

(ii) sodium sulphate solution of concentration 0.1 mol l^{-1}, and

(iii) a solution containing 0.1 mol l^{-1} of calcium orthophosphate, $Ca_3(PO_4)_2$, in water containing a minimum amount of hydrochloric acid.

(c) State with reasons whether or not the concept of solubility product is applicable to the following substances: sodium chloride, benzoic acid, lead(II) chloride and lead(IV) chloride. (L)

14 (a) (i) Write an equation for the equilibrium which may be presumed to exist between solid silver chloride and dissolved silver chloride in a saturated solution.

(ii) Write an expression for the solubility product of silver chloride and state the conditions under which this applies.

(iii) The solubility of silver chloride in water at 25 °C is 1×10^{-5} mol/l. Calculate the solubility product of silver chloride and state the units in which it is expressed.

(b) The solubility product of lead(II) bromide at 25 °C is 3.9×10^{-5} mol^3/l^3. Calculate the solubility in g/l at 25 °C of lead(II) bromide.

(c) **Outline** a practicable method for determining the solubility of a named sparingly soluble ionic compound in water at 25 °C. (AEB)

15 (a) Define the term *pH of a solution*, and calculate the pH values of:

(i) 0.002 M hydrochloric acid,

(ii) 0.002 M sodium hydroxide solution.

$K_w = [H^+][OH^-] = 1.0 \times 10^{-14}$ mol^2 dm^{-6} at 298 K.

(b) Describe what is observed, outline underlying related physico-chemical principles and indicate the pH change to be expected when

(i) hydrochloric acid,

(ii) solid ammonium chloride,

(iii) solid silver chloride
are added respectively to separate samples of ammonia solution,
$NH_3 + H_2O \rightleftharpoons NH_4^+ + OH^-$

(c) (i) If the acidity constant (acid ionisation constant) of ethanoic (acetic) acid at 298 K is 1.8×10^{-5} mol dm^{-3}, calculate the pH of 2.00 dm^3 of a solution containing 0.400 mole of the acid in water at that temperature.

(ii) If 0.200 mole of sodium ethanoate (acetate) is added to the above solution of ethanoic acid, calculate the pH, assuming that any volume change is negligible.

Would you expect the addition of 0.1 mole of hydrochloric acid to affect this pH greatly? Give reasons for your answer. (SUJB)

16 What is the definition of pH? Describe how you would measure the pH of an unknown solution (*not* using an indicator).

Discuss and evaluate the following statements.

(a) More hydrogen will be evolved when excess zinc is added to 50 cm^3 of hydrochloric acid of concentration 0.100 mol dm^{-3} than when excess zinc is added to an equal volume of ethanoic (acetic) acid of the same concentration.

(b) The same volume of a standard solution of sodium hydroxide is needed to neutralise equal volumes of hydrochloric acid and ethanoic acid of the same concentration.

Calculate the pH of the following solutions:

(i) nitric acid of concentration 0.025 mol dm^{-3},

(ii) a saturated solution of 4-hydroxybenzoic acid which contains 6.50 g dm^{-3} at 25 °C (data book),

(iii) a buffer solution made by mixing 60 cm^3 of ethanoic acid of concentration 0.100 mol dm^{-3} with 40 cm^3 of sodium hydroxide of the same concentration. (OC)

17 (a) Describe briefly how you would attempt to verify experimentally the value of the enthalpy change in the following reaction:
$H^+(aq) + OH^-(aq) = H_2O(l)$; $\Delta H = -55.8$ kJ mol^{-1}

(b) What can you predict about the effect of a rise in temperature on the degree of ionisation of water? Explain the basis of your prediction.

(c) Place the following four aqueous solutions in order of increasing pH, giving reasons for your answer: M NH_4Cl; 10^{-3} M NaOH; 10^{-2} M HCl; 10^{-1} M CH_3COONa. (O)

18 (a) Define pH. (b) Write an expression for the ionic product of water. (c) At 0 °C, the ionic product of water is 1.15×10^{-15} mol^2 l^{-2}. At 60 °C, the value of the ionic product is 9.60×10^{-14} mol^2 l^{-2}. What is the hydrogen ion concentration in pure water at (i) 0 °C, and at (ii) 60 °C? (d) What is the pH of pure water at 0 °C? (e) What can you deduce about the sign of ΔH for the reaction
$H_2O(l) = H^+(aq) + OH^-(aq)$? (JMB)

19 (a) Sodium hydroxide solution was added to 25 cm^3 of an aqueous solution of ethanoic acid (acetic acid) of concentration 0.1 mol l^{-1}, and the pH was measured at intervals giving the following results:

cm^3 of sodium hydroxide solution	0	4.0	8.0	12.0	16.0	18.0	20.0	22.0	22.5	23.0	23.5	24.0	28.0
pH	2.8	3.5	4.0	4.5	5.1	5.5	5.8	7.0	9.0	10.5	11.0	11.4	12.3

Plot these results.

Use this graph to determine the following:

(i) the pH at the end point; account for this value,

(ii) the concentration in mol l^{-1} of the sodium hydroxide solution,

(iii) the dissociation constant, K_a, of ethanoic acid.

(b) Name an indicator suitable for this titration, and give its colour in acid and alkaline solutions.

(c) Explain why ethanoic acid is called a *weak* acid. (AEB)

20 Explain what is meant by *solubility product.*

The solubility of anhydrous calcium iodate(v) in water at 298 K is 3.07 g l^{-1}. Calculate (a) its solubility product, (b) its solubility, (in g l^{-1}), in an aqueous solution containing 0.1 mol l^{-1} of sodium iodate(v).

Explain, with full experimental details, how you would determine the solubility of calcium iodate(v) in water at 298 K, emphasising any precautions you would have to take.

(A_r(Ca) = 40; A_r(I) = 127; A_r(O) = 16.) (WJEC)

21 Explain fully what is meant by the term *solubility product*, indicating the limitations on its use, and derive an expression for the solubility product of the salt A_3B_2 given that its solubility in water is *s* mol dm^{-3}.

(a) Calculate the solubility of silver iodide (g dm^{-3}) in water given that the numerical value of its solubility product is 1.0×10^{-16}.

(AgI = 235)

(b) Calculate the solubility product of lead iodide, stating the units in which it is expressed, given that the solubility in water is 6.04×10^{-1} g dm^{-3}.

(PbI_2 = 461)

(c) If solid potassium iodide (which is freely soluble in water) is added in minute quantities at a time to a solution which is 0.02 M in both lead(II) nitrate, $Pb(NO_3)_2$ and silver nitrate, $AgNO_3$,

(i) which of the cations is precipitated first?

(ii) what will be its concentration (mol dm^{-3}) when the second cation begins to be precipitated? (SUJB)

10 Oxidation and reduction

10.1 Electron transfer

10.1.1 BASIC PRINCIPLES

Oxidation and reduction are complementary changes that occur whenever an oxidising agent (or 'oxidant') and a reducing agent (or 'reductant') are allowed to react together. Originally, *oxidation* was defined as the combination of a substance with oxygen, and *reduction* was regarded as a reduction in oxygen content, but the processes are today interpreted in terms of electrons. As an example, consider the thermite reaction between iron(III) oxide and aluminium:

$$\underset{\text{OXIDANT}}{Fe_2O_3} + \underset{\text{REDUCTANT}}{2Al} = 2Fe + Al_2O_3$$

(reduction: $Fe_2O_3 \rightarrow 2Fe$; oxidation: $2Al \rightarrow Al_2O_3$)

As the iron(III) oxide is reduced to iron, so the aluminium is oxidised to aluminium oxide. Always, in an oxidation–reduction (or 'redox') reaction, the oxidant becomes reduced while the reductant becomes oxidised.

The essential change that aluminium suffers in its oxidation is a loss of electrons:

$$Al = Al^{3+} + 3e^-$$

The electrons are accepted by iron(III) ions in their reduction to iron:

$$Fe^{3+} + 3e^- = Fe$$

'Oxidation' can therefore be defined as electron loss and 'reduction' as electron gain. Oxidation–reduction is essentially a process of *electron transfer* from reducing agent to oxidising agent, and compares with neutralisation which, on the Brønsted–Lowry theory, is a process of *proton transfer* from acid to base.

We can now see why neither oxidation nor reduction can occur in isolation. Electrons must originate from one species, which thereby becomes oxidised, and must be transferred to another species, which consequently becomes reduced.

10.1.2 OXIDATION NUMBERS (oxidation states)

An *oxidation number* can be assigned to an element in any chemical species to indicate the extent to which it is oxidised or reduced. An oxidation number of zero is given to the elements themselves, and they are considered to be in neither an oxidised nor a reduced condition. In compounds, a positive oxidation number indicates that an element is in an oxidised state; the higher the oxidation number, the greater is the extent of oxidation. Conversely, a negative oxidation number shows that an element is in a reduced state. The following paragraphs show how oxidation numbers are computed.

Simple ions

If an element is present in the form of simple ions, its oxidation number is merely the ionic charge. Thus, in iron(II) chloride, $Fe^{2+}(Cl^-)_2$, iron has an oxidation number of +2 and chlorine −1.

Molecules

To establish the oxidation number of a particular element in a molecule we write down the formal charge on every atom of that element and then calculate the average. The *formal charge* is based on the assumption that the shared pair of electrons that constitutes the covalent bond is located at the more electronegative atom (§4.6.2). For example, in the hydrogen chloride molecule, in which the electron pair is displaced towards the chlorine atom, we assume that the electrons are located *entirely* at the chlorine atom.

$$\overset{\delta^+}{\mathrm{H}}\!-\!\!:\!\overset{\delta^-}{\mathrm{Cl}} \quad \text{approximates to} \quad \overset{+}{\mathrm{H}} \;\; :\!\overset{-}{\mathrm{Cl}}$$

In effect, we assume that the molecule approximates to the ionic state represented by H^+Cl^-. Thus, the formal charge on the hydrogen atom is +1 and that on the chlorine atom −1. Since, in this particular molecule, there is only one atom of each element, hydrogen has an oxidation number of +1 and chlorine −1.

In the same way that the algebraic sum of the ionic charges of an electrovalent compound is zero, so the algebraic sum of the formal charges on the atoms of a covalent compound is also zero.

As another example, consider dinitrogen oxide, N_2O, which may be formulated as N≡N=O (with delocalised bonds), where --- represents a delocalised π bond, or as N≡N→O. In computing the oxidation numbers of nitrogen and oxygen in this compound we need to remember that the nitrogen atoms have an identical electronegativity, while the electronegativity of oxygen is greater than that of nitrogen. The N=N and N—O bonds must be considered separately from the delocalised π bond.

N=N and N—O bonds The electrons constituting the N=N bond are formally considered to reside at the midway point between the two atoms, so that the charge on each is zero. However, polarisation of the N—O bond leads to the central nitrogen atom acquiring a charge of +1 and the oxygen atom −1:

$$\underset{}{N}{=}\underset{+1}{N}{-}\underset{-1}{O}$$

Delocalised π bond This bond should be treated as a covalent bond involving three atoms. The electron pair is considered to be effectively at the oxygen atom; thus, oxygen gains a charge of −1, and each nitrogen atom $+\frac{1}{2}$:

$$\underset{+\frac{1}{2}}{N}\cdots\underset{+\frac{1}{2}}{N}\cdots\underset{-1}{O}$$

To summarise, the terminal nitrogen atom has a formal charge of $+\frac{1}{2}$, the central nitrogen atom $+1\frac{1}{2}$, and the oxygen atom −2. The oxidation number of oxygen in this compound is −2, and that of nitrogen the mean of $+\frac{1}{2}$ and $+1\frac{1}{2}$, which is +1.

Covalent ions

Covalent ions are treated in essentially the same way as molecules. Consider, for example, the sulphate ion, in which a sulphur atom forms four covalent bonds and two delocalised π bonds with oxygen atoms,

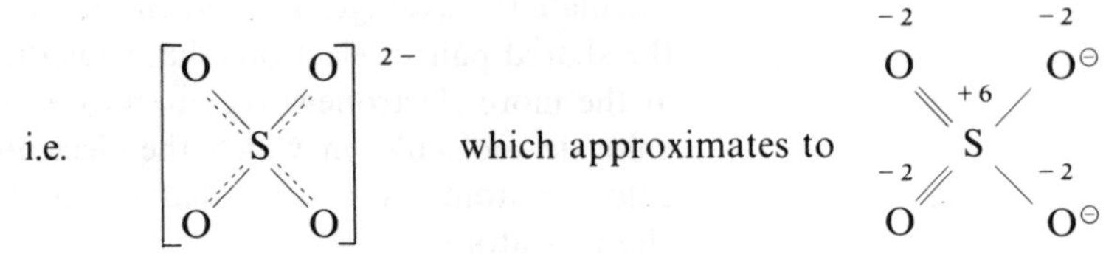

As with all covalent ions, the algebraic sum of the formal charges on the atoms is equal to the charge on the ion.

Uses of oxidation numbers

Oxidation numbers are used in inorganic chemical nomenclature (§1.2). They are also valuable in that if the hydrides, oxides and oxoacids of an element are arranged in order of oxidation number, the chemistry of the element can be seen as part of a logical pattern. Consider the case of nitrogen.

Oxidation number of nitrogen

+5	N_2O_5	HNO_3	NO_3^-
+4*	N_2O_4, NO_2		
+3	N_2O_3	HNO_2	NO_2^-
+2*	NO		
+1	N_2O		
0*	N_2		
−1	NH_2OH		
−2	N_2H_4		
−3*	NH_3	NH_4^+	

(OXIDATION ↑ upwards; REDUCTION ↓ downwards)

* Denotes a relatively stable oxidation state.

The chart helps us to understand why so many substances can appear when nitric acid functions as an oxidising agent. With a feeble reducing agent, such as copper, nitric acid is not extensively reduced and the principal product is nitrogen dioxide, but with a more powerful reductant, e.g. zinc, the nitrogen attains a lower oxidation state and the product may be dinitrogen oxide or even the ammonium ion. Hydrazine, N_2H_4, is a most unlikely product because it is too reactive: reactions tend to stop at relatively stable oxidation states.

The reduction of nitrous acid, in contrast to nitric acid, cannot possibly give nitrogen dioxide because NO_2 is *above* HNO_2 in the chart. Likely products are nitrogen oxide or the ammonium ion, depending on the power of the reducing agent.

The oxidation of ammonia, represented by upward movement on the chart, normally gives nitrogen. If, however, the conditions are severe the ammonia is oxidised further, to nitrogen oxide (§15.2.3).

The difference between the oxidation numbers of an element in reactant and product is equal to the number of electrons transferred in a redox reaction. Consider, for example, the oxidation of hydroxylamine, NH_2OH, to dinitrogen oxide. The difference in oxidation number, $-1-(+1)=-2$, represents the number of electrons lost by a molecule of hydroxylamine in its oxidation. The same figure can be derived from the ionic half-equation for the oxidation,

i.e. $NH_2OH = \frac{1}{2}N_2O + \frac{1}{2}H_2O + 2H^+ + 2e^-$

10.1.3 OXIDISING AGENTS AND REDUCING AGENTS

Oxidising agents

Any chemical species (i.e. atom, molecule or ion) that can gain electrons and become reduced is capable of functioning as an oxidising agent. The more readily electrons are accepted, the more powerful is the oxidising action. Species such as H_2O, HO^- and PO_4^{3-}, in which all the elements are in a relatively stable oxidation state, do not readily accept electrons and are of no practical value as oxidising agents. A measure of the oxidising power of a species in aqueous solution, in the presence of its reduction product, is provided by its standard electrode potential, $E^\ominus$ (§2.5.3). The more positive the $E^\ominus$ value, the more powerful is the oxidising action.

Non-metals, such as oxygen and the halogens, serve as oxidising agents because they receive electrons:

$$\tfrac{1}{2}O_2 + 2H^+ + 2e^- = H_2O \qquad E^\ominus = +1.23\ \text{V}$$
$$\tfrac{1}{2}F_2 + e^- = F^- \qquad E^\ominus = +2.87\ \text{V}$$

Fluorine is the most powerful oxidising agent known. Oxidising power decreases from top to bottom of group 7B, and iodine, although it can be reduced to iodide ions, is not normally regarded as an oxidising agent.

Certain acids and ions that contain an element in a high and relatively unstable oxidation state also function as oxidising agents. Examples are concentrated nitric acid, concentrated sulphuric acid, and cerium(IV), dichromate and permanganate ions. Nitric acid may be reduced to nitrogen dioxide (§15.2.9) and sulphuric acid to sulphur dioxide (§16.2.6).

The cerium(IV) ion is reduced to the more stable cerium(III) ion, and the dichromate ion, which contains chromium(VI), is reduced in acidic solution to the much more stable chromium(III) ion (§18.5.4).

The permanganate ion contains manganese(VII). It is reduced to the manganese(II) ion in strongly acidic solution, and to manganese(IV) oxide in weakly acidic or basic solution (§18.6.5).

Reducing agents

In principle, a reducing agent is a chemical species that is capable of losing electrons, thereby becoming oxidised. Useful reducing agents must have a strong tendency to lose electrons; in water their standard electrode potentials must have negative or low positive values (below approximately +0.80 V). The more negative the $E^\ominus$ value, the more powerful is the reducing action.

Metals that are near the top of the electrochemical series satisfy this condition, for they are strongly electropositive elements, i.e. they readily form positive ions by the loss of electrons. Sodium, zinc, iron and tin are all well-known laboratory reductants.

Ions and molecules that contain an element in a low and relatively unstable oxidation state may also be used. Of particular importance are the tin(II) and iron(II) ions (which become oxidised to tin(IV) and iron(III) ions respectively), sulphur dioxide and the sulphite ion (which on oxidation give the more stable sulphate ion, §16.2.5).

Substances that act as both oxidants and reductants

If a substance can be both oxidised and reduced, then it will function as either a reducing agent or an oxidising agent, depending upon what reagent is added. When mixed with an oxidant more powerful than itself the substance will become oxidised and thus act as a reductant; conversely, in the presence of a more powerful reductant it will behave as an oxidant. Examples are hydrogen peroxide and sulphur dioxide.

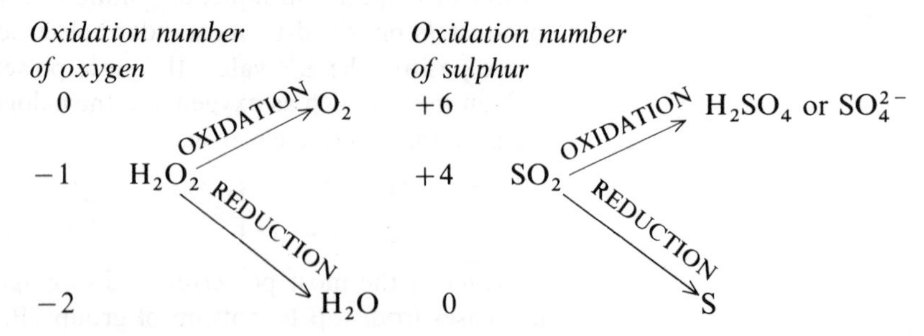

Most nitrogen containing compounds (see chart, p. 326) can act as both oxidising agents and reducing agents. The only exceptions are nitric acid and ammonia and their respective derivatives.

10.1.4 REDOX EQUATIONS

Many redox equations are difficult to balance by inspection and should be constructed by adding together two ionic half-equations. The procedure will be illustrated by reference to the reaction between potassium permanganate and iron(II) sulphate in the presence of dilute sulphuric acid.

The reductant, Fe^{2+}, is oxidised to Fe^{3+} according to the following ionic half-equation:

$$Fe^{2+} = Fe^{3+} + e^- \qquad (1)$$

The oxidant, MnO_4^-, is reduced to Mn^{2+}, and in order to write the ionic half-equation we begin by setting down these formulae:

$$MnO_4^- \longrightarrow Mn^{2+}$$

Then we balance the elements, remembering that the four atoms of oxygen in the permanganate ion finally appear in four molecules of water. The eight hydrogen atoms required for this purpose originate from the sulphuric acid as eight hydrogen ions:

$$MnO_4^- + 8H^+(aq) \longrightarrow Mn^{2+} + 4H_2O$$

We can now see the necessity for the sulphuric acid. (Hydrochloric acid cannot be used instead because it is oxidised by MnO_4^- to chlorine; neither can nitric acid, for it is an oxidant in its own right.)

Finally, we must obtain a charge balance. As things stand, there is a net charge of +7 on the left and +2 on the right. Adding five electrons to the left-hand side will provide a balance:

$$MnO_4^- + 8H^+(aq) + 5e^- = Mn^{2+} + 4H_2O \qquad (2)$$

The ionic equation for the entire reaction is obtained by adding together the ionic half-equations so that the electrons cancel out. (In a redox reaction, electrons are neither gained from the environment nor released to it.) In the present example this is achieved by multiplying equation (1) by five before adding it to equation (2):

$$MnO_4^- + 8H^+(aq) + 5Fe^{2+} = Mn^{2+} + 5Fe^{3+} + 4H_2O$$

For many purposes, such as volumetric analytical calculations, this equation is sufficient, but to perform calculations on reacting masses the stoicheiometric equation is required. This is derived from the ionic equation by inserting *spectator ions*, i.e. ions that play no part in the reaction:

$$KMnO_4 + 4H_2SO_4 + 5FeSO_4 \longrightarrow MnSO_4 + \tfrac{5}{2}Fe_2(SO_4)_3 + 4H_2O$$

The fraction is removed by multiplying throughout by two:

$$2KMnO_4 + 8H_2SO_4 + 10FeSO_4 \longrightarrow 2MnSO_4 + 5Fe_2(SO_4)_3 + 8H_2O$$

A final check shows that the spectator ions, K^+ and SO_4^{2-}, are out of balance. Two more potassium ions and one more sulphate ion are required on the right-hand side, so that the complete stoicheiometric equation is as follows:

$$2KMnO_4 + 8H_2SO_4 + 10FeSO_4 = 2MnSO_4 + 5Fe_2(SO_4)_3 + K_2SO_4 + 8H_2O$$

Stoicheiometric equations for redox reactions can be converted into ionic equations by the reverse of this procedure. Consider, for example, the reaction between potassium permanganate and hydrochloric acid:

$$2KMnO_4 + 16HCl = 2MnCl_2 + 2KCl + 5Cl_2 + 8H_2O$$

The equation can be rewritten so as to show all the ions:

$$2K^+ + 2MnO_4^- + 16H^+(aq) + 16Cl^- = 2Mn^{2+} + 4Cl^- + 2K^+ + 2Cl^- + 5Cl_2 + 8H_2O$$

Cancellation of spectator ions then gives the ionic equation:

$$2MnO_4^- + 16H^+(aq) + 10Cl^- = 2Mn^{2+} + 5Cl_2 + 8H_2O$$

10.1.5 DISPROPORTIONATION

In most redox reactions there is electron transfer between two substances, namely a reducing agent and an oxidising agent, but in a few cases there is an electron transfer within a single substance. Some atoms (or molecules or ions) of the substance serve as an electron source and become oxidised, while others accept electrons and are reduced. The phenomenon is known as *disproportionation.*

Because electrons are not transferred elsewhere, the extent to which a substance is oxidised, as shown by the change in oxidation number, is equal to the extent to which it is reduced. Consider, for example, the reaction that occurs when chlorine dissolves in cold, dilute sodium hydroxide solution:

$$Cl_2 + 2NaOH = NaCl + NaClO + H_2O$$

Chlorine (oxidation number = 0) is partly reduced to the chloride ion (oxidation number = -1) and partly oxidised to the hypochlorite ion, in which the oxidation number of chlorine is $+1$. Similar reactions occur when other non-metals dissolve in alkalis.

Most thermal decomposition reactions involve disproportionation. When, for example, ammonium nitrate decomposes on heating, the ammonium ion is oxidised and the nitrate ion is reduced. Four electrons are lost by the former and gained by the latter so that the product, of both processes, is dinitrogen oxide.

10.1.6 REDOX TITRATIONS

Oxidising and reducing agents, like acids and bases, can be titrated against one another. Special indicators, called *redox indicators*, are often required to detect the end-point. In the same way that indicators for acid–base titrations are themselves acids or bases, so redox indicators are oxidising or reducing agents. In oxidising solutions they become oxidised and display one colour, while in reducing solutions they switch to the reduced state and display another colour. The calculation of results is exactly the same as for acid–base titrations (§1.1.3).

Reducing agents are commonly estimated by titration with potassium permanganate (§18.6.5) or potassium dichromate (§18.5.4). Potassium permanganate is unusual in that it does not require the use of a redox indicator.

For the estimation of oxidising agents direct titration by a reducing agent is unsatisfactory, and two indirect approaches are recommended.

Titration in the reduced state

The oxidising agent (e.g. an iron(III) salt) is completely reduced, perhaps with zinc amalgam and dilute sulphuric acid, and then titrated with a standard solution of an oxidant such as potassium permanganate.

Iodimetry

An excess of potassium iodide is added to a known amount of the oxidising agent in solution. Iodine is liberated quantitatively and is titrated by a solution of sodium thiosulphate:

$$\underset{\text{thiosulphate ion}}{I_2 + 2S_2O_3^{2-}} = \underset{\text{tetrathionate ion}}{2I^- + S_4O_6^{2-}}$$

The end of the titration is marked by the disappearance of the brown colour of the iodine. It is usual to add freshly prepared starch solution near the end-point to intensify the remaining colour by the formation of a dark blue iodine–starch complex.

Sodium thiosulphate pentahydrate is unacceptable as a primary standard because it suffers from a variable water content. Solutions should be standardised against a reliable oxidant such as potassium iodate or potassium dichromate. Both compounds, when treated with an excess of potassium iodide, liberate iodine quantitatively:

$$IO_3^- + 6H^+(aq) + 5I^- = 3I_2 + 3H_2O$$

$$Cr_2O_7^{2-} + 14H^+(aq) + 6I^- = 3I_2 + 2Cr^{3+} + 7H_2O$$

Iodimetric titrations are well suited to the estimation of hypochlorites, hydrogen peroxide, carbon monoxide (§14.2.2) and copper(II) salts (§18.10.2).

10.2 Electrolysis

10.2.1 ELECTRODE REACTIONS

Substances that are electrical conductors in the fused state or in solution are known as *electrolytes*. Electrolytes comprise acids, bases and salts. They are chemically changed by the passage of electricity, the action being known as *electrolysis*. Electrolysis is performed in an *electrolytic cell* or *voltameter* (Fig. 10.1).

On electrolysis, negatively charged ions (*anions*) migrate to the anode

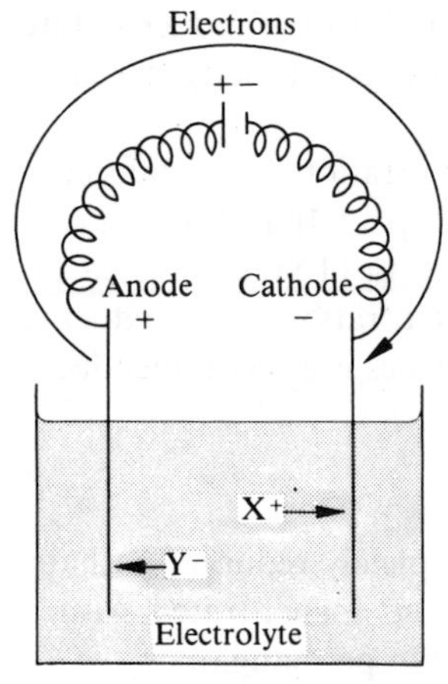

Fig. 10.1 An electrolytic cell.

and positively charged ions (*cations*) to the cathode. Chemical reaction at the anode always results in electron loss and is thus termed *anodic oxidation*, while that at the cathode requires electrons and is known as *cathodic reduction*. Electrons released at the anode travel through the external circuit to the cathode. The one-way movement of electrons through the wire and the two-way movement of ions in the electrolyte together constitute the 'electric current' in this circuit.

There are three types of anode reaction, all of which result in electron loss.

(i) The anode may dissolve. This is what happens, for example, when a potential difference is applied to copper electrodes immersed in a solution of copper(II) sulphate. The copper anode dissolves,

$$Cu(s) = Cu^{2+}(aq) + 2e^-$$

and a corresponding mass of copper is deposited on the cathode by the reverse reaction.

(ii) An anion from the electrolyte may be discharged. When fused sodium chloride is electrolysed with inert electrodes, chlorine is liberated at the anode through the discharge of chloride ions:

$$2Cl^- = Cl_2 + 2e^-$$

(iii) An ion in solution may suffer alteration of charge. When, for example, an acidic solution of an iron(II) salt is electrolysed, anodic oxidation occurs to give the corresponding iron(III) salt:

$$Fe^{2+}(aq) = Fe^{3+}(aq) + e^-$$

Although this may seem unlikely, as positive ions are reacting at the positive electrode, it does not violate the basic principle of electron loss to the anode.

Corresponding to these three types of anode reaction are three types of cathode reaction, all involving electron gain.

(i) An element at the cathode may dissolve. Certain non-metallic elements, e.g. O_2 when bubbled over a platinum cathode, are capable of gaining electrons and entering solution as negative ions.

(ii) A cation from the electrolyte may be discharged. This is by far the commonest type of cathode reaction, and accounts for the use of electrolysis in electroplating and the electrolytic purification of metals,

e.g. $Sn^{2+}(aq) + 2e^- = Sn(s)$

(iii) An ion in solution may undergo alteration of charge. When, for example, tin(IV) chloride solution is electrolysed, tin(IV) ions are reduced to tin(II):

$$Sn^{4+}(aq) + 2e^- = Sn^{2+}(aq)$$

(A continuation of the process will result in the electrodeposition of tin.)

Although these are the only primary reactions that can take place at an electrode, secondary changes may also occur. Oxygen or hydrogen, for instance, common products at anode and cathode respectively, may bring about further oxidation and reduction. More than one primary reaction may occur at an electrode, especially if a high voltage is applied.

10.2.2 FARADAY'S LAW OF ELECTROLYSIS

Michael Faraday in 1832 proposed two laws of electrolysis. They both related to the masses of elements liberated on electrolysis, and are today combined in the following single law. 'One mole of atoms is liberated during electrolysis by zF coulombs of electricity, where F is the Faraday constant ($9.648\,70 \times 10^4$ C mol^{-1}) and z is the charge number of the ion that is discharged.'

The Faraday constant represents one mole of electrons, as will be seen by considering the discharge of singly charged silver ions:

$$Ag^+ + e^- = Ag$$

The equation shows that the liberation of one mole of silver atoms requires one mole of electrons, and according to Faraday's law this is F coulombs.

For any circuit,

$$\underset{\text{(coulombs)}}{\text{quantity of electricity } (Q)} = \underset{\text{(amperes)}}{\text{current } (I)} \times \underset{\text{(seconds)}}{\text{time } (t)}$$

Provided that I and t are known, we can use Faraday's law to calculate the mass of an element, m grams, that is liberated on electrolysis. From Faraday's law,

zF coulombs liberate A_r grams of an element,

where A_r = relative atomic mass.

$\therefore$ It coulombs liberate $\dfrac{A_r \times I \times t}{z \times F}$ grams

i.e. $m = \dfrac{A_r \times I \times t}{z \times F}$

Furthermore, if we have two voltameters containing different elements connected in series, and we know how much element 1 is liberated in one cell, we can readily calculate how much element 2 is liberated in the other cell. I and t are constant under these conditions,

$\therefore$ for cell 1 $\quad m_1 \propto \dfrac{A_{r_1}}{z_1}$

and for cell 2 $\quad m_2 \propto \dfrac{A_{r_2}}{z_2}$

$$\therefore \quad \frac{m_1}{m_2} = \frac{A_{r_1}}{z_1} \times \frac{z_2}{A_{r_2}}$$

10.3

Extraction of metals (A non-thermodynamic treatment)

10.3.1 ORES

It would be astonishing if metallic sodium occurred naturally. It would be equally surprising if it occurred as sodium azide, NaN_3, despite the fact that nitrogen is a common element, because the compound is unstable. Even if sodium azide had been produced when the earth was formed, it would long ago have decomposed. As is well known, sodium occurs naturally as sodium chloride, sodium carbonate and various sodium aluminosilicates; all are highly stable compounds.

Mineral substances seldom contain anions other than carbonate, chloride, oxide, phosphate, silicate, sulphate or sulphide. They may well be *reactive* towards certain reagents – Na_2CO_3 and $CaCO_3$ are good examples – but they are *stable*, in the sense that a relatively large amount of energy is released in their formation from elements. An approximate guide to this energy change is provided by the standard enthalpy of formation, $\Delta H_f^{\ominus}$. A study of such values will thus indicate which compounds of an element are most likely to occur as ores. Table 10.1 lists some common compounds of calcium in decreasing order of standard enthalpy of formation, and it will be seen that only the top four occur naturally.

Table 10.1 Standard enthalpies of formation of various compounds of calcium

Compound	$\Delta H_f^{\ominus}$/kJ mol^{-1}	Ore
$Ca_3(PO_4)_2$	−4 126	phosphorite
$CaSO_4$	−1 433	anhydrite
CaF_2	−1 214	fluorspar
$CaCO_3$	−1 207	calcite, limestone, etc
$Ca(OH)_2$	−987	none
$Ca(NO_3)_2$	−937	none
$CaCl_2$	−795	none
CaO	−635	none
CaS	−483	none
CaC_2	−62.8	none

Generally, the standard enthalpies of formation of carbonates are high, and many metals have an ore of this kind (Table 10.2).

Table 10.2 Standard enthalpies of formation of various compounds

Element	$\Delta H_f^{\ominus}$/kJ mol^{-1}				Ores
	Carbonate	Chloride	Oxide	Sulphide	
sodium	−1 131	−411	−416 (Na_2O)	−373	rock salt, NaCl trona, Na_2CO_3
aluminium	—	−695	−1 669	−509	bauxite, Al_2O_3
zinc	−812	−416	−348	−203	calamine, $ZnCO_3$ zinc blende, ZnS
iron	−748 ($FeCO_3$)	−405 ($FeCl_3$)	−822 (Fe_2O_3)	−178 (FeS_2)	siderite, $FeCO_3$ haematite, Fe_2O_3 pyrites, FeS_2
lead	−700	−359 ($PbCl_2$)	−219 (PbO)	−94.3	cerrusite, $PbCO_3$ galena, PbS
silver	−506	−127	−30.6	−31.8	horn silver, AgCl argentite, Ag_2S metallic silver
gold	—	−118 ($AuCl_3$)	+80.8 (Au_2O_3)	—	metallic gold

Table 10.2 illustrates the principle that $\Delta H_f^\ominus$ values depend on the reactivity of the metal. Highly reactive metals, such as sodium and aluminium, release a considerable amount of enthalpy on forming compounds, whereas other metals release less. Noble metals, notably gold, may *require* enthalpy for their conversion into compounds: $\Delta H_f^\ominus$ for gold(III) oxide has a *positive* value. Certain compounds of gold are thus less stable than the metal itself, which is why gold occurs native; i.e. uncombined.

The compound of a metal for which $\Delta H_f^\ominus$ is greatest may not be commercially the most important ore. Since a lot of enthalpy has been lost in forming the ore, a lot of enthalpy must be supplied to decompose it; and it may well be advantageous to seek another ore for which the enthalpy requirement is less. Sodium, for instance, is extracted from sodium chloride ($\Delta H_f^\ominus = -411$ kJ mol^{-1}) rather than sodium carbonate ($\Delta H_f^\ominus = -1\,131$ kJ mol^{-1}).

10.3.2 METALS FROM ORES

The conversion of an ore to a metal, known as *smelting*, is a reduction process. Energy is required, and may be supplied by heating, adding a reducing agent, or applying an electrical potential difference. A combination of heat and reducing agent is often used.

Heat alone is insufficient to decompose ores of all but the least reactive metals, for although most compounds (e.g. oxides, sulphides and chlorides) become less stable on heating, the temperatures needed to decompose them are unacceptably high. For example, although silver oxide decomposes at 340 °C (613 K) and mercury(II) oxide at 500 °C (773 K), zinc oxide can be broken down only at about 1 900 °C (~2 200 K) and aluminium oxide, calcium oxide and indeed most other metal oxides at over 2 000 °C (~2 300 K).

A reducing agent is therefore necessary in most cases, and the one usually chosen on both technical and economic grounds is carbon, in the form of coke. Carbon will reduce many metal oxides to the metal at temperatures below 2 000 °C (~2 300 K); it is less effective with sulphides and quite useless with chlorides. Other reducing agents used in special cases include magnesium or sodium (for $TiCl_4$) and hydrogen (for the oxides of molybdenum and tungsten).

Electrolysis (cathodic reduction) provides the most powerful method of reducing metal compounds to the metal, for virtually any cathode potential can be applied. All the most reactive metals are produced in this way.

Reduction of oxides

Metal oxides are reduced by carbon in accordance with the following general equation:

$$MO + C = M + CO$$

At room temperature carbon can reduce few metal oxides, partly because the necessary activation energy is lacking, but principally because the enthalpy released as carbon changes into carbon monoxide is insufficient. However, as the temperature rises, carbon monoxide becomes *more* stable, while metal oxides become *less* stable. There is, therefore, for every metal oxide, a certain temperature at which the enthalpy released in the conversion of carbon to carbon monoxide is exactly equal to that required for the reduction of the metal oxide to the metal. At higher temperatures than this, there is a net release of enthalpy and the reaction is able to occur. (A rigorous treatment of this subject, in terms of free energy changes, is given in §2.4.3.)

Iron and zinc are the most notable metals manufactured by the carbon reduction of oxides (§18.7.1 and §18.11.1). However, in the upper, cooler region of a blast furnace carbon monoxide is the principal reductant. This is because, at temperatures below 700 °C (973 K), carbon monoxide is a more powerful reducing agent than carbon itself. The gas reduces some iron(III) oxide to spongy iron and the rest to iron(II) oxide which, in the lower, hotter region of the furnace, is reduced to the metal by carbon.

Carbon is unsuitable for the reduction of oxides of calcium and aluminium, partly because temperatures above 2000 °C (~2300 K) are required, and partly because calcium and aluminium react with carbon to form ionic carbides CaC_2 and Al_4C_3 which prevent isolation of the metals. Most transition elements form carbides of a different kind. Often, as with iron, a small proportion of the carbide improves the properties of the metal; but if, as with titanium, the carbide spoils the metal then carbon cannot be used in the extraction.

Reduction of sulphides

Sulphides are far more difficult than oxides to reduce with carbon. This is partly because carbon fails to form a lower sulphide, corresponding to CO, and partly because CS_2, unlike CO_2, *requires* enthalpy for its formation. Hydrogen, too, is an unsatisfactory reducing agent because the standard enthalpy of formation of hydrogen sulphide is only −20.2 kJ mol^{-1}. Usually, as in the case of zinc, a sulphide ore is first converted to the oxide by roasting it in air (§18.11.1).

Reduction of chlorides

Chlorides are even more resistant than sulphides to reduction by carbon, because of the difficulty of forming tetrachloromethane at high temperatures. One possible reagent for reducing chlorides of the less reactive metals (e.g. tin, lead, silver and copper) is hydrogen, but in practice such metals are best obtained by other routes. The method of chloride reduction is largely confined to the production of titanium by the Kroll process (§18.3.1).

Electrolytic reduction

The electrolytic extraction of metals is expensive, but provides the only commercially viable means of obtaining highly reactive elements, notably sodium, magnesium, calcium and aluminium, whose compounds are difficult to reduce by other methods.

Water must be absent, since if it were present hydrogen would be released at the cathode in preference to the metal. Sodium, magnesium and calcium are all made by the electrolysis of their fused chlorides (§12.2.1). Chlorides have the advantage of relatively low melting temperatures. (Oxides, by contrast, have high lattice enthalpies and are almost infusible, while oxosalts tend to decompose to give oxides.) The temperature is kept as low as possible to minimise costs, reduce the reactivity of the chlorine produced at the anode, and limit the loss of metal through vaporisation.

Aluminium is obtained from its oxide rather than its chloride. (Aluminium chloride is covalent and therefore a non-conductor of electricity.) The difficulty posed by the high melting temperature of aluminium oxide is overcome by dissolving the compound in molten cryolite (§13.2.1).

10.4 Oxidation and reduction in organic chemistry

Simple organic compounds fall into a number of *homologous series*, such as alcohols, aldehydes and carboxylic acids. The word *homologous*, meaning 'of corresponding structure', denotes that all members of a given series share the same functional group (§1.3). All alcohols, for instance, possess the hydroxyl group, HO, as can be seen from the formulae of the first eight members of the homologous series:

CH_3OH	$CH_3CH_2CH_2CH_2OH$
CH_3CH_2OH	$CH_3CH(OH)CH_2CH_3$
$CH_3CH_2CH_2OH$	$(CH_3)_2CHCH_2OH$
$CH_3CH(OH)CH_3$	$(CH_3)_3COH$

As is evident from this example, a homologous series comprises both unbranched and branched chain compounds. Notice, also, that there is a structural difference of CH_2 between any particular *homologue* (i.e. member) and the compounds immediately above and below it in the series.

The properties of an organic compound are dominated by its functional group rather than the hydrocarbon skeleton. Because all members of a homologous series have the same functional group, they have essentially the same chemical properties and can be studied together. There is, however, a decline of reactivity on ascending a homologous series, i.e. as relative molecular mass increases. The decrease is especially large between the first two members of a series, but becomes progressively smaller betweeen the high homologues. Physical properties, such as melting temperature, boiling temperature, density and solubility in water, either increase or decrease in a corresponding manner. For example, for any homologous series, a graph of melting temperature against relative molecular mass rises fairly steeply at first, but gradually flattens out and becomes almost level when there are approximately a hundred carbon atoms in the molecule.

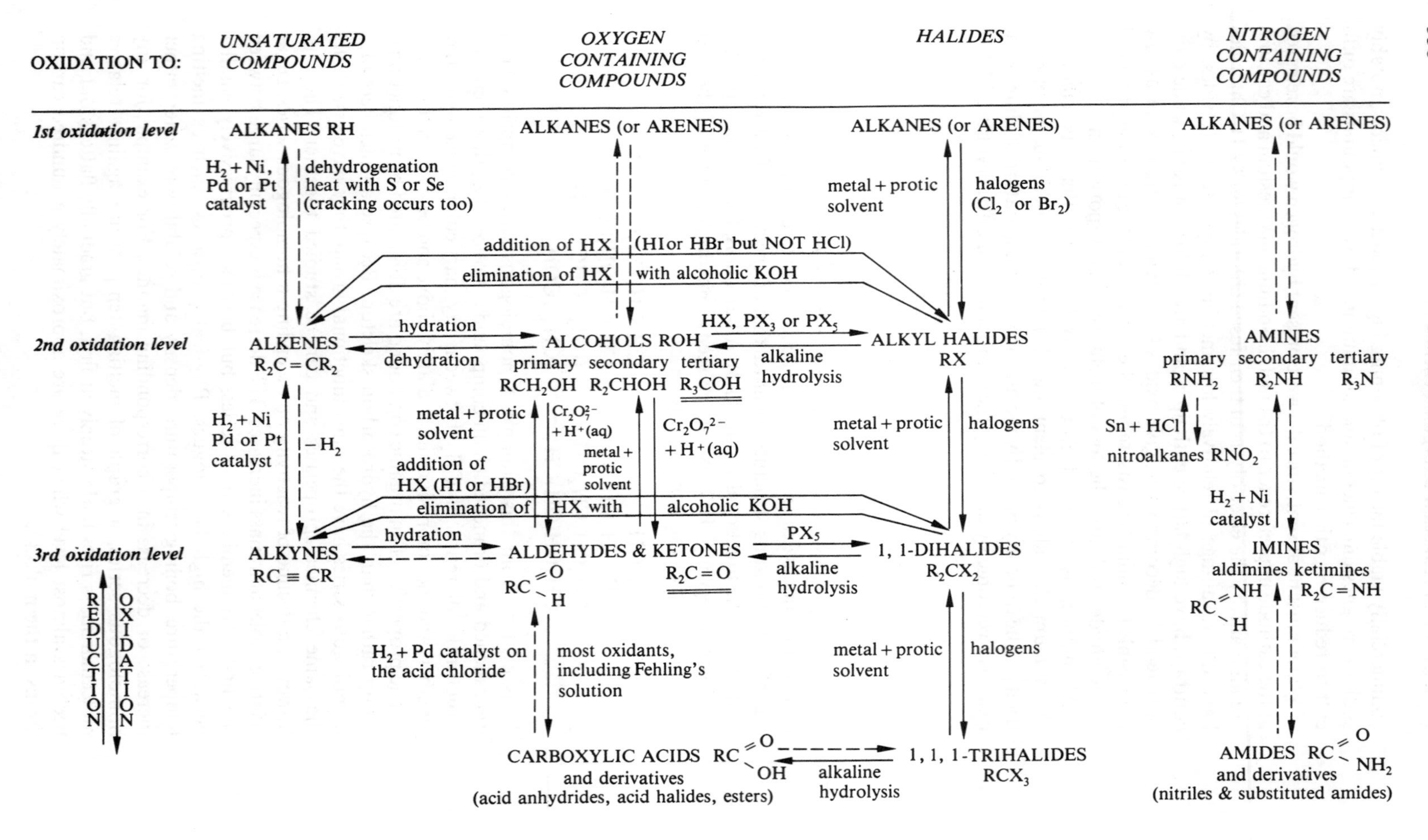

Fig. 10.2 Oxidation sequences in organic chemistry.
Notes
1. Dotted arrows are used to denote changes which, though theoretically possible are difficult to achieve in practice.
2. In general an oxygen containing compound is convertible to the corresponding nitrogen compound by treatment with ammonia. The reverse change, except in the case of amines, is achieved by hydrolysis (primary amines ⟶ alcohols with nitrous acid).
3. This chart is relevant only if the hydrocarbon skeleton of a molecule remains unchanged on reaction. If there is an increase or decrease in chain length, or a rearrangement, *this chart does not apply.*

10.4.1 OXIDATION LEVELS

The oxidation number concept does not work well in organic chemistry, since the oxidation number of carbon varies from one member to another of a homologous series, e.g.

Alkane	Oxidation number of carbon
methane, CH_4	-4
ethane, C_2H_6	-3
propane, C_3H_8	$-2\frac{2}{3}$
butane, C_4H_{10}	$-2\frac{1}{2}$

If, say, ethane could be reduced to methane, or if methane could be oxidised to ethane, these oxidation numbers would be meaningful, but such reactions are impractical.

However, the idea of oxidation numbers can be modified to provide us with a useful tool. We can argue that since the hydrocarbon skeleton of a molecule does not usually participate in reactions we may neglect it, and consider only the carbon atom bearing the functional group. Furthermore, since C—H and C—C single bonds usually remain intact we can neglect these also, and define the *oxidation level of the carbon atom* as the sum of its other bonds, i.e. the number of outer shell electrons used to form bonds other than C—H and C—C.

Figure 10.2 shows the principal homologous series of organic compounds arranged in order of oxidation level. Changes in a *downwards* direction on this chart represent oxidation. These changes require the use of oxidising agents such as potassium permanganate or potassium dichromate. Changes made *upwards* represent reduction and require the use of reducing agents such as sodium or zinc. Changes *across* the chart are made at the same oxidation level and require neither oxidants nor reductants but reagents such as water, ammonia, acids or bases.

EXAMINATION QUESTIONS ON CHAPTER 10

1 (a) The following equations represent oxidation/reduction reactions occurring in aqueous solution. By considering each reaction in terms of electron transfer explain which species is *oxidised* and which is *reduced.*

(i) $2FeCl_2 + Cl_2 \longrightarrow 2FeCl_3$

(ii) $I_2 + 2Na_2S_2O_3 \longrightarrow 2NaI + Na_2S_4O_6$

State, and account for, the colour change observed in each case.

(b) By considering the ionic half equations:

$Ce^{4+} + e^- \longrightarrow Ce^{3+}$

$C_2O_4^{2-} \longrightarrow 2CO_2 + 2e^-$

$Fe^{2+} \longrightarrow Fe^{3+} + e^-$

calculate the volume of cerium(IV) sulphate (of concentration 0.2 mol/l) required to oxidise 25 cm³ of iron(II) ethanedioate (iron(II) oxalate) solution of concentration 0.6 mol/l.

(c) Given the following standard electrode potentials at 25 °C
$Pb^{2+} + 2e^- \rightleftharpoons Pb \quad E^\ominus = -0.126$ V
$Zn^{2+} + 2e^- \rightleftharpoons Zn \quad E^\ominus = -0.763$ V
calculate the e.m.f. of the cell:
$Zn(s)|Zn^{2+}(aq,1\ \text{M})||Pb^{2+}(aq,1\ \text{M})|Pb(s)$ (AEB)

2 Discuss the meanings which are or have been given to the terms *oxidation* and *reduction*, using as examples (a) d-block elements, and (b) organic compounds.

What is meant by *redox potential*? Explain how redox potentials can be used to forecast the direction of chemical change. (L)

3 (a) State Faraday's laws of electrolysis. Describe an experiment to illustrate the second law.

(b) Calculate the time in minutes necessary for a current of 10 A to deposit 1 g of copper from an aqueous solution of copper(II) sulphate.

(c) Show why the ratio of the masses of copper and sodium deposited, under the appropriate conditions, by the same quantity of electricity is 1.38.

(d) Calculate the charge on an electron given that the Avogadro constant is 6.02×10^{23}/mol. (AEB)

4 (a) Outline briefly how the concept of oxidation and reduction has developed into that of electron-transfer reactions and say why such reactions are of importance in the study of chemistry.

(b) Certain standard electrode potentials measured at 25 °C are listed below. Comment on the selection of examples and on the values given. Explain briefly how the electrode potential arises and with the aid of *one* illustrative example how such an *electromotive force of a half-cell* may be measured.

Zn^{2+}/Zn	−0.763 V
Fe^{2+}/Fe	−0.440 V
$H^+/H_2/Pt$	0 V
Cu^{2+}/Cu	+0.337 V
$Fe^{2+}, Fe^{3+}/Pt$	+0.771 V
$Cr^{3+}, Cr_2O_7^{2-}, H^+/Pt$	+1.33 V

Use these values to decide what happens in the following examples.

(i) Iron filings are placed in a solution containing the following ions in their standard state:

$Cu^{2+}, Fe^{2+}, Fe^{3+}, H^+, Zn^{2+}$

Deduce what happens and write equations for the reactions which occur.

(ii) Calculate the standard electromotive force of the cell

$Zn/Zn^{2+}/Cu^{2+}/Cu$

and compare it with that for the cell

$Cu/Cu^{2+}/Zn^{2+}/Zn$

For the first cell write equations for the electrode processes and the cell reaction. Comment on what happens in the cell and explain where the available energy goes when the reaction occurs in a beaker and not in an electrical cell.

(iii) When the following ions (as salts) in their standard states are introduced into a solution which also contains an excess of dilute sulphuric acid, deduce what happens.

Fe^{2+}, Fe^{3+}, Cr^{3+}, $Cr_2O_7^{2-}$

Write separate ion-electron equations for the reactants and deduce the final equation for the reaction.

If 5.6 g of pure iron wire is dissolved, in the absence of air, in excess of dilute sulphuric acid to which 5.6 g of potassium dichromate is added, which one of the solid reactants would be in excess and by how much?

($Fe = 56$, $K_2Cr_2O_7 = 294$) (SUJB)

5 What general principles are involved in the extraction of metals from purified ores? Illustrate your answer by reference to the extraction of iron, zinc, titanium, sodium and aluminium.

(Details of chemical plant are not required.)

1.77 g of a mixture of iron and zinc were dissolved in excess dilute sulphuric acid. When reaction was complete, the resulting solution was titrated with 0.02 M potassium permanganate; 40.0 cm^3 were decolorised. Calculate the percentage of iron in the mixture. (JMB)

6 (a) In general a metal is produced by reduction of one of its compounds which **either** occurs naturally **or** can be produced from a naturally-occurring compound.

Illustrate the above statement by giving an **outline** of the reduction process used for the production of **one** metal from **each** of the following groups of metals:

(i) uranium, chromium, titanium,

(ii) zinc, iron, copper.

(b) Explain why aluminium cannot be produced by electrolysis of the molten chloride or by reduction of its oxide with carbon.

(c) Given the following enthalpy changes of formation:

MgO -602 kJ/mol

Al_2O_3 $-1\,700$ kJ/mol

calculate the enthalpy change for the reaction:

$$Al_2O_3 + 3Mg \longrightarrow 2Al + 3MgO$$

Does your answer indicate whether magnesium will reduce aluminium oxide? Give an explanation. (AEB)

7 For **each** of the following processes used in the isolation of elements from their compounds, describe **one** suitable example and state why the process described is the preferred method:

(a) electrolysis,

(b) reduction with carbon,

(c) reduction with a metal. (AEB)

2 Detailed chemistry

11 Hydrogen

11.1

Introduction

Three isotopes of hydrogen exist: protium, ^1_1H, deuterium, ^2_1H or D, and tritium, ^3_1H or T. The ratio of protium to deuterium in nature is approximately 6 000 : 1, and tritium occurs in extremely small quantities. The mixture of isotopes normally encountered is referred to as 'hydrogen' and is given the symbol H. The hydrogen atom has a $1s^1$ configuration.

Under normal conditions hydrogen is a gas, consisting of diatomic molecules, H_2. Because of its low relative molecular mass, $H_2 = 2.016$, hydrogen has very low melting and boiling temperatures. For protium, the melting temperature is -259 °C (14 K) and the boiling temperature -252.5 °C (20.7 K).

11.1.1 THE BONDING OF HYDROGEN

Hydrogen, in its compounds, is always monovalent. It achieves this condition by one of three electronic processes, as follows:

(i) the loss of its electron to form the hydrogen ion, H^+;
(ii) the gain of one electron to form the hydride ion, H^-;
(iii) the sharing of its electron with one from another atom to form a single covalent bond.

The hydrogen ion

The isolated hydrogen ion, which is in fact a proton, occurs in the hydrogen discharge tube. In aqueous or other solutions, however, the proton never occurs alone, because it is strongly attracted and bound to polar solvent molecules. For example, in aqueous solution, protons are coordinately bonded to water molecules,

i.e.
$$H_2\ddot{O}\!: \longrightarrow H^+ = \left[H_2O{-}H \right]^+$$

The oxonium ion, H_3O^+, which may be described as a monohydrated proton, is formed when compounds known as *protic acids* are dissolved in water,

e.g.
$$HCl + H_2O \rightleftharpoons H_3O^+ + Cl^-$$
$$H_2SO_4 + H_2O \rightleftharpoons H_3O^+ + HSO_4^-$$

There is, however, considerable evidence to indicate that the oxonium ion is extensively hydrated in solution, i.e. $H_3O^+(aq)$. This hydrated ion is known as the *hydrogen ion*, and is denoted by either $H^+(aq)$ or H^+. Protons in solution may therefore be represented by H_3O^+, $H^+(aq)$ or H^+, whichever is convenient, but it should be remembered that in these cases H^+ does not represent an isolated proton. The term 'proton' is often used in discussions of acid–base reactions, although it is doubtful whether free protons ever exist in such reactions.

The proton also occurs coordinately bonded to other molecules, e.g. to ammonia as the NH_4^+ ion, to methanol as $CH_3OH_2^+$ and to ethoxyethane (ether) as $(C_2H_5)_2OH^+$.

The hydride ion

Hydrogen has a low electron affinity (-78 kJ mol^{-1}) and is reduced to the hydride ion, H^-, only by the reactive s-block elements (§12.2.3).

Covalent bonds

Most of the p-block elements form covalent bonds with hydrogen, e.g. NH_3, H_2S and HCl.

Enthalpy changes during the reactions of hydrogen

Overall change	Steps	Enthalpy change of each step/kJ mol^{-1}	Overall enthalpy change/kJ mol^{-1}
(a) $\frac{1}{2}H_2(g) = H(g)$	$\frac{1}{2}H_2(g) = H(g)$	+218	+218
(b) $\frac{1}{2}H_2(g) = H^+(g) + e^-$	$\frac{1}{2}H_2(g) = H(g)$	+218	
	$H(g) = H^+(g) + e^-$	+1 310	+1 528
(c) $\frac{1}{2}H_2(g) = H^+(aq) + e^-$	$\frac{1}{2}H_2(g) = H(g)$	+218	
	$H(g) = H^+(g) + e^-$	+1 310	
	$H^+(g) + aq = H^+(aq)$	−1 200	+328
(d) $\frac{1}{2}H_2(g) + e^- = H^-(g)$	$\frac{1}{2}H_2(g) = H(g)$	+218	
	$H(g) + e^- = H^-(g)$	−70	+140

Table 11.1 The enthalpy changes accompanying the reactions of hydrogen

The enthalpy changes accompanying the reactions of hydrogen to form covalent compounds depend on the other elements. From (b) in Table 11.1, we see that the enthalpy required to form isolated protons, $H^+(g)$, is far in excess of that required for the other changes. Thus, on enthalpy considerations, the overall changes (a), (c) and (d) are more likely to occur than change (b).

11.2 The chemistry of hydrogen

At ordinary temperatures hydrogen has a low reactivity. To bring about a reaction with hydrogen, a high temperature or the presence of a catalyst is usually required. The cause lies in the strength of the bond in the hydrogen molecule. (The H—H bond dissociation enthalpy is 436 kJ mol^{-1}.) For hydrogen to react this bond must be broken, and the breaking process presents an enthalpy barrier which must be overcome before the

H_2

reaction can proceed. Heat or the presence of a catalyst either ruptures or weakens the H—H bond and therefore speeds up the reactions of hydrogen.

The important reactions of hydrogen with other elements and with various compounds are discussed in the appropriate chapters of this book.

11.2.1 THE BINARY COMPOUNDS OF HYDROGEN

Hydrogen forms binary compounds, known as *hydrides*, with most of the elements. Notable exceptions are the noble gases. The term 'compound', however, is perhaps an overstatement when considering the d-block elements, but the s- and p-block elements do form well-defined compounds with hydrogen.

The two principal types of hydride which exist may be designated by means of the periodic table.

The s-block hydrides

Except for beryllium and magnesium, the s-block elements form ionic hydrides containing the hydride ion, H^- (§12.2.3).

The p-block hydrides

All the elements of groups 4B, 5B, 6B and 7B form covalent mononuclear hydrides, i.e. hydrides which contain one atom of a p-block element (Table 11.2).

Table 11.2 The formulae of the mononuclear hydrides of groups 4B, 5B, 6B and 7B

Group	4B	5B	6B	7B
formula†	XH_4	XH_3	H_2X	HX
examples	CH_4, SnH_4	NH_3, PH_3	H_2O, H_2S	HF, HI

† X represents one atom of the group element. Notice the conventional manner of writing the formulae.

The X—H bond dissociation enthalpy tends to increase with decreasing radius of atom X. Therefore the stability of the covalent hydrides increases from group 4B to group 7B, and decreases from top to bottom of each group. The hydrides of the metallic elements, e.g. bismuth and lead, are highly unstable, while those of the head elements of these groups are comparatively stable.

Some elements form polynuclear hydrides as well as mononuclear hydrides. These contain more than one atom of a p-block element. Carbon is particularly outstanding in this respect, and forms alkanes, alkenes, alkynes and arenes (Chapter 14). The remaining p-block elements, notably silicon, germanium, nitrogen, phosphorus, oxygen and sulphur, form a limited number of polynuclear hydrides, e.g. Si_nH_{2n+2}, Ge_nH_{2n+2}, N_2H_4, HN_3, P_2H_4, H_2O_2 and H_2S_n. Boron forms two series of polynuclear hydrides, with the formulae B_nH_{n+4} and B_nH_{n+6}. Many of the p-block hydrides are discussed in more detail in the appropriate groups.

11.2.2 HYDRIDE COMPLEXES

Hydride complexes contain hydride ions, H^-, coordinated to an atom of another element. The best known examples are the tetrahydroborate(III) ion (ASE: tetrahydridoborate(III)), $[BH_4]^-$, and the tetrahydrido-aluminate(III) ion, $[AlH_4]^-$. Both may be regarded as complexes formed between 'BH_3' or 'AlH_3' and the hydride ion,

i.e. $BH_3 + H^- = [BH_4]^-$

$AlH_3 + H^- = [AlH_4]^-$

The tetrahydroborate(III) ion is tetrahedral, and the aluminium complex is probably of the same shape.

$Li[AlH_4]$ and $Na[BH_4]$

Sodium tetrahydroborate(III), $Na[BH_4]$, and lithium tetrahydrido-aluminate(III), $Li[AlH_4]$, are important reducing agents. They function as reductants by acting as sources of the hydride ion. Lithium tetrahydrido-aluminate(III) in ethoxyethane solution is an excellent reducing agent, and is widely used in organic chemistry (§14.10.6). In inorganic chemistry it can be used to prepare p-block hydrides from the corresponding halides,

e.g. $4PCl_3 + 3Li[AlH_4] = 4PH_3 + 3LiCl + 3AlCl_3$

Lithium tetrahydridoaluminate(III) reacts violently with water:

$$Li[AlH_4] + 4H_2O = LiOH + Al(OH)_3(aq) + 4H_2$$

Sodium tetrahydroborate(III), by contrast, is stable in cold water, but is also an excellent reducing agent.

11.2.3 THE POSITION OF HYDROGEN IN THE PERIODIC TABLE

On the basis of its electronic configuration and chemical properties, hydrogen could reasonably be placed in group 1A, group 7B or, with imagination, group 4B of the periodic table. Some of the arguments for and against placing hydrogen in these groups are set out in Table 11.3.

As indicated, there are several arguments against putting hydrogen in any of these groups. Many of the alleged similarities, too, are rather artificial because hydrogen does not closely resemble any other element. It should be regarded as unique and not placed in any of the periodic groups.

Table 11.3 Comparison of hydrogen with the elements of groups 1A, 4B and 7B

	Properties in which hydrogen displays a resemblance to a periodic group	Properties in which hydrogen displays a difference from a periodic group
group 1A	has an ns^1 configuration forms a monopositive ion	is less reactive than the alkali metals is a non-metal forms diatomic molecules forms covalent bonds forms an anion has a higher electronegativity and a higher ionisation enthalpy than the alkali metals gives an ion, H^+, which does not have an independent existence
group 4B	has a half-full outer shell (cf. C) forms covalent bonds is a non-metal (cf. C and Si)	forms diatomic molecules (contrast C, Si, etc) has an electronegativity which does not fit in with the group trend, i.e. $x_H = 2.1$, $x_C = 2.5$, $x_{Si} = 1.8$ etc
group 7B	is a diatomic non-metal forms covalent bonds has one electron less than the following noble gas in the periodic table forms a mononegative ion	forms cationic species is far less reactive than the halogens has a lower electronegativity than the halogens has a much lower electron affinity than the halogens

EXAMINATION QUESTIONS ON CHAPTER 11

1 Hydrogen resembles both the alkali metals and the halogens in its chemical behaviour. Survey the evidence which leads to this conclusion.

Use the compounds which you have described to illustrate the changes in bond-type of the hydrides of elements in a period of the periodic table.
(JMB)

2 Give examples of the use of lithium tetrahydridoaluminate(III) and sodium tetrahydroborate(III) as reducing agents in inorganic and organic chemistry. Describe briefly the bonding in the hydrides of sodium, silicon and chlorine. Comment on the trend in the type of bond across the second period (Li—Ne).

Describe the reactions of the chlorides of magnesium and of silicon with water at room temperature. How far are the reactions of these chlorides typical of the reactions of chlorides of metals and non-metals in general? (Chapters 12 and 14.) (JMB)

12 Groups 1A and 2A

12.1 Introduction

The elements of group 1A, often known as the 'alkali metals' are lithium (Li), sodium (Na), potassium (K), rubidium (Rb), caesium (Cs) and francium (Fr). Group 2A comprises the elements beryllium (Be), magnesium (Mg), calcium (Ca), strontium (Sr), barium (Ba) and radium (Ra). Calcium, strontium, barium and radium are commonly known as the 'alkaline-earth metals'. Little is known about francium since it is radioactive and its isotopes have short half-lives; it is formed from actinium by radioactive decay.

$$^{227}_{89}Ac = ^{223}_{87}Fr + ^{4}_{2}He \qquad (§3.3.1)$$

Radium is also radioactive, but its isotopes are more stable than those of francium and its chemistry is well known. Both francium and radium are commonly excluded from general discussions of the s-block elements, but their properties may be deduced from the trends that occur in each group. Some important properties of the s-block elements are given in Tables 12.1 and 12.2.

All the metals

Table 12.1
Some properties of the elements of group 1A

	Lithium	Sodium	Potassium	Rubidium	Caesium	Francium
character	metal	metal	metal	metal	metal	metal
allotropes	none	none	none	none	none	none
atomic number	3	11	19	37	55	87
relative atomic mass	6.941	22.989 8	39.102	85.467 8	132.905 5	223†
outer electronic configuration	$2s^1$	$3s^1$	$4s^1$	$5s^1$	$6s^1$	$7s^1$
valency	1	1	1	1	1	1
melting temperature/°C	180	97.8	63.7	38.9	28.7	27
boiling temperature/°C	1330	890	774	688	690	—
enthalpy of atomisation/ kJ mol^{-1}	161.0	109.0	90.0	85.8	78.7	—
$E^{\ominus}$/V	−3.04	−2.71	−2.92	−2.92	−2.92	—

† Mass number of the most stable isotope.

12.1.1 THE STRUCTURE OF THE ELEMENTS

The melting temperatures and enthalpies of atomisation of the group 2A elements are generally higher than those of the group 1A elements. This is a consequence of the greater strength of the metallic bonding in the former. Broadly speaking, the higher the ratio of delocalised bonding electrons to atomic radius, the greater is the strength of the metallic bonding. This ratio is small for group 1A elements, with their relatively large atoms and with only one electron per atom available for metallic bonding. For a given group 2A element the ratio is considerably larger than for the adjacent group 1A element (e.g. Mg and Na) because, for the group 2A element, there are two electrons per atom available for metallic bonding and the atomic radius is smaller.

As the atomic radius increases from the top to the bottom of the groups the strength of the metallic bonding decreases. Thus, the melting temperatures and atomisation enthalpies of the group 1A elements, which all crystallise with a body centred cubic lattice, decrease from lithium to caesium. A corresponding decrease is observed in group 2A, but is irregular. This is due to another factor which affects metallic bond strength, namely the crystal structure of the metal. Beryllium and magnesium possess a hexagonal close packed structure, and barium crystallises in a body centred cubic lattice. Calcium and strontium change structure when heated from face centred cubic to hexagonal close packed and finally body centred cubic.

Be and its compounds
Soluble Ba compounds

Table 12.2
Some properties of the elements of group 2A

	Beryllium	**Magnesium**	**Calcium**	**Strontium**	**Barium**	**Radium**
character	metal	metal	metal	metal	metal	metal
allotropes	none	none	two	two	none	none
atomic number	4	12	20	38	56	88
relative atomic mass	9.012 18	24.305	40.08	87.62	137.34	226†
outer electronic configuration	$2s^2$	$3s^2$	$4s^2$	$5s^2$	$6s^2$	$7s^2$
valency	2	2	2	2	2	2
melting temperature/°C	1280	650	850	768	714	700
boiling temperature/°C	2477	1110	1487	1380	1640	1140
enthalpy of atomisation/ kJ mol^{-1}	321	150	193	164	176	130
$E^{\ominus}$ /V	−1.85	−2.38	−2.87	−2.89	−2.90	−2.92

† Mass number of the most stable isotope.

12.1.2 BONDING AND VALENCY

The dominant feature of the s-block elements is the formation of ions by the loss of their outer s electrons; group 1A elements form monopositive ions, and group 2A elements dipositive ions.

Ionisation enthalpies

The ionisation enthalpies decrease from the top to the bottom of each group. Thus, in group 1A, lithium shows the least tendency to ionise and consequently forms a few covalent compounds (§12.2.8), but for the remaining elements of the group covalent bond formation is rare. Likewise, in group 2A, beryllium has a rather high ionisation enthalpy and tends to utilise its $2s^2$ electrons in forming two covalent bonds rather than simple Be^{2+} ions. Indeed, beryllium differs considerably from the other elements of group 2A (§12.2.9).

The covalent character which is evident in many lithium and magnesium compounds arises from distortion of the anions by the small Li^+ and Mg^{2+} cations (§4.6.1).

Standard electrode potentials

The ease with which ions are formed under anhydrous conditions is accurately reflected by the ionisation enthalpies of the appropriate elements. However, when ions are formed in aqueous solution they are hydrated (§4.9.1). Under such conditions ionisation enthalpies are not always a reliable guide to the ease of ion formation, and instead we use standard electrode potentials, $E^{\ominus}$, which by definition, are a measure of the tendency for the following change to occur:

$$M^{n+}(aq) + ne^- = M(s)$$

The more negative the value of $E^{\ominus}$, the greater is the ease with which the *reverse* change, i.e. the formation of hydrated ions from the metal, occurs.

The $E^{\ominus}$ values in Table 12.1 show that the ease with which the alkali metals form hydrated ions in solution decreases in the order Li > K = Rb = Cs > Na. Thus, lithium shows the strongest tendency to form hydrated ions, even though it has the highest ionisation enthalpy in group 1A.

In group 2A the order of standard electrode potentials is Ra > Ba > Sr > Ca > Mg > Be (Table 12.2), which is in accordance with the ionisation enthalpies. This is because for these elements the most important factor is the ionisation enthalpy, which decreases from beryllium to radium.

12.2 The s-block elements

12.2.1 THE OCCURRENCE AND ISOLATION OF THE ELEMENTS

Group 1A

The chief sources of the alkali metals, of which only sodium and potassium are relatively abundant, are dried-up salt lakes. Lithium occurs principally as a complex aluminosilicate.

The elements are isolated by electrolysis of their fused halides. Aqueous solutions cannot be used because the alkali metals all react with water. Sodium, an important metal, is isolated on a large scale by the *Downs process*. The electrolyte is a mixture of sodium chloride and calcium chloride, which melts at approximately 600 °C (873 K). Sodium chloride alone melts at 801 °C (1 074 K); thus by using the mixed electrolyte the cell may be operated at a considerably lower temperature (§8.4.1). In this context the calcium chloride is called a 'flux' because it makes the sodium chloride easier to fuse.

Sodium is used in the manufacture of tetraethyllead and titanium (§18.3.1). Because of its excellent thermal and electrical conductivities it is also used as a heat exchange medium in nuclear reactors and in sodium-filled electrical power lines.

Group 2A

Magnesium, the most important element of this group, is extracted by the electrolysis of fused magnesium chloride. The chloride is obtained by heating a mixture of magnesium oxide and coke in a stream of chlorine:

$$MgO + C + Cl_2 = MgCl_2 + CO$$

Magnesium oxide is produced by heating the carbonate ore or the hydroxide, which is precipitated from sea water with calcium hydroxide:

$$Mg^{2+} + Ca(OH)_2(s) = Mg(OH)_2(s) + Ca^{2+}$$

The remaining elements of group 2A are also prepared by the electrolysis of their fused chlorides.

12.2.2 THE REACTIONS OF THE s-BLOCK ELEMENTS

Most reactions of the s-block elements may be regarded as redox reactions in which the metal, acting as the reductant, becomes oxidised to M^+ ions for group 1A elements or M^{2+} ions for group 2A elements.

Reaction with other elements

The s-block elements, particularly the alkali metals, are extremely reactive and combine with many non-metals. The freshly cut metals have a bright lustre which rapidly disappears as they react with the atmosphere. Most of the s-block elements, except beryllium and magnesium, are therefore stored under liquid paraffin to prevent contact with the atmosphere. Beryllium and magnesium tarnish very slowly because a thin protective film of oxide forms on the metal surface (cf. aluminium).

The reactivity of the s-block elements towards non-metals, such as oxygen (§12.2.4) or the halogens, usually increases from top to bottom of the groups, as the formation of metal ions becomes easier. In some cases, particularly with nitrogen and carbon, the reverse is true. For example, lithium, alone in group 1A, combines directly with nitrogen to form an ionic nitride:

$$6Li + N_2 = 2Li_3N$$

Only in this case is the lattice enthalpy high enough, because of the small size of the lithium ion, to compensate for the energy absorbed during the formation of ions. Lithium nitride is therefore an exothermic compound ($\Delta H_f^{\ominus}[Li_3N] = -198\ kJ\ mol^{-1}$) and is stable. Except for beryllium, all the group 2A elements form ionic nitrides, because of the higher lattice enthalpies associated with the dipositive metal ions.

Only lithium and sodium in group 1A combine directly with carbon to form ionic acetylides, Li_2C_2 and Na_2C_2 respectively. All the group 2A metals combine with carbon to give ionic carbides.

Reaction with water

With the exceptions of beryllium and magnesium, the s-block elements reduce cold water to hydrogen and become oxidised to their hydroxides:

$$\underset{\text{group 1A}}{2M + 2H_2O = 2M^+(aq) + 2HO^- + H_2}$$

$$\underset{\text{group 2A}}{M + 2H_2O = M^{2+}(aq) + 2HO^- + H_2}$$

The group 2A metals are less reactive towards water than the corresponding group 1A elements. Magnesium reacts slowly with cold water but more rapidly with boiling water or steam to form the oxide:

$$Mg + H_2O = MgO + H_2$$

In reactions with water, hydrated metal ions are formed from elements, and we might therefore expect standard electrode potentials to serve as an indication of reactivity. However, this is not the case, because the reactions increase in vigour from top to bottom of each s-block group.

Reaction with ammonia

The alkali metals and calcium, strontium and barium react with liquid ammonia to form hydrogen and metal amides (§15.2.3). Metal amides are ionic and react readily with water,

e.g. $NaNH_2 + H_2O = NaOH + NH_3$

The alkali metals also form amides when heated in ammonia, but the group 2A metals, when similarly treated, produce the nitride or hydride,

e.g. $3Mg + 2NH_3 = Mg_3N_2 + 3H_2$

$3Ca + 2NH_3 = 3CaH_2 + N_2$

12.2.3 THE HYDRIDES OF THE s-BLOCK ELEMENTS

s-Block hydrides

Except for beryllium, the s-block elements combine directly with hydrogen at temperatures in the range 300–700 °C (573–973 K) to give hydrides. Magnesium hydride, MgH_2, is formed only under high pressure.

$$\underset{\text{group 1A}}{2M + H_2 = 2MH}$$

$$\underset{\text{group 2A}}{M + H_2 = MH_2}$$

The hydrides of beryllium and magnesium are covalent and polymeric. The remaining hydrides of the s-block elements are ionic and contain the hydride ion, H^-. The alkali metal hydrides possess a structure similar to that of sodium chloride (Fig. 6.9), with H^- ions replacing Cl^- ions, but the group 2A metal hydrides have a less regular structure.

The presence of the hydride ion in these ionic hydrides can be demonstrated by the electrolysis of the compounds when dissolved in fused alkali metal halides. Hydrogen is liberated at the anode, by oxidation of hydride ions,

i.e. $2H^- = H_2 + 2e^-$

Lithium hydride is sufficiently stable to be fused without decomposition, but the other ionic hydrides decompose into their respective elements below their melting temperatures.

Ionic hydrides are vigorously hydrolysed by water:

$$H^- + H_2O = H_2 + HO^-$$

Because the hydride ion is readily oxidised to hydrogen, ionic hydrides are powerful reducing agents,

e.g. $H^- + CO_2 = HCOO^-$
$H^- + 2CO = HCOO^- + C$

12.2.4 THE OXIDES OF THE s-BLOCK ELEMENTS

Beryllium forms one oxide, BeO, which is essentially covalent. The remaining s-block elements form more than one oxide, all of which are ionic (Table 12.3).

Table 12.3 The oxides of the s-block elements

	Oxides (containing the oxide ion, O^{2-})	**Peroxides (containing the peroxide ion, O_2^{2-})**	**Hyperoxides (containing the hyperoxide ion, O_2^-)**
group 1A	M_2O	M_2O_2	MO_2
group 2A	MO	MO_2	none
comments	all the metals	not Be	

Oxides

Lithium oxide, Li_2O, and the oxides of group 2A are prepared by heating the elements in oxygen, or by thermal decomposition of the carbonates, nitrates or hydroxides. Sodium oxide, Na_2O, is obtained by heating sodium in a limited supply of oxygen, but the oxides of the heavier alkali metals can be prepared only by indirect methods, such as heating the metal with its nitrate,

e.g. $10K + 2KNO_3 = 6K_2O + N_2$

The oxides are strongly basic. Those of group 1A are more basic than those of group 2A, and basic properties increase with descent of the groups.

The oxides of the s-block elements, except that of beryllium, react with water to form the corresponding hydroxide,

e.g. $Na_2O + H_2O = 2NaOH$
$CaO + H_2O = Ca(OH)_2$

Magnesium oxide reacts slowly, but the remaining oxides of these two groups are extremely reactive towards water.

Peroxides

The peroxides of the alkali metals are formed by passing oxygen into solutions of the metals in liquid ammonia:

$$2M + O_2 = M_2O_2$$

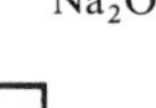

Na_2O_2

Sodium peroxide may also be prepared by heating sodium in an excess of oxygen. The hydrated peroxides of magnesium, calcium and strontium are obtained as white precipitates by adding hydrogen peroxide to cold alkaline solutions of their salts,

e.g. $Ca^{2+} + H_2O_2 + 8H_2O = CaO_2{\cdot}8H_2O(s) + 2H^+(aq)$

BaO_2

Barium peroxide is prepared by direct combination of the elements at 500 °C (773 K). Higher temperatures should be avoided, because above 800 °C (1 073 K) the peroxide decomposes,

i.e. $2BaO_2 = 2BaO + O_2$

The peroxide ion is rather large. Consequently:
(i) it is easily polarised by cations, especially if their radii are small;
(ii) stable crystal lattices can be formed only with large cations.

Lithium peroxide readily decomposes, because the small lithium ion strongly polarises the peroxide ion and the crystal lattice is unstable. The remaining peroxides become increasingly resistant to heat from top to bottom of the groups, because the polarising power of the cation decreases and more stable lattices can be formed.

The peroxides of the alkali metals react readily with acids and with water at 0 °C (273 K) to produce hydrogen peroxide,

e.g. $Na_2O_2 + 2HCl = 2NaCl + H_2O_2$
$Na_2O_2 + 2H_2O = 2NaOH + H_2O_2$

At higher temperatures oxygen is formed by decomposition of the hydrogen peroxide:

$$2Na_2O_2 + 2H_2O = 4NaOH + O_2$$

Oxygen is also liberated from peroxides by treatment with acidic oxides,

e.g. $2Na_2O_2 + 2CO_2 = 2Na_2CO_3 + O_2$

Hyperoxides (formerly called 'superoxides')

All hyperoxides

Hyperoxides are highly coloured compounds, in contrast to the oxides and peroxides, which are colourless. The hyperoxides of potassium, rubidium and caesium are produced by burning the elements in air or oxygen,

e.g. $K + O_2 = KO_2$

Sodium hyperoxide, NaO_2, is prepared by heating sodium or its peroxide with oxygen at 450 °C (723 K) and a pressure of 300 atmospheres (3×10^4 kPa). The highly unstable lithium hyperoxide, LiO_2, is believed to be formed when oxygen is passed into a solution of lithium in liquid ammonia at −78 °C (195 K).

The hyperoxide ion, O_2^-, resembles the peroxide ion in being stable only in combination with large cations. The stability of hyperoxides therefore increases from lithium to caesium.

Hyperoxides react with water and acids at 0 °C (273 K),

e.g. $2KO_2 + 2H_2O = 2KOH + H_2O_2 + O_2$

$2KO_2 + 2HCl = 2KCl + H_2O_2 + O_2$

As with peroxides, the hydrogen peroxide decomposes at higher temperatures to give oxygen. Oxygen is also formed with acidic oxides,

e.g. $4KO_2 + 2CO_2 = 2K_2CO_3 + 3O_2$

12.2.5 THE HYDROXIDES OF THE s-BLOCK ELEMENTS

All group 1A hydroxides

The hydroxides of sodium, potassium, rubidium and caesium are extremely soluble in water, but lithium hydroxide is only sparingly soluble. The hydroxides of the group 2A elements increase in solubility from magnesium to barium; for example, magnesium hydroxide is just soluble enough to give an alkaline reaction to indicators, whereas a saturated solution of barium hydroxide is approximately 0.1 molar.

The hydroxides may be prepared by dissolving the metals or their oxides in water, although this method is unsuitable for magnesium hydroxide and beryllium hydroxide. Because of their low solubility, the group 2A hydroxides may be precipitated from solutions of their salts by an alkali metal hydroxide,

e.g. $MgCl_2 + 2NaOH = Mg(OH)_2(s) + 2NaCl$

The hydroxides of the s-block elements except beryllium function as strong bases because they contain the hydroxide ion. Base strength increases as the attractive forces between metal ions and hydroxide ions decrease. Consequently, the hydroxides of the alkali metals are stronger bases than those of the group 2A metals, and base strength increases from top to bottom of each group.

Only the hydroxides of sodium to caesium are stable to heat; all the others decompose,

e.g. $2LiOH = Li_2O + H_2O$

$Ca(OH)_2 = CaO + H_2O$

The hydroxides of sodium and potassium are white, deliquescent solids that dissolve in water with the production of heat. In the laboratory these hydroxides are used for such purposes as volumetric analysis, for the preparation of ammonia and phosphine, for absorbing acid gases and for drying gases. In industry sodium hydroxide is used in the manufacture of soap, aluminium, petroleum and dyestuffs, and as a degreasing agent.

12.2.6 THE CARBONATES OF THE s-BLOCK ELEMENTS

The carbonates of these elements are ionic and contain the carbonate ion, CO_3^{2-}. Carbonates of the alkali metals other than lithium are readily soluble in water, and are sufficiently stable to be fused without much decomposition. Lithium carbonate is sparingly soluble in water, and when heated decomposes before melting:

$$Li_2CO_3 = Li_2O + CO_2$$

The carbonates of the group 2A metals are insoluble in water and are thermally unstable. They all decompose below their melting temperatures, but the thermal stability increases from magnesium carbonate to barium carbonate as the size of the cation increases (§14.2.3).

Sodium carbonate

In the laboratory, sodium carbonate may be prepared by saturating aqueous sodium hydroxide with carbon dioxide:

$$NaOH + CO_2 = NaHCO_3$$

The carbonate is formed either by heating the hydrogencarbonate, or by adding the stoicheiometric quantity of sodium hydroxide:

$$NaHCO_3 + NaOH = Na_2CO_3 + H_2O$$

In addition to the anhydrous salt, sodium carbonate forms a monohydrate, $Na_2CO_3 \cdot H_2O$, and a decahydrate, $Na_2CO_3 \cdot 10H_2O$. Under normal atmospheric conditions the decahydrate effloresces to give the monohydrate. Anhydrous sodium carbonate is commonly used in the laboratory as a primary standard for the titration of strong acids. When titrated to a methyl orange end-point (§9.2.5) sodium carbonate is decomposed to carbon dioxide:

$$Na_2CO_3 + 2H^+(aq) = 2Na^+ + H_2O + CO_2$$

i.e. $\quad 1 \text{ mol } Na_2CO_3 \equiv 2 \text{ mol } H^+(aq)$

However, when titrated to a phenolphthalein end-point (§9.2.6) the carbonate is converted into the hydrogencarbonate:

$$Na_2CO_3 + H^+(aq) = NaHCO_3 + Na^+$$

i.e. $\quad 1 \text{ mol } Na_2CO_3 \equiv 1 \text{ mol } H^+(aq)$

Anhydrous sodium carbonate is used in the manufacture of glass, soap, paper and detergents, and for softening hard water.

Potassium carbonate

K_2CO_3

Potassium carbonate is not used as a primary standard for standardising acids because, unlike the sodium salt, it is deliquescent. It is made by passing carbon dioxide into potassium hydroxide solution.

12.2.7 THE HYDROGENCARBONATES OF THE s-BLOCK ELEMENTS

Hydrogencarbonates, containing the HCO_3^- ion, are prepared by treating aqueous solutions or suspensions of carbonates with either carbon dioxide or the stoicheiometric amount of a strong acid:

$$CO_3^{2-} + H_2O + CO_2 = 2HCO_3^-$$

$$CO_3^{2-} + H^+(aq) = HCO_3^-$$

The hydrogencarbonates of sodium, potassium, rubidium and caesium can be obtained in the crystalline state by evaporation of their solutions. They are thermally unstable and decompose at 100 °C (373 K) into the carbonate, carbon dioxide and water,

e.g. $2NaHCO_3 = Na_2CO_3 + CO_2 + H_2O$

The hydrogencarbonates of lithium and the group 2A elements exist in solution only, and decompose into the carbonates when their solutions are heated or evaporated,

e.g. $Ca(HCO_3)_2 = CaCO_3 + CO_2 + H_2O$

12.2.8 THE ANOMALOUS PROPERTIES OF LITHIUM

In many respects lithium strongly resembles magnesium but differs from the other alkali metals (Table 12.4). This is an example of a diagonal relationship (§3.2.5).

The reason for these anomalies is the small size of the lithium ion, Li^+, which results in:

(i) high lattice enthalpy of the compounds. This is responsible for their low solubility in water, and also for the formation of a nitride.

(ii) anion distortion. This causes some compounds to have covalent characteristics, e.g. solubility in organic solvents. It is also responsible for the thermal instability of the hydroxide and salts containing oxoanions, e.g. the nitrate and carbonate.

Lithium also forms several true covalent compounds, e.g. Li—CH_3.

Lithium	Sodium, potassium, rubidium and caesium	Magnesium
combines directly with nitrogen to form a nitride, Li_3N	do not combine directly with nitrogen gas	forms a nitride, Mg_3N_2, by direct combination
combines directly with oxygen to form an oxide, Li_2O	form peroxides and hyperoxides	with oxygen forms only an oxide, MgO
carbonate and hydroxide decompose into Li_2O on heating	carbonates and hydroxides are thermally stable	carbonate and hydroxide give MgO on heating
lithium nitrate gives $Li_2O + NO_2 + O_2$ on heating	nitrates produce stable nitrites (NO_2^-) and O_2 on heating	magnesium nitrate gives $MgO + NO_2 + O_2$ on heating
lithium hydrogencarbonate exists only in solution	solid hydrogencarbonates exist	magnesium hydrogencarbonate exists only in solution
fluoride, carbonate, phosphate, ethanedioate and hydroxide are sparingly soluble	corresponding salts are more soluble	salts show similar solubilities to those of lithium
chloride, bromide and iodide are soluble in organic solvents	corresponding halides are less soluble in organic solvents	halides show similar solubilities to those of lithium

Table 12.4 A comparison of lithium with the other group 1A metals and with magnesium

12.2.9 THE ANOMALOUS PROPERTIES OF BERYLLIUM

Beryllium differs from the other elements of group 2A in two major respects. First, it reacts with alkali metal hydroxides to form a complex anion:

$$Be + 2HO^- + 2H_2O = H_2 + \underset{\text{tetrahydroxoberyllate(II) ion}}{[Be(OH)_4]^{2-}}$$

Second, many compounds of beryllium are essentially covalent and quite unlike the corresponding compounds of the other elements of the group. The formation of the Be^{2+} ion requires a large amount of energy, i.e. 2666 kJ mol^{-1}, so beryllium tends to use its outer electrons to form covalent bonds. Alternatively, in accordance with Fajans' rules (§4.6.1), the small Be^{2+} ion is highly polarising and likely to distort anions to produce bonds with a high degree of covalent character.

The oxide and hydroxide of beryllium are essentially covalent and amphoteric. Beryllium does not form a hydrogencarbonate, and the carbonate, $BeCO_3$, is highly unstable and can be prepared only in an atmosphere of carbon dioxide. A basic carbonate is precipitated from a solution of a beryllium salt by adding a soluble carbonate. The halides of beryllium are covalent but ionise in water; cf. aluminium halides (§13.2.6),

$$\text{e.g.} \quad BeF_2 + 4H_2O = [Be(H_2O)_4]^{2+} + 2F^-$$

In the hydrated beryllium ion there are four molecules of water coordinated tetrahedrally around the central ion,

i.e. $\left[Be(H_2O)_4 \right]^{2+}$ (tetrahedral: H_2O above; H_2O, OH_2, H_2O below Be)

Solutions of the $[Be(H_2O)_4]^{2+}$ ion, like those of the $[Al(H_2O)_6]^{3+}$ ion, are acidic because of hydrolysis:

$$[Be(H_2O)_4]^{2+} + H_2O \rightleftharpoons [Be(OH)(H_2O)_3]^{+} + H_3O^{+}$$

In much of its other chemistry beryllium resembles aluminium rather than the remaining members of group 2A (Table 12.5). This is another example of a diagonal relationship.

Table 12.5 A comparison of beryllium with the other group 2A metals and with aluminium

Beryllium	**Magnesium, calcium, strontium and barium**	**Aluminium**
forms a complex ion with alkalis	no reaction with alkalis	forms a complex ion with alkalis
oxide and hydroxide are amphoteric	oxides and hydroxides are exclusively basic	oxide and hydroxide are amphoteric
anhydrous halides are covalent and ionise in water	halides are ionic	chloride, bromide and iodide are covalent and ionise in water: AlF_3 is ionic
beryllium chloride is dimeric in the vapour phase, i.e. Be_2Cl_4	chlorides are ionic	aluminium chloride is dimeric in the vapour phase, i.e. Al_2Cl_6
beryllium fluoride forms a complex with fluoride ions, $[BeF_4]^{2-}$	no fluoro-complexes	aluminium fluoride forms a complex with fluoride ions, i.e. $[AlF_6]^{3-}$

EXAMINATION QUESTIONS ON CHAPTER 12

1 (a) Radon (Rn) is produced naturally by *α-decay* of an isotope of radium, $^{226}_{88}Ra$, which has a *half-life* of 1522 years. In turn radon decays by α-emission giving element X.

(i) Explain what is meant by *α-decay* and *half-life*.

(ii) Write balanced equations for both these nuclear reactions and hence deduce the mass number, atomic number, and the position in the periodic table, of element X.

(b) The alkaline earth elements of the periodic table are (in order of increasing atomic number) beryllium, magnesium, calcium, strontium, barium and radium. From your knowledge of the trends in the chemistry of these elements deduce what you can about:

(i) the solubility of the hydroxide of radium and a likely value for the pH of an aqueous solution of approximate concentration 0.1 mol l^{-1},

(ii) the solubilities of the chloride, carbonate and sulphate of radium,

(iii) the action of heat on the carbonate and on the nitrate of radium.

(c) Give the structure of radium hydride. (AEB)

2 (a) At 1000 °C, the dissociation pressures of carbon dioxide in equilibrium with the carbonates of lithium, sodium and potassium are 90, 19 and 8 mmHg respectively.

(i) Which of these carbonates is thermally the most stable at 1000 °C?

(ii) What is the solid product formed by thermally decomposing the least stable of the carbonates?

(b) Which of the three alkali metal carbonates is least soluble in water at room temperature?

(c) What product is obtained by bubbling carbon dioxide into a cold aqueous solution of sodium carbonate?

(d) Write an equation describing the mild thermal decomposition of the product formed in (c).

(e) Give two examples of the way in which the chemistry of lithium resembles that of magnesium. (JMB)

3 The elements in Group II of the periodic table (alkaline earth metals) are, in alphabetical order, barium (Ba), beryllium (Be), calcium (Ca), magnesium (Mg), radium (Ra) and strontium (Sr).

(a) Arrange these elements in order of increasing atomic numbers.

(b) Write down the electronic configurations of any two of the above elements *other than beryllium (Be)*, stating in each case the name of the element, e.g. Be would be $1s^2\ 2s^2$.

(c) Indicate on a diagram how you would expect successive ionisation energies of magnesium to vary with the number of electrons removed.

(d) (i) What type of chemical bonding is generally found in alkaline earth metal compounds?

(ii) What experiment would you carry out in order to demonstrate the presence of this type of bonding in alkaline earth metal compounds? Briefly indicate the results which you would expect to obtain.

(e) How does the solubility in water of the alkaline earth metal sulphates vary with the relative atomic mass of the metal?

(f) The chlorides of the alkaline earth metals are known to form various hydrates, e.g. $CaCl_2{\cdot}2H_2O$ and $CaCl_2{\cdot}6H_2O$. Suppose you were given a specimen of hydrated strontium chloride, $SrCl_2{\cdot}xH_2O$, and asked to determine the value of x. Outline briefly how you would do this. (L)

4 The s-block of the periodic table contains Group IA, the alkali metals, and Group IIA, the alkaline earths. Give an account of these two groups of elements and their compounds, paying particular attention to similarities and differences within the groups. (L)

5 The following question refers to the group of elements $_{12}Mg$, $_{20}Ca$, $_{38}Sr$, $_{56}Ba$ and $_{88}Ra$.

(a) Give the electronic configurations of the first **four** members of this group.

(b) Which of these metals would you expect to:

(i) have the highest solubility for the sulphate,

(ii) have an unstable nucleus,

(iii) have the lowest first ionisation energy,

(iv) give a crimson flame coloration?

(c) Describe the extraction of a *named* metal by the process of electrolysis. Indicate the purity of the metal obtained and the reasons for this method of extraction.

(d) Magnesium oxide is recognised as having the same crystal structure as sodium chloride.

(i) What are the species present in this structure?

(ii) What is the coordination number of the magnesium species?

(iii) Give **two** further properties showing the chemical similarity between the elements magnesium and sodium. How can this similarity be explained? (AEB)

6 By consideration of the trends in the properties of the Group I elements and their compounds, deduce possible answers to the following questions concerning the element francium (Fr, atomic number 87).

(a) Which noble gas would have the same electronic configuration as the francium ion?

(b) Give the formula of the compound formed between francium and hydrogen.

(c) Write down the equation for the reaction of francium with water.

(d) What further reaction would take place if the solution obtained in (c) were exposed to the atmosphere?

(e) Why would the compound formed between francium and chlorine be soluble in water but insoluble in benzene? (JMB)

7 'The properties of the first member of a group of elements in the periodic table are not typical of the group as a whole.' Discuss this with reference to the chemistry of the elements of Groups I (Li–Cs) and II (Be–Ba). You should include in your answer specific properties which differentiate lithium and beryllium from other members of their respective groups as well as the reasons for the differentiation. (JMB)

13 Group 3B

13.1 Introduction

Group 3B comprises: boron (B), aluminium (Al), gallium (Ga), indium (In) and thallium (Tl) (Table 13.1).

Al powder

Table 13.1 Some properties of the elements of group 3B

	Boron	Aluminium	Gallium	Indium	Thallium
character	non-metal	metal	metal	metal	metal
allotropes	three	none	none	none	none
atomic number	5	13	31	49	81
relative atomic mass	10.81	26.981 5	69.72	114.82	204.37
outer electronic configuration	$2s^2\ 2p^1$	$3s^2\ 3p^1$	$4s^2\ 4p^1$	$5s^2\ 5p^1$	$6s^2\ 6p^1$
valencies	3	3	1, 3	1, 3	1, 3
melting temperature/°C	2300	660	29.8	157	304
boiling temperature/°C	3930	2470	2400	2000	1460

13.1.1 BONDING AND VALENCY

The elements of this group each have three outer electrons (i.e. $ns^2\ np^1$) and show a valency of three (+3 oxidation state). A monovalent state becomes progessively more stable as atomic number increases, and is particularly important for thallium.

Boron

The formation of B^{5-} anions is energetically unfavourable, and boron is therefore covalently bonded in all of its compounds. It forms compounds with many non-metals, e.g. BF_3 and B_2O_3. In most cases the covalency of boron is three, and is reached by promoting an electron from the 2s orbital into a vacant 2p orbital to produce an atom with three unpaired electrons,

i.e. 2s [⇅] 2p [↑| |] —promotion→ 2s [↑] 2p [↑|↑|]
ground state → excited state

Many compounds of trivalent boron act as Lewis acids by accepting a lone pair of electrons into the remaining vacant p orbital. In this way boron attains its maximum covalency or coordination number of four

(§4.5). The remaining elements of group 3B can use vacant d orbitals in their outermost shell and accept further electron pairs to increase their coordination number to six.

Aluminium, gallium, indium and thallium

These elements are metallic and can lose their outer electrons to form tripositive ions. However, few compounds are known that contain simple (i.e. non-complexed) cations. For example, the only simple ionic compounds of aluminium are the anhydrous fluoride, AlF_3, the oxide, Al_2O_3, the perchlorate, $Al(ClO_4)_3$, and the carbide, Al_4C_3.

Many of the compounds of these elements are either covalent or else contain complexed cations.

Covalency

The non-complexed ions have a high surface charge density, which tends to promote anion distortion to give bonds which are best described as polar covalent (§4.6). To support this argument we can consider the energy required to form the trivalent cations from the atoms. The removal of three electrons requires the absorption of a large amount of energy; consequently the atoms tend to use their outer electrons to form covalent bonds. Before this can happen, one of the paired s electrons must be promoted into a vacant p orbital (cf. boron).

Coordination

The high surface charge density of the ions favours complex formation because it increases the attraction of the ions for polar species capable of acting as ligands (§18.1.2). Of particular importance is the hexaaqua-aluminium(III) ion, $[Al(H_2O)_6]^{3+}$, which occurs in many hydrated aluminium salts and solutions of aluminium compounds. The ion is stabilised by the high hydration enthalpy of the Al^{3+} ion, which largely offsets the ionisation enthalpy required in its formation,

i.e. $$Al(g) = Al^{3+}(g) + 3e^- \qquad \Delta H = +5\,143 \text{ kJ mol}^{-1}$$

$$Al^{3+}(g) + 6H_2O = [Al(H_2O)_6]^{3+} \qquad \Delta H = -4\,690 \text{ kJ mol}^{-1}$$

Other important complexes of aluminium include the octahedral hexafluoroaluminate(III) ion, $[AlF_6]^{3-}$, and the tetrahedral tetrachloroaluminate(III) ion, $[AlCl_4]^-$, which are formed by reaction between aluminium halides and the appropriate halide ions in the absence of water. The chloro-complex and the corresponding bromo-complex, $[AlBr_4]^-$, occur as intermediates in the Friedel–Crafts reaction. The tetrahydridoaluminate(III) complex ion is discussed in §11.2.2.

13.2 Boron and aluminium

Boron and aluminium are the most important members of group 3B and will be considered in more detail.

13.2.1 THE OCCURRENCE AND ISOLATION OF ALUMINIUM

Aluminium is the most abundant metallic element, and its compounds are widely distributed in the earth's crust. Examples include clays, mica, feldspars, cryolite, i.e. sodium hexafluoroaluminate(III), $Na_3[AlF_6]$, and the oxide, Al_2O_3. The anhydrous forms of the oxide, i.e. emery, corundum and ruby, are used as abrasives or gem stones, but the hydrated form, bauxite, $Al_2O_3(aq)$, is the principal ore from which aluminium is obtained.

Impure bauxite is purified by dissolution in aqueous sodium hydroxide. This produces a solution of sodium aluminate and sodium silicate; the latter arises from the presence of silica and silicates in the ore.

$$Al_2O_3(aq) + 2HO^- + 7H_2O = 2[Al(OH)_4(H_2O)_2]^-$$

Insoluble impurities, chiefly iron(III) oxide, are removed by filtration. Carbon dioxide is then blown through the filtrate and pure aluminium hydroxide is precipitated and filtered off. It is heated to produce the pure oxide.

$$2[Al(OH)_4(H_2O)_2]^- + CO_2 + H_2O = 2[Al(OH)_3(H_2O)_3](s) + CO_3^{2-}$$

(§13.2.4)

$$2[Al(OH)_3(H_2O)_3] = Al_2O_3 + 9H_2O$$

The pure oxide is dissolved in molten cryolite at 700–1 000 °C (973–1 273 K) and electrolysed. Aluminium is liberated at the cathode and oxygen at the anode:

$$Al^{3+} + 3e^- = Al$$

$$2O^{2-} = O_2 + 4e^-$$

Low voltages, approximately 6 V, must be used to avoid decomposition of the cryolite. The cryolite thus acts as a solvent for the oxide. Currents in the region of 30 000 A are required, and a cheap source of electricity, such as hydroelectric power, is necessary for economic viability. The cell is operated on a continuous basis. Periodically, molten aluminium is removed from the floor of the cell, where it collects, and fresh aluminium oxide is added.

13.2.2 THE REACTIONS OF BORON AND ALUMINIUM

Aluminium is a silvery white metal with a low density. It is reasonably reactive, but in air a thin, impervious oxide film forms on the surface of the metal, which protects it from prolonged attack. A thicker, more protective film of the oxide can be applied to aluminium by the process of *anodising*, in which the metal is made anodic in an elctrolyte of either sulphuric acid (5–80%) or chromic acid (3%). The oxide film so produced is hydrated and may be dyed for decorative purposes.

Pure aluminium is of limited use because of its softness, but its lightweight alloys, especially those with magnesium, are used extensively in aircraft and ship construction, and for window frames and motor-car components.

Generally, the reactions of boron are much slower than those of aluminium (Table 13.2).

Table 13.2 The products of some reactions involving boron and aluminium

Reagent	**Boron**	**Aluminium**
oxygen, when heated	B_2O_3	Al_2O_3
non-oxidising acids, e.g. HCl or dilute H_2SO_4	no reaction	$[Al(H_2O)_6]^{3+}$ (§13.2.4) + H_2
hot concentrated H_2SO_4	H_3BO_3 (slow reaction)	$Al_2(SO_4)_3 + SO_2$
HNO_3	H_3BO_3 (slow reaction)	no reaction; Al is passivated
alkalis	borates (slow reaction)	aluminates + H_2
H_2O at 700 °C (973 K)	B_2O_3	Al_2O_3
sulphur, when heated	B_2S_3	Al_2S_3
halogens	boron halides, BX_3 (except BI_3)	aluminium halides, AlX_3

13.2.3 THE OXYGEN COMPOUNDS OF BORON

Boron oxide

Boron oxide, B_2O_3, is formed by the dehydration of boric acid, H_3BO_3, at 700 °C (973 K):

$$2H_3BO_3 = B_2O_3 + 3H_2O$$

Boron oxide is weakly acidic and with water slowly reforms boric acid.

Oxoacids

If a hot, concentrated solution of disodium tetraborate, $Na_2B_4O_7$, is acidified and cooled, white flaky crystals of boric acid (also known as orthoboric acid), H_3BO_3, separate out:

$$Na_2B_4O_7 + 2HCl + 5H_2O = 4H_3BO_3 + 2NaCl$$

This is the commonest oxoacid of boron, but two others can be obtained from it by the action of heat, namely polymeric metaboric acid, $(HBO_2)_n$, and tetraboric acid, $H_2B_4O_7$.

In crystalline boric acid the planar trigonal molecules are hydrogen bonded together (Fig. 6.22). Boric acid is a weak monoprotic acid, and also behaves as a Lewis acid (§9.2.9) by accepting an electron pair from a water molecule:

$$B(OH)_3 \cdot OH_2 \rightleftharpoons [B(OH)_4]^- + H^+(aq) \qquad pK_a = 9.24$$

The acid strength is considerably enhanced by the addition of certain polyhydroxy compounds such as 1,2,3-propanetriol. In the presence of these compounds boric acid behaves as a strong monoprotic acid and can be titrated by an alkali, with phenolphthalein as the indicator. This effect arises from the formation of complexes, i.e.

$$2\ \begin{matrix} | \\ -C-OH \\ | \\ -C-OH \\ | \end{matrix} + B(OH)_3 = \left[\begin{matrix} -C-O & & O-C- \\ | & B & | \\ -C-O & & O-C- \end{matrix} \right]^- + H^+(aq) + 3H_2O$$

part of 1,2,3-propanetriol

13.2.4 THE OXYGEN COMPOUNDS OF ALUMINIUM

Aluminium oxide (alumina)

Two forms of anhydrous aluminium oxide are known: α-Al_2O_3, which is hard, resistant to hydration and inert to acids or alkalis, and γ-Al_2O_3, which readily hydrates and dissolves slowly in acids or alkalis.

Aluminium hydroxide and the hexaaquaaluminium(III) ion

Solutions of aluminium salts contain the hydrated aluminium(III) ion, $[Al(H_2O)_6]^{3+}$, in which the six water molecules are octahedrally co-ordinated around the Al^{3+} ion,

i.e. $$[Al(H_2O)_6]^{3+}$$

The small, highly charged aluminium ion at the centre of the complex strongly attracts the electrons of the Al—O bonds towards itself, which in turn causes the electrons of the O—H bonds to move towards the oxygen atoms,

i.e. $$Al^{3+} \leftarrow O(H)(H)$$

(the arrows indicate movement of electrons)

These electron movements result in:

(i) the charge of the Al^{3+} ion being delocalised over the whole of the complex,

(ii) a greater partial positive charge residing on the hydrogen atom of a coordinated water molecule than on a non-coordinated molecule,

(iii) a weakening of the O—H bond in a coordinated water molecule compared with that in a non-coordinated molecule. Coordinated water molecules therefore attract bases and lose protons more readily than non-coordinated water molecules.

The hexaaquaaluminium(III) ion functions as an acid by donating protons to bases. In aqueous solution, for example, where solvent water functions as a base, oxonium ions are formed and the solution is acidic:

$$[Al(H_2O)_6]^{3+} + H_2O \rightleftharpoons [Al(OH)(H_2O)_5]^{2+} + H_3O^+ \quad pK_1 = 4.9$$

$$[Al(OH)(H_2O)_5]^{2+} + H_2O \rightleftharpoons [Al(OH)_2(H_2O)_4]^+ + H_3O^+$$

Both reactions are regarded as examples of hydrolysis because water molecules are split up. In the presence of species that are stronger bases than water, e.g. ammonia and the ions CO_3^{2-}, HO^-, S^{2-} and CN^-, a third proton is lost and a neutral, insoluble complex is formed,

e.g. $[Al(OH)_2(H_2O)_4]^+ + HO^- \rightleftharpoons H_2O + \underset{\text{hydrated aluminium hydroxide}}{[Al(OH)_3(H_2O)_3](s)}$

These strong bases act both by removal of oxonium ions, which displaces the previous equilibria over to the right, and also by direct proton removal from the cationic complexes. This explains why, when a carbonate is added to an aluminium salt, carbon dioxide is produced and a white gelatinous precipitate of aluminium hydroxide is formed:

$$2[Al(OH)_2(H_2O)_4]^+ + CO_3^{2-} \rightleftharpoons 2[Al(OH)_3(H_2O)_3](s) + H_2O + CO_2(g)$$

The formula of aluminium hydroxide is sometimes written as $Al(OH)_3(aq)$ for simplicity. The sulphide ion and the cyanide ion also precipitate aluminium hydroxide from solutions of the aluminium(III) ion, and produce hydrogen sulphide and hydrogen cyanide respectively:

$$2[Al(OH)_2(H_2O)_4]^+ + S^{2-} \rightleftharpoons 2[Al(OH)_3(H_2O)_3](s) + H_2S(g)$$

$$[Al(OH)_2(H_2O)_4]^+ + CN^- \rightleftharpoons [Al(OH)_3(H_2O)_3](s) + HCN(g)$$

The hydrated ions of other trivalent metals, e.g. $[Cr(H_2O)_6]^{3+}$ (§18.5.3) and $[Fe(H_2O)_6]^{3+}$ (§18.7.3) behave in a similar manner to the hexaaquaaluminium(III) ion in solution and when treated with carbonate ions.

In the presence of excess hydroxide ions, further protons are removed from aluminium hydroxide to produce soluble aluminate(III) ions:

$$[Al(OH)_3(H_2O)_3] + HO^- \rightleftharpoons [Al(OH)_4(H_2O)_2]^- + H_2O$$

$$[Al(OH)_4(H_2O)_2]^- + HO^- \rightleftharpoons [Al(OH)_5(H_2O)]^{2-} + H_2O$$

$$[Al(OH)_5(H_2O)]^{2-} + HO^- \rightleftharpoons [Al(OH)_6]^{3-} + H_2O$$

The addition of acid displaces the equilibria to the left, by removing hydroxide ions so that aluminium hydroxide is ultimately reprecipitated. This explains the use of carbon dioxide in the preparation of pure aluminium hydroxide for the manufacture of aluminium (§13.2.1). Aluminium hydroxide in turn is soluble in acids, and if the concentration of acid is high enough the hexaaquaaluminium(III) ion is formed.

The hexaaquaaluminium(III) ion and aluminate(III) ions are also formed when aluminium dissolves in acids and alkalis, respectively:

$$2Al + 6H^+(aq) + 12H_2O = 2[Al(H_2O)_6]^{3+} + 3H_2$$

$$2Al + 2HO^- + 10H_2O = 2[Al(OH)_4(H_2O)_2]^- + 3H_2$$

Freshly prepared aluminium hydroxide is readily soluble in both acids and alkalis, but this property is rapidly lost on standing because the hydroxide changes into a hydrated form of aluminium oxide which is less reactive towards these reagents.

13.2.5 THE HALIDES OF BORON

Boron halides

The trihalides BF_3, BCl_3 and BBr_3 are prepared by direct combination of the elements. Boron trifluoride may also be obtained by heating boron oxide with either hydrofluoric acid or a mixture of an ionic fluoride and concentrated sulphuric acid:

$$B_2O_3 + 3CaF_2 + 3H_2SO_4 = 3CaSO_4 + 3H_2O + 2BF_3$$

The iodide is prepared by reacting iodine with $Na[BH_4]$.

Each of the boron trihalide molecules is trigonal planar in shape (§4.8), and functions as a powerful Lewis acid by accepting a lone pair of electrons into the vacant p orbital in the outermost shell,

```
          F                 ┌   F   ┐ -
          |  F⁻             |   |   |
e.g.      B↙         =      | F—B—F |
         / \                |   |   |
        F   F               └   F   ┘
                        tetrafluoroborate(III) ion

             H
             |
       F    N—H          F  H
       |   / |           |  |
       B↙    H     =  F—B—N—H
      / \                |  |
     F   F               F  H
```

The latter compound is called ammonia-boron trifluoride(1/1), where (1/1) denotes the mole ratio of the constituents.

Boron trifluoride is a most useful and powerful Lewis acid, and is used as a catalyst in organic reactions. It reacts with a large excess of water to produce boric acid and tetrafluoroboric(III) acid:

$$4BF_3 + 3H_2O = 3HBF_4 + H_3BO_3$$

In contrast to boric acid, tetrafluoroboric(III) acid is a strong acid, being completely ionised in solution. With smaller amounts of water boron trifluoride forms unstable hydrates, e.g. $BF_3{\cdot}H_2O$.

The chloride, bromide and iodide of boron are vigorously hydrolysed by water in a reaction which is believed to involve the coordination of a water molecule, followed by the elimination of a hydrogen halide,

e.g. $$\mathrm{Cl_3B \leftarrow OH_2} = \mathrm{Cl{-}BCl_2{-}OH_2} = \mathrm{Cl_2B{-}OH} + \mathrm{HCl}$$

(written as $HBOCl_2$)

This reaction is followed by two other similar steps:

$$HBOCl_2 + H_2O = H_2BO_2Cl + HCl$$
$$H_2BO_2Cl + H_2O = H_3BO_3 + HCl$$

The overall reaction is:

$$BCl_3 + 3H_2O = H_3BO_3 + 3HCl$$

The complex ions $[BCl_4]^-$, $[BBr_4]^-$ and $[BI_4]^-$ are less stable than the tetrafluoroborate(III) ion, and exist only in combination with large cations of low polarising power.

13.2.6 THE HALIDES OF ALUMINIUM

The halides of aluminium are prepared by direct combination of the elements or by heating aluminium in a stream of hydrogen halide:

$$\left.\begin{array}{l} 2Al + 3X_2 = 2AlX_3 \\ 2Al + 6HX = 2AlX_3 + 3H_2 \end{array}\right\} X = F, Cl, Br \text{ or } I$$

AlF_3

The fluoride is essentially ionic, i.e. $Al^{3+}(F^-)_3$, and does not react with water or dissolve in organic solvents. By contrast, the chloride, bromide and iodide of aluminium are covalent in the anhydrous state, ionise in water, and dissolve in organic solvents such as benzene.

$AlCl_3$ and $AlBr_3$

Anhydrous aluminium chloride possesses a complex polymeric structure in the solid state. At approximately 200 °C (473 K) it sublimes readily to form a vapour that contains dimeric molecules,

i.e. 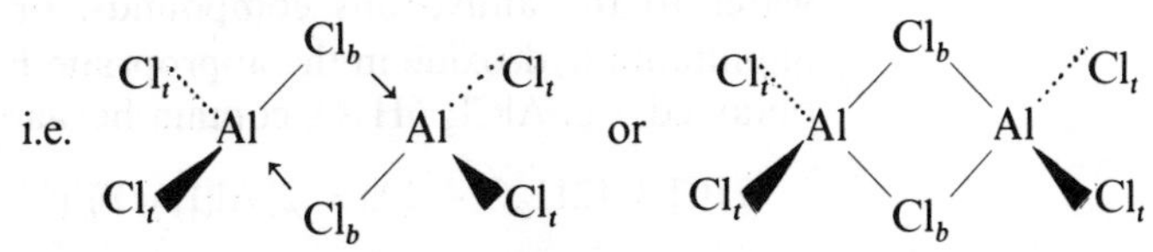

In the dimer each aluminium atom is bonded to four chlorine atoms. The four terminal chlorine atoms (Cl_t) lie at the corners of a rectangular plane which contains the two aluminium atoms. The two bridging chlorine atoms (Cl_b) are situated one above and one below this plane. Each bridging chlorine atom has a covalency of two, which may be envisaged as comprising a normal covalent bond and a coordinate bond, although we cannot distinguish between them (§4.5). Above 400 °C (673 K), the dimeric Al_2Cl_6 molecules break down reversibly to form monomer molecules which are trigonal planar in shape,

i.e. $$Al_2Cl_6 \underset{\substack{200\text{–}400\ °C \\ (473\text{–}673\ K)}}{\overset{400\ °C\ (673\ K)}{\rightleftharpoons}} 2\ AlCl_3$$ (Al bonded to three Cl atoms, trigonal planar)

Al_2Cl_6 molecules are also present in solutions of anhydrous aluminium chloride in benzene, as may be shown by measuring the depression of the freezing temperature (§8.4.1).

In the solid state the anhydrous bromide and iodide of aluminium contain discrete dimeric molecules, i.e. Al_2Br_6 and Al_2I_6 respectively. They behave in a similar manner to the chloride when heated.

The anhydrous aluminium halides, except for the fluoride, behave as powerful Lewis acids. For example, in benzene solution the aluminium halide dimers are split by electron pair donors, such as trimethylamine and ether, to form four coordinate complexes,

e.g. $$Al_2Cl_6 + 2(CH_3)_3N = 2(CH_3)_3N\text{-}AlCl_3$$
$$Al_2Cl_6 + 2(C_2H_5)_2O = 2(C_2H_5)_2O\text{-}AlCl_3$$
$$Al_2Cl_6 + 2Cl^- = 2[AlCl_4]^-$$

For convenience in writing equations, aluminium halides are often represented by the simple formula AlX_3.

The anhydrous chloride, bromide and iodide of aluminium ionise in water,

e.g. $$AlCl_3(s) + 6H_2O = [Al(H_2O)_6]^{3+} + 3Cl^-$$

During this reaction, which is exothermic and irreversible, aluminium–chlorine covalent bonds are broken. The necessary energy is provided by the hydration enthalpy of the small aluminium(III) ion. The hydrated aluminium ions undergo hydrolysis in solution (§13.2.4), but since all the water molecules coordinated to the Al^{3+} ion are not affected we often refer to the reaction as 'partial hydrolysis'. By contrast, the halides of boron are said to be 'completely hydrolysed', because every water molecule that participates in the reaction is cleaved (§13.2.5).

The white hydrated halides of aluminium can be obtained by adding water to the anhydrous compounds, or by dissolving aluminium or aluminium hydroxide in the appropriate hydrohalic acid. The crystals so obtained, e.g. $AlCl_3{\cdot}6H_2O$, contain hexaaquaaluminium(III) ions:

$$6HCl + 12H_2O + 2Al = 2[Al(H_2O)_6]^{3+} + 6Cl^- + 3H_2$$

Anhydrous aluminium halides cannot be prepared by heating the hydrated salts because hydrolysis occurs to give aluminium hydroxide,

e.g. $$[Al(H_2O)_6]^{3+} + 3Cl^- = [Al(OH)_3(H_2O)_3] + 3HCl(g)$$

Unlike the anhydrous aluminium halides, the hydrated compounds do not function as Lewis acids.

13.2.7 OTHER IMPORTANT COMPOUNDS OF ALUMINIUM

Aluminium sulphate and alums

White crystals of aluminium sulphate, $Al_2(SO_4)_3{\cdot}18H_2O$, are prepared by dissolving aluminium, its oxide or its hydroxide in moderately concentrated sulphuric acid.

Alums have the empirical formula $M^IM^{III}(SO_4)_2{\cdot}12H_2O$, where M^I represents NH_4 or a monovalent metal, e.g. Na, K or Rb (not Li), and M^{III} represents a trivalent metal, e.g. Al, Cr, Mn or Fe. The general formula $M^I_2SO_4{\cdot}M^{III}_2(SO_4)_3{\cdot}24H_2O$, which they are often given, reflects their method of preparation (see below). They form octahedrally shaped crystals and are isomorphous with one another, i.e. the arrangement of the constituent ions, $[M^I(H_2O)_6]^+$, $[M^{III}(H_2O)_6]^{3+}$ and SO_4^{2-} within the crystal is the same irrespective of their identity. Alums are therefore known as *double salts*, because they consist essentially of two salts within one crystal. In solution they behave as simple mixtures of the constituent salts.

Alums are prepared by evaporation of a solution containing equimolar amounts of the two sulphates, $M^I_2SO_4$ and $M^{III}_2(SO_4)_3$. Because they are isomorphous, it is possible to form an *overgrowth* of one alum on another, i.e. to grow crystals of one alum around those of another. Many of the common alums still retain their trivial names,

e.g. $(NH_4)_2SO_4{\cdot}Fe_2(SO_4)_3{\cdot}24H_2O$ iron alum

$K_2SO_4{\cdot}Cr_2(SO_4)_3{\cdot}24H_2O$ chrome alum

$K_2SO_4{\cdot}Al_2(SO_4)_3{\cdot}24H_2O$ alum

13.2.8 DIAGONAL RELATIONSHIPS BETWEEN BORON AND SILICON

We have seen (§3.2.5) that the head element of a particular group of the periodic table may resemble the second element of the next group to it.

Table 13.3 A comparison of boron with the remaining elements of group 3B and with silicon

Boron	Aluminium, gallium, indium and thallium	Silicon
non-metal with high melting temperature; non-conductor of electricity	metals with typical metallic properties; melting temperatures are rather low compared with those of most other metals	non-metal with high melting temperature; non-conductor of electricity
oxide, B_2O_3, is covalent, polymeric, glassy and acidic	oxides are ionic and amphoteric or basic	oxide, SiO_2, is covalent, polymeric, glassy and acidic
halides are covalent and completely hydrolysed by water	some halides are ionic; hydrolysis, if it occurs, is partial	halides are all covalent and completely hydrolysed by water
forms a series of volatile hydrides	do not form series of hydrides	forms a series of hydrides
forms binary compounds with metals, e.g. CaB_6	form alloys with metals	forms binary compounds with metals, e.g. Mg_2Si

Boron in many respects resembles silicon rather than the other elements of group 3B (Table 13.3).

EXAMINATION QUESTIONS ON CHAPTER 13

1 Discuss the relative acidity of solutions of $AlCl_3$, $FeCl_2$ and $FeCl_3$, given the ionic radii Al^{3+}, 0.050 nm; Fe^{2+}, 0.076 nm; Fe^{3+}, 0.065 nm.

The hexaaquaaluminium(III) ion yields a species having the general formula $[Al(OH)_y(H_2O)_x]^{charge\ z}$ when treated with a base. State and justify a simple algebraic expression connecting (i) x and y, (ii) y and z.

Show why the particular species obtained is related to the strength of the base added, using H_2O, CO_3^{2-}, and OH^- as examples of bases.

How are the foregoing principles related to the observations made when 2 M sodium hydroxide is added to solutions of (a) $AlCl_3$, (b) $FeCl_2$, (c) $FeCl_3$? (JMB)

2 (a) Starting with aluminium metal **outline**, giving equations and essential experimental conditions, the laboratory preparation of:

(i) aluminium chloride hexahydrate,

(ii) anhydrous aluminium chloride.

(b) What is the action on anhydrous aluminium chloride of:

(i) water,

(ii) excess aqueous alkali.

Give equations.

(c) Using vapour density measurements it is observed that the relative molar mass of aluminium chloride will vary according to the temperature used. Explain this observation.

(d) Aluminium chloride finds use in preparative organic chemistry. Give **two** examples to illustrate this statement and explain the part played by the aluminium chloride. Indicate the conditions necessary for its effective use. (Chapter 14.) (AEB)

3 (a) Describe, with the aid of a diagram, how you would prepare a sample of anhydrous aluminium chloride in the laboratory.

(b) Aluminium chloride vapour has a density of 2.28×10^{-3} g cm^{-3} at 1070 K and 1 atmosphere pressure.

(i) Calculate the apparent relative molecular mass of the sample under these conditions.

(ii) Account for your answer to (i) by discussing the structures of and bonding in aluminium chloride under these conditions and calculate the relative amounts of any species present.

(iii) Describe, giving reasons, how the apparent value of the relative molecular mass will be influenced by increases in temperature and in pressure.

(Gas constant $R = 82.06$ cm^3 atm mol^{-1} K^{-1}) (JMB)

14 Group 4B, including organic chemistry

14.1 Introduction

Group 4B comprises carbon (C), silicon (Si), germanium (Ge), tin (Sn) and lead (Pb). The increase in metallic character from top to bottom of the group, which results from the decreasing ionisation enthalpies of the elements, is pronounced. Carbon and silicon are distinctly non-metallic, tin and lead are typical metals, while germanium possesses the properties of both metals and non-metals. Some important properties of the elements are given in Table 14.1.

Table 14.1 Some properties of the elements of group 4B

	Carbon	Silicon	Germanium	Tin	Lead
character	non-metal	non-metal	semi-metal	metal	metal
allotropes	diamond, graphite	none	none	α-tin, β-tin, γ-tin	none
atomic number	6	14	32	50	82
relative atomic mass	12.011	28.086	72.59	118.69	207.2
outer electronic configuration	$2s^2\ 2p^2$	$3s^2\ 3p^2$	$4s^2\ 4p^2$	$5s^2\ 5p^2$	$6s^2\ 6p^2$
valencies	4	4	2†, 4	2†, 4	2, 4†
melting temperature/°C	3730‡ (sublimes)	1410	937	232§	327
boiling temperature/°C	4830‡	2360	2830	2270§	1744

† Denotes a relatively unstable state.
‡ Data for graphite.
§ Data for β-tin.

14.1.1 THE STRUCTURE OF THE ELEMENTS

Carbon

Two allotropes of carbon exist, namely diamond and graphite (§6.4.3). At ordinary temperatures graphite is the more stable form:

$$\text{C(diamond)} = \text{C(graphite)} \qquad \Delta H^{\ominus} = -1.9\ \text{kJ mol}^{-1}$$

The type of allotropy shown by carbon is referred to as *monotropy*, and is characterised by the following points.

(i) There is no definite temperature at which one allotrope changes into another.
(ii) One allotrope is more stable than the other(s). Unstable forms are said to be *metastable*.

(iii) The change from a metastable form to the stable allotrope is usually very slow.

Diamond is the metastable form because it possesses a higher energy than graphite. Diamond does not change into graphite at room temperature and pressure because of the very high activation energy associated with the moving of atoms within a solid structure. Because of the relatively large distance between the layers in graphite, the density of graphite, 2.25 g cm^{-3}, is lower than that of diamond, 3.51 g cm^{-3}. The change from graphite to diamond, which is endothermic and accompanied by a decrease in volume, is therefore favoured by high temperatures and high pressures.

Charcoals, soot and lampblack, collectively known as *amorphous carbon*, are microcrystalline forms of graphite. Because of its finely divided state, amorphous carbon has an enormous surface area and readily absorbs solutes from solutions, or large volumes of gases and liquids. For maximum effectiveness the amorphous carbon should be activated by heating it in steam at 1 000 °C (1 273 K).

Silicon and germanium

These two elements exist in one structural form only, which is similar to diamond.

Tin

Tin shows the type of allotropy referred to as *enantiotropy*, which is characterised by the following points.

(i) The change from one allotrope to another occurs at a definite temperature, known as the *transition temperature*.

(ii) At the transition temperature between two allotropes, both forms have equal stability.

(iii) Below the transition temperature one allotrope is stable, while above this temperature the other form is stable.

(iv) The allotropes are easily interconverted.

The three allotropes of tin and the transition temperatures are:

$$\underset{\text{(grey tin)}}{\alpha\text{-tin}} \underset{}{\overset{13.2\ ^\circ\text{C (286.4 K)}}{\rightleftharpoons}} \underset{\text{(white tin)}}{\beta\text{-tin}} \overset{161\ ^\circ\text{C (434 K)}}{\rightleftharpoons} \underset{\text{(rhombic tin)}}{\gamma\text{-tin}}$$

The change from β-tin to α-tin is slow above temperatures of −50 °C (223 K) unless some α-tin is already present. The α form has the diamond type of structure, but the β and γ allotropes have metallic structures. In adopting both non-metallic and metallic structures tin is an unusual element.

Lead

Lead exists in one metallic form only.

14.1.2 BONDING AND VALENCY

The elements of group 4B each possess four outer electrons in an ns^2np^2 configuration, and may bond ionically or covalently.

Ionic bonding

Carbon can gain four electrons per atom to form the carbide ion, C^{4-}. The metallic elements of the group, tin and lead, may form M^{4+} ions by the loss of all four outer electrons or M^{2+} ions by the loss of the two p electrons.

Covalent bonding

The formation of four covalent bonds is the dominant feature of the chemistry of carbon and silicon, and is common for the other elements of the group. The oxidation state may vary from -4 to $+4$ for carbon, although the remaining elements nearly always have positive oxidation numbers of 2 or 4. In addition, germanium, tin and lead commonly show a covalency of two by using their unpaired p electrons.

The stability of the tetravalent and divalent states for germanium, tin and lead is shown below.

	Ge	Sn	Pb
	increasing stability of divalent state →		
	← increasing stability of tetravalent state		
more stable state:	4	4	2

Thus germanium and tin in the divalent (+2 oxidation) state act as reducing agents, because they tend to lose electrons and switch to the more stable tetravalent (+4 oxidation) state. For ionic tin(II) compounds,

$$Sn^{2+} \longrightarrow Sn^{4+} + 2e^-$$

For lead the opposite is true, and lead(IV) compounds act as oxidising agents:

$$Pb^{4+} + 2e^- \longrightarrow Pb^{2+}$$

Maximum covalency

With the exception of carbon, the elements of group 4B have vacant d orbitals in their outermost shells which can be used to form coordinate bonds with donors. In such cases the elements have their maximum covalency of six,

e.g. $SiF_4 + 2F^- = [SiF_6]^{2-}$

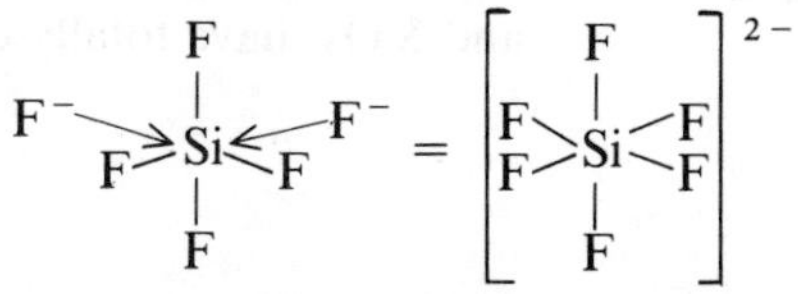

hexafluorosilicate(IV) ion
(octahedral)

Examples of other six coordinate complexes include: $[GeCl_6]^{2-}$, $[SnBr_6]^{2-}$, $[PbCl_6]^{2-}$ and $[Sn(OH)_6]^{2-}$. The last may be compared with the aluminate ion (§13.2.4). Complexes of the type $[SiX_6]^{2-}$, where X represents Cl, Br or I, do not exist because of the difficulty of accommodating six of these large halogen atoms around a silicon atom.

14.1.3 THE UNIQUE PROPERTIES OF CARBON

Carbon differs in many respects from the remaining members of the group.

Maximum covalency

The maximum covalency of carbon is four, due to the absence of d orbitals in the outermost shell. This accounts for many of the differences which occur between carbon compounds and the corresponding compounds of the other group 4B elements. For example, the tetrahalides of carbon do not react with water, in contrast to silicon tetrachloride (§14.2.7).

Bond strength

The C—C bond is considerably stronger than the Si—Si, Ge—Ge, Sn—Sn and Pb—Pb bonds. This accounts for the strong tendency of carbon to form molecules containing chains of carbon atoms covalently bonded together, a property referred to as *catenation*. Carbon also forms exceptionally strong bonds with other elements, such as hydrogen, oxygen, sulphur, nitrogen and the halogens. These two factors account for the immense number of compounds that are studied under the heading of 'organic chemistry'. By contrast, the weaker Si—Si and Si—H bonds result in the silanes being relatively few in number and far more reactive than the alkanes. The Si—O bond is stronger than the Si—Si bond, and the chemistry of silicon is characterised by chains and rings of alternating silicon and oxygen atoms, in the same way that organic chemistry features chain and ring structures held together by strong bonds between carbon atoms.

Multiple bonding

Carbon is the only element of group 4B which is able to form stable multiple bonds with itself or other elements, e.g. C=C, C≡C, C≡N, C=O and C=S. Consequently there are no compounds of silicon and the heavier elements which correspond to the alkenes, alkynes, aldehydes, ketones or nitriles. Even compounds with similar stoicheiometric formulae, e.g. CO_2 and SiO_2, have totally different structures.

14.2 The inorganic chemistry of carbon and silicon

Although many differences exist between carbon and silicon, it is instructive to compare their chemistry. The remaining elements of the group form a family and are conveniently discussed together (§14.12). Most compounds of carbon are classically considered under the heading of 'organic chemistry' (§14.3–§14.11), leaving a relatively small number of compounds, i.e. oxides, tetrahalides, cyanides and carbonates, together with the properties of the element itself, to be studied in 'inorganic chemistry'.

14.2.1 THE REACTIONS OF CARBON AND SILICON

Amorphous carbon is more reactive than graphite because its broken crystal structure allows easy access of reactants. Diamond is less reactive than graphite.

Table 14.2 The products of some reactions of carbon and silicon

Reagent	Carbon	Silicon
oxygen at 700°C (973 K)	CO_2 or CO (§14.2.2)	SiO_2
fluorine	CF_4	SiF_4
chlorine	no reaction	$SiCl_4$
sulphur at 900°C (1173 K)	CS_2	SiS_2
heated metals	metal carbides	metal silicides
oxidising acids, e.g. HNO_3	CO_2	partly oxidised to SiO_2
hydrofluoric acid	no reaction	H_2SiF_6
alkalis	no reaction	silicates and hydrogen

14.2.2 THE OXIDES OF CARBON

Carbon dioxide

Carbon dioxide, CO_2, a gas at room temperature, is formed whenever carbon or its compounds are burned in an excess of oxygen or air. It is more conveniently produced by:

(i) the thermal decomposition of carbonates or hydrogencarbonates (§14.2.3);

(ii) the action of a dilute acid on a carbonate or a hydrogencarbonate,

$$CO_3^{2-} + 2H^+(aq) = CO_2 + H_2O$$

$$HCO_3^- + H^+(aq) = CO_2 + H_2O$$

(iii) fermentation (§14.9.3).

The carbon dioxide molecule is linear and may be represented as:

$$O{=}C{=}O \qquad \text{bond length} = 0.116\text{ nm}$$

The bonding is more complex than this, however, owing to extensive delocalisation of π electrons which produces bonds that are intermediate in character between C=O (bond length, 0.122 nm) and C≡O (bond length, 0.110 nm).

At temperatures in excess of 1 700 °C (1 973 K) carbon dioxide decomposes into carbon monoxide and oxygen:

$$2CO_2 = 2CO + O_2$$

When heated with reactive metals, e.g. sodium or magnesium, reduction occurs:

$$2Na + 2CO_2 = Na_2CO_3 + CO$$

$$2Mg + CO_2 = 2MgO + C$$

Carbon dioxide is weakly acidic and reacts with strongly basic oxides or hydroxides to form carbonates:

$$Na_2O + CO_2 = Na_2CO_3$$

$$Ca(OH)_2 + CO_2 = CaCO_3 + H_2O$$

The formation of insoluble calcium carbonate from a saturated solution of calcium hydroxide (*lime water*) is used as a test for carbon dioxide. The calcium carbonate dissolves if an excess of carbon dioxide is used, owing to the formation of the soluble hydrogencarbonate:

$$CaCO_3 + H_2O + CO_2 = Ca(HCO_3)_2$$

Carbon monoxide

CO

Carbon monoxide is produced whenever carbon or its compounds are burned in a deficiency of oxygen. In the laboratory it may be prepared by:

(i) passing carbon dioxide over carbon at 700 °C (973 K);

(ii) dehydrating methanoic acid or ethanedioic acid with either concentrated sulphuric acid or phosphorus(V) oxide:

$$HCOOH = CO + H_2O$$

$$(COOH)_2 = CO + CO_2 + H_2O$$

The carbon dioxide may be removed by passing the gases through sodium hydroxide solution.

Carbon monoxide is a colourless, odourless, toxic gas. The bonding in the molecule may be represented as $C \overset{\leftarrow}{\equiv} O$ or $C \equiv O$, which indicates one σ bond and two identical π bonds. The electronic configurations of carbon and oxygen are:

C $1s^2\ 2s^2\ 2p^2$ O $1s^2\ 2s^2\ 2p^4$

The end-on and sideways overlap of singly filled 2p orbitals on the carbon and oxygen atoms produces a σ bond and a π bond (§4.3). The second π bond is formed by the overlap of the doubly filled 2p orbital on oxygen and the vacant 2p orbital on carbon. This is a coordinate bond, but once formed it is indistinguishable from the other π bond in the molecule. The bonding in carbon monoxide is therefore similar to that in nitrogen, since both molecules have the same number of electrons. Carbon monoxide can act as a ligand, by donating the lone pair of electrons on the carbon atom.

Carbon monoxide is weakly acidic, and reacts with sodium hydroxide

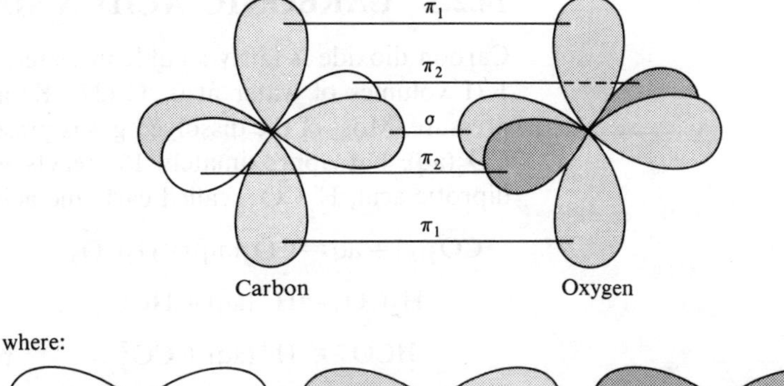

Fig. 14.1 The bonding in carbon monoxide. Bond π_1 is identical to bond π_2.

where:

represents a vacant 2p orbital | represents a singly filled 2p orbital | represents a full 2p orbital

at 200 °C (473 K) and 10 atmospheres (1 000 kPa) pressure to form sodium methanoate:

$$NaOH + CO = HCOONa$$

Despite being the anhydride of methanoic acid, carbon monoxide does not form this acid with water. At room temperature it is slightly soluble to give a neutral solution, and at increased temperatures an oxidation–reduction reaction occurs to give carbon dioxide and hydrogen:

$$CO + H_2O = CO_2 + H_2$$

Carbon monoxide is readily oxidised, and burns in air with a characteristic blue flame:

$$2CO + O_2 = 2CO_2$$

It reduces many heated metallic oxides to metals (§10.3.2).

The reaction with diiodine pentaoxide, I_2O_5, occurs quantitatively at 90 °C (363 K) and may be used to analyse gas mixtures containing carbon monoxide. The liberated iodine is titrated with sodium thiosulphate:

$$I_2O_5 + 5CO = I_2 + 5CO_2$$

$$5 \text{ mol } CO \equiv 1 \text{ mol } I_2 \equiv 2 \text{ mol } Na_2S_2O_3$$

Carbon monoxide is also oxidised by chlorine and by heated sulphur:

$$CO + Cl_2 = O{=}CCl_2$$ carbonyl chloride (phosgene)

$$CO + S = O{=}C{=}S$$ carbonyl sulphide

Solutions of copper(I) chloride in ammonia or hydrochloric acid readily absorb carbon monoxide.

14.2.3 CARBONIC ACID AND ITS SALTS

Carbon dioxide is fairly soluble in water, 1 volume of CO_2 dissolving in 1.71 volumes of water at 0 °C (273 K) and one atmosphere (100 kPa) pressure. Most of the dissolved gas is present as a weakly bound hydrate, $CO_2(aq)$, but approximately 1% reacts with water to produce a weak diprotic acid, H_2CO_3, called carbonic acid:

$$CO_2(g) + aq \rightleftharpoons CO_2(aq) \rightleftharpoons H_2CO_3$$

$$H_2CO_3 \rightleftharpoons H^+(aq) + HCO_3^- \qquad pK_1 \approx 3.7$$

$$HCO_3^- \rightleftharpoons H^+(aq) + CO_3^{2-} \qquad pK_2 = 10.32$$

Carbonic acid is highly unstable and has not been isolated in the pure state.

Two series of salts derived from carbonic acid are known, namely the carbonates and the hydrogencarbonates, and of particular importance are those of the s-block elements. The carbonate ion, which is present in these compounds, is planar, and because of delocalised π bonding the carbon–oxygen bonds are equivalent (Fig. 14.2(a)). Crystalline hydrogencarbonates of the alkali metals are of interest because they contain chains of HCO_3^- ions linked together by hydrogen bonds (Fig. 14.2(b)).

Fig. 14.2
(a) The structure of the carbonate ion. Typical C—O distance is 0.129 nm.
(b) Part of the chain structure in crystalline hydrogencarbonates.

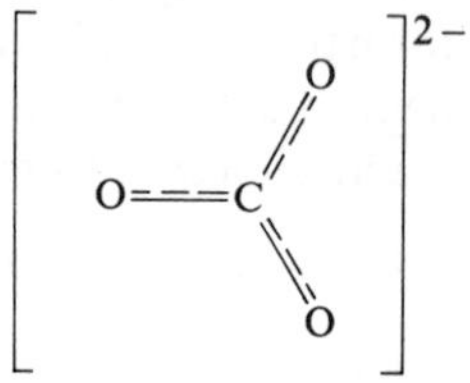

(a)

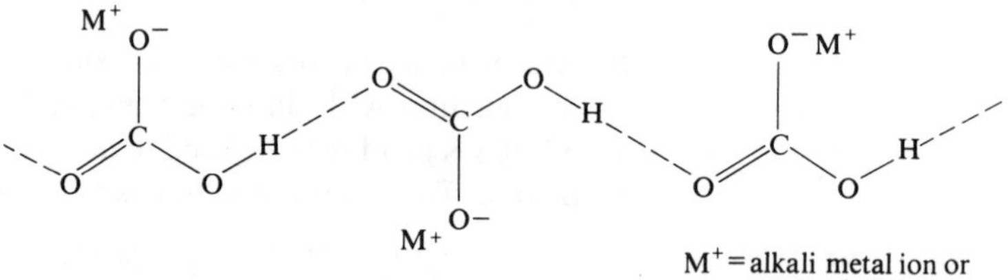

M⁺ = alkali metal ion or ammonium ion

(b)

Carbonates are insoluble in water, with the exception of the ammonium salt and those of the alkali metals. The hydrogencarbonates of the alkali metals (except for lithium), and ammonium hydrogencarbonate, are the only ones which can be obtained in the crystalline state. All the others decompose on evaporation of the solution,

e.g. $Ca(HCO_3)_2 = CaCO_3 + CO_2 + H_2O$

Because of salt hydrolysis (§9.2.8) both carbonates and hydrogencarbonates form alkaline solutions, i.e. $CO_3^{2-} + H_2O \rightleftharpoons HCO_3^- + HO^-$

$$HCO_3^- + H_2O \rightleftharpoons H_2CO_3 + HO^-$$

Thus, 0.1 M solutions of sodium carbonate and sodium hydrogencarbonate have pH values of 11.5 and 8.5 respectively.

All metal carbonates are thermally unstable and decompose into the metal oxide and carbon dioxide on heating. Normally, decomposition occurs below the melting temperature of the carbonate, but alkali metal carbonates, except lithium carbonate, can be fused without appreciable decomposition. Hydrogencarbonates are less stable than the corresponding carbonates and decompose below 100 °C (373 K),

e.g. $CaCO_3 = CaO + CO_2$

$$2NaHCO_3 = Na_2CO_3 + CO_2 + H_2O$$

The carbonate and hydrogencarbonate ions are readily distorted by the electrical field associated with cations. The greater the distortion the lower is the thermal stability (§12.2.6). Thus, the group 2A carbonates become increasingly stable from magnesium to barium as the surface charge density of the cation decreases. Group 2A hydrogencarbonates, and lithium hydrogencarbonate, are stable only in solution because the close approach of a cation, as a prerequisite to forming a crystalline salt, is sufficient to decompose the hydrogencarbonate ion.

14.2.4 THE PRECIPITATION REACTIONS OF CARBONATES AND HYDROGENCARBONATES

Soluble carbonates react with solutions of metal salts to form various products.

Precipitation of a hydroxide or oxide

The hydrated trivalent ions Al^{3+}, Fe^{3+} and Cr^{3+} are acidic in solution and act as acids towards the carbonate ion,

e.g. $2[Al(OH)_2(H_2O)_4]^+ + CO_3^{2-}$

$$= 2[Al(OH)_3(H_2O)_3] + CO_2 + H_2O \quad (\S 13.2.4)$$

In many cases the precipitated hydroxide changes rapidly into a hydrated oxide.

Precipitation of a normal carbonate

The divalent ions Ca^{2+}, Sr^{2+} and Ba^{2+} do not greatly polarise neighbouring water molecules and are neutral or weakly acidic in solution. In these cases the normal carbonate is precipitated by adding a soluble carbonate:

$$M^{2+} + CO_3^{2-} = MCO_3(s) \quad (M = Ca, Sr \text{ or } Ba)$$

Precipitation of a basic carbonate

In between these two extremes are other hydrated divalent metal ions, notably Cu^{2+}, Mg^{2+}, Fe^{2+} and Pb^{2+}. When a soluble carbonate is added to solutions of these ions a precipitate of a basic carbonate is obtained which contains the metal ions, hydroxide ions and carbonate ions. For example, malachite, $CuCO_3 \cdot Cu(OH)_2$ or $Cu_2CO_3(OH)_2$, a naturally occurring basic copper carbonate, possesses a structure with a regular arrangement of Cu^{2+}, CO_3^{2-} and HO^- ions, and white lead, $2PbCO_3 \cdot Pb(OH)_2$ or $Pb_3CO_3(OH)_2$, consists of an array of Pb^{2+} and CO_3^{2-} ions interleaved with layers of $Pb(OH)_2$.

The term 'basic' indicates that the compound is intermediate between the normal salt and the base, i.e. the hydroxide. Basic salts other than carbonates are known, and many of these contain regular arrays of metal ions, hydroxide ions and the relevant anions, e.g. hydroxyapatite, $3Ca_3(PO_4)_2 \cdot Ca(OH)_2$ or $Ca_5(PO_4)_3(OH)$, and basic lead(II) chloride, $PbCl_2 \cdot Pb(OH)_2$ or $PbCl(OH)$.

The hydrogencarbonate ion is a weaker base than the carbonate ion and is less efficient at ionising coordinated water molecules in hydrated cations. Therefore, some metal ions, which form basic carbonates when treated with the carbonate ion, do not form basic salts with the hydrogencarbonate ion. The precipitate obtained is usually a carbonate because most hydrogencarbonates are unstable, particularly if heated,

e.g. $$Mg^{2+} + 2HCO_3^- \xrightarrow{50\ °C\ (323\ K)} MgCO_3(s) + H_2O + CO_2$$

14.2.5 THE OXIDES OF SILICON

Silicon forms a highly unstable monoxide, SiO, and a highly stable dioxide, SiO_2, which is a colourless glassy solid with a high melting temperature. Silicon dioxide exists in three forms (i.e. it is *trimorphic*). In order of decreasing stability at room temperature they are quartz, cristobalite and tridymite. Structurally, each form consists of an extended three-dimensional array in which each silicon atom is surrounded tetrahedrally by four oxygen atoms (§6.4.3). The formula SiO_2 does not, therefore, represent a molecule, and for this reason the name *silica* is often preferred to silicon dioxide (cf. carbon dioxide). Silica adopts this structure because, unlike carbon, silicon cannot form multiple bonds.

The high stability of silica is indicated by its high enthalpy of formation: $\Delta H_f^\ominus[SiO_2(\text{quartz})] = -859\ kJ\ mol^{-1}$. The high value is due to:

(i) the relatively low atomisation enthalpy of silicon, which is a consequence of the weakness of the Si—Si bond;
(ii) the high bond dissociation enthalpy of the Si—O bond.

Silica is insoluble in water and chemically rather inert. It does, however, function as a weakly acidic oxide and reacts, for example, with fused alkali metal hydroxides or carbonates to produce silicates:

$$2NaOH + SiO_2 = Na_2SiO_3 + H_2O$$
$$Na_2CO_3 + SiO_2 = Na_2SiO_3 + CO_2$$

The only other common reagent to attack silica is hydrofluoric acid, with which it forms silicon tetrafluoride and hexafluorosilicic(IV) acid (§14.2.7):

$$SiO_2 + 4HF = SiF_4 + 2H_2O$$

$$SiF_4 + 2HF = H_2SiF_6$$

14.2.6 THE TETRAHALOMETHANES

CCl_4

CBr_4

The fluoride and chloride are stable compounds, in contrast to the bromide and iodide which are unstable because of the difficulty of accommodating four large halogen atoms around a small carbon atom.

Tetrachloromethane is prepared by reacting chlorine with either methane (§14.7.3) or carbon disulphide:

$$CS_2 + 3Cl_2 = CCl_4 + S_2Cl_2$$

$$2S_2Cl_2 + CS_2 = CCl_4 + 6S$$

In the laboratory tetrachloromethane is used as a solvent, although it should be handled with care because it is extremely toxic. Chemically, tetrachloromethane is rather inert; for example, it is not hydrolysed by water despite the reaction being thermodynamically favourable (§14.2.7):

$$CCl_4(l) + 2H_2O(l) = CO_2(g) + 4HCl(g) \qquad \Delta G^\ominus = -233.6 \text{ kJ mol}^{-1}$$

14.2.7 SILICON TETRAHALIDES

$SiCl_4$

All four tetrahalides of silicon are known, i.e. SiX_4, where X represents F, Cl, Br or I. In addition, several catenated halides containing Si—Si bonds are known, e.g. Si_6Cl_{14}, Si_2F_6 and Si_2I_6, but these are less important than the tetrahalides. Silicon tetrafluoride is conveniently prepared by the action of hydrofluoric acid on silica, while the other halides are formed by direct combination of the heated elements.

Silicon tetrahalides have a tetrahedral structure. In contrast to the corresponding halides of carbon they are rapidly hydrolysed by water, which indicates that a favourable pathway exists for the reaction. The reaction is stepwise, each stage comprising the formation of a coordinate bond between a water molecule and the silicon atom, followed by the elimination of a molecule of hydrogen halide.

E.g.

$$SiX_4 + H_2O = X_4Si \leftarrow O{<}^{H}_{H} = X_3Si{-}OH + HX$$

The final product to be expected from the hydrolysis is $Si(OH)_4$, but this has not been isolated since it undergoes condensation polymerisation to form hydrated silica, $SiO_2(aq)$.

In forming a coordinate bond with a water molecule, the silicon atom uses one of its vacant 3d orbitals. Carbon has no d orbitals in its outermost shell and cannot coordinate water molecules in this manner, which accounts for the kinetic stability of its halides. The hydrolysis of silicon tetrafluoride differs from that of the other halides because hydrofluoric acid attacks any unhydrolysed fluoride to produce hexafluorosilicic(IV) acid, which is a strong acid containing the hexafluorosilicate(IV) ion (§14.1.2):

$$SiF_4 + 2H_2O = SiO_2(aq) + 4HF$$
$$SiF_4 + 2HF = 2H^+(aq) + [SiF_6]^{2-}$$

14.2.8 SILANES (silicon hydrides)

All silanes

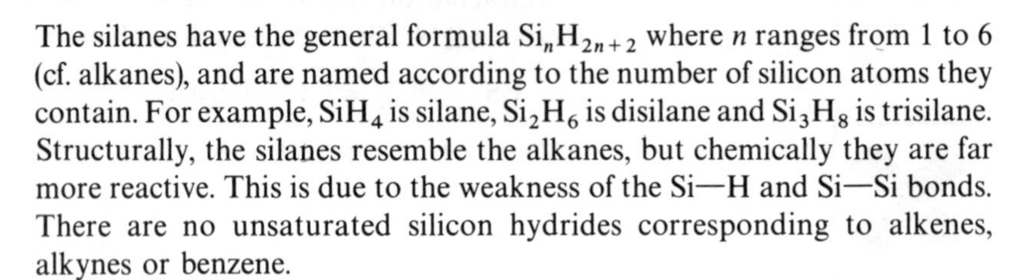

The silanes have the general formula Si_nH_{2n+2} where n ranges from 1 to 6 (cf. alkanes), and are named according to the number of silicon atoms they contain. For example, SiH_4 is silane, Si_2H_6 is disilane and Si_3H_8 is trisilane. Structurally, the silanes resemble the alkanes, but chemically they are far more reactive. This is due to the weakness of the Si—H and Si—Si bonds. There are no unsaturated silicon hydrides corresponding to alkenes, alkynes or benzene.

14.3 The C—C bond in alkanes and other organic compounds

Data on the C—C bond

covalent bond length/nm 0.154
bond dissociation enthalpy/kJ mol^{-1} at 298 K 348

The number of possible organic compounds is almost limitless; so large, that if only one molecule of each were to be made the resulting mass would be far greater than that of the earth. That is why organic chemistry is usually studied separately. The reason for this large number lies in the strength of the C—C covalent bond. Formed by an extensive overlap of s and p orbitals, the bond is stable below about 450 °C (723 K) and enables carbon atoms to join together in chains of virtually any length.

With a few notable exceptions, such as ethene, ethyne and benzene, all organic compounds with more than one carbon atom possess the C—C bond, but it is best studied in the *alkanes* – saturated aliphatic (open-chain) hydrocarbons of general formula C_nH_{2n+2}. In these, there are no other bonds, apart from C—H, to influence the chemistry.

14.3.1 NOMENCLATURE OF ALKANES

Unbranched chain alkanes

The first four alkanes, CH_4, CH_3CH_3, $CH_3CH_2CH_3$ and $CH_3CH_2CH_2CH_3$, are called methane, ethane, propane and butane. For higher members, the name consists of a term representative of the number of carbon atoms, followed by '-ane', thus:

C_5H_{12}	pentane	C_9H_{20}	nonane
C_6H_{14}	hexane	$C_{10}H_{22}$	decane
C_7H_{16}	heptane	$C_{11}H_{24}$	undecane
C_8H_{18}	octane	$C_{12}H_{26}$	dodecane

Alkyl radicals

From each alkane an *alkyl radical* is derived by the loss of a hydrogen atom. Thus, from methane is derived the methyl radical, CH_3—, and from ethane the ethyl radical, CH_3CH_2—, often written as C_2H_5—. From propane are derived two radicals, namely propyl (formerly *n*-propyl), $CH_3CH_2CH_2$—, and isopropyl, $(CH_3)_2CH$—. (The ASE name of the latter is 1-methylethyl.)

There are four radicals derived from butane, with structures and names as follows:

	IUPAC name	ASE name
$CH_3CH_2CH_2CH_2$—	butyl (formerly *n*-butyl)	butyl
$(CH_3)_2CHCH_2$—	isobutyl	2-methylpropyl
$CH_3CH_2CH(CH_3)$—	*sec*-butyl	1-methylpropyl
$(CH_3)_3C$—	*tert*-butyl	1,1-dimethylethyl or dimethylethyl

It is not suggested that radicals are obtained directly from alkanes by loss of hydrogen, or that in normal circumstances they are capable of a separate existence, but such groups form part of many molecules. In general discussions an alkyl radical is represented by R.

Branched chain alkanes

There is only one way in which one carbon and four hydrogen atoms may be arranged, and that is to give a methane molecule. Likewise, there is only one possible form of ethane and one of propane. But for butane, C_4H_{10}, there are two ways, I and II, in which the atoms may be arranged without violating bonding requirements. This is an example

$CH_3CH_2CH_2CH_3$ (I) $\qquad$ $(CH_3)_2CHCH_3$ (II)

of *isomerism* (§5.2.1). To distinguish between them, the unbranched form, I, is called butane, while the branched chain version, II, is called isobutane (ASE: 2-methylpropane). The prefix *iso* is often used to denote a branched chain ending in two methyl groups, thus:

$$\begin{matrix} CH_3 \diagdown & \\ & CH- \\ CH_3 \diagup & \end{matrix}$$

There are three isomers of pentane, C_5H_{12}, with structures and names as follows:

	IUPAC name	ASE name
$CH_3CH_2CH_2CH_2CH_3$	pentane	pentane
$(CH_3)_2CHCH_2CH_3$	isopentane	2-methylbutane
$CH_3-C(CH_3)_2-CH_3$	neopentane	2,2-dimethylpropane

As the relative molecular mass increases, so the number of possible isomers increases greatly. There are, for example, five hexanes, nine heptanes and 18 octanes. It is difficult to find trivial names for them all, especially names that mean anything. (Ideally, the name should reflect the structure.) For this reason, the higher members of the series must be named in accordance with the following IUPAC rules.

(i) The compound is named as an alkyl substituted alkane, by prefixing the designations of the side chains to the name of the *longest chain* present in the formula. For example, compound III is a

$$\boxed{CH_3\underset{\displaystyle CH_3}{\underset{|}{C}}HCH_2CH_2CH_2CH_3}$$

III

methylhexane. (A box has been drawn around the longest chain.) The name is written as one word, and without a hyphen.

(ii) Rule (i) by itself is insufficient, as will be seen by considering IV and V. They are clearly both methylhexanes; but we cannot have two

$$CH_3\underset{\displaystyle CH_3}{\underset{|}{C}}HCH_2CH_2CH_2CH_3 \qquad CH_3CH_2\underset{\displaystyle CH_3}{\underset{|}{C}}HCH_2CH_2CH_3$$

IV V

compounds with the same name. The carbon atoms of the longest chain are therefore numbered from one end to the other, the direction being chosen so as to give the lowest possible numbers to the side chains. Form IV thus becomes 2-methylhexane (observe the hyphen),

$$\begin{array}{l} \overset{1}{C}H_3\overset{2}{C}H\overset{3}{C}H_2\overset{4}{C}H_2\overset{5}{C}H_2\overset{6}{C}H_3 \\ \quad\;\; | \\ \quad\; CH_3 \end{array}$$

IV

$$\begin{array}{l} \overset{1}{C}H_3\overset{2}{C}H_2\overset{3}{C}H\overset{4}{C}H_2\overset{5}{C}H_2\overset{6}{C}H_3 \\ \qquad\quad\;\; | \\ \qquad\quad\; CH_3 \end{array}$$

V

and form V 3-methylhexane. The names 5-methylhexane and 4-methylhexane, obtained by numbering the chain in the opposite direction, are wrong. (There are no such names as these. Turn round a molecule of '5-methylhexane' and it at once becomes 2-methylhexane.)

With more complex structures confusion may arise over the meaning of the phrase 'lowest numbers'. The correct name is always that which contains the lowest number on the occasion of the first difference. Hence, compound VI is 2,**3**,5-trimethylhexane; not 2,**4**,5-trimethylhexane. (The sums of the locants are irrelevant.)

$$\begin{array}{l} \overset{6}{C}H_3\overset{5}{C}H\overset{4}{C}H_2\overset{3}{C}H-\overset{2}{C}H\overset{1}{C}H_3 \\ \quad\;\; | \qquad\quad | \qquad\; | \\ \quad CH_3 \quad CH_3 \; CH_3 \end{array}$$

VI

(iii) The presence of identical radicals is indicated by an appropriate multiplying prefix: di-, tri-, tetra-, penta-, hexa-, etc. For example,

$$\begin{array}{l} \qquad\qquad CH_3 \\ \qquad\qquad\; | \\ CH_3CH_2CCH_2CH_3 \\ \qquad\qquad\; | \\ \qquad\qquad CH_3 \end{array}$$

VII

compound VII is called 3,3-dimethylpentane. (Not 3,3-methylpentane, and not 3-dimethylpentane.)

(iv) If two or more side chains of different nature are present, they are cited in alphabetical order. Consider compound VIII. Since *e* comes

$$\begin{array}{l} \qquad\qquad\qquad\; CH_3 \\ \qquad\qquad\qquad\quad | \\ \qquad\qquad\qquad\; CH_2 \quad\; CH_3 \\ \qquad\qquad\qquad\quad | \qquad\quad\; | \\ \overset{7}{C}H_3\overset{6}{C}H_2\overset{5}{C}H_2\overset{4}{C}H-\!\!-\overset{3}{C}-\overset{2}{C}H_2\overset{1}{C}H_3 \\ \qquad\qquad\qquad\qquad\qquad\;\; | \\ \qquad\qquad\qquad\qquad\quad\; CH_3 \end{array}$$

VIII

before *m*, the name is 4-ethyl-3,3-dimethylheptane, even though the presence of two methyl groups causes us to write 'dimethyl' in accordance with rule (iii).

(v) If two or more side chains are in equivalent positions, the one which is cited first in the name is assigned the lower number. For example,

$$\overset{8}{C}H_3\overset{7}{C}H_2\overset{6}{C}H_2\underset{\displaystyle CH_3}{\overset{5}{C}H}-\underset{\displaystyle \underset{\displaystyle CH_3}{CH_2}}{\overset{4}{C}H}-\overset{3}{C}H_2\overset{2}{C}H_2\overset{1}{C}H_3$$

IX

compound IX is 4-ethyl-5-methyloctane; not 5-ethyl-4-methyloctane.

14.3.2 FORMATION OF THE C—C BOND

Wurtz reaction

When an alkyl halide, RX, is treated with sodium (or certain other metals, e.g. copper) in the presence of an *aprotic solvent* (i.e. one that does not liberate protons by ionisation), usually dry ethoxyethane (ether) or cyclohexane, an alkane (RR) is formed together with a sodium halide:

$$R\boxed{X + 2Na + X}R \xrightarrow{\text{in dry ether}} RR + 2NaX$$

The formation of a sodium halide, with a high lattice enthalpy, drives the reaction to the right. For example:

$$\underset{\text{bromoethane}}{CH_3CH_2Br} + 2Na + BrCH_2CH_3 = \underset{\text{butane}}{CH_3CH_2CH_2CH_3} + 2NaBr$$

The solvent must be aprotic, since in the presence of water and other *protic solvents* (i.e. ones that liberate protons by ionisation) the alkyl halide is reduced to the alkane RH.

A purification problem arises if a mixture of alkyl halides is used in a *mixed Wurtz reaction* to prepare an alkane with an odd number of carbon atoms,

e.g. $CH_3I + CH_3CH_2I + 2Na = CH_3CH_2CH_3 + 2NaI$

Apart from this reaction, simple Wurtz reactions will occur with each of the alkyl halides:

$$2CH_3I + 2Na = CH_3CH_3 + 2NaI$$

$$2CH_3CH_2I + 2Na = CH_3CH_2CH_2CH_3 + 2NaI$$

Such preparations are not normally attempted, although a *Wurtz–Fittig reaction*, which comprises the reaction of an alkyl halide and an aryl halide (§14.8) with sodium, may be quite successful,

e.g. $C_6H_5Br + 2Na + C_2H_5Br = C_6H_5C_2H_5 + 2NaBr$

bromobenzene (Br on benzene ring) ethylbenzene (C_2H_5 on benzene ring) (60% yield)

There is no difficulty in purifying the product, for ethylbenzene, which is a liquid, is easily separated from the by-products butane (a gas) and biphenyl, $C_6H_5 \cdot C_6H_5$, (a solid).

Alkyl halide with potassium or sodium cyanide (§14.8.4)

A C—C bond is formed when an alkyl halide undergoes nucleophilic substitution with the cyanide ion:

$$\mathrm{RX + CN^- \xrightarrow[\text{aqueous alcohol}]{\text{reflux in}} \underset{\text{nitrile}}{RCN} + X^-}$$

Friedel–Crafts reaction (§14.7.4)

This is the name given to the electrophilic substitution of arenes with alkyl or acyl halides. A carbon atom becomes linked to another which forms part of a benzene ring:

$$\underset{\text{benzene}}{C_6H_6} + \underset{\text{alkyl halide}}{RX} \xrightarrow{AlCl_3 \text{ catalyst}} \underset{\text{homologue of benzene}}{C_6H_5R} + HX$$

$$C_6H_6 + \underset{\text{acyl chloride}}{RCOCl} \xrightarrow{AlCl_3} \underset{\text{aromatic ketone}}{C_6H_5COR} + HCl$$

Addition reactions of alkenes (§14.4.4)

The addition of any reagent to an alkene converts the C=C bond into a C—C bond,

$$\text{e.g.}\quad \underset{\text{ethene}}{CH_2{=}CH_2} + H_2 \xrightarrow[\text{finely divided Ni catalyst}]{140\ ^\circ C\ (413\ K)} \underset{\text{ethane}}{CH_3CH_3}$$

14.3.3 GENERAL PROPERTIES OF THE ALKANES

All alkanes

Methane ('natural gas'), ethane, propane and the butanes are gaseous under normal conditions. Then come liquids of gradually increasing boiling temperature, and finally, above $C_{16}H_{34}$, solids of gradually increasing melting temperature. (Both melting temperature and boiling temperature are related to relative molecular mass.)

The alkanes have mild, pleasant smells, and are combustible to give carbon dioxide and water. Being non-polar, they are almost completely insoluble in water but dissolve in less polar solvents.

14.3.4 CHEMICAL PROPERTIES OF THE C—C BOND

Because of its high bond dissociation enthalpy, the C—C bond is a difficult one to break. In most organic reactions, particularly between simple molecules, the product contains the same number of carbon atoms as the reactant, and furthermore they are arranged in the same way: *rearrangement* is uncommon. Usually, the C—C bond can be broken only by heat energy (*pyrolysis*) or under severe oxidising conditions, although in some situations, often where there are neighbouring oxygen atoms, the bond may be broken more easily.

Thermal cleavage

Cracking

When an alkane is heated above about 435 °C (708 K) its molecules *crack* to give smaller ones. The principle is widely applied in utilising the relatively long chain alkanes which are obtained by distilling crude oil (see below) and which would otherwise be of little use.

The principal reaction leads to the formation of a shorter chain alkane and an alkene,

e.g.
$$\underset{\text{butane}}{CH_3CH_2CH_2CH_3} \xrightarrow[\text{(708 K)}]{435\ ^\circ C} \begin{cases} \underset{\text{propene}}{CH_3CH{=}CH_2} + \underset{\text{methane}}{CH_4} \ (50\%) \\ \underset{\text{ethene}}{CH_2{=}CH_2} + \underset{\text{ethane}}{CH_3CH_3} \ (38\%) \end{cases}$$

Important side reactions are isomerisation (see below) and *dehydrogenation*, i.e. the elimination of hydrogen. The latter leads to the formation of alkenes of the original chain length:

$$\underset{\text{butane}}{CH_3CH_2CH_2CH_3} \xrightarrow[\text{(708 K)}]{435\ ^\circ C} \underset{\text{1-butene}}{CH_3CH_2CH{=}CH_2} + \underset{\text{2-butene}}{CH_3CH{=}CHCH_3} + \underset{\text{hydrogen}}{H_2} \ (12\%)$$

Dehydrogenation occurs to a lesser extent with the higher alkanes, i.e. alkanes with a relatively high number of carbon atoms.

In oil refining, mixtures of fairly long chain alkanes are subjected to cracking. Mixtures of shorter chain alkanes are formed, suitable as vehicle fuels, together with ethene, propene and the butenes. These gaseous alkenes can be separated by liquefaction followed by fractional distillation. They are extremely useful, and form the basis of the whole petrochemical industry.

Two main processes are in use, namely *thermal cracking* and *catalytic cracking*. Thermal cracking requires relatively high temperatures (475–600 °C, i.e. 748–873 K) and, like most reactions conducted above 500 °C (773 K), has a mechanism which includes free radicals. A *free radical* is defined as an atom or group of atoms possessing at least one unpaired electron. (A free radical should not be confused with a *radical*, which is merely a group of atoms which occurs repeatedly in several different compounds.)

Thermal cracking proceeds by a *chain reaction*, so called because the reaction of one molecule triggers off the reaction of another, and so on, until all the starting material has been consumed. In every chain reaction there are essentially three stages: (i) initiation; (ii) propagation; and (iii) termination. If, as here, the reaction is propagated by free radicals, it is known as a *free radical chain reaction.*

We shall consider the cracking of butane.

Initiation Free radicals are formed – fairly fast above 450 °C (723 K) – by homolytic fission (§14.7.3) of the C—C bond, which is weaker than the C—H bond:

$$CH_3CH_2CH_2CH_3 \longrightarrow \underset{\text{ethyl free radicals}}{2CH_3CH_2\cdot}$$

$$\text{or} \quad CH_3CH_2CH_2CH_3 \longrightarrow \underset{\text{propyl and methyl free radicals}}{CH_3CH_2CH_2\cdot + CH_3\cdot}$$

A dot is used to symbolise an unpaired electron.

Propagation There are two propagation reactions, in each of which one free radical is consumed and another is formed. Both reactions give, in addition to a free radical, a molecule of the final product.

In the first reaction an alkyl free radical from the initiation stage, represented by R·, abstracts an atom of hydrogen from a butane molecule:

$$R\cdot + CH_3CH_2CH_2CH_3 \longrightarrow \underset{\text{methane, ethane or propane}}{RH} + \underset{\textit{sec}\text{-butyl free radical}}{CH_3CH_2\dot{C}HCH_3}$$

$$\text{or} \quad R\cdot + CH_3CH_2CH_2CH_3 \longrightarrow RH + \underset{\text{butyl free radical}}{CH_3CH_2CH_2\dot{C}H_2}$$

Observe the formation of alkanes of chain length below that of the original alkane.

In the second propagation reaction the butyl or *sec*-butyl free radical undergoes homolytic fission at the β-position, i.e. at the bond next but one to C·, to give an alkene and a further methyl or ethyl free radical which propagates the chain:

$$CH_3 \div CH_2 \div \dot{C}HCH_3 \longrightarrow CH_3\cdot + CH_2{=}CHCH_3$$

$$CH_3CH_2 \div CH_2 \div \dot{C}H_2 \longrightarrow CH_3CH_2\cdot + CH_2{=}CH_2$$

Curved half arrows show the movement of single electrons.

Termination When the concentration of the original alkane has dropped to a low value, two free radicals may join together to form a molecule,

e.g. $CH_3\cdot + CH_3CH_2\cdot \longrightarrow CH_3CH_2CH_3$

Catalytic cracking, by contrast, has an ionic mechanism. An acidic oxide, usually a mixture of silica and alumina, is used to promote the formation of carbon-containing ions. The process can be performed at a relatively low temperature (400–500 °C, i.e. 673–773 K), and is particularly suited to the cracking of *naphthenes*, i.e. cyclopentane and cyclohexane derivatives.

Cracking in the presence of steam is a recent development. The steam cracking of naphtha (§14.3.5) to yield ethene and propene is of great importance in view of the increasing demand for plastics. Naphtha is mixed with steam and heated for about a second at 800 °C (1 073 K). The emergent gas mixture is then chilled and fractionated. As well as lower alkenes, lower alkanes and 1,3-butadiene are obtained.

Decarboxylation of carboxylic acids

The decarboxylation of (i.e. removal of CO_2 from) a carboxylic acid to give a hydrocarbon ruptures a C—C bond. The reaction is used in both aliphatic and aromatic syntheses:

$$RCOOH \longrightarrow RH + CO_2$$

Heat alone is inadequate, but since carbon dioxide is an acidic oxide its elimination may be achieved by heating with a base. If dry sodium ethanoate is heated with soda lime (a mixture of sodium hydroxide and calcium oxide), methane is obtained in nearly theoretical yield:

$$CH_3COONa + NaOH \xrightarrow{\text{heat dry}} CH_4 + Na_2CO_3$$

The methane may be collected over water.

However, the yields are not nearly so good from the higher carboxylic acids, as dehydrogenation occurs to give unsaturated hydrocarbons and free hydrogen. This is a common fault of preparative methods that involve heating organic compounds in the dry state.

Oxidative cleavage

Combustion

Complete oxidation of alkanes occurs on combustion, and gives carbon dioxide, water and a substantial amount of heat energy, e.g.

$$CH_4(g) + 2O_2(g) = CO_2(g) + 2H_2O(l) \qquad \Delta H^\ominus = -887.1 \text{ kJ mol}^{-1}$$

This is the basis of the use of alkanes as fuels both for heating purposes and in the internal combustion engine.

Incomplete combustion in a limited supply of air leads to the formation of water and carbon monoxide or, with less air, to *carbon black*, a pure form of soot which is used as a pigment for paints and inks and as a filler for tyre rubber.

Oxidation by powerful oxidants

Powerful oxidants, especially concentrated nitric acid and acidified potassium permanganate solution, are able to break C—C bonds at moderate temperatures. One notable example concerns the oxidative cleavage of ketones. The molecule may break on either side of the carbonyl group to give mixtures of carboxylic acids:

$$RCH_2 \overset{}{\underset{1}{\vdots}} CO \underset{2}{\vdots} CH_2R' \longrightarrow RCOOH + R'CH_2COOH \quad \text{(cleavage 1)}$$

$$\text{or} \quad RCH_2COOH + R'COOH \quad \text{(cleavage 2)}$$

Cleavage in special situations

In certain situations, often when each of the carbon atoms is joined to an oxygen atom, a C—C bond can be broken with remarkable ease.

$(COOH)_2$
and Salts

Oxidation of ethanedioic acid (oxalic acid)
Ethanedioic acid and the ethanedioate ion undergo oxidative cleavage by the permanganate ion in dilute solution to give carbon dioxide:

$$\begin{matrix} COO^- \\ | \\ COO^- \end{matrix} = 2CO_2 + 2e^-$$

The reaction proceeds quantitatively and is used for the estimation of potassium permanganate (or of ethanedioates, with standardised $KMnO_4$).

Alkaline cleavage of trihaloaldehydes or ketones
Aqueous alkali readily cleaves the C—C bond in compounds such as CCl_3CHO (trichloroethanal or 'chloral') and CCl_3COCH_3(1,1,1-trichloropropanone):

$$CCl_3CHO + HO^- = \underset{\text{trichloromethane}}{CHCl_3} + HCOO^-$$

$$CCl_3COCH_3 + HO^- = CHCl_3 + CH_3COO^-$$

This is the second stage of the *haloform reaction* (§14.7.5).

Hofmann reaction
The term 'Hofmann reaction' is given to the chemical change whereby an amide is converted into an amine with one less carbon atom, by treatment with bromine and alkali:

$$R{-}C(=O){-}NH_2 \xrightarrow{Br_2/HO^-} RNH_2$$

$$\text{e.g.} \quad \underset{\text{ethanamide}}{CH_3CONH_2} \longrightarrow \underset{\text{methylamine}}{CH_3NH_2}$$

The reaction is thus a degradation, and is commonly called the 'Hofmann degradation of amides'. It comprises four separate reactions.

Isomerisation

When an alkane is heated it can not only crack but can also undergo *isomerisation*, i.e. it can form an isomer of the original alkane. It is a general rule that such rearrangement leads to chain branching. Butane, for instance, isomerises to 2-methylpropane; not completely, for the reaction is reversible and an equilibrium mixture is established. Such changes are catalysed by Lewis acids, e.g. aluminium bromide.

14.3.5 OIL REFINING

Crude oil occurs in various parts of the world, notably the Middle East, Mexico, USA, South America, North Africa, Russia and the North Sea. It consists mainly of hydrocarbons, with small amounts of sulphur, oxygen and nitrogen compounds. The hydrocarbons are mainly alkanes, with a wide range of relative molecular masses. There are also some cycloalkanes, known in the industry as *naphthenes*, and varying proportions of aromatic hydrocarbons.

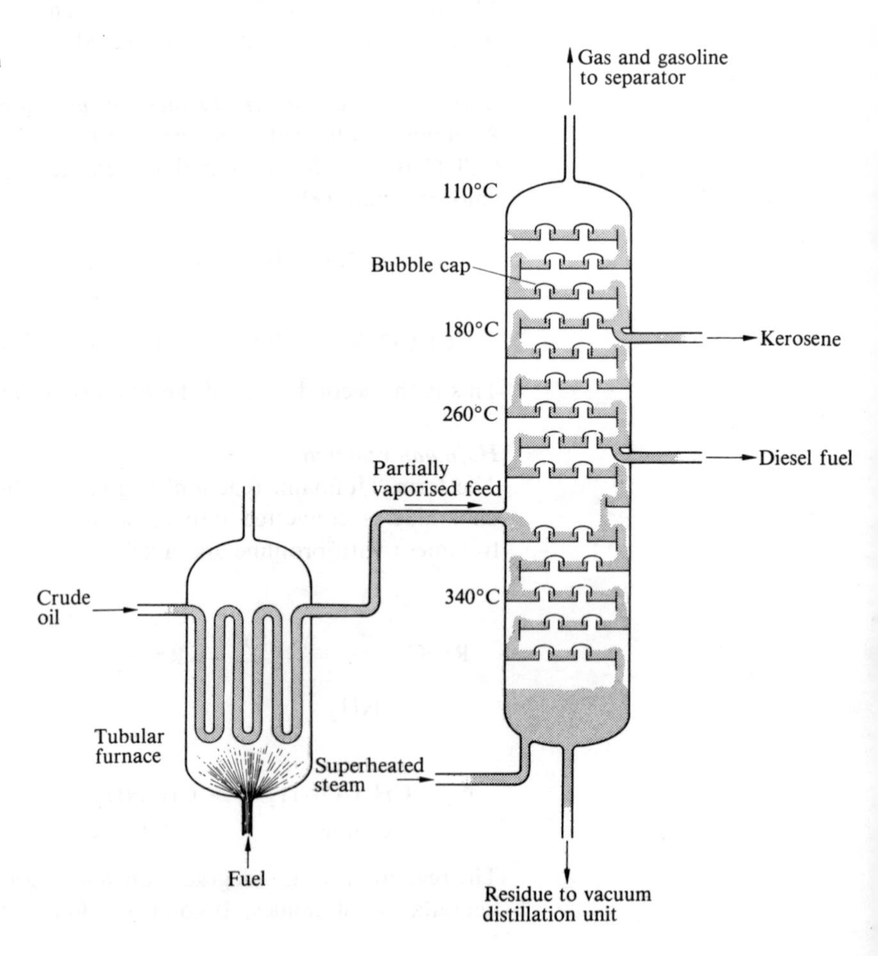

Fig. 14.3 The primary distillation of crude oil. (Adapted from a chart published by the Shell Refining Company Limited, Shell Centre, London SE1.)

Primary distillation

Crude oil is first of all fractionally distilled. The process is often complex and varies from one refinery to another, but the following fractions, listed in increasing order of boiling temperature, are usually collected: gases; light gasoline; naphtha, or heavy gasoline; kerosene; diesel fuel; residue.

The technique consists of heating the crude oil in a tubular furnace, so that it is partially vaporised, and then injecting it into the side of a fractionating column which is heated by superheated steam. The various fractions are drawn off at different points of the column, Fig. 14.3.

The fractionating column contains a series of trays and bubble caps, designed to ensure that the rising vapours come into intimate contact with refluxing liquid. The products from the primary distillation are further treated before they are ready for use. Again, the scheme varies from one plant to another, and the information that follows serves only as a guide.

Gases

Initially in solution in the crude oil, gases – mainly propane and butane – are separated from the gasoline after the primary distillation. They are readily liquefied by applying pressure at ordinary temperatures, and in this form they are sold (e.g. 'Calor gas') for industrial and domestic heating purposes.

Light gasoline

With relatively little treatment this is suitable as motor fuel, but in practice such *straight run petrol* is always blended with petrol from other sources, as follows.

(i) Straight run petrol that has been *thermally reformed*, i.e. subjected to a high temperature (500–600 °C, i.e. 773–873 K) and a pressure of 20–70 atm (2 000–7 000 kPa), so that straight chain molecules re-arrange to become branched.

(ii) Petrol originating from the naphtha fraction on *catalytic reforming* (see below).

(iii) Cracked *light distillate* obtained from the tarry residue (see below).

All petrols sold for use in motor cars contain *additives*, the principal one being tetraethyllead, $(C_2H_5)_4Pb$, which improves the combustion characteristics.

Naphtha (Heavy gasoline)

A certain amount of this fraction, especially if it is rich in aromatics, may be refined and sold as *white spirit*, which is used as a paint thinner and in dry cleaning. Most of it, however, is catalytically reformed to give petrol. Catalytic reforming is a development of thermal reforming and has the same aim, namely to give branched chain and aromatic hydrocarbons.

Kerosene

This fraction is merely refined to give the familiar domestic 'paraffin'. The principal outlet, however, is as a fuel for jet aircraft.

Diesel fuel
The diesel fuel fraction is also little changed.

Residue
Residual hydrocarbons from the base of the primary distillation unit flow to a vacuum distillation unit, from which there are three products: (i) light distillate; (ii) lubricating oil; (iii) residue.

The light distillate is catalytically cracked to give petrol, diesel fuel and gases. The petrol and diesel fuel are blended with the straight run products, while the gases – a mixture of alkanes and alkenes of low relative molecular mass – are separated by liquefaction and distillation.

In practice, three lubricating oil fractions are taken – light, medium and heavy. They are refined in various ways according to the uses to which they are to be put: obviously, oils for medicinal purposes (e.g. 'liquid paraffin') must be treated more drastically than machine oils. The main processing is *dewaxing*, in which paraffin wax is precipitated by a suitable solvent, such as 2-butanone. Apart from producing a useful product, this treatment is necessary to prevent the oil from solidifying in the cold.

The tarry residue from the vacuum distillation is not distilled further. Part of it is sold as *bitumen* for road making, and part is used as fuel oil. For this purpose its viscosity must be reduced, either by blending with diesel oil from the catalytic cracking units or by cracking in a plant called a *visbreaker*.

14.4 The C=C bond in alkenes

Data on the C=C bond

covalent bond length/nm 0.134
bond dissociation enthalpy/kJ mol^{-1} at 298 K 612

A comparison of the data above with that in §14.3 shows that the carbon–carbon double bond, C=C, is considerably shorter and stronger than the carbon–carbon single bond. Double bonds are found in the *alkenes*, formerly known as the *olefins*. Together with the alkynes (§14.5), they are referred to as *unsaturated hydrocarbons*. The word 'unsaturated' in this context means unsaturated with hydrogen: both alkenes and alkynes are capable of adding on hydrogen to give alkanes. Although they do not occur naturally, the lower members of the homologous series are available in large quantities by the cracking of certain petroleum fractions (§14.3.4).

14.4.1 NOMENCLATURE OF ALKENES

Alkenes with one double bond, if they are unbranched, are named by replacing the ending '-ane' of the corresponding alkane by '-ene'.

$$\underset{\text{I}}{CH_3CH{=}CH_2} \qquad \underset{\text{II}}{\overset{4}{C}H_3\overset{3}{C}H_2\overset{2}{C}H{=}\overset{1}{C}H_2}$$

Thus, compound I is called propene. The chain is so numbered as to give the lowest possible numbers to the double bonds. Compound II, for example, is 1-butene; not 3-butene.

$$\underset{\text{III}}{CH_2{=}CH_2} \qquad \underset{\text{IV}}{\overset{4}{C}H_2{=}\overset{3}{C}H{-}\overset{2}{C}H{=}\overset{1}{C}H_2}$$

The trivial name of ethylene is retained by IUPAC for the first member of the homologous series (III). Compounds with two double bonds are referred to as *alkadienes*; those with three double bonds as *alkatrienes*, etc. For example, compound IV is 1,3-butadiene.

For branched chain alkenes, the chain on which the name is based must contain the double bond, even though it may not necessarily be the longest chain in the molecule. For example, compound V is called 2-ethyl-1-butene,

$$\underset{\text{V}}{\overset{4}{C}H_3\overset{3}{C}H_2\overset{2}{C}(CH_2CH_3){=}\overset{1}{C}H_2} \qquad \underset{\text{VI}}{CH_3C(CH_3){=}CH_2}$$

even though the longest carbon chain possesses five atoms. The name of compound VI is 2-methylpropene; the trivial name of isobutene is no longer recognised.

On the system of naming recommended by the Association for Science Education (ASE), alkenes are named by inserting the number or numbers in the name itself. Thus, 1-butene is known as 'but-1-ene', and 1,3-butadiene as 'buta-1,3-diene'. The lowest alkene, compound III, is named systematically as ethene.

14.4.2 FORMATION OF THE C=C BOND

Apart from addition to the carbon–carbon triple bond, C≡C, all the basic methods of forming the C=C bond are elimination reactions. Such reactions often compete with substitution reactions and may be favoured by adjustment of the conditions – principally, by raising the temperature and altering the solvent.

Elimination methods

From $\ \ >\!C(H){-}C(X)\!<$ (X = Cl, Br or I) ***by the elimination of HX*** (§14.8.4)

When an alkyl halide is treated with a base, substitution usually occurs to give an alcohol:

$$RCl + HO^-(aq) = ROH + Cl^-$$

However, at higher temperatures, and with a strong base in concentrated alcoholic solution, dehydrohalogenation occurs to give an alkene:

$$R{-}\underset{H}{\overset{H}{C}}{-}\underset{H}{\overset{Cl}{C}}{-}R' + HO^-(alc) = \underset{H}{\overset{R}{}}\!\!>C{=}C<\!\!\underset{H}{\overset{R'}{}} + Cl^- + H_2O$$

From $-\overset{H}{C}-\overset{OH}{C}-$ ***by the elimination of*** $\boldsymbol{H_2O}$ (§14.9.5)

In the presence of an acid catalyst, e.g. conc. H_2SO_4 or Al_2O_3, alcohols may be dehydrated to give either ethers or alkenes according to the severity of the conditions. Under mild conditions (relatively low temperature and limited acid) a substitution reaction leads to etherification:

$$ROH + ROH \xrightarrow[140\ ^\circ C\ (413\ K)]{conc.\ H_2SO_4} R{-}O{-}R + H_2O$$

Under more drastic conditions, i.e. at a relatively high temperature and with excess acid, an alkene is formed in an elimination reaction:

$$R{-}\underset{H}{\overset{H}{C}}{-}\underset{H}{\overset{OH}{C}}{-}R' \xrightarrow[180\ ^\circ C\ (453\ K)]{conc.\ H_2SO_4} \underset{H}{\overset{R}{}}\!\!>C{=}C<\!\!\underset{H}{\overset{R'}{}} + H_2O$$

From $-\overset{X}{C}-\overset{X}{C}-$ or $-C-\overset{X}{C}-X$ (X = Cl, Br or I) ***by the elimination of*** $\boldsymbol{X_2}$

A double bond can be formed by eliminating two halogen atoms from a suitable dihalide by means of a metal – usually powdered zinc,

$$\left.\begin{array}{ll} \text{e.g.} & CH_3CHBrCH_2Br \\ & \text{a 1,2-dihalide} \\ \text{or} & CH_3CH_2CHBr_2 \\ & \text{a 1,1-dihalide} \end{array}\right\} \xrightarrow[\text{in methanol}]{\text{Zn dust}} CH_3CH{=}CH_2 + ZnBr_2$$

From $-\overset{H}{C}-\overset{H}{C}-$ ***by the elimination of*** $\boldsymbol{H_2}$ (§14.3.4)

The cracking of alkanes at high temperatures is accompanied by dehydrogenation to alkenes.

Addition methods

The partial hydrogenation of an alkyne converts the carbon–carbon triple bond to a double bond, e.g.

$$\underset{\text{ethyne}}{CH\equiv CH} \xrightarrow{+H_2} \underset{\text{ethene}}{CH_2=CH_2}\left[\xrightarrow{+H_2} \underset{\text{ethane}}{CH_3CH_3}\right]$$

The addition may be stopped at the alkene stage by the use of a *Lindlar catalyst*, i.e. palladium, partly poisoned by lead, on calcium carbonate.

Partial halogenation, likewise, results in a double bond. Partial chlorination is barely practicable, but partial bromination and iodination are not difficult to achieve. Other reagents (e.g. HCl) also add across a C≡C bond to give a C=C bond (§14.5).

14.4.3 GENERAL PROPERTIES OF THE ALKENES

All alkenes

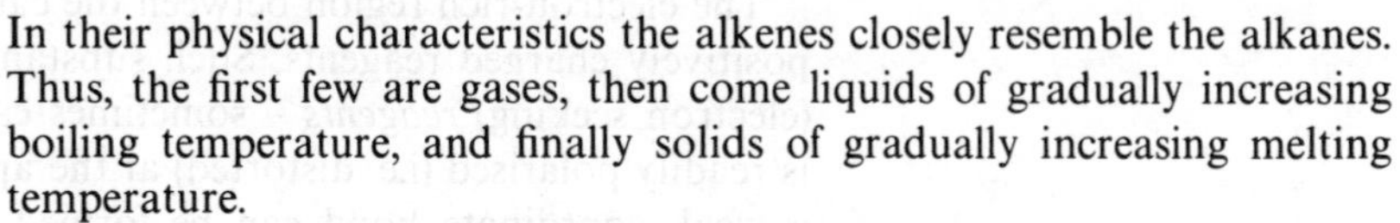

In their physical characteristics the alkenes closely resemble the alkanes. Thus, the first few are gases, then come liquids of gradually increasing boiling temperature, and finally solids of gradually increasing melting temperature.

They have mild, sweet smells, and are non-polar compounds which are insoluble in water. They burn readily to give carbon dioxide and water, although with more yellow, smoky flames than the alkanes. This is a reflection of the higher C:H ratio.

14.4.4 CHEMICAL PROPERTIES OF THE C=C BOND

The C=C bond, in contrast to the C—C bond, is characterised by a high reactivity. Alkenes undergo a considerable number of *addition reactions* across the double bond, as a result of which saturated *adducts*, i.e. addition products, are formed:

$$\underset{\text{alkene}}{>C=C<} + \underset{\text{reagent}}{AB} = \underset{\text{adduct}}{-\overset{A}{\underset{|}{C}}-\overset{B}{\underset{|}{C}}-}$$

Three principal mechanistic routes are possible, namely *electrophilic addition* (A_E), *radical addition* (A_R) and *nucleophilic addition* (A_N). Only electrophilic addition will be discussed here.

Hydrogenation, i.e. the addition of hydrogen in the presence of a nickel or platinum catalyst, is a special case in that it occurs by an adsorption mechanism. The reaction proceeds by the chemisorption (§7.5.3) of hydrogen and the alkene on to the surface of the catalyst, followed by the transfer of hydrogen atoms and desorption of the product:

$$2H_2 + \underset{\text{metal surface}}{\text{////////}} \rightleftharpoons \text{H H} \quad \text{H H (adsorbed)} \xrightleftharpoons{>C=C<}$$

$$\text{H H } {>C=C<} \text{ H H (adsorbed)} \rightleftharpoons \text{H H } {>C-C<} \text{ H H (adsorbed)} \rightleftharpoons \text{H } \overset{>C-C<}{\text{H H}} \text{ H}$$

Electrophilic addition

For such a reaction to occur, the reagent AB must be polar, thus: $\overset{\delta^+}{A}—\overset{\delta^-}{B}$, or at least capable of being polarised under the prevailing conditions. To understand how AB adds across the double bond, it is necessary to have some knowledge of the bond itself. Although we commonly represent it by the symbol C=C for the sake of convenience, the two bonds are far from being identical. As we have seen (§4.3.1), the carbon atoms are joined together partly by a strong σ bond, formed by the overlap of s and p orbitals, and partly by a much weaker π bond, from overlapping p orbitals. It is the latter – weak and accessible – which is broken during addition reactions.

The electron-rich region between the carbon atoms is bound to attract positively charged reagents. Such substances are known as *electrophilic* (electron seeking) *reagents* – sometimes called *electrophiles*. The π bond is readily polarised (i.e. distorted) at the approach of such a reagent, and a weak coordinate bond can be formed by the donation of a pair of electrons from the alkene to the electrophilic reagent. Because π electrons participate in its formation, the resulting complex is called a *π-complex* (Fig. 14.4).

Fig. 14.4 (a) The approach of an electrophile to an alkene. (b) Coordination of π electrons of the C=C bond to the electrophile to give a π-complex. (c) Formation of a carbonium ion.

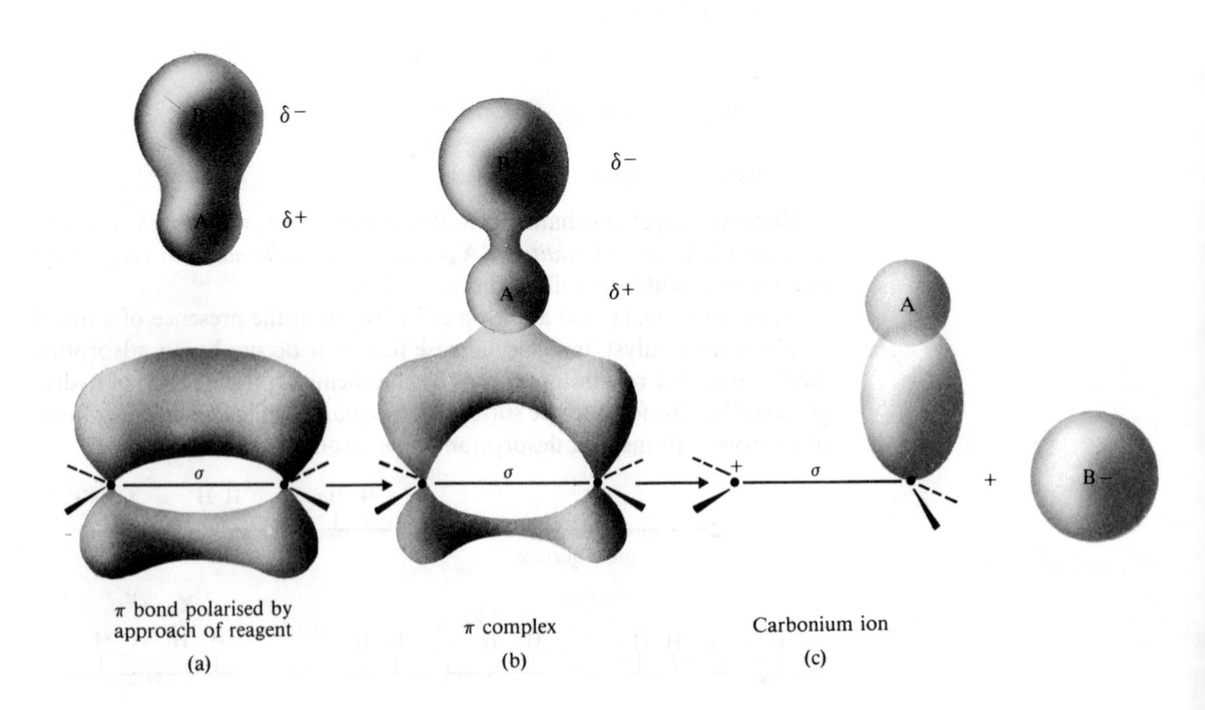

The reaction can be represented by an equation:

$$>C=C< + \overset{\delta^+}{A}-\overset{\delta^-}{B} \longrightarrow >C\overset{\overset{\delta^+\ \ \delta^-}{A-B}}{\overset{\uparrow}{=}}C< \quad \pi\text{-complex}$$

Slowly, the π-complex changes into a *carbonium ion*, i.e. an ion in which positive charge resides on a carbon atom. This happens by the breaking of one bond, A—B, and the formation of another, A—C. At the same time an anion B^- is released into the solvent:

$$>C\overset{\overset{\delta^+\ \ \delta^-}{A-B}}{\overset{\uparrow}{=}}C< \xrightarrow{\text{slowly}} -\overset{A}{\underset{|}{C}}-\overset{\oplus}{C}< + B^-$$

π-complex carbonium ion

During this reaction, the covalent bond A—B in the reagent is effectively broken into two oppositely charged ions, A^+ and B^-. The process is known as *heterolytic cleavage*, or *heterolysis*. There is no evidence that A^+ exists in the free state before bonding to a carbon atom.

Reaction is completed by the rapid coordination of B^- to the carbonium ion:

$$-\overset{A}{\underset{|}{C}}-\overset{\oplus}{C}< + :B^- \xrightarrow{\text{fast}} -\overset{A}{\underset{|}{C}}-\overset{B}{\underset{|}{C}}-$$

For example, when ethene gas is bubbled through a solution of bromine in ethanol at room temperature, the red colour of the bromine is discharged. An addition reaction must take place because hydrogen bromide is not formed, as it would be if a substitution reaction occurred.

$$\underset{\text{ethene}}{CH_2{=}CH_2} + Br_2 = \underset{\text{1,2-dibromoethane}}{CH_2BrCH_2Br}$$

Similar reactions occur with other alkenes. If cyclohexene (a liquid) is used, the red colour again disappears without production of hydrogen bromide.

$$\text{cyclohexene} + Br_2 = \text{1,2-dibromocyclohexane}$$

cyclohexene 1,2-dibromocyclohexane

The way in which molecular bromine, a non-polar substance, behaves as an electrophilic reagent is not immediately obvious. Indeed, pure halogens do not react with pure alkenes in the gas phase. But in the presence of a polar solvent, or certain catalysts such as aluminium bromide, the bromine molecule is polarised and able to form a π-complex with the alkene,

$$\text{e.g.}\quad \underset{\text{permanent dipole}}{\overset{\textit{solvent}}{R{-}\overset{\delta^-}{O}{\div}\overset{\delta^+}{H}}} \qquad \underset{\text{temporary induced dipole}}{\overset{\textit{reagent}}{\overset{\delta^-}{Br}{\div}\overset{\delta^+}{Br}}}$$

Then:

$$CH_2{=}CH_2 + \overset{\delta^-}{Br}{-}\overset{\delta^+}{Br} \xrightarrow{\text{fast}} \underset{\pi\text{-complex}}{\overset{\overset{\delta^-\ \ \delta^+}{Br-Br}\ \uparrow}{CH_2{=}CH_2}}$$

$$\xrightarrow{\text{slow}} \underset{\text{carbonium ion}}{CH_2Br{-}\overset{\oplus}{C}H_2} + Br^- \xrightarrow{\text{fast}} CH_2BrCH_2Br$$

The positive bromine ion, Br^+, is not included in this mechanism. Not only is there no experimental evidence for its existence, but the first ionisation enthalpy of bromine is very high (1 140 kJ mol^{-1} at 298 K) and comparable with the values for noble gases. The Br^- ion, however, *is* involved, as may be proved by conducting the reaction in the presence of chloride ions. Br^- and Cl^- ions compete for the carbonium ion, and the product is a mixture of CH_2BrCH_2Br and CH_2BrCH_2Cl.

When the addition of bromine is performed in solution in ethanol, C_2H_5OH molecules compete with Br^- ions for coordination to the carbonium ions. This leads to the formation of a certain amount of a bromoether, $CH_2BrCH_2OC_2H_5$; cf. the addition of HBrO (see below). To avoid this side reaction, tetrachloromethane or trichloromethane should be used as the solvent.

Chlorine behaves in much the same way as bromine, but iodine does not usually react. With fluorine the reaction is very vigorous, resulting in polysubstitution and breakdown of the alkene into fragments with a smaller number of carbon atoms. (CF_4 is a common product.)

Other electrophilic additions

Bromine atoms in bromine molecules and chlorine atoms in chlorine molecules act as electrophiles because: (i) they can develop partial positive charges; (ii) they can accept a pair of electrons from the alkene; and (iii) bromine and chlorine form strong bonds with carbon atoms. Any other potential electrophile must satisfy the same conditions. The commonest is $\overset{\delta^+}{H}$, belonging to a strong acid such as HBr or H_2SO_4. Weak acids, such as HCN and CH_3COOH, are poor sources of protons and do not react with alkenes. Metal cations, e.g. Na^+, cannot react either, because any products would be unstable.

HF, HCl, HBr and HI

The addition of hydrogen halides follows the expected pattern:

$$CH_2{=}CH_2 + \overset{\delta+}{H}{-}\overset{\delta-}{X} \xrightarrow{\text{fast}} \underset{\pi\text{-complex}}{\overset{\overset{\delta+\ \ \delta-}{H-X}}{\overset{\uparrow}{CH_2{=}CH_2}}}$$

ethene

$$\xrightarrow{\text{slow}} \underset{\text{carbonium ion}}{CH_3{-}\overset{\oplus}{C}H_2} + X^- \xrightarrow{\text{fast}} \underset{\text{ethyl halide}}{CH_3CH_2X}$$

The order of reactivity of the hydrogen halides towards alkenes is $HI > HBr > HCl \gg HF$, which corresponds to the order of their strength as acids. Concentrated aqueous solutions of HI and HBr absorb ethene at room temperature to give iodoethane and bromoethane respectively, but there is no reaction with concentrated hydrochloric acid. Hydrochloric acid attacks only the higher alkenes, and hydrogen fluoride adds to alkenes only under pressure.

The addition of a hydrogen halide to propene (or any other unsymmetrical alkene) could conceivably give two products,

$$\text{e.g.}\quad \underset{\text{propene}}{CH_3CH{=}CH_2} + HBr \begin{cases} \nearrow CH_3CH_2CH_2Br \quad \text{1-bromopropane} \\ \searrow CH_3CHBrCH_3 \quad \text{2-bromopropane} \end{cases}$$

In practice, electrophilic addition leads exclusively to 2-bromopropane. The reason hinges on the difference between the stabilities of primary, secondary and tertiary carbonium ions. In a *primary carbonium ion* the charged carbon atom is attached to *one* alkyl (or aryl) radical; in a *secondary carbonium ion* it is attached to *two* radicals, while in a *tertiary carbonium ion* it is joined to *three.*

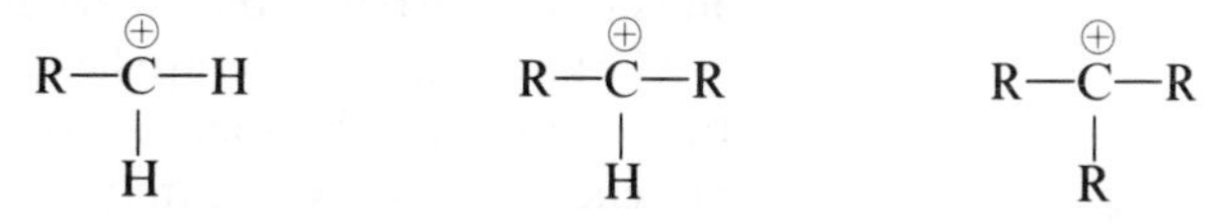

primary carbonium ion secondary carbonium ion tertiary carbonium ion

It is a fundamental law of physics that the stability of any charged system is increased by dispersal of the charge. Thus, an electrical charge on a brass sphere does not remain concentrated at one locality, but becomes evenly distributed over the entire sphere. Carbonium ions, likewise, will gain in stability if the positive charge can be *delocalised*, i.e. shared among two or more carbon atoms. In a primary carbonium ion the scope for delocalisation is limited, because the $\overset{\oplus}{C}$ atom is linked to only one other carbon atom, but in secondary and especially in tertiary carbonium ions dispersal of charge can occur to a greater extent.

$$\underset{\text{primary}}{-\overset{|}{\underset{|}{C}}\rightarrow\overset{\oplus}{\underset{\underset{H}{|}}{C}}-H} \qquad \underset{\text{secondary}}{-\overset{|}{\underset{|}{C}}\rightarrow\overset{\oplus}{\underset{\underset{H}{|}}{C}}\leftarrow\overset{|}{\underset{|}{C}}-} \qquad \underset{\text{tertiary}}{-\overset{|}{\underset{|}{C}}\rightarrow\overset{\oplus}{\underset{\underset{-\underset{|}{C}-}{\uparrow}}{C}}\leftarrow\overset{|}{\underset{|}{C}}-}$$

$$\xrightarrow[\text{Increasing stability and increasing ease of formation}]{\text{Increasing delocalisation of positive charge}}$$

Charge sharing is possible with neighbouring carbon atoms but not with hydrogen atoms because alkyl groups have a *positive inductive effect*, i.e. an electron releasing effect, compared with hydrogen atoms.

Tertiary carbonium ions are thus formed most easily, and are relatively common intermediates in reaction mechanisms. Secondary carbonium ions are formed less readily, and primary carbonium ions are formed only if there is no alternative mechanism.

Let us now reconsider the addition of a hydrogen halide to propene.

$$CH_3CH{=}CH_2 + H^+ \longrightarrow \underset{\text{primary carbonium ion}}{CH_3CH_2\overset{\oplus}{C}H_2} \xrightarrow{Br^-} \underset{\text{1-bromopropane}}{CH_3CH_2CH_2Br}$$

$$CH_3CH{=}CH_2 + H^+ \longrightarrow \underset{\text{secondary carbonium ion}}{CH_3\overset{\oplus}{C}HCH_3} \xrightarrow{Br^-} \underset{\text{2-bromopropane}}{CH_3CHBrCH_3}$$

2-Bromopropane is obtained because its formation proceeds via a secondary carbonium ion, whereas the formation of 1-bromopropane would proceed via a relatively unstable primary carbonium ion.

Other unsymmetrical reagents, e.g. H_2SO_4, act in the same way, a fact which is summarised by *Markownikoff's rule*: 'In the electrophilic addition of protic acids to unsymmetrically substituted alkenes, the hydrogen atom adds to the carbon atom of the double bond bearing the most hydrogen.'

Markownikoff's rule applies only when electrophilic addition occurs. In free radical addition, e.g. the addition of hydrogen bromide in the presence of a peroxide catalyst, the rule is reversed.

H_2SO_4

Chemical reactions occur when alkenes are absorbed by sulphuric acid. The products depend on whether the sulphuric acid is anhydrous or aqueous. In both cases a carbonium ion is formed at first, but various nucleophiles (species with a negative charge) compete for the carbonium ion.

Anhydrous sulphuric acid Alkenes are absorbed at an appreciable rate by anhydrous or highly concentrated sulphuric acid. *Monoalkyl sulphates*, which are half-esters of sulphuric acid (§14.9.5), are formed by the electrophilic addition of H_2SO_4 across the C=C bond,

$$\text{e.g.}\quad CH_2{=}CH_2 + \overset{\delta+}{H}{-}\overset{\delta-}{O}{-}SO_2{-}OH \longrightarrow \underset{\pi\text{-complex}}{\overset{\overset{\delta+\ \ \delta-}{H-O-SO_2-OH}}{\overset{\uparrow}{CH_2{=}CH_2}}}$$

i.e. H_2SO_4

$$\longrightarrow CH_3{-}\overset{\oplus}{C}H_2 + \overset{\ominus}{O}{-}SO_2{-}OH \longrightarrow CH_3CH_2O{-}SO_2{-}OH$$

i.e. HSO_4^- — ethyl hydrogen sulphate

With unsymmetrical alkenes Markownikoff's rule is obeyed,

$$\text{e.g.}\quad CH_3CH{=}CH_2 + H_2SO_4 = \underset{\text{isopropyl hydrogen sulphate}}{(CH_3)_2CHO{-}SO_2{-}OH}$$

Monoalkyl sulphates may be hydrolysed by hot water to give alcohols,

$$\text{e.g.}\quad CH_3CH_2O{-}SO_2{-}OH + H_2O = CH_3CH_2OH + H_2SO_4$$

Aqueous sulphuric acid In the presence of water, and at a high temperature, sulphuric acid converts alkenes into alcohols directly, i.e. not via the monoalkyl sulphate. This is because a carbonium ion is attacked by water in preference to HSO_4^- ions,

$$\text{e.g.}\quad CH_3\overset{\oplus}{C}H_2 + H_2O \longrightarrow \underset{\text{protonated ethanol}}{CH_3\overset{\overset{\oplus}{O}H_2}{\overset{|}{C}}H_2} \longrightarrow \underset{\text{ethanol}}{CH_3CH_2OH} + H^+$$

This forms the basis of the manufacture of alcohols by the acid catalysed hydration of alkenes (§14.4.5). Water alone will not bring about the conversion because its acidity is far too low. The catalyst may be either a protic acid, e.g. H_2SO_4 or H_3PO_4, or a Lewis acid, e.g. WO_3.

Whenever an alcohol is prepared in this way an ether appears as a by-product, because alcohols (like water) can react with carbonium ions,

$$\text{e.g.}\quad CH_3\overset{\cdot\oplus}{C}H_2 + CH_3CH_2OH \longrightarrow \underset{\text{protonated ethoxyethane}}{CH_3\overset{H\overset{\oplus}{O}CH_2CH_3}{\overset{|}{C}}H_2} \longrightarrow \underset{\text{ethoxyethane (diethyl ether)}}{CH_3CH_2{-}O{-}CH_2CH_3} + H^+$$

Chlorine water and bromine water

Chlorine water and bromine water are more than just physical solutions of halogens in water. Some disproportionation (oxidation and reduction) of the halogen occurs to give a hypohalous acid and a hydrohalic acid,

$$\text{e.g.}\quad Cl_2 + H_2O \rightleftharpoons HClO + HCl$$

Consequently, when an alkene is introduced into chlorine water there are four substances competing for attack. Water can be ruled out, and hydrochloric acid does not attack readily (see above): this leaves chlorine and hypochlorous acid. To some extent chlorine adds on,

$$\text{e.g.}\quad CH_2{=}CH_2 + Cl_2 = CH_2ClCH_2Cl$$

but the main reaction appears to be the addition of hypochlorous acid:

$$CH_2{=}CH_2 + HClO = \underset{\substack{\text{2-chloroethanol}\\ \text{(ethylene chlorohydrin)}}}{CH_2ClCH_2OH}$$

Because of the low concentration of hypochlorous acid, HClO molecules are probably not involved in the reaction mechanism. It is more likely that chlorine reacts first, to give chlorocarbonium ions, cf. alkenes with halogens in alcoholic solution. A small proportion of the carbonium ions are attacked by chloride ions to give the dichloride but, if the chloride ion concentration is low, water molecules coordinate preferentially and the chlorohydrin is formed.

When reaction occurs with an unsymmetrical alkene, the *halogen atom* becomes attached to the carbon atom bearing most hydrogen. This is consistent with the theoretical basis of Markownikoff's rule, namely that addition proceeds via the most stable carbonium ion:

$$\underset{\text{propene}}{CH_3CH{=}CH_2} + \overset{\delta^+}{Cl}{-}\overset{\delta^-}{Cl} \xrightarrow{\text{fast}} \underset{\pi\text{-complex}}{CH_3CH{=}CH_2 \leftarrow \overset{\delta^+}{Cl}{-}\overset{\delta^-}{Cl}}$$

$$\xrightarrow{\text{slow}} \underset{\substack{\text{secondary}\\ \text{carbonium ion}}}{CH_3\overset{\oplus}{C}HCH_2Cl} + Cl^- \xrightarrow{H_2O} CH_3\overset{\overset{H_2\overset{\oplus}{O}}{\downarrow}}{C}HCH_2Cl$$

$$\xrightarrow{-H^+} \underset{\text{1-chloro-2-propanol}}{CH_3CH(OH)CH_2Cl}$$

Oxidation

Alkenes are highly susceptible to oxidation by the electrophilic addition of an oxidising agent to the C=C bond. The nature of the product depends very much on the reagent, thus:

$$\text{alkene} \begin{cases} \xrightarrow{O_2 \text{ or a peroxoacid}} \text{epoxide} \\ \xrightarrow{KMnO_4} \text{1,2-diol} \\ \xrightarrow{O_3} \text{ozonide} \end{cases}$$

Secondary changes may occur, so that a variety of final products is possible.

O_2 or peroxoacids Reaction with oxygen occurs at elevated temperatures in the presence of a silver catalyst. The products, known as *epoxides*, are cyclic ethers:

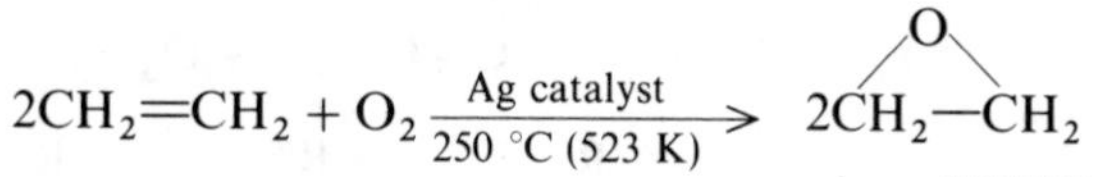

$$2CH_2{=}CH_2 + O_2 \xrightarrow[250\ ^\circ C\ (523\ K)]{\text{Ag catalyst}} 2\,\overset{\large O}{CH_2{-}CH_2}$$

oxirane (IUPAC)

epoxyethane (ASE)

This reaction is of commercial importance (§14.4.5).

Reactions with peroxoacids (e.g. perbenzoic acid, $C_6H_5C(=O){-}O{-}O{-}H$), which likewise give epoxides, may be conducted in inert solvents such as ethoxyethane (diethyl ether) or trichloromethane. If the reaction is performed in aqueous solution the epoxide undergoes acidic hydrolysis to give a *1,2-diol* (see below). The 1,2-diol may subsequently become esterified.

$KMnO_4$ Aqueous potassium permanganate brings about the *hydroxylation* of alkenes, i.e. it results in the addition of two hydroxyl groups across the C=C bond. The products are thus 1,2-diols, commonly called *glycols*,

$$\text{e.g.}\quad \begin{matrix} CH_2 \\ \| \\ CH_2 \end{matrix} \xrightarrow[KMnO_4\ \text{solution}]{\text{cold dilute alkaline}} \begin{matrix} CH_2OH \\ | \\ CH_2OH \end{matrix}$$

1,2-ethanediol

(ethylene glycol)

Yields of 1,2-diols are poor, because they undergo oxidative cleavage to give aldehydes and ketones:

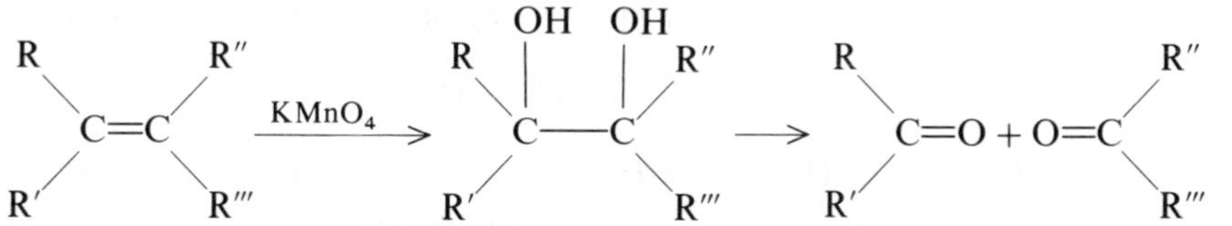

$$RR'C{=}CR''R''' \xrightarrow{KMnO_4} RR'C(OH){-}C(OH)R''R''' \longrightarrow RR'C{=}O + O{=}CR''R'''$$

Ketones formed in this way remain as such, for they are not easily oxidised, but any aldehydes so formed undergo further oxidation by the permanganate to give carboxylic acids.

O_3 In the presence of traces of water, or on contact with glass apparatus, ozone (i.e. trioxygen) becomes polarised:

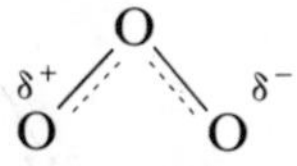

$$\overset{\delta+}{O}{\cdots}O{\cdots}\overset{\delta-}{O}$$

Because of this, it acts as an electrophilic reagent (cf. bromine in similar circumstances) and will attack the double bond of an alkene to give an *ozonide*:

$$\begin{matrix} R & & R'' \\ & C{=}C & \\ R' & & R''' \end{matrix} + O_3 = \begin{matrix} R & O{-}O & R'' \\ & C{-}O{-}C & \\ R' & & R''' \end{matrix}$$

alkene ozonide

The addition of ozone, known as *ozonisation*, is brought about at room temperature by bubbling ozonised oxygen, from an ozoniser, through a solution of an alkene in an inert solvent such as hexane or tetrachloromethane.

Ozonides may be explosive and are seldom isolated. They are usually decomposed by the addition of water:

$$\begin{matrix} R & O{-}O & R'' \\ & C{-}O{-}C & \\ R' & & R''' \end{matrix} + H_2O = \begin{matrix} R & & & R'' \\ & C{=}O + O{=}C & & \\ R' & & & R''' \end{matrix} + H_2O_2$$

Ozonisation, followed by hydrolysis, is known as *ozonolysis.*

R, R′, R″ and R‴ in the formulae shown above can be alkyl groups or hydrogen atoms; consequently, ozonolysis can give two aldehyde molecules, two ketone molecules, or one molecule of aldehyde and one of ketone. Examples are as follows.

$$\begin{matrix} H & & H \\ & C{=}C & \\ H & & H \end{matrix} \xrightarrow{O_3} \begin{matrix} H & O{-}O & H \\ & C{-}O{-}C & \\ H & & H \end{matrix} \xrightarrow{H_2O} 2 \begin{matrix} H & \\ & C{=}O \\ H & \end{matrix} + H_2O_2$$

ethene — ethene ozonide — methanal

$$\begin{matrix} CH_3 & & H \\ & C{=}C & \\ CH_3 & & H \end{matrix} \xrightarrow{O_3} \begin{matrix} CH_3 & O{-}O & H \\ & C{-}O{-}C & \\ CH_3 & & H \end{matrix} \xrightarrow{H_2O} \begin{matrix} CH_3 & & & H \\ & C{=}O + O{=}C & & \\ CH_3 & & & H \end{matrix} + H_2O_2$$

2-methylpropene — 2-methylpropene ozonide — propanone — methanal

The hydrogen peroxide formed on hydrolysis is an oxidising agent and is apt to convert aldehydes into carboxylic acids. This can be minimised by hydrolysing the ozonide in the presence of a reducing agent, e.g. sodium hydrogensulphite or zinc dust, which destroys the peroxide. Alternatively, the ozonide can be subjected to *reductive cleavage* rather than hydrolysis,

e.g. $$\overset{O{-}O}{CH_2{-}O{-}CH_2} + H_2 \xrightarrow[\text{as catalyst}]{\text{Pd on } CaCO_3} 2HCHO + H_2O$$

The ozonolysis of alkenes provides a means of establishing their structures. The products have merely to be separated and identified, and their formulae written down so that the carbonyl groups face each other. Drawing in the double bond and deleting the oxygen atoms then gives the alkene. Consider, for example, an unknown alkene which, on ozonolysis, gives propanal and propanone:

$$\underset{H}{\overset{CH_3CH_2}{\diagdown}}C{=}O \text{ and } O{=}C\overset{CH_3}{\underset{CH_3}{\diagup}} \quad \text{must arise from} \quad \underset{H}{\overset{CH_3CH_2}{\diagdown}}C{=}C\overset{CH_3}{\underset{CH_3}{\diagup}}$$

Addition polymerisation

Addition polymerisation is the principal way of making *plastics*, i.e. organic polymers of high relative molecular mass. It is not the only way, and later (§14.10.14) we shall encounter instances of *condensation polymerisation.*

In the presence of a catalyst, molecules of alkenes can join together to give larger molecules. One molecule of ethene, for instance, can undergo an addition reaction with another:

$$CH_2{=}CH_2 + H|CH{=}CH_2 = CH_3CH_2CH{=}CH_2$$

The product, 1-butene, is the *dimer* of ethene; ethene itself is referred to as the *monomer.*

Three molecules of ethene give one of the *trimer*, 1-hexene:

$$CH_3CH_2CH{=}CH_2 + H|CH{=}CH_2 = CH_3CH_2CH_2CH_2CH{=}CH_2$$

Addition is contrary to Markownikoff's rule because polymerisation does not occur by an electrophilic mechanism.

If a large number of molecules (500 or more) join together, we get a molecule of a *high polymer*, otherwise known as a plastic. Its name is poly(ethene) – commonly polyethylene or polythene – and its formation may be represented by the equation

$$nCH_2{=}CH_2 = CH_3CH_2(CH_2CH_2)_{n-2}CH{=}CH_2$$

In practice polymer chains seldom end in this ideal fashion, and it is customary to write a simplified equation, thus:

$$nCH_2{=}CH_2 = (\text{—}CH_2\text{—}CH_2\text{—})_n$$

Other common monomers are propene, chloroethene (vinyl chloride), propenenitrile (acrylonitrile) and phenylethene (styrene). They may all be regarded as substituted ethenes, $CH_2{=}CHX$. Polymerisation gives substituted polyethenes, possessing chains of $CH_2\text{—}CHX$ units:

$$nCH_2{=}CHX = (\text{—}CH_2\text{—}CHX\text{—})_n$$

Table 14.3 Addition polymers

MONOMER			
Formula	**Name**	**Source**	**Polymerisation conditions**
$CH_2{=}CH_2$	ethene or ethylene	cracking of petroleum fractions	1 000–2 000 atm (100 000–200 000 kPa); 100–300 °C (373–573 K)
$CH_2{=}CH_2$	ethene or ethylene	cracking of petroleum fractions	in solution at 5–30 atm (500–3 000 kPa) with a Ziegler catalyst, e.g. $(C_2H_5)_3Al + TiCl_4$
$CH_3CH{=}CH_2$	propene or propylene	cracking of petroleum fractions	in solution with a Ziegler catalyst
$CH_2{=}CHCl$	chloroethene or vinyl chloride	$CH{\equiv}CH + HCl \xrightarrow[150\ °C\ (423\ K)]{HgCl_2} CH_2{=}CHCl$ $CH_2{=}CH_2 + Cl_2 = CH_2ClCH_2Cl \xrightarrow{500\ °C\ (773\ K)} CH_2{=}CHCl$	emulsion polymerisation with a peroxide catalyst
$C_6H_5CH{=}CH_2$	phenylethene or styrene	from benzene and ethene (§14.4.5)	mass polymerisation with a peroxide catalyst
$CH_2{=}CHCN$	propenenitrile or acrylonitrile	$2CH_3CH{=}CH_2 + 2NH_3 + 3O_2 \xrightarrow{\text{bismuth molybdate, } 450\ °C\ (723\ K)} 2CH_2{=}CHCN + 6H_2O$	in water, with a peroxide catalyst
$CF_2{=}CF_2$	tetrafluoroethene	$CHCl_3 + 2HF = CHClF_2 + 2HCl$ $2CHClF_2 \xrightarrow[(973\ K)]{700\ °C} CF_2{=}CF_2 + 2HCl$	in water, under pressure, with $(NH_4)_2S_2O_8$ catalyst
$CH_2{=}C(CH_3){-}COOCH_3$	methyl 2-methyl-propenoate or methyl methacrylate	$(CH_3)_2C{=}O + HCN \xrightarrow[(293\ K)]{20\ °C} (CH_3)_2C(OH)CN$ $(CH_3)_2C(OH)CN \xrightarrow{CH_3OH,\ H_2SO_4} CH_2{=}C(CH_3){-}C({=}O)OCH_3$	mass polymerisation with a peroxide catalyst
$CH_3COOCH{=}CH_2$	ethenyl ethanoate or vinyl acetate	$CH{\equiv}CH + CH_3COOH \xrightarrow[160\ °C\ (433\ K)]{(CH_3COO)_2Zn} CH_3COOCH{=}CH_2$	emulsion polymerisation with a peroxide catalyst

POLYMER				
Name	**Trade names**	**Outstanding advantages**	**Outstanding disadvantages**	**Main uses**
low density poly(ethene), polyethylene or polythene		cheap and tough, electrically insulating	low melting temperature, partially opaque	squeezy bottles, plastic bags, hosepipes
high density poly(ethene), polyethylene or polythene	Alkathene	higher melting and more transparent than low density poly(ethene)		buckets, washing-up bowls, food boxes
poly(propene) or polypropylene	Propathene Ulstron (fibre)	higher melting and tougher than poly(ethene); no odour		crates, food containers; as fibre, for ropes and blankets
poly(chloroethene) or polyvinyl chloride; PVC		tough and weather resistant; non-flammable		unplasticised, for pipes and gutters: plasticised, for raincoats, upholstery, records, floor tiles
poly(phenylethene) or polystyrene		ideal for injection mouldings	brittle, flammable	toys, pens and other small articles: expanded polystyrene for packaging and ceiling tiles
poly(propenenitrile) or polyacrylonitrile	Acrilan Orlon Courtelle	as fibre, resembles wool		carpets, blankets, sweaters
poly(tetrafluoroethene); PTFE	Fluon Teflon	chemically inert, low coefficient of friction	expensive	non-stick utensils, sleeves for ground glass joints
poly(methyl 2-methyl-propenoate) or polymethyl methacrylate	Perspex Plexiglas	glass-like and tough		baths, aircraft windows, car rear lights
poly(ethenyl ethanoate) or polyvinyl acetate; PVA			low melting temperature	as an emulsion, for emulsion paints and adhesives

The chain length of the product depends on the conditions of polymerisation, notably temperature, pressure, catalyst, and the presence or absence of light. Perhaps the most important factor is the nature of the catalyst. Ethene, for instance, when polymerised in the presence of oxygen, will give poly(ethene) only at pressures above 1 000 atm (100 000 kPa), but with a *Ziegler catalyst*, e.g. a pentyllithium-titanium(IV) chloride complex, the reaction may be conducted at atmospheric pressure.

Addition polymerisation proceeds by a chain reaction which may be propagated by either free radicals or carbonium ions depending on the type of catalyst used. In the presence of a peroxide catalyst, such as dibenzoyl peroxide, $C_6H_5CO{-}O{-}O{-}COC_6H_5$, it occurs by the former route.

All addition polymers are *thermoplastic*, i.e. they soften and become more plastic on heating. Eventually they melt, but return to their original form on cooling. Their plasticity is due to the ability of the long chain molecules to slide over one another; there is little or no *cross-linking* from one chain to another.

The commonest plastics made from single monomers are listed in Table 14.3. Many modern plastics, however, are 'tailor made' for particular purposes by *copolymerising* a mixture of two or more monomers.

14.4.5 CHEMICALS FROM PETROLEUM

The petrochemical industry is heavily dependent on the lower alkenes. These compounds do not occur in crude oil, but are made by the cracking of petroleum fractions (§14.3.5). Alkane molecules are most likely to crack near one end of the hydrocarbon chain, so that the principal alkene is ethene, of which approximately one million tonnes is produced each year in the UK. Next in commercial importance is propene, with an annual production of about 600 000 tonnes, and finally the butenes, i.e. 1-butene, 2-butene and 2-methylpropene.

Ethene

The output is used in approximately the following proportions:

430 000 tonnes – poly(ethene)
200 000 tonnes – chloroethene (vinyl chloride)
160 000 tonnes – 1,2-ethanediol (ethylene glycol)
100 000 tonnes – phenylethene (styrene)
80 000 tonnes – ethanol
30 000 tonnes – other uses

For the manufacture of poly(ethene) see Table 14.3.

Chloroethene

First, 1,2-dichloroethane is produced by reacting gaseous ethene and chlorine together at a temperature of 90–130 °C (363–403 K):

$$CH_2{=}CH_2 + Cl_2 = CH_2ClCH_2Cl$$

At a higher temperature (500 °C; 773 K) this compound undergoes dehydrochlorination to give chloroethene:

$$CH_2ClCH_2Cl \xrightarrow[\text{3 atm (300 kPa)}]{\text{500 °C (773 K)}} CH_2{=}CHCl + HCl$$
over pumice or kaolin

Chloroethene, used entirely for making poly(chloroethene), is stored in liquid form under pressure.

1,2-Ethanediol

Ethene cannot be converted directly to 1,2-ethanediol in good yield, and is produced commercially via oxirane (epoxyethane).This compound is made by oxidising ethene with a large excess of oxygen at a temperature of 230–290 °C (503–563 K), a pressure of 10–30 atm (1 000–3 000 kPa) and in the presence of a silver catalyst:

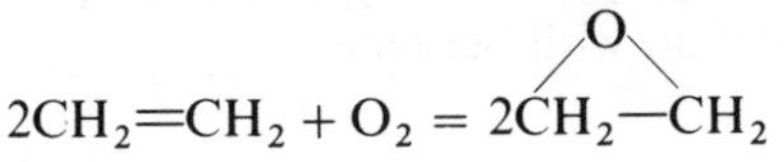

$$2CH_2{=}CH_2 + O_2 = 2\,\overset{O}{CH_2{-}CH_2}$$
oxirane

A certain amount of oxirane is used as a fumigant for preserving food and tobacco, but most is used as a chemical intermediate, especially in the manufacture of 1,2-ethanediol. The hydration is accomplished by heating with steam under pressure:

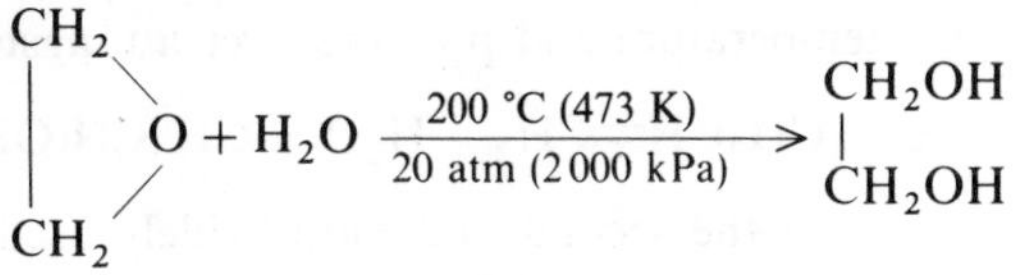

$$(CH_2)_2O + H_2O \xrightarrow[\text{20 atm (2000 kPa)}]{\text{200 °C (473 K)}} CH_2OH{-}CH_2OH$$

1,2-Ethanediol is used as an antifreeze, a hydraulic fluid, and also in the production of polyesters such as Terylene.

Phenylethene

In the presence of aluminium chloride, benzene and ethene undergo a Friedel–Crafts reaction to give ethylbenzene:

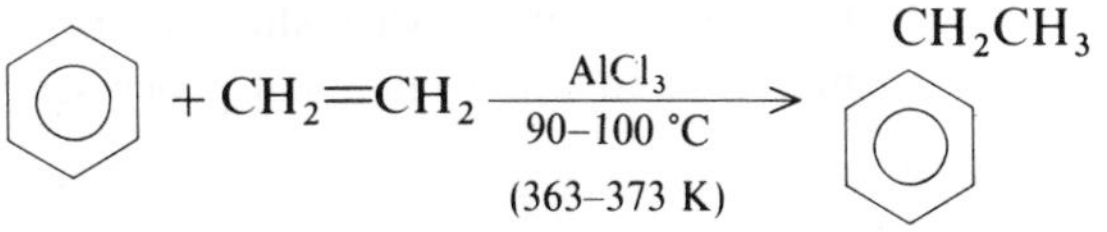

$$C_6H_6 + CH_2{=}CH_2 \xrightarrow[\text{90–100 °C (363–373 K)}]{AlCl_3} C_6H_5CH_2CH_3$$

The catalytic dehydrogenation of ethylbenzene at a high temperature gives phenylethene:

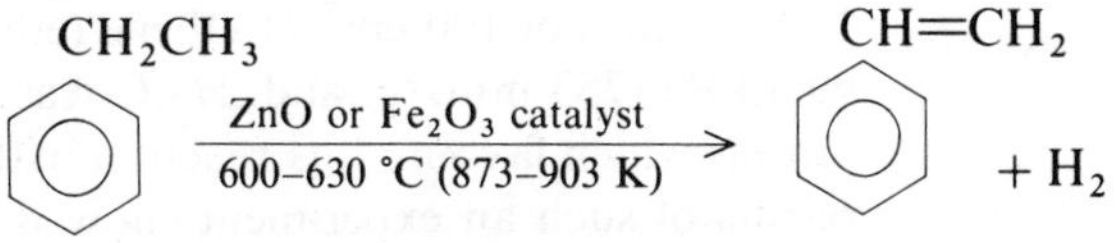

$$C_6H_5CH_2CH_3 \xrightarrow[\text{600–630 °C (873–903 K)}]{ZnO \text{ or } Fe_2O_3 \text{ catalyst}} C_6H_5CH{=}CH_2 + H_2$$

Phenylethene is used for making poly(phenylethene) and various co-polymers.

Ethanol

Although alcoholic beverages are still made by fermentation, industrial alcohol is manufactured by the acid catalysed hydration of ethene. First, ethene is heated with steam to a temperature of 300 °C (573 K). The mixture is then compressed to 70 atm (7 000 kPa) and passed into a reactor containing a catalyst of phosphoric acid absorbed on diatomaceous earth. Partial hydration occurs, and dilute aqueous ethanol is condensed from the gases leaving the reactor.

$$CH_2{=}CH_2 + H_2O = CH_3CH_2OH$$

The product is concentrated by fractional distillation (§8.4.3) to give 'rectified spirit', which is an azeotrope, i.e. a constant boiling mixture, containing approximately 96% ethanol and 4% water. 'Industrial methylated spirit' consists of 95% rectified spirit and 5% methanol, while 'mineralised methylated spirit', sold by pharmacists, contains 90% rectified spirit, 9% methanol, 1% pyridine and a trace of purple dye. 'Absolute alcohol' (99.5% ethanol) is rectified spirit that has been dried by azeotropic distillation with benzene.

Propene

Propene is used principally for the manufacture of poly(propene) (Table 14.3) and 2-propanol. The latter, known commercially as 'isopropanol' or 'IPA', is made from propene by both direct and indirect hydration. In the former process a mixture of propene and steam is passed at a high temperature and pressure over a tungsten oxide catalyst:

$$CH_3CH{=}CH_2 + H_2O = CH_3CH(OH)CH_3$$

In the second and more widely used process propene is absorbed in concentrated sulphuric acid at a temperature no greater than 40 °C (313 K) to give isopropyl hydrogen sulphate which is then hydrolysed by hot water:

$$CH_3CH{=}CH_2 + H_2SO_4 = (CH_3)_2CHO{-}SO_2{-}OH$$

$$(CH_3)_2CHO{-}SO_2{-}OH + H_2O = (CH_3)_2CHOH + H_2SO_4$$

Both ethanol and 2-propanol are widely used as solvents in the manufacture of lacquers, varnishes, printing inks, adhesives, etc, and as starting materials in other chemical processes.

EXAMINATION QUESTIONS ON SECTION 14.4

1 A mixture of 100 cm^3 of ethene (ethylene) and ethane at a pressure of 99.6 kPa (750 mmHg) and 27 °C was treated with bromine, under conditions which favoured its reaction with alkene, but not with alkane. The results of such an experiment showed that 0.285 g of bromine has been used up.

(a) State **two** conditions necessary for the above reaction to occur.

(b) Give the name and structural formula of the product of this reaction.

(c) Give the name and structure of an isomer of the product referred to in (b), and outline **one** method for its preparation.

(d) Calculate the percentage composition by volume of the gaseous mixture.

(e) Write the equations, state the necessary conditions and name the intermediate compounds formed when ethene is converted into (i) butane, (ii) ethanal (acetaldehyde), (iii) a named dicarboxylic acid. (AEB)

2 Alkenes have the general formula $\mathrm{(R)(R')C{=}C(R'')(R''')}$ where R, R′, R″ and R‴ may be either alkyl or aryl groups or hydrogen atoms. The double bond may be cleaved by treatment with ozone, followed by hydrolysis in reducing conditions to give two carbonyl compounds, according to the following equations:

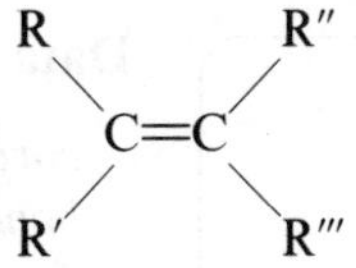

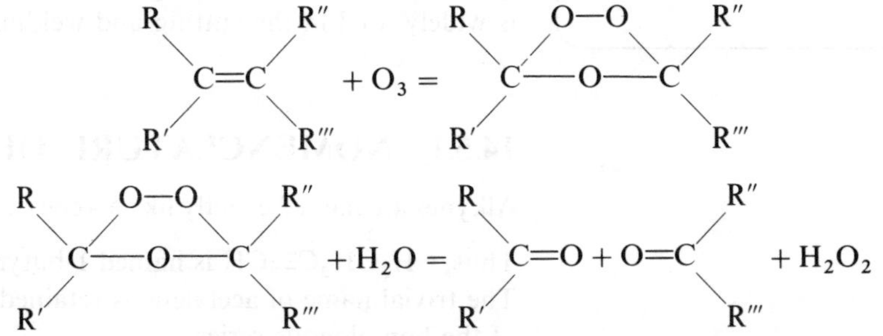

$$\mathrm{(R)(R')C{=}C(R'')(R''') + O_3 = (R)(R')C{-}O{-}C(R'')(R''')} \text{ (with O–O bridging the two C atoms)}$$

$$\mathrm{(R)(R')C{-}O{-}C(R'')(R''')} \text{ (O–O bridged)} \mathrm{ + H_2O = (R)(R')C{=}O + O{=}C(R'')(R''') + H_2O_2}$$

P is an alkene which, on cleavage with ozone followed by hydrolysis in reducing conditions, gave an aldehyde *Q* and a ketone *R*, both of which underwent the triodomethane (iodoform) reaction. Reduction of *R* gave an alcohol which could be resolved into two optically active isomers *S* and *T*. Complete combustion of one mole of *R* required 5.5 moles of oxygen.

(a) Explain why, in the hydrolysis of the ozonide, reducing conditions are necessary.

(b) Identify the compounds *P*, *Q*, *R*, *S* and *T*, explaining fully your reasoning. Give the structures of *S* and *T* and explain in what ways they are different.

(c) An isomer of *P* was found to undergo the same series of reactions to form the same products. Give the structures of *P* and its isomer and explain their existence.

(d) Suggest a method which may be used for adding HOH across the double bond in *P* (or its isomer) and give the structure of the product. (L)

3 (a) Write an equation for the principal reaction which takes place when propene is bubbled into 80% sulphuric acid at room temperature and give the structural formula of the organic product.

(b) Outline the mechanism of the reaction leading to the formation of this product.

(c) How does this mechanism explain the formation of one product rather than another when sulphuric acid reacts with propene?

(d) Write an equation for the reaction which occurs when the product of the reaction between sulphuric acid and propene is added to water and the mixture warmed.

(e) What use is made industrially of reactions similar to those in (a) and (d)? (JMB)

14.5

The C≡C bond in alkynes

Data on the C≡C bond

covalent bond length/nm 0.120
bond dissociation enthalpy/kJ mol^{-1} at 298 K 837

The carbon–carbon triple bond comprises a σ bond and two interacting π bonds (§4.3.2), and is characterised by its great strength. Its bond dissociation enthalpy, shown above, is exceeded only by that of C≡N (890 kJ mol^{-1}) and N≡N (944 kJ mol^{-1}). The only common alkyne is ethyne or 'acetylene', CH≡CH, a gas which burns with a hot flame and which is widely used in the cutting and welding of metals.

14.5.1 NOMENCLATURE OF ALKYNES

Alkynes are named exactly like alkenes, except that the name ends in '-yne'. Thus, $\overset{4}{C}H_3\overset{3}{C}H_2\overset{2}{C}{\equiv}\overset{1}{C}H$ is named 1-butyne (but-1-yne in the ASE system). The trivial name of acetylene is retained by IUPAC for the first member of the homologous series.

14.5.2 FORMATION OF THE C≡C BOND

Elimination methods

The C≡C bond is usually formed by the elimination of two molecules of hydrogen halide from a dihalide,

e.g. $[\underset{\text{ethene}}{CH_2{=}CH_2} \xrightarrow{Br_2}] \underset{\text{1,2-dibromoethane}}{CH_2BrCH_2Br} \xrightarrow[-HBr]{\text{alcoholic KOH}} \underset{\text{bromoethene}}{CH_2{=}CHBr}$

$\xrightarrow[-HBr]{\text{alcoholic KOH}} \underset{\text{ethyne}}{CH{\equiv}CH}$

$[\underset{\text{ethanal}}{CH_3CHO} \xrightarrow{PCl_5}] \underset{\text{1,1-dichloroethane}}{CH_3CHCl_2} \xrightarrow[-HCl]{\text{alcoholic KOH}} \underset{\text{chloroethene}}{CH_2{=}CHCl}$

$\xrightarrow[-HCl]{\text{alcoholic KOH}} \underset{\text{ethyne}}{CH{\equiv}CH}$

Hydrolysis of carbides

The lower alkynes are conveniently prepared by the hydrolysis, with cold water, of ionic metallic carbides:

$$CaC_2 + 2H_2O = CH{\equiv}CH + Ca(OH)_2$$

$$Mg_2C_3 + 4H_2O = \underset{\text{propyne}}{CH_3C{\equiv}CH} + 2Mg(OH)_2$$

Aluminium carbide, however, hydrolyses to give methane.

All alkynes

14.5.3 GENERAL PROPERTIES OF THE ALKYNES

In all their physical properties, the alkynes closely resemble the alkanes and alkenes. Thus, the lower members exist as gases and liquids. Their solubility in water is low, as a result of non-polarity, and they burn readily in air with a smoky yellow flame.

14.5.4 CHEMICAL PROPERTIES OF THE C≡C BOND

The C≡C bond, like the C=C bond, undergoes addition reactions, although often by a different mechanism. Nearly all reagents that add across a C=C bond, and certain others besides, can add to a C≡C bond:

$$-C{\equiv}C- + XY = -\overset{X}{\overset{|}{C}}{=}\overset{Y}{\overset{|}{C}}-$$

Markownikoff's rule (§14.4.4) is obeyed when protic acids attack an unsymmetrical alkyne.

If the reagent XY can add to a C=C bond, the addition continues:

$$-\overset{X}{\overset{|}{C}}{=}\overset{Y}{\overset{|}{C}}- + XY = X-\overset{X}{\overset{|}{C}}-\overset{Y}{\overset{|}{C}}-Y$$

Addition can usually be arrested at the end of the first stage.

Electrophilic addition

Many electrophilic reagents, e.g. H_2SO_4 and HCl, which attack a C=C bond, do not attack a C≡C bond except in the presence of catalysts, usually salts of copper, mercury or nickel. These function by providing cations which are able to disrupt the cylinder of π bonding, with the formation of π-complexes,

e.g. $$\overset{Hg^{2+}}{\overset{\uparrow}{CH{\equiv}CH}}$$

$H_2SO_4 + H_2O$

The hydration of alkynes differs from that of alkenes, both in the conditions of the reaction and the chemical nature of the product.

When ethyne is treated with dilute sulphuric acid at 60 °C (333 K) in the presence of a little mercury(II) sulphate, it hydrates rapidly to give ethanal. The reaction, which at one time was used for the manufacture of ethanal, proceeds via an unstable *enol*, i.e. an unsaturated alcohol with a $-\overset{|}{C}=\overset{\overset{OH}{|}}{C}-$ grouping.

$$\underset{\text{ethyne}}{CH{\equiv}CH} + H_2O \xrightarrow[\text{catalyst}]{HgSO_4 + H_2SO_4} \underset{\substack{\text{ethenol}\\ \text{(vinyl alcohol)}}}{CH_2{=}CHOH} \xrightarrow{\text{rearrangement}} \underset{\text{ethanal}}{CH_3C(=O)H}$$

Hydrogen halides

In their behaviour towards alkynes, hydrogen halides show their usual order of reactivity, which is in accordance with their acid strengths:

$HI > HBr > HCl > HF$

Hydriodic acid reacts readily with ethyne, especially in the presence of a catalyst, to give 1,1-diiodoethane:

$$CH{\equiv}CH \xrightarrow{HI} CH_2{=}CHI \xrightarrow[\text{(Markownikoff's rule)}]{HI} CH_3CHI_2$$

Hydrobromic acid acts in a similar fashion, but not hydrochloric acid, for the addition of HCl to a C=C bond is difficult (§14.4.4).

$$CH{\equiv}CH + \underset{\text{dilute}}{HCl(aq)} \xrightarrow[65\ °C\ (338\ K)]{HgCl_2} CH_2{=}CHCl$$

This reaction is adapted for the manufacture of chloroethene (vinyl chloride) for the plastics industry:

$$CH{\equiv}CH + HCl(g) \xrightarrow[150\ °C\ (423\ K)]{HgCl_2 \text{ on charcoal}} CH_2{=}CHCl$$

Gaseous hydrogen fluoride reacts only under pressure, but liquid anhydrous hydrogen fluoride is more reactive and will give 1,1-difluoroethane:

$$CH{\equiv}CH \xrightarrow{HF} CH_2{=}CHF \xrightarrow{HF} CH_3CHF_2$$

Chlorine

In the presence of a 'halogen carrier' catalyst, such as aluminium chloride or iron(III) chloride, electrophilic addition of chlorine occurs readily but without explosion. Ethyne reacts to give the tetrachloride, although it is possible to isolate the dichloride:

$$CH{\equiv}CH \xrightarrow{Cl_2} \underset{\text{1,2-dichloroethene}}{CHCl{=}CHCl} \xrightarrow{Cl_2} \underset{\text{1,1,2,2-tetrachloroethane}}{CHCl_2CHCl_2}$$

Chlorine water

A molecule of ethyne is able to accept two molecules of hypochlorous acid.

Subsequent loss of water leads to the formation of a dichloroaldehyde:

$$CH{\equiv}CH \xrightarrow{2HClO} CHCl_2CH(OH)_2 \xrightarrow{-H_2O} CHCl_2C\begin{matrix} {/\!/}O \\ \backslash H \end{matrix}$$

dichloroethanal

Polymerisation

Alkynes will polymerise. Any resemblance to the polymerisation of alkenes is no more than superficial, for alkynes do not respond to peroxide catalysts or even Ziegler catalysts. Instead, they polymerise in the presence of the metal salts that catalyse so many of their other reactions. Long chain polymers are not obtained. Ethyne, for example, dimerises in the presence of ammoniacal copper(I) chloride to give 1-buten-3-yne (vinylacetylene):

$$2CH{\equiv}CH \xrightarrow{[Cu(NH_3)_2]^+} \overset{4}{C}H{\equiv}\overset{3}{C}\overset{2}{C}H{=}\overset{1}{C}H_2$$

Radical addition

Halogens

Chlorine and bromine will add to a C≡C bond in the absence of a catalyst. The reactions are accelerated by light, and recent work has shown that they proceed by a free radical mechanism. (Ultraviolet light favours the formation of free radicals (§14.7.3).) The direct chlorination of ethyne is dangerous, for the reaction mixture may explode with the formation of carbon and hydrogen chloride.

Polymerisation

Benzene is formed when ethyne is passed through a red hot tube:

$$3CH{\equiv}CH = C_6H_6 \text{ (benzene ring)}$$

The yield of benzene is low, and the reaction cannot be regarded as a useful method of preparation.

Hydrogenation

The hydrogenation of alkynes to give, first, alkenes and then alkanes has been discussed elsewhere (§14.4.2).

EXAMINATION QUESTION ON SECTION 14.5

1 (a) (i) Draw a well-labelled diagram of the apparatus you would use to prepare a purified specimen of ethyne (acetylene) in the laboratory.

(ii) Write an equation or equations for the reaction involved.

(iii) Name **one** impurity which contaminates the ethyne and explain with an equation how this impurity is removed from the ethyne.

(b) How do ethene and ethyne compare and contrast in their reactions with:

(i) hydrogen,
(ii) gaseous hydrogen bromide followed by aqueous sodium hydroxide,
(iii) an aqueous solution containing diamminosilver(I) ions, $(Ag(NH_3)_2^+(aq))$?
Give equations and essential conditions and name the products for the reactions involved. (SUJB)

14.6
The C⋯C bond in arenes

C⋯C is the symbol used to represent the delocalised carbon–carbon double bond in the benzene ring.

Data on the C⋯C bond

covalent bond length/nm 0.139
bond dissociation enthalpy/kJ mol^{-1} at 298 K 518

Although this bond is most simply studied in *arenes*, i.e. aromatic hydrocarbons such as benzene and methylbenzene, it exists in all aromatic compounds.

Aromatic compounds

Early in the last century, certain naturally occurring substances were discovered that were fundamentally different from aliphatic compounds, in that they contained at least six carbon atoms and could not be degraded readily to compounds with less than six carbon atoms. Such compounds were often fragrant (e.g. benzaldehyde – 'oil of almonds') and were called *aromatic compounds.*

The six carbon atoms forming the basis of the structure are in the form of a regular, planar hexagon known as a *benzene ring*, after the simplest aromatic compound. Aromatic compounds therefore comprise benzene and other compounds with at least one benzene ring. However, the special characteristics of benzenoid aromatic compounds are shared by certain other ring compounds possessing a set of six π electrons and these compounds, too, may also be classed as 'aromatic'. A well-known example is the heterocyclic base, pyridine.

The structure of benzene has already been discussed in detail (§4.4.1). It is important to remember that the ring does not possess three double bonds in definite positions as proposed by Kekulé. The π electrons of the double bonds are not concentrated in three particular localities, but are *delocalised*, i.e. uniformly distributed around the ring. All six carbon–carbon bonds of the ring are identical, as indicated by X-ray diffraction measurements of covalent bond length. This dispersal of electrons leads to great stability; see the calculation of stabilisation energy in §4.4.1. The benzene ring, once formed, is difficult to disrupt, even at high temperatures or by powerful oxidising agents. Indeed, aromatic compounds are often formed at high temperatures. Because of the reduced concentration of electrons, the C⋯C bond is less susceptible than the alkenic C=C bond to addition reactions.

14.6.1 NOMENCLATURE OF ARENES

Homologues of benzene are named systematically as alkylbenzenes, although a few have trivial names which are still recognised by IUPAC,

e.g. CH_3 C_2H_5 CH_3CHCH_3

methylbenzene (toluene) ethylbenzene isopropylbenzene (cumene)

The common aromatic radicals are named as follows:

$C_6H_5—$ phenyl

$C_6H_5—\overset{\alpha}{C}H_2—$ benzyl

Note that a carbon atom attached to a benzene ring is referred to as an α-carbon atom.

The naming of substituted arenes is discussed elsewhere in this book.

14.6.2 FORMATION OF THE C$\overset{...}{—}$C BOND

The benzene ring is formed at a high temperature by the dehydrogenation of *alicyclic compounds*, i.e. cyclic aliphatic compounds. Simple examples are the dehydrogenation of cyclohexane and its derivatives:

$$\text{cyclohexane} \xrightarrow{\text{heat with S or Se}} \text{benzene} + H_2S \quad \text{or} \quad H_2Se$$

The tendency of a stable aromatic ring to be formed is so strong that alicyclic compounds other than those with a six membered ring can be used,

e.g. C_2H_5 (ethylcyclopentane) $\xrightarrow[\text{(ring expansion)}]{\text{heat with S or Se}}$ CH_3 (methylbenzene)

ethylcyclopentane

Acyclic (open-chain) alkanes may also be used, for at high temperatures and in the presence of suitable catalysts they undergo cyclisation to alicyclic hydrocarbons and then dehydrogenation to arenes. This principle is applied in the manufacture of arenes in the petrochemical industry.

C_6H_6

$C_6H_5CH_3$

14.6.3 GENERAL PROPERTIES OF THE ARENES

All the common monocyclic arenes are colourless, mild smelling liquids. They are non-polar and almost completely immiscible with water. They are volatile (benzene boils at 80.1 °C (353.3 K), methylbenzene at 111 °C (384 K) and the dimethylbenzenes at about 140 °C (413 K)), and give rise to flammable vapours which burn with a yellow, sooty flame.

These compounds are major health hazards. Exposure to a high concentration of vapour leads to unconsciousness, damage to the central nervous system, and possibly death. Continual exposure to low concentrations is also risky. Benzene is particularly dangerous, and has a reputation for causing anaemia and leukaemia.

14.6.4 CHEMICAL PROPERTIES OF THE C┈C BOND

The delocalised carbon–carbon double bond is far less reactive than the alkenic double bond. Indeed, the principal reactions of the benzene ring are electrophilic *substitution* reactions of the C—H bond (§14.7.4).

Electrophilic addition reactions to the double bond, so characteristic of the alkenes, are almost totally absent. There is no reaction with hydrogen halides, hypohalous acids or sulphuric acid. (Concentrated sulphuric acid does attack the benzene ring, but by electrophilic substitution of the C—H bond; §14.7.4.) Ozonolysis, however, occurs with difficulty to give ethanedial.

With chlorine or bromine a free radical addition occurs in the presence of sunlight or some other source of ultraviolet radiation,

e.g. $C_6H_6 + 3Cl_2 = C_6H_6Cl_6$

1,2,3,4,5,6-hexachlorocyclohexane
(benzene hexachloride)

The chain reaction is initiated by the homolytic fission of a chlorine molecule (§14.7.3):

$$Cl_2 \xrightarrow{\text{ultraviolet light}} 2Cl\cdot$$

The benzene ring is then attacked in two stages:

$$C_6H_6 + Cl\cdot \longrightarrow C_6H_6Cl\cdot$$

$$C_6H_6Cl\cdot + Cl_2 \longrightarrow C_6H_6Cl_2 + Cl\cdot$$

It can be seen that the second of these steps releases a chlorine free radical that can propagate the chain. Addition continues until six atoms of chlorine have been introduced, after which the chain reaction is terminated in the usual way by the coupling of radicals.

In the presence of a halogen carrier, chlorine and bromine attack the C—H bond of a benzene ring by electrophilic substitution (§14.7.4). There is very little reaction between benzene and iodine.

The hydrogenation of benzene occurs much less readily than that of the alkenes. With Adam's catalyst (PtO_2, reducing with hydrogen to Pt), at 25 °C (298 K), a reaction time of 25 hours is required:

$$C_6H_6 + 3H_2 \xrightarrow{\text{finely divided Pt}} \text{cyclohexane}$$

With nickel as the catalyst, a temperature of 200 °C (473 K) and a pressure of 200 atm (20 000 kPa) is used.

14.7 The C—H bond in alkanes, arenes and C—H acidic compounds

Data on the C—H bond

The C—H covalent bond has a length of 0.109 nm, and a bond dissociation enthalpy that varies considerably with the location of the bond (Table 14.4). The average value is approximately 412 kJ mol^{-1} at 298 K.

With the exception of fully halogenated compounds, nearly all organic compounds possess the C—H bond. The formation and properties of the bond depend very much on its location, and in particular whether it forms part of an aliphatic or aromatic system.

14.7.1 FORMATION OF THE C—H BOND

Both substitution and addition reactions may be used.

Substitution methods

From C—X (where X = halogen); ***the substitution of X by H*** (§14.8.4)
The halogen atom of an aliphatic halide or arylalkyl halide may be substituted by a hydrogen atom with a suitable reducing agent, e.g. metal + protic solvent, or lithium tetrahydridoaluminate(III).

$$\underset{\text{alkyl bromide}}{RBr} \xrightarrow{Zn + HCl} \underset{\text{alkane}}{RH}$$

$$\underset{\text{(chloromethyl)benzene}}{C_6H_5CH_2Cl} \longrightarrow \underset{\text{methylbenzene}}{C_6H_5CH_3}$$

The corresponding reduction of an aryl halide is complicated by the non-lability of a halogen atom when it is attached directly to a benzene ring, and by the fact that drastic reducing conditions could result in the hydrogenation of the benzene ring. Good results may be achieved by the use of a nickel–aluminium alloy in the presence of alkali,

e.g $$\underset{\text{chlorobenzene}}{C_6H_5Cl} \xrightarrow[NaOH]{Ni\text{–}Al} \underset{\text{benzene}}{C_6H_6}$$

From C=O; the substitution of O by 2H

The conversion of $>C=O$ (the carbonyl group of an aldehyde or ketone) to $>CH_2$ requires a powerful reductant; otherwise an alcohol is formed instead $\left(>C=O \rightarrow >C(H)(OH) \right)$. Perhaps the best known reagent is zinc amalgam and boiling concentrated hydrochloric acid: this is the *Clemmensen reduction.* The mechanism is not well understood. The only certain fact is that the reduction does not proceed via an alcohol, for alcohols cannot be reduced to alkanes by means of this reagent. For example,

$$\left.\begin{array}{l} CH_3CH_2CHO \\ \text{propanal} \\ \text{or} \quad CH_3COCH_3 \\ \text{propanone} \end{array}\right\} \xrightarrow{Zn/Hg + \text{conc. } HCl} \underset{\text{propane}}{CH_3CH_2CH_3}$$

An alternative reducing agent for converting $>C=O$ to $>CH_2$ is a mixture of concentrated hydriodic acid and red phosphorus.

From C—OH; the substitution of OH by H

This is a difficult substitution to achieve in one stage. The usual reductants of the type metal + protic solvent are ineffective, but phenols (aromatic hydroxy compounds in which the HO group is attached directly to the benzene ring) may be successfully reduced by distilling them in the dry state with zinc dust,

$$\text{e.g.} \quad \underset{\text{phenol}}{C_6H_5OH} + Zn = \underset{\text{benzene}}{C_6H_6} + ZnO$$

Similar treatment applied to alcohols would result in cracking. Alcohols are best reduced by means of concentrated hydriodic acid and a little red phosphorus at 150 °C (423 K):

$$ROH \xrightarrow[\text{substitution}]{HI} RI \xrightarrow[\text{reduction}]{HI} RH$$

From C—NH₂; the substitution of NH₂ by H

Direct reduction is impossible. However, aromatic primary amines can be *diazotised* with sodium nitrite and hydrochloric acid, and the resulting diazonium salt reduced with, for example, phosphinic acid:

$$\underset{\substack{\text{phenylamine} \\ \text{(aniline)}}}{C_6H_5NH_2} \xrightarrow{NaNO_2 + HCl} \underset{\substack{\text{benzenediazonium} \\ \text{chloride}}}{C_6H_5\overset{\oplus}{N}{\equiv}N\,Cl^-} \xrightarrow{HPH_2O_2} C_6H_6 + N_2(g) + HCl$$

Addition methods

The hydrogenation of C=C and C≡C bonds, by hydrogen in the presence of a nickel, palladium or platinum catalyst, leads to the formation of C—H bonds (§14.4.4).

14.7.2 CHEMICAL PROPERTIES OF THE C—H BOND

The conditions under which a C—H bond reacts, and the mechanisms of the reactions, vary enormously with the location of the bond. The electronegativity difference between carbon and hydrogen is only 0.4 (C = 2.5, H = 2.1 on the Pauling scale), which means that although there is a certain amount of charge separation, $\overset{\delta^-}{C}$—$\overset{\delta^+}{H}$, in the C—H bond it is normally insufficient to permit electrophilic attack at the carbon atom.

With alkanes, and the alkyl groups of alkylaromatic hydrocarbons such as methylbenzene, substitution proceeds by a free radical mechanism, whereby ultraviolet light or visible light at the blue end of the spectrum is used to split the reagent into free radicals which are highly reactive and capable of attacking the hydrocarbon. Ultraviolet light is potentially capable of breaking many covalent bonds and bringing about many reactions.

However, there are two special cases in which the hydrogen atom is particularly susceptible to substitution. One is where the hydrogen atom is directly bonded to a benzene ring. Aromatic compounds, because of their delocalised π electrons (§4.4.1), behave quite differently from alkenes. Electrophilic addition does not occur readily, as this would destroy the aromatic ring with its stable π bonding. Instead, the principal reactions are electrophilic substitutions.

The other special case is C—H acidic compounds. The hydrogen atom of a C—H bond is not normally acidic, but the presence of certain groupings, notably C=O and C≡C, can cause acidity. Aldehydes, ketones and certain alkynes show this effect and are therefore known as *C—H acidic compounds.*

C—H bonds also participate in elimination reactions, namely dehydrohalogenation (§14.8.4), dehydration (§14.9.5) and dehydrogenation (§14.9.5).

14.7.3 THE C—H BOND IN ALKANES AND ALKYLAROMATIC HYDROCARBONS

Halogenation

Generally, alkanes react rapidly with halogens either in ultraviolet light or on heating to give a mixture of compounds. Methane and chlorine, for example, give a mixture of mono-, di-, tri- and tetrachloromethanes:

$$CH_4 + Cl_2 = CH_3Cl + HCl$$
$$CH_3Cl + Cl_2 = CH_2Cl_2 + HCl$$
$$CH_2Cl_2 + Cl_2 = CHCl_3 + HCl$$
$$CHCl_3 + Cl_2 = CCl_4 + HCl$$

It is impossible to stop the reaction at any of the intermediate stages, although the composition of the product can be controlled to some extent by adjusting the methane:chlorine ratio.

From ethane a mixture of nine chlorides is obtained, and the higher alkanes give even more complex mixtures. Their separation on anything but a small scale by chromatographic techniques is virtually impossible, and the reactions are thus of little use.

Alkylaromatic hydrocarbons undergo side chain halogenation at the α-position in essentially the same manner as alkanes, except that the substitution is *stepwise* and can be stopped at any of the intermediate stages. Methylbenzene, for example, can be chlorinated at its boiling temperature in the presence of ultraviolet light to give three chlorides:

$$C_6H_5CH_3 + Cl_2 = C_6H_5CH_2Cl + HCl$$

(chloromethyl)benzene

$$C_6H_5CH_2Cl + Cl_2 = C_6H_5CHCl_2 + HCl$$

(dichloromethyl)benzene

$$C_6H_5CHCl_2 + Cl_2 = C_6H_5CCl_3 + HCl$$

(trichloromethyl)benzene

By stopping the reaction when the increase in mass corresponds to that which is calculated, good yields of (chloromethyl)benzene and (dichloromethyl)benzene can be obtained.

The different halogens have widely different reactivities, the order being $F_2 > Cl_2 > Br_2 > I_2$. Fluorine reacts violently to give highly fluorinated products, some of which arise through the cleavage of carbon–carbon bonds. Chlorination can be performed safely and smoothly in the presence of diffused sunlight or at about 400 °C (673 K) in the dark. Bromine is much less reactive than chlorine, both in ultraviolet light and at high temperatures, and is capable of attacking only tertiary hydrogen atoms (see below). Iodine does not react.

Different substrates, too, have different reactivities. 2-Methylpropane (isobutane) and 2-methylbutane (isopentane), for example, are particularly reactive because they possess tertiary hydrogen atoms. Hydrogen atoms are classified as primary, secondary or tertiary, according to whether they are attached to primary, secondary or tertiary carbon atoms. A *primary carbon atom* is one which is attached to *one* alkyl or aryl group, a *secondary carbon atom* is one which is joined to *two*, and a *tertiary carbon atom* one which is joined to *three*, thus:

```
     H                    R                    R
     |                    |                    |
   R—C—H                R—C—H                R—C—H
     |                    |                    |
     H                    H                    R
```

primary carbon atom / primary hydrogen atoms — secondary carbon atom / secondary hydrogen atoms — tertiary carbon atom / tertiary hydrogen atom

Tertiary hydrogen atoms are the most easily substituted. The order is tertiary > secondary > primary, although the reactivities of all three types vary considerably from one type of halogenation to another.

These facts are explained in terms of a free radical chain reaction. Consider the chlorination of methane.

Initiation Two reactions are conceivable:

$$Cl{-}\!:\!{-}Cl \xrightarrow[\text{or heat}]{\text{ultraviolet light}} 2Cl\cdot \qquad (1)$$

$$\Delta H^{\ominus} = +242 \text{ kJ mol}^{-1}$$

$$H_3C{-}\!:\!{-}H \longrightarrow CH_3\cdot + H\cdot \qquad (2)$$

$$\Delta H^{\ominus} = +426 \text{ kJ mol}^{-1}$$

Because of the lower enthalpy requirement, reaction (1) occurs to the total exclusion of (2).

The chlorine molecule is said to undergo *homolytic fission*, or *homolysis*, because the molecule is broken into fragments which are similar to each other in the sense that they are both free radicals (*homo* = same). Homolysis need not necessarily give identical free radicals. Thus,

$$ICl \longrightarrow I\cdot + Cl\cdot$$

is also an example of homolytic fission.

An alternative to homolytic fission is *heterolytic fission*, or *heterolysis* (*hetero* = different), in which the bond breaks in such a way that one atom takes both of the bonding electrons. The fragments are therefore dissimilar, in that they are ions of opposite charge,

e.g. $Cl{-}\!:\!{-}Cl \longrightarrow Cl^+ + Cl^-$

Heterolytic fission of the halogens is not brought about by ultraviolet light, but requires a polarising influence; cf. halogens with alkenes, §14.4.4.

Propagation

Reaction 1 $\qquad Cl\cdot + CH_4 \longrightarrow CH_3\cdot + HCl$

Reaction 2 $\qquad CH_3\cdot + Cl_2 \longrightarrow CH_3Cl + Cl\cdot$

An alternative to reaction 1, namely

$$Cl\cdot + CH_4 \longrightarrow H\cdot + CH_3Cl$$

is precluded on enthalpy grounds, for the standard enthalpy change on forming the H—Cl bond is -431 kJ mol^{-1}, whereas that for the formation of the C—Cl bond is only -338 kJ mol^{-1}.

The chlorine free radical formed in reaction 2 can attack either another molecule of methane, to give more CH_3Cl, or a molecule of CH_3Cl to give CH_2Cl_2: consequently, the product is inevitably a mixture of chlorinated methanes.

Termination By the combination of any available pairs of free radicals:

$$2Cl\cdot \longrightarrow Cl_2$$

$$CH_3\cdot + Cl\cdot \longrightarrow CH_3Cl$$

$$2CH_3\cdot \longrightarrow CH_3CH_3$$

The fact that traces of ethane have been detected in the product provides powerful evidence to support this theory.

The rapidity of the reaction is due to the fact that chlorine free radicals are highly reactive, and that every time one of them is used up another is formed automatically. Bromination, by contrast, is relatively slow, because bromine free radicals are less reactive than chlorine free radicals.

The radical chlorination of methylbenzene resembles that of methane, with propagation represented by the following equations:

$$Cl\cdot + C_6H_5CH_3 \longrightarrow C_6H_5\dot{C}H_2 + HCl$$

benzyl free radical

$$C_6H_5\dot{C}H_2 + Cl_2 \longrightarrow C_6H_5CH_2Cl + Cl\cdot$$

However, the α C—H bonds in (chloromethyl)benzene and (dichloromethyl)-benzene are much less reactive than those in methylbenzene; consequently, the chlorination of (chloromethyl)benzene does not begin until that of methylbenzene is complete. When radical bromination of methylbenzene is performed (tribromomethyl)benzene is not formed.

See §14.7.4 for the chlorination of methylbenzene in the aromatic ring.

The possibility of a free radical chain reaction taking place can be gauged quite accurately by considering the overall standard enthalpy change, $\Delta H^{\ominus}$, i.e. the algebraic sum of the individual standard enthalpy changes. If the reaction as a whole is exothermic ($\Delta H^{\ominus}$ -ve) it will proceed, otherwise not. (For chemical reactions in general, feasibility is dictated by the standard free energy change, $\Delta G^{\ominus}$; $\Delta H^{\ominus}$ provides no more than a rough guide (§2.4.2).)

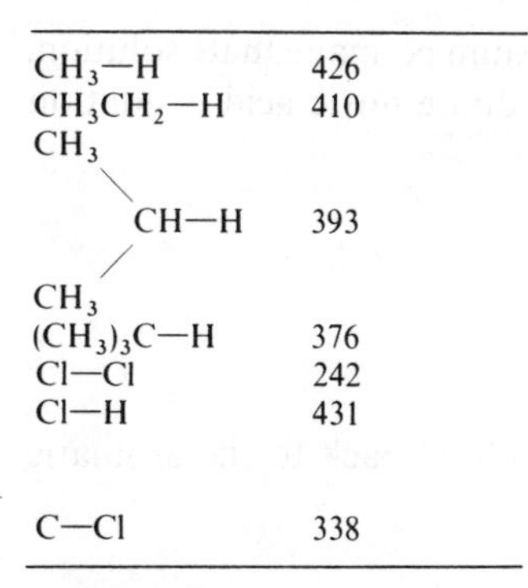

CH_3—H	426
CH_3CH_2—H	410
$(CH_3)_2$CH—H	393
$(CH_3)_3$C—H	376
Cl—Cl	242
Cl—H	431
C—Cl	338

Table 14.4 Bond dissociation enthalpies/kJ mol^{-1}

We can calculate the standard enthalpy change for the conversion of methane to chloromethane, using the bond dissociation enthalpies shown in Table 14.4. To break the Cl—Cl and CH_3—H bonds *requires* $242 + 426$ kJ mol^{-1}, i.e. $\Delta H^{\ominus} = +668$ kJ mol^{-1}. To form the Cl—H and C—Cl bonds *liberates* $431 + 338$ kJ mol^{-1}, i.e. $\Delta H^{\ominus} = -769$ kJ mol^{-1}. The net enthalpy change for the reaction is thus $-769 + 668 = -101$ kJ mol^{-1}. The reaction is exothermic, and capable of proceeding.

Nitration

Nitration means a reaction with nitric acid in which a hydrogen atom becomes substituted by a nitro group, NO_2:

$$RH + HNO_3 = \underset{\text{nitroalkane}}{RNO_2} + H_2O$$

Alkanes may be nitrated either in the liquid phase, with concentrated nitric acid under pressure at 140 °C (413 K), or in the vapour phase at 420 °C (693 K),

$$\text{e.g.}\quad CH_3CH_2CH_3 \xrightarrow[\substack{420\ °C\\(693\ K)}]{HNO_3} \begin{cases} CH_3CH_2CH_2NO_2 & \text{1-nitropropane}\\ CH_3CH(NO_2)CH_3 & \text{2-nitropropane}\\ CH_3CH_2NO_2 & \text{nitroethane}\\ CH_3NO_2 & \text{nitromethane}\end{cases}$$

The nitroethane and nitromethane arise through cracking.

Aliphatic nitration is believed to have a free radical mechanism. Unlike the nitration of benzene, it is not an electrophilic substitution and cannot be brought about by a mixture of concentrated nitric acid and concentrated sulphuric acid.

Oxidation

Alkanes are relatively resistant to attack by oxidising agents, but do respond to acidified permanganate or dichromate solutions on heating. In a free radical chain reaction, an alkane is oxidised to the corresponding alcohol:

$$RH \longrightarrow ROH$$

In practice, the reaction is complicated by the oxidation of the alcohol to give an aldehyde or a ketone. Any aldehyde so formed oxidises to a carboxylic acid. Also, cleavage of C—C bonds occurs to give products of low relative molecular mass, and rearrangement reactions may also occur. The oxidation of alkanes is thus of little value for preparing definite compounds.

Alkyl groups which are attached to a benzene ring are much more susceptible to attack, and the oxidation of alkylaromatic hydrocarbons is of both industrial and laboratory importance. Methyl groups are oxidised according to the general scheme

$$—CH_3 \longrightarrow —CH_2OH \longrightarrow —C(=O)H \longrightarrow —C(=O)OH$$

With a powerful oxidant, e.g. alkaline potassium permanganate solution, or acidified sodium dichromate solution, or dilute nitric acid, oxidation is complete,

$$\text{e.g.}\quad C_6H_5CH_3 \longrightarrow C_6H_5COOH$$

Alkyl groups other than CH_3 are usually oxidised back to the aromatic nucleus:

$$C_6H_5(CH_2)_nCH_3 \xrightarrow{KMnO_4} C_6H_5COOH$$

With certain less powerful oxidants, intermediate oxidation products can be obtained,

$$\text{e.g.}\quad C_6H_5CH_3 \xrightarrow[300\ ^\circ C\ (573\ K)]{SeO_2} C_6H_5CHO \quad \text{benzaldehyde}$$

14.7.4 THE C—H BOND IN THE BENZENE RING

The π electrons of the benzene ring, like those of the C=C bond in alkenes, attract electrophilic reagents: these always possess at least a partial positive charge, and are able to accept an electron pair (from the π bond). However, while an alkene undergoes electrophilic *addition*, an aromatic compound suffers electrophilic *substitution*. The two reactions are not as different as they appear to be, and it is useful to begin by revising the subject of electrophilic addition:

$$\underset{\text{alkenic double bond}}{>C{=}C<} + \underset{\text{electrophile}}{\overset{\delta^+}{X}{-}\overset{\delta^-}{Y}} \longrightarrow \underset{\pi\text{-complex}}{>C{=}C< \cdot \overset{\delta^+}{X}{-}\overset{\delta^-}{Y}} \xrightarrow{\text{slow}} \underset{\text{carbonium ion}}{>C(X){-}\overset{\oplus}{C}<} + Y^- \xrightarrow{\text{fast}} \underset{\text{product}}{>C(X){-}C(Y)<}$$

Electrophilic substitution in aromatics is closely similar to this, the only real difference being that in the final stage of the reaction the anion Y^-, instead of adding to the carbonium ion, abstracts a proton from the carbonium ion so as to reform the highly stable aromatic ring:

$$\underset{\text{benzene}}{C_6H_6} + \underset{\text{electrophile}}{\overset{\delta^+}{X}{-}\overset{\delta^-}{Y}} \longrightarrow \underset{\pi\text{-complex}}{C_6H_6 \cdot \overset{\delta^+}{X}{-}\overset{\delta^-}{Y}} \longrightarrow \underset{\substack{\text{carbonium ion}\\(\sigma\text{-complex})}}{[C_6H_6X]^{\oplus}} + Y^- \xrightarrow{-H^+} \underset{\text{products}}{C_6H_5X + HY}$$

The π-complex is similar to that which is formed with alkenes, in that the electrophile is loosely bound by the acceptance of a pair of π electrons. The carbonium ion, however, is a little different from the carbonium ions that we have encountered before, in that the positive charge is delocalised over five carbon atoms. It is usually called a *σ-complex*, reflecting the fact that X is now joined to a carbon atom by a σ bond. The slowest and hence the rate determining stage of the reaction is usually the formation of the σ-complex (cf. addition to alkenes).

Because of the delocalisation of π electrons, the benzene ring is less susceptible than the alkenic double bond to electrophilic attack. Hydrogen bromide, hydrogen iodide, chlorine water and bromine water – well-known reagents for alkenes – do not attack benzene. With the exception of sulphuric acid, the few electrophiles which do react with the C—H bonds of the benzene ring require the presence of catalysts to increase their *electrophilicity*, i.e. their electrophilic character. Aromatic nitration, for example, is effected by nitric acid in the presence of sulphuric acid, which converts weakly electrophilic HNO_3 molecules into strongly electrophilic NO_2^+ cations. Chlorine and bromine substitute in the benzene ring only in the presence of Lewis acid catalysts (e.g. anhydrous $AlCl_3$), which polarise the halogen molecules and so increase their electrophilicity. In *Friedel–Crafts reactions*, aluminium chloride serves a similar purpose, by increasing the polarisation and hence the electrophilicity of molecules of an alkyl halide or acyl halide. This information is summarised in Table 14.5. It must be emphasised that the reagents and catalysts shown in Table 14.5 attack not only benzene itself, but also the C—H bonds of any other aromatic compound.

Table 14.5 Catalysts for electrophilic substitution in the benzene ring

Process	Reagent	Catalyst	Electrophile
nitration	HNO_3	H_2SO_4	NO_2^+
chlorination	Cl_2	anhydrous $AlCl_3$ or $FeCl_3$	$\overset{\delta+}{Cl}—Cl\cdots\overset{\delta-}{AlCl_3}$†
bromination	Br_2	anhydrous $AlBr_3$ or $FeBr_3$	$\overset{\delta+}{Br}—Br\cdots\overset{\delta-}{AlBr_3}$†
Friedel–Crafts alkylation	RX	anhydrous $AlCl_3$	$\overset{\delta+}{R}—X\cdots\overset{\delta-}{AlCl_3}$†
Friedel–Crafts acylation	RCOX	anhydrous $AlCl_3$	$(R)(X)\overset{\delta+}{C}=O\cdots\overset{\delta-}{AlCl_3}$†
sulphonation	H_2SO_4	none required	SO_3, i.e. $S^{\delta+}$ bonded to $O^{\delta-}$, $^{\delta-}O$, $O^{\delta-}$ †

† These species act as electrophiles because they possess atoms with a partial positive charge.

Some other reagents, notably nitrous acid and iodine, although incapable of attacking benzene, will react with certain aromatic compounds in which the ring is activated by a suitable substituent group, such as HO, NH_2, NHR or NR_2. Such groups possess a lone pair of electrons (on O or N) which can interact with the π electrons of the ring, so increasing the electron density of the ring and its attraction for electrophiles. This is known as a 'mesomeric effect' (§14.8.4).

Nitration

Nitric acid can participate in aromatic substitution reactions because the benzene ring is highly resistant to oxidation. (Any attempt at a similar reaction with an alkene would lead to oxidative cleavage.) The electrophile is the *nitryl cation*, NO_2^+, formed by the protonation of nitric acid:

$$H^+ + H{-}\ddot{O}{-}NO_2 \rightleftharpoons H_2\overset{\oplus}{O}{-}NO_2 \rightleftharpoons H_2O + NO_2^+$$

Nitric acid is capable of self-protonation,

$$2HNO_3 \rightleftharpoons H_2\overset{\oplus}{O}{-}NO_2 + NO_3^-$$
$$\rightleftharpoons H_2O + NO_2^+$$

but the change occurs only to a small extent and nitric acid is thus a weak nitrating agent when used alone.

Nitric acid is best protonated by a strong acid. Concentrated sulphuric acid serves the purpose, and is the usual catalyst for aromatic nitration:

$$HNO_3 + H_2SO_4 \rightleftharpoons H_2NO_3^+ + HSO_4^-$$
$$H_2NO_3^+ + H_2SO_4 \rightleftharpoons H_3O^+ + NO_2^+ + HSO_4^-$$

In accepting a proton, the nitric acid molecule behaves as a base!

Evidence in support of this theory is provided by the depression of freezing temperature that is observed when nitric acid is dissolved in sulphuric acid. Values which are four times those expected on the basis of a covalent structure are found, suggesting that one molecule of nitric acid gives rise to four ions. Salts containing the NO_2^+ ion are known (§15.2.9).

A mixture of concentrated nitric acid and concentrated sulphuric acid, known as 'nitrating acid', converts benzene into nitrobenzene, a yellow oily liquid with a boiling temperature of 211 °C (484 K):

$$C_6H_6 + HNO_3 \xrightarrow[55\ ^\circ C\ (328\ K)]{conc.\ H_2SO_4} C_6H_5NO_2 + H_2O$$

In detail:

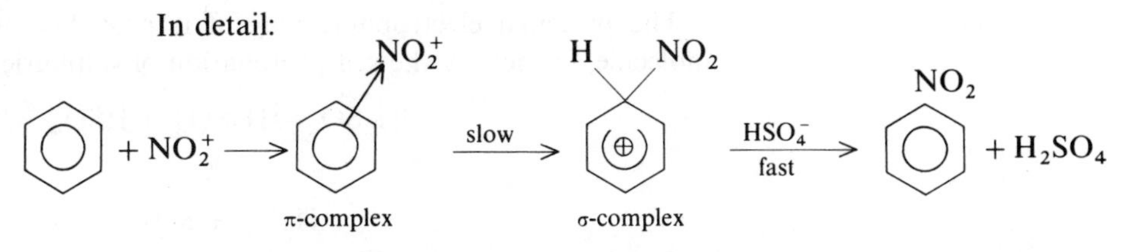

Further nitration may occur to give 1,3-dinitrobenzene, a yellow solid with a melting temperature of 90 °C (363 K). The nitro group already in position is *1,3-directing*, i.e. it directs all other incoming groups into the 3-position. Dinitration is relatively difficult, for NO_2, like all 1,3-directing groups, deactivates the benzene ring by an electron withdrawing effect.

$$C_6H_5NO_2 + \underset{\substack{\text{fuming} \\ \text{(i.e. 95\% } HNO_3)}}{HNO_3} \xrightarrow[100\ ^\circ C\ (373\ K)]{\text{conc. } H_2SO_4} \underset{\text{1,3-dinitrobenzene}}{C_6H_4(NO_2)_2} + H_2O$$

TNT

Only a little 1,3,5-trinitrobenzene is formed.

Methylbenzene is easier to nitrate than benzene, because the methyl group has an electron releasing effect; a *positive inductive effect*, +I. Because the methyl group is 1,2- and 1,4-directing, the ultimate product is methyl-2,4,6-trinitrobenzene, a high explosive commonly called trinitrotoluene (TNT):

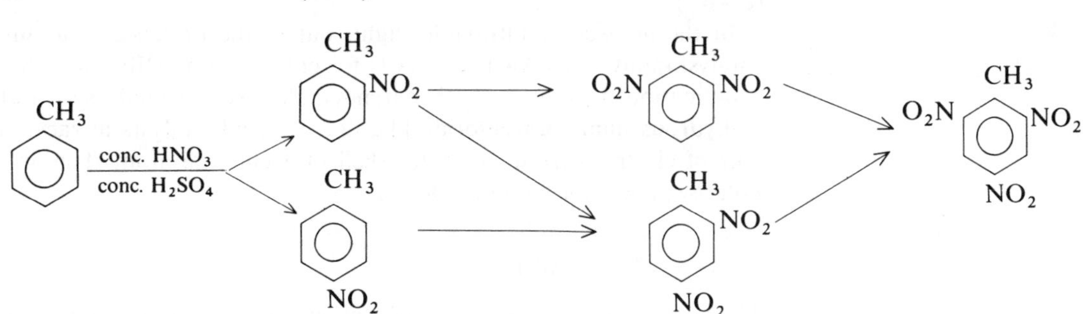

Sulphonation

Benzene undergoes substitution with concentrated sulphuric acid at 80 °C (353 K) to give benzenesulphonic acid, a colourless, highly soluble solid with a melting temperature of 50–51 °C (323–324 K).

$$C_6H_6 + H_2SO_4 \rightleftharpoons C_6H_5SO_3H + H_2O$$

Methylbenzene and the other alkylbenzenes behave in a similar way:

$$C_6H_5CH_3 \xrightarrow{\text{conc. } H_2SO_4} \text{2-}CH_3C_6H_4SO_3H + \text{4-}CH_3C_6H_4SO_3H$$

2- and 4-methylbenzenesulphonic acids

The principal electrophilic reagent is probably the sulphur trioxide molecule, formed by the self-protonation of sulphuric acid:

$$2H_2SO_4 \rightleftharpoons H_3SO_4^+ + HSO_4^-$$
$$\updownarrows$$
$$H_3O^+ + SO_3$$

$$C_6H_6 + \overset{\delta-}{O}{=}S^{\delta+}(\overset{\delta-}{O})_2 \underset{}{\overset{\text{fast}}{\rightleftharpoons}} \underset{\pi\text{-complex}}{C_6H_6 \cdot SO_3} \overset{\text{fast}}{\rightleftharpoons} \underset{\sigma\text{-complex}}{[C_6H_6SO_3^-]^{\oplus}} \overset{\text{slow}}{\rightleftharpoons} C_6H_5SO_3^-H^+$$

This mechanism is different from that of nitration in two ways. First, the slowest (hence, rate determining) stage is the abstraction of a proton. Second, the reaction is reversible. The reverse change, *desulphonation*, may be accomplished by refluxing the sulphonic acid with dilute sulphuric acid or hydrochloric acid, or by treating it with superheated steam.

Halogenation

In the presence of ultraviolet light benzene reacts with both chlorine and bromine in free radical addition reactions to give hexahalocyclohexanes (§14.6.4).

In the *absence* of ultraviolet light, but in the presence of a suitable catalyst (anhydrous $AlCl_3$ or $FeCl_3$ for chlorination; $AlBr_3$ or $FeBr_3$ for bromination), benzene and other arenes undergo electrophilic substitution. Anhydrous aluminium chloride, like all Lewis acid catalysts, attracts a lone pair of electrons from the outer shell of a chlorine atom. This in turn polarises the chlorine molecule:

$$\overset{\delta+}{Cl} \rightarrow \overset{\delta-}{Cl}\!: \longrightarrow \overset{\delta-}{AlCl_3}$$

The partial negative charge is distributed among the four chlorine atoms attached to the aluminium atom, although it is often written, for convenience, on the aluminium atom itself. This convention is adopted in the mechanism which follows.

The positively charged chlorine atom can attack the benzene ring in the usual manner:

$$C_6H_6 + \overset{\delta+}{Cl}{-}Cl\cdots\overset{\delta-}{AlCl_3} \longrightarrow \underset{\pi\text{-complex}}{C_6H_6 \cdot \overset{\delta+}{Cl}{-}Cl\cdots\overset{\delta-}{AlCl_3}} \longrightarrow \underset{\sigma\text{-complex}}{[C_6H_6Cl]^{\oplus}} + [AlCl_4]^-$$

$$\xrightarrow[\text{abstracts } H^+]{[AlCl_4]^-} C_6H_5Cl + HCl + AlCl_3$$

Chlorination and bromination may be conducted at room temperature. To ensure that only monosubstitution occurs, equimolar proportions of benzene and the halogen are used: with excess reagent a mixture of 1,2- and 1,4-dihalobenzenes is formed:

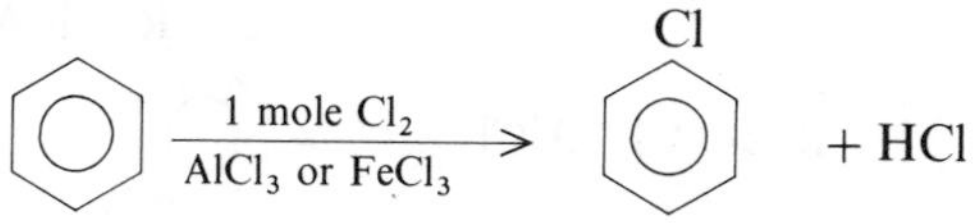

$$C_6H_6 \xrightarrow[\text{AlCl}_3 \text{ or FeCl}_3]{\text{1 mole Cl}_2} C_6H_5Cl + HCl$$

chlorobenzene, boiling temperature 132 °C (405 K)

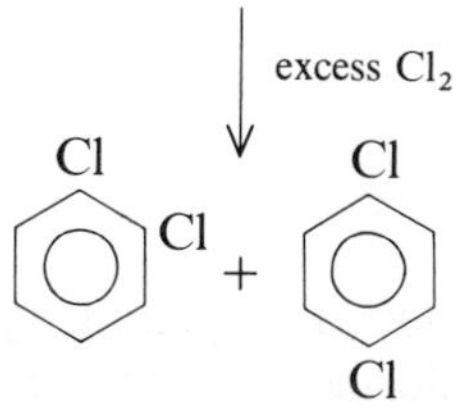

$$C_6H_5Cl \xrightarrow{\text{excess Cl}_2} \text{1,2-}C_6H_4Cl_2 + \text{1,4-}C_6H_4Cl_2$$

The catalytic halogenation of methylbenzene follows the same course:

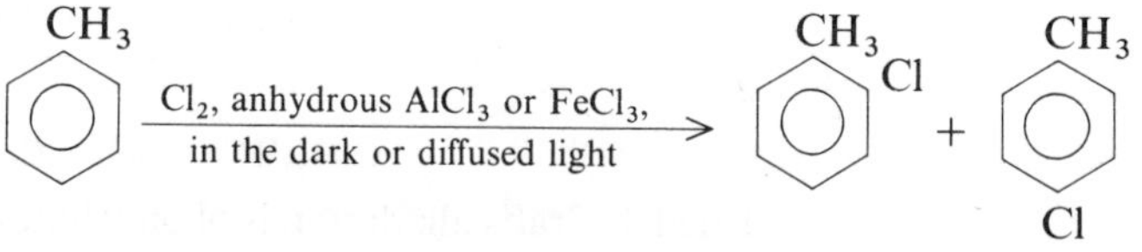

$$C_6H_5CH_3 \xrightarrow[\text{in the dark or diffused light}]{\text{Cl}_2\text{, anhydrous AlCl}_3 \text{ or FeCl}_3\text{,}} \text{2-}ClC_6H_4CH_3 + \text{4-}ClC_6H_4CH_3$$

2- and 4-chloromethylbenzenes

Phenol and phenylamine, in which the benzene ring is activated by an electron donating group (HO or NH_2), undergo *tri*halogenation when they are treated with chlorine water or bromine water at room temperature,

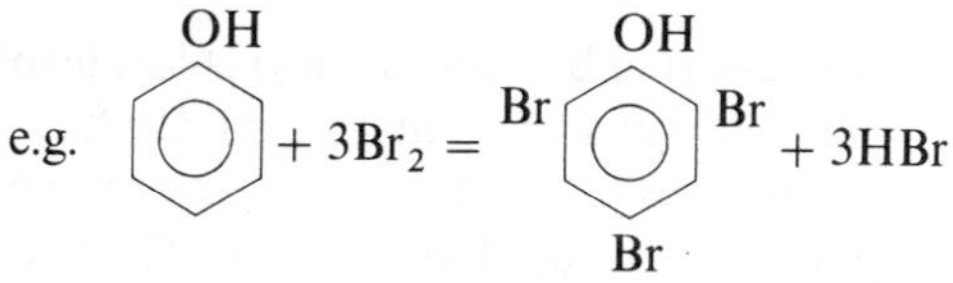

$$\text{e.g.} \quad C_6H_5OH + 3Br_2 = C_6H_2Br_3OH + 3HBr$$

2,4,6-tribromophenol

The products precipitate as white crystalline solids immediately the reactants are mixed together. No catalyst is required.

Friedel–Crafts alkylation

The introduction of an alkyl group into the benzene ring by means of an alkyl halide or other alkylating agent is referred to as a *Friedel–Crafts reaction.* A Lewis acid catalyst is needed and, while many compounds have been used for this purpose, aluminium chloride usually gives the best yield. The reaction is often conducted in the presence of excess hydrocarbon which acts as a solvent but, should additional solvent be required, carbon disulphide or nitrobenzene may be used. At the optimum temperature of 20 °C (293 K) a reaction time of about 12 hours is required.

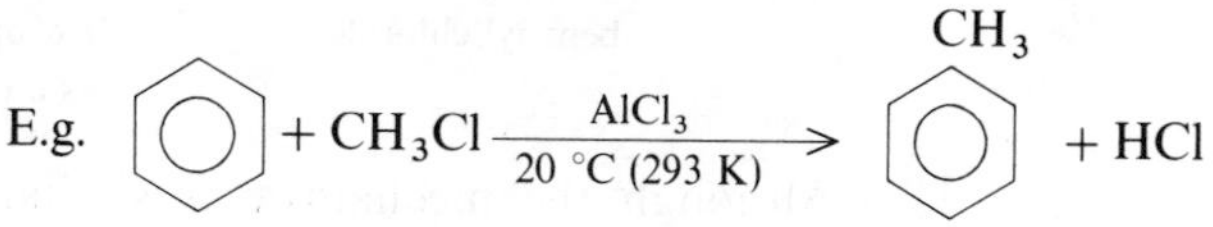

$$\text{E.g.} \quad C_6H_6 + CH_3Cl \xrightarrow[\text{20 °C (293 K)}]{\text{AlCl}_3} C_6H_5CH_3 + HCl$$

The mechanism of the reaction is similar to that of chlorination. The function of the catalyst is to enhance the polarisation and thus the electrophilicity of the alkyl halide.

$$C_6H_6 + \overset{\delta+}{R} \rightarrow Cl\text{---}\overset{\delta-}{AlCl_3} \longrightarrow \overset{\delta+\quad\delta-}{R\text{–}Cl\cdots AlCl_3}\text{(π-complex)} \longrightarrow \text{σ-complex} + [AlCl_4]^-$$

π-complex σ-complex

$$\xrightarrow[\text{abstracts } H^+]{[AlCl_4]^-} C_6H_5R + HCl + AlCl_3$$

Other alkylating agents are alcohols and alkenes (§14.4.5). Aluminium chloride is again the preferred catalyst, although it is less effective with alcohols and more of it is needed,

e.g. $$C_6H_6 \xrightarrow[C_2H_5OH + 1 \text{ mole } AlCl_3]{CH_2{=}CH_2 + 0.2 \text{ mole } AlCl_3} C_6H_5C_2H_5$$

Friedel–Crafts alkylation is of only limited importance in the laboratory as it suffers from several disadvantages.

Friedel–Crafts acylation

An acyl group, $R\text{—}C{\overset{O}{\lessgtr}}$ or $Ar\text{—}C{\overset{O}{\lessgtr}}$, may be introduced into the benzene ring by using an acyl chloride (often called an 'acid chloride') or an acid anhydride. As in Friedel–Crafts alkylations, aluminium chloride is usually used as a catalyst, and an excess of the hydrocarbon or carbon disulphide may be used as a solvent. The reaction, which provides the principal route to aromatic ketones, proceeds at room temperature, although a reaction time of about 12 hours is necessary to ensure a good yield.

$$C_6H_6 + CH_3COCl \xrightarrow[\text{of } AlCl_3]{1.1 \text{ moles}} C_6H_5COCH_3 + HCl$$

ethanoyl chloride acetophenone (formed as a complex with $AlCl_3$)

$$C_6H_6 + C_6H_5COCl \xrightarrow{AlCl_3} C_6H_5COC_6H_5 + HCl$$

benzoyl chloride benzophenone (as a complex)

Although the mechanism is similar to that of the Friedel–Crafts alkylation, there is some uncertainty about the origin of the electrons which are attracted by the catalyst. Clearly, a lone pair of electrons in the outer shell of the chlorine atom could be involved, as previously, but oxygen donates electrons more readily than chlorine and it is considered likely that electrons of the C=O bond are attracted:

$$\mathrm{R(Cl)C{=}O} + \mathrm{AlCl_3} \rightleftharpoons \mathrm{R(Cl)\overset{\delta+}{C}{=}\overset{\delta-}{O}\text{---}\overset{\delta-}{A}lCl_3}$$

Partial negative charge, shared among O and Cl atoms, is usually written on Al as before.

Electrophilic substitution then occurs in the usual way:

$$\mathrm{C_6H_6} + \mathrm{R(Cl)\overset{\delta+}{C}{=}O\text{---}\overset{\delta-}{A}lCl_3} \longrightarrow \underset{\pi\text{-complex}}{\mathrm{C_6H_6 \rightarrow \overset{\delta+}{C}(R)(Cl){=}O\text{---}\overset{\delta-}{A}lCl_3}} \longrightarrow \underset{\sigma\text{-complex}}{[\mathrm{C_6H_6(COR)}]^{\oplus}} + [\mathrm{AlCl_4}]^-$$

$$\xrightarrow[\text{abstracts } \mathrm{H^+}]{[\mathrm{AlCl_4}]^-} \mathrm{C_6H_5COR} + \mathrm{HCl} + \mathrm{AlCl_3}$$

The ketone, like the reagent, coordinates to the aluminium chloride:

$$\mathrm{Ar(R)C{=}O} + \mathrm{AlCl_3} \rightleftharpoons \mathrm{Ar(R)C{=}O\text{---}AlCl_3}$$

This means that an unusually large amount of catalyst is needed, since up to one mole of aluminium chloride per mole of acyl chloride is gradually removed during the reaction. When the reaction is complete, the ketone is recovered from its aluminium chloride complex after acidification of the cooled mixture. Aluminium salts dissolve in the aqueous phase, while the ketone, which remains in the organic phase, is isolated by distilling off the solvent.

When an acid anhydride is used as the acylating agent, a ketone and a carboxylic acid are formed. Because both these compounds react with a mole of aluminium chloride, at least two moles of the catalyst must be used,

e.g. $$\mathrm{C_6H_6} + \underset{\text{ethanoic anhydride}}{\mathrm{(CH_3CO)_2O}} \xrightarrow[\mathrm{AlCl_3}]{\text{2.4 moles of}} \mathrm{C_6H_5COCH_3} + \mathrm{CH_3COOH}$$

Aromatic disubstitution

The reactions quoted above show that disubstitution yields predominantly either a 1,3-compound or a mixture of 1,2- and 1,4-isomers. It is the group already present that influences disubstitution; not the incoming group.

Although various rules have been proposed to help in predicting the directive influence of a substituent group, none is entirely satisfactory. The best approach lies in classifying substituent groups according to their structure.

Class (*i*) Those in which the substituent atom (i.e. the one which is attached to the ring) is further joined by a multiple bond to a more electronegative atom, e.g.

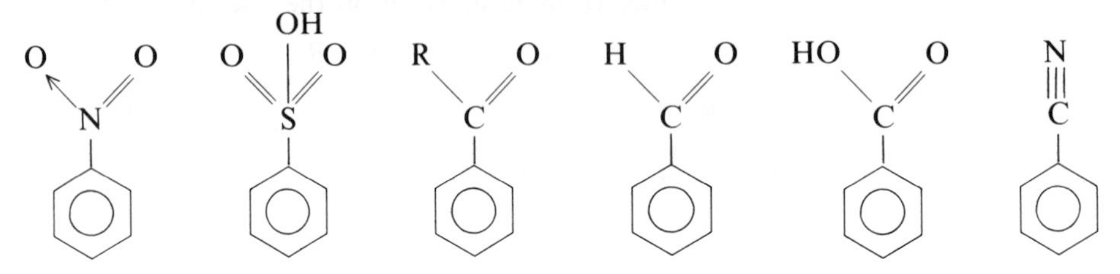

All such groups are 1,3-directing.

Class (*ii*) Those which consist of a single atom, or groups XY, XY_2, etc, in which there is no multiple bonding, e.g.

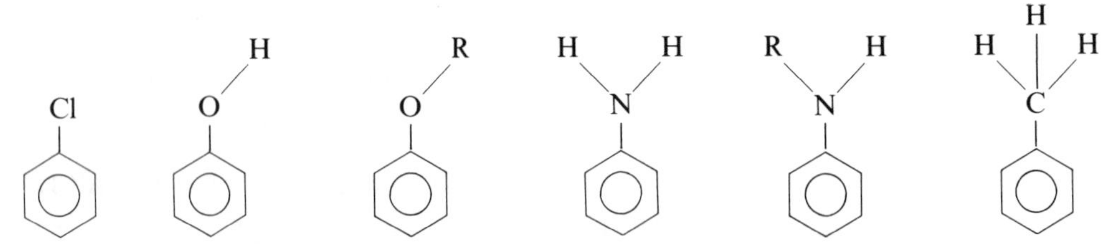

All the groups illustrated here are both 1,2- and 1,4-directing. However, a few such groups, notably CCl_3, $\overset{+}{N}H_3$ and $\overset{+}{N}R_3$, are 1,3-directing.

14.7.5 C—H ACIDIC COMPOUNDS

Aldehydes and ketones

All aldehydes and ketones contain the carbonyl group, C=O. The electronegativity of oxygen is considerably greater than that of carbon, so that the π electrons of the C=O bond are displaced towards the oxygen atom. The electron shift in this bond is transmitted to some extent to adjoining bonds in the molecule, thus:

Positions γ β α

Note: curved arrows show the displacement of bonding pairs of electrons.

The displacement is particularly large at the α position, but dies away as the distance from the C=O bond increases.

Because of its polarisation, each α C—H bond is weakened and can lose its hydrogen atom relatively easily in the form of a proton. Consequently, α hydrogen atoms can readily be substituted by atoms of chlorine, bromine or iodine. In the presence of alkali, all the hydrogen atoms at an α position become replaced,

e.g. $CH_3COCH_3 + 3Br_2 + 3HO^- = CBr_3COCH_3 + 3Br^- + 3H_2O$
1,1,1-tribromopropanone

Trihalogenation occurs because monohalogenated and dihalogenated aldehydes and ketones are more susceptible to halogenation than the original aldehydes and ketones.

In the presence of acid, only monohalogenation occurs. This is because, under these conditions, the monosubstituted product is *less* readily halogenated than the original compound,

e.g. $CH_3COCH_3 + Br_2 = CH_2BrCOCH_3 + HBr$
1-bromopropanone

Haloform reaction

Halogenation under alkaline conditions is used in the *haloform reaction* for preparing the haloforms, i.e. chloroform, $CHCl_3$, bromoform, $CHBr_3$, and iodoform, CHI_3 (§14.8.1). Three reactants are required.

(i) Ethanal or a methyl ketone. A $CH_3C(=O)—$ group in an aldehyde or ketone is essential.
(ii) The appropriate halogen, i.e. chlorine, bromine or iodine.
(iii) An alkali, e.g. sodium hydroxide.

The reaction proceeds in two stages.

Substitution The $CH_3C(=O)—$ grouping becomes halogenated to $CX_3C(=O)—$, where X = Cl, Br or I,

e.g. $CH_3CHO + 3Cl_2 + 3NaOH = CCl_3CHO + 3NaCl + 3H_2O$

In the case of propanone (see above) only *one* CH_3 group becomes substituted.

Alkaline cleavage The trihalogenated aldehyde or ketone is immediately attacked by the alkali to give a haloform and a salt of a carboxylic acid,

e.g. $CCl_3CHO + NaOH = CHCl_3 + HCOONa$

Ethanol and secondary alcohols possessing a methyl group also yield haloforms upon treatment with halogen and alkali, not because they themselves undergo the haloform reaction, but because they become oxidised by the halogen or the hypohalite ion to aldehydes or methyl ketones:

$$C_2H_5OH + NaClO = CH_3CHO + NaCl + H_2O$$
$$CH_3CH(OH)R + NaClO = CH_3COR + NaCl + H_2O$$

These principles are utilised in the *iodoform test* in organic qualitative analysis. The test is performed by placing approximately 1 cm^3 of the unknown compound in a test tube and then adding about 3 cm^3 of a solution of iodine in aqueous potassium iodide. This is followed by sodium hydroxide solution, dropwise, until the brown colour of the iodine almost disappears. Finally, the solution is cooled. A yellow crystalline precipitate of iodoform is positive for the presence of ethanal, a methyl ketone, ethanol, or a secondary alcohol with a methyl group.

Carboxylic acids

Carboxylic acids do not react with halogens under alkaline conditions, but do so, largely at the α-position, in the presence of red phosphorus, e.g.

$$\underset{\text{propanoic acid}}{CH_3CH_2COOH} + Br_2 \xrightarrow{\text{red phosphorus}} \underset{\text{2-bromopropanoic acid}}{CH_3CHBrCOOH} + HBr$$

Further halogenation may occur but in a stepwise manner, because the halogenation of carboxylic acids increases their resistance to attack under these conditions. The chlorination of ethanoic acid at its boiling temperature is particularly well known.

$$CH_3COOH + Cl_2 \xrightarrow{\text{red phosphorus}} CH_2ClCOOH + HCl$$

$$CH_2ClCOOH + Cl_2 = CHCl_2COOH + HCl$$

$$CHCl_2COOH + Cl_2 = CCl_3COOH + HCl$$

Each stage is complete before the next begins.

The reaction may be stopped at either of the intermediate stages with a good yield of chloroethanoic acid or dichloroethanoic acid. Contrast this with the chlorination of methane, which proceeds by a free radical mechanism and invariably gives a mixture of products.

Alkynes

1-Alkynes, containing the grouping ≡CH, are weak protic acids. Ethyne (acetylene) has a pK_a value of about 26, which shows that it is a weaker acid than ethanol (p$K_a \approx 16$) but much stronger than methane (p$K_a > 40$). Such acidity is remarkable in a hydrocarbon, and results from the relative remoteness of the electrons of the C≡C bond from those of the C—H bond. The latter can approach the carbon atom more closely than usual. This leads to charge separation, $\equiv\overset{\delta^-}{C}\div\overset{\delta^+}{H}$, and the loss of the hydrogen atom as a proton to a suitable acceptor.

Copper

A reddish-brown precipitate of copper(I) acetylide (ASE: copper(I) dicarbide) can be obtained by passing ethyne through an ammoniacal solution of copper(I) chloride:

$$CH{\equiv}CH + 2[Cu(NH_3)_2]^+ = CuC{\equiv}CCu + 2NH_4^+ + 2NH_3$$

Ammonia acts as the proton acceptor.

The general reaction is:

$$RC{\equiv}CH \xrightarrow{\text{ammoniacal CuCl}} RC{\equiv}CCu(s)$$

This is used in the laboratory as a test for the presence of the $\equiv$CH grouping.

Silver

Silver acetylide (ASE: silver dicarbide) can be prepared by the action of ethyne on ammoniacal silver nitrate. It forms as a white precipitate:

$$CH{\equiv}CH + 2[Ag(NH_3)_2]^+ = AgC{\equiv}CAg + 2NH_4^+ + 2NH_3$$

For a general case,

$$RC{\equiv}CH \xrightarrow{\text{ammoniacal } AgNO_3} RC{\equiv}CAg(s)$$

EXAMINATION QUESTIONS ON SECTION 14.7

1 Discuss the haloform reaction and its importance in organic chemistry, emphasising and, as far as possible, explaining all of its unusual features. (OC)

2 Under suitable conditions toluene may be converted to (a) benzyl chloride, and (b) chlorotoluenes. In each case state appropriate reagents and conditions for the conversion, outline the reaction mechanism and give an equation. (JMB)

3 (a) With full experimental details, describe the laboratory preparation of a pure sample of **either** bromoethane **or** any other halogenoalkane (alkyl halide).

(b) State the conditions required for bromine to react with:

(i) ethane,

(ii) benzene to give monobromobenzene,

(iii) benzene to give $C_6H_6Br_6$,

(iv) methylbenzene (toluene) to give bromo-2-methylbenzene (*o*-bromotoluene),

(v) methylbenzene to give (bromomethyl)benzene (benzyl bromide).

N.B. Descriptions of the preparations are **not** required.

(c) Write structural formulae for the bromo-compounds produced in (b).

(d) Compare the reactions (if any) of silver nitrate solution with bromoethane, bromobenzene and (bromomethyl)benzene. (AEB)

4 A sample of phenylethanone (acetophenone), $C_6H_5{\cdot}CO{\cdot}CH_3$, was prepared as follows:

(i) 20 g of ethanoyl chloride (acetyl chloride) was slowly added to a mixture of 75 g of dry benzene and anhydrous aluminium(III) chloride (aluminium chloride) (catalyst) in a round bottom flask fitted with a reflux water condenser.

(ii) The mixture was then heated in a water bath at 50 °C for one hour.

(iii) After cooling, the mixture was shaken with a large volume of water, after which the lower layer was discarded.

(iv) The top layer was successively washed with dilute aqueous sodium hydroxide and water and then left in contact with anhydrous calcium chloride for 24 hours.

(v) Finally, the sample was separated from the calcium chloride and fractionally distilled. 15 g of phenylethanone was collected as the fraction boiling between 197 and 202 °C.

(a) Draw a labelled diagram of the apparatus used in stage (i).

(b) Why must the reagents be dry during stages (i) and (ii)?

(c) Write an equation for the reaction in stage (ii) explaining how the aluminium chloride facilitates the reaction.

(d) Why should stages (i) and (ii) be carried out in a fume cupboard?

(e) What substances are extracted during stage (iii)? Write equations for reactions which occur at this stage.

(f) Name the organic compound in the top layer with phenylethanone at stage (iv). Why is this substance still present?

(g) Explain how impurities are removed during treatment of the top layer in stage (iv).

(h) What is the theoretical yield of phenylethanone?
(H = 1, C = 12, O = 16, Cl = 35.5)

(i) What is the percentage yield of phenylethanone in this preparation?

(SUJB)

14.8

The C—halogen bond in organic halides

Data on C—halogen bonds

	C—F	C—Cl	C⩦Cl (C_6H_5Cl)	C—Br	C—I
covalent bond length/nm	0.138	0.177	0.169	0.193	0.214
bond dissociation enthalpy/kJ mol^{-1} at 298 K	484	338	457	276	238

Carbon–halogen bonds are encountered in a wide variety of organic halides, e.g.

(i) *alkyl halides*, i.e. haloalkanes, e.g. chloromethane, CH_3Cl.

(ii) *aryl halides*, i.e. aromatic halides in which a halogen atom is attached directly to the benzene ring, e.g. chlorobenzene, C_6H_5Cl.

(iii) *arylalkyl halides*, i.e. aromatic halides in which one or more halogen atoms appear in an aliphatic side chain, e.g. (chloromethyl)benzene, $C_6H_5CH_2Cl$.

(iv) *1,1-dihalides*, i.e. dihalides in which the two halogen atoms reside on the same carbon atom, e.g. 1,1-dichloroethane, CH_3CHCl_2.

(v) *1,2-dihalides*, i.e. dihalides in which the two halogen atoms are attached to adjoining carbon atoms, e.g. 1,2-dichloroethane, CH_2ClCH_2Cl.

(vi) *acyl halides*, otherwise known as *acid halides*, e.g. ethanoyl chloride, CH_3COCl.

(vii) *polyhalides*, e.g. trichloromethane (chloroform), $CHCl_3$.

14.8.1 NOMENCLATURE OF ORGANIC HALIDES

The commonest types of systematic names listed by IUPAC are as follows.

(i) *Substitutive names* – the only ones recognised by ASE – are based on the names of the alkanes or arenes from which the halides are derived. For example, the compound CH_3Cl is derived from methane by the substitution of an atom of chlorine for one of hydrogen, and is given the substitutive name of chloromethane.

(ii) *Radicofunctional names*, e.g. methyl chloride for CH_3Cl, are drawn from the names of the hydrocarbon radicals which are present in the molecules. Table 14.6 shows some common radicofunctional names, together with the equivalent substitutive names.

Table 14.6 Nomenclature of organic halides

Formula	**Substitutive name**	**Radicofunctional name**
CH_3Cl	chloromethane	methyl chloride
CH_3CH_2Cl	chloroethane	ethyl chloride
$CH_3CH_2CH_2Cl$	1-chloropropane	propyl chloride
$CH_3CHClCH_3$	2-chloropropane	isopropyl chloride
$CH_2{=}CHCl$	chloroethene	vinyl chloride
CH_2Cl_2	dichloromethane	methylene dichloride
CH_3CHCl_2	1,1-dichloroethane	ethylidene dichloride
CH_3CCl_3	1,1,1-trichloroethane	ethylidyne trichloride
C_6H_5Cl (benzene ring with Cl)	chlorobenzene	phenyl chloride†
$C_6H_5CH_2Cl$ (benzene ring with CH_2Cl)	(chloromethyl)benzene	benzyl chloride
$C_6H_5CHCl_2$ (benzene ring with $CHCl_2$)	(dichloromethyl)benzene	benzylidene dichloride
$C_6H_5CCl_3$ (benzene ring with CCl_3)	(trichloromethyl)benzene	benzylidyne trichloride†

† These names are not in current use. For $C_6H_5CCl_3$, the trivial name of benzotrichloride is widely used, although this is not recognised by IUPAC.

For the *haloforms*, trihalides of the type CHX_3, the trivial names fluoroform, chloroform, bromoform and iodoform are retained by IUPAC, but not by ASE.

For the naming of acyl halides see §14.10.1.

14.8.2 FORMATION OF THE C—HALOGEN BOND

Both substitution and addition reactions may be used.

Substitution methods

From C—H; the substitution of H by Cl, Br or I

Alkanes undergo free radical chlorination or bromination in ultraviolet light at room temperature, or in the dark on heating to about 400 °C (673 K),

e.g. $CH_4 + Cl_2 = CH_3Cl + HCl$ (§14.7.3)

Similar reactions occur when methylbenzene is treated with chlorine or bromine under these conditions (§14.7.3).

The substitution of H by halogen occurs more readily, but by different mechanisms, in aldehydes, ketones and carboxylic acids, where the C—H bond is activated by a neighbouring C=O bond,

e.g. $CH_3COCH_3 + 3Cl_2 + 3NaOH = CCl_3COCH_3 + 3NaCl + 3H_2O$ (§14.7.5)

The C—H bond in the benzene ring is subject to electrophilic substitution with chlorine or bromine in diffused light and the presence of a halogen carrier,

e.g. $C_6H_6 + Cl_2 \xrightarrow{FeCl_3} C_6H_5Cl + HCl$ (§14.7.4)

From C—OH; the substitution of HO by Cl, Br or I

Alcohols can be converted into alkyl halides by treatment with a reactive inorganic halide, usually a hydrogen halide or a phosphorus halide,

e.g. $C_2H_5OH + PCl_5 = C_2H_5Cl + POCl_3 + HCl$ (§14.9.5)

Phenols do not react with hydrogen halides, and even with phosphorus halides there is little aryl halide formation (§14.9.5).

The HO group in carboxylic acids responds to phosphorus halides but not to hydrogen halides,

e.g. $CH_3C(=O)OH + PCl_5 = CH_3C(=O)Cl + POCl_3 + HCl$ (§14.10.7)

From C—NH₂; the substitution of NH₂ by Cl, Br or I (§14.11.8)

Aromatic (but not aliphatic) primary amines can be diazotised with nitrous acid at 5° C (278 K). Suitable treatment of the diazonium salt will result in the substitution of chlorine, bromine or iodine atoms.

From C=O; the substitution of O by 2Cl or 2Br

Aldehydes and ketones react with phosphorus pentachloride to give 1,1-dichlorides,

$$\text{e.g.}\quad \underset{\text{ethanal}}{CH_3C(=O)H} + PCl_5 = \underset{\text{1,1-dichloroethane}}{CH_3CHCl_2} + POCl_3$$

Yields are quite good from aldehydes, but less so from ketones.

Phosphorus pentabromide may be used to make 1,1-dibromides, but has a tendency to dissociate into phosphorus tribromide and bromine, and the latter can bring about substitution in the alkyl group of the carbonyl compound.

Addition methods

From C═C

Alkenes undergo electrophilic addition with chlorine and bromine to give 1,2-dihalides, and with hydrogen bromide and hydrogen iodide to give alkyl halides (§14.4.4).

From C≡C (§14.5.4)

Alkynes, like alkenes, respond to addition with halogens and hydrogen halides. Usually two molecules of the reagent react, but only one in the case of iodine or hydrogen chloride because these reagents are unable to attack a C═C bond.

14.8.3 GENERAL PROPERTIES OF ORGANIC HALIDES

CCl_4

All organic halides

Some, e.g. C_6H_5Cl

The introduction of a halogen atom into a hydrocarbon raises its relative molecular mass and hence its boiling temperature. Of the alkyl halides, only three are gaseous at ordinary temperatures: CH_3Cl, CH_3Br and C_2H_5Cl. The rest, below C_{16}, are liquids. Because of the relationship between relative molecular mass and boiling temperature, the iodides have higher boiling temperatures than the bromides, which in turn are higher boiling than the chlorides.

Although the halides are strongly polar compounds – the negative inductive effect of the halogen atom underlies much of their chemistry – they are immiscible with water because of the absence of hydrogen bonding. Their densities are mostly greater than 1 g cm^{-3}, and with water they form a separate *lower* phase.

The lower organic halides are fire resistant and some, notably bromochlorodifluoromethane, are used in fire extinguishers.

14.8.4 CHEMICAL PROPERTIES OF THE C—HALOGEN BOND

The bond is most easily studied in alkyl halides. Like alcohols, they are classified as primary, secondary or tertiary compounds:

RCH_2X	R_2CHX	R_3CX
primary	secondary	tertiary

Alkyl halides may undergo *nucleophilic substitution reactions* with a wide range of reagents, and *elimination reactions* (dehydrohalogenation) to give alkenes.

Nucleophilic substitution reactions

Alkyl halides are usually prepared from alcohols by reaction with hydrogen halides,

e.g $C_2H_5OH + HBr \rightleftharpoons C_2H_5Br + H_2O$

The reaction is reversible, but if bromoethane is refluxed with water for a few minutes, and dilute nitric acid and silver nitrate solution are then added, there is no precipitate of silver bromide, showing that hydrolysis with water alone is slow. If, however, the experiment is repeated with sodium hydroxide solution, a turbidity of silver bromide results. Replacement of the sodium hydroxide by any other soluble hydroxide produces the same result, suggesting that bromoethane is attacked more rapidly by the HO^- ion than the H_2O molecule.

Now, bromine is a more electronegative element than carbon, which causes a permanent charge separation in the C—Br bond:

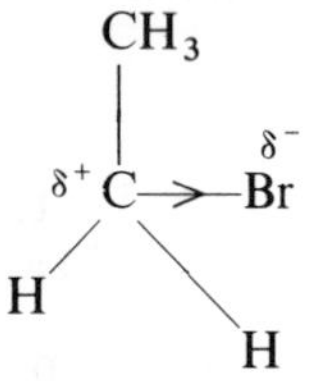

$$\begin{array}{c} CH_3 \\ | \\ {}^{\delta+}C \rightarrow Br^{\delta-} \\ H \quad H \end{array}$$

This is known as a negative inductive effect, $-I$ (p. 304).

It is logical to suppose that the HO^- ion will approach and perhaps attack the positively charged carbon atom:

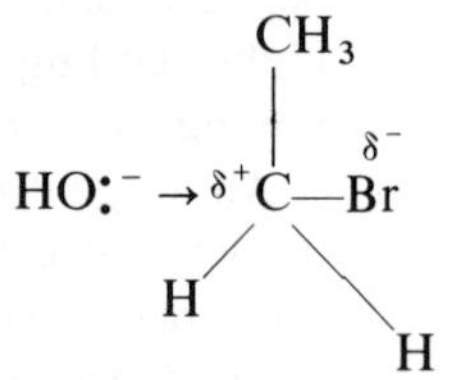

$$\begin{array}{c} CH_3 \\ | \\ HO\!:^- \rightarrow {}^{\delta+}C—Br^{\delta-} \\ H \quad H \end{array}$$

HO^- in this context is referred to as a *nucleophilic reagent*, or *nucleophile*. (Because of its negative charge, it seeks out a centre of positive charge, cf. an atomic nucleus.)

As the HO^- ion approaches the carbon atom, the electrons of the C—Br bond are driven nearer and nearer to the bromine atom. The bond becomes increasingly polarised, i.e. less covalent and more electrovalent. Eventually, the HO^- ion coordinates on to the carbon atom, through a lone pair of electrons in the outer shell of the oxygen atom, and at about the same time the electrons of the C—Br bond move completely to the bromine atom to give a separate Br^- ion.

CH_3 / HO:$^-$ C—Br / H H	CH_3 / HO:$^-$ C—Br / H H	CH_3 / HO→C :Br$^-$ / H H
HO$^-$ ion approaching C atom	HO$^-$ ion close to C; C—Br bond strongly polarised	HO$^-$ ion coordinated on to C atom; Br$^-$ ion leaving

Alkyl iodides are the most reactive alkyl halides. The reactivity order of organic halides in general is

$$\text{iodides} > \text{bromides} > \text{chlorides}$$

At first sight we might expect chlorides to be the most reactive. The C—Cl bond is more polarised than C—Br or C—I, because of the relatively high electronegativity of chlorine, and this leads to a strong attraction for nucleophiles. However, and this is the more important consideration, the C—I bond is the most easily polarised *at the approach of nucleophiles.* Consequently, the bond is easily broken with the loss of an iodide ion. (The C—I bond has a low bond dissociation enthalpy.) I^- is said to be a good *leaving group.*

Other nucleophilic substitution reactions

In aqueous solution, HO^- is a good nucleophile, and hence a good *entering group,* for three reasons: (i) it has a negative charge, and is attracted to $C^{\delta+}$; (ii) the oxygen atom has a lone pair of electrons available for coordination; and (iii) a strong bond can be established with carbon. (The C—O bond dissociation enthalpy is high.) Any other anion that satisfies these conditions will behave as an entering group. Examples are CN^-, RO^-, $RCOO^-$ and H^-. Certain molecules, too, in which there is charge separation, e.g. H_2O, NH_3 and amines, will also react, although not as readily as nucleophiles with a full negative charge.

Certain common anions, notably Cl^-, SO_4^{2-}, CO_3^{2-} and PO_4^{3-}, do not act as nucleophiles because of their reluctance to donate lone pairs of electrons. Generally, strong Lewis bases make the best nucleophiles.

CN^-

Organic cyanides, called *nitriles,* are formed when alkyl halides are refluxed with sodium cyanide or potassium cyanide in aqueous alcohol:

$$RX + K^+(C{\equiv}N)^- = R{-}C{\equiv}N + K^+X^-$$

$$\text{e.g.}\quad CH_3CH_2Br + KCN = \underset{\text{propanenitrile}}{CH_3CH_2CN} + KBr$$

All cyanides

Warning Sodium cyanide and potassium cyanide are extremely poisonous.

This reaction is important in synthetic organic chemistry, for it represents one of the very few ways of increasing the length of a carbon chain. Nearly always, a reaction product contains the same number of carbon atoms as the starting material, but by reacting potassium cyanide with an alkyl halide an extra carbon atom may be introduced. For example, we could convert an alcohol ROH to a higher homologue RCH_2OH by one of the following routes:

$$ROH \xrightarrow{HBr} RBr \xrightarrow{KCN} RCN \begin{cases} \xrightarrow{\text{Mendius reduction}} RCH_2NH_2 \xrightarrow{HNO_2} \\ \xrightarrow{\text{hydrolysis}} RCOOH \xrightarrow{Li[AlH_4]} \end{cases} RCH_2OH$$

As this scheme suggests, nitriles are important intermediates in the synthesis of primary amines and carboxylic acids. The Mendius reduction, to primary amines, is brought about by hydrogen and a nickel catalyst, or by a metal and a protic solvent (e.g. Zn + dilute H_2SO_4), or by lithium tetrahydridoaluminate(III) in dry ether. The hydrolysis of nitriles, to carboxylic acids, may be conducted in either acidic or basic solution, e.g. with concentrated hydrochloric acid or 30% aqueous sodium hydroxide. With water alone the reaction is slow.

RO^-

Ethers are obtained, in a reaction known as the *Williamson synthesis*, e.g.

$$\underset{\text{sodium ethoxide}}{C_2H_5Br + C_2H_5O^-Na^+} \xrightarrow[\text{alcoholic solution}]{\text{reflux in}} \underset{\text{ethoxyethane}}{C_2H_5{-}O{-}C_2H_5} + NaBr$$

The product of this particular reaction is a *simple ether*, so called because the two hydrocarbon radicals are identical. The Williamson synthesis is most valuable for the preparation of *mixed ethers*, in which the radicals are dissimilar, because there is no other means of making such compounds, e.g.

$$\underset{\text{2-bromopropane}}{CH_3CHBrCH_3} + C_2H_5O^-Na^+ = \underset{\text{2-ethoxypropane}}{(CH_3)_2CH{-}O{-}C_2H_5} + NaBr$$

$RCOO^-$

Alkyl halides react rapidly with the silver salts of carboxylic acids to give esters,

$$\text{e.g.}\quad C_2H_5Br + CH_3C\begin{matrix} \diagup\!\!\diagup O \\ \diagdown O^-Ag^+ \end{matrix}$$

$$\xrightarrow[\text{alcoholic solution}]{\text{reflux in}} \underset{\text{ethyl ethanoate}}{CH_3C\begin{matrix} \diagup\!\!\diagup O \\ \diagdown OC_2H_5 \end{matrix}} + AgBr$$

The cheaper sodium salts can be used instead, but silver salts are preferred because the silver ion, Ag^+, catalyses the reaction.

A variation on the more usual esterification of carboxylic acid and alcohol, this method has the merit of giving almost 100% yield of ester and not just an equilibrium mixture.

Metals

Sodium, zinc and other metals which lie towards the top of the electrochemical series behave very much like nucleophilic reagents, because of the ease with which they can ionise and donate electrons (§14.10.6). In the presence of protic solvents, e.g. dilute HCl or H_2SO_4, such metals can attack alkyl halides and reduce them to alkanes,

$$\text{e.g.}\quad CH_3CH_2Br \xrightarrow{\text{Zn + dilute HCl}} CH_3CH_3$$

An alternative reducing agent is lithium tetrahydridoaluminate(III) in dry ether.

NH_3 (ammonolysis)

Ammonia is a weaker nucleophile than anions such as HO^- and CN^-, and reaction with an alkyl halide occurs only on heating in a sealed tube to a temperature of 100 °C (373 K). Amine salts are formed, in equilibrium with primary amines and hydrogen halides:

$$H_3N^{\bullet} + \overset{|}{C}{}^{\delta+} \!\!-\!\!\!\times\!\!- Br^{\delta-} \longrightarrow H_3N\!: \; \overset{|}{C}{}^{\delta+} \!\!-\!\!\!\times\!\!- Br^{\delta-} \longrightarrow \left[H_3N\!\rightarrow\!\overset{|}{C} \right]^+ \overset{\times}{:}Br^-$$

$$\text{e.g.}\quad C_2H_5Br + NH_3(\text{alcoholic}) \xrightarrow[(373\ K)]{100°\ C} C_2H_5NH_3^+Br^- \;\; \text{ethylammonium bromide}$$

$$\rightleftharpoons C_2H_5NH_2 + HBr \;\; \text{(ethylamine)}$$

This preparation, like all sealed tube reactions, is hazardous because of the risk of explosion. A safety shield is essential. The reaction should not be performed in schools and colleges.

The product, when cooled to room temperature, is entirely an amine salt, from which the free amine can be liberated by double decomposition with a base,

$$\text{e.g.}\quad C_2H_5NH_3^+Br^- + NaOH = C_2H_5NH_2 + NaBr + H_2O$$

Good yields of amines can be obtained only from primary alkyl halides: secondary and tertiary halides (except for isopropyl halides) undergo an elimination reaction to give alkenes.

The ammonolysis of alkyl halides is known as the *Hofmann method* of preparing amines. It suffers from the disadvantage that it invariably gives a mixture of four compounds, namely a primary amine, a secondary amine, a tertiary amine and a quaternary ammonium salt, e.g. $C_2H_5Br \xrightarrow{NH_3} C_2H_5NH_2$ (ethylamine), $(C_2H_5)_2NH$ (diethylamine), $(C_2H_5)_3N$ (triethylamine) and $[(C_2H_5)_4N]^+Br^-$ (tetraethylammonium bromide)

The reason for this is that amines, like ammonia, can act as nucleophiles and react with the alkyl halide, e.g.

$$C_2H_5NH_2\text{:} + CH_3CH_2^{\delta+}\text{—}Br^{\delta-} \longrightarrow [C_2H_5NH_2 \rightarrow CH_2CH_3]^+ Br^-$$

i.e. $[(C_2H_5)_2NH_2]^+Br^-$

The composition of the product can be controlled to some extent by adjusting the ratio of alkyl halide to ammonia. (A large excess of ammonia favours the primary amine.) Primary, secondary and tertiary amines can be separated by fractional distillation, unless their boiling temperatures are very close together, or by chromatographic techniques.

Special cases

Aryl halides

Aryl halides, such as chlorobenzene, bromobenzene and iodobenzene, are remarkably unreactive compounds. For example, from our knowledge of the chemistry of alkyl halides we should expect chlorobenzene to be hydrolysed to phenol on refluxing with aqueous alkali:

$$C_6H_5Cl + NaOH(aq) = C_6H_5OH + NaCl$$

However, this reaction does not occur, except at a high temperature (300 °C; 573 K) and pressure (150–200 atm; 15 000–20 000 kPa). Likewise, the reaction with ammonia, to give phenylamine, occurs only at 200 °C (473 K) and a pressure of 50 atm (5 000 kPa), in the presence of a copper(I) oxide catalyst. With many other nucleophiles that react readily with alkyl halides, e.g. CN^-, RO^- and $RCOO^-$, there is no reaction at all.

The halogen atom in chlorobenzene and related compounds is said to be *non-labile*, i.e. not readily substituted. The cause lies partly in the geometry of the molecule and partly in the nature of the bonding.

Steric hindrance Because of the presence of the benzene ring, nucleophilic reagents are hindered in their approach to the carbon atom of the C—Cl bond.

Mesomeric effect The principal cause of low reactivity is the *mesomeric effect*, whereby a lone pair of electrons in the outer shell of the chlorine atom interacts with the π electrons of the benzene ring, as symbolised by Fig. 14.5(a).

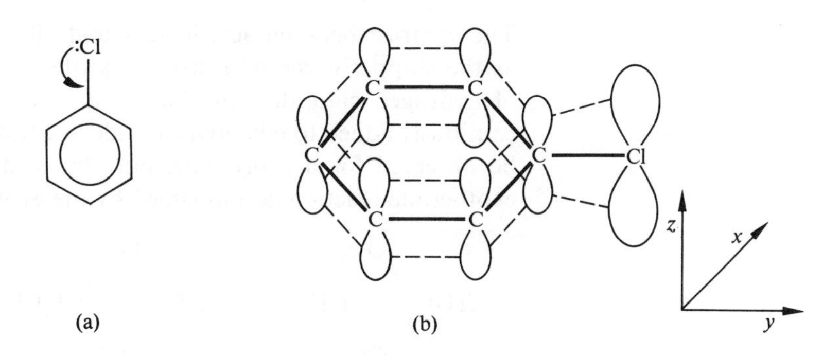

Fig. 14.5 The bonding in chlorobenzene.

Chlorine has the electronic configuration: $1s^2\ 2s^2\ 2p^6\ 3s^2\ 3p_x^2\ 3p_y^2\ 3p_z^1$. The unpaired 3p electron forms a σ bond with a carbon atom of the benzene ring, while the other 3p electrons constitute two lone pairs. One of these 3p orbitals lies in the same plane as the σ-framework of the benzene ring (§4.4.1), but the other is perpendicular to it and interacts to a certain extent with the $2p_z$ orbital of the carbon atom (Fig. 14.5(b)), thus introducing a certain amount of double bond character into the carbon–chlorine bond. This interaction does not destroy the stable aromatic character of the benzene ring system, but it does mean that the C—Cl bond in chlorobenzene is shorter and stronger than the corresponding bond in an alkyl chloride. It is therefore difficult to remove chlorine from chlorobenzene.

A second consequence of the mesomeric effect is that the partial positive charge on the carbon atom of the C—Cl bond is reduced, so that the atom holds less attraction for nucleophilic reagents.

By no means all aromatic halides are unreactive. (Chloromethyl)benzene, $C_6H_5CH_2Cl$, possesses a labile halogen atom because in this compound the chlorine atom and the π-system are prevented from interacting by an intervening saturated carbon atom. A halogen atom is non-labile only if it is attached to a carbon atom which is doubly bonded or which forms part of a benzene ring.

Other aromatic compounds which display the mesomeric effect include phenol and phenylamine:

Consequently these compounds, like chlorobenzene, do not undergo nucleophilic substitution reactions. Furthermore, because the electron availability at oxygen or nitrogen is reduced, they are reluctant to act as nucleophilic reagents.

Acyl halides, e.g. ethanoyl chloride, $CH_3C(=O)Cl$

The contrast between acyl halides and alkyl halides is obvious as soon as the stopper is removed from a bottle of ethanoyl chloride, for fumes of hydrogen chloride immediately appear in moist air. When water is cautiously added to ethanoyl chloride in a test tube, two liquid layers may be observed for a short time, but almost at once there is a vigorous, exothermic reaction accompanied by the evolution of hydrogen chloride.

$$CH_3C(=O)Cl + H_2O = CH_3C(=O)OH + HCl$$

The extreme lability of the chlorine atom indicates that the carbon atom to which it is attached has a particularly high partial positive charge. It arises because the polarisation of the C—Cl bond is reinforced by a similar effect involving the π electrons of the C=O bond. (Oxygen, like chlorine, is a highly electronegative element.)

$$\overset{\delta+}{\text{—C}}(=\overset{\delta-}{O})\text{—}\overset{\delta-}{Cl}$$

Reactions with nucleophiles thus occur readily, but they are not necessarily nucleophilic substitution reactions. Addition–elimination reactions are also possible, i.e. nucleophilic addition across the C=O bond, followed by the elimination of HCl, and with most reagents the mechanism is of this kind. It is therefore customary to regard acyl halides primarily as carbonyl compounds (§14.10.10).

Elimination of hydrogen halide

When an alkyl halide is boiled with an ethanolic solution of potassium hydroxide, it does not become hydrolysed to an alcohol but undergoes dehydrohalogenation (elimination of HX) to give an alkene,

e.g. $CH_3CH_2Br + HO^-(\text{alcoholic}) = CH_2{=}CH_2 + H_2O + Br^-$

H^+ (a proton) and X^- (a halide ion) are nearly always eliminated from adjoining carbon atoms, a process known as *1,2-elimination.*

Only one alkene (ethene) may be obtained from the ethyl halides and each of the propyl halides (propene), but from many of the higher halides there are two possible products. 2-Chlorobutane, for instance, could conceivably give 1-butene or 2-butene:

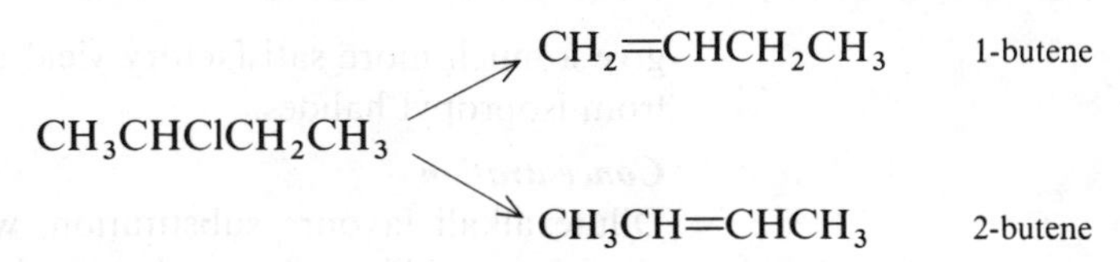

In fact the product is mainly 2-butene. It is a general rule – *Saytzeff's rule* – that the hydrogen atom is eliminated from the carbon atom with the fewer hydrogen atoms.

Substitution versus elimination

Whenever an alkyl halide is treated with an alkali there are two competing reactions:

(i) substitution,

e.g. $CH_3CH_2Br + HO^- = CH_3CH_2OH + Br^-$

Here, the HO^- ion behaves as a nucleophile, coordinating to a carbon atom.

(ii) elimination of hydrogen halide,

e.g. $CH_3CH_2Br + HO^- = CH_2{=}CH_2 + H_2O + Br^-$

Here, the HO^- ion behaves as a base, abstracting a proton from the position next to that of the halogen atom.

The ratio of substitution to elimination is not fixed, but greatly depends on the conditions – namely the nature of the reagent, the nature of the solvent, the concentration of the reagent and the temperature.

Reagent

In theory, strong bases favour elimination while strong nucleophiles favour substitution. In practice, however, species, such as HO^- and RO^-, which are strong bases are also strong nucleophiles and so tend to promote both types of reaction. Tertiary amines are an exception, in that they are moderately strong bases but weak nucleophiles and thus encourage elimination rather than substitution.

Solvent

For tertiary halides the use of water favours substitution, while less polar solvents, such as ethanol, lead to the formation of alkenes. For primary and secondary halides the influence of the solvent is less clear.

An objection to the use of alcoholic alkali for elimination lies in the production of ethyl ethers, $C_2H_5{-}O{-}R$. These compounds are formed by neutralisation of the alcohol by the alkali, to give a metallic alkoxide, followed by a Williamson synthesis with the alkyl halide,

e.g. $$C_2H_5OH + NaOH \rightleftharpoons C_2H_5O^-Na^+ + H_2O$$

$$C_2H_5O^-Na^+ + C_2H_5Cl = C_2H_5{-}O{-}C_2H_5 + NaCl$$

Chloroethane with alcoholic alkali gives almost no ethene, and while the yield is a little better with the bromide and iodide, the main product in all cases is ethoxyethane. The higher members of the homologous series give a much more satisfactory yield of alkene, e.g. 80% yield of propene from isopropyl halides.

Concentration

Dilute alkali favours substitution, while concentrated alkali promotes elimination. This is the result of a change of mechanism.

Temperature

At low temperatures only substitution can occur. Elimination reactions are prevented from taking place because the activation energy (§7.4.4) is relatively high. An increase of temperature, provided that it is sufficient to supply this activation energy, will have a proportionately greater effect on the rate of elimination than substitution.

To summarise, an ideal reagent for substitution is dilute aqueous calcium hydroxide (a relatively weak base), at the lowest temperature consistent with a reasonable rate of reaction. Another reagent, often recommended, is wet silver oxide. A suitable reagent for elimination is a hot, concentrated, alcoholic solution of potassium hydroxide.

EXAMINATION QUESTIONS ON SECTION 14.8

1 (a) Draw a well-labelled diagram of the apparatus you would use to prepare bromoethane (ethyl bromide), b.p. 38 °C, in the laboratory. Your diagram should indicate the reagents and conditions to be used.

(b) Write an equation (or equations) for the reaction involved.

(c) Name **four** impurities which might contaminate the initial sample of bromoethane.

(d) Describe how you would purify your initial sample of bromoethane.

(e) Explain with appropriate equations how **two** of the impurities you mentioned in part (c) are removed by the purification process. (SUJB)

2 Deduce the structures of the compounds **A** to **G** in the scheme below, giving your reasons. (*hv* represents irradiation with light.)

$$\underset{\mathbf{A}}{C_7H_6Cl_2} \xrightarrow[\text{Boil}]{\text{NaOH(aq)}} \underset{\mathbf{B}}{C_7H_7ClO}$$

$$\mathbf{A} \xrightarrow{Cl_2,\ hv} \mathbf{C} \qquad \mathbf{B} \xrightarrow{CrO_3} \mathbf{D}$$

$$\underset{\mathbf{C}}{C_7H_5Cl_3} \xrightarrow[\text{Boil}]{Na_2CO_3\text{(aq)}} \underset{\mathbf{D}}{C_7H_5ClO}$$

$$\mathbf{C} \xrightarrow{Cl_2,\ hv} \mathbf{E} \qquad \mathbf{D} \xrightarrow{KMnO_4} \mathbf{F}$$

$$\underset{\mathbf{E}}{C_7H_4Cl_4} \xrightarrow[\text{Boil}]{Na_2CO_3\text{(aq)}} \underset{\mathbf{F}}{C_7H_5ClO_2}$$

$$\mathbf{F} \xrightarrow[\text{conc } HNO_3]{\text{conc } H_2SO_4} \underset{\mathbf{G}}{C_7H_4ClNO_4}$$

G is a mixture of only **two** compounds. (OC)

3 State the reagents and give the equations summarising the preparation of each of the following from bromoethane:

(a) ethanol,
(b) ethyl ethanoate (*acetate*),
(c) ethoxypropane,
(d) 1-propylamine (two steps are needed). (JMB)

4 (a) Describe, giving essential practical details, how you would prepare, in the laboratory, a pure sample of bromoethane (ethyl bromide), starting from ethanol.

(b) Under what different conditions will bromoethane react with the following reagents: (i) silver oxide, (ii) potassium cyanide, (iii) sodium hydroxide?

Give the structural formula(e) of the organic product(s) from each reaction.

(c) Choose **one** of the products formed in (b) and outline an experiment you would perform in order to identify it. (AEB)

5 This question concerns the preparation and chemical properties of a dibromoalkane of the type $R_1 \cdot CHBr \cdot CHBr \cdot R_2$ where R_1, R_2 may be alkyl or aryl groups or H.

(a) Give the name of a compound having the above structure.

(b) Name the reagents you would use to prepare the compound in the laboratory and write an equation for the reaction.

(c) Indicate any necessary conditions, e.g. of temperature and pressure.

(d) Give a labelled sketch of the apparatus you would use.

(e) Describe briefly how the preparation would be carried out and how the product would be purified.

(f) Describe briefly how you would convert the compound into a dicarboxylic acid. Write equations and give reaction conditions for the stages involved.

(g) Give the name of a dibromo compound which is isomeric with the compound named in (a). Write equations to show the stages involved in its formation from a hydrocarbon. (AEB)

14.9

The C—O bond in alcohols, phenols and ethers

Data on the C—O bond

covalent bond length/nm 0.143
bond dissociation enthalpy/kJ mol^{-1} at 298 K 360

The carbon–oxygen single bond is encountered in alcohols, phenols and ethers. It also occurs, together with the carbon–oxygen double bond, in carboxylic acids, carboxylic acid anhydrides and carboxylic acid esters, but consideration of such compounds will be left until §14.10 because their chemistry is governed primarily by the C=O bond.

An *alcohol* is a compound whose molecules possess at least one *hydroxyl group*, HO. Alcohols may be aliphatic, e.g. ethanol, alicyclic, e.g. cyclohexanol, or aromatic with the HO group in a side chain, e.g. phenylmethanol. Compounds of the last kind are referred to as *alkaryl alcohols.*

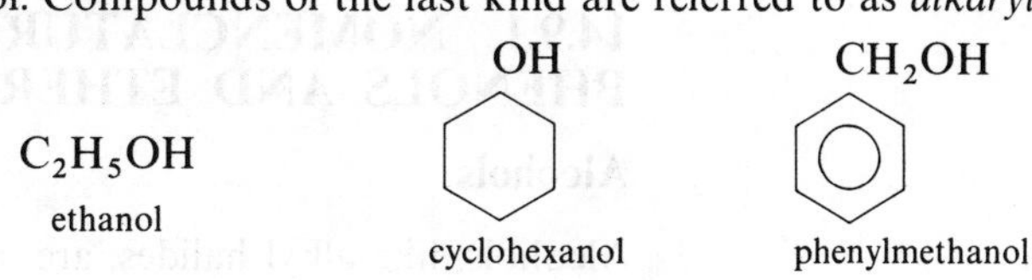

Alcohols may be classified as primary, secondary or tertiary, depending upon whether the hydroxyl group is attached to a primary, secondary or tertiary carbon atom (§14.7.3).

```
     H                R                 R
     |                |                 |
  R—C—OH           R—C—OH            R—C—OH
     |                |                 |
     H                H                 R
  primary          secondary         tertiary
```

Thus, ethanol and phenylmethanol are primary alcohols, while cyclohexanol is secondary.

An alcohol with one hydroxyl group is described as *monohydric*, one with two hydroxyl groups is *dihydric*, and so on. A dihydric alcohol is said to be a *diol* (the common name is 'glycol'), a trihydric alcohol is called a *triol*, etc. Examples are as follows:

```
                          CH2OH
                            |
   CH2OH                   CHOH
     |                      |
   CH2OH                  CH2OH
```

1,2-ethanediol (ethylene glycol), a dihydric alcohol

1,2,3-propanetriol (glycerol), a trihydric alcohol

With very few exceptions, alcohols with more than one hydroxyl group are stable only if those groups are attached to different carbon atoms. Thus, 1,2-ethanediol is stable, but 1,1-ethanediol is unstable, except in aqueous solution, and decomposes spontaneously with the elimination of water between HO groups. This process always leads to the establishment of a C=O bond, i.e. to the formation of an aldehyde or ketone.

```
         OH                      O
        /                       //
CH3—C—OH  ——>  CH3—C          + H2O     (§14.10.6)
        \                       \
         H                       H
  1,1-ethanediol          ethanal
```

A *phenol* is an aromatic hydroxy compound in which at least one HO group is attached directly to the benzene ring. The simplest compound is phenol itself, C_6H_5OH. Like alcohols, phenols are said to be monohydric, dihydric or trihydric according to the number of hydroxyl groups in the molecule.

An *ether* is a compound whose molecules consist of two alkyl or aryl radicals joined together via an atom of oxygen.

14.9.1 NOMENCLATURE OF ALCOHOLS, PHENOLS AND ETHERS

Alcohols

Alcohols, like alkyl halides, are usually referred to by *substitutive names*

in which the final 'e' of the name of the parent hydrocarbon is replaced by 'ol'; e.g. methanol for CH_3OH. To distinguish between isomers, the number of the carbon atom bearing the HO group is introduced *before* the name, e.g. 2-propanol for $CH_3CH(OH)CH_3$.
(**Note.** ASE recognises only substitutive names for alcohols, but requires the number to be inserted into the name, e.g. propan-2-ol for $CH_3CH(OH)CH_3$.)

When HO is not the principal group (§1.3), it is indicated by the prefix 'hydroxy'. The compound $CH_3CH(OH)CH_2CH_2CHO$ is called 4-hydroxypentanal because, according to the IUPAC rules of priority, the aldehyde group takes precedence over the hydroxyl group and must be cited as a suffix.

Diols and triols are named as such,

e.g. $HOCH_2CH_2CH_2CH_2OH$
1,4-butanediol (ASE: butane-1,4-diol)

Older, *radicofunctional names*, e.g. methyl alcohol for CH_3OH, are still allowed by IUPAC, and certain *trivial names* are retained, e.g. ethylene glycol for $HOCH_2CH_2OH$. *Industrial names*, such as isopropanol for $CH_3CH(OH)CH_3$, are not permitted by IUPAC.

Table 14.7 summarises the various names that may be encountered for C_1–C_4 saturated monohydric alcohols.

Table 14.7 Nomenclature of alcohols

Formula	Substitutive name	Radicofunctional name	Industrial name
CH_3OH	methanol	methyl alcohol	methanol
CH_3CH_2OH	ethanol	ethyl alcohol	ethanol
$CH_3CH_2CH_2OH$	1-propanol	propyl alcohol	*n*-propanol
$CH_3CH(OH)CH_3$	2-propanol	isopropyl alcohol	isopropanol
$CH_3CH_2CH_2CH_2OH$	1-butanol	butyl alcohol	*n*-butanol
$(CH_3)_2CHCH_2OH$	2-methyl-1-propanol	isobutyl alcohol	isobutanol
$CH_3CH(OH)CH_2CH_3$	2-butanol	*sec*-butyl alcohol	*sec*-butanol
$(CH_3)_3COH$	2-methyl-2-propanol	*tert*-butyl alcohol	*tert*-butanol

Phenols

Phenols, like alcohols, may be given substitutive names by replacing the last letter 'e' of the name of the parent arene by 'ol', 'diol', etc, but IUPAC recommends the retention of trivial names for simple compounds. ASE accepts trivial names only for phenol and methyl-substituted phenols.

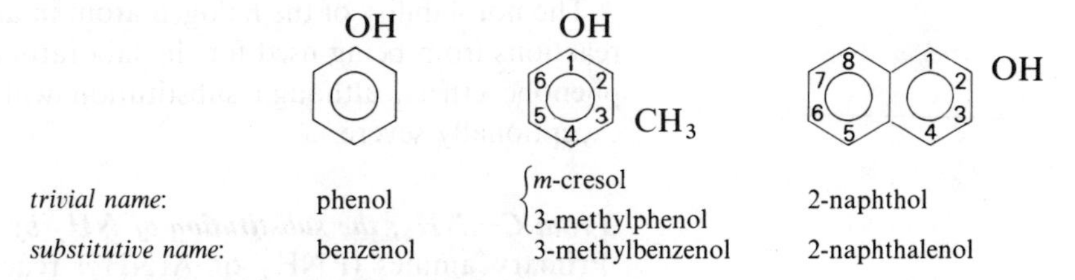

The ASE variation of inserting the number in the name, e.g. naphthalen-2-ol, should again be noted.

Ethers

IUPAC allows the use of either substitutive or radicofunctional names, although only the former are recommended by ASE.

	$CH_3—O—CH_2CH_3$	$CH_3CH_2—O—CH_2CH_3$
substitutive name:	methoxyethane	ethoxyethane
radicofunctional name:	ethyl methyl ether	diethyl ether

14.9.2 FORMATION OF THE C—O BOND

Both substitution and addition reactions may be used.

Substitution methods

From C—H; the substitution of H by HO

The oxidation of alkanes to alcohols, $RH \rightarrow ROH$, is usually very difficult. Methane is exceptional, in that it may be oxidised with oxygen under drastic conditions to provide a 30% yield of methanol:

$$2CH_4 + O_2 \xrightarrow[\substack{130\text{–}200\ \text{atm}\ (13\,000\text{–}20\,000\ \text{kPa}) \\ \text{metallic catalyst}}]{400\ °\text{C}\ (673\ \text{K})} 2CH_3OH$$

If other alkanes are treated in this way, they crack to yield products with fewer carbon atoms than the reactants.

Arenes cannot be converted directly into phenols.

From C—X (X = halogen)***; the substitution of X by HO or RO*** (§14.8.4)

Alkyl halides can readily be converted to alcohols or ethers by nucleophilic substitution reactions. Alcohols are formed on refluxing with aqueous alkali,

e.g. $C_2H_5Br + NaOH = C_2H_5OH + NaBr$

The method is of limited use because alkyl halides are normally prepared from alcohols in the first place. However, the refluxing of an alkyl halide with a sodium or potassium alkoxide in alcoholic solution provides a valuable route to ethers, especially mixed ethers,

e.g. $C_2H_5Br + CH_3ONa = C_2H_5—O—CH_3 + NaBr$

The non-lability of the halogen atom in an aryl halide prevents similar reactions from being used for the laboratory preparation of phenols and phenolic ethers, although substitution will occur if the conditions are exceptionally severe.

From C—NH$_2$; the substitution of NH$_2$ by HO (§14.11.7)

Primary amines (RNH_2 or $ArNH_2$) react with nitrous acid to yield diazonium salts which undergo hydrolysis to give alcohols or phenols. Aliphatic diazonium salts are highly unstable, so that on treatment with nitrous acid a primary aliphatic amine appears to be converted directly to an alcohol,

e.g. $C_2H_5NH_2 + HNO_2 = C_2H_5OH + N_2 + H_2O$

Except in the case of ethylamine, the yield of alcohol is low and the method is of little preparative value.

From C—SO₃H; the substitution of SO₃H by HO

Heating a sodium arylsulphonate with solid sodium hydroxide is one of the main ways of making phenols,

e.g. $C_6H_5SO_3Na$ (sodium benzenesulphonate) $\xrightarrow{\text{fuse with NaOH}}$ C_6H_5ONa $\xrightarrow{\text{HCl}}$ C_6H_5OH

Alkylsulphonic acids do not undergo this reaction.

Substitution of HO by RO (§14.9.5)

The *etherification* of alcohols by dehydration under acidic conditions is one of the most important ways of making simple ethers,

$$\text{e.g.}\quad 2C_2H_5OH \xrightarrow[130\ ^\circ C\ (403\ K)]{\text{conc. } H_2SO_4} C_2H_5{-}O{-}C_2H_5 + H_2O$$

This is a nucleophilic substitution reaction of an alcohol, and is often accompanied by the elimination of water to give an alkene. An excess of alcohol is used to lessen alkene formation.

Substitution of RO by HO

In a reaction which is essentially the reverse of etherification, an ether may be cleaved in the presence of a strong acid, usually hydriodic acid, to give an alcohol,

e.g. $C_2H_5{-}O{-}C_2H_5 + HI = C_2H_5OH + C_2H_5I$ (§14.9.5)

Alcohols are formed also by the hydrolysis of esters,

$$\text{e.g.}\quad CH_3C(=O)OC_2H_5 + H_2O \rightleftharpoons CH_3C(=O)OH + C_2H_5OH$$ (§14.10.11)

Although these reactions are superficially similar to each other, they proceed by totally different mechanisms.

Addition methods

From C=C (§14.4.4)

The hydration of alkenes is the major industrial way of making alcohols and ethers. Water alone is inadequate, and an acid catalyst is always used,

$$\text{e.g.}\quad CH_2{=}CH_2 + H_2O \xrightarrow[\text{70 atm (7000 kPa)}]{H_3PO_4\text{ catalyst, 300 °C (573 K)}} CH_3CH_2OH$$

The oxidation of alkenes by reagents such as potassium permanganate gives a low yield of 1,2-diols,

$$\text{e.g.}\quad CH_2{=}CH_2 \xrightarrow{\text{cold alkaline }KMnO_4} HOCH_2CH_2OH$$

From C=O (§14.10.5)
The principal feature of the C=O bond is the readiness with which it undergoes nucleophilic addition reactions. If we represent the reagent by $H^+Nu{:}^-$, where $Nu{:}^-$ stands for a nucleophilic anion, addition proceeds as follows:

$$>C{=}O + Nu{:}^- \longrightarrow >C(Nu)O^{\ominus} \xrightarrow{H^+} >C(Nu)OH$$

Thus, in all such reactions, the doubly bonded oxygen atom is converted into an HO group.

Of particular interest is the hydrogenation of aldehydes and ketones to primary and secondary alcohols respectively by means of a metal and a protic solvent,

$$\text{e.g.}\quad \underset{\text{ethanal}}{CH_3CHO} \xrightarrow{\text{metal + acid}} \underset{\text{ethanol}}{C_2H_5OH}$$

Alternative reducing agents are lithium tetrahydridoaluminate(III) and sodium tetrahydroborate(III).

14.9.3 MANUFACTURE OF ALCOHOLS AND PHENOL

Methanol is manufactured in the UK by the catalytic addition of hydrogen to carbon monoxide:

$$CO + \underset{\text{(in excess)}}{2H_2} \underset{\substack{\text{50 atm (5000 kPa)} \\ \text{copper catalyst}}}{\overset{\text{270 °C (543 K)}}{\rightleftharpoons}} CH_3OH$$

Ethanol (commonly called 'alcohol') and *2-propanol* are made by the acid catalysed hydration of ethene and propene respectively (§14.4.5).

Ethanol is also produced commercially by *fermentation.* This term is used to describe various chemical reactions brought about by the catalytic action of *enzymes*, i.e. proteins which are present in the living cells of plants and animals. *Yeasts* and *bacteria* may be used as sources of enzymes. Both are single celled micro-organisms, although they differ from each other in that whereas yeast cells have a nucleus, bacteria cells, which are much smaller, do not.

Fermentation causes *carbohydrates*, which are essentially polyhydric alcohols, to be broken down into simple alcohols. In the making of alcoholic beverages, an enzyme called 'zymase', which is present in yeast cells, converts the sugars of a fruit juice into ethanol and carbon dioxide:

$$\underset{\text{glucose or fructose}}{C_6H_{12}O_6} \xrightarrow{\text{zymase}} 2C_2H_5OH + 2CO_2$$

Fermentation proceeds best in the warm (~30 °C (303 K)) and ceases when the concentration of alcohol reaches ~14%. Alcoholic beverages stronger than this ('spirits') are made by distilling the fermented liquor.

Some ethanol is prepared industrially by the fermentation of molasses, i.e. the brown syrupy residue produced in the refining of cane sugar. Most industrial alcohol, however, is manufactured by the hydration of ethene.

The lower alcohols are used as solvents and in the preparation of other organic compounds, notably esters.

Phenol is made by the oxidation of isopropylbenzene. The manufacture by the fusion of sodium benzenesulphonate with sodium hydroxide is no longer worked in the UK.

Phenol is used for the manufacture of phenol–methanal plastics (Bakelite), dyes, drugs (especially aspirin) and selective weed killers.

CH_3OH and many phenols

C_4H_9OH

Phenols

All alcohols and ethers

14.9.4 GENERAL PROPERTIES OF ALCOHOLS, PHENOLS AND ETHERS

In contrast to hydrocarbons, the simplest of which are gases, all the common alcohols are liquids. Table 14.8 provides an interesting comparison of the boiling temperatures of alcohols and alkanes of comparable relative molecular mass.

Table 14.8 Boiling temperatures of alcohols and alkanes

Alkane	Relative molecular mass	Boiling temperature/ °C (K)	Alcohol	Relative molecular mass	Boiling temperature/ °C (K)
ethane	30	−88.6 (184.6)	methanol	32	64.5 (337.7)
propane	44	−42.2 (231.0)	ethanol	46	78.5 (351.7)
butane	58	−0.5 (272.7)	1-propanol	60	97.2 (370.4)
pentane	72	36.3 (309.5)	1-butanol	74	117 (390)

Because boiling temperature is a function of relative molecular mass, it is reasonable to assume that alcohols are associated through hydrogen bonding, in exactly the same way as water (§6.6.1).

$$\underset{R}{\overset{\delta^-}{O}}-\overset{\delta^+}{H}\cdots\underset{R}{\overset{\delta^-}{O}}-\overset{\delta^+}{H}\cdots\underset{R}{\overset{\delta^-}{O}}-\overset{\delta^+}{H}\cdots\underset{R}{\overset{\delta^-}{O}}-\overset{\delta^+}{H}$$

The dotted lines symbolise hydrogen bonds.

Because of the hydrogen bonding that can exist between alcohol and water molecules the lower aliphatic alcohols, namely methanol, ethanol, 1-propanol, 2-propanol and 2-methyl-2-propanol, are completely miscible with water in all proportions. Alkanes, by contrast, are immiscible. The remaining butyl alcohols and the higher members of the homologous series are only partially miscible, with solubility falling to a very low level at around $C_7H_{15}OH$ (1-heptanol = 0.09 g per 100 g of water at 18 °C (291 K)).

Phenols, too, show hydrogen bonding. Phenol itself is a solid with a melting temperature of 40.9 °C (314.1 K), while methylbenzene, with a similar relative molecular mass, is a liquid. Phenol is soluble in water, although aromatic hydrocarbons are not. When phenol is shaken with cold water a two-phase liquid system is obtained, the lower of which is a solution of water in phenol and the upper, phenol in water.

Ethers cannot associate through hydrogen bonding. Consequently, the boiling temperatures of ethers are close to those of alkanes of similar relative molecular mass, but widely different from those of comparable alcohols. Compare Table 14.9 with Table 14.8.

Table 14.9 Boiling temperatures of ethers

Ether	Relative molecular mass	Boiling temperature/ °C (K)
methoxymethane	46	−24.8 (248.4)
methoxyethane	60	7.0 (280.2)
ethoxyethane	74	34.5 (307.7)

Hydrogen bonding is possible, however, between ether and water molecules; thus, the water solubility of ethers is comparable with that of alcohols. For example, the solubilities of ethoxyethane and 1-butanol (isomers of formula $C_4H_{10}O$) are 7.5 and 7.9 g per 100 g of water respectively, at 20 °C (293 K).

14.9.5 CHEMICAL PROPERTIES OF THE C—O BOND

The chemistry of alcohols, phenols and ethers can be studied under three main headings: (i) acidic and basic character; (ii) nucleophilic substitution reactions; (iii) elimination reactions. We shall consider each in turn.

Acidic and basic character

Alcohols and phenols resemble water in that they can donate protons and also accept protons, i.e. they have both acidic and basic character. Ethers, by contrast, are purely basic. We must consider this aspect first, because a knowledge of it is crucial to an understanding of the rest of the chemistry.

Acidity

Alcohols and especially phenols act as weak protic acids, ionising in the presence of water to give a proton and an anion:

$$R—O\!:\!\!-\!H \rightleftharpoons RO^- + H^+$$

alkoxide ion

$$Ar—O\!:\!\!-\!H \rightleftharpoons ArO^- + H^+$$

phenoxide ion

The acidity of hydroxy compounds varies considerably, depending partly on the ease with which a proton can be lost, and partly on the stability of the anion that is formed (§9.2.10). Alcohols are extremely weak acids: the pK_a values of methanol and ethanol are 15.5 and ~16 respectively (cf. 15.7 for water). Aqueous solutions of alcohols are neutral to litmus and other acid–base indicators. Their salts can be made only by adding metal to alcohol; sodium ethoxide, for example, is produced in a vigorous reaction between ethanol and sodium at room temperature:

$$C_2H_5OH + Na = C_2H_5O^-Na^+ + \tfrac{1}{2}H_2$$

Sodium ethoxide remains in solution, but can be isolated as a white solid after evaporation of unreacted ethanol.

Only the alkali metals (group 1A) are sufficiently reactive to dissolve readily in alcohols, although alkoxides can be prepared from group 2A metals (notably magnesium), and also aluminium.

Metallic alkoxides, being salts of strong bases and weak acids, are extensively hydrolysed in solution (§9.2.8), which is the reason why they cannot be made by neutralisation methods.

E.g. $C_2H_5ONa + H_2O \rightleftharpoons C_2H_5OH + NaOH$

Alkoxides are useful compounds in the laboratory, especially in the Williamson synthesis (§14.8.4).

Phenols are much more acidic than alcohols. Phenol itself – formerly called 'carbolic acid' – has a pK_a value of 10.00. Phenols give acidic solutions with water, which turn litmus red, and their sodium and potassium salts can be made not only from the metals but also from their hydroxides,

e.g. $C_6H_5OH + NaOH \rightleftharpoons C_6H_5O^-Na^+ + H_2O$

sodium phenoxide

Phenol can be liberated from an alkaline solution by the addition of a strong acid,

e.g. $C_6H_5O^-Na^+ + HCl = C_6H_5OH + NaCl$

Most phenols are weaker than carboxylic acids. (Ethanoic acid has a pK_a value of 4.76.) They are also weaker than carbonic acid, and will therefore not liberate carbon dioxide from carbonates or hydrogencarbonates. However, the presence of a nitro group on the aromatic ring may increase the acidity of a phenol to such an extent that a reaction with sodium hydrogencarbonate solution is possible,

e.g. $C_6H_4(OH)(NO_2) + NaHCO_3 = C_6H_4(O^-Na^+)(NO_2) + H_2O + CO_2$

The nitro group has a tendency to withdraw electrons from the aromatic ring. As a result, electrons are drawn into the ring from the HO group. Polarisation of the O—H bond is increased, and loss of a proton is facilitated.

Sodium phenoxides, like sodium alkoxides, react readily with alkyl halides in the Williamson synthesis,

e.g. $C_6H_5O^-Na^+ + CH_3I = C_6H_5OCH_3 + NaI$

methoxybenzene

Ethers, because they lack an O—H bond, are not acidic.

Basicity

Alcohols, phenols and ethers, like water, can function as Lewis bases by donating a lone pair of electrons from the outer shell of the oxygen atom. If the lone pair is donated to a proton the alcohol or ether is said to be a 'base'; if it is donated to some other Lewis acid the alcohol or ether is referred to as a 'nucleophile'. We shall consider each in turn.

Alcohols as bases In the same way that water molecules are converted into oxonium ions in the presence of protic acids, so alcohols and ethers give substituted oxonium ions, RH_2O^+ and R_2HO^+ respectively:

$$H_2O: + H^+ \rightleftharpoons H_2O^{\oplus}\rightarrow H \quad \text{i.e.} \quad H_3O^+$$

oxonium ion (protonated water)

$$RHO: + H^+ \rightleftharpoons RHO^{\oplus}\rightarrow H \quad \text{i.e.} \quad RH_2O^+$$

protonated alcohol

$$R_2O: + H^+ \rightleftharpoons R_2O^{\oplus}\rightarrow H \quad \text{i.e.} \quad R_2HO^+$$

protonated ether

Although it is conventional to write the positive charge on the central oxygen atom, the charge does not reside there but is distributed over the ion as a whole.

The ability to donate a lone pair of electrons increases in the order:

$$\mathrm{H_2\ddot{O}} < \mathrm{R{\rightarrow}\ddot{O}{-}H} < \mathrm{R{\rightarrow}\ddot{O}{\leftarrow}R}$$

water alcohol ether

because the +I effect of the alkyl groups increases the electron availability on the oxygen atom. Even so, alcohol and ether salts formed in this way are unstable, and exist only in acidic solution.

Alcohols as nucleophiles Alcohols and, to a lesser extent, ethers can react with a considerable variety of Lewis acids. Coordination of alcohol is often followed by the elimination of another species. Phenols seldom act as nucleophiles because of the mesomeric effect (§14.8.4).

Carbonyl compounds Alcohols react with a wide range of carbonyl compounds, i.e. substances whose molecules possess the carbonyl group, C=O. Examples are aldehydes, carboxylic acids, carboxylic acid anhydrides and carboxylic acid chlorides. In most cases there occurs an addition–elimination reaction, in which the nucleophilic addition of alcohol across the C=O bond is followed by the loss of a small molecule such as H_2O (§14.10.5).

PCl_3, PBr_3 and PI_3 Water hydrolyses phosphorus trichloride to phosphonic acid and hydrogen chloride (§15.2.6). The reaction proceeds by the coordination of water to PCl_3, followed by the elimination of HCl. In a similar fashion, alcohols convert phosphorus trichloride to phosphonic acid and alkyl chlorides:

$$3ROH + PCl_3 = 3RCl + H_2PHO_3$$

Phosphorus trichloride is not a highly recommended reagent for the preparation of alkyl chlorides, as the yields are low. The main products are esters, $P(OR)_3$, of the unknown acid H_3PO_3, i.e. $P(OH)_3$. Phosphorus tribromide and phosphorus triiodide, however, give good yields of alkyl bromides and iodides. The reagents are usually made *in situ* from red phosphorus and the appropriate halogen.

PCl_5 and PBr_5 Phosphorus pentachloride reacts vigorously with alcohols at room temperature to give good yields of alkyl chlorides, although only one chlorine atom of the reagent can be used:

$$ROH + PCl_5 = RCl + POCl_3 + HCl$$

The reaction, which has a mechanism similar to that of the reaction with phosphorus trichloride, serves not only as a means of making alkyl chlorides, but also as a test for the presence of a hydroxyl group in an organic compound. (The substance under test must be dried beforehand.) Phosphorus pentabromide, likewise, is excellent for the preparation of alkyl bromides. Phosphorus pentaiodide does not exist (§15.2.6).

Phenols, when treated with phosphorus halides, give low yields of aryl halides. The main products are esters of phosphorus-containing acids.

$SOCl_2$ Thionyl chloride (§16.2.7) is often suggested as a reagent for the preparation of alkyl chlorides:

$$ROH + SOCl_2 = RCl + SO_2 + HCl$$

It has an advantage in that both the by-products are gaseous, but the reagent is relatively unreactive and must be used in excess.

Nucleophilic substitution reactions

Because the C—O bond is polarised in the same way as the C—halogen bond, we might expect alcohols and ethers to take part in nucleophilic substitutions, thus:

$$\underset{\text{nucleophile}}{Nu\!:^-} + \overset{\delta+}{C}\!-\!\overset{\delta-}{OH} = Nu\rightarrow C + HO^-$$

However, except under acidic conditions such reactions do not occur, because the C—O bond, with its relatively high bond dissociation enthalpy, is too difficult to break. HO^- and RO^- are said to be *poor leaving groups*. (The best leaving groups, e.g. I^- and Br^-, form bonds with carbon that are easily broken.)

In the presence of a protic acid the substrate is no longer an alcohol (or ether) but a protonated alcohol (or ether), and the equation must be modified as follows:

$$Nu\!:^- + C\!-\!\overset{\oplus}{O}H_2 = Nu\rightarrow C + H_2O$$

Reaction is facilitated for two reasons.

(i) The positive charge on the protonated alcohol (or ether) increases the attraction for nucleophiles.

(ii) When an alcohol becomes protonated, the proton attracts a lone pair of electrons *away* from the oxygen atom. The oxygen atom therefore becomes electron deficient, and draws towards itself the electrons of the C—O bond. In other words, the C—O bond becomes increasingly polarised, and the loss of H_2O (not HO^-) is encouraged.

$$Nu\!:^- + C\!-\!\overset{H^+}{\underset{H}{O}} \longrightarrow Nu\rightarrow C + H_2O$$

Lewis acid catalysts, such as aluminium oxide, can also be used.

Phenols rarely participate in nucleophilic substitution reactions because of the mesomeric effect (§14.8.4).

Esterification of alcohols with inorganic acids

An *ester* is a compound made in an *esterification* reaction between an acid and an alcohol:

$$\text{acid} + \text{alcohol} \rightleftharpoons \text{ester} + \text{water}$$

Perhaps the best known esters are those of carboxylic acids: they will be discussed in §14.10.11. We are concerned here with esters of inorganic acids, notably hydrogen halides and sulphuric acid. (Esters of hydrogen halides are alkyl halides.)

HI, HBr and HCl Alcohols react with hydrogen halides to give alkyl halides and water:

$$ROH + HX \rightleftharpoons RX + H_2O$$

The reactivity order of the reagents is as follows:

$$HI > HBr > HCl$$

Hydriodic acid and hydrobromic acid react readily to give alkyl iodides and bromides respectively. Hydrochloric acid reacts at a reasonable rate only with tertiary alcohols and phenylmethanol. Other alcohols must be saturated with hydrogen chloride gas and then heated in a sealed tube for reaction to occur. Anhydrous zinc chloride, which acts as a Lewis acid catalyst, is usually included.

In the preparation of alkyl bromides, hydrobromic acid is often replaced by a mixture of concentrated sulphuric acid (which supplies protons) and sodium bromide (which supplies bromide ions). 1-Bromobutane, for example, can be prepared by heating together a mixture of 1-butanol, concentrated sulphuric acid and sodium bromide:

$$\underset{\text{1-butanol}}{CH_3CH_2CH_2CH_2OH} + H^+ \longrightarrow \underset{\text{protonated 1-butanol}}{CH_3CH_2CH_2CH_2\overset{\oplus}{O}H_2} \xrightarrow{Br^-} \underset{\text{1-bromobutane}}{CH_3CH_2CH_2CH_2Br} + H_2O$$

The presence of sulphuric acid introduces complications. Because it is a dehydrating agent it partially converts alcohols into ethers and alkenes (see below), and for this reason its use is not recommended with secondary and tertiary alcohols. Also, it is an oxidising agent, and may oxidise primary and secondary alcohols to carboxylic acid esters and ketones respectively. Esters of sulphuric acid may be formed too, but such compounds are relatively involatile and are unlikely to contaminate the final product.

The same technique cannot be used for iodides, owing to the ease with which the iodide ion is oxidised to iodine by concentrated sulphuric acid. Instead, constant boiling hydriodic acid is used. The yield of alkyl iodide may be poor, because hydriodic acid can reduce such compounds to alkanes.

Alcohols vary considerably in their reactivity towards hydrogen halides. Besides the decline of reactivity that is always associated with the ascent of a homologous series, there is also a considerable difference between the reactivity of primary, secondary and tertiary alcohols:

tertiary > secondary > primary

As we saw above, phosphorus trihalides and pentahalides are often used instead of hydrogen halides for converting alcohols into alkyl halides. The choice of reagent is dictated partly by consideration of yield and partly by ease of purifying the product.

H_2SO_4 Sulphuric acid is diprotic and thus gives rise to two series of esters,

e.g.
$$(HO)_2SO_2 \xrightarrow{CH_3OH} (CH_3O)(HO)SO_2 \xrightarrow{CH_3OH} (CH_3O)_2SO_2$$

sulphuric acid — methyl hydrogen sulphate — dimethyl sulphate

Although alkyl hydrogen sulphates can be made in fairly good yield by heating the appropriate alcohol with concentrated sulphuric acid on a boiling water bath, attempts to prepare dialkyl sulphates by direct esterification lead, except in the case of methanol and ethanol, to etherification of the alcohol and the formation of alkenes (see below).

Concentrated sulphuric acid with phenols has the effect of sulphonating the aromatic ring.

Etherification of alcohols

When an alcohol is heated in the presence of a protic acid (usually concentrated sulphuric acid) or a Lewis acid (usually aluminium oxide) it undergoes etherification through the loss of a molecule of water between two molecules of the alcohol,

e.g.
$$2C_2H_5OH \xrightarrow[\text{or } Al_2O_3 \text{ at } 250\ ^\circ C\ (523\ K)]{\text{conc. } H_2SO_4 \text{ at } 130\ ^\circ C\ (403\ K)} C_2H_5{-}O{-}C_2H_5 + H_2O$$

The temperature is rather critical, for at lower temperatures esters are formed (e.g. ethyl hydrogen sulphate), and at higher temperatures the principal product is an alkene (e.g. ethene).

Etherification should be regarded as a nucleophilic substitution reaction between two molecules of alcohol, one of which, in the protonated state, acts as the substrate, while the other behaves as the nucleophilic reagent.

$$R(H)\ddot{O}: \xrightarrow{\text{approach}} {-}C{-}\overset{\oplus}{O}H_2 \xrightarrow{\text{coordination}} R(H)\overset{\oplus}{O}{\rightarrow}C{-} + H_2O \longrightarrow R{-}O{-}C{-} + H^+$$

ROH acts as a nucleophile — protonated alcohol (or alcohol–Lewis acid complex) — protonated ether — ether — proton

Secondary and, in particular, tertiary alcohols are dehydrated so easily that dilute sulphuric acid is adequate.

Cleavage of ethers

Etherification is a reversible reaction, and in the presence of a strong acid catalyst an ether may revert to the alcohol or alcohols from which it is derived,

$$\text{e.g.}\quad C_2H_5{-}O{-}C_2H_5 + H_2O \xrightarrow[\text{heat under pressure}]{\text{dilute } H_2SO_4} 2C_2H_5OH$$

When concentrated hydriodic acid – the most usual reagent – is used, the product is a mixture of an alcohol and an alkyl iodide,

$$\text{e.g.}\quad C_2H_5{-}O{-}C_2H_5 + HI \xrightarrow{\text{reflux}} C_2H_5OH + C_2H_5I$$

Elimination reactions of alcohols

Elimination of water

When an alcohol is heated with an acid catalyst, etherification is always accompanied by a competing reaction leading to the formation of an alkene. One molecule of water is lost from one molecule of the alcohol in a 1,2-elimination reaction (cf. 1,2-elimination of HX from an alkyl halide),

$$\text{e.g.}\quad CH_3CH_2OH \xrightarrow[\substack{\text{or conc. } H_3PO_4 \text{ at } 210\ °C\ (483\ K) \\ \text{or } Al_2O_3 \text{ at } 350\ °C\ (623\ K)}]{\text{conc. } H_2SO_4 \text{ at } 180\ °C\ (453\ K)} CH_2{=}CH_2 + H_2O$$

An excess of acid is necessary to ensure an adequate rate of reaction.

The reaction occurs by a three-stage mechanism.

(i) The alcohol becomes protonated by the acid catalyst, immediately the two compounds are mixed together at room temperature:

$$-\overset{\overset{\large H}{|}}{\underset{|}{C}}-\overset{\overset{\large OH}{|}}{\underset{|}{C}}- + H^+ \rightleftharpoons -\overset{\overset{\large H}{|}}{\underset{|}{C}}-\overset{\overset{\large \overset{\oplus}{O}H_2}{|}}{\underset{|}{C}}-$$

(ii) On heating, the protonated alcohol decomposes. Water is driven off, and a carbonium ion remains:

$$-\overset{\overset{\large H}{|}}{\underset{|}{C}}-\overset{\overset{\large \overset{\oplus}{O}H_2}{|}}{\underset{|}{C}}- \rightleftharpoons -\overset{\overset{\large H}{|}}{\underset{|}{C}}-\overset{\oplus}{\underset{|}{C}}- + H_2O$$

(iii) Finally, the carbonium ion decomposes with the loss of a proton to give an alkene:

$$-\overset{\overset{\large H}{|}}{\underset{|}{C}}-\overset{\oplus}{\underset{|}{C}}- \rightleftharpoons \;\rangle C{=}C\langle\; + H^+$$

It will be noticed that the acid catalyst is regenerated in the final stage, which is a feature of all catalysed reactions. It will also be seen that all stages of the reaction are reversible. The reverse change, i.e. the acid catalysed hydration of alkenes to alcohols, is particularly important in the chemical industry as a means of manufacturing alcohols (§14.4.5).

Secondary and, in particular, tertiary alcohols form alkenes most easily. Secondary alcohols respond to orthophosphoric acid at about 140 °C (413 K), and tertiary alcohols at about 100 °C (373 K).

Table 14.10 The action of concentrated sulphuric acid on ethanol at various temperatures

Temperature/ °C (K)	Type of reaction	Mechanism of reaction	Product
20 (293)	neutralisation	protonation of the alcohol	unstable oxonium salt, $C_2H_5\overset{\oplus}{O}H_2\ HSO_4^-$
100 (373)	esterification	nucleophilic substitution (HSO_4^- as nucleophile)	ethyl hydrogen sulphate, $C_2H_5OSO_2OH$
130 (403)	etherification	nucleophilic substitution (C_2H_5OH as nucleophile)	ethoxyethane, $C_2H_5{-}O{-}C_2H_5$
180 (453)	alkene formation	elimination of water	ethene, $CH_2{=}CH_2$

The formation of alkenes, rather than ethers, is favoured by high temperatures (Table 14.10). Etherification has a considerably lower activation energy than alkene formation, so that while the former occurs readily at about 130 °C (403 K) the latter does not. At higher temperatures the rate of etherification is increased to some extent, but the rate of alkene formation is increased proportionately more because many more molecules have the activation energy necessary for them to react.

In the elimination of water from secondary or tertiary alcohols Saytzeff's rule (§14.8.4) is followed, i.e. the thermodynamically more stable alkene is formed through the loss of a hydrogen atom from the adjoining carbon atom which possesses the *lower* number of hydrogen atoms,

e.g. $CH_3CH(OH)CH_2CH_3$ (2-butanol) $\xrightarrow{\text{conc. } H_2SO_4}$ $CH_3CH{=}CHCH_3$ (2-butene (main product)) + $CH_2{=}CHCH_2CH_3$ (1-butene (minor product))

Elimination of hydrogen

The oxidation of alcohols is the most important way of preparing aldehydes and ketones. The reaction should be regarded essentially as a dehydrogenation, in which two hydrogen atoms are lost from the same position on the chain. One is a hydrogen atom attached directly to carbon, while the other forms part of the HO group. Thus, primary alcohols are oxidised to aldehydes, and secondary alcohols to ketones.

Tertiary alcohols, however, cannot be oxidised in this way, because there is no hydrogen atom on the carbon atom which carries the HO group. Nevertheless, in the presence of acidic oxidising agents, they readily undergo oxidative cleavage, probably via alkenes.

$$CH_3{-}C(H)(H){-}OH \xrightarrow[-H_2]{\text{oxidation}} CH_3C({=}O)H$$

ethanol (*primary alcohol*) → ethanal

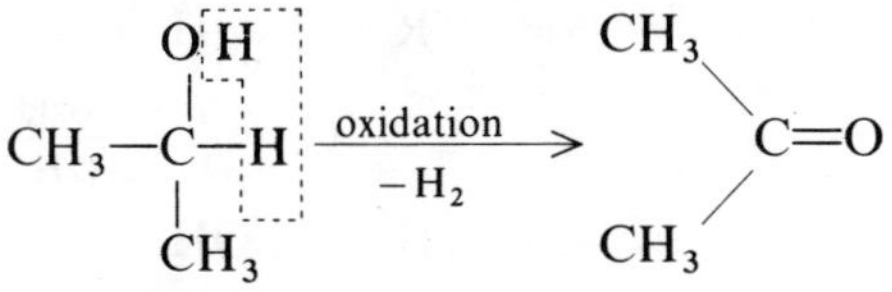

$$CH_3-\underset{CH_3}{\overset{OH}{C}}-H \xrightarrow[-H_2]{\text{oxidation}} (CH_3)_2C{=}O$$

2-propanol *secondary alcohol* → propanone

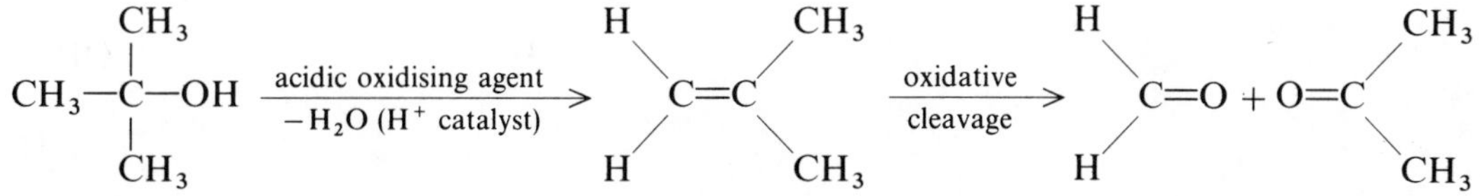

$$CH_3-\underset{CH_3}{\overset{CH_3}{C}}-OH \xrightarrow[-H_2O\ (H^+\ \text{catalyst})]{\text{acidic oxidising agent}} H_2C{=}C(CH_3)_2 \xrightarrow[\text{cleavage}]{\text{oxidative}} H_2C{=}O + O{=}C(CH_3)_2$$

2-methyl-2-propanol *tertiary alcohol* → 2-methylpropene → methanal + propanone

The elimination of hydrogen can be achieved catalytically, by passing the alcohol vapour over a metal that can act as a hydrogen acceptor. Copper or silver is highly effective, at temperatures of 400–450 °C (673–723 K).

Alternatively, an oxidising agent such as acidified sodium dichromate solution or alkaline potassium permanganate solution may be used to eliminate hydrogen in the form of water. Dichromate is usually preferred because, unlike permanganate, it will not normally attack any C=C bonds that there may be in the molecule.

In aqueous solution, aldehydes and ketones form hydrates by the nucleophilic addition of water across the C=O bond (§14.10.6):

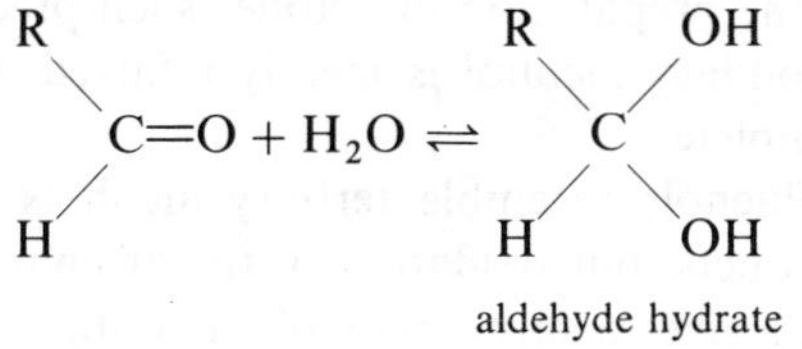

$$RHC{=}O + H_2O \rightleftharpoons RHC(OH)_2$$

aldehyde hydrate

$$R_2C{=}O + H_2O \rightleftharpoons R_2C(OH)_2$$

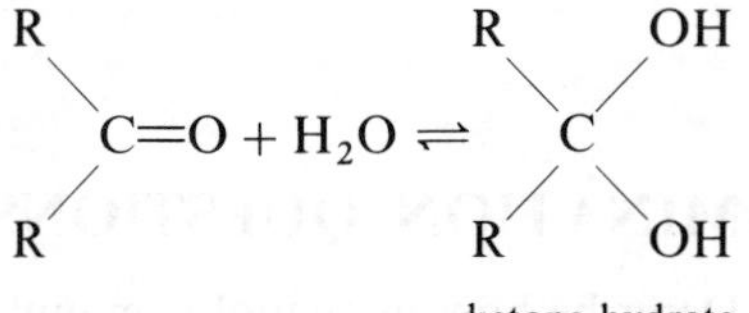

ketone hydrate

Aldehyde hydrates bear a structural resemblance to primary and secondary alcohols in that they possess the $\geq C(H)(OH)$ grouping. They can therefore lose two atoms of hydrogen from the same carbon atom and become oxidised to carboxylic acids:

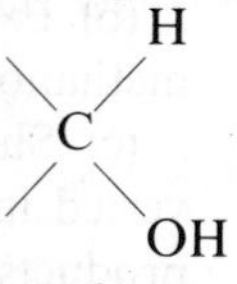

$$\mathrm{R(H)C(OH)_2} \xrightarrow[-(2H^+ + 2e^-)]{\text{oxidation}} \mathrm{R{-}C(=O)OH}$$

For example,

$$\mathrm{CH_3C(=O)H} \longrightarrow \mathrm{CH_3C(=O)OH}$$

Ketone hydrates, on the other hand, do not possess the $\mathrm{>C(H)OH}$ grouping and are not susceptible to oxidation.

Thus, if a primary alcohol is refluxed with an oxidising agent, the product is a carboxylic acid rather than an aldehyde. To obtain the aldehyde, it is vital that the aldehyde is distilled off as it is formed. This is possible because aldehydes are more volatile than the alcohols from which they are formed or the acids to which they oxidise,

e.g.	$C_2H_5OH \longrightarrow$	$CH_3CHO \longrightarrow$	CH_3COOH
boiling temperature/°C (K)	78.5 (351.7)	20.8 (294.0)	118 (391)

In making ethanal by the oxidation of ethanol, the reaction mixture is maintained at about 50 °C (323 K). At this temperature ethanal distils off rapidly, while most of the other liquids remain in the distillation flask. In the preparation of ketones such precautions are unnecessary, and the secondary alcohol is merely refluxed with the oxidant until reaction is complete.

Phenols resemble tertiary alcohols in yielding no simple oxidation products, but oxidation of the aromatic ring occurs very easily. Phenol itself gives a tarry mass often containing 1,4-benzoquinone, especially if chromic acid is used as the oxidant.

EXAMINATION QUESTIONS ON SECTION 14.9

1 (a) Describe how methanol is manufactured from a hydrocarbon (one method), and explain carefully the reasons for the physical and chemical conditions under which the process is carried out.

(b) Explain fully one chemical test you would use to distinguish between methanol and ethanol.

(c) Show by writing balanced equations how methanol may be converted into ethanol. Give essential conditions but no details of how the products are isolated or purified. (O)

2 Show, by means of equations, how (a) ethanol, and (b) ethane-1,2-diol

(ethylene glycol) are obtained on an industrial scale. How would you obtain anhydrous ethanol from a 50% aqueous solution?

Give **two** reactions which demonstrate the presence of a hydroxyl group in ethanol.

How would you demonstrate that phenol is more acidic than cyclohexanol ($C_6H_{11}OH$)? (OC)

3 (a) (i) Write the structural formula for each of the four isomeric alcohols C_4H_9OH.

(ii) Which one of these isomers would be *optically active*? Explain briefly what is meant by this term.

(iii) Write the structural formulae of the oxidation products (if any) when each of the isomeric alcohols is treated with acidified potassium dichromate(VI) solution.

(b) (i) Outline **two** different ways of introducing an O—H group into a benzene ring.

(ii) Give **three** ways in which phenol differs chemically from phenylmethanol (benzyl alcohol). (AEB)

4 (a) For each of parts (i) to (vii) below **one or more** of the alternatives A, B, C and D listed below are correct. Decide which of the alternatives is (are) correct and write the appropriate letters in answer to the question.

A $CH_3 \cdot CH_2 \cdot CH_2 \cdot CH_2OH$
B $(CH_3)_3COH$
C $(CH_3)_2CH \cdot CH_2OH$
D $CH_3 \cdot CH(OH) \cdot CH_2 \cdot CH_3$

(i) Which will turn acidified potassium dichromate(VI) solution green on warming?

(ii) Which will have optically active forms?

(iii) Which will give hydrogen chloride gas on treatment with phosphorus pentachloride?

(iv) Which will give a yellow precipitate on warming with iodine and alkali?

(v) Which could be oxidised to a ketone?

(vi) Which could be oxidised to a carboxylic acid?

(vii) Which could form an ester with ethanoyl (acetyl) chloride?

(b) Write the systematic names for each of the compounds A, B, C and D.

(c) Write the structural formula of the principal organic substance produced when:

(i) ethane-1,2-diol is refluxed with excess acidified potassium manganate (VII) (potassium permanganate),

(ii) tetrahydrofuran,

$$\begin{array}{ccc} CH_2 & — & CH_2 \\ | & & | \\ CH_2 & & CH_2 \\ & \diagdown \; \diagup & \\ & O & \end{array}$$

is warmed with excess concentrated hydriodic acid,

(iii) ethane-1,2-diol is treated with excess sodium. (SUJB)

14.10
The C=O bond in organic carbonyl compounds

Data on the C=O bond

covalent bond length/nm 0.122
bond dissociation enthalpy/kJ mol^{-1} at 298 K 743

The simplest compounds to contain the carbon–oxygen double bond, which is often called the *carbonyl group*, are *aldehydes* and *ketones*. In aldehydes at least one of the bonds to the carbonyl carbon atom is attached to an atom of hydrogen; in ketones both bonds to the carbonyl carbon atom are attached to alkyl or aryl radicals.

H(H)C=O	R(H)C=O	R(R)C=O	R(R′)C=O
methanal	higher aldehyde	simple ketone	mixed ketone

The formula of an aldehyde may be written as R—C(=O)—H or as RCHO, but not as RCOH, for this implies the presence of a hydroxyl group.

Other carbonyl compounds, shown below, possess additional bonds which considerably affect their chemistry.

R—C(=O)—OH	R—C(=O)—O—C(=O)—R	R—C(=O)—Cl
carboxylic acid	carboxylic acid anhydride	carboxylic acid chloride

R—C(=O)—$O^{\ominus}M^+$	R—C(=O)—OR′	R—C(=O)—NH_2
carboxylic acid salt ($M^+ = Na^+, K^+$, etc)	carboxylic acid ester	carboxylic acid amide

Carboxylic acids contain the carboxyl group, —C(=O)—OH, which is a combination of the *carb*onyl group and the hydr*oxyl* group; hence the name. Acids with one carboxyl group (such as ethanoic acid and benzoic acid) are said to be *monocarboxylic*; those with two such groups are *dicarboxylic*, and so on. Examples of dicarboxylic acids are as follows:

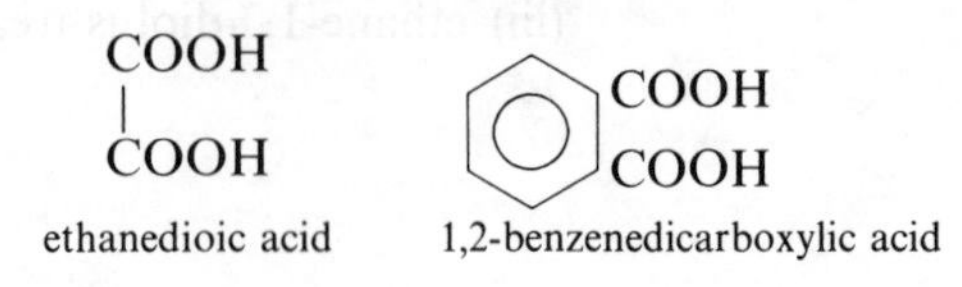

ethanedioic acid 1,2-benzenedicarboxylic acid

A *carboxylic acid anhydride* is derived from the parent acid, in theory but not always in practice, by the elimination of one molecule of water between two molecules of the acid, thus:

$$\text{R—C(=O)—O}\boxed{\text{H}} + \boxed{\text{OH}}\text{—C(=O)—R} \longrightarrow \text{R—C(=O)—O—C(=O)—R} + H_2O$$

acid — acid anhydride

Certain (but by no means all) dicarboxylic acids form cyclic anhydrides by the elimination of a molecule of water from *one* molecule of the acid,

e.g. $C_6H_4(COOH)_2 \longrightarrow C_6H_4(CO)_2O + H_2O$

1,2-benzenedicarboxylic acid — 1,2-benzenedicarboxylic anhydride

An *acid chloride* is derived from an acid, in theory and in practice, by the substitution of the hydroxyl group by an atom of chlorine. Carboxylic acid chlorides are often referred to as *acyl chlorides*, because RC(=O)— (or ArC(=O)—) is termed an 'acyl' group.

An *ester* is the product of *esterification* between an acid and an alcohol (or phenol):

acid + alcohol $\rightleftharpoons$ ester + water

Here we are concerned only with the esters of carboxylic acids. Esters of inorganic acids are considered in §14.9.5.

An *amide* is derived from an acid by the replacement of the hydroxyl group by an amino group, NH_2. Thus, the formula of a carboxylic acid amide is $RCONH_2$.

14.10.1 NOMENCLATURE OF CARBONYL COMPOUNDS

Aldehydes

In the IUPAC system, the name of an aliphatic aldehyde is formed from that of the corresponding hydrocarbon by replacing the terminal 'e' by '-al'. Dialdehydes are named as 'dials'. Aromatic aldehydes are named by adding the suffix '-carbaldehyde' to the name of the ring system.

Trivial names, which relate to the names of the carboxylic acids to which aldehydes are oxidised, are approved by IUPAC but not by ASE.

Table 14.11 Nomenclature of aldehydes

Formula	Substitutive name	Trivial name
$HCHO$	methanal	formaldehyde†
CH_3CHO	ethanal	acetaldehyde
CH_3CH_2CHO	propanal	propionaldehyde
$CH_3CH_2CH_2CHO$	butanal	butyraldehyde
$(CH_3)_2CHCHO$	2-methylpropanal	isobutyraldehyde
$OHCCHO$	ethanedial	glyoxal
C_6H_5CHO	benzenecarbaldehyde	benzaldehyde

† An aqueous solution of formaldehyde is often called 'formalin'.

Ketones

Aliphatic ketones are usually given substitutive names by adding '-one', '-dione', etc to the name of the corresponding hydrocarbon, with elision of the final 'e'. An older system of radicofunctional names, in which the ketone $R'COR''$ is named by citing the radicals R' and R'' in alphabetical order, followed by the word 'ketone', is still approved by IUPAC but not by ASE.

IUPAC also recognises the trivial name, acetone, for CH_3COCH_3.

Table 14.12 Nomenclature of aliphatic ketones

Formula	Substitutive name†	Radicofunctional name
CH_3COCH_3	propanone	dimethyl ketone
$CH_3COCH_2CH_3$	2-butanone	ethyl methyl ketone
$CH_3CH_2COCH_2CH_3$	3-pentanone	diethyl ketone
$CH_3COCOCH_3$	2,3-butanedione	

† ASE recommends inserting locants in the name, e.g. pentan-3-one, and omitting locants from 2-butanone and 2,3-butanedione.

To name an aromatic ketone on the IUPAC system, we look for the acid corresponding to the acyl group which is present in the compound, and change its name ending from '-ic acid' to '-ophenone'. For example, $C_6H_5COCH_3$, which contains the acetyl group CH_3CO, derived from acetic acid, is called acetophenone.

ASE disregards this system and recommends the use of substitutive names. Thus, $C_6H_5COC_6H_5$, which is derived from diphenylmethane, $C_6H_5CH_2C_6H_5$, is called diphenylmethanone. (The IUPAC name is benzophenone.) An exception concerns $C_6H_5COCH_3$, which is related to the hydrocarbon $C_6H_5CH_2CH_3$. The hydrocarbon is called ethylbenzene, but could logically be named phenylethane, and from the latter is obtained the ASE name of phenylethanone for the ketone.

Carboxylic acids

Although IUPAC recommends the retention of trivial names for simple aliphatic acids, substitutive names derived from those of the corresponding hydrocarbons by the change of ending from '-e' to '-oic acid' are gaining wide acceptance and are the only ones recommended by ASE.

Table 14.13 Nomenclature of aliphatic carboxylic acids

Formula	Substitutive name	Trivial name
$HCOOH$	methanoic acid	formic acid
CH_3COOH	ethanoic acid	acetic acid
CH_3CH_2COOH	propanoic acid	propionic acid
$CH_3CH_2CH_2COOH$	butanoic acid	butyric acid
$(CH_3)_2CHCOOH$	2-methylpropanoic acid	isobutyric acid
$HOOC{-}COOH$	ethanedioic acid	oxalic acid
$HOOC{-}(CH_2)_4{-}COOH$	hexanedioic acid	adipic acid

Aromatic acids may be given substitutive names by means of the suffix '-carboxylic acid'.

Table 14.14 Nomenclature of aromatic acids

Formula	Substitutive name†	Trivial name
COOH	benzenecarboxylic acid	benzoic acid
COOH, CH_3 (ring positions 1–6)	2-methylbenzenecarboxylic acid	*o*-toluic acid
COOH, COOH	1,2-benzenedicarboxylic acid	phthalic acid
COOH, COOH	1,3-benzenedicarboxylic acid	isophthalic acid
COOH, COOH	1,4-benzenedicarboxylic acid	terephthalic acid

†ASE recommends inserting locants in the name, e.g. benzene-1,2-dicarboxylic acid.

Acid anhydrides

Such compounds are named after their parent acids by replacing the word 'acid' by 'anhydride'. Thus, $(CH_3CO)_2O$ is called ethanoic anhydride or acetic anhydride.

Acid chlorides

Acid chlorides are given radicofunctional names, in which the name of the acyl radical is followed by the word 'chloride'. Acyl radicals (RCO or ArCO) may have trivial or systematic names taken from those of the carboxylic acids from which they are derived (Table 14.15).

Table 14.15 The naming of acid chlorides

Radical	Trivial name	Systematic name	Acid chloride	Trivial name	Systematic name
CH_3CO	acetyl	ethanoyl	CH_3COCl	acetyl chloride	ethanoyl chloride
CH_3CH_2CO	propionyl	propanoyl	CH_3CH_2COCl	propionyl chloride	propanoyl chloride
C_6H_5CO	benzoyl	benzenecarbonyl	C_6H_5COCl	benzoyl chloride	benzenecarbonyl chloride

Salts and esters

These compounds, also, are named after the parent acids,

e.g. CH_3COONa sodium ethanoate or sodium acetate
$CH_3COOC_2H_5$ ethyl ethanoate or ethyl acetate

Amides

The names of amides are derived from those of the parent acids, by replacement of the ending '-oic acid' or '-ic acid' by '-amide',

e.g. CH_3CONH_2 { ethanamide (from 'ethanoic acid') / acetamide (from 'acetic acid') }

An ending of the type '-carboxylic acid' must be replaced by '-carboxamide',

e.g. $C_6H_5CONH_2$ { benzamide (from 'benzoic acid') / benzenecarboxamide (from 'benzenecarboxylic acid') }

N-Substituted amides are named in accordance with the same rules,

e.g. $CH_3CONHC_2H_5$ *N*-ethylethanamide or *N*-ethylacetamide

However, *N*-phenyl substituted amides, because they are derived from aniline (phenylamine), may be named in trivial fashion as *anilides.* (ASE does not permit this.) For example,

$CH_3CONHC_6H_5$ acetanilide or *N*-phenylethanamide or *N*-phenylacetamide

N-Substituted amides which are derived from aromatic amines other than phenylamine must be named systematically.

14.10.2 FORMATION OF THE C=O BOND

Like the C=C bond, the C=O bond is established principally by elimination reactions. The addition of ozone across a C=C bond, or of sulphuric acid and water across a C≡C bond, may also be used.

Elimination methods

From $\geq C-C(OH)\langle H$ ***by the elimination of*** H_2

Both in the laboratory and the chemical industry the oxidation of alcohols is the most important way of making aldehydes and ketones. In the laboratory an oxidising agent is used, while in industry the favoured technique is dehydrogenation (§14.9.5).

From $-CH_2-CCl_2-$ ***by the action of alkali***

On treatment with aqueous alkali, 1,1-dihalides give rise to aldehydes or ketones,

$$\text{e.g.}\quad C_6H_5CHCl_2 + 2NaOH = C_6H_5CHO + 2NaCl + H_2O$$

The method is important in the manufacture of benzaldehyde, but is of limited use for the preparation of other aldehydes and ketones because most 1,1-dihalides have to be made from aldehydes and ketones by the action of phosphorus pentachloride or pentabromide.

Addition methods

From C=C (§14.4.4)

The ozonolysis of alkenes, i.e. ozonisation followed by hydrolysis, results in the cleavage of the C=C bond and the formation of aldehydes and ketones:

$$\begin{matrix} H \\ R \end{matrix}\!\!>C=C<\!\!\begin{matrix} R' \\ R'' \end{matrix} \xrightarrow{O_3} \begin{matrix} H \\ R \end{matrix}\!\!>C\overset{O-O}{\underset{}{-O-}}C<\!\!\begin{matrix} R' \\ R'' \end{matrix} \xrightarrow{H_2O} \begin{matrix} H \\ R \end{matrix}\!\!>C=O + O=C<\!\!\begin{matrix} R' \\ R'' \end{matrix}$$

The reaction is important not so much as a method of preparing aldehydes and ketones, but as a means of locating the position of a double bond.

From C≡C (§14.5.4)

Alkynes are readily hydrated to aldehydes and ketones by the action of warm dilute sulphuric acid in the presence of mercury(II) sulphate,

$$\text{e.g.}\quad CH{\equiv}CH \xrightarrow[HgSO_4]{H_2SO_4,\ H_2O} CH_2{=}CHOH \xrightarrow{\text{rearrangement}} CH_3CHO$$

Until 1940 ethanal was manufactured in the UK in this way.

14.10.3 INTERCONVERSION OF CARBONYL COMPOUNDS

Many carbonyl compounds can be converted into one another, and carboxylic acids, acid anhydrides, acid chlorides, esters and amides are usually prepared in this way. Such reactions do not appear to include the C=O bond, although, as we shall see, the bond is very much involved in the reaction mechanisms.

Carboxylic acids

The two main routes to carboxylic acids are: (i) the oxidation of aldehydes (§14.10.6); and (ii) the hydrolysis of acid derivatives. These include acid anhydrides (§14.10.9), acid chlorides (§14.10.10), esters (§14.10.11) and amides (§14.10.12).

Acid anhydrides

Although certain dicarboxylic acids can be converted to their anhydrides on treatment with a dehydrating agent, most acid anhydrides must be made by heating the acid chloride with the sodium salt of the same acid:

$$\left.\begin{array}{l} R{-}C({=}O){-}Cl \\ + \\ R{-}C({=}O){-}ONa \end{array}\right\} \xrightarrow[\text{distillation}]{\text{dry}} R{-}C({=}O){-}O{-}C({=}O){-}R + NaCl$$

The reaction should be regarded as a nucleophilic substitution at the C—Cl bond, with $RCOO^-$ acting as the nucleophile.

If an acid chloride is heated with the sodium salt of a different acid, the product is a *mixed anhydride*.

Acid chlorides (acyl chlorides)

In the same way that an alcohol may readily be converted to an alkyl chloride by means of phosphorus trichloride, phosphorus pentachloride or thionyl chloride, so the same reagents may be used for converting a carboxylic acid to its acid chloride (§14.10.7).

Esters

Although direct esterification between carboxylic acids and alcohols is usually employed (§14.10.7), it is also possible to make esters from acid anhydrides, acid chlorides and from carboxylates (§14.8.4).

Amides

In a study of the chemistry of amides, the following sequence of reactions should be borne in mind:

$$\underset{\text{alkyl halide}}{RX} \xrightarrow[\text{NaCN}]{\text{KCN or}} \underset{\text{nitrile}}{R{-}C{\equiv}N} \underset{-H_2O\ (P_4O_{10})}{\overset{+H_2O\ (H^+)}{\rightleftharpoons}} \underset{\text{amide}}{R{-}C({=}O)NH_2}$$

$$\underset{(180\ ^\circ C;\ 453\ K)}{\underset{-H_2O}{\overset{+H_2O\ (H^+\ \text{or}\ HO^-)}{\rightleftharpoons}}} \underset{\text{ammonium salt}}{R{-}C({=}O)ONH_4} \overset{+H_2O}{\rightleftharpoons} \underset{\text{acid}}{R{-}C({=}O)OH} + NH_3$$

There are thus two principal routes to an amide. Either a nitrile can be partially hydrolysed – and it is difficult to stop the reaction at the amide stage – or a carboxylic acid can be treated with ammonia and the resulting ammonium salt heated at 180 °C (453 K) to bring about partial dehydration.

To improve the yield it is better to use an acid derivative rather than the carboxylic acid itself. Acid anhydrides, acid chlorides and esters may all be used.

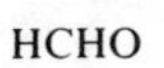

HCHO

CH_3CHO and C_6H_5CHO

14.10.4 GENERAL PROPERTIES OF CARBONYL COMPOUNDS

Aldehydes and ketones

Aliphatic aldehydes, such as methanal and ethanal, have pungent, acrid odours, whereas ketones, such as propanone, are pleasant smelling. All the common aldehydes and ketones are liquids, except for methanal which is a gas. Their boiling temperatures are somewhat higher than those of comparable alkanes, because of the association that results from charge

All aldehydes and ketones

separation within the carbonyl group: $\overset{\delta+}{C}=\overset{\delta-}{O}$. (Oxygen is much more electronegative than carbon.)

Aldehyde and ketone molecules form hydrogen bonds with water molecules; thus the lower aldehydes and ketones are soluble in water. As in other homologous series, solubility decreases with increasing chain length, so that whereas ethanal is completely miscible with water, propanal is soluble only to the extent of 20 g per 100 g of water at 20 °C (293 K).

Table 14.16 Comparison of the boiling temperatures and water solubilities of aldehydes, ketones and alkanes

Compound	Relative molecular mass	Boiling temperature/ °C (K)	Solubility/g per 100 g of water at 20°C (293 K)
CH_3CH_2CHO	58	48.8 (322.0)	20.0
CH_3COCH_3	58	56.2 (329.4)	∞
$CH_3CH_2CH_2CH_3$	58	−0.5 (272.7)	0.037

HCOOH

CH_3COOH and HCOOH

CH_3COOH

Carboxylic acids

Because their molecules contain hydroxyl groups, it is only to be expected that carboxylic acids will show hydrogen bonding, yet they do not form long chain structures. Instead, *dimerisation* occurs. The shape of the molecule, particularly the carboxylic acid grouping, allows two molecules to join together rather like two pieces of a jigsaw puzzle. A cyclic dimer is formed, with the hydrogen atom of one carboxyl group attracting the carbonyl oxygen atom of the second:

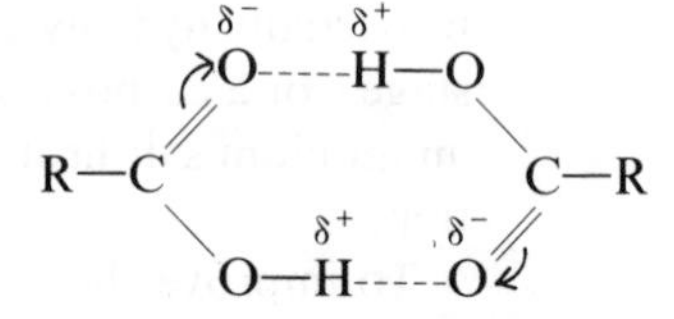

Evidence for this is provided by relative molecular mass measurements in suitable solvents, such as benzene.

The lower carboxylic acids – even methanoic acid – are all pungent smelling liquids, with boiling temperatures much higher than would be expected from their relative molecular masses. Up to and including C_2H_5COOH they are completely miscible with water. Thereafter they are only partially miscible, and the solubility decreases sharply with increasing chain length.

$(CH_3COO)_2Pb$

Salts

All common salts of carboxylic acids are crystalline solids with a high solubility in water. (Lead(II) ethanoate is one of the few soluble salts of lead.) Sodium and potassium salts are alkaline in aqueous solution, owing to hydrolysis. The sodium salts of long chain acids, e.g. sodium octadecanoate (sodium stearate), $C_{17}H_{35}COONa$, are known as 'soaps'.

$(CH_3CO)_2O$

Acid anhydrides

Methanoic anhydride, $(HCO)_2O$, does not exist. (The anhydride of methanoic acid is carbon monoxide.) Anhydrides of other carboxylic acids look and smell very much like their parent acids, but owing to the greater number of carbon atoms their solubility in water is considerably lower. Ethanoic anhydride, for instance, has a solubility in the cold of only 13.6 g per 100 g of water; an excess of anhydride will form a separate *lower* phase (density = 1.087 g cm^{-3} at 20 °C (293 K)). Ethanoic acid, by contrast, is infinitely soluble in water.

C_6H_5COCl

CH_3COCl and C_6H_5COCl

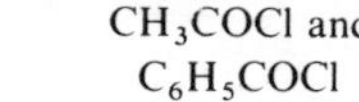

CH_3COCl

Acid chlorides

The acid chloride of methanoic acid, HCOCl, is highly unstable and decomposes spontaneously into carbon monoxide and hydrogen chloride. The other carboxylic acids, however, give stable acid chlorides which are colourless liquids with unpleasant acrid smells. They fume in moist air, owing to hydrolysis to the parent carboxylic acid and hydrogen chloride. Acid chlorides are immiscible with water – they form a separate lower phase – but rapidly undergo hydrolysis when an aqueous mixture is shaken. (Aromatic acid chlorides, such as benzoyl chloride, are hydrolysed by water much more slowly.)

Many esters

Esters

Hydrogen bonding is impossible between the molecules of an ester, and the lower esters are a family of volatile liquids with boiling temperatures close to those of comparable hydrocarbons but far removed from those of carboxylic acids. They have fruity odours, and are in fact responsible for the natural flavours of many fruits. Butyl ethanoate is commonly referred to as 'pear drops'. Esters of low relative molecular mass are fairly soluble in water because hydrogen bonding can occur between molecules of ester and those of water, cf. ethers, aldehydes and ketones.

Esters of 1,2,3-propanetriol (glycerol) and the long chain fatty acids constitute animal and vegetable oils and fats. They have a special importance in that on hydrolysis with sodium hydroxide solution they form soaps (§14.10.13).

Amides

The simplest amide, that of methanoic acid, is a liquid with a boiling temperature of 193 °C (466 K). Other amides are solids. (Ethanamide has a melting temperature of 82.3 °C (355.5 K).) The high melting and boiling temperatures of amides, compared with those of carboxylic acids, suggest the existence of association due to hydrogen bonding:

$$\begin{array}{ccccc} H_2N & O\text{----}H-NH & & O\text{----}H-NH & O \\ \diagdown\,C\,\diagup\!\!\diagup & & \diagdown\,C\,\diagup\!\!\diagup & & \diagdown\,C\,\diagup\!\!\diagup \\ | & & | & & | \\ R & & R & & R \end{array}$$

Hydrogen bonding with water molecules leads to the lower members of the homologous series having a high solubility in water. Methanamide is miscible in all proportions, and ethanamide dissolves to the extent of 97.5 g per 100 g of water at 20 °C (293 K).

14.10.5 CHEMICAL PROPERTIES OF THE C=O BOND

The C=O bond can take part in a great many chemical reactions, although, as we shall see later, its reactivity depends very much on the type of molecule in which it occurs. All its reactions are essentially addition reactions. If we represent the reagent by HA, where H stands for hydrogen and A for a suitable group of atoms, we may write a general equation as follows:

$$>C=O + H-A = \;>C<^{OH}_{A}$$

The adduct thus contains two functional groups, one of which is always HO. In other words the product is always, in part, an alcohol.

Superficially, addition to the C=O bond resembles addition to the C=C bond. Both reactions are accompanied by loss of the double bond. However, the C=O bond is quite different from the C=C bond in that there is considerable polarisation. Oxygen, because it is more electronegative than carbon, exerts a negative inductive effect, and the π electrons of the double bond are displaced towards the oxygen atom, thus:

$$>\overset{\delta+}{C}\overset{\curvearrowright}{=}\overset{\delta-}{O}$$

Consequently, reagents that add to the C=O bond are totally different from those that add to the C=C bond. Bromine, hydrogen bromide, bromine water and sulphuric acid, well known for their reactions with alkenes, do not take part in addition reactions with carbonyl compounds. Any reagent that attacks the carbonyl carbon atom must be a nucleophile, i.e. a chemical species with an atom possessing a negative charge ($-$ or δ^-) and a lone pair of electrons (§9.2.9). *Thus, we are concerned with nucleophilic addition to the C=O bond, whereas, as we saw in §14.4.4, there is electrophilic addition to the C=C bond.*

The nucleophile may be an anion or a molecule, and the reaction may be uncatalysed or acid catalysed. There are thus four basic reaction mechanisms to consider.

I Uncatalysed addition of a nucleophilic anion

Let Nu:$^-$ represent the nucleophilic anion. (The dots symbolise a lone pair of electrons.) Coordination of Nu:$^-$ to the carbonyl carbon atom is accompanied by coordination of the carbonyl oxygen atom to a proton. (In theory, other positively charged species could join the oxygen atom, but combination with a proton gives stable products.) The question immediately arises – which species attacks first? Do we get nucleophilic attack on carbon, followed by electrophilic attack on oxygen (route 1) or vice versa (route 2)?

route 1 $$\overset{\delta+}{\mathrm{C}}=\overset{\delta-}{\mathrm{O}} + \mathrm{Nu:}^- \longrightarrow \mathrm{Nu{-}C{-}O}^{\ominus} \xrightarrow{+\mathrm{H}^+} \mathrm{Nu{-}C{-}OH}$$

alkoxide ion

route 2 $$\overset{\delta+}{\mathrm{C}}=\overset{\delta-}{\mathrm{O}} + \mathrm{H}^+ \longrightarrow \overset{\oplus}{\mathrm{C}}\mathrm{{-}OH} \xrightarrow{+\mathrm{Nu:}^-} \mathrm{Nu{-}C{-}OH}$$

carbonium ion

In uncatalysed reactions the first route is considered to be the more likely. Alkoxide ions, because they contain electronegative oxygen bearing a negative charge, are more stable and are thus more readily obtained than carbonium ions.

II Acid catalysed addition of a nucleophilic anion

In highly acidic conditions route 2 is the more probable. It will be seen that protons catalyse the reaction by converting the partial positive charge on the carbonyl carbon atom into a full positive charge, thereby increasing the attraction for nucleophiles.

III Uncatalysed addition of a nucleophilic molecule

In many reactions a nucleophilic molecule, which we can represent by H—Nu:, takes the place of the anion Nu:$^-$ and the proton H$^+$. (Examples are H_2O and NH_3.) In such reactions, the initial attack of the nucleophile leads to the formation of an intermediate which, because it possesses both positive and negative charges, may be referred to as a *zwitterion.* This is converted to the final adduct by the transfer of a proton.

$$\overset{\delta-}{\mathrm{Nu}}{-}\overset{\delta+}{\mathrm{H}} \downarrow \overset{\delta+}{\mathrm{C}}=\overset{\delta-}{\mathrm{O}} \longrightarrow \mathrm{H{-}}\overset{\oplus}{\mathrm{Nu}}\mathrm{{-}C{-}O}^{\ominus} \xrightarrow{-\mathrm{H}^+} \mathrm{Nu{-}C{-}O}^{\ominus} \xrightarrow{+\mathrm{H}^+} \mathrm{Nu{-}C{-}OH}$$

coordination of the nucleophile — zwitterion — alkoxide ion

During the transfer of the proton an alkoxide ion is momentarily formed. *Uncatalysed addition across the C=O bond always proceeds via an alkoxide ion.*

IV Acid catalysed addition of a nucleophilic molecule

Molecular nucleophiles are relatively weak, because they possess only a partial negative charge, and may require the assistance of an acid catalyst. In such circumstances a proton attacks first, at the oxygen atom, followed by the molecule H—Nu: to the carbon atom. The reaction is completed by regeneration of the proton catalyst:

$$\overset{\delta+}{>C}=\overset{\delta-}{O} + H^+ \longrightarrow >\overset{\oplus}{C}-OH \xrightarrow{+H-Nu:} >C(\overset{\oplus}{Nu}-H)(OH) \xrightarrow{-H^+} >C(Nu)(OH)$$

carbonium ion

Acid catalysed addition across the C=O bond always proceeds via a carbonium ion.

To produce stable end-products, addition reactions across the C=O bond are often followed by the loss of water or other small molecules such as HCl or C_2H_5OH (Table 14.17). Such reactions are known as *addition–elimination* or *condensation* reactions. Almost all the reactions of carboxylic acids and their derivatives are of this type.

Table 14.17 Addition–elimination reactions of carbonyl compounds

Carbonyl compound	Molecule eliminated
aldehyde, RCHO	H_2O
ketone, R_2CO	H_2O
carboxylic acid, RCOOH	H_2O
acid anhydride, $(RCO)_2O$	RCOOH
acid chloride, RCOCl	HCl
ester, RCOOR′	R′OH
amide, $RCONH_2$	NH_3

Reactivity of carbonyl compounds

The rates of carbonyl addition reactions vary tremendously, depending partly on the nature of the nucleophile and partly on that of the carbonyl compound itself. We have discussed (§9.2.9) the requirements of a good nucleophile, noting particularly that its lone pair of electrons should readily be donated, and that it should form a strong bond with a carbon atom. The nature of the carbonyl compound is perhaps even more important, for such compounds range in reactivity from carboxylic acid salts, which undergo no addition reactions at all, to aldehydes and acid chlorides which are highly reactive. The order of reactivity is as follows:

acid chloride > acid anhydride > aldehyde > ketone ≫ ester > amide > acid ≫ salt

To understand this, we must realise that there are three effects which can influence reactivity.

Inductive effect (§9.2.10) The high reactivity of aldehydes compared with ketones is accounted for very largely by the existence of *two* inductive effects in a ketone. Consider the following compounds:

$H_2C{=}O$ (methanal) $\quad CH_3(H)C{=}O$ (ethanal) $\quad (CH_3)_2C{=}O$ (propanone)

Because the methyl group has a +I effect, it reduces the positive charge on the carbonyl carbon atom and renders it less susceptible to nucleophilic attack. The reactivity order of these three compounds is thus methanal > ethanal > propanone.

Alkyl groups usually exert a +I effect. That of the ethyl group is slightly greater than that of the methyl group; thus propanal is slightly less reactive than ethanal, and 3-pentanone is less reactive than propanone.

The inductive effect explains also the high reactivity of acid chlorides:

$R{-}C({=}O){-}Cl$

The chlorine atom has a −I effect, i.e. it draws electrons away from the carbonyl carbon atom and so increases its positive charge.

Mesomeric effect The generally low reactivity of carboxylic acids and their derivatives, in comparison with aldehydes and ketones, is accounted for by the mesomeric effect. This is closely similar to the mesomeric effect in chlorobenzene (§14.8.4). In the same way that in chlorobenzene a p

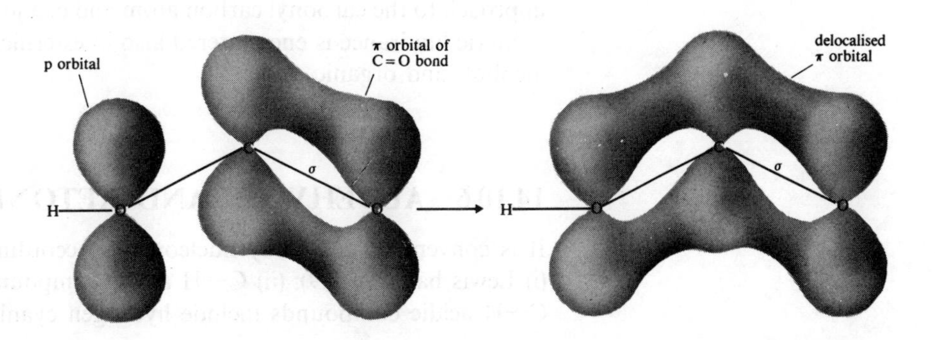

Fig. 14.6 Mesomeric effect in the carboxyl group.

orbital of the chlorine atom interacts with the π orbital of the benzene ring, so in carboxylic acids a p orbital of the oxygen atom of the HO group interacts with the π orbital of the C=O bond. Hence a delocalised π orbital is established, covering the carbonyl carbon atom and both oxygen atoms.

Electrons in the delocalised π orbital tend to shield the carbon atom from attack by nucleophilic reagents. Carboxylic acids are therefore relatively unreactive in addition–elimination reactions, and we say that these compounds have a low *carbonyl activity.*

The 2p orbital on the hydroxyl oxygen atom is occupied by a lone pair of electrons, and for most purposes it is sufficient to argue that this lone pair of electrons is drawn towards the carbonyl carbon atom, thereby reducing its partial positive charge and its attraction for nucleophiles. The effect is symbolised in the following way:

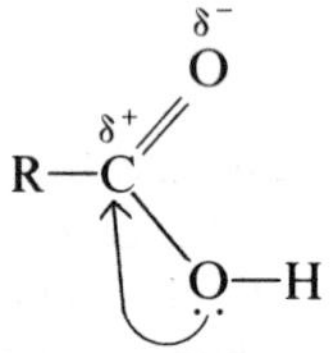

$$R-\overset{\delta+}{C}(=\overset{\delta-}{O})-\ddot{O}-H$$

The mesomeric effect is present in acid derivatives, such as esters, amides and acid chlorides:

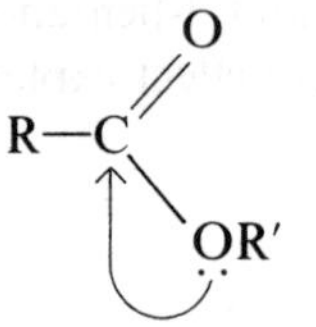

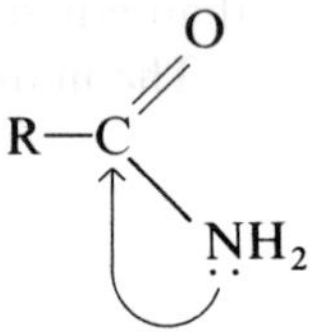

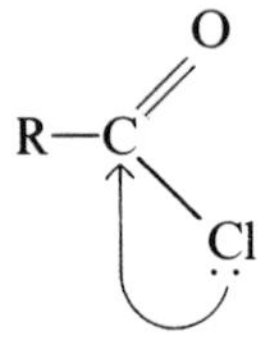

$$R-C(=O)-\ddot{O}R' \qquad R-C(=O)-\ddot{N}H_2 \qquad R-C(=O)-\ddot{Cl}$$

Consequently, esters and amides, like carboxylic acids, have a relatively low carbonyl activity. Acid chlorides, however, are highly reactive because the negative inductive effect of the chlorine atom more than compensates for the mesomeric effect.

Steric hindrance One of the reasons for the reactivity difference between aldehydes and ketones is that the latter are *sterically hindered.* In an aldehyde molecule the functional group is joined on one side to a hydrogen atom which, being very small, permits ready access of reagents. A ketone, however, has its carbonyl group shielded by hydrocarbon groups on both sides. A reagent, especially if it is bulky, is mechanically hindered in its approach to the carbonyl carbon atom and can join it only with difficulty.

Steric hindrance is encountered also in esterification reactions between alcohols and organic acids.

14.10.6 ALDEHYDES AND KETONES

It is convenient to classify nucleophiles according to whether they are: (i) Lewis bases (§9.2.9); (ii) C—H acidic compounds; or (iii) cryptobases. C—H acidic compounds include hydrogen cyanide and many carbonyl compounds such as aldehydes, ketones and esters with an acidic α-hydrogen atom (§14.7.5). Cryptobases are reagents with hidden basic characteristics. The best known examples are complex aluminium hydrides and borohydrides.

Reaction with Lewis bases

The reagents that concern us here include water, alcohols, sodium hydrogensulphite, ammonia, and derivatives of ammonia both organic and inorganic. While many of these compounds react at a reasonable rate in neutral or weakly alkaline conditions, there are some – notably alcohols and 2,4-dinitrophenylhydrazine – which are too feebly basic to do so. In such cases a strong acid (HCl or H_2SO_4) is used as a catalyst. The concentration of acid catalyst is critical, for the acid affects not only the carbonyl compound but also the reagent. Too high a concentration converts the reagent into an unreactive form,

e.g. $:NH_2R + H^+ \rightleftharpoons \overset{\oplus}{N}H_3R$ (no lone pair on N)

ammonia derivative — substituted ammonium ion

Certain reagents which are amphoteric, notably hydroxylamine, function best in strongly alkaline solution.

H_2O

Aldehydes in aqueous solution exist in equilibrium with 1,1-diols, commonly called *aldehyde hydrates*,

e.g. $CH_3C(=O)H + H_2O \rightleftharpoons CH_3C(OH)_2H$

1,1-ethanediol (acetaldehyde hydrate)

Mechanism III, §14.10.5. (To master the subject, the mechanism of this and other reactions in this section should be worked out in detail.)

In the case of methanal, equilibrium lies almost entirely to the right-hand side; for ethanal, the extent of hydrate formation at room temperature has been estimated at 58%. Ketones do not undergo extensive hydration into $R_2C(OH)_2$ because the carbonyl carbon atom is insufficiently positively charged to attract H_2O – a weak nucleophile.

Most 1,1-diols are unstable and cannot be isolated because H_2O, which is a highly stable molecule, can readily be formed from the diol via an alkoxide ion:

$$R\text{—}C(=O)H + H_2O \rightleftharpoons R\text{—}C(OH)(\overset{\ominus}{O})H + H^+ \rightleftharpoons R\text{—}C(OH)_2H$$

alkoxide ion

By contrast, monohydric alcohols and 1,2-diols are stable, for they can only be dehydrated via a carbonium ion (§14.9.5), a much more difficult route.

H_2O + oxidant

Aldehydes are readily oxidised in aqueous solution to give carboxylic acids. The reactions involve the elimination of hydrogen from aldehyde hydrates, and are thus similar to the oxidation of alcohols by means of dichromate solution (§14.9.5).

$$\underset{\text{aldehyde}}{R{-}C({=}O){-}H} \xrightarrow{+H_2O} \underset{\text{aldehyde hydrate}}{R{-}C(OH)(H){-}OH} \xrightarrow{-(2H^+ + 2e^-)} \underset{\text{carboxylic acid}}{R{-}C({=}O){-}OH}$$

$$\text{cf.} \quad \underset{\text{primary alcohol}}{R{-}C(H)(H){-}OH} \xrightarrow{-(2H^+ + 2e^-)} \underset{\text{aldehyde}}{R{-}C({=}O){-}H}$$

Ketones are unable to take part in this reaction for they do not possess a hydrogen atom on the carbonyl carbon atom. (With a powerful oxidant, such as HNO_3, ketones undergo oxidation with cleavage of C—C bonds (§14.3.4).)

Mild oxidants, such as Fehling's solution or Tollens' reagent, which do not attack ketones, are useful in the laboratory for distinguishing between aldehydes and ketones. Fehling's solution is a mixture of copper(II) sulphate and an alkaline solution of potassium sodium 2,3-dihydroxybutanedioate (potassium sodium tartrate, 'Rochelle salt'). It functions as an oxidant because it contains copper(II) as a complex 2,3-dihydroxybutanedioate, which can be reduced to copper(I) in the form of copper(I) oxide:

$$RCHO + 2Cu^{2+} + 5HO^- = RCOO^- + Cu_2O + 3H_2O$$

Fehling's solutions I and II (the $CuSO_4$ and the alkaline solution of Rochelle salt) are mixed together in roughly equal proportions, the aldehyde is added and the mixture boiled. The dark blue colour of the solution gradually fades and a reddish brown precipitate of copper(I) oxide appears as redox occurs.

Tollens' reagent is an ammoniacal solution of silver nitrate, prepared by adding one or two drops of sodium hydroxide solution to silver nitrate solution and then adding aqueous ammonia until the precipitate of silver oxide dissolves. The reagent contains the diamminesilver(I) ion, and can oxidise an aldehyde because silver(I) in this form may be readily reduced to silver(0), i.e. metallic silver.

$$RCHO + 2[Ag(NH_3)_2]^+ + 3HO^- = RCOO^- + 2Ag + 4NH_3 + 2H_2O$$

The reaction is usually conducted in such a way that silver is deposited on the walls of a test tube. For this purpose, the test tube must be chemically clean and the silver must be liberated slowly. This entails using a very dilute solution and a moderate temperature – no more than 50 °C (323 K).

Unlike Fehling's solution, which can be stored without deterioration, Tollens' reagent must be prepared at the time it is required. On standing, silver nitride, Ag_3N, is formed, which is highly explosive ('fulminating silver').

H_2O + acid (polymerisation)

Aldehyde hydrates readily undergo etherification to give polyethers in the presence of an acid catalyst; cf. the etherification of alcohols in the presence of acid (§14.9.5). For the simplest possible case,

$$\mathrm{R{-}C(=O){-}H} + \mathrm{H_2O} \rightleftharpoons \mathrm{R{-}C(OH)(OH){-}H}$$

$$\mathrm{HO{-}CHR{-}\boxed{OH + H}O{-}CHR{-}OH} = \mathrm{HO{-}CHR{-}O{-}CHR{-}OH} + \mathrm{H_2O}$$

Many molecules of aldehyde hydrate may be involved, and a variety of chain and ring structures can be formed.

For example, the evaporation of an acidic solution of methanal gives a long chain polymer commonly called paraformaldehyde:

$$\mathrm{HOCH_2\,\boxed{OH + \mathit{n}H}\,OCH_2\,\boxed{OH + H}\,OCH_2OH} = \mathrm{HOCH_2(OCH_2)_\mathit{n}OCH_2OH} + (n+1)\mathrm{H_2O}$$

poly(methanal)
(paraformaldehyde)

Paraformaldehyde

Paraformaldehyde is a white amorphous powder, which may be readily depolymerised on heating to 180–200 °C (453–473 K). For this reason it is often used as a convenient source of methanal.

Ketones, because they do not form hydrates, do not polymerise in this way.

ROH

If an aldehyde is allowed to react with an alcohol in the presence of anhydrous mineral acid (usually HCl), nucleophilic addition occurs by mechanism IV (§14.10.5) to give a *hemiacetal*:

$$\mathrm{R{-}C(=O){-}H} + \mathrm{R'OH} \rightleftharpoons \mathrm{R{-}C(OH)(OR'){-}H}$$

(cf. the addition of H_2O)

$$\text{e.g.}\quad CH_3C(=O)H + CH_3OH \rightleftharpoons CH_3C(OH)(H)-OCH_3$$

1-methoxyethanol

A further reaction occurs readily to give the *acetal*; but this second reaction is an etherification and not an addition:

$$R-C(OH)(H)-OR' + R'OH = R-C(OR')(H)-OR' + H_2O$$

Because they are diethers, acetals are unreactive compounds.

Ketones cannot give *ketals* by direct reaction with alcohols as there is insufficient positive charge on the carbonyl carbon atom. Ketals can, however, be prepared indirectly.

$NaHSO_3$

When an aldehyde or a methyl ketone is shaken at room temperature with a saturated aqueous solution of sodium hydrogensulphite, a white crystalline precipitate is obtained of a so-called 'bisulphite addition compound' – in reality, a sodium salt of a hydroxysulphonic acid. The reaction is uncatalysed; HSO_3^- adds first, to carbon, followed by the transfer of H^+ to oxygen (mechanism I, §14.10.5),

$$\text{e.g.}\quad (CH_3)_2C{=}O + Na^+HSO_3^- \rightleftharpoons (CH_3)_2C(\overset{\ominus}{O}Na^+)(SO_3H)$$

$$\rightleftharpoons (CH_3)_2C(OH)(SO_3^{\ominus}\,Na^+)$$

sodium 2-hydroxy-2-propanesulphonate
(acetone sodium bisulphite)

A proton is transferred in the final stage of the mechanism because the SO_3H group is more acidic than the HO group and has a greater tendency to ionise.

Hydrogensulphite addition compounds can be hydrolysed back to the compounds from which they are derived by treatment with sodium carbonate solution or hydrochloric acid. Because of this, and because they are formed only by aldehydes and ketones, hydrogensulphite addition compounds can be used in the purification of aldehydes and ketones. (The impure aldehyde or ketone is mixed with sodium hydrogensulphite solution, and the resulting precipitate is filtered off and washed. Subsequent treatment with sodium carbonate or hydrochloric acid liberates the aldehyde or ketone, which can be recovered by distillation or ether extraction.)

Aldehydes undergo the reaction more easily than ketones, especially the higher ketones, a fact which is utilised in *Schiff's reagent* for distinguishing between aldehydes and ketones. This reagent is a solution of the dyestuff rosaniline ('magenta'), which has been decolorised by the passage of sulphur dioxide. Aldehydes will restore the colour to Schiff's reagent at room temperature, as they can abstract sulphur dioxide by an addition reaction, but ketones react only very slowly.

RNH₂

Primary amines, RNH_2, react readily with aldehydes and ketones by an uncatalysed nucleophilic addition reaction:

$$\underset{\text{aldehyde or ketone}}{>C=O} + \underset{\text{primary amine}}{RNH_2} \rightleftharpoons >C(OH)(NHR)$$

mechanism III, §14.10.5

The adducts are unstable – so much so that they cannot be isolated – and undergo spontaneous loss of water to give more stable products. (This is an addition–elimination reaction.)

There are two possible ways in which water could be eliminated:

$$H{-}\overset{|}{\underset{|}{C}}{-}C(OH)(NHR) = {-}\overset{|}{C}{=}C(NHR) + H_2O \qquad (1)$$

$$>C(OH)(NHR) = \underset{\text{Schiff's base}}{>C=NR} + H_2O \qquad (2)$$

From our knowledge of the dehydration of alcohols to alkenes we might expect reaction (1), but in practice reaction (2) occurs because it is easier for a proton to be lost from a nitrogen atom than a carbon atom. (The reason is that the N—H bond is more polarised and hence more easily broken than the C—H bond.)

Wherever possible, in the elimination of water, a hydrogen atom leaves a nitrogen atom rather than a carbon atom.

The products of these reactions are therefore substituted imines, usually known as *Schiff's bases* or *azomethines.* A well-known example is *N*-benzylidenephenylamine, made by warming together benzaldehyde and phenylamine:

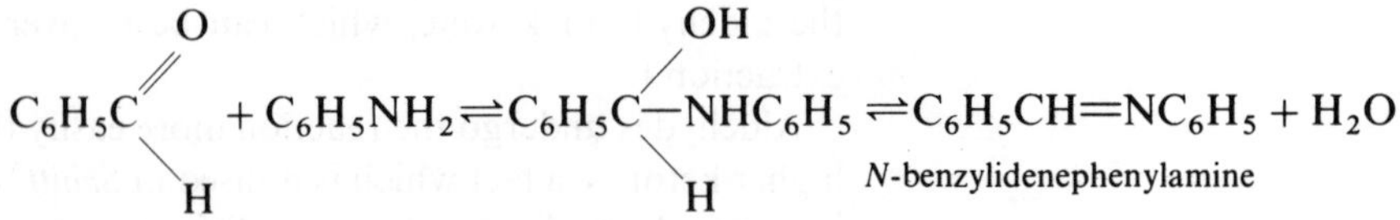

$$C_6H_5C(=O)H + C_6H_5NH_2 \rightleftharpoons C_6H_5C(OH)(H)-NHC_6H_5 \rightleftharpoons C_6H_5CH{=}NC_6H_5 + H_2O$$

N-benzylidenephenylamine

Schiff's bases may be reduced to secondary amines by the addition of hydrogen across the double bond. Note the reversible nature of these reactions. Schiff's bases, like many other derivatives of aldehydes and ketones, can be hydrolysed back to the parent compounds.

NH_2OH

Hydroxylamine, NH_2OH, undergoes addition–elimination reactions with aldehydes and ketones to give stable solids called *oximes*, which crystallise well, have sharp melting temperatures, and can therefore be used in the identification of aldehydes and ketones. E.g.

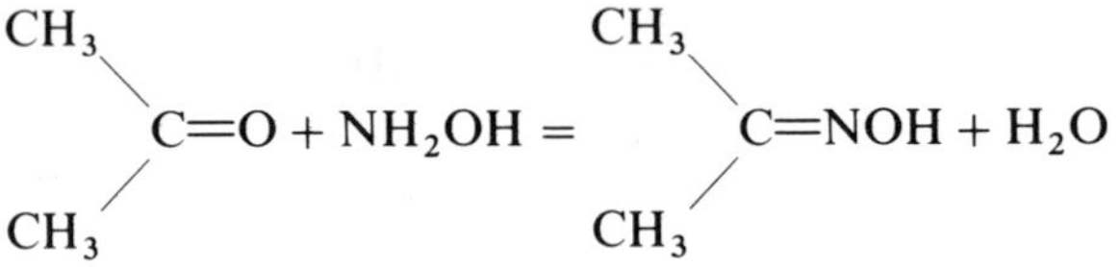

$$(CH_3)_2C{=}O + NH_2OH = (CH_3)_2C{=}NOH + H_2O$$

propanone oxime
$\theta_{c,m} = 61\ °C$ ($T_m = 334$ K)

Although the reaction can be conducted in acidic solution, it is more usual to work under alkaline conditions. Alkali converts hydroxylamine into a salt, which is completely dissociated in solution:

$$NH_2OH + NaOH = Na^+\overset{\ominus}{N}HOH + H_2O$$

The hydroxylamide ion is a better nucleophile than the hydroxylamine molecule because of the full negative charge on the nitrogen atom, and attacks the aldehyde or ketone by mechanism I, §14.10.5.

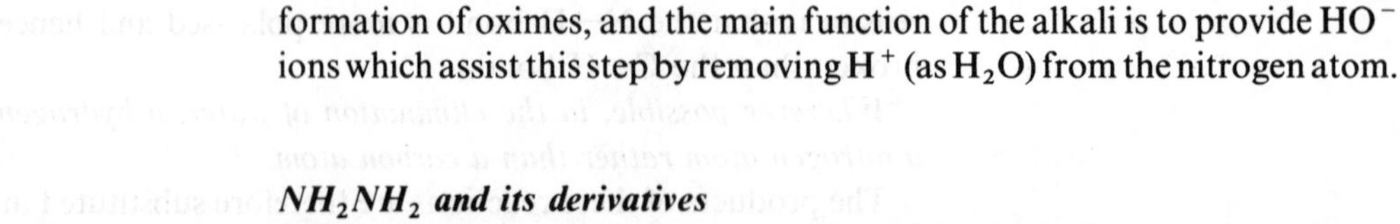

$$>C{=}O + \overset{\ominus}{N}HOH \rightleftharpoons >C(NHOH)\overset{\ominus}{O} \xrightarrow{H^+ \text{(from } H_2O)} >C(NHOH)OH \rightleftharpoons >C{=}NOH + H_2O$$

adduct — oxime

However, dehydration of the adduct is the rate determining step in the formation of oximes, and the main function of the alkali is to provide HO^- ions which assist this step by removing H^+ (as H_2O) from the nitrogen atom.

NH_2NH_2 *and its derivatives*

Hydrazine, NH_2NH_2, reacts readily with aldehydes and ketones in addition–elimination reactions (mechanism III, §14.10.5) to give *hydrazones*:

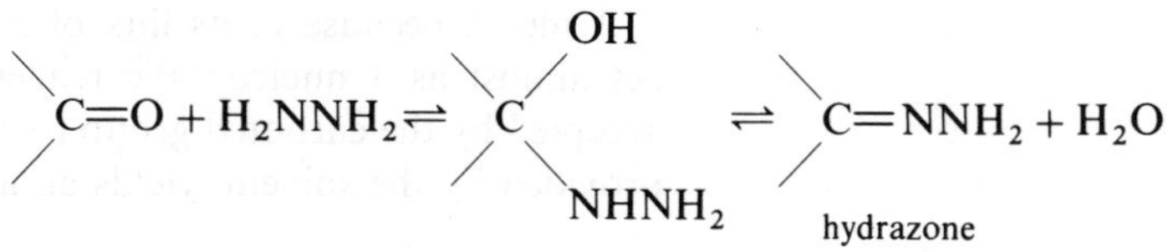

$$\rangle C{=}O + H_2NNH_2 \rightleftharpoons \rangle C\langle^{OH}_{NHNH_2} \rightleftharpoons \rangle C{=}NNH_2 + H_2O$$

hydrazone

Propanone, for example, gives propanone hydrazone, $(CH_3)_2C{=}NNH_2$.

The use of hydrazine in characterising aldehydes and ketones is limited by the fact that most hydrazones are liquids. In addition, a further reaction with the aldehyde or ketone can occur to give an *azine*:

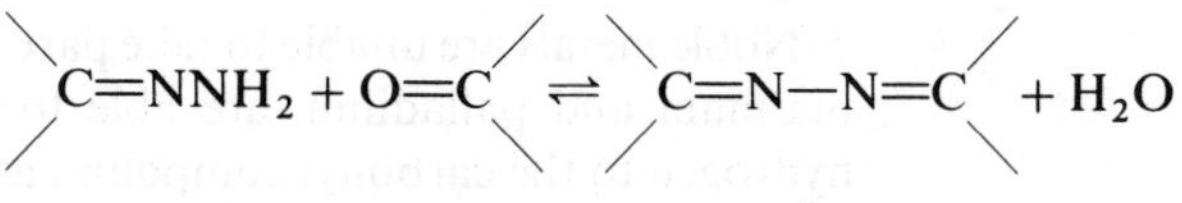

$$\rangle C{=}NNH_2 + O{=}C\langle \rightleftharpoons \rangle C{=}N{-}N{=}C\langle + H_2O$$

azine

To overcome this difficulty, it is necessary to use a substituted hydrazine, such as phenylhydrazine or 2,4-dinitrophenylhydrazine,

e.g. $$CH_3(H)C{=}O + H_2NNHC_6H_5 = CH_3(H)C{=}NNHC_6H_5 + H_2O$$

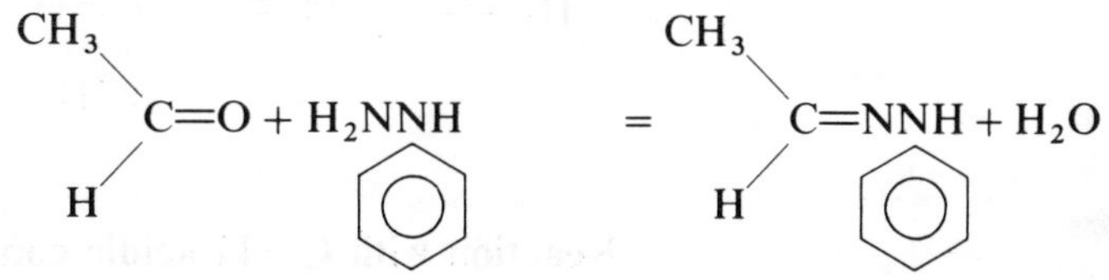

ethanal phenylhydrazone
$\theta_{c,m} = 101\ °C\ (T_m = 374\ K)$

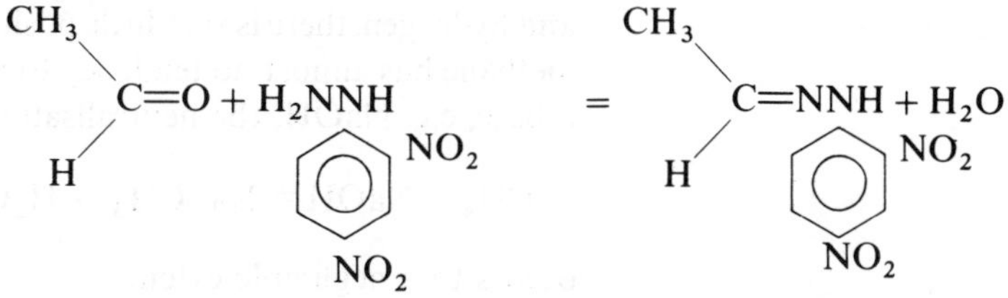

$$CH_3(H)C{=}O + H_2NNHC_6H_3(NO_2)_2 = CH_3(H)C{=}NNHC_6H_3(NO_2)_2 + H_2O$$

ethanal-2,4-dinitrophenylhydrazone
$\theta_{c,m} = 168\ °C\ (T_m = 441\ K)$

2,4-Dinitrophenylhydrazine requires the use of an acid catalyst (mechanism IV, §14.10.5).

2,4-Dinitrophenylhydrazine in dilute sulphuric acid (*Brady's reagent*) is used as a general test for aldehydes and ketones. Nearly all such compounds, at room temperature, give an immediate orange precipitate of the 2,4-dinitrophenylhydrazone. The derivative can be filtered off, recrystallised from ethanol, dried, and its melting temperature determined. The result, when compared with tabulated values in reference books, provides a guide to the identity of the original aldehyde or ketone.

Metals

Aldehydes and ketones can be reduced to their corresponding primary or secondary alcohols by using a metal and a protic solvent,

e.g. $$CH_3COCH_3 \xrightarrow{Zn + HCl} CH_3CH(OH)CH_3$$

A metal, because of its flux of readily available electrons (§6.5.1), can act almost as a nucleophilic reagent. Electrons from the metal can be accepted by the carbonyl group to give a dianion (I), which, with protons provided by the solvent, yields an alcohol (II).

$$\mathrm{>C{=}O + 2e^- \longrightarrow \underset{(I)}{>\overset{\ominus}{C}{-}\overset{\ominus}{O}} \xrightarrow{2H^+} \underset{(II)}{>C(H){-}OH}}$$

Noble metals are unable to take part in such reactions, but some, notably platinum and palladium, are able to *transfer* electrons from molecular hydrogen to the carbonyl compound and thereby bring about its catalytic hydrogenation, e.g.

$$\mathrm{H_2 \xrightarrow{2e^-} Pt \xrightarrow{2e^-} (CH_3)_2C{=}O \longrightarrow 2H^+ + (CH_3)_2\overset{\ominus}{C}{-}\overset{\ominus}{O} \longrightarrow H{-}C(CH_3)_2{-}OH}$$

Reaction with C—H acidic compounds

Very few C—H compounds are acidic. Consider methane, for example. Because of the small difference between the electronegativities of carbon and hydrogen, there is very little positive charge on the hydrogen atoms and methane has almost no tendency to ionise. (Its pK_a value is over 40.) With a base, e.g. NaOH, the neutralisation:

$$CH_4 + NaOH \rightleftharpoons Na^+CH_3^- + H_2O$$

occurs to a negligible extent.

But some C—H compounds, because of their structure, are slightly acidic. A simple example is provided by hydrogen cyanide, where polarisation of the C≡N bond induces a corresponding electronic displacement in the C—H bond:

$$\overset{\delta+}{H}{-}C{\equiv}\overset{\delta-}{N}$$

There is sufficient positive charge on the hydrogen atom to cause appreciable ionisation in water ($pK_a = 9.40$). With a base, e.g. NaOH, neutralisation occurs:

$$HCN + NaOH \rightleftharpoons Na^+CN^- + H_2O$$

The anion $:CN^-$, with a lone pair of electrons, is a nucleophile and can react with aldehydes and ketones. Pure HCN, by contrast, is not a nucleophile.

Other C—H acidic compounds include aldehydes and ketones with an α-hydrogen atom (§14.7.5).

HCN

A mixture of sodium cyanide and ethanoic acid reacts with aldehydes and ketones at room temperature to give hydroxynitriles, commonly known as *cyanohydrins.* The reaction is reversible, and yields at equilibrium are better with aldehydes than with ketones.

$$\text{E.g.}\quad CH_3C(=O)H + HCN \rightleftharpoons CH_3C(OH)(H){-}CN \qquad \text{mechanism I, §14.10.5}$$

2-hydroxypropanenitrile (acetaldehyde cyanohydrin)

Provided that a basic catalyst is included to form CN^- ions, the reaction may also be accomplished by means of liquid or aqueous hydrogen cyanide.

Cyanohydrins are useful intermediates in the preparation of α-hydroxy-acids,

$$\text{e.g.}\quad CH_3CH(OH)CN \xrightarrow{\text{acidic hydrolysis}} CH_3CH(OH)COOH$$

2-hydroxypropanoic acid (lactic acid)

Aldehydes or ketones with an α-hydrogen atom (aldol addition and aldol condensation)

When ethanal is treated with calcium hydroxide solution at room temperature, it dimerises to give a syrupy liquid commonly called aldol:

$$2CH_3CHO \xrightarrow{Ca(OH)_2} CH_3CH(OH)CH_2CHO$$

3-hydroxybutanal (aldol)

Any reaction of this type, in which an aldehyde or a ketone (or a mixture of such compounds) possessing an α-hydrogen atom dimerises in the presence of a base, is known as an *aldol addition.*

Ethanal has acidic hydrogen atoms at the α position (§14.7.5), and when treated with a base (usually the hydroxide of a metal in group 1A or 2A) is partially converted into a *carbanion*, literally 'carbon anion', i.e. an ion in which negative charge resides on a carbon atom:

$$H{-}\overset{\overset{\delta+}{H}}{\underset{H}{C}}{-}C(=\overset{\delta-}{O})H + HO^- \rightleftharpoons H{-}\overset{\ominus}{\underset{H}{C}}{-}C(=O)H + H_2O$$

The nucleophilic carbanion attacks a second molecule of ethanal, and the addition is completed by a proton from the water (mechanism I, §14.10.5).

$$(CH_3)(H)C{=}O + \overset{\ominus}{C}H_2CHO \rightleftharpoons (CH_3)(H)C(CH_2CHO)(\overset{\ominus}{O}) \xrightleftharpoons{+H^+} (CH_3)(H)C(CH_2CHO)(OH)$$

With calcium hydroxide as the catalyst, the reaction stops mainly at this stage, although with sodium hydroxide, which is a stronger base, repeated aldol additions occur to give a long chain polymer which appears as a sticky brown resin.

In most aldol additions it is possible to isolate the product provided that the temperature is kept at about 20 °C (293 K). However, on warming, or in the presence of an acid, the product undergoes the elimination of water to give an unsaturated aldehyde or ketone,

$$\text{e.g.}\quad CH_3CH(OH)CH_2CHO \xrightarrow{\text{warm}} \underset{\text{2-butenal (crotonaldehyde)}}{CH_3CH{=}CHCHO} + H_2O$$

An aldol addition followed by the elimination of water is known as an *aldol condensation.*

Reaction with cryptobases

The word 'cryptic' means 'hidden'; thus, a *cryptobase* is a chemical species with hidden basic characteristics. A simple example is the tetrahydrido-aluminate(III) ion, $[AlH_4]^-$. This ion is *not* a Lewis base, for it does not possess a lone pair of electrons and cannot form coordinate bonds. But it can serve as a source of the hydride ion, H^-, which *is* a base. The species $[AlH_4]^-$ is therefore referred to as a cryptobase.

Cryptobases can be useful for adding certain anions, notably H^- and R^- (an alkyl anion) to the C=O bond. Suppose, for example, that we wished to introduce H^- into an aldehyde or a ketone. What reagent could we use? Not sodium hydride or calcium hydride, because these compounds are insoluble in organic solvents and are decomposed by water. But we could use lithium tetrahydridoaluminate(III), $Li[AlH_4]$, which can serve as a source of hydride ions and which is soluble in ether.

Li[AlH₄] and Na[BH₄]

Both these reagents readily reduce virtually all aldehydes and ketones to their respective primary and secondary alcohols in high yield. Sodium tetrahydroborate(III), $Na[BH_4]$, is often preferred to $Li[AlH_4]$ because it can be used in ethanol or water rather than ether.

The reaction mechanism involves the transfer of the true base, H^-, from the cryptobase to the carbonyl compound:

$$\underset{[AlH_4]^-\text{ ion}}{H_3Al^{\ominus}{-}H} + \underset{\text{aldehyde or ketone}}{>C{=}O} \longrightarrow AlH_3 + \underset{\text{alkoxide ion}}{H{-}\overset{|}{\underset{|}{C}}{-}O^{\ominus}}$$

All four hydrogen atoms of the $[AlH_4]^-$ ion are used, because alkoxide ion coordinates to the aluminium hydride molecule:

$$H{-}\overset{|}{\underset{|}{C}}{-}O^{\ominus} + AlH_3 \longrightarrow [H{-}\overset{|}{\underset{|}{C}}{-}O{-}AlH_3]^-$$

The ion formed in this way is also a cryptobase and reacts with another molecule of aldehyde or ketone. The process is repeated twice more, so that ultimately the lithium tetrahydridoaluminate(III) is converted into a lithium tetraalkoxyaluminate(III):

$$\mathrm{Li[AlH_4]} \longrightarrow \mathrm{Li[Al(O{-}\overset{|}{\underset{|}{C}}{-}H)_4]}$$

The alcohol is obtained on hydrolysis with dilute sulphuric acid.

Aldehydes without an α-hydrogen atom (Cannizzaro reaction)
We have seen that if an aldehyde possesses an α-hydrogen atom it undergoes the aldol addition when it is treated with alkali. If it does not possess an α-hydrogen atom it may instead take part in a *Cannizzaro reaction*, which is a disproportionation of two molecules of the aldehyde. One molecule is oxidised to the corresponding carboxylic acid anion, while the other is reduced to the corresponding primary alcohol,

$$\text{e.g.}\quad \underset{\text{methanal}}{2\mathrm{HCHO}} + \underset{\substack{50\%\\\text{solution}}}{\mathrm{NaOH}} \xrightarrow[(303\ \mathrm{K})]{30\ ^\circ\mathrm{C}} \underset{\substack{\text{sodium}\\\text{methanoate}}}{\mathrm{HCOONa}} + \underset{\text{methanol}}{\mathrm{CH_3OH}}$$

$$2\,\underset{\text{benzaldehyde}}{\mathrm{C_6H_5CHO}} + \underset{\substack{60\%\\\text{solution}}}{\mathrm{KOH}} \xrightarrow[(333\ \mathrm{K})]{60\ ^\circ\mathrm{C}} \underset{\substack{\text{potassium}\\\text{benzoate}}}{\mathrm{C_6H_5COOK}} + \underset{\text{phenylmethanol}}{\mathrm{C_6H_5CH_2OH}}$$

It is a feature of such reactions that a strong base is required. Under such conditions, the aldehyde hydrate is partially converted into an alkoxide ion which is a cryptobase, capable of transferring H^- to a molecule of aldehyde.

14.10.7 CARBOXYLIC ACIDS

Because of the structure of the carboxyl group, $-\mathrm{C}\!\begin{smallmatrix}\nearrow \mathrm{O}\\ \searrow \mathrm{OH}\end{smallmatrix}$, carboxylic acids display two sets of reactions, namely those of the C—OH bond (cf. alcohols), and those of the C=O bond (cf. aldehydes and ketones).

Reactions of the C—OH bond (including the O—H bond)

When we discussed alcohols (§14.9.5), we did so under three main headings: (i) acidic and basic character; (ii) nucleophilic substitution reactions; and (iii) elimination reactions. It is logical to consider carboxylic acids under the same three headings.

Acidic and basic character

Acidity The acidic character of carboxylic acids is due to the presence of the hydroxyl group, which allows ionisation to occur in the presence of water:

$$R{-}C(=O){-}O{-}H \rightleftharpoons R{-}C(=O){-}O^{\ominus} + H^+$$

carboxylate ion

Methanoic acid and ethanoic acid have pK_a values of 3.75 and 4.76 respectively, showing that carboxylic acids are more strongly acidic than alcohols and phenols. (The corresponding values for methanol and phenol are 15.5 and 10.00.) The increased acidity is due to the negative inductive effect of the carbonyl oxygen atom, which, transmitted through the molecule, increases the polarisation of the O—H bond:

$$R{-}C(=O){-}\overset{\delta-}{O}{-}\overset{\delta+}{H}$$

In this way the hydrogen atom acquires an increased positive charge and can more easily break away as a proton in the presence of water (§9.2.10).

Carboxylic acids can be neutralised by bases to give salts,

e.g. $CH_3COOH + NaOH \rightleftharpoons CH_3COO^-Na^+ + H_2O$

sodium ethanoate

$CH_3COOH + NH_3(aq) \rightleftharpoons CH_3COO^-NH_4^+$

ammonium ethanoate

They react with strongly electropositive metals with the production of hydrogen,

e.g. $2CH_3COOH + 2Na = 2CH_3COONa + H_2$

Because they are stronger acids than carbonic acid, they will liberate carbon dioxide from carbonates and hydrogencarbonates,

e.g. $2CH_3COOH + Na_2CO_3 = 2CH_3COONa + H_2O + CO_2(g)$

They are, however, much weaker than the principal mineral acids. Halogenated acids are stronger than the unsubstituted acids, due to the negative inductive effect of the halogen atoms (§9.2.10).

Basicity Carboxylic acids act as Lewis bases in their reactions with phosphorus trichloride:

$$3RCOOH + PCl_3 \xrightarrow{\text{heat}} 3RCOCl + H_2PHO_3$$

The reaction is analogous to that with alcohols (§14.9.5).

Phosphorus trichloride converts carboxylic acids to acid chlorides in reasonably good yield, and is a satisfactory reagent for the purpose provided that the acid chloride has a boiling temperature well below 200 °C (473 K) – the decomposition temperature of phosphonic acid – so that the two compounds can be separated by distillation.

Phosphorus pentachloride reacts in a closely similar fashion to the trichloride:

$$RCOOH + PCl_5 = RCOCl + POCl_3 + HCl \qquad \text{(cf. §14.9.5)}$$

Reactions are vigorous, yields of acid chloride are good, and the products can readily be separated by fractional distillation provided that their boiling temperatures are not too close together. (The boiling temperature of $POCl_3$ is 107 °C (380 K).)

Another common reagent is thionyl chloride:

$$RCOOH + SOCl_2 \xrightarrow{\text{heat}} RCOCl + SO_2 + HCl \qquad \text{(cf. §14.9.5)}$$

Thionyl chloride is the least reactive of these three reagents and must be used in excess to obtain a good yield. The acid chloride and excess reagent are subsequently separated by fractional distillation. The use of thionyl chloride is thus restricted to the preparation of those acid chlorides which boil well above its boiling temperature of 77 °C (350 K).

Nucleophilic substitution reactions

Difficult to achieve with alcohols, nucleophilic substitution reactions are virtually impossible with carboxylic acids. There is, for example, no reaction with HCl or HBr.

Elimination of hydrogen

Methanoic acid bears a structural resemblance to primary and secondary alcohols in that the molecule possesses a carbon atom bearing both a hydrogen atom and an HO group. Dehydrogenation can therefore occur on heating:

$$\mathrm{HCOOH} \xrightarrow[\text{160 °C (433 K)}]{\text{heat under pressure}} O{=}C{=}O + H_2$$

methanoic acid — carbon dioxide

$$\text{cf.} \quad \mathrm{RCH_2OH} \xrightarrow[\text{Cu catalyst}]{\text{450 °C (723 K)}} \mathrm{RCHO} + H_2$$

primary alcohol — aldehyde

Alternatively, methanoic acid (like primary and secondary alcohols) may be treated with an oxidising agent so that hydrogen is eliminated in the form of water. Such reactions occur readily, and both methanoic acid and the methanoates are powerful reducing agents. They will decolorise acidified potassium permanganate solution in the cold, and reduce silver nitrate solution and Tollens' reagent to silver.

Acids other than methanoic acid, because they lack the

$$\ce{>C(H)(OH)}$$

grouping, do not give off hydrogen on heating and do not function as reducing agents.

Reactions of the C=O bond

Carboxylic acids take part in nucleophilic addition reactions across the C=O bond. As with aldehydes and ketones, some of the reactions do not need a catalyst whereas others require the presence of acid (see mechanisms, §14.10.5).

Carboxylic acids have a relatively low carbonyl activity, due mainly to the mesomeric effect (§14.10.5). The inductive effect of the alkyl group also plays a part, and methanoic acid, in which the inductive effect is absent, is somewhat more reactive than the other acids. Many reagents that react with aldehydes and ketones, for example hydrogen cyanide, hydroxylamine and 2,4-dinitrophenylhydrazine, do not attack carboxylic acids. The principal reagents that do attack carboxylic acids are alcohols, ammonia, primary and secondary amines, and lithium tetrahydridoaluminate(III).

The products of nucleophilic addition can never be isolated, for they are unstable 1,1-diols. As we should expect, they immediately stabilise themselves by the elimination of a molecule of water between the two hydroxyl groups:

$$\mathrm{R{-}C({=}O)OH} + \mathrm{HNu} \longrightarrow \mathrm{R{-}C(OH)_2{-}Nu} \longrightarrow \mathrm{R{-}C({=}O)Nu} + \mathrm{H_2O}$$

Derivatives of carboxylic acids behave in a similar manner. Always, nucleophilic addition across the C=O bond is followed, to produce a stable product, by the elimination of a small molecule, such as HCl or C_2H_5OH (Table 14.17). The effect of such elimination is to reform the C=O bond: in this way, acids and their derivatives are converted into one another, and the C=O bond *appears* to play no part in the reaction.

Many of the addition–elimination reactions of carboxylic acids and their derivatives are reversible, so that equilibrium mixtures may be established. A well-known example concerns esterification:

$$\mathrm{R{-}C({=}O)OH} + \mathrm{R'OH} \rightleftharpoons \mathrm{R{-}C({=}O)OR'} + \mathrm{H_2O}$$

By contrast, if an acid chloride is allowed to react with an alcohol, the reaction proceeds almost entirely to completion, even when the hydrogen chloride that is produced is retained in the system:

$$R{-}C(=O){-}Cl + R'OH \rightleftharpoons R{-}C(=O){-}OR' + HCl$$

The position of equilibrium depends on the difference between the reactivities of the carbonyl compounds (see order of reactivities, §14.10.5). If the compounds are of comparable reactivity (e.g. acid and ester), reactions proceed with roughly equal ease in both directions, but if the compounds have very different reactivities (e.g. acid chloride ≫ ester), this is not the case and the position of equilibrium lies far to one side.

ROH

Carboxylic acids react with alcohols in the presence of an acid catalyst to give esters. Nucleophilic addition of alcohol across the C=O bond (mechanism IV, §14.10.5) is at once followed by loss of water:

$$R{-}C(=O){-}OH + R'OH \xrightleftharpoons{H^+} R{-}C(OH)_2{-}OR' \rightleftharpoons R{-}C(=O){-}OR' + H_2O$$

Esterification is a reversible reaction, for the reason mentioned above. If, for example, ethanoic acid and ethanol are refluxed together in equimolar proportions, equilibrium is established when approximately two-thirds of the acid and alcohol have been transformed (§8.2.1).

If an involatile acid is reacted with an involatile alcohol, equilibrium can be destroyed by distilling off the water as it is formed, but this technique cannot be used for the lower acids and alcohols. In such cases the yield of ester may be increased by using an excess of alcohol: in this way as much as possible of the more expensive acid is esterified.

The acids commonly used as catalysts are concentrated sulphuric acid and anhydrous hydrogen chloride. The technique consists of refluxing the mixture of carboxylic acid, alcohol and acid catalyst until equilibrium has been established – usually about half an hour – and then distilling off the ester. The concentration of catalyst is critical, for too much acid protonates not only the carboxylic acid but also the alcohol, to give RH_2O^+, which, because of its positive charge, will not attack a carbonyl carbon atom.

The esterification of carboxylic acids by secondary alcohols is slower than by primary alcohols due to steric hindrance. Because of its relatively large bulk, a molecule of a secondary alcohol can approach a protonated acid only with difficulty. If an alcohol contains both primary and secondary hydroxyl groups, e.g. 1,2,3-propanetriol, $CH_2OH{\cdot}CHOH{\cdot}CH_2OH$, the primary HO groups always esterify more readily than the secondary ones.

Curiously, tertiary alcohols, which would be expected to present the most steric hindrance, react at much the same rate as primary alcohols In such cases esterification proceeds by an entirely different mechanism.

***NH*₃**

At room temperature, ammonia behaves as a base and neutralises carboxylic acids to give ammonium salts:

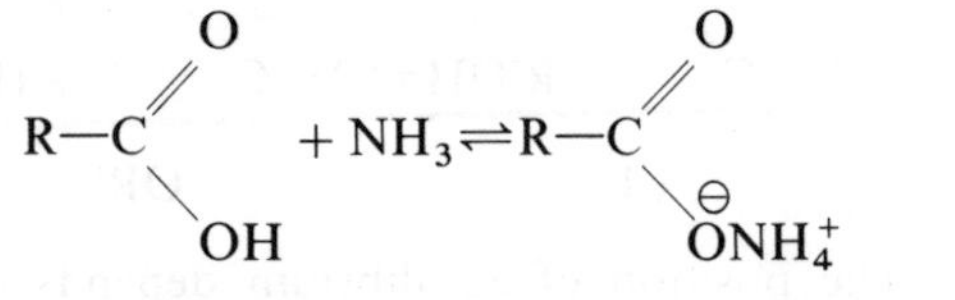

$$R{-}C(=O){-}OH + NH_3 \rightleftharpoons R{-}C(=O){-}O^{\ominus}NH_4^+$$

At 150–200 °C (423–473 K) ammonia behaves as a nucleophile, and amides are formed in an addition–elimination reaction:

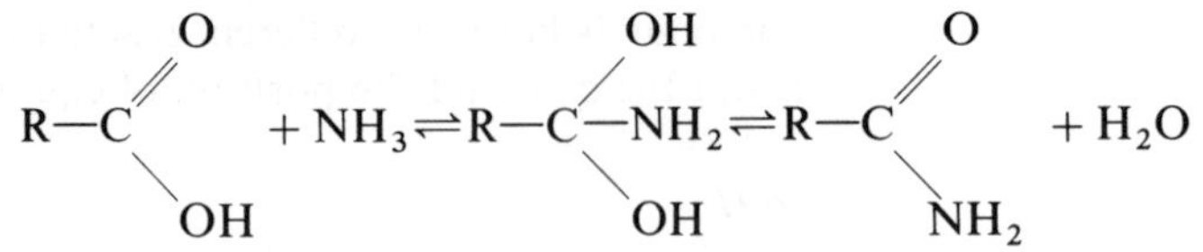

$$R{-}C(=O){-}OH + NH_3 \rightleftharpoons R{-}C(OH)_2{-}NH_2 \rightleftharpoons R{-}C(=O){-}NH_2 + H_2O$$

Amides of high boiling acids may therefore be prepared by passing a current of gaseous ammonia through the carboxylic acid maintained at 150–200 °C (423–473 K). Lower amides, notably ethanamide, are made by distilling the corresponding ammonium salt in the dry state at about 180 °C (453 K). Dissociation occurs into ammonia and the carboxylic acid, which recombine, at this temperature, to give the amide. Yields by this latter method are poor.

RNH*₂ *and R*₂*NH

Primary and secondary amines, like ammonia, react with carboxylic acids to give mono- and disubstituted amides respectively:

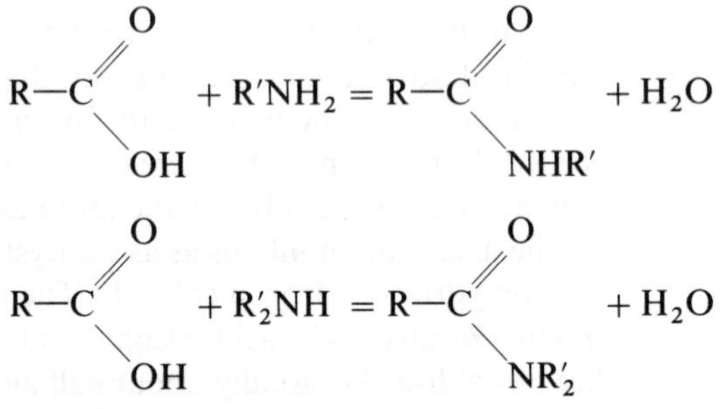

$$R{-}C(=O){-}OH + R'NH_2 = R{-}C(=O){-}NHR' + H_2O$$

$$R{-}C(=O){-}OH + R'_2NH = R{-}C(=O){-}NR'_2 + H_2O$$

The reactions are of little importance, for substituted amides are best prepared from acid chlorides or esters.

***Li*[*AlH*₄]**

Carboxylic acids are readily reduced to primary alcohols with lithium tetrahydridoaluminate(III). This is the only way there is of reducing carboxylic acids. Sodium tetrahydroborate(III) is ineffective.

14.10.8 SALTS OF CARBOXYLIC ACIDS

Carboxylate ions, because of their negative charge, have an extremely low carbonyl activity and do not participate in addition–elimination reactions.

Their principal property is their ability to donate a lone pair of electrons, so that they can act as nucleophiles and ligands. They behave as nucleophiles in substitution reactions with alkyl halides,

$$\text{e.g.}\quad CH_3COOAg + C_2H_5I = CH_3COOC_2H_5 + AgI \qquad (\S 14.8.4)$$

Similar reactions with acyl halides give rise to acid anhydrides,

$$\text{e.g.}\quad CH_3COONa + CH_3COCl = (CH_3CO)_2O + NaCl$$

Carboxylate ions act as ligands in coordinating to certain transition metal cations, notably Fe^{3+} and Cr^{3+}, to give complex ions of the type $[M_3^{III}(RCOO)_6O]^+$. The colour of the complex depends upon the particular carboxylate ion, a feature that may be used in identifying such ions. Neutral solutions of ethanoates, in the cold, when treated with neutral iron(III) chloride solution, give a blood red coloration; when boiled, this changes to a brown precipitate of a basic iron(III) ethanoate. Methanoates behave similarly.

14.10.9 ACID ANHYDRIDES

Carboxylic acid anhydrides are characterised by addition–elimination reactions, i.e. nucleophilic addition across a C=O bond, followed by the elimination of a molecule of carboxylic acid. The C—O bond, by contrast, is unreactive; cf. ethers.

```
      O                  Nu                    Nu
     //                 /⊖                    /
R—C                R—C—O               R—C
     \                  \⊕                    \\
      O + HNu ⇌          O→H    ⇌              O
     /                  /                      OH
R—C                R—C                       /
     \\                 \\              +R—C
      O                  O                    \\
                                               O
```

The mechanism of addition is a little different from usual, in that the proton attaches itself to the ether oxygen atom rather than the carbonyl oxygen atom. This gives an adduct which, on cleavage, gives two molecules of carbonyl compounds. The energy required to break a C—O bond in the final stage is available because a comparable bond is being formed (C—O → C=O).

Acid anhydrides have a high carbonyl activity, and their chemistry is much more extensive than that of carboxylic acids. In the scope of their reactions, they compare with aldehydes and ketones rather than carboxylic acids.

H_2O

Acid anhydrides can be hydrolysed to give the parent acids,

$$\text{e.g.}\quad (CH_3CO)_2O + H_2O = 2CH_3COOH$$

The reaction is slow to start at room temperature because of the immiscibility of acid anhydrides and water, but occurs rapidly on warming or shaking **and can be violent**.

ROH and ArOH

Unlike carboxylic acids, acid anhydrides will form esters with phenols as well as with alcohols. Because a molecule of carboxylic acid is eliminated, only half the molecule of acid anhydride is effectively used.

E.g. $(CH_3CO)_2O + C_2H_5OH = CH_3COOC_2H_5 + CH_3COOH$

Most of these esterifications occur readily when the reactants are warmed together. In difficult cases an acid or base catalyst may be used.

NH_3, RNH_2 and R_2NH

The reactions between acid anhydrides and ammonia, or primary or secondary amines, occur readily and are often used for the preparation of amides and substituted amides. Once again only half the acid anhydride is utilised.

E.g. $(CH_3CO)_2O + NH_3 = CH_3CONH_2 + CH_3COOH \xrightarrow{NH_3} CH_3COONH_4$

Li[AlH₄]

Like carboxylic acids, acid anhydrides can be reduced to primary alcohols with lithium tetrahydridoaluminate(III) in dry ether. Ethanoic anhydride, for example, gives ethanol.

14.10.10 ACID CHLORIDES (acyl chlorides)

Acid chlorides possess the $-C\begin{smallmatrix}\nearrow O\\ \searrow Cl\end{smallmatrix}$ grouping and may therefore display nucleophilic substitution reactions typical of the C—Cl bond and nucleophilic addition–elimination reactions typical of the C=O bond.

Most common nucleophiles, including H_2O, ROH, NH_3, RNH_2 and R_2NH, are able to attack both C—Cl and C=O bonds, and on reacting with acid chlorides would give identical products by either mode of attack. For example, the reaction between an acid chloride and an alcohol, which yields an ester and hydrogen chloride, could be formulated as a nucleophilic substitution:

$$R'(H)\ddot{O}: + \overset{\delta+}{C}(R)(=\overset{\delta-}{O})-\overset{\delta-}{Cl} \longrightarrow R'(H)\overset{\oplus}{O}\rightarrow C(R)=O + Cl^- \xrightarrow{Cl^-\text{ abstracts }H^+} R'-O-C(R)=O + HCl$$

protonated ester ... ester

Alternatively, the alcoholysis could be represented as addition–elimination:

$$R{-}C(=O){-}Cl + R'OH \rightleftharpoons R{-}C(OH)(Cl){-}OR' \longrightarrow R{-}C(=O){-}OR' + HCl$$

Which mechanism is correct? Although we cannot be certain, we think that it is probably the latter. Nucleophilic substitution is unlikely because the chloride ion is not a good base for abstracting a proton in the final stage.

Hydrolysis, ammonolysis and aminolysis are likely to have a similar mechanism to alcoholysis. *All occur by nucleophilic addition across the C=O bond, followed by the elimination of HCl.*

Because of the inductive effect of the chlorine atom, which reinforces that of the oxygen atom, acid chlorides have an exceptionally high reactivity. They are the most reactive of the carbonyl compounds, and will take part in a wide range of reactions, even with reagents whose nucleophilicity is low.

H_2O

When water is cautiously added to an aliphatic acid chloride at room temperature, two liquid phases may be observed for a short time, but almost at once there is a vigorous, exothermic reaction accompanied by the production of some hydrogen chloride. A carboxylic acid and most of the hydrogen chloride remain in solution,

e.g. $CH_3COCl + H_2O = CH_3COOH + HCl$

This reaction can be violent.

Aromatic acid chlorides, such as benzoyl chloride, are more resistant to hydrolysis and may require a strong acid to act as catalyst. Alternatively, hydrolysis can be conducted in the presence of a base, which provides the powerful nucleophile HO^- (cf. hydrolysis of esters, §14.10.11).

ROH and ArOH

Aliphatic acid chlorides, like acid anhydrides, readily undergo esterification with both alcohols and phenols. Aromatic acid chlorides are less reactive, but respond to the use of acid catalysts. The esterification of an aromatic acid chloride by a phenol is best performed in the presence of aqueous alkali, a reaction known as the *Schotten–Baumann reaction*. The alkali has the effect of converting a weakly nucleophilic phenol molecule into a strongly nucleophilic phenoxide ion. Aliphatic acid chlorides are rapidly hydrolysed by aqueous alkali and cannot be esterified in this way.

NH_3, RNH_2 and R_2NH

Concentrated aqueous ammonia readily converts acid chlorides to amides at room temperature,

$$\text{e.g.}\quad CH_3C(=O)Cl + 2NH_3 = CH_3C(=O)NH_2 + NH_4Cl$$

Primary and secondary amines, likewise, react with acid chlorides to give substituted amides,

$$\text{e.g.}\quad CH_3C(=O)Cl + 2C_2H_5NH_2 = CH_3C(=O)NHC_2H_5 + C_2H_5\overset{\oplus}{N}H_3Cl^-$$

N-ethylethanamide — ethylammonium chloride

To avoid the loss of one mole of amine as the amine salt, reactions with aromatic acid chlorides (which are relatively resistant to hydrolysis) may be conducted in the presence of dilute aqueous sodium hydroxide,

$$\text{e.g.}\quad C_6H_5C(=O)Cl + C_6H_5NH_2 \xrightarrow{NaOH(aq)} C_6H_5C(=O)NHC_6H_5 + \underset{(\text{as NaCl})}{HCl}$$

benzoyl chloride — phenylamine — N-phenylbenzamide

This is another type of Schotten–Baumann reaction.

The ethanoylation (acetylation) and benzoylation of amines is of great importance in their identification. In the same way that aldehydes and ketones are characterised by the melting temperatures of their 2,4-dinitrophenylhydrazones, so primary and secondary amines are characterised by the melting temperatures of their ethanoyl and benzoyl derivatives.

14.10.11 ESTERS

Esters have a $—C(=O)O—C—$ grouping and take part in addition–elimination reactions, i.e. nucleophilic addition across the C=O bond, followed by the elimination of a molecule of alcohol. The C—O bond is unreactive; cf. ethers.

H_2O; H^+ catalyst (acidic hydrolysis)

On hydrolysis, esters revert to the carboxylic acids and alcohols from which they are derived. The reaction with water alone is very slow, partly because the H_2O molecule is a weak nucleophile and partly because most esters have a low solubility in water, and the reaction may be catalysed by means of a strong acid:

$$R{-}C(=O){-}OR' + H_2O \overset{H^+}{\rightleftharpoons} R{-}C(OH)(OH){-}OR' \rightleftharpoons R{-}C(=O){-}OH + R'OH$$

The acid catalysed addition of water occurs in accordance with mechanism IV (§14.10.5).

Acidic hydrolysis is the exact reverse of acid catalysed esterification, and is of limited use as it leads to the establishment of an equilibrium mixture.

HO⁻ (alkaline hydrolysis)
Alkaline hydrolysis is of much more practical importance, for the reaction goes to completion. Here, the nucleophile is not the H_2O molecule but the much more effective HO^- ion:

$$R{-}C(=O){-}OR' + Na^+HO^- \rightleftharpoons R{-}C(O^\ominus Na^+)(OH){-}OR' \longrightarrow R{-}C(=O){-}O^\ominus Na^+ + R'OH$$

e.g. $CH_3COOC_2H_5 + NaOH = CH_3COONa + C_2H_5OH$

The reaction goes to completion because the very low reactivity of carboxylate salts in addition reactions causes the final step to be irreversible.

In the laboratory, hydrolysis is conducted by refluxing the ester with aqueous alkali until a homogeneous mixture is obtained. This may well take several hours. A mutual solvent for the ester and the aqueous alkali (e.g. ethanol or diethylene glycol) is often included. After hydrolysis is complete, the alcohol is distilled off and the residue then acidified with a strong acid to liberate the carboxylic acid. If the latter is a liquid it may be distilled off or extracted with ether; if solid, it is filtered off.

The alkaline hydrolysis of esters of long chain carboxylic acids (*fatty acids*) is known as *saponification*, for the products – sodium salts of the long chain acids – are soaps (§14.10.13).

NH_3

$$R{-}C(=O){-}OR' + NH_3 \rightleftharpoons R{-}C(OH)(NH_2){-}OR' \rightleftharpoons R{-}C(=O){-}NH_2 + R'OH$$

e.g. $CH_3COOC_2H_5 + NH_3 = CH_3CONH_2 + C_2H_5OH$

The ammonolysis of esters, unlike that of carboxylic acids, occurs readily, for not only are esters more reactive than acids but there is no competing reaction to give an ammonium salt.

This is, perhaps, the best method of preparing amides. The reaction is performed merely by shaking the ester with aqueous or alcoholic ammonia at room temperature.

RNH_2 and R_2NH

The aminolysis of esters with primary or secondary amines is closely related to the previous reaction and gives rise to substituted amides.

Metals

Aliphatic (but not aromatic) esters may be reduced by sodium and ethanol in the *Bouveault–Blanc reduction.* Two alcohols are obtained; one represents the reduction product of the parent acid, while the other is the alcohol that was used in making the ester,

$$\text{e.g.}\quad CH_3CH_2C(=O)OCH_3 \xrightarrow{Na + C_2H_5OH} CH_3CH_2CH_2OH + CH_3OH$$

methyl propanoate (from propanoic acid and methanol) — 1-propanol — methanol

$Li[AlH_4]$

Lithium tetrahydridoaluminate(III) in dry ether reduces an ester RCOOR′ to give a mixture of two alcohols, RCH_2OH and R′OH; cf. the Bouveault–Blanc reduction.

14.10.12 AMIDES

We might expect amides, with their $-C(=O)NH_2$ grouping, to undergo reactions of the C=O bond (cf. aldehydes, carboxylic acids, etc) plus reactions of the $C-NH_2$ bond (cf. primary amines).

In practice, the C=O bond is dominant. Except in the reaction with nitrous acid, the $C-NH_2$ bond contributes little towards the chemistry of amides.

Reactions of the C=O bond

Amides have a relatively low carbonyl activity, comparable with that of carboxylic acids. They take part in a limited number of addition–elimination reactions, i.e. nucleophilic addition across the C=O bond, followed by the elimination of ammonia.

H_2O; H^+ catalyst (acidic hydrolysis)

Amides and substituted amides can be hydrolysed to give the parent carboxylic acid and ammonia or an amine:

$$\mathrm{R{-}C({=}O)NH_2 + H_2O \rightleftharpoons R{-}C(OH)(OH)NH_2 \rightleftharpoons R{-}C({=}O)OH + NH_3}$$

$$\mathrm{R{-}C({=}O)NHR' + H_2O \rightleftharpoons R{-}C(OH)(OH)NHR' \rightleftharpoons R{-}C({=}O)OH + R'NH_2}$$

In both theory and practice, the hydrolysis of amides is analogous to that of esters. The reaction is very slow with water alone, but proceeds at a reasonable rate in the presence of an acid catalyst. Because amides are less reactive than esters, concentrated acid may be required.

$$\text{E.g.}\quad \mathrm{CH_3CONH_2 + H_2O \xrightarrow[\text{boil}]{H_2SO_4\,(aq)} CH_3COOH + NH_3}$$

Ammonia is retained by the sulphuric acid as ammonium hydrogensulphate.

HO$^-$ (alkaline hydrolysis)
Like esters, amides can also be hydrolysed under alkaline conditions, in which case the nucleophile is the HO^- ion, rather than the H_2O molecule, and the organic product is a salt rather than a carboxylic acid:

$$\mathrm{R{-}C({=}O)NH_2 + Na^+HO^- \rightleftharpoons R{-}C(\overset{\ominus}{O}Na^+)(OH)NH_2 \longrightarrow R{-}C(\overset{\ominus}{O}Na^+){=}O + NH_3}$$

$$\text{e.g.}\quad \mathrm{CH_3CONH_2 + NaOH \xrightarrow{\text{boil}} CH_3COONa + NH_3}$$

The hydrolysis of amides under more gentle conditions may be achieved, should this be necessary, by means of nitrous acid at 5 °C (278 K) (see below).

Metals
Amides may be reduced to primary amines by means of sodium and ethanol:

$$\mathrm{RCONH_2 \xrightarrow{Na + C_2H_5OH} RCH_2NH_2}$$

The reduction of an amide bears a superficial resemblance to the Hofmann reaction ($RCONH_2 \rightarrow RNH_2$), for in both cases a primary amine is formed. The Hofmann reaction, however, is accompanied by a decrease in the number of carbon atoms from n to $n-1$ (§14.3.4).

Li[AlH$_4$]
Lithium tetrahydridoaluminate(III) is very efficient for converting amides

into primary amines, and is recommended in preference to sodium and ethanol.

Elimination of water

Amides can be dehydrated to nitriles by means of various Lewis acids,

$$\text{e.g.}\quad \underset{\text{ethanamide}}{CH_3CONH_2} \xrightarrow[\text{heat}]{P_4O_{10}} \underset{\text{ethanenitrile}}{CH_3CN} + H_2O$$

Although phosphorus(V) oxide is the best known dehydrating agent for this purpose, thionyl chloride or boron trifluoride is generally preferable.

The reaction is the exact opposite of those just described under reactions of the C=O bond in that it is an elimination rather than an addition reaction. It is the reverse of the nucleophilic addition reaction in which water (in the presence of an acid catalyst) converts a nitrile into an amide.

Reactions of the C—NH$_2$ bond (including the N—H bond)

Acidic and basic character

Amides in aqueous solution are neutral to litmus, but have a weakly defined amphoteric character.

Acidity Amides are slightly acidic for the same reason that carboxylic acids are acidic, i.e. the inductive effect of the carbonyl oxygen atom, which promotes polarisation of the N—H bonds. The acidity of amides is not as great as that of acids because nitrogen is less electronegative than oxygen and polarisation of the N—H bonds is relatively small.

Basicity Like ammonia and the amines, amides are weakly basic because they can donate a lone pair of electrons from the outer shell of the nitrogen atom to a proton. With hydrogen chloride, for example, they form chlorides which are extensively hydrolysed in aqueous solution,

$$\text{e.g.}\quad \underset{\text{ethanamide}}{CH_3C(=O)NH_2} + HCl \rightleftharpoons \underset{\text{ethanoylammonium chloride}}{CH_3C(=O)\overset{\oplus}{N}H_3Cl^-}$$

$$\text{cf.}\quad \underset{\text{ethylamine}}{CH_3CH_2NH_2} + HCl \rightleftharpoons \underset{\text{ethylammonium chloride}}{CH_3CH_2\overset{\oplus}{N}H_3Cl^-}$$

The basic character of amides is considerably less than that of ammonia and amines, for two reasons. First, there is the mesomeric effect, whereby the lone pair of electrons on the nitrogen atom is drawn towards the carbonyl carbon atom:

$$R-C(=O)-\ddot{N}H_2$$

This reduces the ability of the nitrogen atom to coordinate to a proton. Second, when an amide reacts with an acid not all the protons join the NH_2 group; as with carboxylic acids, most of them join the carbonyl oxygen atom:

$$R-C(=O)NH_2 + H^+ \rightleftharpoons R-\overset{\oplus}{C}(OH)NH_2$$

Amides as nucleophiles

The chemistry of amines is dominated by their ability to act as nucleophiles (§14.11.7). With amides, however, because the electron density at the nitrogen atom is reduced by the mesomeric effect, this property is almost entirely absent. Only the reaction with nitrous acid proceeds as it does with primary amines:

$$RCONH_2 + HNO_2 \xrightarrow[(273\text{–}278\ K)]{0\text{–}5\ °C} RCOOH + N_2 + H_2O$$

14.10.13 FATS, OILS AND SOAPS

Animal and vegetable oils and fats are all esters of 1,2,3-propanetriol (glycerol) and long chain carboxylic acids, known as *fatty acids*. Examples of saturated fatty acids are as follows:

$CH_3(CH_2)_{14}COOH$ — hexadecanoic acid (palmitic acid)

$CH_3(CH_2)_{16}COOH$ — octadecanoic acid (stearic acid)

Almost all naturally occurring fatty acids have even numbers of carbon atoms.

Oils and fats are *triglycerides*, so called because all three hydroxyl groups of the glycerol molecule are esterified. An example is provided by tristearin (glyceryl tristearate):

$$\begin{array}{l} CH_2OCO(CH_2)_{16}CH_3 \\ | \\ CHOCO(CH_2)_{16}CH_3 \\ | \\ CH_2OCO(CH_2)_{16}CH_3 \end{array} \quad \text{tristearin}$$

Tristearin is a solid and is a constituent of stearin, a fat found in animals and some plants, and the chief constituent of tallow and suet.

Tristearin is a *simple glyceride*, in that all three fatty acid residues are identical. Most natural glycerides, however, are *mixed glycerides*, based on two or even three different fatty acids. Generally, a preponderance of saturated fatty acids gives a fat, while a preponderance of unsaturated fatty acids leads to an oil. In the manufacture of margarine, unsaturated oils are partially hydrogenated to give semi-solid fats. Reaction is conducted at 150–200 °C (423–473 K) in the presence of finely divided nickel, which is afterwards recovered by filtration.

Soaps comprise the sodium and potassium salts of long chain fatty acids. The raw materials used in soap making are low grade oils and fats, such as coconut oil, olive oil, whale oil, and cattle and mutton tallow. In the traditional *kettle process*, the oil or fat is heated with sodium hydroxide solution in a large kettle for several hours:

$$\begin{array}{lcl} CH_2OCOR & & RCOONa & & CH_2OH \\ | & & & & | \\ CHOCOR' & + 3NaOH = & + R'COONa & + & CHOH \\ | & & & & | \\ CH_2OCOR'' & & + R''COONa & & CH_2OH \\ & & \text{'soap'} & & \text{1,2,3-propanetriol} \end{array}$$

Potassium hydroxide may be used instead of sodium hydroxide for making liquid soap, soap flakes and some granulated soaps.

When saponification is complete, a large quantity of sodium chloride is added to 'salt out' the soap. It is then boiled with salt solution to remove residual alkali and 1,2,3-propanetriol, and afterwards with water to remove the salt. 1,2,3-Propanetriol, obtained from the liquor by distillation, is a valuable by-product. It is used mainly for making synthetic resins and explosives.

In recent years the kettle process has been largely superseded by continuous hydrolysis by steam (not alkali) in a stainless steel column called a hydrolyser.

Soap is a *detergent*, i.e. cleansing agent, and functions as such because one part of the anion – the carboxylate group – is water attracting, while the other part – the hydrocarbon chain – attracts grease and oil. Greasy matter is thus converted into an emulsion with water, and can be removed by rinsing.

14.10.14 POLYESTERS AND POLYAMIDES

Unlike polyalkenes and related polymers, which are made by addition polymerisation, often from a single monomer, polyesters and polyamides are made by *condensation polymerisation* in addition–elimination reactions, usually between two compounds.

Polyethylene terephthalate ('Terylene', 'Trevira')

Whenever a monocarboxylic acid reacts with a monohydric alcohol a simple ester is produced, but if the acid and alcohol are both difunctional the product is a *linear polyester*, which may be suitable for forming filaments and hence fibres. If the acid has a functionality of two and the alcohol has a functionality greater than two, or vice versa, the product is a *polyester resin*, whose molecules are cross-linked in three dimensions. Such materials are used in the manufacture of paints.

Terylene is a well-known example of a linear polyester. It is made from a dicarboxylic acid, 1,4-benzenedicarboxylic acid, and a dihydric alcohol, 1,2-ethanediol. For the remainder of this discussion trivial names will be used.

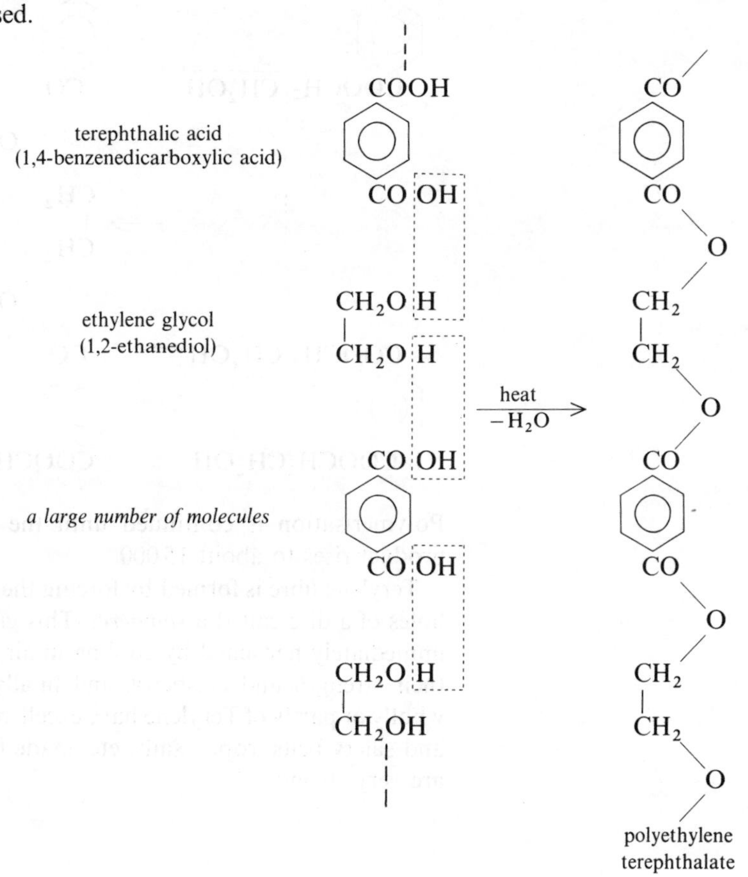

terephthalic acid (1,4-benzenedicarboxylic acid): $HOOC-C_6H_4-COOH$

ethylene glycol (1,2-ethanediol): $HOCH_2CH_2OH$

a large number of molecules

$$\cdots HOOC-C_6H_4-CO{:}OH + H{:}OCH_2CH_2O{:}H + HO{:}OC-C_6H_4-CO{:}OH + H{:}OCH_2CH_2OH \cdots \xrightarrow[-H_2O]{\text{heat}} \cdots -CO-C_6H_4-CO-O-CH_2-CH_2-O-CO-C_6H_4-CO-O-CH_2-CH_2-O- \cdots$$

polyethylene terephthalate

Direct esterification is extremely difficult because of the insolubility of terephthalic acid in ethylene glycol, and in practice the manufacture of Terylene begins with the esterification of terephthalic acid by *methanol*:

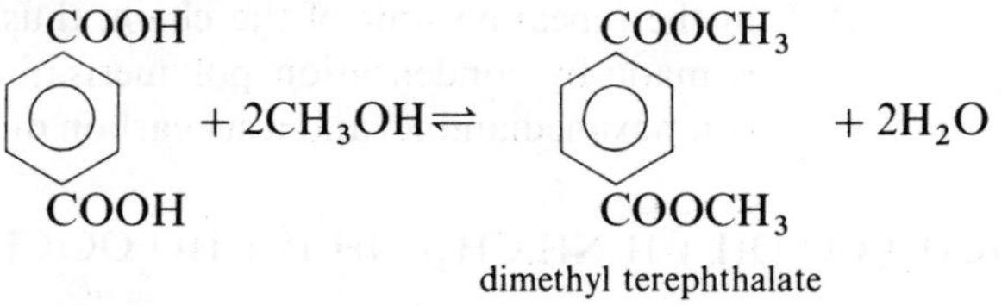

$$HOOC-C_6H_4-COOH + 2CH_3OH \rightleftharpoons CH_3OOC-C_6H_4-COOCH_3 + 2H_2O$$

dimethyl terephthalate

Dimethyl terephthalate is then reacted with ethylene glycol at about 230 °C (503 K) in the presence of a metal oxide catalyst:

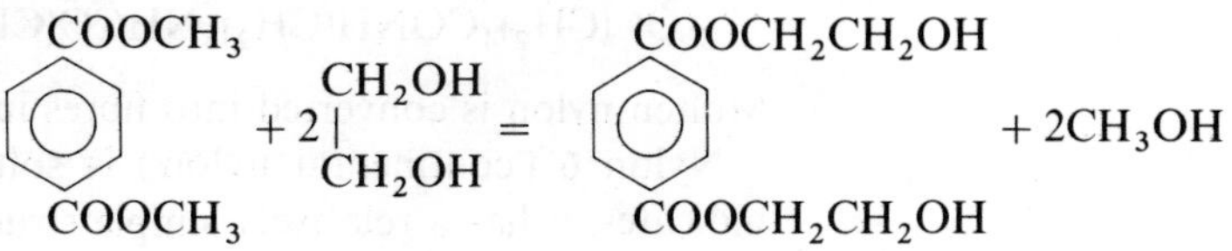

$$CH_3OOC-C_6H_4-COOCH_3 + 2\,\begin{matrix}CH_2OH\\|\\CH_2OH\end{matrix} = HOCH_2CH_2OOC-C_6H_4-COOCH_2CH_2OH + 2CH_3OH$$

On heating to a temperature of 280 °C (553 K), under reduced pressure, the dihydroxydiethyl terephthalate undergoes condensation polymerisation with the loss of ethylene glycol:

$$\begin{array}{c} COOCH_2CH_2OH \\ C_6H_4 \\ COOCH_2\,\vdots\,CH_2OH \\ \\ COOCH_2\,\vdots\,CH_2OH \\ C_6H_4 \\ COOCH_2CH_2OH \end{array} \longrightarrow \begin{array}{c} COOCH_2CH_2OH \\ C_6H_4 \\ CO \\ O \\ CH_2 \\ | \\ CH_2 \\ O \\ CO \\ C_6H_4 \\ COOCH_2CH_2OH \end{array} + \begin{array}{c} CH_2OH \\ | \\ CH_2OH \end{array}$$

Polymerisation is continued until the relative molecular mass of the product rises to about 15 000.

Terylene fibre is formed by forcing the molten polymer through the fine holes of a disc called a *spinneret*. This gives very fine filaments which are immediately hardened by cooling in air, then stretched so as to improve their strength and elasticity, and finally spun into fibre. Textiles made wholly or partly of Terylene have excellent durability and crease resistance, and safety belts, ropes, sails, etc, made from a special grade of Terylene, are very strong.

Nylon

The term *nylon* covers numerous linear polyamides. The commonest are nylon 66 and nylon 6. The numbers refer to the numbers of carbon atoms in the repeating unit of the chain; thus, nylon 66 is so called because it is made by condensation polymerisation between a six carbon diamine (1,6-hexanediamine) and a six carbon dicarboxylic acid (hexanedioic acid):

$$HOOC(CH_2)_4CO\,\boxed{OH + H}\,NH(CH_2)_6NH\,\boxed{H + HO}\,OC(CH_2)_4CO\,\boxed{OH + H}\,NH(CH_2)_6NH_2$$

$$\downarrow \text{heat, 280 °C (553 K)}; -H_2O$$

$$-OC(CH_2)_4CONH(CH_2)_6NHCO(CH_2)_4CONH(CH_2)_6NH- \quad \text{etc.}$$

Molten nylon is converted into fibres in the same way as Terylene.

Nylon 6 ('continental nylon') is softer than nylon 66. As its name indicates, it has a relatively simple structure, reflecting the fact that it is

$$-[(CH_2)_5CONH]-_n \qquad -[CO(CH_2)_4CONH(CH_2)_6NH]-_n$$

nylon 6 nylon 66

made from only one substance, namely caprolactam, which is the cyclic amide of caproic acid (hexanoic acid).

The polymerisation of caprolactam to nylon 6 is accomplished by heating the material at 250 °C (523 K) for about four hours with a trace of water:

$$n \text{ caprolactam} \xrightarrow{\text{heat}} -[(CH_2)_5CONH]-_n$$

caprolactam

Nylon textiles are too common to need description. Not so well known is the fact that many light engineering components, such as gear wheels, are moulded from nylon. They are tough, have a low coefficient of friction, and retain their shape with use.

14.10.15 POLYPEPTIDES AND PROTEINS

Polypeptides and proteins are compounds of natural origin which resemble nylon in the sense that they have a polyamide structure. Proteins occur in the form of muscle, skin, hide, hair, nail, hoof, etc. Animal cells in general are composed largely of protein, and certain plant cells (e.g. soya beans) are also rich in this type of material.

Polypeptides and proteins are composed of amino-acid molecules chemically joined together; on hydrolysis, either with mineral acids or with enzymes, they revert to mixtures of amino-acids. An *amino-acid* is a compound whose molecules possess both an amino group, NH_2, and a carboxyl group, COOH. Naturally occurring amino-acids, i.e. those which account for proteins, are termed α-amino-acids to denote that the amino group is attached to the α-carbon atom:

$$R-\overset{NH_2}{\underset{H}{\overset{|\alpha}{\underset{|}{C}}}}-COOH$$

Examples of α-amino-acids are glycine, in which R = H; alanine, in which $R = CH_3$; and phenylalanine, in which $R = C_6H_5CH_2$.

$$H-\overset{NH_2}{\underset{H}{\overset{|}{\underset{|}{C}}}}-COOH \qquad CH_3-\overset{NH_2}{\underset{H}{\overset{|}{\underset{|}{C}}}}-COOH \qquad C_6H_5CH_2-\overset{NH_2}{\underset{H}{\overset{|}{\underset{|}{C}}}}-COOH$$

glycine alanine phenylalanine

Some amino-acids are more complicated than these, in that they may contain HO groups, atoms of sulphur or halogens, ring structures containing nitrogen, or two amino groups or carboxyl groups.

Whenever a carboxylic acid, RCOOH, and a primary amine, $R'NH_2$, react together, the product is an *N*-substituted amide:

$$RCO\boxed{OH + H}NHR' = RCONHR'$$

If the carboxyl and amino groups both belong to the same molecule, i.e. if an amino-acid is used, a substituted amide is still obtained:

$$H_2N—R—CO\boxed{OH + H}NH—R—COOH$$

$$= H_2N—R—CONH—R—COOH$$

A compound of this kind, derived from two amino-acid molecules, is called a *dipeptide*. If three amino-acid molecules are used, the product is called a *tripeptide*, and so on. If the polymer chain is derived from a large number of amino-acid molecules the material is termed a polypeptide or a protein.

Polypeptides and proteins have closely related structures, and the distinction between them is a fine one. There are, however, two important differences, as follows.

(i) Protein molecules are always hydrated, but polypeptides are not. Water molecules are held to a protein chain by hydrogen bonding, and the water forms an integral part of the structure. When a protein is dehydrated it changes, irreversibly, into a polypeptide.

(ii) Because of their extremely complicated structures, it is not possible to synthesise proteins. Polypeptides, by contrast, can be prepared in the laboratory, although it must be stressed that at present only one or two have been made in this way. For all practical purposes, polypeptides are obtainable only from natural sources.

From the equation shown above we can see that peptide and protein chains consist of organic radicals joined together by —CONH—, commonly referred to as a *peptide link*. Essentially, this is an amide group; thus proteins, like nylon, are linear polyamides. Protein chains can be held in close proximity to one another by hydrogen bonding between the NH and CO groups of neighbouring chains. The same applies to nylon, which is why many proteins resemble nylon in having a fibrous character.

EXAMINATION QUESTIONS ON SECTION 14.10

1 (a) Name **two** reagents which will add to alkenes and **two** which will add to aldehydes. Write an equation for each reaction.

(b) Compare the electronic structure of ethene (ethylene) with that of methanal (formaldehyde).

(c) Make a mechanistic comparison between the addition reactions of alkenes and the addition reactions of aldehydes, illustrating your answer with suitable reactions. (OC)

2 When compound **A**, $C_7H_{14}O_2$, is refluxed with concentrated aqueous ammonia, it forms two products, **B**, C_3H_8O, and **C**, C_4H_9NO.

When **C** is treated with bromine and then warmed with aqueous sodium hydroxide it forms **D**, C_3H_9N.

D reacts with dilute hydrochloric acid to form the ionic compound **E**.

When **E** is treated with aqueous sodium nitrate(III) (sodium nitrite) it yields a compound **F**, isomeric with **B**.

B and **F** give isomeric compounds **G** and **H**, C_3H_6O, on warming with aqueous acidified potassium manganate(VII) (potassium permanganate).

Both **G** and **H** give precipitates with 2,4-dinitrophenylhydrazine reagent.

G has no reaction with an aqueous solution containing diammino-silver(I) ions (Tollens' reagent), whereas **H** produces a mirror on the sides of the test tube and is converted to **I**.

Write the names and structural formulae for the substances **A** to **I**. (SUJB)

3 Discuss the chemistry of phenylethanone, $C_6H_5COCH_3$, by considering the following:

(a) a method of synthesis,

(b) addition and addition-elimination reactions of the carbonyl group,

(c) a method of preparing benzenecarboxylic (*benzoic*) acid from phenyl-ethanone. (JMB)

4 (a) Write the structural formulae of all the aldehydes and ketones of molecular formula C_4H_8O which do not contain a carbon to carbon double bond $\left(\rangle C{=}C\langle \right)$. List the isomers under the headings *aldehydes*, and *ketones*.

(b) Write equations for the reaction of (i) an aldehyde, (ii) a ketone, with sodium hydrogensulphite. **Outline** the preparation of **one** of the products. State the value of this reaction in practical chemistry.

(c) For an aldehyde and a ketone which are isomeric, give:

(i) **two** reactions, other than the reaction in (b), which are common,

(ii) **two** reactions which are different. (AEB)

5 What is a polymer?

(a) Outline the synthesis of a nylon and give the structure of the final product.

(b) Show how benzene-1,4-dicarboxylic acid (*terephthalic acid*) and ethane-1,2-diol react to form a polymer. In what way would the products differ if a small quantity of ethanol were added to the reactants?

(c) A natural polymer is made up from molecules of the general formula $RCHNH_2COOH$. Suggest a basic structure for this polymer.

When R is CH_3CHOH the molecule is *threonine*.

Predict what reaction might take place when *threonine* is treated with (i) nitrous acid, (ii) phosphorus pentachloride. (JMB)

6 (a) Explain what conclusions can be deduced about the structural formula of ethanoic acid (acetic acid), $C_2H_4O_2$, from each of the following results of practical tests carried out on a sample of the pure acid.

(i) Reaction with sodium metal displaces one hydrogen atom per molecule.

(ii) Reaction with phosphorus pentachloride gives fumes of hydrogen chloride.

(iii) When heated with chlorine, in direct sunlight, three hydrogen atoms per molecule are displaced by chlorine atoms.

(iv) It gives no reaction with 2,4-dinitrophenylhydrazine.

(v) Determination of the relative molecular mass of ethanoic acid by the depression of freezing point of benzene gives a value of approximately 120 instead of the expected value of 60.

(b) Giving full experimental details, describe the preparation of a pure sample of ethyl ethanoate (ethyl acetate). (AEB)

7 (a) Describe briefly the preparation of aminoethanoic acid, starting from ethanoic acid.

(b) Write equations for **two** typical reactions of aminoethanoic acid.

(c) Name and give **one** example of the natural polymers which are made from substances such as aminoethanoic acid. What is the *name* and *structure* of the characteristic functional group in such polymers?

(d) Which of the two formulae below best represents aminoethanoic acid? Give the reason for your choice.

$$H_2NCH_2COOH \qquad H_3\overset{+}{N}CH_2CO\overset{-}{O}$$

(e) Give **one** reaction which would enable you to distinguish between ethanoic acid and ethanamide. (O)

8 (a) State how you would convert ethanoic (acetic) acid into (i) ethanoyl (acetyl) chloride and (ii) ethanoyl (acetic) anhydride, giving the essential conditions for the reactions, but no details of the apparatus or of the processes whereby the products are purified.

(b) Compare the behaviour of ethanoyl chloride and ethanoic anhydride towards (i) water, (ii) phenylamine (aniline) and (iii) phenol.

(c) Describe briefly a simple test tube experiment you would use to distinguish ethanoyl chloride from benzoyl chloride. (O)

9 Describe in outline how benzene may be converted into benzoic acid.

Benzoic acid exists as a dimer in benzene whilst in water it is a monomer. Why is this?

Give equations to show how ethyl benzoate reacts with (a) ammonia, (b) lithium tetrahydroaluminate ($LiAlH_4$), and (c) nickel and hydrogen at 100 °C and 100 atm. pressure. Draw the structural formulae of the products. (OC)

10 Addition and addition-elimination reactions are a common feature of the chemistry of carbonyl compounds. Give an account of the general mechanisms of these reactions. Discuss the similarities and differences which exist between these two types of reaction by considering the reactions between:

(a) ethanal and lithium tetrahydridoaluminate,

(b) propanone and hydrazine,

(c) ethanoyl (*acetyl*) chloride and ammonia,

(d) ethanoic (*acetic*) acid and ethanol.

Show how the principles of these reactions find application in the production of (i) Terylene, (ii) nylon 66. (JMB)

11 Indicate the reagents and conditions by which the following conversions may be effected:

(a) ethanol to ethyl ethanoate (ethyl acetate) using ethanol as the only organic starting material,

(b) methyl benzoate to benzene,

(c) ethanal (acetaldehyde) to ethoxyethane (diethyl ether),

(d) methylbenzene (toluene) to benzyl alcohol ($C_6H_5{\cdot}CH_2OH$).

You are not expected to discuss the purification of either intermediates or the final product. (SUJB)

12 (a) What do you observe when an acidified solution of 2,4-dinitrophenylhydrazine and a solution of an aldehyde are warmed together and then cooled?

(b) Name the mechanism of this reaction.

(c) Write equations for the reaction of benzenecarbaldehyde (*benzaldehyde*) with the following:

(i) 2,4-dinitrophenylhydrazine,

(ii) hydroxylamine.

(d) Benzenecarbaldehyde (*benzaldehyde*) can give two different products **A**, $C_7H_8N_2$, and **B**, $C_{14}H_{12}N_2$, with hydrazine (NH_2NH_2). Write structures for **A** and **B**. (JMB)

13 (a) Write the structural formula of each of the following derivatives of ethanoic acid (acetic acid):

(i) ethanoic anhydride (acetic anhydride),

(ii) ethyl ethanoate (ethyl acetate),

(iii) ethanoyl chloride (acetyl chloride),

(iv) ethanamide (acetamide).

(b) Each of the derivatives in (a) can be *hydrolysed*.

(i) Explain what is meant by *hydrolysis*.

(ii) Give the conditions for the hydrolysis of each derivative.

(iii) Write the derivatives in order of ease of hydrolysis, putting the easiest first.

(c) Stating your reasons, arrange the following in order of increasing acid strength (i.e. putting the least acidic first): ethanoic acid; monobromoethanoic acid; monochloroethanoic acid; monoiodoethanoic acid. (AEB)

14 When pentan-2-ol, $CH_3CH_2CH_2CH(OH)CH_3$, is heated with an acidified aqueous solution of potassium dichromate(VI), the colour of the solution changes from orange to green and the alcohol is oxidised to the corresponding ketone.

(a) What is the oxidation state of the chromium species responsible for the green colour?

(b) Name one other reagent which could have been used instead of potassium dichromate(VI) to oxidise the alcohol to the ketone.

(c) What is the structural formula of the ketone formed by oxidising pentan-2-ol?

(d) Show, by means of an equation in each case, how this ketone reacts with (i) 2,4-dinitrophenylhydrazine (ii) hydroxylamine.

(e) How might you use the 2,4-dinitrophenylhydrazine derivatives of two isomeric ketones in order to distinguish between them? (JMB)

15 The reaction of propanone (acetone) with hydrogen cyanide is imperceptibly slow in the absence of a suitable catalyst. What substance would you use to catalyse this reaction? How does the catalyst function?
What is the reaction of propanone with:
(a) 2,4-dinitrophenylhydrazine,
(b) lithium tetrahydridoaluminate ($LiAlH_4$),
(c) aqueous ammoniacal silver nitrate?
What mass of tri-iodomethane (iodoform) is obtained by treating 1.00 g of propanone with an excess of iodine and alkali? (OC)

16 (a) Describe briefly a method for the preparation of propanone from propene.
(b) Under what conditions does propanone react with hydrogen cyanide? Name and give the structural formula of the product. Write equations to show the reaction of this product with dilute aqueous sodium hydroxide solution.
(c) Treatment of propanone with iodine and aqueous sodium hydroxide solution gives triiodomethane. What mass of triiodomethane may be obtained from 1 g of propanone?
(d) Write down the structural formula of an isomer of propanone. Describe and explain a simple experiment which would enable you to differentiate between these two isomers. (O)

17 A sample of pure ethanoyl chloride (acetyl chloride) was prepared as follows:
(i) 0.3 moles of phosphorus(III) chloride were added drop by drop to 0.2 moles of glacial ethanoic (acetic) acid. The mixture was heated on a water bath and the impure ethanoyl chloride was condensed and collected in a receiver surrounded by an ice/water mixture and with a side-arm attached to a calcium chloride tube.
(ii) The crude product was purified by fractional distillation to give 6.28 g of ethanoyl chloride.
(a) Write an equation for the reaction of phosphorus(III) chloride (phosphorus trichloride) with glacial ethanoic acid.
(b) Draw a labelled diagram of the apparatus used in stage (i) to obtain the impure ethanoyl chloride.
(c) Why was the receiver surrounded by an ice/water mixture?
(d) Why did the receiver have a calcium chloride side-arm tube?
(e) Name **three** impurities which may contaminate the impure ethanoyl chloride obtained in stage (i).
(f) Organic liquids are usually purified by first washing with aqueous reagents. Explain with an appropriate equation why this is not possible with ethanoyl chloride.
(g) What reagent was in excess during stage (i)?
(h) Calculate the theoretical yield of ethanoyl chloride in grams ($H = 1$, $C = 12$, $O = 16$, $Cl = 35.5$).
(i) What is the percentage yield of ethanoyl chloride in the preparation described above?

(j) Write the structural formula of the principal organic substance produced when ethanoyl chloride reacts with:
(i) methylamine,
(ii) anhydrous sodium ethanoate (sodium acetate),
(iii) benzene and anhydrous aluminium chloride. (SUJB)

18 (a) Describe, giving equations and essential conditions, how a solution of ethanoic acid may be obtained starting from (i) ethanol, (ii) ethanoyl chloride.
(b) Explain briefly:
(i) why compound **A** exhibits geometrical isomerism but compound **B** does not,

$$HOOCCH{=}CHCOOH \qquad HOOCCH_2CH_2COOH$$

A **B**

(ii) why compound **C** exhibits optical isomerism but compound **D** does not,

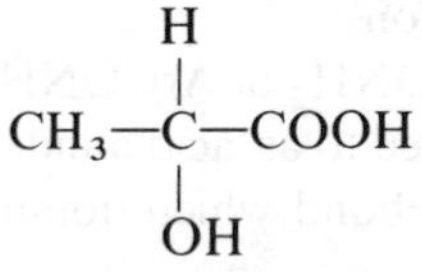

$$CH_3{-}\overset{\overset{H}{|}}{\underset{\underset{OH}{|}}{C}}{-}COOH \qquad CH_3{-}\overset{\overset{H}{|}}{\underset{\underset{COOH}{|}}{C}}{-}COOH$$

C 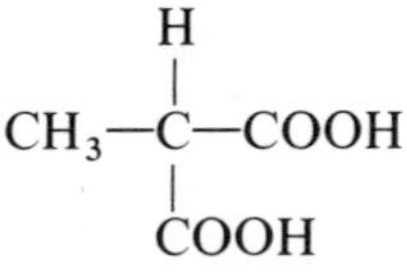 **D**

(c) **Predict**, giving your reasons, whether compound **E** will exhibit geometrical isomerism.

H_2C bonded in a three-membered ring to two CHCOOH groups (CHCOOH—CHCOOH)

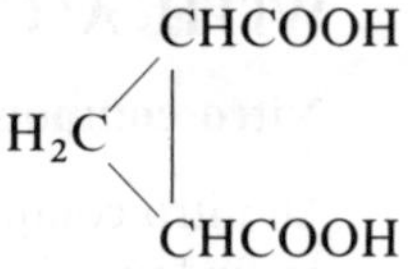

E (O)

19 Starting with bromomethane and benzene and no other organic materials, give a reaction scheme (i.e. reagents, conditions of reactions, and structures of intermediate compounds) for the formation of X.

$$X = C_6H_5{-}\underset{\underset{O}{\|}}{C}{-}\underset{\underset{H}{|}}{N}{-}C_6H_4{-}CH_3$$

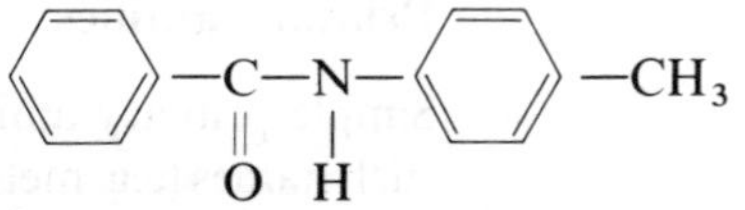

State what you would see, and name the products formed, when X is treated with hot concentrated hydrochloric acid. (AEB)

14.11 The C—N bond in nitro compounds, amines and diazonium salts

Data on the C—N bond

covalent bond length/nm 0.147
bond dissociation enthalpy/kJ mol^{-1} at 298 K 305

Of the many compounds whose molecules contain a carbon–nitrogen single bond, the most important are the nitro compounds, the amines, the diazonium salts and the amides.

Nitro compounds possess the nitro group, NO_2. They must not be confused with nitrites, i.e. esters of nitrous acid, with which they are isomeric.

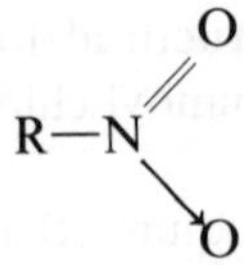

$$R{-}N{\overset{O}{\underset{O}{\lessgtr}}}$$

nitro compound

$$R{-}O{-}N{=}O$$

nitrite

Amines are derivatives of ammonia, in which one or more hydrogen atoms of the ammonia molecule have been substituted by alkyl or aryl groups. They are classified as primary, secondary and tertiary amines, according to the number of such groups.

Diazonium salts are electrovalent compounds consisting of a diazonium ion, $Ar\overset{\oplus}{N}{\equiv}N$, in association with an anion, usually Cl^-. The best known example is benzenediazonium chloride, $C_6H_5\overset{\oplus}{N}{\equiv}N\ Cl^-$. Aliphatic diazonium salts are highly unstable and decompose rapidly, even in cold dilute aqueous solution.

Amides, $RCONH_2$ or $ArCONH_2$, are derivatives of carboxylic acids and are often referred to as 'acid amides'. They possess, in addition to the C—N bond, a C=O bond which dominates their chemistry (§14.10.12).

14.11.1 NOMENCLATURE OF COMPOUNDS WITH A C—N BOND

Nitro compounds

All nitro compounds have substitutive names which are formed from the prefix 'nitro-' and the name of the appropriate alkane or arene,

e.g. CH_3NO_2 nitromethane; $C_6H_5NO_2$ nitrobenzene

Primary amines

Simple primary amines have radicofunctional names which, unlike other such names (e.g. methyl alcohol and methyl chloride), are written as a single word. Most of the differences between IUPAC and ASE recommendations are due to variations in the names of radicals, rather than to any fundamental difference in the system of naming. Examples are as follows:

	$CH_3CH_2CH_2NH_2$	$(CH_3)_2CHNH_2$
IUPAC:	propylamine	isopropylamine
ASE:	propylamine	(1-methylethyl)amine

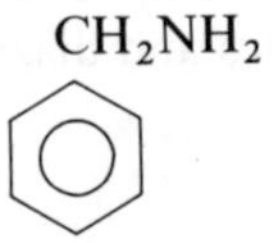

	$CH_3CH_2CH(NH_2)CH_3$	$C_6H_5CH_2NH_2$
IUPAC:	*sec*-butylamine	benzylamine
ASE:	(1-methylpropyl)amine	(phenylmethyl)amine

Certain trivial names are retained by IUPAC but not by ASE,

e.g.

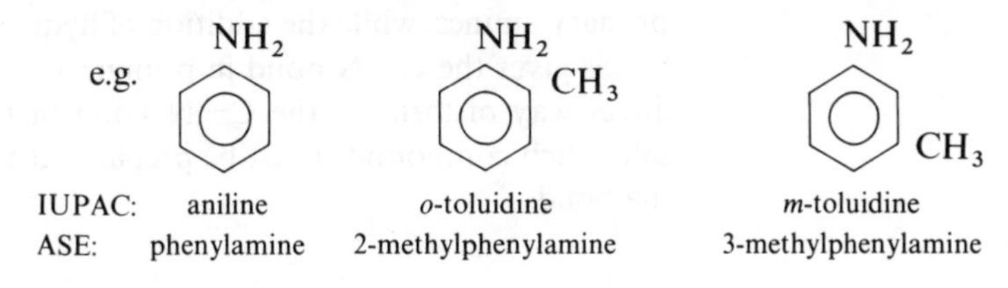

IUPAC:	aniline	*o*-toluidine	*m*-toluidine
ASE:	phenylamine	2-methylphenylamine	3-methylphenylamine

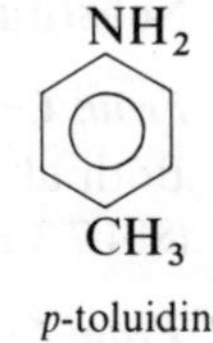

p-toluidine
4-methylphenylamine

Primary diamines are given substitutive names by addition of the suffix '-diamine' to the name of the parent hydrocarbon,

e.g. $NH_2CH_2CH_2NH_2$

IUPAC: 1,2-ethanediamine
ASE: ethane-1,2-diamine

Additive names, e.g. ethylenediamine for $NH_2CH_2CH_2NH_2$, are allowed by IUPAC but not recognised by ASE.

Secondary and tertiary amines

Symmetrical secondary and tertiary amines are named by adding the prefix 'di-' or 'tri-' to the name of the radical,

e.g. $(C_2H_5)_2NH$ — diethylamine; $(C_2H_5)_3N$ — triethylamine

Diazonium salts

These compounds are named by adding the suffix '-diazonium' and the name of the anion to that of the parent arene,

e.g.

benzenediazonium bromide 4-methylbenzenediazonium chloride

14.11.2 FORMATION OF THE C—N BOND

Both substitution and addition reactions are available. Substitution may be used to establish the C—N bond in the form of nitro compounds and primary amines, while the addition of hydrogen across the C=N or C≡N bonds gives the C—N bond in primary or secondary amines. There is no direct way of forming the C—N bond in tertiary amines or diazonium salts; such compounds must be prepared from others that already possess the bond.

Substitution methods

From C—H; the substitution of H by NO_2

Both alkanes and arenes react with nitric acid to give nitro compounds (§14.7.3 and 14.7.4).

From C—Cl, C—Br or C—I; the substitution of halogen by NH_2 (§14.8.4)

Alkyl halides undergo nucleophilic substitution with ammonia to give primary amines. Further reactions invariably occur because amines, like ammonia, act as nucleophiles and attack the remaining alkyl halide. The final product is thus a mixture of primary, secondary and tertiary amines and a quaternary ammonium salt.

Aryl halides, because of the non-lability of their halogen atoms, participate in such reactions only under drastic conditions.

Addition methods

From C≡N

This bond is present in nitriles, RCN (§14.8.4). On reduction, either by catalytic hydrogenation or by means of a metal plus protic solvent, nitriles give primary amines as the principal product. Because of a side reaction, however, appreciable quantities of secondary amines are also formed, e.g.

$$\underset{\text{ethanenitrile}}{CH_3CN} \xrightarrow[\text{or } Na + C_2H_5OH]{H_2/Ni \text{ catalyst}} CH_3CH_2NH_2 \text{ and } (CH_3CH_2)_2NH$$

The use of lithium tetrahydridoaluminate(III) as reductant avoids the formation of secondary amines.

14.11.3 INTERCONVERSION OF C—N COMPOUNDS

The preparation of amines by the ammonolysis of halides or alcohols is of limited use, because the reactions are difficult to perform and lead to a mixture of products whose separation may present problems. Consequently, it is common practice to establish the C—N bond in nitro compounds or in amides (§14.10.3) and then prepare other C—N compounds from these starting materials.

Primary amines

Both in the laboratory and the chemical industry aromatic primary amines are usually prepared by the reduction of nitro compounds,

e.g. $C_6H_5NO_2 \xrightarrow{Sn + HCl} C_6H_5NH_2$ (§14.11.6)

The method is not well suited to the preparation of aliphatic primary amines because of the difficulty of making aliphatic nitro compounds.

Amides may be converted into primary amines either by reduction (§14.10.12) or by treatment with bromine and alkali (§14.3.4),

e.g. $CH_3CONH_2 \xrightarrow[\text{reduction}]{Li[AlH_4]} CH_3CH_2NH_2$

$CH_3CONH_2 \xrightarrow[\text{Hofmann reaction}]{Br_2 + KOH} CH_3NH_2$

Diazonium salts

The diazotisation of aromatic primary amines is achieved by treatment with sodium nitrite and dilute hydrochloric acid (or sulphuric acid) at a temperature of 0–5 °C (273–278 K) (§14.11.7).

E.g. $C_6H_5NH_2 + NaNO_2 + 2HCl = C_6H_5\overset{\oplus}{N}{\equiv}N\ Cl^- + NaCl + 2H_2O$

14.11.4 GENERAL PROPERTIES OF COMPOUNDS WITH A C—N BOND

Most aliphatic amines

Amines

Aliphatic amines are colourless compounds with a fishy, ammoniacal smell. Because the charge separation in the N—H bond is less than in the O—H bond, amines are more volatile than alcohols of comparable relative molecular mass, e.g.

	Boiling temperature/ °C (K)		Boiling temperature/ °C (K)
CH_3NH_2	−6.3 (266.9)	CH_3OH	64.5 (337.7)
$CH_3CH_2NH_2$	16.6 (289.8)	CH_3CH_2OH	78.5 (351.7)

So volatile are the methylamines and ethylamines that (like ammonia) they irritate the sensitive tissues of the nose and eyes.

Hydrogen bonding between amine and water molecules is responsible for the lower amines having a high solubility in water.

$C_6H_5NH_2$
and most
aromatic amines

Most aromatic amines are crystalline solids, although the lower ones, notably phenylamine, are oily liquids with a density greater than 1 g cm^{-3}. They have a sweet, sickly smell and are highly poisonous.

The lower members can be steam distilled. This is because they have a low solubility in water, are steam-volatile, and are not hydrolysed by hot water.

Diazonium salts

Generally, aromatic diazonium salts are only marginally more stable than their aliphatic counterparts. They are normally prepared in cold aqueous solution and used at once, for they decompose slowly even at 0 °C (273 K).

14.11.5 CHEMICAL PROPERTIES OF COMPOUNDS WITH A C—N BOND

The electronegativity difference between carbon and nitrogen leads to a permanent charge separation of the carbon–nitrogen bond, $\overset{\delta+}{C}—\overset{\delta-}{N}$. A similar effect in the carbon–halogen bond causes alkyl halides to take part in many nucleophilic substitution reactions, and we must consider the possibility of such reactions with amines, nitro compounds and diazonium salts:

$$Nu\!:^- + \;\rangle C—NH_2 = Nu—C\langle \; + NH_2^-$$

$$Nu\!:^- + \;\rangle C—NO_2 = Nu—C\langle \; + NO_2^-$$

$$Nu\!:^- + \;\rangle C—\overset{\oplus}{N}\equiv N = \;\rangle C—Nu + N_2$$

However, the nitrogen atom is small and the C—N bond is not readily polarised any further. As a result, NH_2^- and NO_2^- ions are poor leaving groups, and *amines and nitro compounds do not undergo substitution by nucleophilic reagents.* The only C—N compounds which do react in this way are diazonium salts. Even so, aromatic diazonium salts do not react readily except in the presence of a catalyst.

14.11.6 NITRO COMPOUNDS

Nitro compounds are attacked readily only by reducing agents. Reduction is often accomplished by a non-noble metal (e.g. Fe, Zn, Sn) in acidic solution, but can also be achieved electrolytically or by catalytic hydrogenation.

$C_6H_5NO_2$

In the laboratory, nitrobenzene can rapidly be reduced to phenylamine by refluxing with tin and hydrochloric acid:

$$C_6H_5NO_2 + 6H^+ + 6e^- = C_6H_5NH_2 + 2H_2O \quad (1)$$

$$Sn = Sn^{4+} + 4e^- \quad (2)$$

Multiplication of the half-equation (1) by 2 and (2) by 3, followed by addition, gives the ionic equation for the complete reaction:

$$2C_6H_5NO_2 + 12H^+ + 3Sn = 2C_6H_5NH_2 + 4H_2O + 3Sn^{4+}$$

In acidic solution the phenylamine forms as an involatile amine salt. When reduction of nitrobenzene is complete, alkali is added to liberate the free amine:

$$C_6H_5NH_3^+ + HO^- = C_6H_5NH_2 + H_2O$$

Phenylamine can then be extracted from the mixture by steam distillation.

14.11.7 AMINES

The chemistry of amines is dominated by the lone pair of electrons in the outer shell of the nitrogen atom. Donation of this lone pair to a proton accounts for the basic properties of amines, while donation to another species, usually a carbon atom, is responsible for amines acting as nucleophiles in their reactions with alkyl halides, alcohols, carbonyl compounds and nitrous acid.

Amines as bases

All amines, aliphatic and aromatic, resemble ammonia in that their molecules are able to coordinate to a proton,

e.g. $$\text{H—N(R)(H):} + H^+ \rightleftharpoons [\text{H—N(R)(H)}\rightarrow\text{H}]^+$$

primary amine — monosubstituted ammonium ion

Secondary and tertiary amines, similarly, give rise to di- and trisubstituted ammonium ions respectively.

Ionisation occurs in water, with the result that aqueous solutions of amines are alkaline:

$$RNH_2 + H_2O \rightleftharpoons RNH_3^+ + HO^-$$

cf. $$NH_3 + H_2O \rightleftharpoons NH_4^+ + HO^-$$

Amines, like ammonia, are weak bases, for their ionisation in water is far from complete. Because α (i.e. the degree of ionisation) is low, K_b values are relatively low and pK_b values relatively high (§9.2.1).

Neutralisation reactions occur with acids at room temperature to give amine salts. If, for example, a bottle of aqueous methylamine is held close to one of concentrated hydrochloric acid, a dense white smoke of methylammonium chloride is observed:

$$CH_3NH_2 + HCl = CH_3\overset{\oplus}{N}H_3Cl^-$$

cf. $$NH_3 + HCl = NH_4^+Cl^-$$

If hydrochloric acid is added to phenylamine, and the resulting mixture is cooled, crystals of phenylammonium chloride are deposited:

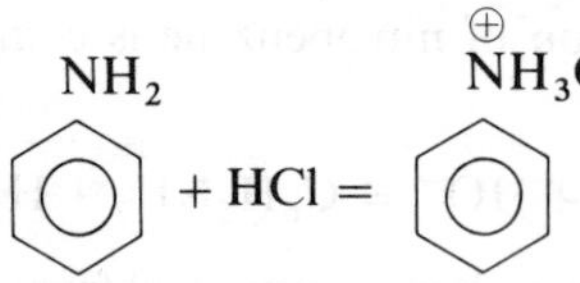

$$C_6H_5NH_2 + HCl = C_6H_5\overset{\oplus}{N}H_3Cl^-$$

The term *amine salt* is used to describe the mono-, di- and trisubstituted ammonium salts made by dissolving amines in acids, and also the quaternary ammonium salts, i.e. tetrasubstituted ammonium salts, made by the nucleophilic attack of tertiary amines on alkyl halides (§14.8.4). All such compounds are named as substituted ammonium salts,

e.g. $[(CH_3)_2NH_2]^+Cl^-$ dimethylammonium chloride
$[(CH_3)_4N]^+Cl^-$ tetramethylammonium chloride

If different hydrocarbon radicals are present, they are cited in alphabetical order,

e.g. $$[CH_3CH_2CH_2{-}N(CH_2CH_3)(H){-}CH_3]^+ Cl^-$$

ethylmethylpropylammonium chloride

Should the name of the parent base end in '-ine', rather than '-amine', the cation name is formed by adding the suffix '-ium' to that of the base, with elision of the final 'e',

e.g $[C_6H_5NH_3]^+Cl^-$
IUPAC: anilinium chloride
ASE: phenylammonium chloride

As may be expected from their ionic structure, amine salts are crystalline solids with a moderate solubility in water. They are extensively hydrolysed in aqueous solution and give an acidic reaction if they are derived from strong acids (§9.2.8).

Amines as nucleophiles

Amines are able to react with a considerable variety of Lewis acids. Often, coordination of the amine precedes the elimination of another species. In some of these reactions, e.g. with nitrous acid and with carbonyl compounds, amines resemble alcohols. They are, however, better than alcohols at donating a lone pair of electrons, and will react with certain compounds (e.g. alkyl halides) that are not attacked by alcohols.

Alkyl halides

In nucleophilic substitution reactions with alkyl halides, primary amines give secondary amines, secondary amines give tertiary amines, and tertiary amines give quaternary ammonium salts (§14.8.4).

Carbonyl compounds

Primary and secondary amines undergo nucleophilic addition–elimination reactions with carbonyl compounds. With aldehydes and ketones, primary amines give Schiff's bases (§14.10.6). With carboxylic acids and their derivatives (acid anhydrides, acid chlorides and esters), primary and secondary amines give *N*-substituted amides (e.g. §14.10.10).

Nitrous acid on primary amines

Nitrous acid is an unstable compound and must be generated *in situ* from sodium nitrite and hydrochloric acid:

$$NaNO_2 + HCl = HNO_2 + NaCl$$

The amine under test is first dissolved in excess dilute hydrochloric acid. The solution is cooled to 5 °C (278 K) by means of ice, and sodium nitrite then added in small portions with stirring and continued cooling.

An excess of hydrochloric acid is needed to form the nitrosyl cation, NO^+:

$$\text{H}{-}\ddot{\text{O}}{-}\text{N}{=}\text{O} + \underset{\text{(from HCl)}}{\text{H}^+} \rightleftharpoons \underset{\text{protonated nitrous acid}}{\text{H}_2\overset{\oplus}{\text{O}}{-}\text{N}{=}\text{O}} \rightleftharpoons \text{H}_2\text{O} + \underset{\text{nitrosyl cation}}{\overset{\oplus}{\text{N}}{=}\text{O}}$$

The primary amine coordinates to the nitrosyl cation to give a species which changes, by a series of reactions, into a diazonium ion, $R{-}\overset{\oplus}{N}{\equiv}N$,

$$\text{R}(\text{H})_2\text{N}{:} + \overset{\oplus}{\text{N}}{=}\text{O} = \text{R}{-}\overset{\oplus}{\text{N}}{\equiv}\text{N} + \text{H}_2\text{O}$$

Chloride ions are also present in solution; thus, at this point we have a diazonium chloride, $R{-}\overset{\oplus}{N}{\equiv}N\ Cl^-$. *Aliphatic* diazonium salts are highly unstable and cannot be isolated. They are immediately hydrolysed to give alcohols:

$$R{-}\overset{\oplus}{N}{\equiv}N\ Cl^- + H_2O = ROH + N_2 + HCl \qquad (§14.11.8)$$

The overall reaction between an aliphatic primary amine and nitrous acid can therefore be represented as follows:

$$RNH_2 + HNO_2 = ROH + N_2 + H_2O$$

In effect, the amino group, NH_2, is replaced by the hydroxyl group, HO. Nitrous acid is used for this purpose throughout the whole of organic chemistry; see, for example, the conversion of an amide, $RCONH_2$, to a carboxylic acid, RCOOH, (§14.10.12).

Aromatic diazonium salts are more stable than their aliphatic counterparts and remain in solution at 5 °C (278 K). The overall reaction can be summarised as follows:

$$ArNH_2 + NaNO_2 + 2HCl = Ar\overset{\oplus}{N}{\equiv}N\ Cl^- + NaCl + 2H_2O$$

Hydrolysis, to phenols, occurs at an appreciable rate only on warming (§14.11.8).

Alcohols are not the only compounds to be formed when aliphatic primary amines are treated with nitrous acid. The product, even at 0 °C (273 K), is likely to be a mixture of an alcohol, alkene, alkyl nitrite, alkyl chloride and other compounds. These substances arise mainly through reactions between the highly reactive diazonium ion and various nucleophiles in the solution,

$$\text{e.g.}\quad R{-}\overset{\oplus}{N}{\equiv}N + NO_2^- = RONO + N_2$$

14.11.8 AROMATIC DIAZONIUM SALTS

Although the formula of the aryldiazonium ion is usually written $Ar{-}\overset{\oplus}{N}{\equiv}N$, the positive charge is not located entirely on the penultimate nitrogen atom but is shared among both nitrogen atoms and the aromatic ring. (The formula $\overbrace{Ar{-}N{\equiv}N}^{\oplus}$ may be considered more appropriate.) This accounts for the relative stability of aromatic diazonium ions; in aliphatic ones there is less scope for delocalisation of charge.

Most of the reactions of diazonium salts are nucleophilic substitutions. They may take place with or without the loss of nitrogen.

I The sharing of positive charge with the adjoining carbon atom leads to nucleophilic substitution with the loss of molecular nitrogen:

$$Nu\!:^- + \;\rangle\overbrace{C{-}N{\equiv}N}^{\oplus} = \;\rangle C{-}Nu + N_2$$

II The sharing of positive charge with the terminal nitrogen atom leads to nucleophilic substitution with the retention of nitrogen:

$$\rangle\overbrace{C{-}N{\equiv}N}^{\oplus} + Nu\!:^- = \;\rangle C{-}N{=}N{-}Nu$$

While a few reactions of type I take place satisfactorily without a catalyst, most need to be catalysed by a copper(I) compound. Reactions of type II are known as *coupling reactions.* They occur with phenols and aromatic amines, and give rise to brightly coloured azo dyes.

Uncatalysed nucleophilic substitution with loss of nitrogen

HO⁻

Diazonium salts react with water to give phenols; slowly at room temperature, but more rapidly on warming,

$$\text{e.g.}\quad C_6H_5\overset{\oplus}{N}\equiv N\ Cl^- + H_2O = C_6H_5OH + N_2 + HCl$$

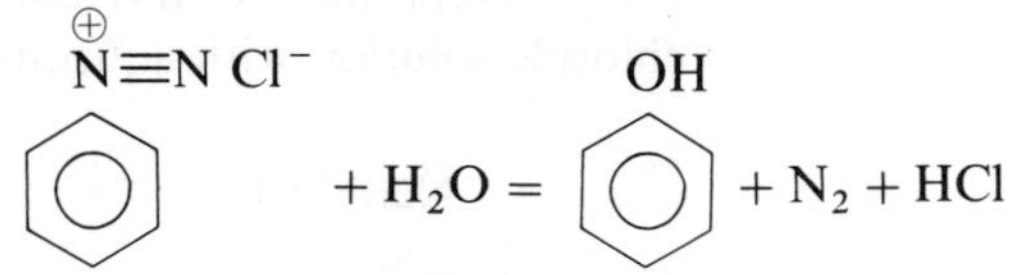

Competing reactions occur (principally the formation of an aryl chloride), the yield of phenol is low, and the method is used only in special circumstances, e.g. when it is impracticable to make the phenol by other means.

RO⁻

On boiling with methanol or ethanol, diazonium salts react in the same way as they do with water, ethers being formed instead of phenols:

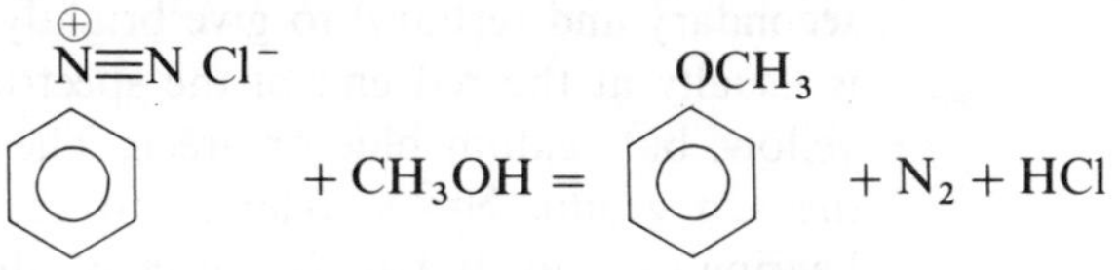

$$C_6H_5\overset{\oplus}{N}\equiv N\ Cl^- + CH_3OH = \underset{\text{methoxybenzene}}{C_6H_5OCH_3} + N_2 + HCl$$

At the same time a small amount of benzene is formed by reduction of the diazonium salt.

Catalysed nucleophilic substitution with loss of nitrogen (Sandmeyer reactions)

Although N_2 is a much better leaving group than, say, Cl^- or HO^- (compare the reactivities of $Ar\overset{\oplus}{N}\equiv N\ Cl^-$ and ArCl), a catalyst is often required to encourage its removal and replacement by other groups. Groups that are introduced by means of Sandmeyer reactions include Cl^-, Br^- and CN^- (catalysed by Cu^I), and I^- (self-catalysed).

Cl⁻

In the Sandmeyer reaction for introducing Cl^-, the diazonium chloride, at 0 °C (273 K), is added to a solution of copper(I) chloride in concentrated hydrochloric acid, and the mixture is then boiled until there is no further production of nitrogen,

$$\text{e.g.}\quad C_6H_5\overset{\oplus}{N}\equiv N\ Cl^- \xrightarrow{\text{CuCl in conc. HCl}} C_6H_5Cl + N_2$$

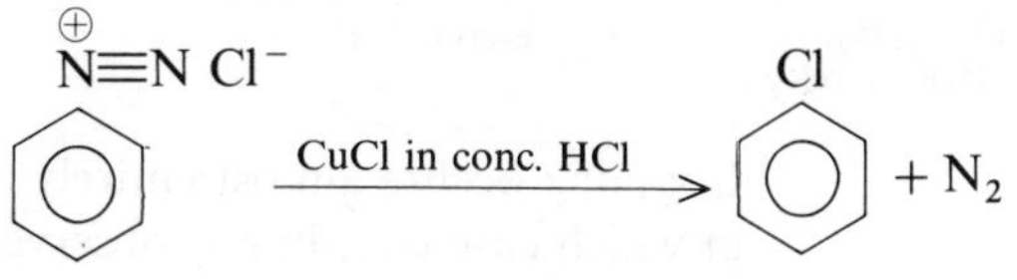

Metallic copper, too, is suitable as a catalyst, probably because of the thin film of copper(I) oxide that exists on its surface.

Br⁻

The bromide ion can be introduced in a similar fashion to the chloride ion, by treating the diazonium bromide with copper(I) bromide dissolved in hydrobromic acid.

I⁻

For the preparation of aryl iodides it is necessary only to boil the diazonium chloride solution with aqueous potassium iodide,

e.g. $C_6H_5N^{\oplus}{\equiv}N\ Cl^- + KI = C_6H_5I + KCl + N_2$

Superficially, the reactions appear to be uncatalysed, but it is likely that they are catalysed by the iodide ion.

Coupling reactions

Diazonium salts couple with phenols and aromatic amines (primary, secondary and tertiary) to give brightly coloured *azo dyes*. Their colour is usually at the red end of the spectrum; i.e. it may be red, orange or yellow, but seldom blue or green. The colour is due to the presence of the azo group, N=N, referred to as a *chromophoric* (literally 'colour bearing') group. It has the effect of absorbing the blue components of white light, so that only the red and yellow ones are transmitted or reflected.

Coupling reactions are conventionally regarded as electrophilic substitution reactions of the phenol or aromatic amine, in which the diazonium ion acts as an electrophile.

Diazonium ions are relatively weak electrophiles because of the low concentration of positive charge on the terminal nitrogen atom. Benzene itself does not react with diazonium ions because the electron density of the ring is insufficient. However, in phenols and amines the electron density is increased by the mesomeric effect (§14.8.4). Consequently, phenols and amines are much more reactive than benzene in all electrophilic substitutions.

$C_6H_5X + \overset{\oplus}{N{\equiv}N{-}Ar} \longrightarrow$ π-complex $\longrightarrow$ σ-complex (H and N=N—Ar on the same ring carbon, ring ⊕) $\xrightarrow{-H^+}$ azo dye ($X{-}C_6H_4{-}N{=}N{-}Ar$)

X = HO, NH$_2$, NHR or NR$_2$

π-complex σ-complex azo dye

Coupling occurs almost entirely at the 4-position, unless this is blocked, in which case coupling is observed at the 2-position.

The pH of the solution is critical. Phenols couple most rapidly in basic solution at a pH of 9–10, because the negative charge of the phenoxide ion, which is present under these conditions, helps to increase the electron density of the ring still further. For example,

$$C_6H_5N^{\oplus}{\equiv}N\ Cl^- + \text{2-naphthol} \xrightarrow[\substack{5\text{–}10\ ^\circ C \\ (278\text{–}283\ K)}]{\text{NaOH solution}} \text{1-phenylazo-2-naphthol (red)} + HCl$$

2-naphthol; 1-phenylazo-2-naphthol (red)

Strongly basic solutions should be avoided.

Best results with amines are obtained in weakly acidic solution. The optimum pH is ~5, achieved by the use of a buffer solution of ethanoic acid and sodium ethanoate. For example,

$$C_6H_5N^{\oplus}{\equiv}NCl^- + C_6H_5N(CH_3)_2 \xrightarrow[\substack{5\text{–}10\ ^\circ C \\ (278\text{–}283\ K)}]{CH_3COOH + CH_3COONa} C_6H_5N{=}NC_6H_4N(CH_3)_2 + HCl$$

4-dimethylaminoazobenzene (yellow)

An excess of acid must be avoided, as this will completely convert the amine into its salt. In this condition the lone pair of electrons on the nitrogen atom is used in bonding to a proton, and is unable to increase the electron density of the ring. In neutral solution coupling occurs at the amine nitrogen atom rather than the 4-position of the ring, and in basic solution there is no reaction.

EXAMINATION QUESTIONS ON SECTION 14.11

1 Describe in *detail* the laboratory preparation of phenylamine (aniline) from nitrobenzene. Your description should include materials, conditions, the recovery and purification of the product and the principles underlying the procedures adopted at each stage of the preparation. (AEB)

2 In the laboratory preparation of phenylamine (aniline) from nitrobenzene the following five stages are used:

(i) nitrobenzene is heated with tin and excess concentrated hydrochloric acid,

(ii) after cooling, excess sodium hydroxide is added,

(iii) the mixture is then steam distilled,

(iv) excess salt is added to the distillate,

(v) the resultant mixture is extracted twice with an equal volume of ether.

(a) State the purpose of each stage and, where possible, give the principle(s) underlying each stage.

(b) Describe the precautions you would take in carrying out the distillation of the ether extract:

(i) to recover the ether,

(ii) to obtain a pure sample of phenylamine.

(c) Describe a chemical test you would use to confirm that the final product is phenylamine. (AEB)

3 Addition of sodium nitrite solution to an ice-cold solution of phenylamine (aniline) in hydrochloric acid yields a solution of benzenediazonium chloride. What happens when sodium nitrite is added to a cold solution of l-aminopropane in hydrochloric acid? How do you account for this?

Starting from benzenediazonium chloride solution, how would you obtain (a) chlorobenzene, (b) an azo dyestuff, and (c) benzene?

How may 1,3,5-tribromobenzene be obtained from suitable starting materials? (OC)

4 (a) For each of parts (i) to (x) below **only one** of the alternatives **A**, **B**, **C**, **D**, **E** is correct. Answer each part by giving the appropriate letter.

A $CH_3{\cdot}CH_2{\cdot}NH_2$ **B** $CH_3{\cdot}CO{\cdot}NH_2$ **C** $C_6H_5NO_2$
D $C_6H_5{\cdot}NH_2$ **E** $C_6H_5N_2Cl$

(i) Which is a strong electrolyte?

(ii) Which dissolves in dilute hydrochloric acid, but not in water?

(iii) Which is insoluble in water, acid and alkali?

(iv) Which is a colourless liquid at 20 °C when pure?

(v) Which has the highest vapour pressure at 15 °C?

(vi) Which is explosive when pure?

(vii) Which yields a product with one less carbon atom when treated with bromine and potassium hydroxide solution?

(viii) Which is the most ready to combine with a proton?

(ix) Which gives ammonia on warming with aqueous sodium hydroxide solution?

(x) Which evolves nitrogen on treatment with nitric(III) acid (nitrous acid) at 5 °C?

(b) How and under what conditions does ethanamide (acetamide) react with:

(i) bromine and sodium hydroxide,

(ii) dilute hydrochloric acid? (SUJB)

5 Nitrobenzene was reduced to aniline and the aniline converted into benzenediazonium chloride by its reaction with aqueous sodium nitrite and hydrochloric acid below 10 °C.

(a) State a reagent or mixture of reagents suitable for the reduction of nitrobenzene to aniline.

(b) Write an equation for the conversion of aniline into benzenediazonium chloride.

(c) When the aqueous solution of benzenediazonium chloride is heated to 50 °C and then cooled, a low-melting organic solid is formed. (i) Name the organic product. (ii) Write an equation for its formation. (iii) Suggest one simple chemical test by which the product can be distinguished from aniline.

(d) Give one example of the use of benzenediazonium chloride in the preparation of azo-compounds, by writing out the structures of (i) the other reagent, (ii) the product. (JMB)

6 (a) For each of the transformations (i) to (viii) in the reaction scheme below choose the most appropriate reaction conditions from the list **A** to **I**. (You need only write the number of the reaction and append to it the appropriate letter of the reaction conditions.)

$$
\begin{array}{ccccc}
 & & CH_3NH_2 & & CH_3COOH \\
 & & \uparrow\text{(iii)} & & \uparrow\text{(iv)} \\
CH_3COONH_4 & \xrightarrow{\text{(i)}} & CH_3CONH_2 & \xrightarrow{\text{(ii)}} & CH_3CN \\
 & \swarrow\text{(vii)} & & \searrow\text{(viii)} & \downarrow\text{(v)} \\
CH_3COONa & & & & CH_3CH_2NH_2 \\
 & & & & \downarrow\text{(vi)} \\
 & & & & CH_3CH_2OH
\end{array}
$$

A Distil with dry phosphorus(V) oxide (phosphorus pentoxide).
B Add dilute hydrochloric acid and sodium nitrate(III) (sodium nitrite) solution.
C Heat strongly in a sealed tube.
D Add bromine, then warm with aqueous sodium hydroxide.
E Treat with lithium tetrahydridoaluminate(III) (lithium aluminium hydride) ($LiAlH_4$) in ether.
F Reflux with dilute hydrochloric acid.
G Boil with aqueous sodium hydroxide.
H Reflux with excess glacial ethanoic acid (acetic acid).
I Bubble hydrogen gas through the solution.

(b) The compound $HOC_6H_4NH_2$ is very soluble in dilute hydrochloric acid and in dilute sodium hydroxide solution, but it is insoluble in water. Explain this with appropriate equations.

(c) Write the structural formula of the principal organic substance produced when phenylamine (aniline) reacts with:
(i) ethanoyl (acetyl) chloride,
(ii) bromine water,
(iii) iodomethane. (SUJB)

7 Give the structural formula, indicating the characteristic functional group, of (a) a *carboxylic acid amide*, (b) a *primary aliphatic amine*, and (c) a *primary aromatic amine*.

Describe **two** ways in which the chemical properties of amides differ from those of amines and **two** ways in which they resemble those of amines.

How may your example in (a) be converted into an amine,
in (b) be converted into an amide,
and in (c) be converted into a nitrile? (OC)

14.12

Germanium, tin and lead

14.12.1 THE OCCURRENCE AND ISOLATION OF THE ELEMENTS

Germanium is obtained by reduction of germanium(IV) oxide with hydrogen at red heat:

$$GeO_2 + 2H_2 = Ge + 2H_2O$$

Impure tin(IV) oxide ore, known as cassiterite or tinstone, is reduced by heating with coke. This produces crude tin which is refined by heating in a furnace with a sloping hearth until the tin melts and flows away from less fusible impurities, e.g. iron.

$$SnO_2 + 2C = Sn + 2CO$$

Lead is obtained from galena, PbS, the most abundant lead ore.

14.12.2 THE REACTIONS OF GERMANIUM, TIN AND LEAD

Table 14.18 The products of some reactions of germanium, tin and lead

Reagent	Germanium	Tin	Lead
air or oxygen on heating	GeO_2	SnO_2	PbO or Pb_3O_4
chlorine	$GeCl_4$	$SnCl_4$	$PbCl_2$
heated sulphur	GeS_2	SnS_2	PbS
concentrated HNO_3	GeO_2	SnO_2	$Pb(NO_3)_2$
concentrated HCl	no reaction	$SnCl_2$	$PbCl_2$
concentrated H_2SO_4	GeO_2	$Sn(SO_4)_2$	$PbSO_4$
alkalis	$[Ge(OH)_6]^{2-}$	$[Sn(OH)_6]^{2-}$	$[Pb(OH)_6]^{4-}$
heated metals	alloy formation	alloy formation	alloy formation

We notice from their reactions (Table 14.18) that germanium and tin, in contrast to lead, have a distinct preference for the tetravalent state. The reaction of lead with dilute hydrochloric acid or cold concentrated sulphuric acid is slow, because the insoluble lead salts which are formed tend to coat the metal surface. Attack by hot concentrated sulphuric acid is more rapid because lead sulphate is soluble under these conditions. Warm concentrated hydrochloric acid is effective because it dissolves lead(II) chloride to form the tetrachloroplumbate(II) ion, (§14.12.4).

14.12.3 THE OXIDES OF GERMANIUM, TIN AND LEAD

The oxides of these elements are:

GeO	SnO	PbO	(divalent or +2 oxidation state)
GeO_2	SnO_2	PbO_2	(tetravalent or +4 oxidation state)
		Pb_3O_4	(mixed oxide)

The following general comments can be made.

(i) All the oxides are amphoteric.
(ii) Basic character increases from germanium to lead.
(iii) Basic character is more pronounced in the divalent state than the tetravalent state.
(iv) Lead, alone in this group, forms a mixed oxide.

Germanium(II) oxide

Germanium(II) oxide is readily oxidised to germanium(IV) oxide, GeO_2, by oxygen. It dissolves in alkalis to produce the germanate(II) ion, GeO_2^{2-} or possibly $[Ge(OH)_4(aq)]^{2-}$, and forms germanium(II) salts with acids.

Germanium(IV) oxide

Germanium(IV) oxide is a white powder which can be prepared by heating germanium in oxygen. A hydrated form is obtained by dissolving germanium in concentrated nitric acid, or by hydrolysing germanium(IV) chloride with an excess of water:

$$Ge + O_2 = GeO_2$$

$$Ge + 4HNO_3 = GeO_2(aq) + 4NO_2 + 2H_2O$$

$$GeCl_4 + 2H_2O \rightleftharpoons GeO_2(aq) + 4HCl$$

At room temperature germanium(IV) oxide is ionic, i.e. $Ge^{4+}\ (O^{2-})_2$, with a similar structure to that of rutile (Fig. 6.15), but above 1 033 °C (1 306 K) it has a silica type of structure. In forming an oxide which has both ionic and covalent structures germanium exhibits both metallic and non-metallic character.

Germanium(IV) oxide dissolves in an excess of concentrated hydrochloric acid to form $GeCl_4$ and $[GeCl_6]^{2-}$ (§14.12.4). With alkalis germanate(IV) species are produced, which are oxoanions containing germanium, e.g. $[GeO(OH)_3]^-$ or $[Ge(OH)_6]^{2-}$. Like all germanium(IV) compounds, the oxide does not possess any reducing or oxidising properties.

Tin(II) oxide

Tin(II) oxide is prepared by heating tin(II) ethanedioate:

$$(COO)_2Sn = SnO + CO + CO_2$$

The hydrated oxide, SnO(aq), which is produced as a white precipitate by adding a limited quantity of alkali to a tin(II) salt, can be dehydrated at 120 °C (393 K) to give the blue-black anhydrous compound.

Tin(II) oxide dissolves readily in acids or alkalis to form tin(II) salts or stannate(II) ions, possibly $[Sn(OH)_6]^{4-}$,

e.g. $$SnO + 2HCl = SnCl_2 + H_2O$$

$$SnO + 4HO^- + H_2O = [Sn(OH)_6]^{4-}$$

A solution of the stannate(II) ion is a powerful reducing agent.

Tin(IV) oxide

Tin(IV) oxide can be prepared by methods similar to those used for germanium(IV) oxide. The hydrated forms of the oxide, which vary widely in their water content, readily become anhydrous on heating. Tin(IV) oxide is ionic, i.e. $Sn^{4+}(O^{2-})_2$, with the rutile type of structure (Fig. 6.15).

Tin(IV) oxide dissolves in acids or alkalis, particularly fused alkalis,

e.g. $$SnO_2 + 2H_2SO_4 = Sn(SO_4)_2 + 2H_2O$$

$$SnO_2 + 2HO^- + 2H_2O = [Sn(OH)_6]^{2-}$$

The octahedral hexahydroxostannate(IV) ion, $[Sn(OH)_6]^{2-}$, comprises six hydroxide ions coordinated to a tin(IV) ion, and may be compared with the aluminate(III) ion (§13.2.4). Like all tin(IV) compounds, the oxide does not function as a reductant or an oxidant.

The reaction of tin with concentrated nitric acid possibly involves the formation of hydrated tin(IV) ions, which undergo extensive hydrolysis because of the high surface charge density of the tin(IV) ion (cf. aluminium).

$$(x-4)H_2O + Sn + 4NO_3^- + 8H^+(aq) = [Sn(H_2O)_x]^{4+} + 4NO_2$$

$$[Sn(H_2O)_x]^{4+} + H_2O \rightleftharpoons [Sn(OH)(H_2O)_{x-1}]^{3+} + H_3O^+$$

$$[Sn(OH)(H_2O)_{x-1}]^{3+} + H_2O \rightleftharpoons [Sn(OH)_2(H_2O)_{x-2}]^{2+} + H_3O^+$$

The loss of two more hydrogen ions would produce hydrated tin(IV) hydroxide, i.e. $[Sn(OH)_4(H_2O)_{x-4}]$. However, there is no evidence for the existence of this species, presumably because it rapidly changes into the hydrated oxide:

$$[Sn(OH)_4(H_2O)_{x-4}] \rightleftharpoons SnO_2(aq) + (x-2)H_2O$$

With an excess of alkali, further ionisation would produce anionic species, e.g. $[Sn(OH)_6]^{2-}$

Lead(II) oxide

PbO

Lead(II) oxide exists in two different structural forms, litharge (red) and massicot (yellow). The stable form at room temperature is litharge, but the change from massicot to litharge is very slow under these conditions. Lead(II) oxide may be prepared by the thermal decomposition of lead(IV) oxide or other lead(II) compounds, e.g. the carbonate, hydroxide or nitrate. At temperatures below 550 °C (823 K) litharge is formed, but above 650 °C (923 K) the product is massicot.

Lead(II) oxide is the most basic group 4B oxide and dissolves in acids to produce lead(II) salts, although the reaction may be slow if the resulting salt is insoluble and forms a coating on the surface of the oxide particles. Lead(II) oxide also dissolves slowly in solutions of alkali to produce anionic plumbate(II) species, possibly $[Pb(OH)_6]^{4-}$; cf. tin(II) oxide.

The only hydroxide formed by the elements of this group appears to be lead(II) hydroxide, $Pb(OH)_2$, which is produced as a white precipitate by the addition of limited alkali to a solution of a lead(II) salt:

$$Pb^{2+}(aq) + 2HO^- = Pb(OH)_2$$

Lead(II) hydroxide is amphoteric and possesses no reducing or oxidising properties.

PbO_2

Lead(IV) oxide

A brown precipitate of lead(IV) oxide, PbO_2, is formed by:

(i) oxidising the lead(II) ion with the hypochlorite ion, ClO^-, in alkaline solution:

$$Pb^{2+} + 2HO^- + ClO^- = PbO_2 + H_2O + Cl^-$$

(ii) treating trilead tetraoxide, Pb_3O_4, with dilute nitric acid:

$$Pb_3O_4 + 4HNO_3 = 2Pb(NO_3)_2 + PbO_2 + 2H_2O$$

(iii) the electrolytic oxidation of a lead(II) salt:

$$Pb^{2+} + 2HO^- = PbO_2 + 2H^+(aq) + 2e^- \quad \text{(at the anode)}$$

Lead(IV) oxide is similar in structure to rutile (Fig. 6.15). Above 300 °C (573 K) it decomposes into lead(II) oxide and oxygen:

$$2PbO_2 = 2PbO + O_2$$

Lead(IV) oxide is a powerful oxidising agent. For example, it oxidises sulphur dioxide and warm concentrated hydrochloric acid:

$$PbO_2 + SO_2 = PbSO_4$$

$$PbO_2 + 4HCl = PbCl_2 + 2H_2O + Cl_2$$

At 0 °C (273 K) lead(IV) oxide dissolves in concentrated hydrochloric acid to produce the unstable lead(IV) chloride, $PbCl_4$, and with fused alkalis forms the hexahydroxoplumbate(IV) ion, $[Pb(OH)_6]^{2-}$ (cf. SnO_2).

Pb_3O_4

Trilead tetraoxide (dilead(II) lead(IV) oxide)

This orange-red compound is manufactured by heating lead(II) oxide in air at temperatures close to 340 °C (613 K):

$$6PbO + O_2 \underset{\text{above 550 °C (823 K)}}{\overset{\text{300–550 °C (573–823 K)}}{\rightleftharpoons}} 2Pb_3O_4$$

It is a mixed oxide which, as its name suggests, behaves chemically (but not physically) as a 2:1 mixture of lead(II) oxide and lead(IV) oxide. For example, with nitric acid the 'lead(II) part' dissolves, leaving a precipitate of lead(IV) oxide:

$$\underset{(\text{'}PbO_2\cdot 2PbO\text{'})}{Pb_3O_4} + 4HNO_3 = PbO_2 + 2Pb(NO_3)_2 + 2H_2O$$

Warm concentrated hydrochloric acid is oxidised to chlorine by the 'lead(IV) part':

$$\underset{(\text{'}PbO_2\cdot 2PbO\text{'})}{Pb_3O_4} + 8HCl = 3PbCl_2 + 4H_2O + Cl_2$$

Structurally, the oxide contains a regular array of Pb^{2+}, Pb^{4+} and O^{2-} ions in the ratio 2:1:4.

Trilead tetraoxide under the commercial name of 'red lead' is used in anti-rust paints.

14.12.4 THE HALIDES OF GERMANIUM, TIN AND LEAD

Germanium(II) chloride

Germanium(II) chloride is prepared by passing germanium(IV) chloride vapour over heated germanium:

$$Ge + GeCl_4 = 2GeCl_2$$

Like all germanium(II) compounds it is a powerful reducing agent.

Germanium(IV) chloride

$GeCl_4$

Germanium(IV) chloride is prepared by passing chlorine over heated germanium, or a mixture of germanium(IV) oxide and carbon:

$$Ge + 2Cl_2 = GeCl_4$$

$$GeO_2 + C + 2Cl_2 = GeCl_4 + CO_2$$

In many respects germanium(IV) chloride resembles tin(IV) chloride. It is reversibly hydrolysed by water to a white precipitate of hydrated germanium(IV) oxide, and forms the hexachlorogermanate(IV) ion, $[GeCl_6]^{2-}$, with ionic chlorides.

Germanium also forms a fluoro-complex, $[GeF_6]^{2-}$, but not the corresponding bromo- and iodo-complexes, presumably because it is not possible to accommodate these large halogen atoms around a germanium atom.

Tin(II) chloride

$SnCl_2$

The hydrate, $SnCl_2{\cdot}2H_2O$, is prepared by dissolving tin in hydrochloric acid, followed by evaporation of the solution:

$$Sn + 2HCl + 2H_2O = SnCl_2{\cdot}2H_2O + H_2$$

The anhydrous compound is prepared by passing hydrogen chloride over heated tin. It cannot be prepared by heating the hydrate because hydrolysis occurs and a basic tin(II) chloride is formed:

$$SnCl_2{\cdot}2H_2O = \underset{\text{tin(II) chloride hydroxide}}{SnCl(OH)} + H_2O + HCl$$

Tin(II) chloride is molecular (bond angle 95°) with a lone pair of electrons on the tin atom (§4.8). Aqueous solutions of tin(II) chloride appear milky due to the formation, by hydrolysis, of insoluble basic chlorides,

$$\text{e.g.} \quad SnCl_2 + H_2O \rightleftharpoons SnCl(OH) + HCl$$

The addition of hydrochloric acid clears the solution by displacing the equilibrium to the left and then forming the trichlorostannate(II) ion and the tetrachlorostannate(II) ion:

$$SnCl_2 + 2Cl^- \rightleftharpoons [SnCl_3]^- + Cl^- \rightleftharpoons [SnCl_4]^{2-}$$

There is no evidence for the existence of simple hydrated tin(II) ions in solution because of the strong tendency for hydrolysis to occur and produce complex ions containing tin(II).

Tin(II) chloride reduces many substances, e.g. mercury(II) and iron(III) compounds:

$$2HgCl_2 + SnCl_2 = SnCl_4 + Hg_2Cl_2$$

$$Hg_2Cl_2 + SnCl_2 = SnCl_4 + 2Hg$$

$$2Fe^{3+} + 2Cl^- + SnCl_2 = SnCl_4 + 2Fe^{2+}$$

Tin(II) chloride also reduces nitrobenzene to phenylamine.

$SnCl_4$

Tin(IV) halides

Tin(IV) halides are prepared by methods similar to those used for germanium(IV) chloride. Tin(IV) fluoride appears to be ionic but the other tin(IV) halides are covalent, with tetrahedral molecules.

Tin(IV) chloride is a colourless liquid which is reversibly hydrolysed by water to hydrated tin(IV) oxide:

$$SnCl_4 + 2H_2O \rightleftharpoons SnO_2(aq) + 4HCl$$

The white fumes emitted from tin(IV) chloride in moist air are a mixture of hydrogen chloride and a 'smoke' of hydrated tin(IV) oxide produced in this hydrolysis reaction. The reaction with water is thought to include an ionic stage (cf. aluminium), because the addition of small amounts of water to tin(IV) chloride results in the formation of an ionic compound with the formula $SnCl_4{\cdot}5H_2O$. Undoubtedly this compound contains hydrated tin(IV) ions, although their exact nature is not known. Hydrated ions of the type $[Sn(H_2O)_x]^{4+}$ are also formed in solution by the reaction:

$$SnCl_4 + xH_2O = [Sn(H_2O)_x]^{4+} + 4Cl^-$$

but these undergo extensive hydrolysis to produce hydrated tin(IV) oxide (§14.12.3). The ease of hydrolysis of tin(IV) halides decreases from the fluoride to the iodide.

The addition of hydrochloric acid to tin(IV) chloride produces the octahedral hexachlorostannate(IV) ion:

$$2Cl^- + SnCl_4 = [SnCl_6]^{2-}$$

The ammonium salt, $(NH_4)_2[SnCl_6]$, is used in dyeing. The other tin(IV) halides form similar halo-complexes, i.e. $[SnF_6]^{2-}$, $[SnBr_6]^{2-}$ and $[SnI_6]^{2-}$, when treated with the appropriate hydrohalic acid.

The tin(IV) halides do not possess oxidising or reducing properties.

All lead(II) halides

Lead(II) halides

The lead(II) halides are sparingly soluble in cold water and are prepared by metathesis,

i.e. $Pb^{2+}(aq) + 2X^-(aq) = PbX_2(s)$ (X = F, Cl, Br or I)

They are colourless compounds, with the exception of lead(II) iodide, which is yellow. Lead(II) halides are appreciably soluble in hot water and in concentrated solutions of the hydrohalic acids. For example, lead(II) chloride dissolves in hydrochloric acid to form the tetrachloroplumbate(II) complex:

$$PbCl_2 + 2Cl^- \rightleftharpoons [PbCl_4]^{2-}$$

The lead(II) halides are ionic, but their structures become increasingly covalent from the fluoride to the iodide (§4.6.1). They do not function as reductants or oxidants.

PbF_4 and $PbCl_4$

Lead(IV) halides

Bromine and iodine are unable to oxidise lead(II) to lead(IV). Therefore the only lead(IV) halides to exist are the essentially ionic fluoride and the covalent chloride. Lead(IV) chloride is prepared by passing chlorine through a solution of lead(II) chloride in concentrated hydrochloric acid at 0 °C (273 K). Oxidation to lead(IV) chloride occurs and the hexachloroplumbate(IV) ion is formed:

$$PbCl_2 + Cl_2 = PbCl_4$$

$$PbCl_4 + 2Cl^- = [PbCl_6]^{2-}$$

The addition of ammonium chloride precipitates the ammonium salt of this ion, $(NH_4)_2[PbCl_6]$, which is filtered off and treated with concentrated sulphuric acid to liberate lead(IV) chloride as a yellow oil:

$$(NH_4)_2[PbCl_6] + H_2SO_4 = PbCl_4 + 2HCl + (NH_4)_2SO_4$$

Lead(IV) chloride is reversibly hydrolysed by water (cf. $SnCl_4$) and decomposes into lead(II) chloride and chlorine on warming.

$Pb(NO_3)_2$

14.12.5 OTHER COMMON COMPOUNDS OF LEAD

The only common water soluble lead(II) salts are the nitrate, $Pb(NO_3)_2$, and the ethanoate, $(CH_3COO)_2Pb{\cdot}3H_2O$. In solution the hydrated lead(II) ion undergoes hydrolysis to produce polymeric cations such as $[Pb_4(OH)_4]^{4+}$.

14.12.6 THE HYDRIDES OF GERMANIUM, TIN AND LEAD

Germanium forms a series of hydrides, known as *germanes*, with the general

All hydrides of Ge, Sn and Pb

formula Ge_nH_{2n+2}. Germanes containing up to nine germanium atoms are known. They are surprisingly stable to hydrolysis, even with 30% alkali, but react rapidly with oxygen,

e.g. $GeH_4 + 2O_2 = GeO_2 + 2H_2O$

The only hydride of tin, stannane, SnH_4, is prepared by reducing tin(IV) chloride with lithium tetrahydridoaluminate(III):

$$SnCl_4 + Li[AlH_4] = SnH_4 + LiCl + AlCl_3$$

It is thermally unstable but resistant to hydrolysis.

The very unstable plumbane, PbH_4, is formed in small quantities when a magnesium–lead alloy is dissolved in an acid.

PROBLEMS – ORGANIC CHEMISTRY

1 A substance, X, has the following composition by mass:

carbon	52.2%
hydrogen	13.0%
oxygen	34.8%

(a) What is the empirical formula of X?

(b) When completely vaporised, 0.023 g of X is found to occupy a volume of 11.2 cm^3 (corrected to s.t.p.). What is the relative molecular mass of X?

(c) What is the molecular formula of X?

(d) Write down two different structural formulae which are consistent with your answer to (c).

(e) X reacts with iodine in aqueous, alkaline solution to produce a yellow precipitate of triiodomethane (*iodoform*). Write the structural formula of substance X and show by means of equations how the triiodomethane is formed. (JMB)

2 (a) (i) Draw a well-labelled diagram of the apparatus you would use to determine the percentages of carbon and hydrogen in a *solid* organic compound.

(ii) Indicate the weighings you would make and how you would obtain the percentages of carbon and hydrogen.

(b) The compound, X, contains *only* carbon, hydrogen and fluorine. When 0.32 g of X is burnt in an excess of oxygen, 0.44 g of carbon dioxide and 0.09 g of water are formed.

(i) What is the empirical formula of X? Show your working.

(ii) The relative molecular mass of X is 64. What is its molecular formula?

(iii) Write the structural formula of all the possible compounds which X could be and give the systematic name for each formula.

(iv) Indicate those structures which are structural isomers, those which are geometrical isomers and those (if any) which are optical isomers.

(H = 1, C = 12, O = 16, F = 19) (SUJB)

3 An aromatic compound A with a relative molecular mass of 250 contained 33.6% carbon, 2.40% hydrogen and 64.0% bromine by mass. On refluxing A with aqueous sodium hydroxide, half of the bromine was

removed and the resulting organic compound, B, was converted into C by the action of a mild oxidising agent. C gave a crystalline derivative with 2,4-dinitrophenylhydrazine and was readily oxidised, even by exposure to the air, to a monobasic acid, D. The latter contained the same number of carbon atoms as C, and 2.01 g of it neutralised 100 cm^3 of an aqueous solution of sodium hydroxide containing 0.100 moles per dm^3.

On further investigation it was found that A gave two isomers E and F when a radical *X* was substituted into the aromatic ring in place of one hydrogen atom.

Deduce the structural formulae of the compounds A, B, C, D, E and F, explaining fully your reasoning. (O)

4 A colourless liquid, of relative molecular mass 106.5, which fumes in moist air, reacts *readily* with water to produce a mixture of hydrochloric acid and a monobasic carboxylic acid. The compound contains 45.1% carbon and 6.6% hydrogen. It also contains oxygen and chlorine. 1.00 g of the compound under suitable conditions gives 1.35 g of silver chloride. Give its possible structural formulae, and outline the method you would use to prepare one of the compounds so shown from the corresponding carboxylic acid.

Describe **briefly**, but giving essential experimental details, how you would prepare (chloromethyl)benzene (benzyl chloride) from methylbenzene (toluene).

Compare the reactivity of the halogen in chlorobenzene, (chloromethyl) benzene and benzenecarbonyl chloride (benzoyl chloride) giving reasons for any differences in behaviour.
(A_r(H) = 1; A_r(C) = 12; A_r(O) = 16; A_r(Cl) = 35.5; A_r(Ag) = 108.)
(WJEC)

5 A liquid X contains by mass 31.5% of carbon, 5.3% of hydrogen and 63.2% of oxygen. Its solution in water liberates carbon dioxide from sodium carbonate and from the resulting solution a substance of formula $C_2H_3O_3Na$ can be obtained. With phosphorus pentachloride, X gives hydrogen chloride and a compound of molecular formula $C_2H_2OCl_2$.

State and explain the information which may be obtained from each of these facts. Give the molecular formula of X and devise a synthesis of it from ethene (ethylene). (L)

6 (a) Identify **X** and **Y** from the following information:

(i) **X** contains a halogen; when 1.85 g of **X** was vaporised 448 cm^3 of vapour were obtained at s.t.p.,

(ii) when **X** was treated with a hot alcoholic solution of potassium hydroxide, **Y** was formed; **Y** did not contain a halogen,

(iii) on vigorous oxidation with alkaline potassium manganate(VII) (potassium permanganate) **Y** yielded an acid of relative molecular mass 74.

(b) Distinguishing between primary, secondary and tertiary alcohols, give the structural formulae of all the alcohols which have the molecular formula $C_4H_{10}O$.

Describe a laboratory test to distinguish between a primary, a secondary and a tertiary alcohol. (AEB)

7 An organic compound **A** has relative molecular mass 150 and contains 72.00% carbon, 6.67% hydrogen and 21.33% oxygen.

A reacts with hot, aqueous sodium hydroxide solution to give a solution which contains **B**, which is ionic, and **C**, C_2H_6O. On addition of hydrochloric acid to the mixture in solution, crystals of **D**, $C_7H_6O_2$, are precipitated. **D** displaces carbon dioxide from an aqueous solution of sodium hydrogen carbonate.

Reaction of **A** with lithium tetrahydrido aluminate gives two products, **C**, C_2H_6O, and **E**, C_7H_8O. **E** can be converted into **D** by reaction with acidified potassium manganate(VII) solution.

When **A** is treated with a mixture of concentrated nitric and sulphuric acids, and the mixture is then cooled and poured into a mixture of ice and water, a solid **F**, $C_9H_9O_4N$, is obtained.

Identify all the compounds **A** to **F** inclusive, giving your reasons. Give the structural formula of each compound and write equations for **four** of the changes involved. (O)

8 The organic compound X, which contains carbon, hydrogen and oxygen only, was found to have a relative molecular mass of about 85. When 0.43 g of X is burnt in excess oxygen, 1.10 g of carbon dioxide and 0.45 g of water are formed.

($H = 1$, $C = 12$, $O = 16$)

(a) What is the empirical formula of X?

(b) What is the molecular formula of X?

(c) Write an equation for the complete combustion of X.

(d) X undergoes a condensation reaction with hydroxylamine and also with 2,4-dinitrophenylhydrazine. Write the structural formulae of **five** possible non-cyclic compounds which X could be and give the systematic names for each of the five formulae.

(e) What is meant by a condensation reaction?

(f) Write an equation for the reaction of **one** of the five possible compounds for X with hydroxylamine.

(g) Look closely at the five structures you have written for X.

(i) Which will give a yellow precipitate on warming with a mixture of iodine and alkali?

(ii) Which will reduce diamminosilver(I) ions? (SUJB)

EXAMINATION QUESTIONS – ORGANIC CHEMISTRY

1 How would you carry out the following transformations? State essential conditions and reagents.

$$CH_2{=}CH(CH_2)_4CO_2C_2H_5 \begin{cases} \xrightarrow{(a)} CH_2{=}CH(CH_2)_4CH_2OH \\ \xrightarrow{(b)} CH_3(CH_2)_5CH_2OH \\ \xrightarrow{(c)} CH_2{=}CH(CH_2)_4CONHMe \\ \xrightarrow{(d)} HO_2C(CH_2)_4CO_2H \\ \xrightarrow{(e)} CH_3CHBr(CH_2)_4CO_2C_2H_5 \\ \xrightarrow{(f)} OCH(CH_2)_4CO_2C_2H_5 \end{cases}$$

Indicate the mechanisms of reactions (c) and (e). (OC)

2 Write equations for the reactions, name the organic products formed, and state what you would observe when:

(a) benzoyl chloride is added slowly to cooled aqueous ammonia, and the product isolated and distilled with phosphorus(V) oxide (phosphorus pentoxide),

(b) benzene is added slowly to a mixture of concentrated nitric and sulphuric acids at a temperature below 45 °C and the resulting mixture is poured into water,

(c) propan-2-ol is mixed with a little potassium iodide solution, and a solution of sodium chlorate(I) (sodium hypochlorite) is added,

(d) ethanamide (acetamide) is treated with bromine, followed by dilute potassium hydroxide solution and the resulting mixture is run into hot concentrated potassium hydroxide solution,

(e) benzaldehyde is shaken with concentrated potassium hydroxide solution and, after standing overnight, water is added. The mixture is then extracted with ether and the aqueous solution is acidified. (AEB)

3 Illustrate, with one example in each case, the uses of the following substances in organic chemistry, mentioning the conditions that are necessary, and giving the equations for the reactions:

(a) hydrogen cyanide,
(b) sodium nitrite,
(c) bromine,
(d) sodium tetrahydridoborate(III), $NaBH_4$,
(e) aluminium chloride,
(f) nickel. (O)

4 (a) State **two** chemical properties which are associated with **each of** the groups represented by $-CH_2OH$ and $-COOH$.

(b) Compare the behaviour of aminoethane (ethylamine), ethanamide (acetamide) and aminoethanoic acid (glycine) in their reactions with aqueous solutions of:

(i) sodium hydroxide,
(ii) hydrochloric acid.

(c) Outline the preparation of bromobenzene from benzene. Compare the reactions of bromobenzene and bromoethane (ethyl bromide) with:

(i) ammonia in alcoholic solution,
(ii) sodium hydroxide in aqueous solution,
(iii) sodium in ethereal solution. (AEB)

5 (a) For each of parts (i) to (viii) below **one or more** of the alternatives **A**, **B**, **C**, **D** and **E** listed below are correct. Decide which of the alternatives is (are) correct and write the appropriate letters in answer to the question.

A hexane
B hex-2-ene
C hex-1-yne
D hex-2-yne
E cyclohexane

(i) Which would decolorise bromine in the absence of sunlight?

(ii) Which would give a yellow precipitate with a solution containing diamminosilver(I) ions?

(iii) Which would react with chlorine only when heated or exposed to sunlight?

(iv) Which have three or more carbon atoms arranged linearly?

(v) Which are unsaturated compounds?

(vi) Which of the compounds have an empirical formula of CH_2?

(vii) Which of the compounds have geometrical isomers?

(viii) Which would absorb 2 moles of hydrogen per mole in the presence of a nickel catalyst?

(b) Outline, giving equations, the reactions by which hex-1-ene may be converted to:

(i) hex-1-yne,

(ii) hexane,

(iii) hexan-2-ol. (SUJB)

6 Describe, giving essential reagents and conditions, how phenol can be prepared in the laboratory, starting from nitrobenzene.

(a) How would you confirm that the product is a phenol?

(b) What are the products of the reaction of phenol with (i) ethanoic (*acetic*) anhydride and (ii) benzenecarbonyl (*benzoyl*) chloride? Give equations for these reactions. (JMB)

7 Give **one** example of **each** of the following organic processes giving necessary reaction conditions and equations: nitration, cracking, reduction, oxidation, sulphonation, decarboxylation, aromatic nuclear halogenation. (WJEC)

8 Deduce the structures of the compounds **A** to **G** below, giving the reason for selecting each structure.

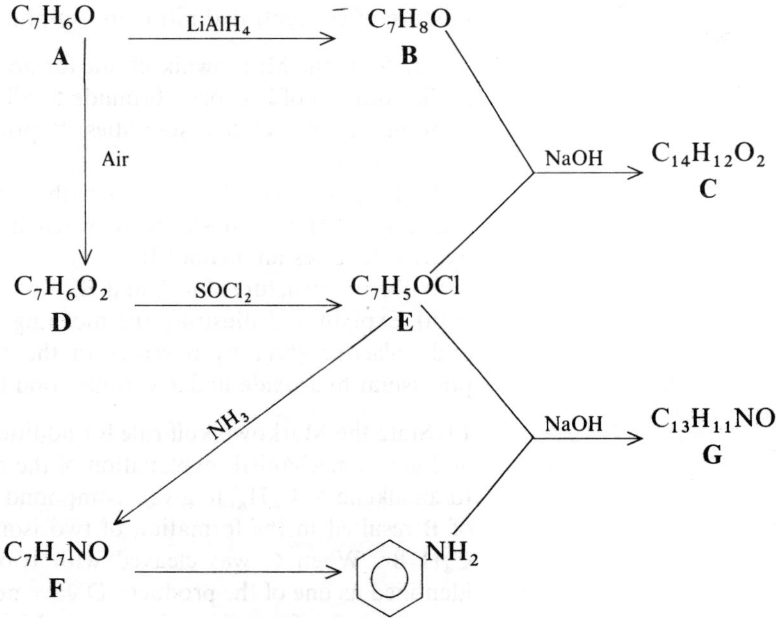

How would you show that the compound **G** was identical with one of the same formula from the laboratory stock-room? (L)

9 Starting in each case from either benzene or methylbenzene (toluene), outline **one** preparation of each of the following compounds:
$PhCH(OH)CH_3$; $PhCH(NH_2)CH_3$; $PhCH_2CH_2NH_2$; $PhCH_2CO_2H$; $PhCO_2H$. (OC)

10 (a) **Outline** with essential experimental conditions how the following conversions can be carried out:
(i) $CH_3CH_2OH \longrightarrow CH_3CH_2CH_2OH$
(ii) $CH_3CH_2OH \longrightarrow CH_3CH(OH)CH_3$
(iii) $CH_3CH_2OH \longrightarrow CH_3CH_2OCH_2CH_3$
(iv) $CH_3CHO \longrightarrow CH_3CH(OH)COOH$
(v) $CH_3CH_2COOH \longrightarrow CH_3COOH$
(b) What chemical tests would you perform to show that conversions (ii) and (iv) had taken place?
(c) Which of the organic substances above exhibit *optical isomerism*? (AEB)

11 Describe how you would carry out the following conversions, giving the essential conditions for the reactions but no details of the apparatus or of the separation and purification of the final products:
(a) $C_2H_5OH \longrightarrow CH_3CH_2CH_2NH_2$;
(b) $C_2H_5OH \longrightarrow CH_3CONH_2$;
(c) $C_6H_5COOH \longrightarrow C_6H_5NH_2$. (O)

12 Explain what is meant by *electrophilic* and *nucleophilic* reagents, and give **two** examples of each. Discuss the mechanisms of the following reactions: (a) bromine and an alkene such as ethene (ethylene), (b) bromine and an alkane such as hexane, C_6H_{14}, (c) bromine and benzene, (d) a mixture of concentrated nitric and sulphuric acids with benzene. (L)

13 (a) State the Markownikoff rule for predicting the direction of electrophilic addition of hydrogen bromide to alkenes. Explain in detail the rule in terms of the relative stabilities of primary, secondary, and tertiary carbonium ions.
(b) Propene (propylene) reacts with hydrogen bromide to give a substance A, C_3H_7Br. Substance A, when heated with aqueous potassium hydroxide, gives an alcohol B.
(i) Derive structures for A and B.
(ii) Explain and illustrate the meaning of the terms *base*, *nucleophile*, and *inductive effect* by referring to the reactions of substance A with potassium hydroxide under various conditions. (JMB)

14 State the Markownikoff rule for addition reactions of HBr to alkenes, and give a mechanistic explanation of the rule. Bromine water was added to an alkene A, C_4H_8, to give a compound B. Acid-catalysed dehydration of B resulted in the formation of two isomeric bromoalkenes C and D, C_4H_7Br. When C was cleaved with trioxygen (*ozone*), methanal was identified as one of the products. D gave no methanal with trioxygen, but propanone was formed in this reaction. Identify compounds A, B, C and D, and explain the observations described above. (JMB)

15 (a) State two properties of carbonium ions.

(b) What is the formula of the carbonium ion formed when propene reacts with hydrogen bromide to form 2-bromopropane?

(c) State whether this carbonium ion is a primary, secondary or tertiary species.

(d) How does carbonium ion theory explain the fact that the principal product of the reaction between propene and hydrogen bromide is 2-bromopropane rather than 1-bromopropane?

(e) What reagents and conditions would be used to prepare propan-2-ol from 2-bromopropane?

(f) In what way would you alter the conditions in (e) in order to prepare propene rather than propan-2-ol from 2-bromopropane? (JMB)

16 (a) For each of parts (i) to (x) below **only one** of the alternatives **A**, **B**, **C**, **D** or **E** is correct. Answer each part by giving the appropriate letter.

A ethandioic acid (oxalic acid)
B methanoic acid (formic acid)
C ethanoic acid (acetic acid)
D propanoic acid (propionic acid)
E benzoic acid

(i) Which has the lowest boiling point?
(ii) Which is the strongest acid?
(iii) Which forms acid salts?
(iv) Which has a sodium salt, which, on heating with soda lime, gives ethane?
(v) Which decolourises potassium manganate(VII) (potassium permanganate) solution only on warming?
(vi) Which has a calcium salt, which, on heating, gives propanone (acetone)?
(vii) Which gives a buff precipitate with neutral iron(III) chloride solution?
(viii) Which gives **two** oxides of carbon on heating with concentrated sulphuric acid?
(ix) Which is the strongest reducing agent?
(x) Which is the least soluble in cold water?

(b) Outline the production of ethanoic (acetic) acid from crude petroleum via ethene (ethylene).

(c) When the dibasic acid, **F**, is heated it gives **G** ($C_5H_{10}O_2$) which is optically active. Draw **one** possible structure for each of **F** and **G**.

(SUJB)

17 The following represent important examples of different types of organic reactions:

A $C_6H_5\text{-}COONa + NaOH \rightarrow C_6H_6 + Na_2CO_3$
B $CH_3\text{-}CHO + NH_3 \rightarrow CH_3\text{-}CH(OH)\text{-}NH_2$
C $CH_3\text{-}CO\text{-}CH_3 + NH_2OH \rightarrow (CH_3)_2\text{-}C{=}NOH + H_2O$
D $CH_3\text{-}CH_2\text{-}I + H_2O \rightarrow CH_3\text{-}CH_2\text{-}OH + HI$
E styrene (phenylethene) $\xrightarrow[\text{(lauroyl peroxide)}]{\text{dodecanoyl peroxide}}$ polystyrene (polyphenylethene)
F $C_6H_6 + CH_3\text{-}Cl \rightarrow C_6H_5\text{-}CH_3 + HCl$

(a) Explaining your reasons, place each of these reactions in the appropriate category from the following list:
(i) polymerisation, (ii) decarboxylation,
(iii) substitution (displacement), (iv) condensation,
(v) addition, (vi) oxidation, (vii) hydrolysis.
(b) Give **one** example, other than any of those above, of each of the following:
(i) a condensation reaction,
(ii) a substitution reaction,
(iii) a polymerisation reaction.
(c) Give the conditions of reaction **F**.
(d) What is the function of dodecanoyl peroxide in reaction **E**?
(e) State, giving a reason, whether a similar reaction to **B** takes place with propanone (acetone). (AEB)

18 (a) Draw structural formulae which show the spatial arrangement of the atoms in the molecules of (i) ethyne (acetylene), (ii) propanone (acetone), (iii) dichlorobenzenes, (iv) butenedioic acids, $C_2H_2(COOH)_2$, and (v) 2-hydroxypropanoic (lactic) acids. In the case of (iv) and (v) above, explain why there is more than one structure.
(b) Explain the following:
(i) The carbonyl group in a carboxylic acid is less reactive than that in an aldehyde or ketone,
(ii) a dilute aqueous solution of aminoethanoic acid (glycine) is a much weaker electrolyte than an aqueous solution of ethanoic (acetic) acid of equivalent concentration. (O)

19 By means of equations and essential experimental conditions, outline how you would prepare a sample of the substance $CH_3CH(OH)COOH$ starting with ethanal, CH_3CHO.

Describe experiments which you could undertake to show that the substance you have prepared:
(a) contains the elements carbon and hydrogen, (b) behaves as a secondary alcohol, (c) is acidic in nature.

What would be the effect of a portion of your sample on the plane of polarisation of light? Explain your answer. (JMB)

20 Explain the following terms used in organic chemistry, illustrating **each** by **one** specific example and giving brief experimental details in each case:
(a) alkylation, (b) acylation, (c) decarboxylation, (d) diazotisation, (e) reduction, (f) ozonolysis.

C_6H_{12} is a straight-chain alkene. Describe how ozonolysis can assist in the determination of its molecular structure. (WJEC)

21 (a) Describe how you would prepare a pure sample of nitrobenzene from benzene in the laboratory. Your account should include:
(i) the reagents used,
(ii) the conditions of the reaction,
(iii) the equation of the reaction,
(iv) the method of purifying your sample of nitrobenzene.
(b) How and under what conditions does benzene react with:

(i) chlorine,
(ii) sulphuric acid,
(iii) chloromethane? (SUJB)

EXAMINATION QUESTIONS – GROUP 4B INORGANIC CHEMISTRY

1 Survey the chemistry of the Group IV elements (C-Pb) by giving:
(a) a summary of the physical and chemical properties of the elements,
(b) brief descriptions of preparative routes to the chlorides and oxides,
(c) a discussion of group trends in valencies and bond types of the chlorides and oxides,
(d) a discussion of the special properties of carbon and the ways in which its chemistry differs from the other members of the group. (JMB)

2 (a) **Outline** the preparation of:
(i) the dioxides of carbon and silicon,
(ii) the tetrachlorides of carbon and silicon.
(b) How do the electronic structures of the elements carbon and silicon (atomic numbers C = 6, Si = 14) and the structures of the dioxides and tetrachlorides help to explain that, at ordinary temperature and pressure,
(i) carbon dioxide is a gas, while silicon(IV) oxide (silicon dioxide) is a solid of high melting point,
(ii) carbon tetrachloride does not react with water, while silicon tetrachloride reacts vigorously with cold water giving steamy fumes and a gelatinous precipitate?
(c) For the systems

$$Sn^{4+}(aq) + 2e^- \rightleftharpoons Sn^{2+}(aq); \quad E^\ominus = +0.15 \text{ volt}$$

and

$$PbO_2(s) + 4H^+(aq) + 2e^- \rightleftharpoons Pb^{2+}(aq) + 2H_2O(l); \quad E^\ominus = +1.46 \text{ volts}$$

state, with an explanation, which of the following reactions is more likely to occur:
(i) $PbO_2(s) + 4H^+(aq) + Sn^{2+}(aq) \rightarrow Pb^{2+}(aq) + Sn^{4+}(aq) + 2H_2O(l)$
(ii) $Sn^{4+}(aq) + Pb^{2+}(aq) + 2H_2O(l) \rightarrow Sn^{2+}(aq) + PbO_2(s) + 4H^+(aq)$
(AEB)

3 Give an account, with examples where appropriate, of the chemistry of the elements of Group IV (C—Pb) of the periodic table by considering:
(a) the atypical nature of carbon, comparing its properties with those of silicon,
(b) changes in the stability of oxidation states of the elements,
(c) changes in ionic and covalent character of their compounds.
Show, by means of equations and stating essential conditions only, how you could prepare tin(IV) iodide and lead(II) iodide, starting from the respective Group IV elements. (JMB)

4 Four of the elements in Group IV of the periodic table are:
carbon, atomic number 6
silicon, atomic number 14

tin, atomic number 50
lead, atomic number 82

(a) Write the electronic structure of each atom.

(b) Explain in terms of their electronic configurations why tin shows typical metallic behaviour whereas carbon does not.

(c) Tin reacts with dry chlorine to form a colourless liquid. 0.001 25 mol of this liquid, when added to water, produced a solution containing hydrochloric acid which was neutralised by sodium hydroxide of concentration 0.100 mol l^{-1}.

(i) Give the name and formula of the colourless liquid.

(ii) Give an equation for the reaction with cold water.

(iii) Calculate the volume of sodium hydroxide solution needed to neutralise the hydrochloric acid formed.

(d) The element germanium (atomic number 32) comes between silicon and tin in the group.

(i) Suggest one method for preparing germanium(IV) chloride.

(ii) Predict the colour of germanium(IV) oxide. How would you expect this oxide to react with excess alkali?

(e) What would be the product obtained if an aqueous solution of tin(II) chloride were added to (i) excess dilute alkali, (iii) mercury(II) chloride solution? (AEB)

5 The elements carbon to lead in Group IV of the periodic table illustrate differences and similarities between elements in a particular group.

By reference to carbon and lead **only**, illustrate this statement by considering each of the following and explaining any differences.

(a) Methods of formation and structures of the dioxides.

(b) The type of bonds that the elements form.

(c) The relative stabilities of the lower and higher oxides.

(d) The relative stabilities of the tetrachlorides.

(e) The formation of a large number of hydrides of carbon.

In each case make a prediction for germanium, remembering that germanium is between carbon and lead in the periodic table. (L)

6 Write an account of the Group IV elements (C, Si, Sn, Pb) by giving:

(a) a description of the physical and chemical properties of the elements themselves;

(b) preparative routes from the elements themselves to their (i) oxides, (ii) chlorides,

(c) a commentary on the group trends in bond type in their (i) oxides, (ii) chlorides,

(d) a discussion on the way in which the chemistry of carbon differs from that of the other members of the group. (SUJB)

7 Illustrate the transition from non-metallic to metallic character of carbon, silicon and tin by describing the structure of, and giving one chemical property of, any **one** type of compound of these elements.

One of the products of the reaction between magnesium silicide (Mg_2Si) and sulphuric acid is a gas, X.

A 0.620 g sample of X occupied 224 cm^3. When hydrolysed, this sample yielded 1 568 cm^3 of hydrogen and a residue (SiO_2) which, after it had been strongly heated, weighed 1.200 g. (All gas volumes corrected to s.t.p.)

What is the molecular formula of X?

Write an equation for the hydrolysis of X. (OC)

8 Tin(IV) iodide (stannic iodide) is prepared by direct combination of the elements.

Add 2 g of granulated tin to a solution of 6.35 g of iodine in 25 cm^3 of tetrachloromethane (carbon tetrachloride) in a 100 cm^3 flask fitted with a reflux condenser. Reflux gently until reaction is complete. Filter through pre-heated funnel, wash residue with 10 cm^3 of hot tetrachloromethane and add washings to the filtrate. Cool in ice until orange crystals of tin(IV) iodide separate. Filter off the crystals and dry. The melting point of tin(IV) iodide is 144 °C.

(Sn = 119, I = 127)

(a) Write an equation for the reaction.

(b) Suggest reasons why tetrachloromethane is used.

(c) Which reactant is in excess?

(d) Calculate the maximum theoretical yield of product.

(e) Sketch the apparatus used initially.

(f) How would you know when reaction is complete?

(g) Sketch the filtration apparatus, explain why washing is necessary and suggest reasons for the procedure adopted.

(h) How would you obtain a further crop of crystals?

(i) Sketch the apparatus for determining the melting point of the product and say very briefly how you would measure it.

(j) What is the principle underlying the technique of using mixed melting points in identifying compounds?

(k) Comment on the structure and colour of tin(IV) iodide. (SUJB)

9 Locate the element germanium (Ge) in Group IV of the periodic table.

(a) Write down the electronic configuration of a germanium *atom*.

(b) Suggest one method (reagents and conditions only) for preparing germanium(IV) chloride.

(c) By comparison with the chlorides of the other elements in the same group predict (i) the colour and (ii) the physical state at s.t.p. of germanium(IV) chloride.

(d) What, if anything, would you expect to happen if germanium(IV) chloride were added to water?

(e) Give the formula and state the shape of a *complex ion* which you would expect to be formed between germanium(IV) and fluoride ions.

(f) Suggest one method by which germanium could be prepared, starting from germanium(IV) oxide, GeO_2. (JMB)

10 Write an account of the common oxides of the Group IV elements carbon, silicon, tin and lead, referring especially to their structure, physical properties and acid-base behaviour.

Mention **two** characteristics which you would expect germanium dioxide, GeO_2, to possess remembering that germanium comes between silicon and tin in Group IV. (L)

11 (a) Describe with essential experimental details a laboratory preparation of carbon monoxide.

(b) Draw a structural formula to indicate the type of bonding believed to be present in this gas.

(c) How, and under what conditions, does carbon monoxide react with (i) chlorine, (ii) iodine(V) oxide (iodine pentoxide), (iii) sodium hydroxide?

(d) 50.0 cm^3 of a mixture of carbon monoxide, carbon dioxide and hydrogen were exploded with 25.0 cm^3 of oxygen. After explosion, the volume measured at the original room temperature and pressure was 37.0 cm^3. After treatment with potassium hydroxide solution the volume was reduced to 5.0 cm^3. Calculate the percentage composition by volume of the original mixture. (AEB)

12 State two industrial uses of tin.

Tin is said to be enantiotropic; explain what this means.

Write down the electronic configurations of (a) Sn^{2+}, (b) Sn^{4+}.

How would you convert a solution containing tin(II) into a solution containing tin(IV)?

Pure tin was reacted with chlorine to give a colourless liquid that fumed in moist air. A solution of 0.1629 g of the liquid in water was equivalent to 24.00 cm^3 of exactly 0.1 M sodium hydroxide solution using methyl orange as the indicator. Calculate the percentage purity of the liquid tin compound. (JMB)

13 Give a comparative account of the chemistry of the oxides and chlorides of carbon, silicon, germanium, tin and lead, paying particular attention to:

(a) the formation, composition and stability of oxides,

(b) the formation, hydrolytic behaviour and stability of chlorides.

(JMB)

14 (a) Give two characteristic properties of (i) metallic oxides, (ii) non-metallic oxides, (iii) metallic chlorides, (iv) non-metallic chlorides.

(b) Illustrate the general transition from non-metallic to metallic properties exhibited by the Group IV elements C, Si, Sn and Pb by considering the structures and properties of their oxides and chlorides. (O)

Group 5B

15.1 Introduction

Group 5B comprises: nitrogen (N), phosphorus (P), arsenic (As), antimony (Sb) and bismuth (Bi). There are many fundamental differences between nitrogen and the other elements of the group, mostly arising from the inability of nitrogen to utilise d orbitals in bonding (§3.2.4). Some important properties of the elements are shown in Table 15.1.

15.1.1 THE STRUCTURE OF NITROGEN AND PHOSPHORUS

Nitrogen

At room temperature nitrogen is an unreactive colourless gas consisting of diatomic molecules, N≡N (§4.3.3). Its low reactivity stems from two

Table 15.1 Some properties of the elements of group 5B

As and P

P

	Nitrogen	Phosphorus	Arsenic	Antimony	Bismuth
character	non-metal	non-metal	semi-metal	semi-metal	metal
allotropes	none	white, red, black	non-metallic and metallic	non-metallic and metallic	none
atomic number	7	15	33	51	83
relative atomic mass	14.006 7	30.973 8	74.921 6	121.75	208.980 6
outer electronic configuration	$2s^2\ 2p^3$	$3s^2\ 3p^3$	$4s^2\ 4p^3$	$5s^2\ 5p^3$	$6s^2\ 6p^3$
valencies	3	3, 5	3, 5	3, 5	3, 5†
melting temperature/°C	−210	44.2 (white) 590 (red)	—	630	271
boiling temperature/°C	−196	280 (white)	613‡ (sublimes)	1380	1560

† Denotes a relatively unstable state. ‡ Data for non-metallic form.

factors: (i) the lack of polarity in the N_2 molecule; (ii) the high bond dissociation enthalpy,

i.e. $N_2(g) = 2N(g) \qquad \Delta H^\ominus = +944\ \text{kJ mol}^{-1}$

It is therefore difficult for charged species to attack the molecule, and difficult to break the N≡N bond. Other molecules containing triple bonds, e.g. alkynes and carbon monoxide, are far more reactive than nitrogen because their bonds are polar and weaker than the N≡N bond.

Phosphorus

Unlike nitrogen, phosphorus is unable to form multiple bonds between its atoms; cf. silicon and carbon. Hence phosphorus cannot form P_2 molecules analogous to N_2. Like carbon, phosphorus exhibits the type of allotropy known as monotropy. The allotropes are usually named according to their colour; *white* phosphorus (sometimes *yellow* phosphorus), *red* phosphorus and *black* phosphorus. The two principal allotropes are the white and red forms (Table 15.2).

Table 15.2 Some properties of white and red phosphorus

	White	Red
appearance and colour	yellowy white waxy solid	red-violet powder
density/g cm^{-3}	1.82	2.34
melting temperature	44.2 °C (317.4 K)	590°C (863 K) (under pressure)
solubility in CS_2	soluble	insoluble
ignition temperature in air	~35 °C (308 K)	~260°C (533 K)
electrical conduction	none	none

White phosphorus (P(white))

This is the least stable and by far the most reactive allotrope. It is normally stored under water to protect it from atmospheric oxidation. Because of a complex oxidation process, it emits a faint green light (*phosphorescence*) in air at room temperature.

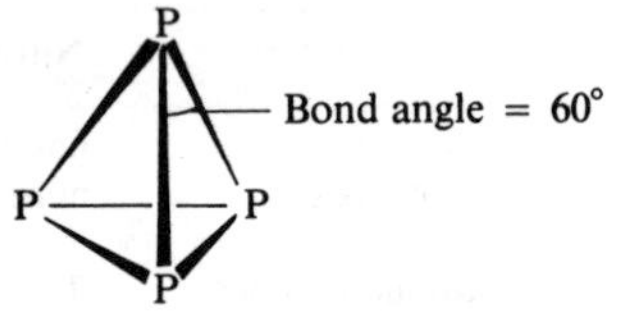

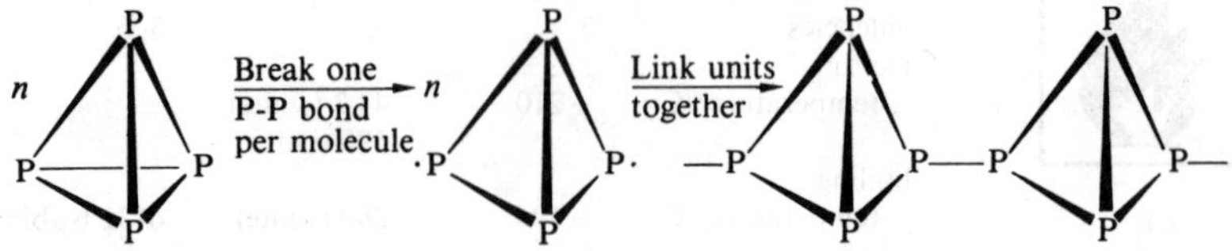

Fig. 15.1
(a) The P_4 molecule of white phosphorus. The phosphorus atoms are located at the corners of a tetrahedron.
(b) The suggested structure of red phosphorus and its formation from white phosphorus. Cross-linking also occurs between the chains.

White phosphorus consists of P_4 molecules in which each phosphorus atom uses its 3p orbitals to form three single bonds (Fig. 15.1(a)). On this basis a bond angle of 90° would be expected, but the value of 60° in the molecule means that considerable strain is present in the P—P bonds. For this reason the P—P bonds are easily broken, a fact which explains the high reactivity of white phosphorus. The low melting temperature of white phosphorus and its solubility in carbon disulphide are accounted for by the relatively weak van der Waals' forces which exist between the P_4 molecules.

Molecules of P_4 are also present in the liquid and vapour phases of phosphorus.

Red phosphorus (P(red))
Red phosphorus is more stable and far less reactive than the white allotrope. Nevertheless, at room temperature, the change from white to red phosphorus is very slow, taking several years to reach completion. The interconversion of white and red phosphorus is indicated below.

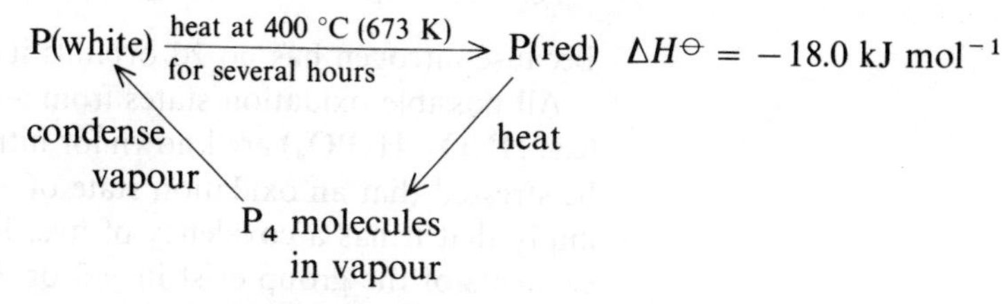

Red phosphorus is polymeric (Fig. 15.1(b)), which accounts for its insolubility, low volatility and in part for its low reactivity. The relief of strain as a P—P bond in a P_4 molecule is broken to form a polymer chain unit also contributes to the low reactivity of red phosphorus.

Black phosphorus
This is the least reactive and most stable allotrope of phosphorus. It is prepared by heating white phosphorus at 350 °C (623 K) for eight days.

15.1.2 BONDING AND VALENCY

Each element of this group has five outer electrons (i.e. $ns^2\ np^3$) and can enter into ionic or covalent bonding. Electrovalency involves either the loss or gain of three electrons per atom, but various covalencies are possible.

Ionic bonding

Nitrogen forms the nitride ion, N^{3-},

e.g. $6Li + N_2 = 2(Li^+)_3N^{3-}$

The nitride ion is an exceedingly strong base and is rapidly hydrolysed by water:

$$N^{3-} + 3H_2O = NH_3 + 3HO^-$$

The remaining elements of the group form anions with increasing difficulty as the non-metallic character decreases. For example, Ca_3P_2 and Mg_3P_2 contain the phosphide ion, P^{3-}, and the arsenide ion, As^{3-}, is believed to exist in Mg_3As_2, but antimony and bismuth do not form the simple anions Sb^{3-} and Bi^{3-}.

Antimony and bismuth are sufficiently metallic to form tripositive ions by the loss of their outer p electrons. Ions of charge +5 are not formed.

Covalent bonding

The elements of this group reach a covalency of three by using their unpaired p electrons. Apart from nitrogen they also show a covalency of five. This is reached by the promotion of an electron from the outer s orbital to a vacant d orbital to give an atom with five unpaired electrons with which to form five covalent bonds,

$$\text{i.e.}\quad ns^2\ \ np^3 \xrightarrow{\text{energy}} ns^1\ \ np^3\ \ nd^1$$

Because nitrogen has no 2d orbitals it cannot show a covalency of five.

All possible oxidation states from −3 (e.g. NH_3, Li_3N, Ca_3P_2) to +5 (e.g. HNO_3, H_3PO_4) are known for nitrogen and phosphorus, but it must be stressed that an oxidation state of +5 for nitrogen (§10.1.2) does not imply that it has a covalency of five. In their compounds the remaining elements of the group exist in +3 or +5 oxidation states only.

Nitrogen and, to a limited extent, phosphorus reach a covalency or *coordination number* of four by donation of the outer lone pair of s electrons. In this way are formed the ammonium ion, NH_4^+, and the phosphonium ion, PH_4^+ (§15.2.4).

Except for nitrogen, which has no available d orbitals, the elements of the group may accept an electron pair from a donor and thus reach a covalency or coordination number of six,

e.g. $PF_5 + F^- \rightarrow [PF_6]^-$

hexafluorophosphate(V) ion

Other examples include $[AsF_6]^-$, $[SbCl_6]^-$ and $[Sb(OH)_6]^-$.

15.2 Nitrogen and phosphorus

Nitrogen and phosphorus, the most important members of this group, will be discussed together. The chemistry of arsenic, antimony and bismuth is described in §15.3.

15.2.1 THE OCCURRENCE AND ISOLATION OF THE ELEMENTS

Industrially, nitrogen is obtained by the fractional distillation of liquefied air. In the laboratory, pure nitrogen is obtained by *carefully* heating a solution containing ammonium and nitrite ions:

$$NH_4^+ + NO_2^- = N_2 + 2H_2O$$

Phosphorus is extracted from various phosphate-containing minerals, the most important of which are apatite, $3Ca_3(PO_4)_2{\cdot}CaF_2$, and hydroxy-apatite, $3Ca_3(PO_4)_2{\cdot}Ca(OH)_2$. When heated to 1 500 °C (1 773 K) with silica and coke in an electric furnace the following reactions occur:

$$2Ca_3(PO_4)_2 + 6SiO_2 = 6CaSiO_3 + P_4O_{10}$$

$$P_4O_{10} + 10C = 10CO + P_4$$

White phosphorus is obtained by condensing the vapours that distil from the furnace.

15.2.2 THE REACTIONS OF NITROGEN AND PHOSPHORUS

Some important reactions of these elements are summarised in Table 15.3.

Table 15.3 The products of some reactions of nitrogen and phosphorus

Reagent	Nitrogen	White phosphorus	Red phosphorus
hydrogen	NH_3	no reaction	no reaction
oxygen	a small amount of NO is formed reversibly at 3000 °C (3273 K)	ignites at approximately 35 °C (308 K) to form P_4O_6 and P_4O_{10}	ignites at approximately 260 °C (533 K) to form P_4O_6 and P_4O_{10}
halogens (F_2, Cl_2, Br_2, I_2)	no reaction	PX_3 and PX_5	PX_3 and PX_5 (reactions are less vigorous than with P(white))
sulphur	no reaction	complex mixture of sulphides	complex mixture of sulphides
heated metals	ionic nitrides or interstitial nitrides	phosphides	phosphides
alkalis, e.g. NaOH, KOH	no reaction	$PH_3 + PH_2O_2^-$	no reaction
oxidising acids, e.g. HNO_3	no reaction	H_3PO_4	H_3PO_4

15.2.3 AMMONIA

The preparation of ammonia

Ammonia is manufactured in the *Haber process* by the direct combination of nitrogen and hydrogen:

$$\tfrac{1}{2}N_2 + \tfrac{3}{2}H_2 \rightleftharpoons NH_3 \qquad \Delta H^{\ominus} = -46.2 \text{ kJ mol}^{-1}$$

The formation of ammonia is favoured by high pressures and low temperatures (§8.3.3). However, at low temperatures equilibrium is established very slowly, even in the presence of a catalyst, and increased temperatures are necessary to secure a reasonable rate of reaction. Typical pressures for the synthesis of ammonia are 100 to 1 000 atmospheres (10^4–10^5 kPa), with a temperature of approximately 550 °C (823 K) and a catalyst of finely divided iron containing, as a promoter, a small amount of aluminium oxide. (A promoter increases the efficiency of a catalyst.)

In the laboratory, ammonia is usually prepared by heating an ammonium salt with an alkali:

$$NH_4^+ + HO^- = NH_3 + H_2O$$

Water is removed by passing the ammonia over calcium oxide. Other drying agents, e.g. concentrated sulphuric acid or calcium chloride, cannot be used because they react with ammonia.

Ammonia can also be obtained by the hydrolysis of ionic nitrides, e.g. Mg_3N_2 or Li_3N.

NH_3

The properties of ammonia

Ammonia is a colourless gas with a characteristic odour. Because of hydrogen bonding its boiling temperature, −33.4 °C (239.8 K), is higher than expected (§6.6). The molecule is pyramidal (§4.8). On the nitrogen atom there is a lone pair of electrons that can be donated to other species. For this reason ammonia is a Lewis base, a property which dominates its chemistry.

Aqueous solutions of ammonia

Ammonia is extremely soluble in water, e.g. at 0 °C (273 K) and one atmosphere (100 kPa) pressure, one volume of water dissolves approximately 1 300 volumes of ammonia. A saturated solution, which contains about 35% of ammonia, is often referred to as '880 ammonia' because it has a density of 0.880 g cm^{-3}. The high solubility results from the ability of ammonia to hydrogen bond with water,

i.e. $NH_3(g) + H_2O(l) \rightleftharpoons$

```
        ---H
            \
             O
            / ·.
        ---H    ·.
        ---H      H              H---
            \      \            /
             O -- H — N -- H — O
            /        /
        ---H        H
        ---H       ·
            \    ·
             O ·
            /
        ---H
```

Most of the ammonia is present in solution in this hydrated form, i.e. $NH_3(aq)$. However, a small proportion of the dissolved ammonia ionises thus:

$$NH_3(aq) + H_2O \rightleftharpoons NH_4^+(aq) + HO^-(aq) \qquad pK_b = 4.75$$

A one molar solution of ammonia is about 0.0042 molar in NH_4^+ and HO^- ions. In view of the limited extent of this ionisation it is best to refer to solutions of ammonia as *aqueous ammonia* or *ammonia solution* rather than 'ammonium hydroxide'. Because of the relatively low concentration of hydroxide ions, aqueous ammonia is regarded, on the Arrhenius theory (§9.2), as a weak base or weak alkali.

The basic properties of ammonia

The ammonia molecule has an affinity for protons, and is thus regarded as a base on the Brønsted–Lowry theory (§9.2.7). The ammonium ion, NH_4^+, is formed when ammonia reacts with covalent compounds containing acidic hydrogen atoms, e.g. hydrogen chloride, or aqueous solutions of acids:

$$NH_3(g) + HCl(g) = NH_4^+Cl^-(s)$$

$$NH_3(g) + H^+(aq) = NH_4^+(aq)$$

The basic character of ammonia towards water is fairly limited because water is a weak acid.

The addition of aqueous ammonia to a solution of a metal salt often produces a precipitate of the metal hydroxide, but this may not always be so, for the following reasons:

(i) the hydroxide may be reasonably soluble;
(ii) the hydroxide ion concentration may not be high enough to exceed the solubility product of the metal hydroxide (§9.3.1);
(iii) with excess ammonia the metal ion may form a soluble ammine complex (§18.1.2), since ammonia is a Lewis base.

The oxidation of ammonia

At high temperatures ammonia reacts with oxygen. In the absence of a catalyst nitrogen is formed, but in the presence of platinum the product is nitrogen oxide.

$$4NH_3 + 3O_2 = 2N_2 + 6H_2O$$

$$4NH_3 + 5O_2 = 4NO + 6H_2O$$

The latter reaction is of considerable importance in the production of nitric acid (§15.2.9).

Ammonia is a weak reducing agent and therefore stable towards many oxidising agents. Chlorine, however, oxidises ammonia to nitrogen, although nitrogen trichloride (§15.2.5) is produced if an excess of chlorine is used.

$$2NH_3 + 3Cl_2 = N_2 + 6HCl$$

Similarly, the hypochlorite ion, ClO^-, oxidises ammonia to nitrogen:

$$2NH_3 + 3ClO^- = N_2 + 3Cl^- + 3H_2O$$

Ammonia also reduces some heated metal oxides to the metal,

e.g. $2NH_3 + 3CuO = 3Cu + N_2 + 3H_2O$

Liquid ammonia

Liquid ammonia, which is prepared by compressing the gas, closely resembles water in being an excellent ionising solvent and in undergoing self-ionisation:

$$2NH_3 \rightleftharpoons NH_4^+ + NH_2^- \quad pK \approx 30 \text{ at } -50\ °C\ (223\ K)$$

$$2H_2O \rightleftharpoons H_3O^+ + HO^- \quad pK = 15.74 \text{ at } 25\ °C\ (298\ K)$$

A system of acids and bases exists in liquid ammonia analogous to that in water. For example, in liquid ammonia acids furnish the ammonium ion, NH_4^+ (cf. H_3O^+), and bases the amide ion, NH_2^- (cf. HO^-). Thus a neutralisation reaction occurs between ammonium chloride (an acid) and sodium amide (a base):

$$NH_4^+Cl^- + Na^+NH_2^- = Na^+Cl^- + 2NH_3$$

i.e. $$NH_4^+ + NH_2^- = 2NH_3$$

cf. $$H_3O^+ + HO^- = 2H_2O$$

Liquid ammonia dissolves the alkali metals and, to a lesser extent, calcium, strontium and barium to produce deep blue solutions which are excellent reducing agents. Evaporation of the ammonia from freshly prepared solutions results in the recovery of the alkali metal or, for the group 2A metals, metal ammines such as $Ca(NH_3)_6$. On standing, the blue solutions slowly become colourless as metal amides are formed,

e.g. $$2Na + 2NH_3 = 2NaNH_2 + H_2$$

$$Ca + 2NH_3 = Ca(NH_2)_2 + H_2$$

Transition metal ions, e.g. Fe^{3+} and Cu^{2+}, catalyse the formation of amides. Metal amides contain the strongly basic amide ion, which readily removes a proton from a water molecule:

$$NH_2^- + H_2O = NH_3 + HO^-$$

The deep blue solutions are excellent electrical conductors and reducing agents, for they contain *solvated electrons*, i.e. free electrons associated with molecules of solvent,

e.g. $$Na = Na^+ + e^-$$

$$e^- + nNH_3(l) = e^-(NH_3)_n$$

Ammonium salts

Ammonia combines with protic acids to form ionic ammonium salts which contain the tetrahedral ammonium ion, NH_4^+. In structure and solubility ammonium salts closely resemble the corresponding compounds of potassium and rubidium. This is due to similarities in ionic radii: NH_4^+ 0.143 nm, K^+ 0.133 nm and Rb^+ 0.148 nm. Like the salts of potassium and rubidium, ammonium salts are generally soluble in water but form anhydrous crystals.

All ammonium salts are thermally unstable. The decomposition products depend on the nature of the anion in the salt.

(i) With anions derived from non-oxidising acids, ammonia and the acid or its decomposition products are obtained, e.g. $$NH_4Cl \rightleftharpoons NH_3 + HCl$$

$$(NH_4)_2CO_3 = 2NH_3 + H_2O + CO_2$$

(ii) Anions derived from oxidising acids oxidise the ammonium ion to nitrogen or one of its oxides,

e.g. $$NH_4NO_2 = N_2 + 2H_2O$$

$$NH_4NO_3 = N_2O + 2H_2O$$

$$(NH_4)_2Cr_2O_7 = N_2 + 4H_2O + Cr_2O_3$$

The first two reactions may occur explosively and should not be attempted.

When ammonium salts are treated with alkalis ammonia is liberated:

$$NH_4^+ + HO^- = NH_3 + H_2O$$

Ammonium salts may be analysed by boiling with alkali and dissolving the expelled ammonia in a known excess of a standard solution of an acid. The amount of acid that is not neutralised is then determined by titration with a standard solution of an alkali.

$$1 \text{ mol } H^+(aq) \equiv 1 \text{ mol } NH_3 \equiv 17 \text{ g } NH_3$$

PH_3

15.2.4 PHOSPHINE

Phosphine, PH_3, is prepared by warming white phosphorus with a concentrated solution of an alkali metal hydroxide:

$$P_4 + 3HO^- + 3H_2O = PH_3 + 3\underset{\text{phosphinate ion}}{PH_2O_2^-}$$

The reaction is an example of disproportionation (§10.1.5). Pure phosphine is stable in air at room temperature, but when prepared by this method it immediately ignites due to the presence of spontaneously flammable impurities, notably diphosphane, P_2H_4, and phosphorus vapour. The impurities can be condensed by passing the impure phosphine gas through a freezing mixture. Phosphine is also formed by the hydrolysis of ionic phosphides, or by reducing phosphorus trichloride with lithium tetrahydridoaluminate(III):

$$Ca_3P_2 + 6H_2O = 2PH_3 + 3Ca(OH)_2$$

$$4PCl_3 + 3Li[AlH_4] = 4PH_3 + 3LiCl + 3AlCl_3$$

The phosphine molecule is pyramidal, with a bond angle of 93°. Hydrogen bonding does not occur with phosphine (Fig. 6.21), and as a result its boiling temperature, −87.4 °C (185.8 K), and solubility in water are much lower than those of ammonia. Solutions of phosphine are unstable and slowly decompose into phosphorus and hydrogen.

Phosphine is a much weaker base than ammonia, because phosphorus and hydrogen have the same electronegativity. Consequently, there is no accumulation of negative charge on the phosphorus atom, and the tetrahedral phosphonium ion, PH_4^+, is far less stable than the ammonium ion. This is shown by the fact that phosphonium iodide, one of the most stable phosphonium salts, dissociates above 60 °C (333 K):

$$PH_4I \rightleftharpoons PH_3 + HI$$

Water is a stronger base than phosphine and therefore readily decomposes the phosphonium ion:

$$PH_4^+ + H_2O = PH_3 + H_3O^+$$

Phosphine is a stronger reducing agent than ammonia because the P—H bond is readily oxidised to P—OH. For example, phosphine ignites readily in air at 150 °C (423 K):

$$PH_3 + 2O_2 = H_3PO_4$$

Liquid phosphine displays none of the properties of liquid ammonia.

15.2.5 THE HALIDES OF NITROGEN

NCl_3, NBr_3 and NI_3

The trihalides of nitrogen are prepared by oxidising ammonia with an excess of the appropriate halogen:

$$NH_3 + 3X_2 = NX_3 + 3HX \qquad (X = F, Cl, Br\ or\ I)$$

Nitrogen trifluoride is a stable gas at room temperature and does not react with water or alkalis. The trichloride is an unstable, highly reactive yellow liquid, which is rapidly hydrolysed by water to ammonia and hypochlorous acid:

$$NCl_3 + 3H_2O = NH_3 + 3HClO$$

The reaction is stepwise and involves the formation of a coordinate N—H bond in which the nitrogen atom is the donor:

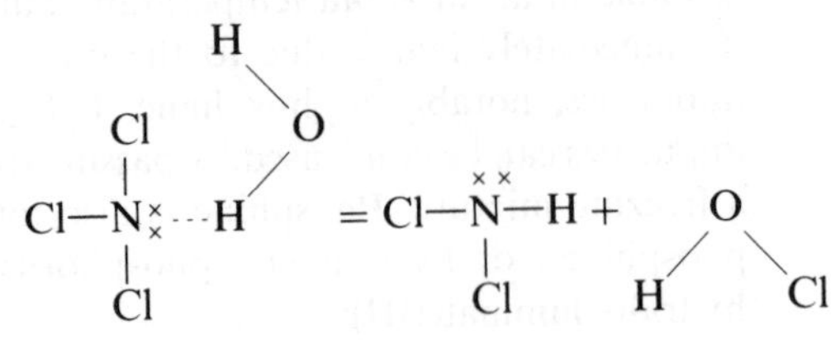

This is followed by:

$$NHCl_2 + H_2O = NH_2Cl + HClO$$

and $NH_2Cl + H_2O = NH_3 + HClO$

The hydrolysis should be contrasted with similar reactions of other non-metal halides (§13.2.5) where the central atom, unlike nitrogen, can act as an electron *acceptor* towards a water molecule.

Nitrogen tribromide, NBr_3, is known in solution in trichloromethane, but decomposes quite rapidly at room temperature. Nitrogen triiodide has been prepared only as an addition compound with ammonia, $NI_3{\cdot}NH_3$. It is a black crystalline solid which detonates on shock.

Nitrogen forms another fluoride, dinitrogen difluoride, N_2F_2, which exhibits geometrical isomerism (§5.2.2):

cis isomer — *trans* isomer

(x x) represents a lone pair of electrons

15.2.6 THE HALIDES OF PHOSPHORUS

Halides of phosphorus are known in both the trivalent and pentavalent states.

PCl_3, PBr_3 and PI_3

Phosphorus trihalides (phosphorus(III) halides)

Phosphorus trihalides are prepared by reacting the appropriate halogen with an excess of phosphorus in the absence of oxygen and water. For example, when a slow stream of chlorine is passed over white phosphorus, the chlorine burns in the phosphorus vapour and phosphorus trichloride distils over. The tribromide and triiodide are prepared by slowly adding bromine or iodine to red phosphorus.

Phosphorus trihalides have a pyramidal structure (§4.8) and, with the exception of the fluoride, are vigorously hydrolysed to phosphonic acid by cold water,

e.g. $PCl_3 + 3H_2O = H_2PHO_3 + 3HCl$

The reaction is believed to involve the coordination of a water molecule, using a vacant 3d orbital on the phosphorus atom, followed by the elimination of hydrogen chloride; cf. the hydrolysis of BCl_3 and $SiCl_4$ (§13.2.5 and §14.2.7).

$$Cl_3P \leftarrow O(H)H = Cl_3P{-}O(H)H = Cl_2P(OH) + HCl$$

(written as $HPOCl_2$)

This reaction is followed by two further steps to give H_3PO_3, i.e. $P(OH)_3$, which immediately rearranges into phosphonic acid, H_2PHO_3:

$$HPOCl_2 + H_2O = H_2PO_2Cl + HCl$$

$$H_2PO_2Cl + H_2O = H_2PHO_3 + HCl$$

Similar reactions occur with other molecules that contain hydroxyl groups, e.g. alcohols and carboxylic acids (§14.9.5 and §14.10.7).

Phosphorus trichloride is readily oxidised by oxygen or chlorine:

$$2\,\mathrm{PCl_3} + \mathrm{O_2} = 2\,\mathrm{O{=}PCl_3}$$

phosphoryl chloride

$$\mathrm{PCl_3} + \mathrm{Cl_2} \rightleftharpoons \mathrm{PCl_5}$$

PCl_5 and PBr_5

Phosphorus pentahalides (phosphorus(V) halides)

Phosphorus pentaiodide, PI_5, is unknown. This is probably because of the difficulty of accommodating five large iodine atoms around the smaller phosphorus atom. The chloride is prepared by reacting phosphorus trichloride with an excess of chlorine:

$$\mathrm{PCl_3(g)} + \mathrm{Cl_2(g)} \rightleftharpoons \mathrm{PCl_5(g)} \qquad \Delta H^{\ominus} = -93\ \mathrm{kJ\ mol^{-1}}$$

The reaction is reversible, and the formation of the pentachloride is favoured by an excess of chlorine and a low temperature (§8.3.3). Similar principles apply to the formation of the pentabromide, but, because of the difficulty of handling fluorine, the fluoride is prepared by the reaction between phosphorus pentachloride and calcium fluoride at 300–400 °C (573–673 K):

$$2\mathrm{PCl_5} + 5\mathrm{CaF_2} = 2\mathrm{PF_5} + 5\mathrm{CaCl_2}$$

At room temperature phosphorus pentafluoride is a gas which consists of trigonal bipyramidal molecules (§4.8), while the chloride and bromide are colourless ionic solids of constitution $[PCl_4]^+ [PCl_6]^-$ and $[PBr_4]^+ Br^-$ respectively. The $[PX_4]^+$ ions are tetrahedral, and the $[PCl_6]^-$ ion is octahedral. In the vapour state, or in solutions in non-polar solvents, phosphorus pentachloride exists as trigonal bipyramidal molecules.

The pentafluoride is a powerful Lewis acid and forms complexes with bases such as ethers, amines and the fluoride ion (§15.1.2). Phosphorus pentachloride and pentabromide dissociate on heating to form the corresponding trihalide and halogen, but phosphorus pentafluoride is thermally stable.

Phosphorus pentachloride reacts with water, producing first phosphoryl chloride and then orthophosphoric acid:

$$\mathrm{PCl_5} + \mathrm{H_2O} = \mathrm{POCl_3} + 2\mathrm{HCl}$$

$$\mathrm{POCl_3} + 3\mathrm{H_2O} = \mathrm{H_3PO_4} + 3\mathrm{HCl}$$

Phosphorus pentachloride is useful for replacing the hydroxyl groups of molecules by chlorine atoms,

e.g.

$$\mathrm{(HO)_2SO_2} + 2\mathrm{PCl_5} = \mathrm{Cl_2SO_2} + 2\mathrm{POCl_3} + 2\mathrm{HCl}$$

sulphuric acid — sulphuryl chloride

For other examples see §14.9.5 and §14.10.7.

15.2.7 THE OXIDES OF NITROGEN

Oxides of nitrogen are known in all oxidation states from +1 to +5 (Table 15.4).

Table 15.4 The oxides of nitrogen

Oxidation number of nitrogen	Formula	Name	$\Delta H_f^{\ominus}$/kJ mol^{-1}
+1	N_2O	dinitrogen oxide	+81.6
+2	NO	nitrogen oxide	+90.4
+3	N_2O_3	dinitrogen trioxide	+92.9
+4	NO_2	nitrogen dioxide	+33.9
+4	N_2O_4	dinitrogen tetraoxide	+9.7
+5	N_2O_5	dinitrogen pentaoxide	+15†

† Data for formation of the gaseous compound.

The oxides become increasingly acidic as the oxidation number of the nitrogen atom increases. Thus, N_2O and NO are very weakly acidic, but N_2O_5 is strongly acidic. They are all endothermic compounds, due principally to the high bond dissociation enthalpy of the nitrogen molecule.

Dinitrogen oxide

Dinitrogen oxide is formed by the reaction between hydroxylammonium ions and nitrite ions in solution:

$$HONH_3^+ + NO_2^- = N_2O + 2H_2O$$

Dinitrogen oxide is a colourless gas, which is collected over warm water since it is appreciably soluble in cold water.

The dinitrogen oxide molecule is linear, and involves delocalised π bonding, i.e. N⩶N⋯O.

Dinitrogen oxide is unreactive to most reagents at room temperature. On heating, however, it decomposes into its elements:

$$2N_2O = 2N_2 + O_2$$

Hot substances may cause this decomposition and burn in the oxygen so produced. For example, a glowing splint (hot carbon) relights in dinitrogen oxide, just as it does in oxygen, and heated phosphorus burns easily in the gas.

$$C + 2N_2O = 2N_2 + CO_2$$

$$P_4 + 10N_2O = P_4O_{10} + 10N_2$$

Dinitrogen oxide may be distinguished from oxygen by the facts that no reaction occurs with nitrogen oxide, and that an equal volume of nitrogen is produced on passing the gas over heated copper:

$$Cu + N_2O = CuO + N_2$$

NO

Nitrogen oxide

Impure nitrogen oxide gas is formed by reacting copper with concentrated nitric acid that has been diluted with an equal volume of water:

$$3Cu + 8HNO_3 = 3Cu(NO_3)_2 + 4H_2O + 2NO$$

The pure gas is prepared by the reduction, in acidic solution, of nitrate ions by iron(II) ions or nitrite ions by iodide ions:

$$3Fe^{2+} + NO_3^- + 4H^+(aq) = 3Fe^{3+} + NO + 2H_2O$$

$$2NO_2^- + 2I^- + 4H^+(aq) = I_2 + 2NO + 2H_2O$$

Commercially, nitrogen oxide is produced by the oxidation of ammonia (§15.2.9).

The nitrogen oxide molecule is often described as 'odd' because it possesses an odd number of electrons. One electron is therefore unpaired. Like the nitrogen molecule, which has fourteen electrons and a triple bond between its atoms, the nitrogen oxide molecule has fifteen electrons and, essentially, a triple bond between the nitrogen and oxygen atoms. The extra or odd electron is easily removed to yield the nitrosyl cation, NO^+, which is isoelectronic with the nitrogen molecule (§4.3.3). Salts containing the nitrosyl cation are known; e.g. nitrosyl tetrafluoroborate(III), $NOBF_4$. At low temperatures nitrogen oxide dimerises to $(NO)_2$.

Nitrogen oxide acts as a ligand (§4.5.1) by donating the lone pair of electrons on the nitrogen atom, and forms complexes such as the $[Fe(H_2O)_5(NO)]^{2+}$ ion, which is produced in the brown ring test, and the so-called 'nitroprusside' ion, $[Fe(CN)_5(NO)]^{2-}$ (§5.1.1).

Nitrogen oxide is readily oxidised. For example, nitrogen dioxide is rapidly formed with oxygen, and in the presence of a charcoal catalyst the halogens, except for iodine, yield nitrosyl halides:

$$2NO + O_2 = 2NO_2$$

$$2NO + X_2 = 2NOX \quad (X = F, Cl \text{ or } Br)$$

The reaction with the permanganate ion in acidic solution is quantitative and may be used to estimate the gas in a mixture:

$$3MnO_4^- + 4H^+ + 5NO = 3Mn^{2+} + 5NO_3^- + 2H_2O$$

Above 1 000 °C (1 273 K) nitrogen oxide decomposes into its elements. Thus burning magnesium, which is intensely hot, forms its oxide and nitride when plunged into nitrogen oxide:

$$2NO = N_2 + O_2$$

$$5Mg + N_2 + O_2 = Mg_3N_2 + 2MgO$$

N_2O_3

Dinitrogen trioxide

This highly unstable compound is obtained as a blue liquid when equimolar quantities of nitrogen oxide and dinitrogen tetraoxide condense together at −20 °C (253 K):

$$2NO + N_2O_4 \rightleftharpoons 2N_2O_3$$

Reaction with water or alkali yields respectively nitrous acid, HNO_2, or the nitrite ion, NO_2^-; thus dinitrogen trioxide is the anhydride of nitrous acid.

$$H_2O + N_2O_3 = 2HNO_2$$
$$N_2O_3 + 2HO^- = 2NO_2^- + H_2O$$

NO_2 and N_2O_4

Nitrogen dioxide and dinitrogen tetraoxide

Nitrogen dioxide is prepared in the laboratory by heating a nitrate. The lead salt is commonly used because it is anhydrous:

$$2Pb(NO_3)_2 = 2PbO + 4NO_2 + O_2$$

Nitrogen dioxide and oxygen are separated by passing the gases through a cooled U-tube. Nitrogen dioxide dimerises to dinitrogen tetraoxide (see below) and condenses to a yellow liquid. Alternatively, nitrogen dioxide can be obtained by the reduction of concentrated nitric acid by many metals,

e.g. $$Cu + 4HNO_3 = Cu(NO_3)_2 + 2H_2O + 2NO_2$$

Nitrogen dioxide is a toxic brown gas with an unpleasant smell. The molecule is angular with delocalised π bonding. It possesses one unpaired electron which is localised mainly on the nitrogen atom.

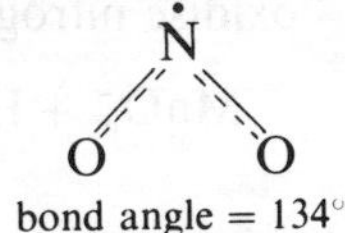

bond angle = 134°

Dinitrogen tetraoxide, when pure, is a colourless liquid and has a boiling temperature of 21 °C (294 K). It possesses no unpaired electrons. The molecule contains a weak N—N bond which is formed by the pairing up of odd electrons on two nitrogen dioxide molecules,

i.e. $O_2N\cdot\cdot NO_2$ = $O_2N{:}NO_2$ or $O_2N{-}NO_2$

The dioxide and tetraoxide exist in a temperature-dependent equilibrium:

$$N_2O_4 \rightleftharpoons 2NO_2 \qquad \Delta H^\ominus = +58.1 \text{ kJ mol}^{-1}$$

Low temperatures favour the tetraoxide. At temperatures in excess of 150 °C (423 K) nitrogen dioxide dissociates into nitrogen oxide and oxygen. These changes are summarised below:

$$N_2O_4 \rightleftharpoons 2NO_2 \rightleftharpoons 2NO + O_2$$

dissociation begins at −11 °C (262 K) and is complete at 140 °C (413 K)

dissociation begins at 150 °C (423 K) and is complete at 600 °C (873 K)

Nitrogen dioxide is highly reactive. It is a mixed anhydride and dissolves in water at 0 °C (273 K) to form a mixture of nitric and nitrous acids:

$$2NO_2 + H_2O = HNO_3 + HNO_2$$

oxidation number of N +4 +5 +3

At room temperature nitrous acid rapidly disproportionates and the overall reaction is:

$$H_2O + 3NO_2 = 2HNO_3 + NO$$

In the presence of air nitrogen oxide is oxidised to nitrogen dioxide, which reacts with further water to give nitric acid as the ultimate product.

With alkalis, nitrogen dioxide forms a mixture of nitrite ions and nitrate ions:

$$2NO_2 + 2HO^- = NO_3^- + NO_2^- + H_2O$$

Nitrogen dioxide is a powerful oxidising agent, and heated elements such as magnesium, iron, copper, carbon, phosphorus and sulphur burn in the gas to produce their oxides and nitrogen. Most other reducing agents convert it to nitrogen oxide,

e.g.

$$H_2S + NO_2 = NO + H_2O + S$$

$$SO_2 + H_2O + NO_2 = NO + H_2SO_4$$

Powerful oxidants, such as the permanganate ion in acidic solution, oxidise nitrogen dioxide to the nitrate ion:

$$MnO_4^- + H_2O + 5NO_2 = 5NO_3^- + 2H^+ + Mn^{2+}$$

N_2O_5

Dinitrogen pentaoxide

Dinitrogen pentaoxide is obtained as a white crystalline solid by dehydrating nitric acid with phosphorus(V) oxide at low temperatures:

$$4HNO_3 + P_4O_{10} = 4HPO_3 + 2N_2O_5$$

In the crystalline state it is ionic, i.e. $NO_2^+ \, NO_3^-$, but N_2O_5 molecules exist in the vapour phase. Above 0 °C (273 K) it decomposes:

$$2N_2O_5 = 2N_2O_4 + O_2$$

Dinitrogen pentaoxide is the anhydride of nitric acid:

$$N_2O_5 + H_2O = 2HNO_3$$

15.2.8 THE OXIDES OF PHOSPHORUS

Phosphorus(III) oxide

Phosphorus(III) oxide is a white solid, prepared by heating white phosphorus in a slow stream of air:

$$P_4 + 3O_2 = P_4O_6$$

Phosphorus(V) oxide, which is produced at the same time, is condensed to a solid by passing the emergent gases through a U-tube at about 150 °C (423 K). Phosphorus(III) oxide passes on and is collected in an ice cooled receiver.

Phosphorus(III) oxide exists as P_4O_6 molecules in the vapour phase and in solution in organic solvents (Fig. 15.2(b)).

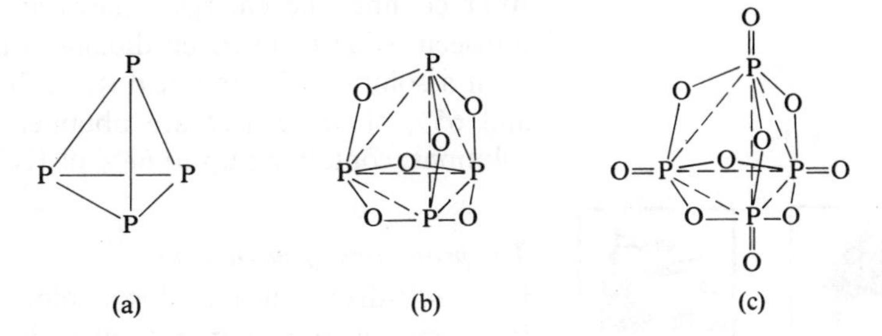

Fig. 15.2 The structural relationship between molecules of (a) P_4, (b) P_4O_6 and (c) P_4O_{10}. (Dotted lines in (b) and (c) outline the P_4 tetrahedra and do not represent bonds.)

Phosphorus(III) oxide dissolves in cold water to form phosphonic acid, H_2PHO_3:

$$P_4O_6 + 6H_2O = 4H_2PHO_3$$

It is readily oxidised and is thus a reducing agent,

e.g. $$P_4O_6 + 2O_2 = P_4O_{10}$$

$$P_4O_6 + 6Cl_2 = 4POCl_3 + O_2$$

P_4O_{10}

Phosphorus(V) oxide

This is the product of heating phosphorus in excess air or oxygen:

$$P_4 + 5O_2 = P_4O_{10}$$

It is a white powder. P_4O_{10} molecules (Fig. 15.2(c)) exist in the vapour phase and in one of numerous crystalline forms.

Phosphorus(V) oxide has a strong affinity for water, with which it reacts to form the acids HPO_3 and H_3PO_4 (§15.2.10). Consequently it is used as a desiccating agent, but when exposed to the air its drying capacity is rapidly lost because a glassy layer of acids forms on the particles. Phosphorus(V) oxide is also used as a dehydrating agent.

15.2.9 THE OXOACIDS OF NITROGEN AND THEIR SALTS

Nitric acid

Pure nitric acid is obtained by heating sodium nitrate or potassium nitrate with concentrated sulphuric acid:

$$NaNO_3 + H_2SO_4 = NaHSO_4 + HNO_3$$

The acid is manufactured by passing a mixture of ammonia and dry dust-free air over a platinum or platinum–rhodium catalyst at 900 °C (1 173 K). The reaction which occurs is sufficiently exothermic to maintain the catalyst at the working temperature.

$$4NH_3(g) + 5O_2(g) = 4NO(g) + 6H_2O(g)$$
$$\Delta H^{\ominus} = -226.4 \text{ kJ per mol of } NH_3$$

After cooling, the emergent gases are mixed with more air to oxidise nitrogen oxide to nitrogen dioxide, which is then absorbed in water to form a solution of nitric acid (§15.2.7). Solutions containing between 50 and 65% of nitric acid are obtained in this way. More concentrated solutions, containing up to 68% of HNO_3, are obtained by distillation.

HNO_3

The properties of nitric acid

Pure, anhydrous nitric acid is a colourless liquid of density 1.50 g cm^{-3}. It is composed of planar molecules which are linked together by hydrogen bonds,

i.e.

O=N(–O)–O–H······O=N(–O)–O–H·······

······ represents a hydrogen bond

– – – represents a delocalised π bond

Pure nitric acid does not display any acidic properties; for example, it does not decompose carbonates, but due to self-ionisation it is a weak conductor of electricity:

$$2HNO_3 \rightleftharpoons NO_2^+ + NO_3^- + H_2O$$

Salts containing the nitryl cation are known; e.g. $NO_2^+\ ClO_4^-$ and $NO_2^+\ NO_3^-$. The nitryl cation is also formed when pure nitric acid or concentrated nitric acid is mixed with concentrated sulphuric acid (§14.7.4).

In the laboratory nitric acid is usually encountered as an aqueous solution. Concentrated nitric acid (density 1.42 g cm^{-3}) is a constant boiling mixture (122 °C (395 K)) containing about 68% of the pure acid. In dilute solution nitric acid is approximately 93% ionised; therefore it is a strong acid, which possesses the usual acid properties:

$$HNO_3 + H_2O \rightleftharpoons H_3O^+ + NO_3^-$$

Aqueous nitric acid is an oxidant, a property which is derived from the nitrate ion. The possible reduction products of the nitrate ion and the relevant ionic half-equations are as follows:

(a) $NO_3^- + 2H^+(aq) + e^- = H_2O + NO_2$

(b) $NO_3^- + 4H^+(aq) + 3e^- = 2H_2O + NO$

(c) $2NO_3^- + 10H^+(aq) + 8e^- = 5H_2O + N_2O$

(d) $NO_3^- + 8H^+(aq) + 6e^- = 2H_2O + HONH_3^+$

(e) $NO_3^- + 10H^+(aq) + 8e^- = 3H_2O + NH_4^+$

Generally, with concentrated nitric acid and weak reductants, reaction (a) predominates, but with the dilute acid and strong reductants other products are possible as shown below.

Non-metals Often nitrogen oxide is produced, together with an oxoacid of the non-metal,

e.g. $$S + 2HNO_3 = H_2SO_4 + 2NO$$

$$4P + 10HNO_3 + H_2O = 4H_3PO_4 + 5NO + 5NO_2$$

Metals In their reactions with nitric acid, metals can be roughly divided into six categories.

(i) Calcium and magnesium are the only metals capable of liberating hydrogen from very dilute (1%) nitric acid.
(ii) Metals close to, or below, hydrogen in the electrochemical series, e.g. Pb, Cu and Ag, are weak reductants and generally liberate nitrogen dioxide from concentrated nitric acid, or nitrogen oxide from the dilute acid.
(iii) Powerful reducing agents, e.g. Mg or Zn, reduce dilute nitric acid to dinitrogen oxide, the hydroxylammonium ion or the ammonium ion, depending on reaction conditions, such as temperature and acid concentration. Not surprisingly, a mixture of these products is often obtained.
(iv) Weak metals, e.g. Sn, Sb and As, reduce concentrated nitric acid to nitrogen dioxide and form their hydrated oxides (§14.12.3).
(v) Iron, chromium and aluminium are rendered *passive*, i.e. unreactive, by concentrated nitric acid. The passivity is due to a thin film of oxide on the surface of the metal, and until this is removed either by scraping or by means of a reducing agent the metal will not dissolve readily in any acid.
(vi) Noble metals, e.g. Au and Pt, are not attacked.

Cations Iron(II) ions are oxidised to iron(III) (§15.2.7). Tin(II) compounds are oxidised to tin(IV) compounds, and reduce nitric acid to the hydroxylammonium ion and the ammonium ion.

Anions Sulphide ions and iodide ions are weak reductants and reduce the acid to nitrogen dioxide:

$$S^{2-} + 2NO_3^- + 4H^+(aq) = S + 2NO_2 + 2H_2O$$

$$2I^- + 2NO_3^- + 4H^+(aq) = I_2 + 2NO_2 + 2H_2O$$

All nitrates

Ionic nitrates

All metal nitrates are soluble in water and are prepared by dissolving metals, oxides, hydroxides or carbonates in nitric acid. With the exception of those of the alkali metals, lead and ammonium they are all hydrated, and contain the planar nitrate ion which involves delocalised π bonding,

i.e. 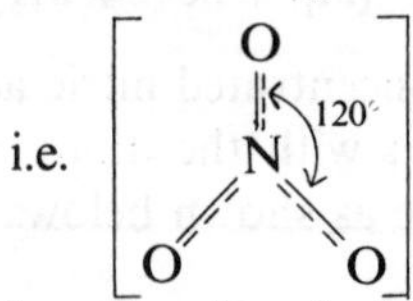

Although the *hydrated* nitrates of some metals are ionic, e.g. $Cu(NO_3)_2 \cdot 3H_2O$, the *anhydrous* compounds are covalent, e.g. $Cu(NO_3)_2$. Covalent nitrates cannot be prepared by dehydration of the hydrated salts.

In contrast to its behaviour in acidic solution, the nitrate ion in neutral or alkaline solution is only a weak oxidising agent. Under these conditions, Devarda's alloy (Al, 45%; Cu, 50%; Zn, 5%) quantitatively reduces the nitrate ion to ammonia, and is used for estimating nitrates in the absence of ammonium salts:

$$NO_3^- + 6H_2O + 8e^- = NH_3 + 9HO^-$$

Nitrates are thermally unstable and decompose on heating in one of three ways.

(i) Nitrites are formed from the nitrates of group 1A metals except lithium,

e.g. $2NaNO_3 = 2NaNO_2 + O_2$

(ii) Oxides, oxygen and nitrogen dioxide are produced from the nitrates of other metals, including lithium. Some metal oxides are thermally unstable, e.g. Ag_2O and HgO, and in such cases the metal is produced:

$$2AgNO_3 = 2Ag + 2NO_2 + O_2$$

(iii) Dinitrogen oxide is formed by heating ammonium nitrate (§15.2.3).

Nitrous acid

Nitrous acid is a weak acid ($pK_a = 3.34$). It is unstable and readily disproportionates at room temperature:

$$3HNO_2 = HNO_3 + 2NO + H_2O$$

oxidation number of N +3 +5 +2

A pale blue solution containing the acid is obtained by adding dilute hydrochloric acid to a cold solution of a nitrite, or by dissolving dinitrogen trioxide in water (§15.2.7). The salts of nitrous acid contain the nitrite ion, NO_2^-, and are more stable than the acid itself. The common nitrites, i.e. $NaNO_2$ and KNO_2, are prepared by heating the corresponding nitrates either alone or with a reducing agent such as lead, or by dissolving an equimolar mixture of nitrogen oxide and nitrogen dioxide in a solution of the appropriate hydroxide:

$$NaNO_3 + Pb = NaNO_2 + PbO$$

$$NO + NO_2 + 2HO^- = 2NO_2^- + H_2O$$

In acidic solution the nitrite ion is a powerful oxidant, and oxidises iodide ions to iodine, hydrogen sulphide to sulphur, and ammonia to nitrogen. The reduction products of the nitrite ion include nitrogen oxide, dinitrogen oxide and the ammonium ion. Powerful oxidising agents, such as the permanganate ion or chlorine, oxidise the nitrite ion to the nitrate ion.

Test to distinguish between nitrites and nitrates The addition of ethanoic acid to a nitrite produces nitrous acid, which decomposes to give brown fumes of nitrogen dioxide in air:

$$CH_3COOH + NaNO_2 = CH_3COONa + HNO_2$$

$$3HNO_2 + O_2 = HNO_3 + 2NO_2 + H_2O$$

Nitrates do not react with ethanoic acid.

15.2.10 THE OXOACIDS OF PHOSPHORUS AND THEIR SALTS

The structures of the simple acids are shown below. In each molecule the phosphorus atom is surrounded by other atoms or groups to give a distorted tetrahedron.

O=P(H)(H)(OH)	O=P(H)(OH)(OH)	O=P(OH)(OH)(OH)
phosphinic acid	phosphonic acid	orthophosphoric acid

The hydrogen atoms of HO groups bonded to a phosphorus atom are ionisable, but hydrogen atoms directly linked to phosphorus are not.

H_3PO_4

Orthophosphoric acid (phosphoric acid)

This is the commonest acid of phosphorus, and is prepared by oxidising red phosphorus with nitric acid or by adding phosphorus(V) oxide to hot water. Orthophosphoric acid is a weak triprotic acid:

$$H_3PO_4 + H_2O \rightleftharpoons H_3O^+ + H_2PO_4^- \qquad pK_1 = 2.15$$

$$H_2PO_4^- + H_2O \rightleftharpoons H_3O^+ + HPO_4^{2-} \qquad pK_2 = 7.21$$

$$HPO_4^{2-} + H_2O \rightleftharpoons H_3O^+ + PO_4^{3-} \qquad pK_3 = 12.36$$

Phosphoric acid is often encountered as 'syrupy phosphoric acid', an 85% aqueous solution which has a consistency of syrup resulting from extensive hydrogen bonding. The acid is rather unreactive and difficult to reduce.

Salts containing the PO_4^{3-} and HPO_4^{2-} ions are insoluble, except for those of the alkali metals and ammonium, but those containing the $H_2PO_4^-$ ion tend to be more soluble. For example, $Ca_3(PO_4)_2$ and $CaHPO_4$ are insoluble, but $Ca(H_2PO_4)_2$ is soluble. Orthophosphates are alkaline in solution because of salt hydrolysis (§9.2.8).

15.3 Arsenic, antimony and bismuth

The elements are obtained by reduction of their oxides with hydrogen or carbon. Some of their reactions are summarised in Table 15.5.

Table 15.5 The products of some reactions of arsenic, antimony and bismuth

Reagent	Arsenic	Antimony	Bismuth
concentrated H_2SO_4	$As_4O_6(aq)$	$Sb_2(SO_4)_3$	$Bi_2(SO_4)_3$
dilute HNO_3	$As_4O_6(aq)$	Sb_4O_6	$Bi(NO_3)_3$
concentrated HNO_3	H_3AsO_4	Sb_2O_5	$Bi(NO_3)_3$
alkalis	AsO_3^{3-}	SbO_3^{3-}	no reaction
oxygen when heated	As_4O_6	Sb_4O_6	Bi_2O_3

All As compounds

15.3.1 THE COMPOUNDS OF ARSENIC, ANTIMONY AND BISMUTH

The hydrides

The hydrides are: arsine, AsH_3, stibine, SbH_3 and bismuthine, BiH_3. Arsine and stibine are prepared by reducing the respective trichlorides with lithium tetrahydridoaluminate(III),

e.g. $$4AsCl_3 + 3Li[AlH_4] = 4AsH_3 + 3LiCl + 3AlCl_3$$

Bismuthine is reportedly formed by dissolving a magnesium–bismuth alloy in a dilute acid.

The hydrides become decreasingly stable from arsine to bismuthine. Like ammonia, they are composed of pyramidal molecules, but show no tendency to form ions of the type MH_4^+.

The chlorides

The trichlorides, which are all covalent, are prepared by direct combination of the elements with a deficiency of chlorine, if necessary, to minimise pentachloride formation. Arsenic trichloride resembles phosphorus trichloride, except that the reaction with water is reversible,

i.e. $$AsCl_3 + 3H_2O \rightleftharpoons H_3AsO_3 + 3HCl$$

Antimony and bismuth trichlorides are also reversibly hydrolysed by water, to give white insoluble basic salts,

e.g. $$BiCl_3 + H_2O \rightleftharpoons BiClO + 2HCl$$

The only known pentachloride is $SbCl_5$, which is prepared by reacting antimony or its trichloride with an excess of chlorine.

The trivalent oxides

These are prepared by direct combination of the elements (Table 15.5). Arsenic(III) oxide and antimony(III) oxide are similar in structure to phosphorus(III) oxide and are thus formulated as As_4O_6 and Sb_4O_6

respectively. Because bismuth(III) oxide contains no discrete molecules it is represented by the stoicheiometric formula, Bi_2O_3.

Arsenic(III) oxide is principally acidic and dissolves in alkalis to form the arsenite ion:

$$As_4O_6 + 12HO^- = 4AsO_3^{3-} + 6H_2O$$

Antimony(III) oxide is distinctly amphoteric, dissolving in alkalis to form the antimonite ion, SbO_3^{3-}, and in concentrated acids to give antimony salts, e.g. $Sb_2(SO_4)_3$. Bismuth(III) oxide, like the hydroxide, $Bi(OH)_3$, is basic.

The pentavalent oxides

The compounds are represented by their stoicheiometric formulae, i.e. As_2O_5, Sb_2O_5 and Bi_2O_5, because their structures are unknown. Arsenic(V) oxide and antimony(V) oxide are prepared by oxidising the elements with concentrated nitric acid and dehydrating the products. Bismuth(V) oxide, which exists only in an impure state, is prepared by oxidising bismuth(III) oxide with peroxodisulphate ions, followed by acidification.

All these oxides are acidic. They readily decompose into the trivalent oxides and oxygen on heating.

EXAMINATION QUESTIONS ON CHAPTER 15

1 (a) Describe the Haber process for the manufacture of ammonia, explaining carefully the reasons for the physical and chemical conditions under which the process is carried out, but omitting details of industrial plant.

(b) How does ammonia react with chlorine?

(c) By means of a clearly labelled schematic diagram **only**, describe the nitrogen cycle. Your diagram should give the principal classes of compounds involved and show how they are interconverted. (O)

2 (a) State how each of the following compounds may be prepared, in each case confining your answer to naming the reactants, stating the conditions and giving an equation: (i) N_2O, (ii) NO, (iii) N_2O_4, (iv) P_2O_5.

(b) Discuss the interaction of each of the above compounds with water. (O)

3 Describe and explain the action of (a) heat, (b) concentrated sulphuric acid, on each of the following substances:

(i) ammonium chloride, (ii) ammonium nitrate, (iii) sodium chlorate.

When 0.2000 g of a mixture containing these three substances only was boiled with excess sodium hydroxide solution, the ammonia evolved was neutralised by 20.00 cm^3 of exactly 0.1 M hydrochloric acid. When, however, the same weight of the mixture was boiled with excess aluminium powder and sodium hydroxide, the ammonia evolved was neutralised by 30.00 cm^3 of exactly 0.1 M hydrochloric acid. Calculate the weights of ammonium chloride, ammonium nitrate and sodium chlorate in 0.2000 g of the mixture. (JMB)

4 Describe and account for the similarities and differences in the chemistry of nitrogen and phosphorus, using (a) the elements, (b) the hydrides, and (c) the chlorides as examples. (JMB)

5 (a) Compare the chemistry of nitrogen with that of phosphorus with particular regard to the structure, bonding and properties of (i) the elements, (ii) the hydrides, (iii) the chlorides.

(b) Use the values of bond enthalpies tabulated below to show that nitrogen is unlikely to form an oxide N_4O_6, analogous to P_4O_6, or an acid of formula H_3NO_3.

Bond type	*Bond enthalpy/kJ mol^{-1}*
N≡N	944
N—N	163
O=O	496
O—O	146
N=O	594
N—O	163
H—H	436
N—H	388
O—H	463

(OC)

6 This question concerns the elements arsenic (atomic number 33), nitrogen (atomic number 7) and phosphorus (atomic number 15).

(a) For what reason are the elements placed in the same group of the periodic table?

(b) Distinguishing between s, p and d electrons, give the electronic configuration of the elements.

(c) Place the elements in order of decreasing electronegativity (i.e. putting the most electronegative element first). Name one other element which is more electronegative than all three.

(d) Give the structures of:
(i) the tetraamminezinc(II) ion,
(ii) dinitrogen tetraoxide,
(iii) solid phosphorus pentachloride.
Outline the preparation of (iii).

(e) Explain why:
(i) phosphorus forms a trichloride and a pentachloride but nitrogen forms only a trichloride,
(ii) ammonia is more basic than phosphine,
(iii) nitrogen is more inert than phosphorus. (AEB)

7 An allotropic form of phosphorus reacts with aqueous copper(II) sulphate to give metallic copper and an acidic solution. In an experiment, 0.93 g of phosphorus reacted with excess of the copper(II) sulphate to give 4.8 g of copper.
(P = 31; Cu = 64)

(a) Calculate the number of moles of copper atoms formed.

(b) Calculate the number of moles of phosphorus atoms used.

(c) Calculate the number of moles of copper atoms formed for one mole of phosphorus atoms.

(d) What is the oxidation number of copper before and after reaction?

(e) Using your answer in (c) and (d), work out the oxidation number of phosphorus before and after reaction.

(f) The acid in the final solution has the empirical formula HPO_x. Find the value of x.

(g) Write a balanced equation for the reaction, using water molecules on the reactants side of the equation and hydrogen ions on the products side.

(h) (i) Explain what is meant by the term allotropic form.

(ii) Explain why phosphorus but not nitrogen can form a pentachloride, XCl_5.

(iii) Name products and write an equation for the action of heat on phosphorus(V) chloride (phosphorus pentachloride).

(iv) Name products and write their formulae for the stage-by-stage neutralisation of the acid H_3PO_4 by sodium hydroxide solution. (SUJB)

8 (a) **Outline** the chemistry involved in the extraction of phosphorus from calcium orthophosphate and state which allotrope is produced.

(b) Give the equation and conditions for the preparation of phosphine from white phosphorus.

(c) Compared with phosphine explain why ammonia:

(i) is more basic,

(ii) has a higher boiling point,

(iii) has a greater bond angle than phosphine.

(d) Give the names or structures of the organic products obtained when concentrated aqueous ammonia reacts with a named halogenoalkane (alkyl halide). (AEB)

9 (a) By writing equations for suitable reactions, illustrate the behaviour of ammonia as (i) a base, (ii) a reducing agent (reductant), (iii) a ligand.

(b) For the elements, nitrogen, phosphorus and arsenic:

(i) give the formulae of their characteristic hydrides,

(ii) state how the thermal stability of the characteristic hydride varies from nitrogen to phosphorus to arsenic.

(c) (i) Give the oxidation number (state) of nitrogen in the following compounds: nitric acid; nitrous acid; ammonia.

(ii) Explain why nitric acid can be reduced but cannot be oxidised; give an example of its reduction.

(iii) Explain, giving one example in each case, why nitrous acid can act both as a reducing agent and as an oxidising agent. (AEB)

10 (a) Compare the chemical behaviour of the trihydrides of nitrogen and phosphorus by considering (i) their thermal stability, (ii) their reactions with oxygen, (iii) their reactions with dilute hydrochloric acid.

(b) How, and under what conditions, does ammonia react with (i) sodium, (ii) iodine, (iii) aqueous copper(II) sulphate solution?

(c) Explain why an aqueous solution of sodium nitrate (of concentration 0.1 mol l^{-1}) has a pH of approximately 7, while an aqueous solution of sodium phosphate(V), Na_3PO_4, of the same concentration has a pH greater than 7.

(d) What is meant by the statement that *phosphorus exhibits polymorphism*? (AEB)

16 Group 6B

16.1 Introduction

Group 6B comprises: oxygen (O), sulphur (S), selenium (Se), tellurium (Te) and polonium (Po). Oxygen, the head element, differs from the remaining members of the group, mainly because of its inability to use d orbitals in bonding (§3.2.4). Some important properties of the elements are given in Table 16.1.

Table 16.1 Some properties of the elements of group 6B

	Oxygen	Sulphur	Selenium	Tellurium	Polonium
character	non-metal	non-metal	non-metal	non-metal	metal
allotropes	oxygen, ozone	α, β and others	α, β and metallic	none	none
atomic number	8	16	34	52	84
relative atomic mass	15.999 4	32.06	78.96	127.60	210†
outer electronic configuration	$2s^2\ 2p^4$	$3s^2\ 3p^4$	$4s^2\ 4p^4$	$5s^2\ 5p^4$	$6s^2\ 6p^4$
valencies	2	2, 4, 6	2, 4, 6	2, 4, 6	2, 4, 6(?)
melting temperature/°C	−218	113(α), 119(β)	217(metallic)	450	254
boiling temperature/°C	−183	445	685	990	960

† Mass number of the most stable isotope.

16.1.1 THE STRUCTURE OF OXYGEN AND SULPHUR

Oxygen

Oxygen exhibits allotropy and exists in two forms, O_2 and O_3, with the systematic names of dioxygen and trioxygen, and the trivial names (which will be used here) of oxygen and ozone, respectively. Although negligible amounts of ozone occur at sea level, it is formed from oxygen in the upper atmosphere by the action of ultraviolet light. In the laboratory ozone is prepared by passing an electrical discharge through oxygen. Sparking should be avoided as this generates heat and decomposes the ozone. At a potential of 20 000 V about 10% of oxygen is converted into ozone to give a mixture known as 'ozonised oxygen'. Pure ozone is obtained as a pale blue gas by the fractional distillation of liquefied ozonised oxygen.

The formation of ozone from oxygen is believed to involve partial atomisation of the latter into free atoms, which then combine with oxygen molecules:

$$O_2 \rightleftharpoons 2O \quad ; \quad O_2 + O \rightleftharpoons O_3$$

The ozone molecule is bent and symmetrical, with delocalised π bonding between the oxygen atoms, i.e.

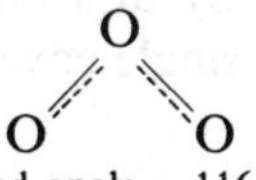

$$O{=}O{=}O \text{ (bent, delocalised)}$$

bond angle = 116.8°
bond length = 0.127 8 nm

Oxygen and ozone in the upper atmosphere form an equilibrium mixture and provide an example of *dynamic allotropy*, although it should be noted that ozonised oxygen is not an equilibrium mixture. Ozone is formed endothermically from oxygen ($\Delta H_f^{\ominus}[O_3] = +142$ kJ mol^{-1}) and thus decomposes exothermically and sometimes explosively to oxygen, particularly if heated. At room temperature the change is slow, especially in the absence of catalysts such as finely divided metals or metal oxides.

Sulphur

Crystalline sulphur exhibits enantiotropy (§14.1.1):

$$\alpha\text{-sulphur} \rightleftharpoons \beta\text{-sulphur}$$

(rhombic sulphur) (monoclinic sulphur)

Below the transition temperature of 95.6 °C (368.8 K) the α-form is stable. The β-form is stable above this temperature up to the melting temperature, 119 °C (392 K). α-Sulphur and β-sulphur differ in the way that the cyclic S_8 molecules, which they both contain, are packed together,

i.e.

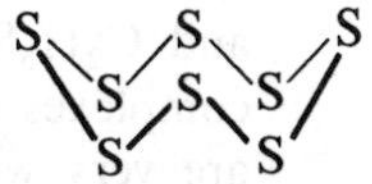

side view of S_8 molecule

Other forms of solid sulphur are known. *Plastic sulphur* contains spiral chains of sulphur atoms, and is obtained by pouring molten sulphur at 200 °C (473 K) into cold water. *Colloidal sulphur* is made by carefully acidifying a solution of sodium thiosulphate (§16.2.6). The rapid cooling of sulphur vapour produces *flowers of sulphur*, which appears to be a mixture of α-sulphur and an amorphous form.

16.1.2 BONDING AND VALENCY

The elements of this group each possess an outer $ns^2\ np^4$ electronic configuration and may participate in ionic or covalent bonding.

Ionic bonding

The elements form dinegative ions by the gain of two electrons, i.e. O^{2-}, S^{2-}, Se^{2-} and Te^{2-}. Many metallic oxides are ionic and contain the oxide ion, O^{2-}, but the remaining elements form ions only in combination with the s-block elements. Consequently, most compounds of sulphur, selenium and tellurium are covalent.

Covalent bonding

By utilising their two unpaired p electrons the elements can form two single covalent bonds or one double bond,

e.g. H–O–H   H–S–H $R_2C{=}O$ $R_2S{=}O$

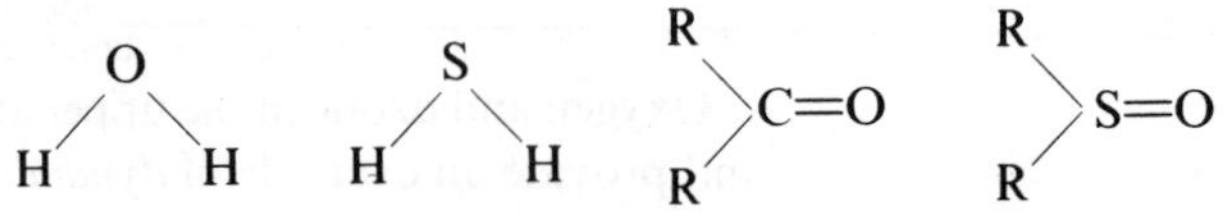

There are several ways in which the elements of this group may possess covalencies greater than two.

(i) A covalently bound oxygen atom can form a coordinate bond by using one of its lone pairs of electrons, thus showing a covalency or coordination number of three. For example,

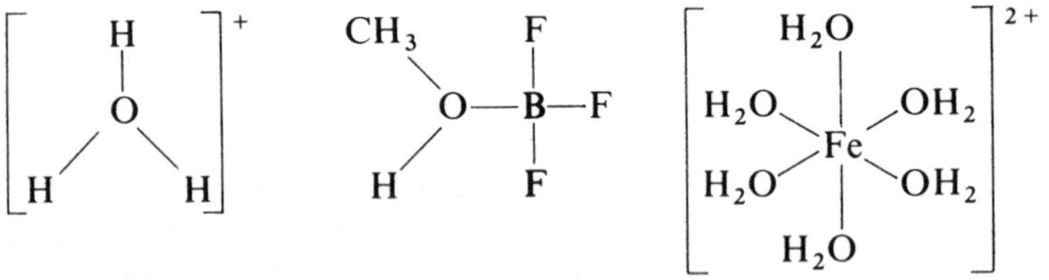

$[H_3O]^+$ $CH_3(H)O{\rightarrow}BF_3$ $[Fe(H_2O)_6]^{2+}$

Sulphur is a much weaker electron pair donor than oxygen and does not form the $[H_3S]^+$ ion. Some sulphur compounds, such as $(CH_3)_2S$ and C_2H_5SH, act as ligands through the sulphur atom and form complexes with transition elements (§18.1.2). Selenium and tellurium are very weak electron pair donors, although some selenium compounds do function as ligands.

(ii) Except for oxygen, the elements of group 6B each possess vacant d orbitals in their outermost shell. By absorbing energy to promote electrons from s or p orbitals into the vacant d orbitals, these elements may participate in four and six covalent bonds as the following scheme for sulphur illustrates.

	3s	3p	3d	number of unpaired electrons (= covalency)
ground state	⇅	⇅ 1 1		2
first excited state	⇅	1 1 1	1	4
second excited state	1	1 1 1	1 1	6

The four and six unpaired electrons of the excited states can then be used in covalent bond formation. Oxygen has no suitable d orbitals and a prohibitive amount of energy would be required for electron promotion into the third shell.

16.2

Oxygen and sulphur

Only oxygen and sulphur, the most important members of this group, will be discussed in detail.

16.2.1 THE OCCURRENCE AND ISOLATION OF OXYGEN AND SULPHUR

Oxygen

Oxygen accounts for approximately 50% of the earth's crust. The atmosphere and water contain 23% and 89% by mass of oxygen respectively, and most of the world's rocks contain combined oxygen.

In the laboratory oxygen may be prepared by one of the following methods.

(i) The decomposition of hydrogen peroxide, either catalytically or by oxidation with the permanganate ion:

$$2H_2O_2 = 2H_2O + O_2 \quad \text{(catalyst of manganese(IV) oxide, } MnO_2\text{)}$$

$$6H^+(aq) + 2MnO_4^- + 5H_2O_2 = 2Mn^{2+} + 5O_2 + 8H_2O$$

(ii) The thermal decomposition of some oxoanions, usually in salts of the alkali metals,

e.g. $$2NO_3^- = 2NO_2^- + O_2 \quad \text{(§15.2.9)}$$

$$2ClO_3^- = 2Cl^- + 3O_2 \quad \text{(catalyst of manganese(IV) oxide, } MnO_2\text{)}$$

(iii) The thermal decomposition of certain metallic oxides,

e.g. $$2HgO = 2Hg + O_2$$

$$2BaO_2 = 2BaO + O_2$$

(iv) The electrolysis of solutions of acids or alkalis. Oxygen is liberated at the anode (§10.2.1).

Industrially, oxygen is obtained by the fractional distillation of liquefied air.

Sulphur

Large quantities of sulphur are obtained from underground deposits of the element in the USA and other parts of the world. Sulphur is also produced in vast quantities from hydrogen sulphide by partial combustion or by reaction with sulphur dioxide:

$$2H_2S + O_2 = 2H_2O + 2S$$

$$2H_2S + SO_2 = 2H_2O + 3S$$

The hydrogen sulphide is obtained from natural hydrocarbon gases – some contain up to 30% of H_2S – or from the sulphur compounds present in crude oil.

16.2.2 THE REACTIONS OF OXYGEN AND SULPHUR

Oxygen (dioxygen)

Oxygen combines with many metals and non-metals to form oxides. In many cases the reaction occurs only on heating and the elements are said to burn. With some elements, for example the majority of the s-block elements, combination with oxygen occurs readily at room temperature. When some of the less reactive metals, e.g. lead or iron, are finely powdered, they react exothermically with oxygen at room temperature, even though a lump of the metal is unaffected under the same conditions. These finely divided metals are said to be *pyrophoric*, and often sufficient heat is liberated during the reaction to set the powder on fire.

Acids and alkalis do not react with oxygen unless, like hydriodic acid (§17.2.4), they are capable of being oxidised.

Many of the other reactions of oxygen are considered in other sections of this book.

Sulphur

Most metals, particularly if heated, react with sulphur to form sulphides. With reactive metals, such as magnesium or aluminium, the reaction is exceptionally vigorous. The non-metals phosphorus, oxygen, carbon and the halogens except for iodine also combine directly with sulphur. For example, sulphur burns in air to form sulphur dioxide and a small amount of sulphur trioxide (§16.2.5).

Oxidising acids, for example concentrated nitric acid or concentrated sulphuric acid, attack sulphur:

$$6HNO_3 + S = 2H_2O + H_2SO_4 + 6NO_2$$

$$2H_2SO_4 + S = 2H_2O + 3SO_2$$

16.2.3 WATER

The shape of the water molecule is discussed in §4.8 and the structure of ice and liquid water in §6.6.1.

Water is thermally very stable, and decomposition into oxygen and hydrogen becomes appreciable only above 1 000 °C (1 273 K). Pure water self-ionises to a very limited extent (§9.2.3) and is therefore only a weak conductor of electricity:

$$H_2O \rightleftharpoons H^+(aq) + HO^-$$

It is amphoteric, and acts as a Brønsted–Lowry acid or base by, respectively, donating or accepting protons.

(i) Water functions as an acid with species that are stronger bases, i.e. stronger proton acceptors, than itself,

e.g. $NH_3 + H_2O \rightleftharpoons NH_4^+ + HO^-$

(ii) Water acts as a base towards compounds that are stronger acids, i.e. stronger proton donors,

e.g. $HCl + H_2O \rightleftharpoons H_3O^+ + Cl^-$

The oxonium ion, H_3O^+, is further hydrated to form the *hydrogen ion*, $H^+(aq)$, (§11.1.1).

Thus, water is an excellent ionising solvent for covalent compounds such as ammonia or hydrogen chloride.

Water possesses weak oxidising and weak reducing properties:

$$2H_2O + 2e^- = H_2 + 2HO^- \qquad E^\ominus = -0.83\ V$$
$$4H^+(aq) + O_2 + 4e^- = 2H_2O \qquad E^\ominus = +1.23\ V$$

Metals with standard electrode potentials that are more negative than −0.83 V can reduce water to hydrogen. In these reactions the metals are oxidised to their oxides or hydroxides. Only the more electropositive s-block elements reduce water at ordinary temperatures, but at higher temperatures metals such as magnesium, zinc or iron are able to reduce steam. Oxidising agents with standard electrode potentials more positive than +1.23 V can, in principle, oxidise water to oxygen, but in practice the reaction is often slow. Fluorine or the hexaaquacobalt(III) ion, however, rapidly oxidises water at room temperature (§17.2.2 and §18.1.2).

16.2.4 HYDROGEN SULPHIDE

H_2S

Hydrogen sulphide is often prepared in the laboratory by the action of dilute hydrochloric acid on iron(II) sulphide:

$$FeS + 2HCl = FeCl_2 + H_2S$$

The product so obtained is contaminated with hydrogen because iron(II) sulphide always contains a small amount of free iron.

Hydrogen sulphide is a colourless gas with an odour of bad eggs. It is highly toxic, more so than hydrogen cyanide, but because of its strong odour it can be detected even in minute concentrations. However, hydrogen sulphide very quickly kills the sense of smell and should therefore be used with extreme caution.

The hydrogen sulphide molecule has a bond angle of 93°. The electronegativity of sulphur is not high enough to enable hydrogen bonding to occur, and as a result hydrogen sulphide shows none of the anomalous properties of water (§6.6). Because the S—H bond is weaker than the O—H bond, hydrogen sulphide is less stable than water. For example, it decomposes into sulphur and hydrogen at approximately 800 °C (1 073 K), and burns in an excess of air:

$$2H_2S + 3O_2 = 2H_2O + 2SO_2$$

It is moderately soluble in cold water to give weakly acidic solutions:

$$H_2S + H_2O \rightleftharpoons H_3O^+ + HS^- \qquad pK_1 = 7.05$$
$$HS^- + H_2O \rightleftharpoons H_3O^+ + S^{2-} \qquad pK_2 = 13.92$$

It is therefore a weak diprotic acid and forms two series of salts with alkalis,

e.g. $H_2S + NaOH = H_2O + NaHS$ (sodium hydrogensulphide)
$H_2S + 2NaOH = 2H_2O + Na_2S$ (sodium sulphide)

Because hydrogen sulphide is readily oxidised to sulphur by various oxidising agents, it is considered to be a good reductant.

e.g. $X_2 + H_2S = 2HX + S$ (X = Cl, Br or I)

$$2MnO_4^- + 6H^+(aq) + 5H_2S = 2Mn^{2+} + 8H_2O + 5S$$

$$2Fe^{3+} + H_2S = 2Fe^{2+} + 2H^+(aq) + S$$

Even sulphur dioxide, which is usually considered to be a reducing agent, oxidises hydrogen sulphide, particularly in the presence of water:

$$SO_2 + 2H_2S = 2H_2O + 3S$$

Aqueous solutions of hydrogen sulphide are slowly oxidised in air:

$$2H_2S + O_2 = 2H_2O + 2S$$

16.2.5 THE OXIDES OF SULPHUR

Sulphur dioxide

SO_2

Sulphur dioxide, SO_2, is a colourless, toxic gas with a characteristic choking odour. It is formed by burning sulphur in air, although a small proportion of sulphur trioxide, SO_3, which is produced simultaneously, gives the gas a white smoky appearance. In the laboratory pure sulphur dioxide is prepared by reacting a sulphite or a hydrogensulphite with dilute hydrochloric acid, or by reducing hot concentrated sulphuric acid with metals:

$$SO_3^{2-} + 2H^+(aq) = SO_2 + H_2O$$

$$HSO_3^- + H^+(aq) = SO_2 + H_2O$$

$$Cu + 2H_2SO_4 = CuSO_4 + 2H_2O + SO_2$$

Copper is a useful metal in this respect, because it does not form hydrogen by reacting with the acid as it becomes diluted by the water produced in the reaction.

Large quantities of sulphur dioxide are manufactured for conversion into sulphuric acid (§16.2.6).

The sulphur dioxide molecule is bent,

i.e.

bond angle = 119.5°
bond length = 0.143 nm
(××) represents a lone pair of electrons

The compound is thermally very stable, as indicated by its high enthalpy of formation; $\Delta H_f^\ominus[SO_2(g)] = -297$ kJ mol^{-1}.

Sulphur dioxide is readily oxidised and is thus a powerful reducing agent. In many cases the oxidation product is the sulphate ion, although with chlorine, sulphuryl chloride, SO_2Cl_2, is formed,

e.g. $$5SO_2 + 2MnO_4^- + 2H_2O = 5SO_4^{2-} + 4H^+(aq) + 2Mn^{2+}(aq)$$

$$SO_2 + Cl_2 = SO_2Cl_2 \quad \text{(with a camphor catalyst)}$$

SO_3

Sulphur trioxide (ASE: sulphur(VI) oxide)

Sulphur trioxide, SO_3, may be prepared either by dehydrating concentrated sulphuric acid with phosphorus(V) oxide, or by heating sulphates:

$$P_4O_{10} + 2H_2SO_4 = 4HPO_3 + 2SO_3$$

$$Fe_2(SO_4)_3 = Fe_2O_3 + 3SO_3$$

Industrially, large quantities of sulphur trioxide are produced for the manufacture of sulphuric acid (§16.2.6).

In the vapour state, above 60 °C (333 K), sulphur trioxide exists as trigonal planar molecules, but in the solid state there are at least three polymeric forms, known as α-, β- and γ-sulphur trioxide.

$$\left[-\overset{O}{\underset{O}{S}}-O-\overset{O}{\underset{O}{S}}-O- \right]_n \quad \beta\text{-}SO_3$$

γ-SO_3

α-SO_3 is believed to consist of cross-linked chains made up of alternating sulphur and oxygen atoms.

Sulphur trioxide is extremely reactive and strongly acidic. It combines exothermically with water, to produce sulphuric acid, and with basic oxides to form sulphates,

e.g. $$SO_3(g) + H_2O(l) = H_2SO_4(l) \qquad \Delta H^\ominus = -130 \text{ kJ mol}^{-1}$$

$$SO_3(g) + MgO(s) = MgSO_4(s) \qquad \Delta H^\ominus = -281 \text{ kJ mol}^{-1}$$

16.2.6 THE OXOACIDS OF SULPHUR AND THEIR SALTS

'Sulphurous acid' and sulphites

Sulphur dioxide dissolves in water to give an acidic solution which has long been called 'sulphurous acid' ('H_2SO_3'). However, there is no evidence for the existence of H_2SO_3 molecules, and solutions of sulphur dioxide are believed to contain hydrated molecules, i.e. $SO_2(aq)$, which ionise:

$$SO_2(aq) + H_2O \rightleftharpoons H^+(aq) + HSO_3^- \qquad pK_1 = 1.92$$

$$HSO_3^- + aq \rightleftharpoons H^+(aq) + SO_3^{2-} \qquad pK_2 = 7.21$$

Thus, in solution sulphur dioxide behaves as a weak diprotic acid and forms two series of salts, one containing the sulphite ion, SO_3^{2-}, and the other the hydrogensulphite ion, HSO_3^-. The common alkali metal salts are prepared by passing sulphur dioxide into solutions of alkalis,

e.g. $2NaOH + SO_2 = Na_2SO_3 + H_2O$ (with an excess of alkali)

$NaOH + SO_2 = NaHSO_3$ (with an excess of SO_2)

All sulphites decompose into sulphur dioxide when treated with strong non-oxidising acids,

i.e. $SO_3^{2-} + 2H^+(aq) = SO_2 + H_2O$

$HSO_3^- + H^+(aq) = SO_2 + H_2O$

Acidic or alkaline solutions of sulphites or hydrogensulphites are powerful reductants. Although in alkaline solution the reducing properties are due to the sulphite ion, in acidic solution they arise from the presence of sulphur dioxide:

$$SO_4^{2-} + 4H^+(aq) + 2e^- = SO_2(aq) + 2H_2O \qquad E^\ominus = +0.17 \text{ V}$$

$$SO_4^{2-} + H_2O + 2e^- = SO_3^{2-} + 2HO^- \qquad E^\ominus = -0.93 \text{ V}$$

The standard electrode potentials (§2.5.3) indicate that the sulphites are stronger reductants in alkaline solutions than sulphur dioxide is in acidic solutions, and the equations show that the oxidation product in both cases is the sulphate ion,

e.g. $SO_3^{2-} + 2Fe^{3+}(aq) + H_2O = SO_4^{2-} + 2Fe^{2+}(aq) + 2H^+(aq)$

$3SO_2 + Cr_2O_7^{2-} + 2H^+(aq) = 3SO_4^{2-} + 2Cr^{3+}(aq) + H_2O$

The last reaction is accompanied by a colour change from orange ($Cr_2O_7^{2-}$ ion) to green (complex chromium(III) ion) and is used as a test for sulphur dioxide. The test is performed by exposing to the gas a filter paper soaked in an acidified dichromate solution.

Sulphur dioxide and sulphites act as oxidants towards more powerful reducing agents, such as hydrogen sulphide (§16.2.4).

White barium sulphite is precipitated from neutral solutions of sulphites by the addition of barium chloride. Unlike barium sulphate, the sulphite is soluble in hydrochloric acid, a property which enables sulphites to be distinguished from sulphates:

$$BaSO_3 + 2H^+(aq) = Ba^{2+} + H_2O + SO_2$$

Sulphuric acid

The contact process

The *contact process*, which is used to manufacture huge quantities of sulphuric acid, involves the catalytic oxidation of sulphur dioxide to sulphur trioxide. Sulphur dioxide is obtained: (i) by burning sulphur in air (§16.2.5), (ii) as a by-product of some metal extraction processes in which a metal sulphide ore is roasted in air (e.g. §18.11.1), or (iii) by combustion of the hydrogen sulphide extracted from natural gas or petroleum.

Impurities, which can poison the catalyst, are removed by washing the gas with water and passing it through an electric field. (Solid particles are electrostatically precipitated.) The purified sulphur dioxide is then mixed with a slight excess of air and passed into the *converter* which contains the catalyst. Here, the oxidation of sulphur dioxide occurs:

$$SO_2(g) + \tfrac{1}{2}O_2(g) \rightleftharpoons SO_3(g) \qquad \Delta H^\ominus = -98 \text{ kJ mol}^{-1}$$

A 'promoted' vanadium catalyst is used, comprising an alkali metal sulphate and either vanadium(V) oxide, V_2O_5, or a vanadate (§18.4.5) on an inert support of silica. The reaction temperature is 440 °C (713 K).

Sulphur dioxide vapour reacts with pure water to form a mist of sulphuric acid which is slow to settle. However, it combines readily with the 2 per cent of water in 98 per cent sulphuric acid without forming a mist. The cooled vapour from the converter is therefore absorbed in this medium, to which water is added continuously to maintain the concentration at 98 per cent.

H_2SO_4

The physical properties of sulphuric acid

Pure sulphuric acid is a colourless, viscous liquid with a density of 1.84 g cm^{-3} and a freezing temperature of 10 °C (283 K). At its boiling temperature, 290 °C (563 K), sulphuric acid decomposes into sulphur trioxide and water. 'Concentrated sulphuric acid' is a constant boiling mixture (§8.4.3) containing approximately 98% of the acid. The high viscosity and high boiling temperature of sulphuric acid are attributed to the presence of extensive hydrogen bonding between neighbouring molecules,

i.e.

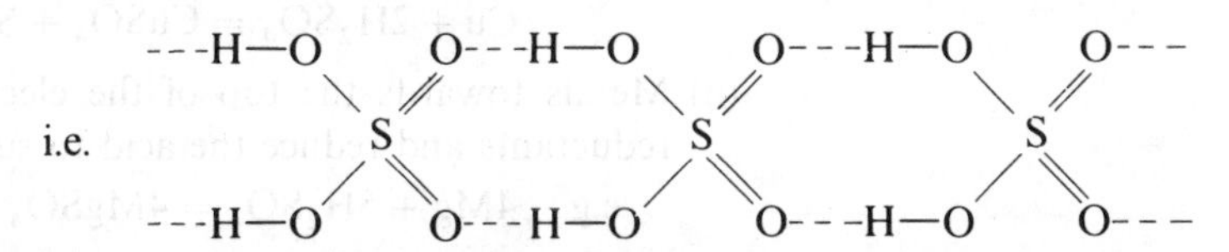

The acidic properties of sulphuric acid

Pure sulphuric acid is a covalent compound. In dilute aqueous solution it is extensively ionised and behaves as a diprotic acid:

$$H_2SO_4 + H_2O \rightleftharpoons H_3O^+ + HSO_4^-$$
$$HSO_4^- + H_2O \rightleftharpoons H_3O^+ + SO_4^{2-}$$

The first ionisation stage is virtually complete, i.e. sulphuric acid is a strong acid, but the ionisation of the hydrogensulphate ion, HSO_4^-, is incomplete; $pK_2 = 1.92$. A dilute solution of sulphuric acid (approximately 2M) therefore contains hydrogen ions and hydrogensulphate ions together with a small proportion of sulphate ions. Dilute sulphuric acid possesses typical acidic properties; e.g. it forms salts with bases, liberates carbon dioxide from carbonates, and reacts with metals above hydrogen in the electrochemical series to produce hydrogen.

The affinity of sulphuric acid for water

When concentrated sulphuric acid is diluted with water a large amount of heat is produced as the ions formed become hydrated. Concentrated sulphuric acid therefore has a strong affinity for water and is used as a drying agent and a dehydrating agent, i.e. a catalyst for the elimination of water. The uses of concentrated sulphuric acid include:

(i) the drying of gases, such as chlorine, oxygen and sulphur dioxide;
(ii) the removal of water of crystallisation from hydrated salts;
(iii) the removal of hydrogen and oxygen in the form of water from compounds which contain no water, e.g. methanoic acid (§14.2.2), alcohols (§14.9) and ethanedioic acid (§14.2.2).

The oxidising properties of sulphuric acid

Hot concentrated sulphuric acid is a weak oxidising agent, but this property is rapidly lost on dilution. The reduction product of the acid may be sulphur dioxide, sulphur or hydrogen sulphide, depending on the strength of the reducing agent with which the acid reacts.

$$2H_2SO_4 + 2e^- = SO_2 + 2H_2O + SO_4^{2-}$$

$$4H_2SO_4 + 6e^- = S + 4H_2O + 3SO_4^{2-}$$

$$5H_2SO_4 + 8e^- = H_2S + 4H_2O + 4SO_4^{2-}$$

The sulphate ion is not a reduction product, but originates from sulphuric acid molecules which act as proton suppliers.

(i) Non-metals or metals low in the electrochemical series are weak reductants and reduce sulphuric acid to sulphur dioxide,

e.g. $$C + 2H_2SO_4 = CO_2 + 2SO_2 + 2H_2O$$

$$Cu + 2H_2SO_4 = CuSO_4 + SO_2 + 2H_2O$$

(ii) Metals towards the top of the electrochemical series are powerful reductants and reduce the acid to sulphur or hydrogen sulphide,

e.g. $$4Mg + 5H_2SO_4 = 4MgSO_4 + 4H_2O + H_2S$$

$$3Zn + 4H_2SO_4 = 3ZnSO_4 + 4H_2O + S$$

(iii) Hydrogen bromide and hydrogen iodide reduce sulphuric acid (§17.2.4).

The displacement reactions of sulphuric acid

Sulphuric acid has a low volatility and can displace more volatile acids from their salts. For example, when concentrated sulphuric acid is added to a crystalline ionic chloride at room temperature the following equilibrium is established:

$$Cl^- + H_2SO_4 \rightleftharpoons HSO_4^- + HCl$$

If the gaseous hydrogen chloride is allowed to escape the equilibrium is disturbed towards the right, i.e. hydrochloric acid is displaced from its salts by a less volatile acid. (The sulphate ion, SO_4^{2-}, is formed only at much higher temperatures; see below.)

A similar reaction occurs with nitrates:

$$NO_3^- + H_2SO_4 \rightleftharpoons HSO_4^- + HNO_3$$

except that nitric acid is less volatile than hydrogen chloride and is lost from the equilibrium mixture only on heating. The reaction between sulphuric acid and ionic bromides and iodides is discussed in §17.2.4.

The reactions of sulphuric acid with organic compounds

Many organic compounds, including alkenes (§14.4.4), alkynes (§14.5.4), alcohols (§14.9.5) and arenes (§14.7.4) react with sulphuric acid.

The salts of sulphuric acid

Sulphuric acid is diprotic and forms two series of salts. Sulphates contain the tetrahedral SO_4^{2-} ion, and hydrogensulphates (formerly bisulphates) contain the HSO_4^- ion which, like the acid, has a distorted tetrahedral structure:

Most sulphates are soluble in water and are prepared by dissolving a metal or its oxide, hydroxide or carbonate in sulphuric acid. Insoluble sulphates, i.e. those of calcium, strontium, barium, lead and mercury(I), are prepared by metathesis (double decomposition) reactions,

e.g. $BaCl_2 + H_2SO_4 = BaSO_4(s) + 2HCl$

Many of the soluble metal sulphates are hydrated, e.g. $NiSO_4{\cdot}6H_2O$, with the water molecules coordinated on to the cation. However, in some cases, e.g. $CuSO_4{\cdot}5H_2O$ and $FeSO_4{\cdot}7H_2O$, water molecules are also attached to the sulphate ion by hydrogen bonds (§6.6.2).

The only common hydrogensulphates are those of the alkali metals, and ammonium hydrogensulphate. The crystalline salts may be prepared by:

$NaHSO_4$ and $KHSO_4$

(i) reacting an alkali and sulphuric acid together in equimolar amounts, followed by evaporation,

e.g. $NaOH + H_2SO_4 = NaHSO_4 + H_2O$

(ii) the evaporation of a solution containing equimolar quantities of a sulphate and sulphuric acid,

e.g. $Na_2SO_4 + H_2SO_4 = 2NaHSO_4$

(iii) the action of warm concentrated sulphuric acid on a chloride (see above).

Hydrogensulphates are acidic. They ionise in solution:

$$HSO_4^- + H_2O \rightleftharpoons H_3O^+ + SO_4^{2-}$$

and they displace hydrogen chloride from chlorides at temperatures above 500 °C (773 K):

$$NaHSO_4 + NaCl \rightleftharpoons Na_2SO_4 + HCl(g)$$

Tests for sulphates

The addition of barium chloride to a solution of a sulphate or a hydrogensulphate produces a white precipitate of barium sulphate, which, unlike barium sulphite, is insoluble in hydrochloric acid. Hydrogensulphates can be distinguished from sulphates because they:

(i) liberate carbon dioxide from soluble carbonates, without the formation of a precipitate (see below);
(ii) react with zinc to form hydrogen (see below).

Some sulphates, e.g. aluminium sulphate, also liberate carbon dioxide from carbonates and produce hydrogen with zinc, but a precipitate of the metal hydroxide is formed too (§13.2.4).

Thiosulphates

Thiosulphates are prepared by boiling a solution of a sulphite with sulphur. Sodium thiosulphate, $Na_2S_2O_3$, is commonly used in the laboratory for the volumetric analysis of iodine (§17.2.2):

$$2S_2O_3^{2-} + I_2 = 2I^- + S_4O_6^{2-} \quad \text{tetrathionate ion}$$

In acidic solution the thiosulphate ion decomposes into sulphur dioxide and a yellow precipitate of sulphur. This is a disproportionation reaction:

$$S_2O_3^{2-} + 2H^+(aq) = H_2O + SO_2 + S$$

oxidation number of S: +2 ... +4 0

$SOCl_2$

16.2.7 THIONYL CHLORIDE (ASE: SULPHUR DICHLORIDE OXIDE)

Thionyl chloride is a colourless liquid which fumes in air because of the rapid hydrolysis reaction which occurs with water vapour:

$$SOCl_2 + H_2O = SO_2 + 2HCl$$

A similar reaction occurs with liquid water and with coordinated water in hydrated metal salts. For this reason anhydrous metal chlorides can be prepared by heating the hydrated salts with thionyl chloride,

e.g. $CrCl_3{\cdot}6H_2O + 6SOCl_2 = CrCl_3 + 6SO_2(g) + 12HCl(g)$

After completion of the reaction excess thionyl chloride is removed by distillation, preferably in a vacuum.

Thionyl chloride is used in organic chemistry for replacing hydroxyl groups by chlorine atoms (§14.9.5).

16.2.8 THE STABILISATION OF HIGH OXIDATION STATES BY OXYGEN

A number of oxides and oxoanions are known which contain an element in an unusually high oxidation state. (For most purposes oxidation state is equal numerically to valency, §10.1.2.) For example,

ionic compounds: $Mn^{IV}O_2$, $Pb^{IV}O_2$;

covalent compounds: $Mn_2^{VII}O_7$, $Cr^{VI}O_3$, $S^{VI}O_3$, $Cl_2^{VII}O_7$, together with their derived oxoanions, $Mn^{VII}O_4^-$, $Cr^{VI}O_4^{2-}$, $Cr_2^{VI}O_7^{2-}$, $S^{VI}O_4^{2-}$ and $Cl^{VII}O_4^-$.

No other element, apart from fluorine, has the same ability as oxygen to stabilise these high oxidation states. There is no single reason for this property and we must discuss ionic and covalent species separately.

Ionic oxides

Oxides such as manganese(IV) oxide, $Mn^{4+}(O^{2-})_2$, possess a high lattice enthalpy because of the attraction between the highly charged cations and the relatively small doubly charged oxide ions. In this case the lattice enthalpy:

i.e. $Mn^{4+}(g) + 2O^{2-}(g) = MnO_2(s) \quad \Delta H^\ominus = -13\,390 \text{ kJ mol}^{-1}$

is high enough to compensate for the large amount of energy required for ion formation:

$$Mn(s) = Mn^{4+}(g) + 4e^- \quad \Delta H^\ominus = +10\,945 \text{ kJ mol}^{-1}$$

$$O_2(g) + 4e^- = 2O^{2-}(g) \quad \Delta H^\ominus = +1\,924 \text{ kJ mol}^{-1}$$

Manganese(IV) oxide is therefore stable, as its enthalpy of formation indicates:

$$Mn(s) + O_2(g) = MnO_2(s) \quad \Delta H^\ominus = -521 \text{ kJ mol}^{-1}$$

Similar arguments apply to lead(IV) oxide and other ionic oxides of metals in relatively high oxidation states.

Covalent oxides and oxoanions

The stabilisation of these species is due to three factors.

The small size of the oxygen atom

Oxygen atoms are relatively small and pack easily around a central atom so that strong covalent bonds can be formed. It is more difficult for larger atoms to do this.

Oxygen forms strong covalent bonds with the atoms of other elements

Generally, for a compound to be stable, the energy produced during covalent bond formation should be in excess of that required to atomise the elements and raise the atoms to the necessary excited states. For example, in the formation of sulphur trioxide, the energy produced during the formation of strong sulphur–oxygen bonds is greater than that required for atomisation of oxygen and sulphur, and excitation of sulphur atoms from the configuration $3s^2\ 3p^4$ to $3s^1\ 3p^3\ 3d^2$ (§16.1.2). As a result, sulphur trioxide is an exothermic compound: $\Delta H_f^{\ominus}[SO_3(g)] = -395\ kJ\ mol^{-1}$. Thus, the +6 oxidation (or hexavalent) state of sulphur is stabilised by oxygen because of the great strength of the sulphur–oxygen bond. Other elements, except for fluorine, form weaker bonds with sulphur and are unable to stabilise the +6 oxidation state. Similar principles apply to the oxides of other elements in very high oxidation states and also to anions; for example, MnO_4^- exists but not MnS_4^-, because the manganese–oxygen bond is stronger than the manganese–sulphur bond.

The high electronegativity of oxygen

An atom in a high oxidation state has a great affinity for electrons and therefore usually possesses oxidising properties. For example, the permanganate ion is an oxidant because of the strong electron attracting tendency of the manganese atom in a +7 oxidation state. In the absence of a reductant, an atom in a high oxidation state will attempt to obtain electrons from the nearest available source, such as the oxygen atoms to which it is bonded in an oxide or an oxoanion. Because of the high electronegativity of oxygen, this transfer of electrons is prevented and the high oxidation state of the element is stabilised.

EXAMINATION QUESTIONS ON CHAPTER 16

1. (a) Name two substances containing sulphur from which sulphur dioxide is prepared industrially. **Outline** the method of production in each case.

(b) Compare the bleaching action of sulphur dioxide with that of chlorine.

(c) Give examples, and explanations, of the following properties of concentrated sulphuric acid:

(i) its dehydrating properties,

(ii) its use as a sulphonating agent.

(d) Why has sulphuric acid a higher boiling point (384 °C) than would be expected for a covalent compound of formula H_2SO_4? (AEB)

2. (a) Compare the properties of water and hydrogen sulphide, giving explanations of any similarities or differences that you quote.

(b) Comment on the following observations.

(i) Oxygen normally exists in the form of diatomic molecules, whereas sulphur normally exists in the form of S_8 molecules.

(ii) Oxygen does not form a hexafluoride analogous to SF_6.

(iii) There are no compounds containing the O^- or S^- ions although the electron affinities of oxygen and sulphur are $-142\ kJ\ mol^{-1}$ and $-200\ kJ\ mol^{-1}$ respectively.

(iv) Oxygen has a higher electronegativity than sulphur although more energy is released when a sulphur atom accepts an electron than when an oxygen atom accepts an electron. (OC)

3. (a) Describe the manufacture of sulphuric acid, explaining carefully the reasons for the chemical and physical conditions used in each stage of the process. Details of the industrial plant are **not** required.

(b) Give two industrial uses of sulphuric acid.

(c) How and under what conditions does sulphuric acid react with (i) potassium bromide, (ii) oxalic acid (ethanedioic acid), (iii) hydrogen sulphide? (O)

4. (a) Describe, with full practical instructions, how you would determine the concentration (g dm^{-3}) of a solution of hydrogen peroxide using the quantitative reactions:

$$H_2O_2 + 2H^+ + 2I^- \rightarrow I_2 + 2H_2O$$
$$I_2 + 2S_2O_3^{2-} \rightarrow 2I^- + S_4O_6^{2-}$$

The usual titrimetric (volumetric) apparatus is available together with aqueous solutions of potassium iodide, sulphuric acid, starch and 0.10 M sodium thiosulphate ($Na_2S_2O_3$).

(b) Show how you would calculate the result from the experimental data in (a).
(H = 1, O = 16)

(c) Sodium thiosulphate can be prepared by boiling sodium sulphate(IV) (sodium sulphite) solution with excess of sulphur. Outline, with equations and reaction conditions, how you would prepare sodium thiosulphate pentahydrate using elemental sulphur as the only sulphur-containing material in your scheme.

(d) Compare, and comment on, the relative volatilities of water and hydrogen sulphide in terms of their respective molar masses
(H = 1, O = 16, S = 32). (SUJB)

5. The following equilibrium is involved in the industrial preparation of sulphuric acid.

$$2SO_2(g) + O_2(g) \rightleftharpoons 2SO_3(g); \qquad \Delta H = -188 \text{ kJ mol}^{-1}.$$

(a) Write the expression for the equilibrium constant K_p for the above equilibrium.

(b) State the effect upon the equilibrium of, and give an explanation for, each of the following changes.

(i) Increase of pressure at constant temperature.

(ii) Increase of temperature at constant pressure.

(c) Give approximate values of the temperature and pressure which are used in the industrial preparation of sulphur(VI) oxide. (JMB)

Group 7B

17.1 Introduction

F_2, Cl_2 and Br_2

I_2

The elements of group 7B, commonly called 'halogens', are fluorine (F), chlorine (Cl), bromine (Br), iodine (I) and astatine (At). Fluorine, the head element of the group, differs appreciably from chlorine, bromine and iodine, which show distinct chemical similarities. Astatine is radioactive and does not occur naturally (§17.2.7). Some properties of the elements are shown in Table 17.1.

Table 17.1 Some properties of the elements of group 7B

	Fluorine	Chlorine	Bromine	Iodine	Astatine
character	non-metal	non-metal	non-metal	non-metal	non-metal
allotropes	none	none	none	none	none
atomic number	9	17	35	53	85
relative atomic mass	18.998 4	34.453	79.904	126.904 5	210†
outer electronic configuration	$2s^2\ 2p^5$	$3s^2\ 3p^5$	$4s^2\ 4p^5$	$5s^2\ 5p^5$	$6s^2\ 6p^5$
valencies	1	1, 3, 5, 7	1, 3, 5, 7	1, 3, 5, 7	1, 3, 5, 7(?)
melting temperature/°C	−220	−101	−7.2	114	302
boiling temperature/°C	−188	−34.7	58.8	184	not known

† Mass number of the most stable isotope.

In general discussions the symbol X is often used to designate a halogen atom. Compounds of the halogens are referred to as 'halides'.

17.1.1 THE STRUCTURE OF THE ELEMENTS

The halogens all consist of diatomic molecules, i.e. F_2, Cl_2, Br_2 and I_2, in which the two halogen atoms are joined by a single covalent bond. The structure of crystalline iodine is shown in Fig. 6.18.

Fluorine is a pale yellow gas, chlorine is a pale green gas, bromine is a dark brown volatile liquid, and iodine is a black shiny solid which sublimes readily on heating to form a violet vapour.

17.1.2 BONDING AND VALENCY

Each element of group 7B has seven outer electrons (i.e. $ns^2\ np^5$) and can participate in ionic or covalent bonding.

Ionic bonding

Halide ions

All the halogens form mononegative ions by accepting one electron into the singly filled outer p orbital. These ions, i.e. fluoride (F^-), chloride (Cl^-), bromide (Br^-) and iodide (I^-), are collectively referred to as 'halide ions'. The halides of metals in low valency (or oxidation) states are mostly ionic, but in high valency states metal halides tend to be polar covalent.

Halogen cations

Iodine, the least electronegative of the common halogens, is present in some of its compounds as a cationic species, e.g. $[(C_5H_5N)_2I]^+$, which contains an I^+ ion stabilised by the coordination of two molecules of pyridine, C_5H_5N.

Covalent bonding

A covalency of one is reached by using the unpaired electron in the outer p orbital to form a single covalent bond. The halides of all non-metals and some metals are covalent, and the halogen atom may have an oxidation state of $+1$ or -1, depending on the electronegativity of the element to which it is bonded. Fluorine, however, is the most electronegative of all the elements, and in compounds always has an oxidation state of -1. Single covalent bonds exist in the halogen molecules, but here the oxidation state of each atom is zero.

Except for fluorine, the halogens show valencies of greater than one by utilisation of the d orbitals in the outermost shell, in a manner similar to that discussed for sulphur (§16.1.2). Thus, by promotion of electrons from the outer p and s orbitals into vacant d orbitals in the same shell, valencies of 3, 5 and 7 are reached, as the following scheme for iodine illustrates:

	5s	5p			5d					number of unpaired electrons (= covalency)
ground state	⇅	⇅	⇅	1						1
first excited state	⇅	⇅	1	1	1					3
second excited state	⇅	1	1	1	1	1				5
third excited state	1	1	1	1	1	1	1			7

The additional energy produced by the formation of the extra covalent bonds usually compensates for the energy absorbed during promotion. Because there are no 2d orbitals, fluorine could achieve higher covalencies only by promotion of electrons from the second shell to vacant orbitals in the third shell. The excessive amount of energy required, however, is not offset by the energy liberated by the formation of additional covalent bonds, and fluorine is therefore restricted to a valency of one. The species in Table 17.2 illustrate the various valencies that the halogens may possess. Notice that valencies of 3, 5 or 7 occur only in combinations of the halogens with elements of higher electronegativity, usually oxygen or fluorine, and the oxidation state is always $+3$, $+5$ or $+7$.

Table 17.2 Examples of the various valency and oxidation states of the halogens

All the halogens can show a covalency or coordination number of two by forming a single covalent bond with one atom and a coordinate bond with another, e.g. aluminium chloride (§13.2.6).

Valency	Oxidation state	Fluorine	Chlorine	Bromine	Iodine
1	−1	HF, F^-	HCl, Cl^-	HBr, Br^-	HI, I^-
1	+1	—	HClO, Cl_2O, ClF	HBrO, Br_2O, BrF	HIO, ICl, IBr
3	+3	—	$HClO_2$, ClF_3	BrF_3	ICl_3
5	+5	—	$HClO_3$	$HBrO_3$, BrF_5	HIO_3, IF_5, I_2O_5
7	+7	—	$HClO_4$, Cl_2O_7	$HBrO_4$	HIO_4, IF_7

17.2 Fluorine, chlorine, bromine and iodine

17.2.1 THE OCCURRENCE AND ISOLATION OF THE ELEMENTS

The halogens are far too reactive to occur in the uncombined state in nature, but compounds of the halogens are reasonably abundant. Fluorspar, cryolite and apatite, CaF_2, $Na_3[AlF_6]$ and $3Ca_3(PO_4)_2{\cdot}CaF_2$ respectively, are the principal sources of fluorine. Chlorides and bromides occur extensively in sea water and dried-up salt lake deposits. Iodine compounds occur in some species of seaweed, but the most important source of iodine is sodium iodate, $NaIO_3$, which constitutes up to 5% of the huge deposits of Chile saltpetre (sodium nitrate) in South America.

The free halogens are usually obtained by the oxidation of halide ions,

i.e. $2X^- = X_2 + 2e^-$

Fluorine

Fluorine is the strongest known oxidising agent and cannot, therefore, be prepared by the chemical oxidation of fluoride ions. Fluoride ions can be oxidised only by electrolysis, which must be conducted under anhydrous conditions because water, if present, would be preferentially oxidised to oxygen. Fluorine is conveniently prepared by dissolving potassium fluoride in anhydrous liquid hydrogen fluoride to make the latter electrically conducting. With a HF to KF ratio of 2:1 the electrolyte melts at around 100 °C (373 K) and fluorine is liberated at the anode during electrolysis.

Chlorine

Chlorine, an important industrial chemical, is manufactured by several processes, including:

(i) the electrolysis of a concentrated solution of sodium chloride;

(ii) the oxidation of hydrogen chloride, obtained as a by-product from chlorination reactions (§14.7.3), with atmospheric oxygen at 400 °C (673 K) in the presence of a copper(II) chloride catalyst:

$$4HCl + O_2 = 2H_2O + 2Cl_2$$

Alternatively, the hydrogen chloride is dissolved in water and the resulting solution of hydrochloric acid electrolysed to yield chlorine at the anode;

(iii) the electrolysis of the fused chlorides of sodium or magnesium (§12.2.1).

In the laboratory, chlorine is usually prepared by the oxidation of chloride ions under acidic conditions. For example, chlorine is produced when manganese(IV) oxide, MnO_2, or lead(IV) oxide, PbO_2, is warmed with either concentrated hydrochloric acid or a mixture of an ionic chloride and concentrated sulphuric acid:

$$MO_2 + 4H^+(aq) + 2Cl^- = M^{2+} + 2H_2O + Cl_2 \qquad (M = Pb \text{ or } Mn)$$

Potassium permanganate oxidises concentrated hydrochloric acid to chlorine

$$2MnO_4^- + 16H^+(aq) + 10Cl^- = 2Mn^{2+} + 8H_2O + 5Cl_2$$

but the reaction is sometimes explosive, possibly because of the formation of unstable oxides of chlorine.

Chlorine may also be prepared in the laboratory by the action of dilute hydrochloric acid on hypochlorites such as bleaching powder, which is a complex mixture (§17.2.2). The reaction may be represented as

$$Ca(ClO)_2 + 4H^+(aq) + 2Cl^- = Ca^{2+} + 2H_2O + 2Cl_2$$

Bromine

Industrially, bromine is recovered from the bromides in sea water by oxidation with chlorine at a pH of about 3.5:

$$2Br^- + Cl_2 = 2Cl^- + Br_2$$

Bromine may be prepared in the laboratory by warming a mixture of potassium bromide, manganese(IV) oxide and concentrated sulphuric acid:

$$MnO_2 + 2H_2SO_4 + 2Br^- = Mn^{2+} + 2H_2O + Br_2 + 2SO_4^{2-}$$

Iodine

Iodine is obtained commercially from the sodium iodate which remains in solution after the crystallisation of sodium nitrate from Chile saltpetre. In the laboratory it is usually prepared by warming a mixture of potassium iodide, manganese(IV) oxide and concentrated sulphuric acid:

$$MnO_2 + 2H_2SO_4 + 2I^- = Mn^{2+} + 2H_2O + I_2 + 2SO_4^{2-}$$

17.2.2 THE REACTIONS OF THE HALOGENS

The halogens are highly reactive, but there is a decrease in reactivity from fluorine to iodine.

Reaction with other elements

Fluorine, because of its high reactivity, is rather difficult to handle. It combines directly with all other elements except nitrogen, helium, neon and argon. In many cases the reaction occurs at room temperature, but with the more noble elements heat is required. For example, xenon tetrafluoride, XeF_4, is obtained as colourless crystals by heating a mixture of fluorine and xenon in the ratio of 5:1 to 400 °C (673 K) under a pressure of 6 atm (600 kPa).

Many elements reach higher oxidation states in combination with fluorine than with the other halogens, as the following examples show:

AgF_2	CoF_3	AsF_5	PtF_6	SF_6	IF_7
$AgCl$	$CoCl_2$	$AsCl_3$	$PtCl_4$	SCl_4	ICl_3

Fluorine, like oxygen, is able to stabilise elements in high oxidation states and for similar reasons, with one additional feature – the weakness of the bond in the F_2 molecule.

Chlorine, especially when heated, combines directly with most of the common metals and non-metals, with the exception of carbon, oxygen, nitrogen, helium, neon, argon and krypton. If more than one product is possible, and the chlorine is present in excess, the higher chloride is usually formed, e.g. PCl_5 (not PCl_3) and $FeCl_3$ (not $FeCl_2$). Bromine and iodine are similar in many respects to chlorine except that they do not react with any of the noble gases; furthermore, iodine does not combine with boron.

The reactions of the halogens with hydrogen to produce hydrogen halides provide an excellent illustration of the decrease in reactivity from fluorine to iodine. Fluorine and hydrogen combine explosively, even at −253 °C (20 K). Chlorine and hydrogen react very slowly at room temperature in the dark, but in the presence of sunlight or ultraviolet light an explosive free radical chain reaction occurs; cf. the reaction between chlorine and methane (§14.7.3).

Initiation

$$Cl_2 \xrightarrow{\text{light}} 2Cl\cdot$$

Chain propagation

$$H_2 + Cl\cdot \longrightarrow HCl + H\cdot$$
$$H\cdot + Cl_2 \longrightarrow HCl + Cl\cdot$$

Chain termination

$$H\cdot + H\cdot \longrightarrow H_2$$
$$Cl\cdot + Cl\cdot \longrightarrow Cl_2$$
$$H\cdot + Cl\cdot \longrightarrow HCl$$

Chlorine and hydrogen can be made to combine non-explosively by passing them over a catalyst of activated charcoal, or by burning a jet of chlorine in hydrogen, or hydrogen in chlorine.

In the presence of ultraviolet light bromine reacts with hydrogen by a non-explosive reversible free radical reaction. In the absence of light, but in the presence of a platinum catalyst, the two elements combine readily at 375 °C (648 K).

Hydrogen and iodine combine at about 400 °C (673 K), especially in the presence of a platinum catalyst. This reaction is a classical example of a reversible reaction (§8.1). Light has no effect upon it.

Reaction with water

Fluorine is a powerful oxidant and oxidises water to oxygen:

$$2F_2 + 2H_2O = 4HF + O_2$$

Side reactions occur, resulting in the formation of small quantities of ozone, hydrogen peroxide and oxygen difluoride, OF_2. Recently, hypofluorous acid, HOF, has been identified as another product of the interaction of fluorine and water at 0 °C (273 K).

Chlorine, bromine and iodine are less powerful oxidants than fluorine and undergo disproportionation with water. Chlorine and bromine are fairly soluble in water to give solutions known as 'chlorine water' and 'bromine water' respectively. Iodine, by contrast, is sparingly soluble. Two equilibria are established in aqueous solutions of halogens,

e.g. $$Cl_2(g) + aq \rightleftharpoons Cl_2(aq)$$

$$Cl_2(aq) + H_2O \rightleftharpoons HCl + HClO \qquad K = 4.2 \times 10^{-4}$$

In the second equilibrium, hydrated chlorine molecules react with water to produce low concentrations of hydrochloric acid and hypochlorous acid, as indicated by the low value of the equilibrium constant. Similar reactions occur in aqueous solutions of bromine and iodine but the concentrations of the corresponding acids are even lower than those produced from chlorine, i.e. $K = 7.2 \times 10^{-9}$ for Br_2 and $K = 2.0 \times 10^{-13}$ for I_2. Hypochlorous acid, HClO, and hypobromous acid, HBrO, are both powerful oxidising agents, which accounts in part for the oxidising properties of chlorine water and bromine water.

Reaction with alkalis

In many respects the reactions of the halogens with alkalis resemble those with water.

Fluorine oxidises concentrated alkalis to oxygen:

$$2F_2 + 4HO^- = O_2 + 4F^- + 2H_2O$$

With dilute alkalis (approximately 2%) oxygen difluoride is formed:

$$2F_2 + 2HO^- = 2F^- + H_2O + OF_2$$

Chlorine, bromine and iodine are far more soluble in solutions of alkalis than they are in water alone, because the acid-forming equilibria are displaced towards the right by reaction with hydroxide ions:

$$X_2(aq) + H_2O \rightleftharpoons HX + HXO \qquad (X = Cl, Br \text{ or } I)$$

$$\begin{array}{cc} HX & HXO \\ \downarrow HO^- & \downarrow HO^- \\ H_2O + X^- & H_2O + XO^- \end{array}$$

The overall reaction is therefore as follows:

$$X_2 + 2HO^- = XO^- + X^- + H_2O$$

The hypohalite ions, XO^-, show an increasing tendency, in the order $IO^- > BrO^- > ClO^-$, to disproportionate in alkaline solution into halide and halate ions:

$$3XO^- = 2X^- + XO_3^-$$

In cold dilute alkaline solution the hypochlorite ion, ClO^-, is reasonably stable, and as a result chlorine dissolves in cold dilute solutions of alkalis to form a chloride and a hypochlorite in a 1:1 mole ratio:

$$Cl_2 + 2HO^- = Cl^- + ClO^- + H_2O$$

At temperatures above 75 °C (348 K) and in concentrated alkali the hypochlorite ion rapidly disproportionates, and under these conditions chlorine dissolves in alkalis to form chloride and chlorate ions:

$$3Cl_2 + 6HO^- = 5Cl^- + ClO_3^- + 3H_2O$$

The hypobromite ion, BrO^-, is stable at 0 °C (273 K), but above this temperature it rapidly disproportionates into bromide ions and bromate ions, BrO_3^-. The hypoiodite ion, IO^-, disproportionates at all temperatures into iodide ions and iodate ions, IO_3^-. Therefore, when bromine or iodine is dissolved in aqueous alkali at room temperature the following overall reactions occur:

$$3Br_2 + 6HO^- = 5Br^- + BrO_3^- + 3H_2O$$

$$3I_2 + 6HO^- = 5I^- + IO_3^- + 3H_2O$$

The hypochlorites of sodium and calcium are of particular importance. An aqueous solution of sodium hypochlorite is commonly used as a bleach or disinfectant, being sold under trade names such as 'Parazone' or 'Domestos'. Bleaching powder is a mixture of a basic hypochlorite and a basic chloride, of formulae $Ca_3(OCl)_2(OH)_4$ and $Ca_2Cl_2(OH)_2 \cdot H_2O$ respectively.

The redox properties of the halogens

The electron affinities, i.e. the enthalpy changes accompanying

$$X(g) + e^- = X^-(g)$$

are not suitable for comparing the oxidising power of the halogens because: (i) reactions of the halogens usually involve molecules, not atoms; (ii) hydrated ions are usually formed, not gaseous ions.

If we consider the more realistic change

$$X_2 + 2e^- + aq = 2X^-(aq)$$

we see that the standard electrode potential, $E^{\ominus}$, (§2.5.3) gives us a realistic guide to oxidising power, for it expresses the tendency of halogen molecules to be reduced to hydrated ions in solution. On this basis the oxidising power of the halogens decreases from fluorine to iodine, as the $E^{\ominus}$ values become less positive.

Fluorine is the most powerful oxidant of all, which explains why fluoride ions can be oxidised only by electrolysis. In aqueous solution fluorine oxidises chloride, bromide and iodide ions to the free halogens:

$$F_2 + 2X^-(aq) = 2F^-(aq) + X_2 \qquad (X = Cl, Br \text{ or } I)$$

These are often called *displacement reactions*, because halogens are displaced from their salts by other halogens which are more powerful oxidising agents. Similarly, chlorine oxidises bromide and iodide ions, and bromine oxidises iodide ions. Iodine is a relatively weak oxidant and does not oxidise other halide ions, except the astatide ion, At^-. The iodide ion is, in fact, a good reductant, because it readily loses its electron to form iodine.

Fluorine is not generally used as an oxidant because it is difficult to handle and its reactions tend to be rather vigorous. Chlorine and bromine do not suffer from these drawbacks and are commonly used as oxidants. In aqueous solution the oxidising action of chlorine arises from chlorine molecules and hypochlorous acid,

i.e. $Cl_2 + 2e^- = 2Cl^-$

$$HClO + H^+(aq) + 2e^- = Cl^- + H_2O$$

Several examples of the use of chlorine and bromine as oxidants occur throughout this book, but for convenience a few illustrations are listed below.

$$X_2 + H_2O + SO_3^{2-} = 2H^+(aq) + SO_4^{2-} + 2X^- \qquad (X = Cl \text{ or } Br)$$

$$4X_2 + 5H_2O + S_2O_3^{2-} = 10H^+(aq) + 2SO_4^{2-} + 8X^- \qquad (X = Cl \text{ or } Br)$$

Although iodine is a much weaker oxidant than the other halogens, it does oxidise hydrogen sulphide and the thiosulphate ion:

$$X_2 + H_2S = 2HX + S \qquad (X = Cl, Br \text{ or } I)$$

$$I_2 + 2S_2O_3^{2-} = 2I^- + S_4O_6^{2-}$$

The latter reaction is used in the laboratory for the quantitative estimation of iodine:

$$1 \text{ mole } I_2 \equiv 2 \text{ moles } Na_2S_2O_3$$

In anhydrous conditions chlorine and, to a lesser extent, bromine may act as oxidants by removing hydrogen from certain molecules,

e.g. $CH_4 + Cl_2 = CH_3Cl + HCl$

Iodine is a weaker oxidant in this respect although it does oxidise ammonia. Fluorine reacts explosively with many organic compounds.

17.2.3 COMPOUNDS OF THE HALOGENS

Most elements form compounds with the halogens, and for the most part these are discussed in more detail in other sections of this book. A few generalisations will be made in this section concerning the halides of metals, and of non-metals excluding hydrogen and oxygen.

Metal halides

Many hydrated metal halides are prepared by dissolving a metal or its oxide, hydroxide or carbonate in the appropriate hydrohalic acid. The anhydrous halides may be obtained by heating the metal in a stream of dry halogen or hydrogen halide. Metals, such as tin or iron, which show two common valency (or oxidation) states generally show the higher state with the free halogen and the lower state with the hydrogen halide.

E.g. $Sn + 2I_2 = SnI_4$

$$Sn + 2HI = SnI_2 + H_2$$

Exceptions to this generalisation are:

$$Fe + I_2 = FeI_2 \quad \text{(not } FeI_3\text{)}$$

$$2Cu + I_2 = 2CuI \quad \text{(not } CuI_2\text{)}$$

Many anhydrous metal halides differ considerably from the corresponding hydrated compounds. Anhydrous aluminium chloride, for example, is covalent but the hydrated salt, $AlCl_3{\cdot}6H_2O$, is ionic (§13.2.6). Anhydrous metal halides, with a few exceptions such as $BaCl_2$, cannot be prepared by heating the hydrated halides because the latter undergo hydrolysis when heated to give either a hydroxide or a basic chloride,

e.g. $AlCl_3{\cdot}6H_2O = Al(OH)_3 + 3HCl + 3H_2O$

$$MgCl_2{\cdot}6H_2O = MgCl(OH) + HCl + 5H_2O$$

If hydrolysis occurs on heating, the water can often be removed by refluxing the hydrated salt with reagents that react with water (§16.2.7). Some anhydrous metal chlorides can be prepared by heating the metal oxide with tetrachloromethane, or by heating a mixture of the metal oxide and carbon in a stream of chlorine,

e.g. $TiO_2 + CCl_4 = TiCl_4 + CO_2$

$$SnO_2 + C + 2Cl_2 = SnCl_4 + CO_2$$

Ionic metal halides possess the usual properties of electrovalent compounds, i.e. they have high melting temperatures and conduct electricity in the fused state or in aqueous solution. By contrast, covalent metal halides have low melting temperatures and do not conduct electricity when molten. They are usually anhydrous, and react with water to produce hydrated metal ions,

e.g. $AlCl_3 + 6H_2O = [Al(H_2O)_6]^{3+} + 3Cl^-$ (§13.2.6)

By a consideration of Fajans' rules (§4.6.1) we can gain an insight into the bonding in metal halides. From top to bottom of a periodic group, as the surface charge density of metal ions decreases, the distortion of any given halide ion becomes less important and the bonding becomes more ionic or less covalent. In group 2A, for example, the halides of beryllium are covalent, those of magnesium are ionic with some covalent characteristics, but those of calcium, strontium and barium are essentially ionic. The effect of cationic charge on halide ion distortion can be seen in the chlorides of the metallic elements of the third period, i.e. NaCl (ionic), $MgCl_2$ (ionic with some covalency) and $AlCl_3$ (covalent).

We must also remember that the ease of distortion of halide ions increases from fluoride to iodide. Thus, the halides of a particular metal become increasingly covalent in character from the fluoride to the iodide. In some cases, e.g. aluminium and tin(IV), the fluoride is ionic but the chloride, bromide and iodide are covalent.

Non-metal halides

Non-metal halides are usually prepared by direct combination of the elements, and many examples are quoted throughout this book. Multivalent elements usually reach their highest valency if an excess of the halogen is used. If the halogen is deficient in quantity then a lower valency results. The formation of the chlorides of phosphorus (§15.2.6) provides a good illustration of this point.

Non-metal halides generally have low melting and boiling temperatures, which increase with relative molecular mass. Generally, the compounds are soluble in non-polar solvents and are hydrolysed by water,

e.g. $BI_3 + 3H_2O = H_3BO_3 + 3HI$

$SiCl_4 + 2H_2O = SiO_2(aq) + 4HCl$

Tetrachloromethane is exceptional in its resistance to hydrolysis (§14.2.6).

17.2.4 HYDROGEN HALIDES

HF

The hydrogen halides are among the most important halogen compounds. In the anhydrous state their names are:

HF	hydrogen fluoride	HBr	hydrogen bromide
HCl	hydrogen chloride	HI	hydrogen iodide

HF, HCl, HBr and HI

Preparation

Direct combination

Direct combination (§17.2.2) is a useful means of preparing hydrogen chloride, but not the other hydrogen halides. The reaction between hydrogen and fluorine is far too violent to be of use, while the reactions between hydrogen and bromine or iodine are reversible and do not give hydrogen bromide or hydrogen iodide in the pure state.

Displacement reactions

Concentrated sulphuric acid displaces hydrogen chloride from crystalline ionic chlorides at room temperature (§16.2.6). At higher temperatures, the hydrogensulphate which is formed reacts with further ionic chloride (§16.2.6). Displacement can also be used for the preparation of hydrogen fluoride from ionic fluorides. Hydrogen bromide and hydrogen iodide, however, are oxidised by concentrated sulphuric acid (see below) and therefore cannot be prepared in this way.

Hydrolysis of phosphorus trihalides

Gaseous hydrogen halides are produced when water is added dropwise to phosphorus trihalides, other than the fluoride:

$$PX_3 + 3H_2O = H_2PHO_3 + 3HX(g)$$

If the water is added in large amounts the hydrogen halide dissolves in it and is not isolated. The method is particularly useful for preparing hydrogen bromide and hydrogen iodide, for which purpose the phosphorus trihalide is usually made *in situ*. Hydrogen bromide may be obtained by adding bromine dropwise to a paste of red phosphorus and water, while hydrogen iodide is conveniently produced by adding water dropwise to a mixture of red phosphorus and iodine:

$$2P + 3X_2 = 2PX_3$$

$$PX_3 + 3H_2O = H_2PHO_3 + 3HX \quad (X = Br \text{ or } I)$$

Any halogen vapour which escapes with the hydrogen halide is removed by passing the gases through a tube packed with moist red phosphorus.

The action of halogens on covalent hydrides

The halogens are reduced to hydrogen halides by hydrogen sulphide:

$$H_2S + X_2 = 2HX + S \quad (X = F, Cl, Br \text{ or } I)$$

Hydrogen chloride is often obtained as a by-product of the reactions between hydrocarbons and chlorine (§14.7.3).

General properties

Table 17.3 shows that both the strength and the ionic character of hydrogen–halogen bonds decrease with increasing size of the halogen atom.

Table 17.3 Some properties of the hydrogen halides

	HF	HCl	HBr	HI
$\Delta H_f^{\ominus}$/kJ mol^{-1}	−269	−92.3	−36.2	+25.9
bond dissociation enthalpy/kJ mol^{-1}	+562	+431	+366	+299
boiling temperature/°C	+20	−85	−67	−35
melting temperature/°C	−83	−115	−88	−51
% ionic character in H—X bond	43	17	13	7
pK_a (in 1 M solutions)	3.25	−7.4	−9.5	−10

In line with the decreasing bond dissociation enthalpy, the thermal stability of the hydrogen halides decreases from the fluoride to the iodide. For example, hydrogen iodide decomposes at 400 °C (673 K), while hydrogen fluoride and hydrogen chloride are quite stable at this temperature.

At room temperature the hydrogen halides are gaseous, although the fluoride is readily liquefied by cooling. The gases are colourless, but in contact with moist air white fumes are formed due to the production of droplets of the hydrohalic acid. Hydrogen bonding occurs in hydrogen fluoride and accounts for its anomalously high melting and boiling temperatures. The effect is absent in the other hydrogen halides.

Acidic properties

The hydrogen halides are extremely soluble in water; for example, at 0 °C (273 K) and one atmosphere (100 kPa) pressure, one volume of water dissolves approximately 500 volumes of hydrogen halide. The high solubility arises because the hydrogen halides ionise in water, i.e. they react with water to form ions:

$$HX(g) + H_2O \rightleftharpoons H_3O^+ + X^- \quad (X = F, Cl, Br \text{ or } I)$$

Solutions of hydrogen halides are therefore acidic and are known generally as 'hydrohalic acids'. Their individual names are:

HF	hydrofluoric acid	HBr	hydrobromic acid
HCl	hydrochloric acid	HI	hydriodic acid

Hydrochloric, hydrobromic and hydriodic acids are almost completely ionised and are therefore strong acids. The p*K*a values (Table 17.3) indicate that acid strength increases as the hydrogen–halogen bond becomes weaker. This is to be expected because the hydrogen–halogen bond is broken during ionisation. By contrast, hydrofluoric acid is a weak acid in dilute solution because of the great strength of the hydrogen–fluorine bond; for example, in 0.1 M solution it is approximately 10% ionised. In solutions of between 5 M and 15 M, however, hydrofluoric acid is a much stronger acid. The reason is that two equilibria exist in aqueous solutions of hydrogen fluoride:

$$HF + H_2O \rightleftharpoons F^- + H_3O^+ \quad (1)$$

$$HF + F^- \rightleftharpoons HF_2^- \quad (K = 5.2\ \text{mol}^{-1}\ \text{dm}^3) \quad (2)$$

At high concentrations of HF equilibrium (2) is important, since it removes fluoride ions from solution and hence displaces equilibrium (1) to the right. Thus, as the solution becomes more concentrated there is an increase in the degree of ionisation of HF and in the hydrogen ion concentration of the solution.

Recent studies of the hydrogendifluoride ion, HF_2^-, suggest that the hydrogen atom lies midway between the two fluorine atoms, and that the H—F distance in the ion (0.113 nm) is considerably greater than in the HF molecule (0.092 nm). It has been proposed that the ion has a 'one electron bond' structure represented thus:

$$[F\text{---}H\text{---}F]^-$$

The ion is *not* formed by hydrogen bonding between the HF molecule and the F^- ion.

Several stable salts containing the hydrogendifluoride ion are known, e.g. KHF_2 and NH_4HF_2. The remaining hydrohalic acids do not readily form the corresponding hydrogendihalide ions, HX_2^-, because of:

(i) the lower surface charge density of chloride, bromide and iodide ions, which causes the attraction of these ions for hydrogen halide molecules to be weak;

(ii) the low concentration of non-ionised hydrogen halide in solution.

The hydrohalic acids show typical acidic properties. For example, they form salts with bases, and react with metals above hydrogen in the electrochemical series to form hydrogen. Hydrofluoric acid and moist (but not dry) hydrogen fluoride rapidly attack glass or silica to form hexafluorosilicic(IV) acid:

$$SiO_2 + 4HF = SiF_4 + 2H_2O$$

$$SiF_4 + 2HF = H_2SiF_6$$

Because hydrogen fluoride is formed when fluorine reacts with water, moist fluorine also attacks glass and silica.

Ease of oxidation

Hydrogen iodide is readily oxidised to iodine and is therefore a good reducing agent. For example, hydriodic acid is oxidised by atmospheric oxygen and various oxidising agents:

$$2HI = 2H^+(aq) + I_2 + 2e^-$$

Acidic solutions of iodides are also readily oxidised, a property which is used as a test for oxidising agents. A positive result is denoted by the liberation of iodine which gives a blue-black coloration with starch.

Hydrogen bromide is more difficult to oxidise than hydrogen iodide, and is therefore a weaker reductant. For example, hydrogen bromide and ionic bromides slowly reduce concentrated sulphuric acid to sulphur dioxide,

$$\text{e.g.} \quad 2HBr + H_2SO_4 = 2H_2O + Br_2 + SO_2$$

but hydrogen iodide and ionic iodides rapidly reduce sulphuric acid not only to sulphur dioxide but also to hydrogen sulphide:

$$8HI + H_2SO_4 = 4H_2O + 4I_2 + H_2S$$

Additional examples of the powerful reducing properties of hydrogen iodide include the reduction of alcohols to alkanes, nitric acid to nitrous acid, and dinitrogen oxide to the ammonium ion.

Hydrogen chloride is unaffected by concentrated sulphuric acid and is oxidised only by strong oxidants such as manganese(IV) oxide or potassium permanganate. Hydrogen fluoride can be oxidised only by electrolysis. Thus, there is a well-defined increase in the ease of oxidation of hydrogen halides from the fluoride to the iodide.

17.2.5 POLYHALIDE IONS

Polyhalide ions carry a single negative charge and comprise three or more halogen atoms. The best known example is the triiodide ion, I_3^-, which is formed by dissolving iodine in an aqueous solution of an ionic iodide:

$$I^- + I_2 \rightleftharpoons I_3^-$$

The stability of this ion is favoured by large non-distorting cations, e.g. K^+, Rb^+ and Cs^+.

17.2.6 SUMMARY OF THE PROPERTIES OF FLUORINE AND ITS COMPOUNDS

Reference has often been made to the fact that fluorine and its compounds differ appreciably from the other halogens and their compounds. Some of the important differences are summarised below.

(i) The bond in the fluorine molecule is weak, but fluorine forms exceptionally strong bonds with most other elements.
(ii) Fluorine has a limited covalency.
(iii) Fluorine is the most powerful oxidant and can be prepared only by electrolysis.
(iv) Hydrogen bonding occurs in hydrogen fluoride but not the other hydrogen halides.
(v) Hydrofluoric acid is a weak acid in dilute solution, but becomes stronger with increasing concentration. The remaining hydrohalic acids are strong at all concentrations.
(vi) Differences exist between the solubilities in water of some metal fluorides and the corresponding chlorides, bromides and iodides, e.g.
   (1) CaF_2 is insoluble, but $CaCl_2$, $CaBr_2$ and CaI_2 are soluble. This is due to the high lattice enthalpy of the fluoride;
   (2) AgF is soluble, but AgCl, AgBr and AgI are insoluble. In this case the high hydration enthalpy of the fluoride ion provides sufficient energy for the AgF lattice to be broken down.

17.2.7 ASTATINE

Astatine is produced by bombarding bismuth isotopes with α-particles:

$$^{207}_{83}Bi + ^{4}_{2}He = ^{211}_{85}At$$

Little is known of astatine because it is radioactive and its isotopes have short half-lives. Much of its chemistry, however, can be predicted from a knowledge of the other halogens, particularly iodine. There is evidence that astatine forms the ions At^-, AtO_3^- and At^+.

EXAMINATION QUESTIONS ON CHAPTER 17

1 **Outline** the chemistry of an industrial process in which chlorine is one of the important products (N.B. detailed diagrams of industrial plant are not required.)

Describe the action of the reagents listed below on (i) sodium chloride, (ii) sodium bromide, and (iii) sodium iodide, giving equations and explanations.

(a) Concentrated sulphuric acid is added separately to each of the solids (i), (ii) and (iii) and the mixture warmed.

(b) To aqueous solutions of (i), (ii) and (iii) aqueous silver nitrate is added, followed by aqueous ammonia.

(c) To aqueous solutions of (i), (ii) and (iii) chlorine water is added followed by tetrachloromethane (carbon tetrachloride); the mixture is then shaken and allowed to settle. (AEB)

2 (a) Outline an industrial method for the manufacture of chlorine from sodium chloride.

(b) Compare the chemical behaviour of the halogens chlorine, bromine and iodine, by referring to their reactions with (i) hydrogen, (ii) aqueous potassium iodide solution.

(c) Compare the action of sodium chloride, sodium bromide and sodium iodide on concentrated sulphuric acid. (AEB)

3 (a) Astatine is a *naturally radioactive* member of the halogens having a *half life* of 54.0 seconds. It is considered to decay according to the following sequence:

$^{216}_{85}At \rightarrow ^{212}_{83}Bi \rightarrow ^{212}_{84}Po \rightarrow ^{208}_{82}Pb$

(i) Explain the terms in italics.

(ii) Name the particles emitted in each stage and indicate their nature and chief properties.

(iii) Give the number of protons, electrons and neutrons in the isotope of lead obtained.

(b) Show by means of an equation how astatine could possibly be obtained from an aqueous solution of potassium astatide.

(c) Give the name of an astatide which might be expected to be insoluble in water.

(d) If astatoethane (ethyl astatide) was treated with aqueous alkali, would you expect the hydrolysis to be faster or slower than that with bromoethane (ethyl bromide)?

Give reasons for your answer.

What would be the products of such a hydrolysis? (AEB)

4 (a) Fluorine and its compounds often have properties noticeably different from those of chlorine and other halogens. Discuss the following, relating the differences where possible to the data given at the end of this section.

(i) Chlorine gas can be prepared from chlorides by 'chemical' methods but fluorine gas has to be prepared by electrolysis.

(ii) The vapour pressures of aluminium fluoride, chloride and bromide reach a value of one atmosphere at 1564 K, 696 K and 530 K respectively.

(iii) The pH of an aqueous solution of sodium fluoride is greater than 7 while that of a similar solution of sodium chloride is 7.

Size of ions/nm F^- 0.133; Cl^- 0.181; Br^- 0.196

Standard electrode potentials/V	$F_2(g)\ 2F^-(aq)$	+2.87
	$Cl_2(aq)\ 2Cl^-(aq)$	+1.36
	$Br_2(aq)\ 2Br^-(aq)$	+1.09

The dissociation constant of hydrofluoric acid is 5.6×10^{-4} mol l^{-1}.

(b) Aqueous copper(II) sulphate solution is blue in colour; when aqueous potassium fluoride is added, a green *precipitate* is formed but, when aqueous potassium chloride is added instead, a bright green *solution* is formed. What do you think is happening in the two cases? (L)

5 Discuss the general trends in the properties of the halogens (fluorine to iodine), and suggest the reason for such trends, limiting yourself to the following:

(a) the reaction of the halogens with (i) water, (ii) aqueous sodium hydroxide, (iii) aqueous alkali-metal halides,

(b) the reaction of alkali-metal halides with sulphuric acid,

(c) the strengths of the halogen hydracids. (WJEC)

6 It is said that the chemistry of fluorine and its compounds differs markedly from that of the other halogens. Discuss this statement by reference to:

(a) the reactions of the halogens with hydrogen and with sodium hydroxide solutions,

(b) the acidity of aqueous solutions of the hydrogen halides,

(c) the solubility of the halides,

(d) the oxidising properties of the halogens. (JMB)

7 (a) Given the following physical data for the halogen hydrides.

	Boiling point in °C	K_a	Bond dissociation energy in kJ/mol
hydrogen fluoride	+19.5	10^{-4}	562
hydrogen chloride	−85.0	10^7	431
hydrogen bromide	−67.0	10^9	366
hydrogen iodide	−35.0	10^{11}	298

(i) Account for the differences in boiling points of the halogen hydrides.

(ii) Place the halogen hydrides in order of **increasing acidity** and account for the order you have suggested.

(iii) Explain why a solution of hydrogen chloride in methyl benzene (toluene), does not conduct electricity, whilst in aqueous solution it behaves as a strong electrolyte.

(b) Phosphorus trichloride and aluminium chloride have important but different functions as reagents in organic chemistry. For each compound give **one** example of the reaction for which it is used.

(c) A sample of xenon tetrafluoride reacts with iodide ion as shown in the equation

$$XeF_4 + 4I^- \longrightarrow 2I_2 + Xe + 4F^-$$

The liberated iodine from such a reaction was reduced by 40.0 cm^3 of sodium thiosulphate(VI) solution of concentration 0.01 mol/l. Calculate the volume at s.t.p. of xenon which was liberated in the reaction. (AEB)

8 (a) Outline the essential features of the process used for obtaining fluorine commercially (details of plant are **not** required).

Name materials, quote conditions, and give chemical equations as appropriate.

(b) Compare, or contrast any **two** of the following:

(i) any action fluorine and chlorine, separately, might have with cold dilute sodium hydroxide,

(ii) any action hydrogen chloride and hydrogen bromide, separately, might have with concentrated sulphuric acid,

(iii) any action chlorine and iodine, separately, might have with sodium thiosulphate ($Na_2S_2O_3$) solution.

Briefly ascribe *reasons* for the differences in behaviour between the halogens and/or their hydrides in each of the reactions selected.

(c) Suggest either (i) a reason for the insolubility of calcium fluoride when the other calcium halides are soluble in water,

or (ii) a structure for F_2O, giving a bond diagram and predicting the shape of the molecule. (SUJB)

9 (a) Outline the laboratory preparation of *each* of the **four** hydrogen halides HF, HCl, HBr and HI.

(b) The boiling points in Kelvin of the hydrogen halides are given below.

Hydrogen halide	Boiling point/K
HF	293
HCl	188
HBr	206
HI	238

Account for this variation in boiling point.

(c) Compare and contrast:

(i) the thermal stabilities of the four hydrogen halides in (b),

(ii) the strengths of the acids formed in aqueous solution. (AEB)

10 Astatine (At, atomic number 85), the fifth member of the halogen group, exists only as short-lived radioactive isotopes and is not found in nature. Use your knowledge of the properties of the halogens, and the trends in those properties, to predict the principal features of the chemistry of astatine *and* its compounds if a stable isotope were discovered. (L)

11 (a) Illustrate the trends in the chemical properties of the halogens F, Cl, Br and I by considering how these elements react with (i) metals, (ii) hydrogen, (iii) aqueous sodium hydroxide.

(b) Discuss the bonding in hydrogen fluoride and explain how this accounts for one physical property and one chemical property of this hydride which may be described as 'abnormal'. (O)

18 d-block transition elements

18.1 General characteristics

18.1.1 INTRODUCTION

Transition elements are strictly defined as those which have partially filled d or f orbitals. A slightly broader definition, which is commonly adopted, includes those elements that have partially filled d or f orbitals in any of their common valency states.

This broad definition embraces some 55 elements, counting up to atomic number 103 (lawrencium). Subdivided into two groups, they are:

(i) the main transition elements (those of the d block) and (ii) the inner transition elements (those of the f block).

The main transition elements fall into three series. The nine elements scandium to copper, whose atoms or ions possess partially filled 3d orbitals, comprise the first series transition elements. Zinc, with the configuration $3d^{10}\ 4s^2$, forms no compounds involving 3d electrons, and is therefore not a transition element. The second transition series (the nine elements yttrium to silver) are characterised by partially filled 4d orbitals in their atoms or ions. Like zinc, cadmium is non-transitional in its characteristics. The third transitional series, arising from the filling of 5d orbitals, includes the nine elements from lanthanum to gold; mercury is excluded.

The inner transition elements fall into two categories: the *lanthanoids* and *actinoids*. The lanthanoids comprise the fourteen elements having partially filled 4f orbitals (elements cerium to lutecium), although lanthanum is commonly included because of its similarity. The actinoids,

Table 18.1 First series transition elements

Element	Outer electronic configuration of atoms	Outer electronic configuration of ions		
scandium	$3d^1\ 4s^2$			Sc^{3+} $3d^0$
titanium	$3d^2\ 4s^2$		Ti^{2+} $3d^2$	Ti^{3+} $3d^1$
vanadium	$3d^3\ 4s^2$		V^{2+} $3d^3$	V^{3+} $3d^2$
chromium	$3d^5\ 4s^1$		Cr^{2+} $3d^4$	Cr^{3+} $3d^3$
manganese	$3d^5\ 4s^2$		Mn^{2+} $3d^5$	Mn^{3+} $3d^4$
iron	$3d^6\ 4s^2$		Fe^{2+} $3d^6$	Fe^{3+} $3d^5$
cobalt	$3d^7\ 4s^2$		Co^{2+} $3d^7$	Co^{3+} $3d^6$
nickel	$3d^8\ 4s^2$		Ni^{2+} $3d^8$	
copper	$3d^{10}\ 4s^1$	Cu^+ $3d^{10}$	Cu^{2+} $3d^9$	
zinc	$3d^{10}\ 4s^2$		Zn^{2+} $3d^{10}$	

Notes

(i) The configurations given are those outside an argon noble gas core.

(ii) Zinc is included to show its non-transitional configuration.

which have partially filled 5f orbitals, comprise the elements thorium to lawrencium; actinium may also be included for convenience.

We shall concentrate on the first series transition elements. The outer electronic configurations of the atoms and some ions of these elements are given in Table 18.1.

In an isolated atom or ion the five 3d orbitals are *degenerate*. In accordance with Hund's rule (§3.1.6), each of these orbitals is singly occupied by an electron before spin pairing occurs. A common way of representing this electronic arrangement is the box notation (Table 18.2).

Table 18.2 The outer electronic configurations of the first series transition elements

	3d					4s
scandium	↑					↑↓
titanium	↑	↑				↑↓
vanadium	↑	↑	↑			↑↓
chromium	↑	↑	↑	↑	↑	↑
manganese	↑	↑	↑	↑	↑	↑↓
iron	↑↓	↑	↑	↑	↑	↑↓
cobalt	↑↓	↑↓	↑	↑	↑	↑↓
nickel	↑↓	↑↓	↑↓	↑	↑	↑↓
copper	↑↓	↑↓	↑↓	↑↓	↑↓	↑
zinc	↑↓	↑↓	↑↓	↑↓	↑↓	↑↓

Notes

(i) All the atoms, except for chromium and copper, have doubly filled 4s orbitals. Because of repulsion between the paired 4s electrons the $3d^4\ 4s^2$ configuration is higher in energy than $3d^5\ 4s^1$. Thus, chromium adopts the latter configuration. The 3d and 4s sub-shells do not differ greatly in energy, and towards the end of the series the 3d level appears to fill preferentially. Thus, copper adopts a $3d^{10}\ 4s^1$ configuration rather than $3d^9\ 4s^2$.

(ii) The 4s and 3d electrons are very close in energy, but when ions are formed it is the 4s electrons that are lost first.

18.1.2 FEATURES OF THE FIRST SERIES TRANSITION ELEMENTS AND THEIR IONS

The transition elements have the following general features in common.

(a) They are all metals, showing many similarities in their physical properties.

(b) They form complexes.

(c) Their complexes are often coloured.

(d) They show a variety of valency states (oxidation states).

(e) Both metals and ions often show catalytic activity.

Metallic properties

Most are dense, hard, lustrous, strong metals with high melting and boiling temperatures. They are good conductors of heat and electricity. The metallic bonding (§6.5.1), which involves both 4s and 3d electrons, undoubtedly accounts for many of these properties. The ready formation of alloys is also a feature of the elements.

Atomic and metallic radii (§3.2.4) all show a general decrease in passing from scandium to copper (Fig. 18.1).

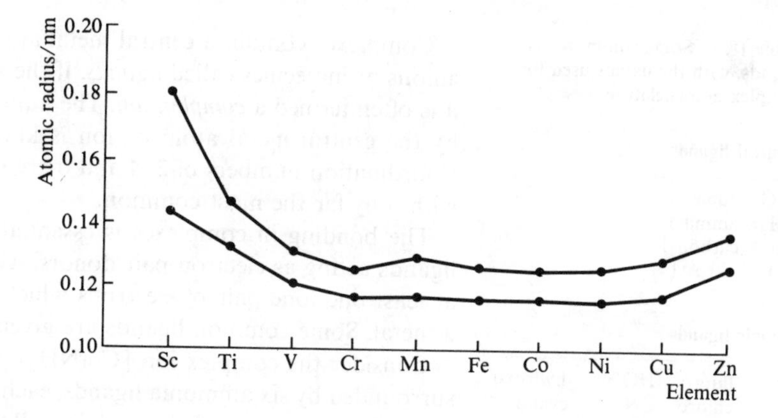

Fig. 18.1 Atomic and metallic radii of the first series transition elements. The upper curve shows metallic radius and the lower curve atomic radius, i.e. covalent bond radius.

Initially, the radius decreases steeply, but levels out after vanadium. Electrons in 3d orbitals are poor *screeners*, i.e. they shield 4s electrons from the nuclear charge in a highly inefficient way. Therefore in passing from scandium to copper a decrease in atomic radius is to be expected, since the increase in nuclear charge will have a greater effect than the addition of an electron. The flattening of the graph after vanadium could be due to there being a certain minimum volume that atoms may occupy

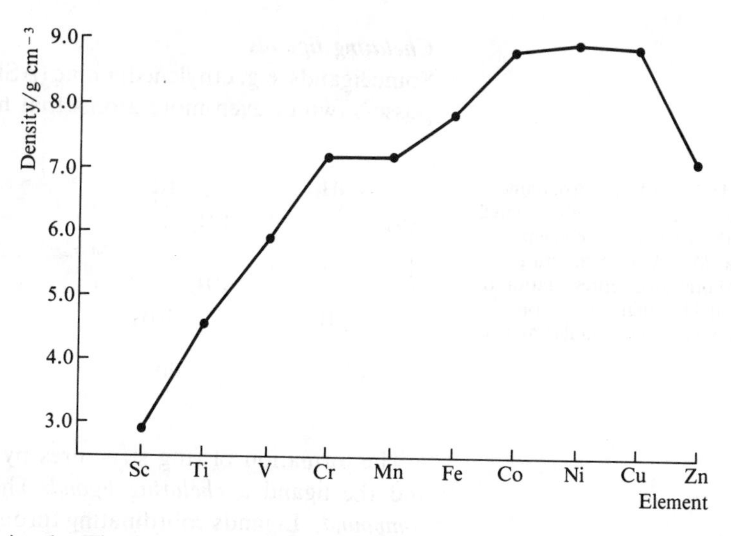

Fig. 18.2 Densities of the first series transition elements.

efficiently. The change in density (Fig. 18.2) is explained by the decreasing volume and increasing atomic mass in passing from scandium to copper. The small variation in metallic radius for the elements (particularly vanadium to copper) explains, in part at least, the ease of alloy formation.

Complex formation

Although the formation of complexes is an important property of transition elements, it is not exclusive to them, for certain non-transition elements, e.g. aluminium and zinc, also form complexes.

Table 18.3 Some common ligands, with the names used for complex nomenclature

Neutral ligands	
H_2O	aqua
NH_3	ammine
CO	carbonyl
NO	nitrosyl

Anionic ligands			
F^-	fluoro	HO^-	hydroxo
Cl^-	chloro	CN^-	cyano
Br^-	bromo	NO_2^-	nitro
I^-	iodo	H^-	hydrido†

†IUPAC, but not ASE, recommends that the term 'hydro' is used in complexes of boron.

Complexes contain a central metal atom or ion linked to one or more anions or molecules called *ligands*. If the species carries an overall charge it is often termed a *complex ion*. The number of coordinate bonds formed by the central metal atom or ion is known as its *coordination number*. Coordination numbers of 2, 4 and 6 are often encountered in complexes, with 6 by far the most common.

The bonding in complexes is essentially dative in character, with the ligands acting as electron pair donors. All ligands must therefore possess at least one lone pair of electrons which is capable of being donated to a metal. Some common ligands are given in Table 18.3.

Consider the complex ion $[Co(NH_3)_6]^{3+}$. This comprises a Co^{3+} ion surrounded by six ammonia ligands, each nitrogen atom donating its lone pair of electrons to the metal ion. Because ammonia molecules are coordinated to the Co^{3+} ion, we say that the Co^{3+} ion is 'complexed by ammonia'.

Water, also, may act as a ligand by using one of the lone pairs on the oxygen atom. Aqueous solutions of simple salts (such as sulphates) contain metal ions complexed with water molecules. For example, a solution of iron(II) sulphate contains the hydrated ions $[Fe(H_2O)_6]^{2+}$, in which the six water molecules are coordinated to an Fe^{2+} ion. These hydrated or complex ions usually persist in the crystalline salts. For example, $FeSO_4 \cdot 7H_2O$ crystals contain $[Fe(H_2O)_6]^{2+}$ complex ions.

Chelating ligands

Some ligands, e.g. ethylenediamine (ASE: ethane-1,2-diamine) (Fig. 18.3), possess two or even more atoms that may bond to the central metal.

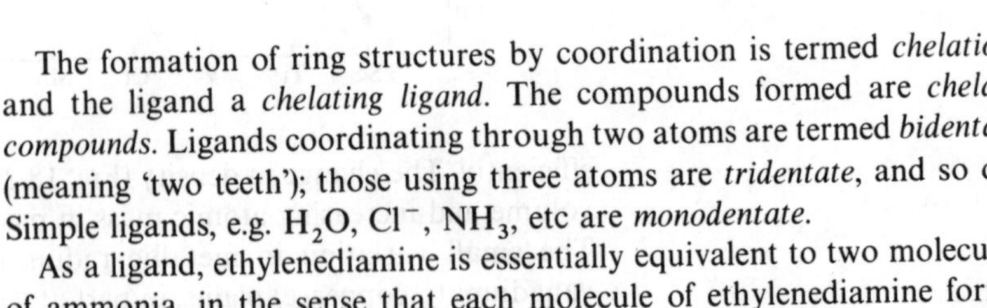

Fig. 18.3 (a) Ethylenediamine molecule (commonly abbreviated en). (b) Mode of attachment to metal M; note ring structure. (c) Abbreviated representation of (b). (d) The ethanedioate ion: This may act in a similar manner to (a).

The formation of ring structures by coordination is termed *chelation*, and the ligand a *chelating ligand*. The compounds formed are *chelate compounds*. Ligands coordinating through two atoms are termed *bidentate* (meaning 'two teeth'); those using three atoms are *tridentate*, and so on. Simple ligands, e.g. H_2O, Cl^-, NH_3, etc are *monodentate*.

As a ligand, ethylenediamine is essentially equivalent to two molecules of ammonia, in the sense that each molecule of ethylenediamine forms two coordinate bonds. Thus, the coordination number in the complex $[Co(en)_3]^{3+}$ is not three but six, because the cobalt(III) ion forms six coordinate bonds.

Charge on complex ions

The charge on a complex ion is delocalised over the whole of the complex.

The charge is the algebraic sum of that on the central ion and the total charge carried by the ligands. When all ligands are neutral, the overall charge of the complex ion is simply that carried by the central ion; hence complex *cations* arise. With anionic ligands, complex *anions* may result. For example, $[Fe(CN)_6]^{4-}$ comprises six cyanide ligands, CN^- (total charge -6), complexed to one Fe^{2+} ion (charge $+2$).

$\therefore$ overall charge $= +2 - 6 = -4$

Likewise, $[CrCl_2(H_2O)_4]^+$ comprises a central Cr^{3+} ion, complexed with two chloride ions (total charge -2) and four neutral water molecules. This is a complex cation (overall charge $= +3 - 2 = +1$), despite its anionic ligands.

The overall complex charge may even be zero. This may happen when a neutral ligand, particularly carbon monoxide, forms a complex with a metal atom, e.g. $[Cr(CO)_6]$ and $[Ni(CO)_4]$. It also happens when the total charge on anionic ligands is equal and opposite to that carried by a central metal ion, e.g. $[CrCl_3(NH_3)_3]$, which represents a chromium(III) ion complexed with three ammonia molecules and three chloride ions.

Nomenclature rules (IUPAC)

The following rules apply to all complexes, whether or not they are derived from transition elements. Any differences between IUPAC and ASE names are noted where they occur.

Formulae of complexes The symbol for the central atom is written first, followed by anionic ligands and neutral ligands in that order. Within each ligand class the order should be alphabetical in terms of the symbol for the donor atom of the ligand. Polyatomic ligands, but not monoatomic ligands, are enclosed in curved brackets, and the formula of the whole complex is enclosed in square brackets.

e.g. $[Cr(H_2O)_6]^{3+}$

$[Al(OH)(H_2O)_5]^{2+}$

$[CoCl(NH_3)_5]^{2+}$ (NH_3 in parentheses, but not Cl)

Naming of complexes Ligands are cited first, followed by the metal.

Ligand names Names of common ligands are given in Table 18.3.

Ligand numbers A Greek prefix: di, tri, tetra, penta or hexa, is used to denote the number of each type of ligand. 'Mono' is not normally used.

Order of citation of ligands Ligands are listed in alphabetical order, the multiplying prefix being ignored. Thus, pentaaqua is cited before dicyano.

Central metal For all complexes the name of the metal follows the names of the ligands. In the case of complex anions the metal name is modified to end in -ate (Table 18.4). Generally, for elements ending in -ium, the anionic name is obtained by replacing this ending by -ate, e.g. chrom*ium* to chrom*ate*.

Oxidation state of the central atom In the Stock notation the oxidation state of the central atom, i.e. the formal charge on the central ion, is

Table 18.4 Names used for some metals in anionic complexes

Element	Name in complex
Ti	titanate
V	vanadate
Cr	chromate
Mn	manganate
Fe	ferrate†
Co	cobaltate
Ni	niccolate† (ASE: nickelate)
Cu	cuprate†
Zn	zincate
Ag	argentate†
Au	aurate†
Hg	mercurate
B	borate
Al	aluminate
Ge	germanate
Sn	stannate†
Pb	plumbate†

† Based on the Latin names of the elements.

indicated by a Roman numeral in parentheses after the name of the complex.

A few examples should help to clarify the above rules. Notice that complex names are written as one word, with no hyphens, and no spacing between the name and the oxidation state of the central atom.

$[Cr(H_2O)_6]^{3+}$	hexaaquachromium(III) ion
$[Al(OH)(H_2O)_5]^{2+}$	pentaaquahydroxoaluminium(III) ion (penta*a*qua before *h*ydroxo)

The salts of these and other complex cations are named accordingly. For example, $[CoCl(NH_3)_5]Cl_2$ is pentaamminechlorocobalt(III) chloride.

$[Fe(CN)_6]^{4-}$	hexacyanoferrate(II) ion
$[BH_4]^-$	tetrahydroborate(III) ion (ASE: tetrahydridoborate(III); Table 18.3)
$[Ni(CO)_4]$	tetracarbonylnickel(0) (The oxidation state of nickel in this complex is zero.)

Stereochemistry (shapes of complexes)

In six coordinate complexes the ligands are arranged octahedrally about the metal, as in $[Co(NH_3)_6]^{3+}$, $[Fe(H_2O)_6]^{2+}$, etc. Figure 18.4 illustrates a convenient method of representing octahedral complexes.

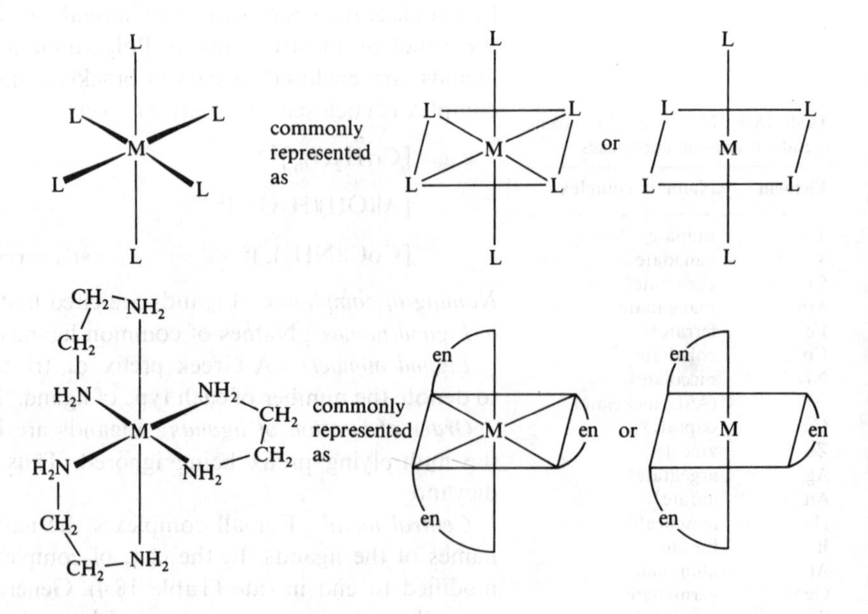

Fig. 18.4 Octahedral coordination with a monodentate ligand L and the bidentate ligand ethylenediamine.

Complexes with coordination number 4 are either square planar or tetrahedral. With coordination number 2, ligands and metal are collinear.

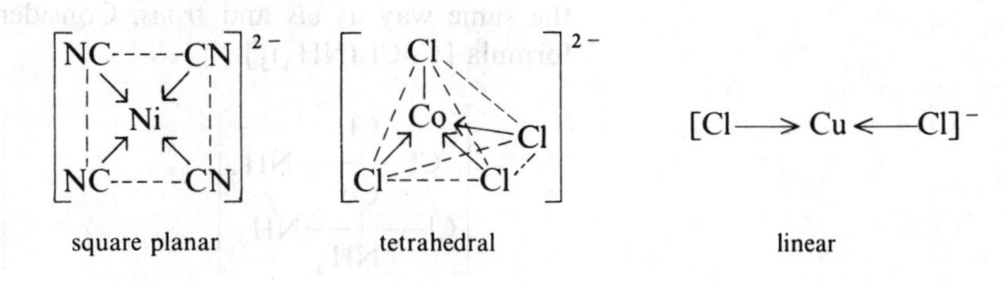

square planar tetrahedral linear

Table 18.5 The geometry of some typical complexes

Octahedral	Square planar	Tetrahedral	Linear
$[Ni(NH_3)_6]^{2+}$	$[Ni(CN)_4]^{2-}$	$[CoCl_4]^{2-}$	$[Cu(NH_3)_2]^+$
$[CoCl_2(NH_3)_4]^+$	$[CuCl_4]^{2-}$ (§18.10.3)	$[Cu(CN)_4]^{3-}$	$[CuCl_2]^-$
$[Fe(CN)_6]^{4-}$		$[Ni(CO)_4]$	$[Ag(NH_3)_2]^+$
$[Fe(CN)_5(NO)]^{2-}$			
$[Cr(CO)_6]$			

Isomerism

Both geometrical and optical isomerism may be encountered in transition element complexes.

Geometrical isomerism This occurs with square planar complexes of the type $ML_2L'_2$, and octahedral complexes of the type $ML_4L'_2$ and $ML_3L'_3$. The best known square planar examples occur with platinum(II) complexes,

e.g. $[PtCl_2(NH_3)_2]$: Cl, NH_3 / Pt / Cl, NH_3 — *cis* isomer; H_3N, Cl / Pt / Cl, NH_3 — *trans* isomer

cis isomer *trans* isomer

For octahedral complexes, *cis–trans* isomerism may occur with mono- and bidentate ligands,

e.g. $[Co(NH_3)_2(H_3N)_2Cl_2]^+$ arrangements: NH_3; H_3N—Cl; Co; H_3N—Cl; NH_3 (*cis* isomer) and NH_3; H_3N—Cl; Co; Cl—NH_3; NH_3 (*trans* isomer)

cis isomer *trans* isomer

Both have the formula $[CoCl_2(NH_3)_4]^+$

Cl; Cl; Co; en; en (*cis* isomer) and Cl; en; Co; en; Cl (*trans* isomer)

cis isomer *trans* isomer

Both have the formula $[CoCl_2(en)_2]^+$

Octahedral complexes of the type $ML_3L'_3$ exist as *facial* and *meridional* isomers, for which the abbreviations *fac* and *mer*, respectively, are used in the same way as *cis* and *trans*. Consider, for example, the isomers of formula $[CoCl_3(NH_3)_3]$.

fac isomer

mer isomer

In the *fac* isomer, all ligands of the same type are adjacent to one another. In the *mer* isomer, two of the three identical ligands are opposite to each other while both are adjacent to the third.

Optical isomerism Optical isomerism (§5.2.2) occurs with complexes such as $[Co(en)_3]^{3+}$ and $[Cr(en)_3]^{3+}$,

e.g.

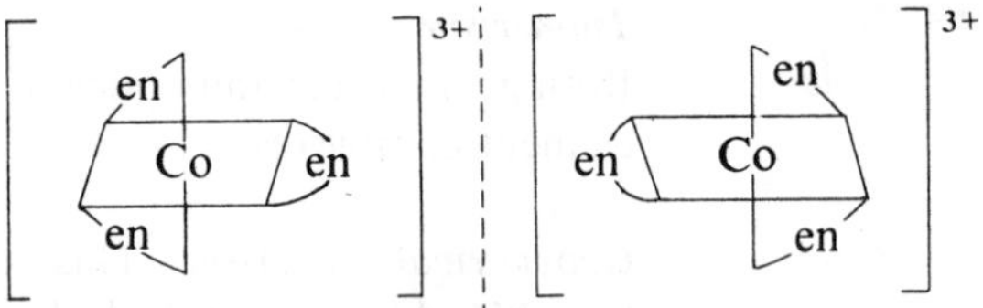

A mirror plane B

Neither A nor B possesses a centre or plane of symmetry. A is non-superimposable on B.

The *cis* (but not the *trans*) isomer of $[CoCl_2(en)_2]^+$ also shows optical activity:

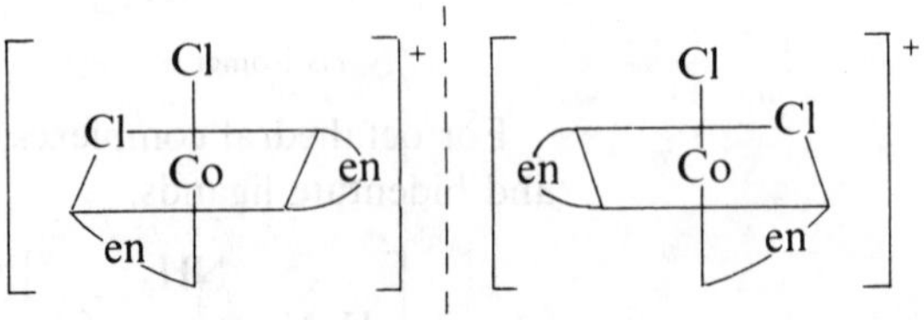

A mirror plane B

Optical activity in simple tetrahedral complexes MLL′L″L‴ may be expected but has not yet been confirmed.

Ionisation isomerism Compounds that have the same composition but are composed of different ions are known as *ionisation isomers*. For example, there are two compounds with the formula $Co(NH_3)_4Cl_2NO_2$. One is $[CoCl_2(NH_3)_4]^+NO_2^-$, and the other $[CoCl(NO_2)(NH_3)_4]^+Cl^-$. Although the latter gives a precipitate of silver chloride with Ag^+ ions, the former does not because it has no free Cl^- ions.

The three different coloured complexes, each with the formula $CrCl_3 \cdot 6H_2O$, may be called ionisation isomers or *hydration isomers*.

$[Cr(H_2O)_6]^{3+} (Cl^-)_3$
grey-blue

$[CrCl(H_2O)_5]^{2+} (Cl^-)_2 \cdot H_2O$
light green

$[CrCl_2(H_2O)_4]^+ Cl^- \cdot 2H_2O$
dark green

Coloured complex ions

Isolated transition element ions are devoid of colour, whereas complexed ions are usually coloured. For example, the pale green colour of hydrated iron(II) sulphate heptahydrate, $FeSO_4 \cdot 7H_2O$, arises from the presence in the crystal of hexaaquairon(II) ions.

The colour of complex ions is associated with partially filled d orbitals and the nature of the metal, and also with the nature of the ligands.

Nature of the metal and d orbital occupation

Table 18.6 Electronic configurations and colour of some hydrated ions, $[M(H_2O)_6]^{n+}$ ($n = 2$ or 3)

M	Electronic configuration	Colour	M	Electronic configuration	Colour
Sc^{3+}	$3d^0$	none	Fe^{3+}	$3d^5$	pale violet
Ti^{3+}	$3d^1$	violet	Fe^{2+}	$3d^6$	pale green
V^{3+}	$3d^2$	green	Co^{2+}	$3d^7$	pink
Cr^{3+}	$3d^3$	grey-blue	Ni^{2+}	$3d^8$	green
Mn^{3+}	$3d^4$	violet	Cu^{2+}	$3d^9$	blue
Mn^{2+}	$3d^5$	pale pink	Cu^+	$3d^{10}$	none
			Zn^{2+}	$3d^{10}$	none

Vacant or fully occupied d orbitals are associated with a lack of colour (Table 18.6). For a particular metal, the colour produced depends on the number of d electrons (compare Fe^{3+} and Fe^{2+}). Different metals produce differently coloured ions, even when those ions possess the same electronic configuration (compare Mn^{2+} and Fe^{3+}).

Nature of the ligand

Complexes undergo *substitution* reactions in which all or some of the ligands are replaced by different ligands. Aqua complexes readily undergo such reactions and many examples will be met. Altering the ligands around an ion often brings about a marked colour change,

e.g. $\underset{\text{blue}}{[CoCl_4]^{2-}} \xleftarrow{\text{excess HCl}} \underset{\text{pink}}{[Co(H_2O)_6]^{2+}} \xrightarrow{\text{excess } NH_3} \underset{\text{red-brown}}{[Co(NH_3)_6]^{2+}}$

The five 3d orbitals of an *isolated* metal ion are degenerate. The approach to the central ion of six ligands in the formation of a complex results in the degeneracy being partially relieved. When this happens, two of the d orbitals are raised in energy, relative to the d orbital energy of the isolated ion, while three are lowered.

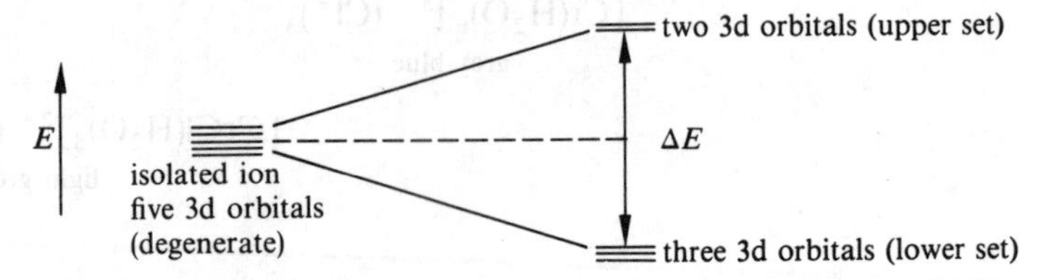

The energy difference (ΔE) between the lower and upper set is equal to

$\Delta E = h\nu$ (§3.1.4)

An electron can be promoted from the lower to the upper set on absorbing radiation of the frequency (ν) corresponding to ΔE in the above equation. For most transition metal complexes this frequency corresponds to light in the visible part of the spectrum. For $[Cu(H_2O)_6]^{2+}$ ΔE corresponds to radiation in the red end of the visible spectrum. Thus, when white light falls on this ion, red light is absorbed in promoting one electron from the lower to the upper set, so that compounds containing the ion appear blue. (White light minus red gives blue.)

$$\xrightarrow{\text{white light}} [Cu(H_2O)_6]^{2+} \xrightarrow[\text{passes on}]{\text{blue light}}$$

Similar arguments apply to other complex ions.

For an electronic transition to occur from the lower to the upper set, there must be at least one electron in the lower set and at least one vacancy in the upper set (to accommodate the electron). Hence, colourless complexes arise from Sc^{3+}, where there is no electron in the lower set, and from Zn^{2+} and Cu^{+}, where there is no vacancy in the upper set.

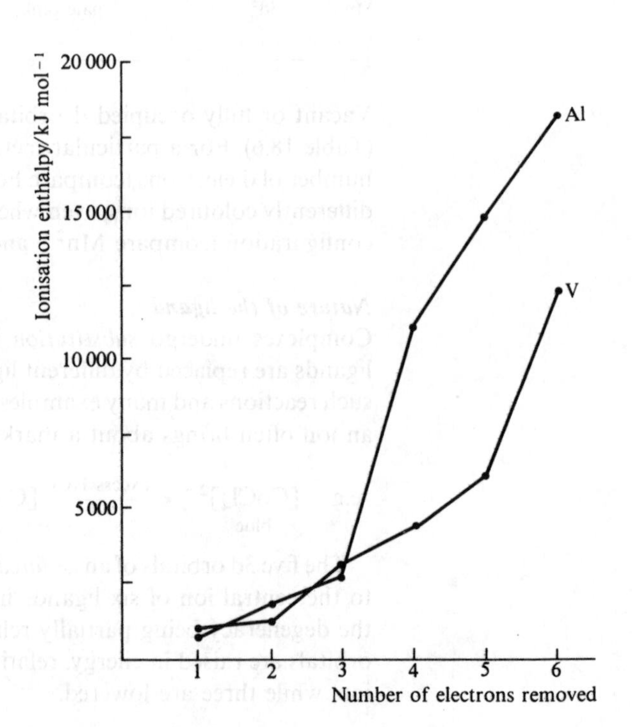

Fig. 18.5 Successive ionisation enthalpies of aluminium and vanadium.

The value of ΔE, and hence the colour, varies from metal to metal, and for a particular metal it depends on the charge on the ion and the nature of the ligands.

Variable valency state (oxidation state)

The successive ionisation enthalpies of an atom control the number of electrons that may be involved in the formation of compounds. For non-transition elements, large jumps occur in ionisation enthalpies after the removal of one electron from a group 1A metal, two electrons from a group 2A metal, and three electrons from a group 3B metal.

The large jump after the removal of the third electron means that Al^{4+} cannot be obtained by chemical means. Aluminium atoms can use only three electrons (3p and $3s^2$) in the formation of ionic or covalent bonds. However, the ionisation enthalpies of a transition metal atom, such as vanadium (Fig. 18.5), increase gradually as electrons are removed from the energetically similar 3d and 4s orbitals. This close similarity in energy means that both 4s and 3d electrons are available for bond formation (ionic or covalent), and accounts in part for the variability of valency or oxidation state. A transition metal atom is therefore able to adjust its valency or oxidation state within limits, depending on its environment,

Table 18.7 Common valency states of the first series transition elements. The commonest states are underlined.

Sc	Ti	V	Cr	Mn	Fe	Co	Ni	Cu	(Zn)
$\underline{3}$	$\underline{4}$	5	6	7	$\underline{3}$	$\underline{3}$	3	$\underline{2}$	$\underline{2}$
	3	$\underline{4}$	$\underline{3}$	6	$\underline{2}$	$\underline{2}$	$\underline{2}$	$\underline{1}$	
	2	3	2	4					
		2		3					
				$\underline{2}$					

i.e. ligands, anions, oxidants, reductants, etc.

The common oxidation states are all positive and numerically equal to the valency states (Table 18.7). With the exception of cobalt and copper, the highest valency state that a metal can possibly exhibit is equal to the number of 4s electrons plus the number of unpaired 3d electrons in the neutral atom.

The higher states (above 4) occur in fluorides, oxides, oxoanions and, to a limited extent, in the chlorides of the metals. (Highly electronegative elements stabilise high oxidation states (§16.2.8).) The bonding in the higher states (5, 6 and 7) is predominantly covalent, since the production of discrete, highly charged ions is energetically impossible in chemical systems. For the lower states (2 and 3) discrete ions can exist (e.g. in the oxides), but complex ion formation is perhaps the outstanding feature.

The mono, di and trivalent states (+1, +2 and +3 oxidation states)
The monovalent state is of importance only for copper and will be discussed later.

The divalent or +2 oxidation state involves both the 4s electrons (one 4s and one 3d for chromium and copper), while of necessity 3d electrons are always used in the trivalent or +3 oxidation state.

Successive ionisation enthalpies show an overall increase from scandium to copper, in keeping with the decreasing atomic radius (Fig. 18.6).

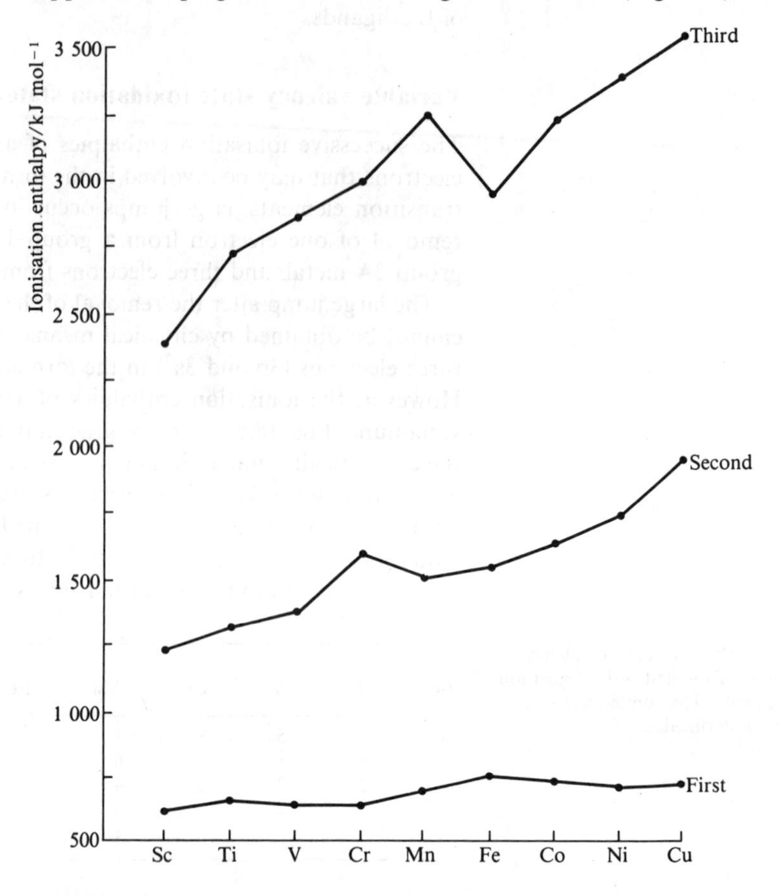

Fig. 18.6 Successive ionisation enthalpies of the transition elements.

The production of M^{3+} ions from elements on the right of the series (Co, Ni, Cu) is energetically expensive, and consequently such elements show a preference for the divalent (+2) state. Attainment of the trivalent (+3) state by metals on the left (Sc, Ti, V, Cr) is easier, and this state is in fact preferred to the divalent one.

In the absence of oxidising agents, the general preference for the divalent (+2) or trivalent (+3) state is as follows:

Sc	Ti	V	Cr	Mn	Fe	Co	Ni	Cu
3	3	3	3	2	2	2	2	2

The divalent state for those elements preferring the trivalent one is thus a reducing condition, although scandium does not form Sc^{2+} ions. Conversely the trivalent state, for those preferring the divalent one, is an oxidising condition. The stability of a particular valency or oxidation state is influenced by the nature of the ligands attached to the ion (e.g. §18.8.3), but the above pattern is often followed.

In aqueous solution, in the absence of other ligands, water usually coordinates to metal ions to give *aqua ions* (hydrated ions). The properties of these ions in solution are worthy of further discussion.

Stability in solution

The divalent state (*+2 oxidation state*) Except for scandium and possibly titanium, all the metals form aqua-complexes (hydrated ions) of general formula $[M(H_2O)_6]^{2+}$, often abbreviated to $M^{2+}(aq)$ or even M^{2+}. Solutions of V^{2+} and Cr^{2+} are strongly reducing, even attacking the solvent water, especially if it is acidified:

$$2[M(H_2O)_6]^{2+} + 2H_2O = 2[M(H_2O)_6]^{3+} + H_2 + 2HO^- \quad (M = V \text{ or } Cr)$$

$[Fe(H_2O)_6]^{2+}$ is oxidised by oxygen and many other oxidants, but the ions of manganese, cobalt, nickel and copper are quite stable in this respect.

The trivalent state (*+3 oxidation state*) Simple $[M(H_2O)_6]^{3+}$ ions, abbreviated $M^{3+}(aq)$ or M^{3+}, exist for all the metals except copper and nickel. The $[Co(H_2O)_6]^{3+}$ ion oxidises water (see below) while $[Mn(H_2O)_6]^{3+}$ is highly unstable. The simple hydrated ions of the remaining metals are stable in solution, although those of titanium and vanadium are readily oxidised by many common oxidants, including air.

The tetravalent state (+4 oxidation state)
This is the commonest state for titanium, and one of the more common for vanadium. The only stable compound of manganese in this state is the insoluble manganese(IV) oxide, MnO_2.

Because of hydrolysis simple hydrated Ti^{4+} and V^{4+} ions do not exist (§18.3.4 and §18.4.4). Manganese(IV) has no aqueous chemistry of note.

The penta, hexa and heptavalent states (+5, +6 and +7 oxidation states)
Valency states of 5 for vanadium, 6 for chromium and 7 for manganese are common. The compounds are oxidising agents, becoming increasingly powerful as the valency increases; e.g. dichromates ($Cr_2O_7^{2-}$) and permanganates (MnO_4^-).

Aqueous solutions of chromate (CrO_4^{2-}) ions, dichromate ions (§18.5.4) and vanadium(V) species (§18.4.5) are stable, but not solutions of the permanganate ion (§18.6.5).

Effect of ligands on oxidation–reduction behaviour
Ligands can affect the stability of a particular valency or oxidation state of a metal and therefore the oxidation-reduction behaviour of the ion. The standard electrode potentials (§2.5.3) for two Co^{2+}/Co^{3+} complexes illustrate this point:

$$[Co(H_2O)_6]^{3+} + e^- = [Co(H_2O)_6]^{2+} \qquad E^\ominus = +1.82 \text{ V}$$

$$[Co(NH_3)_6]^{3+} + e^- = [Co(NH_3)_6]^{2+} \qquad E^\ominus = +0.10 \text{ V}$$

The oxidising nature of $[Co(H_2O)_6]^{3+}$ is reflected in its high $E^\ominus$ value. (The trivalent (+3) state readily changes to the divalent (+2) one.) By comparing it with

$$\tfrac{1}{2}O_2 + 2H^+ + 2e^- = H_2O \qquad E^\ominus = +1.23 \text{ V}$$

we can see why this ion oxidises water to oxygen. The greater stability of the trivalent (relative to the divalent) state with ammonia ligands is indicated by the much lower $E^\ominus$ value. Molecular oxygen, in fact, converts

$[Co(NH_3)_6]^{2+}$ to $[Co(NH_3)_6]^{3+}$. Similarly, the cobalt(II) complex with nitrite ions is easily oxidised to the corresponding cobalt(III) complex. Thus we see that with some ligands, e.g. H_2O and Cl^-, the divalent state is the more stable for cobalt, while with others, e.g. NH_3 and NO_2^-, the trivalent condition is preferred. (See §18.7.2 for other examples.)

Catalytic activity

Heterogeneous catalysis (§7.5.3) is a feature of many transition metals and their compounds, particularly the oxides. The multiplicity of orbitals (3d and 4s) for the first series elements no doubt facilitates bond formation between reactant molecules and the catalyst surface, thereby lowering the activation energy of the reaction. In solution, the ions may function as homogeneous catalysts (§7.5.2) by changing their valency or oxidation states.

Table 18.8 Examples of catalysts taken from the first transition series

Catalyst	Reaction catalysed	Comment
V_2O_5 or vanadate(V)	$2SO_2 + O_2 \rightleftharpoons 2SO_3$	contact process
Fe	$N_2 + 3H_2 \rightleftharpoons 2NH_3$	Haber process
$Cr_2O_3 + ZnO$	$CO + 2H_2 = CH_3OH$	
MnO_2	$2KClO_3 = 2KCl + 3O_2$	
Ti compounds	polymerisation of ethene and propene to poly(ethene) and poly(propene)	Ziegler catalyst
Ni	$>C{=}C< + H_2 = -\overset{H}{\underset{\vert}{C}}-\overset{H}{\underset{\vert}{C}}-$	hydrogenation
Ni	$C_xH_y + xH_2O = xCO + (x + \frac{y}{2})H_2$	steam reforming
Co compounds	alkenes + CO + H_2 = aldehydes and ketones	OXO process
Mn^{2+}	$2MnO_4^- + 5(COO)_2^{2-} + 16H^+ = 2Mn^{2+} + 10CO_2 + 8H_2O$	autocatalysis
Co^{2+}	$2ClO^- = 2Cl^- + O_2$	

Properties of the individual transition elements

The important properties of the first series transition elements and their compounds are discussed in the following sections. Zinc is included for convenience, but it should be remembered that it is not a transition element. The outer electronic configuration and principal valency states of each element are given in the appropriate sections, and some physical properties are shown in Table 18.9.

Table 18.9 Some properties of the first series transition elements

	Sc	Ti	V	Cr	Mn	Fe	Co	Ni	Cu	Zn†
atomic number	21	22	23	24	25	26	27	28	29	30
relative atomic mass	44.9559	47.90	50.9414	51.996	54.9380	55.847	58.9332	58.71	63.546	65.37
density/g cm^{-3}	2.99	4.54	5.96	7.19	7.20	7.86	8.90	8.90	8.92	7.14
melting temperature/°C	1540	1675	1900	1890	1240	1535	1492	1453	1083	420
boiling temperature/°C	2730	3260	3000	2482	2100	3000	2900	2730	2595	907

† Included for comparison.

18.2 Scandium

Outer electronic configuration: $3d^1\ 4s^2$. Principal valency: 3.

Scandium is an uncommon element, which resembles aluminium rather than the other transition elements. It forms halides, ScX_3, salts with acids, and a hydrated ion, $[Sc(H_2O)_6]^{3+}$, that undergoes hydrolysis; cf. $[Al(H_2O)_6]^{3+}$. The oxide of scandium, Sc_2O_3, is more basic than that of aluminium.

18.3 Titanium

Outer electronic configuration: $3d^2\ 4s^2$. Principal valencies: 2, 3, 4.

18.3.1 THE ELEMENT

The principal ores of titanium are rutile, TiO_2, and ilmenite, $FeTiO_3$. Titanium is isolated by the *Kroll process*, in which titanium(IV) chloride is first prepared by heating the ore with carbon in a stream of chlorine:

$$TiO_2 + C + 2Cl_2 = TiCl_4 + CO_2$$

$$2FeTiO_3 + 3C + 7Cl_2 = 2TiCl_4 + 2FeCl_3 + 3CO_2.$$

The titanium(IV) chloride is purified by fractional distillation and then reduced with magnesium or sodium at 800 °C (1 073 K) in an atmosphere of argon:

$$TiCl_4 + 2Mg = 2MgCl_2 + Ti$$

$$TiCl_4 + 4Na = 4NaCl + Ti$$

The atmosphere of argon is necessary because titanium combines readily with oxygen and nitrogen at elevated temperatures.

In its physical properties titanium resembles stainless steel, i.e. it is hard, has a high melting temperature and is resistant to corrosion. It is widely used for the construction of aircraft, space vehicles, nuclear reactors, chemical plants and steam turbines.

Titanium is unreactive at room temperature, but at higher temperatures, approximately 800 °C (1 073 K), it combines readily with most non-metals, particularly oxygen and nitrogen.

18.3.2 THE DIVALENT STATE (+2 oxidation state) $3d^2$

Few compounds of divalent titanium are known. The oxide and chloride are prepared by heating the appropriate titanium(IV) compound with titanium. Titanium(II) compounds are readily oxidised.

18.3.3 THE TRIVALENT STATE (+3 oxidation state) $3d^1$

Titanium(III) compounds display the typical properties of transition element compounds because the 3d orbitals contain one electron.

The violet hydrated titanium(III) ion, $[Ti(H_2O)_6]^{3+}$, is prepared by reducing acidic solutions of titanium(IV) compounds either with zinc or electrolytically. The hexaaquatitanium(III) ion undergoes hydrolysis in solution,

i.e. $[Ti(H_2O)_6]^{3+} + H_2O \rightleftharpoons [Ti(OH)(H_2O)_5]^{2+} + H_3O^+$

and with alkalis hydrated titanium(III) oxide, $Ti_2O_3(aq)$, is formed.

Titanium(III) compounds are powerful reducing agents, and are rapidly oxidised to titanium(IV) species by reagents such as oxygen or the halogens.

18.3.4 THE TETRAVALENT STATE (+4 oxidation state) $3d^0$

This is the commonest state of titanium in the presence of air. It involves either the loss of the four outer electrons, to form the Ti^{4+} ion, or the participation of 3d and 4s electrons in four covalent bonds. Because there are no electrons in the 3d orbitals, titanium(IV) compounds do not show any of the typical characteristics of transition element compounds. For example, they are colourless, and often show a strong resemblance to the corresponding compounds of tin(IV).

Titanium(IV) oxide (titanium dioxide)

Like tin(IV) oxide, titanium(IV) oxide is a white ionic solid, i.e. $Ti^{4+}(O^{2-})_2$, which is amphoteric although rather inert to most reagents. It dissolves slowly in concentrated sulphuric acid to form a compound known as 'titanyl sulphate', $(TiO)SO_4{\cdot}H_2O$, (see below) and with fused alkalis it forms *titanates* of unknown structure but of empirical formula M_2TiO_3. The hexafluorotitanate(IV) ion, $[TiF_6]^{2-}$, is formed when titanium(IV) oxide is dissolved in hydrofluoric acid.

Titanium(IV) chloride (titanium tetrachloride)

$TiCl_4$

Titanium(IV) chloride, $TiCl_4$, the most important halide of titanium, is prepared by heating titanium(IV) oxide with carbon and chlorine or by direct combination of the heated elements. Like tin(IV) chloride, titanium(IV) chloride is covalent. It is a colourless liquid which fumes in moist air and reacts with water to form a white precipitate of the oxide,

i.e. $TiCl_4 + 2H_2O = TiO_2(aq) + 4HCl$

Complex ions, such as the hexachlorotitanate(IV) ion, $[TiCl_6]^{2-}$, are formed when titanium(IV) chloride is dissolved in concentrated hydrochloric acid.

Aqueous chemistry

Because of its high surface charge density, the Ti^{4+} ion distorts coordinated water molecules and hydrolysis occurs in aqueous solution to form hydroxo species (cf.§13.2.4), e.g. $[Ti(OH)_2(H_2O)_4]^{2+}$ (often written TiO^{2+}). Proton loss occurs on the addition of alkali to form the hydrated dioxide, cf. tin (§14.12.3). 'Titanyl sulphate' contains polymeric chains consisting of alternate titanium and oxygen atoms.

18.4
Vanadium

Outer electronic configuration: $3d^3\ 4s^2$. Principal valencies: 2, 3, 4, 5.

18.4.1 THE ELEMENT

Vanadium resembles titanium in its physical and chemical properties at room temperature, but is highly reactive at increased temperatures.

18.4.2 THE DIVALENT STATE
(+2 oxidation state) $3d^3$

This is the least stable state of vanadium. The lavender coloured hydrated vanadium(II) ion, $[V(H_2O)_6]^{2+}$, is formed by reducing acidic solutions of other vanadium compounds with zinc. Vanadium(II) compounds are powerful reducing agents.

18.4.3 THE TRIVALENT STATE
(+3 oxidation state) $3d^2$

Vanadium(III) oxide, V_2O_3, is basic and dissolves in acids to produce the green hydrated vanadium(III) ion, $[V(H_2O)_6]^{3+}$. This ion is also formed by the reduction of compounds of pentavalent or tetravalent vanadium with zinc and hydrochloric acid, or by electrolysis.

Vanadium(III) compounds are easily oxidised to vanadium(IV) or vanadium(V) compounds.

18.4.4 THE TETRAVALENT STATE
(+4 oxidation state) $3d^1$

Because vanadium(IV) compounds possess no marked oxidising or reducing properties, this is one of the commoner states of vanadium.

The dark blue amphoteric vanadium(IV) oxide, VO_2, is formed by reducing vanadium(V) oxide with sulphur dioxide. It dissolves in acids to form the hydrated vanadium(IV) oxide ion, $VO^{2+}(aq)$, which is believed to be $[VO(H_2O)_5]^{2+}$. Complex polymeric vanadate(IV) anions are formed when vanadium(IV) oxide dissolves in alkalis. Vanadium(IV) chloride, VCl_4, which is prepared by heating vanadium in chlorine, resembles titanium(IV) chloride in its chemical properties.

18.4.5 THE PENTAVALENT STATE (+5 oxidation state) $3d^0$

The pentavalent state for vanadium corresponds to the formal loss of all five outer electrons. The resemblance between vanadium and the metallic elements of group 5B, however, is much less marked than between titanium(IV) and tin(IV) compounds. Acidic solutions of vanadium(V) compounds are moderately powerful oxidants.

V_2O_5

Vanadium(V) oxide, V_2O_5, an orange-coloured powder, is prepared by direct combination of the heated elements. It is amphoteric and dissolves readily in alkalis to produce anionic vanadate(V) species, e.g. VO_4^{3-}. In acids the yellow vanadium(V) oxide ion, $VO_2^+(aq)$, is formed.

The only binary halide of vanadium(V) is the fluoride, VF_5. Covalent halides of vanadium(V) containing oxygen are also known, e.g. vanadium(V) oxide trichloride, $VOCl_3$.

Vanadium(V) compounds are used as catalysts in the manufacture of sulphuric acid (§16.2.6).

Summary

The colours of vanadium compounds in the four different states can conveniently be demonstrated by reducing an acidic solution of a vanadium(V) compound with zinc. The colour changes observed are as follows:

valency state	5	4	3	2
species present	$VO_2^+(aq)$	$VO^{2+}(aq)$	$[V(H_2O)_6]^{3+}$	$[V(H_2O)_6]^{2+}$
colour	yellow†	blue†	green	lavender

the colour changes in this sequence as reduction proceeds →

† A green colour is often observed, due to the combined presence of pentavalent and tetravalent species.

18.5 Chromium

Outer electronic configuration: $3d^5\ 4s^1$. Principal valencies: 2, 3, 6.

18.5.1 THE ELEMENT

Pure chromium is obtained either by reducing chromium(III) oxide with aluminium at a high temperature:

$$Cr_2O_3 + 2Al = Al_2O_3 + 2Cr$$

or by the electrolysis of an aqueous solution of chromium(VI) oxide, CrO_3, to which a small quantity of sulphuric acid has been added. It is a white metal with a bluish tinge. The metal dissolves readily in non-oxidising acids, but is passivated (§15.2.9) by oxidising acids such as nitric acid. At elevated temperatures chromium combines directly with many non-metals, e.g. the halogens, oxygen and sulphur, to form chromium(III) compounds.

Chromium is highly resistant to atmospheric corrosion, which accounts for its extensive use in chromium plating and stainless steels. The latter are alloys of iron, chromium and nickel.

18.5.2 THE DIVALENT STATE
(+2 oxidation state) $3d^4$

The turquoise hexaaquachromium(II) ion, $[Cr(H_2O)_6]^{2+}$, is prepared either by dissolving chromium in non-oxidising acids, or by reducing acidic solutions of chromium(III) or chromium(VI) compounds with zinc. Like most chromium(II) species, the hydrated chromium(II) ion is easily oxidised, e.g. by oxygen:

$$4[Cr(H_2O)_6]^{2+} + 2H_2O + O_2 = 4[Cr(H_2O)_6]^{3+} + 4HO^-$$

18.5.3 THE TRIVALENT STATE
(+3 oxidation state) $3d^3$

Chromium(III) compounds possess no marked reducing or oxidising properties, and the trivalent state is therefore considered to be the most stable for chromium. Perhaps the most distinctive feature of chromium(III) is its strong tendency to form complexes which are noted for their kinetic inertness, i.e. the rate of substitution of one ligand by another is very low. Because of this, a large number of chromium(III) complexes have been isolated and studied. Most of them are octahedral, and vary in colour from green to grey-blue depending on the ligands present (e.g. §18.1.2).

In aqueous solution, the hexaaquachromium(III) ion, $[Cr(H_2O)_6]^{3+}$, resembles the hydrated aluminium ion, $[Al(H_2O)_6]^{3+}$, in that the small, highly charged cation (Cr^{3+}) distorts the coordinated water molecules so that hydrolysis occurs (§13.2.4):

$$[Cr(H_2O)_6]^{3+} + H_2O \rightleftharpoons [Cr(OH)(H_2O)_5]^{2+} + H_3O^+ \quad pK_1 = 3.9$$

The addition of bases, e.g. alkalis, ammonia, sulphide ions or carbonate ions, to hydrated chromium(III) ions produces a green precipitate of hydrated chromium(III) oxide, $Cr_2O_3(aq)$, which is often referred to as 'chromium(III) hydroxide'. Anhydrous chromium(III) oxide, Cr_2O_3, is a green powder which is prepared by heating the hydrated oxide or ammonium dichromate (§15.2.3). The hydrated oxide and the anhydrous oxide are amphoteric, dissolving in acids to form the $[Cr(H_2O)_6]^{3+}$ ion, and in alkalis to produce a green solution containing anionic chromate(III) species, believed to be $[Cr(OH)_4(H_2O)_2]^-$, $[Cr(OH)_5(H_2O)]^{2-}$ and $[Cr(OH)_6]^{3-}$.

18.5.4 THE HEXAVALENT STATE
(+6 oxidation state) $3d^0$

CrO_3

Stable compounds of chromium(VI) exist only in combination with oxygen.

Chromium(VI) oxide (chromium trioxide)

Red-brown crystals of chromium(VI) oxide, CrO_3, are precipitated by the addition of cold concentrated sulphuric acid to a saturated solution of a dichromate. Chromium(VI) oxide is extremely soluble in water to give an acidic solution which is known as 'chromic acid', although the exact nature of the species present in the solution is far from certain.

Two well-defined series of salts are derived from chromium(VI) oxide. They are the *chromates*, and the *dichromates*.

Chromates

The chromate ion (ASE: chromate(VI)) is obtained by oxidising chromium(III) salts with the peroxide ion, O_2^{2-}, under alkaline conditions. Sodium peroxide or hydrogen peroxide are convenient reagents:

$$2Cr^{3+}(aq) + 4HO^- + 3O_2^{2-} = 2CrO_4^{2-} + 2H_2O$$

Chromates are also formed by the fusion of a chromium(III) compound with potassium chlorate.

Na_2CrO_4 and K_2CrO_4

Many yellow crystalline salts containing the tetrahedral CrO_4^{2-} ion are known, e.g. Na_2CrO_4, K_2CrO_4 and $BaCrO_4$. Silver chromate, Ag_2CrO_4, is unusual in being a brick-red colour. Only the chromates of the alkali metals, ammonium, magnesium and calcium are soluble in water. The chromate ion is stable in neutral or alkaline solution, but in acidic solution immediately changes into the dichromate ion:

$$\underset{\text{yellow}}{2CrO_4^{2-}} + 2H^+(aq) = \underset{\text{orange}}{Cr_2O_7^{2-}} + H_2O$$

If alkali is added to a solution of a dichromate the reverse change occurs,

i.e. $Cr_2O_7^{2-} + 2HO^- = 2CrO_4^{2-} + H_2O$

The chromate ion is also produced from dichromate solutions by the addition of cations that form soluble dichromates but insoluble chromates:

$$Cr_2O_7^{2-} + 2M^{2+} + H_2O = 2MCrO_4(s) + 2H^+(aq) \quad (M^{2+} = Pb^{2+} \text{ or } Ba^{2+})$$

Dichromates

The dichromate ion (ASE: dichromate(VI)) consists of two tetrahedral CrO_4 units bonded together by a common oxygen atom,

i.e. 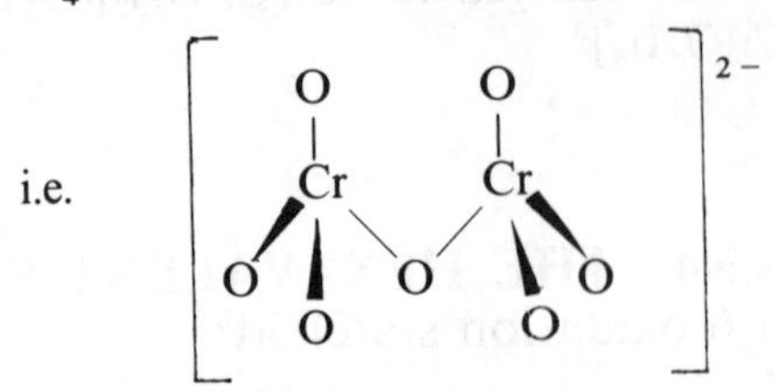

In acidic media it possesses powerful oxidising properties, and the sodium and potassium salts are commonly used for this purpose in the laboratory.

$$Cr_2O_7^{2-} + 14H^+(aq) + 6e^- = 2Cr^{3+} + 7H_2O \qquad E^{\ominus} = +1.33\text{ V}$$

Chromates are weaker oxidising agents than dichromates, as evidenced by the more negative $E^{\ominus}$ value:

$$2CrO_4^{2-} + 5H_2O + 6e^- = Cr_2O_3(aq) + 10HO^- \quad E^{\ominus} = -0.13\ V$$

$Na_2Cr_2O_7$ and $K_2Cr_2O_7$

Sodium dichromate is often used as an oxidant, especially in organic preparations. It is deliquescent. Potassium dichromate is not deliquescent and can be obtained in a high state of purity. Consequently it is used as a primary standard in volumetric analysis; principally for the determination of iron(II) in acidic solution:

$$Cr_2O_7^{2-} + 14H^+(aq) + 6Fe^{2+} = 2Cr^{3+} + 6Fe^{3+} + 7H_2O$$

The dichromate solution is placed in the burette and the acidified iron(II) solution in the conical flask. A *redox indicator* is necessary to detect the end-point of the titration. Such an indicator is barium diphenylamine-sulphonate, a colourless substance which is converted to a deep blue compound by oxidants. Thus, the end-point of the titration is observed as a deep blue colour with the first drop of dichromate solution in excess. The hydrated iron(III) ion can interfere, because it, too, is able to oxidise the indicator, turning it deep blue. However, phosphate complexes of iron(III) do not affect the indicator, and for this reason phosphoric acid is added to the iron(II) solution before starting the titration.

Unlike permanganates, dichromates do not oxidise chloride ions, since the $E^{\ominus}$ value for Cl_2/Cl^- is more positive than that for $Cr_2O_7^{2-}/Cr^{3+}$ (§2.5.3):

$$\tfrac{1}{2}Cl_2(g) + e^- = Cl^- \quad E^{\ominus} = +1.36\ V$$

Dichromates can therefore be used for the titration of iron(II) in solutions containing chloride ions.

In addition to oxidising iron(II), the dichromate ion readily oxidises sulphur dioxide or sulphite ions to sulphate ions, and iodide ions to iodine.

Chromyl chloride (ASE: chromium(VI) dichloride dioxide)

Chromyl chloride, CrO_2Cl_2, the acid chloride of the acid H_2CrO_4, is formed by warming a mixture of a dichromate and a chloride with concentrated sulphuric acid:

$$Na_2Cr_2O_7 + 4NaCl + 3H_2SO_4 = 2CrO_2Cl_2 + 3H_2O + 3Na_2SO_4$$

Chromyl chloride vapour distils off and can be condensed to a red-brown liquid. This reaction can be used as a specific test for ionic chlorides, since neither bromides nor iodides form the corresponding chromyl compounds.

Chromyl chloride is hydrolysed by water to chromic acid and hydrochloric acid. With alkalis it reacts to form chromates:

$$CrO_2Cl_2 + 4HO^- = CrO_4^{2-} + 2Cl^- + 2H_2O$$

Summary

The colours of chromium compounds in the three principal valency states can be demonstrated by reducing an acidified dichromate solution with zinc. The colour changes which are observed are as follows:

valency state	6	3	2
species present	$Cr_2O_7^{2-}$	complexed Cr^{3+} ions	$[Cr(H_2O)_6]^{2+}$
colour	orange	green	turquoise

colour change as reduction proceeds →

Some reactions of $[Cr(H_2O)_6]^{3+}$

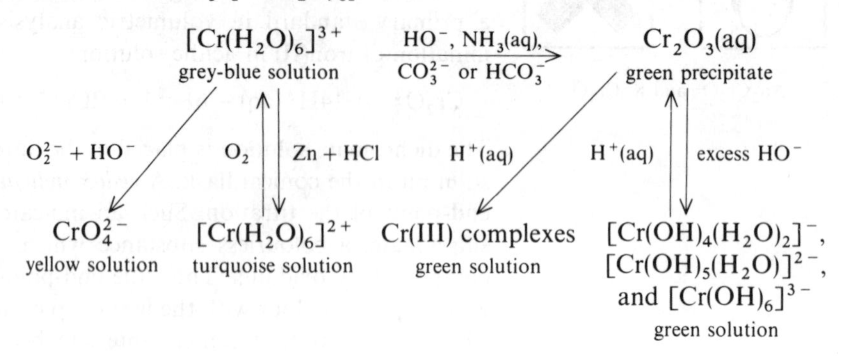

18.6 Manganese

Outer electronic configuration: $3d^5\ 4s^2$. Principal valencies: 2, 4, 6, 7.

18.6.1 THE ELEMENT

Manganese can be obtained by the electrolysis of aqueous manganese(II) salts. It dissolves readily in non-oxidising acids to produce manganese(II) salts, and reacts with various non-metals on heating. It is used principally as an alloying ingredient in steels.

18.6.2 THE DIVALENT STATE (+2 oxidation state) $3d^5$

This is the most stable state for manganese. Neutral or acidic solutions of manganese(II) salts are very pale pink and contain the hexaaqua-manganese(II) ion, $[Mn(H_2O)_6]^{2+}$. The addition of alkali to aqueous manganese(II) salts precipitates white manganese(II) hydroxide, $Mn(OH)_2(aq)$, which is rapidly oxidised by atmospheric oxygen to give brown hydrated manganese(III) oxide, $Mn_2O_3(aq)$. Both the oxide and hydroxide of manganese(II) are basic.

18.6.3 THE TETRAVALENT STATE (+4 oxidation state) $3d^3$

MnO_2

The only important stable compound of tetravalent manganese is the ionic manganese(IV) oxide, MnO_2. The stability of this compound is due to its high lattice enthalpy (§16.2.8). Manganese(IV) oxide is normally en-countered in the anhydrous form as a black powder, which is prepared by heating hydrated manganese(IV) oxide or manganese(II) nitrate:

$$Mn(NO_3)_2 = MnO_2 + 2NO_2$$

A brown hydrated form of manganese(IV) oxide, $MnO_2(aq)$, with a variable water content, is obtained by reducing permanganate ions in alkaline solution (§18.6.5) or by oxidising manganese(II) ions with peroxodisulphate ions, $S_2O_8^{2-}$, or hypochlorite ions, ClO^-:

$$\text{e.g.}\quad Mn^{2+} + S_2O_8^{2-} + 2H_2O = MnO_2(aq) + 4H^+(aq) + 2SO_4^{2-}$$

Because of its high lattice enthalpy manganese(IV) oxide is insoluble in water.

Manganese(IV) oxide is a powerful oxidising agent, particularly under acidic conditions. For example, warm concentrated hydrochloric acid is oxidised to chlorine (§17.2.1).

Manganese(IV) oxide functions as a catalyst in many reactions (Table 18.8).

18.6.4 THE HEXAVALENT STATE (+6 oxidation state) $3d^1$

The deep green tetrahedral manganate ion, MnO_4^{2-} (ASE: manganate(VI)), is the only known species of manganese(VI). It is prepared by fusing a mixture of manganese(IV) oxide and potassium hydroxide with an oxidant such as potassium chlorate or potassium nitrate:

$$3MnO_2 + 6HO^- + ClO_3^- = 3MnO_4^{2-} + 3H_2O + Cl^-$$

Alternatively, the manganate ion can be obtained by reducing permanganate ions with manganese(IV) oxide in a strongly alkaline solution. The reaction is reversible:

$$2MnO_4^- + MnO_2 + 4HO^- \rightleftharpoons 3MnO_4^{2-} + 2H_2O$$

The manganate ion is stable only in alkaline solution. In neutral or acidic media the equilibrium is displaced to the left by the removal of hydroxide ions, and the manganate ion disproportionates into manganese(IV) oxide and permanganate ions.

18.6.5 THE HEPTAVALENT STATE (+7 oxidation state) $3d^0$

$KMnO_4$

Potassium permanganate, $KMnO_4$, a deep purple compound which contains the tetrahedral permanganate ion, MnO_4^- (ASE: manganate(VII)), is the most familiar compound of heptavalent manganese. The permanganate ion can be produced from various other species of manganese, including:

(i) the manganate ion, by disproportionation (see above), by electrolytic oxidation, or by oxidation with chlorine:

$$2MnO_4^{2-} + Cl_2 = 2MnO_4^- + 2Cl^-$$

(ii) the manganese(II) ion, by oxidation, in acidic solution, with powerful oxidants such as sodium bismuthate, $NaBiO_3$, or the periodate ion, IO_4^-,

(iii) the element itself, by anodic oxidation. For this purpose, an aqueous alkali metal carbonate is electrolysed using a manganese anode.

In solution the permanganate ion is unstable:

$$4MnO_4^- + 4H^+(aq) = 4MnO_2(aq) + 2H_2O + 3O_2$$

The reaction occurs slowly in acidic solution but is accelerated by light. For this reason, standard permanganate solutions are stored in dark bottles. In strongly alkaline solution the permanganate ion rapidly decomposes into the manganate ion:

$$4MnO_4^- + 4HO^- = 4MnO_4^{2-} + 2H_2O + O_2$$

The permanganate ion is a powerful oxidising agent. In acidic solution it is reduced to the manganese(II) ion:

$$MnO_4^- + 8H^+(aq) + 5e^- = Mn^{2+} + 4H_2O \qquad E^\ominus = +1.52\ V$$

If the permanganate is present in excess, however, hydrated manganese(IV) oxide is obtained by the oxidation of manganese(II) ions:

$$2MnO_4^- + 3Mn^{2+} + 2H_2O = 5MnO_2(aq) + 4H^+(aq)$$

Neutral or alkaline solutions of permanganates are also oxidants, and under these conditions hydrated manganese(IV) oxide is produced:

$$MnO_4^- + 2H_2O + 3e^- = MnO_2(aq) + 4HO^- \qquad E^\ominus = +1.67\ V$$

In strongly alkaline solution, however, and with an excess of permanganate, the manganate ion is formed:

$$MnO_4^- + e^- = MnO_4^{2-} \qquad E^\ominus = +0.56\ V$$

Aqueous potassium permanganate is widely used as a laboratory oxidant, in both volumetric analysis and preparative chemistry.

Redox titrations with potassium permanganate

Acidic solutions of reductants, notably iron(II) salts, ethanedioates, nitrites and hydrogen peroxide, are rapidly and quantitatively oxidised by potassium permanganate. Under these conditions the manganese(II) ion is the reduction product of the permanganate ion. Solutions of these reductants can therefore be analysed with a standard solution of the permanganate (§10.1.6 and §1.1.3). The acid used is usually sulphuric acid. Hydrochloric acid is unsuitable for the purpose because it is oxidised to chlorine by permanganate ions, and nitric acid is unsuitable since it, also, is an oxidising agent. The end-point of the titration is recognised by the appearance of the first permanent pink coloration in the solution being titrated. Potassium permanganate is not used as a primary standard because it is difficult to obtain in a high state of purity, and its solutions are not stable over long periods of time.

Preparative reactions of potassium permanganate

Alkaline solutions of potassium permanganate are commonly used as oxidants in organic chemistry (§14.4.4 and §14.9.5). Under these conditions the final reduction product of the permanganate ion is hydrated manganese(IV) oxide, although the green manganate ion is often formed as an intermediate.

Some reactions of $[Mn(H_2O)_6]^{2+}$

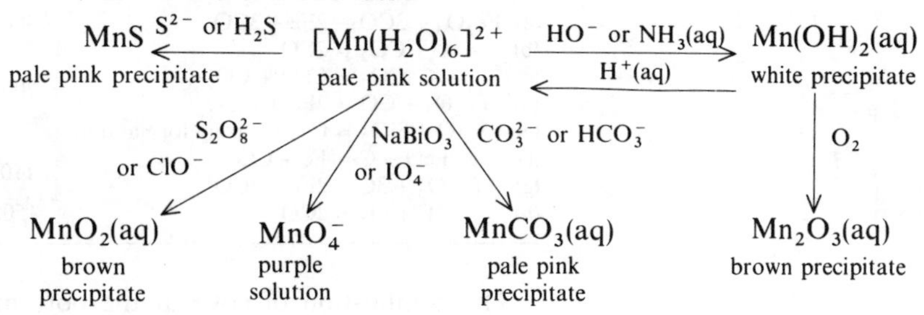

18.7 Iron

Outer electronic configuration: $3d^6\ 4s^2$. Principal valencies: 2, 3.

18.7.1 THE ELEMENT

Pure iron, in contrast to most steels, is rather soft, malleable and ductile. It is fairly reactive and is attacked by moist air to form, ultimately, hydrated iron(III) oxide, $Fe_2O_3(aq)$, i.e. rust. Non-oxidising acids attack the metal to form hydrogen and, in the absence of air, iron(II) salts:

$$Fe + 2H^+(aq) = Fe^{2+} + H_2$$

In the presence of air some iron(II) is oxidised to iron(III):

$$4Fe^{2+} + O_2 + 4H^+(aq) = 4Fe^{3+} + 2H_2O$$

Warm dilute nitric acid dissolves iron to give iron(II) and iron(III) nitrates and oxides of nitrogen, i.e. reduction products of nitric acid, but concentrated nitric acid renders the metal passive. On heating, iron combines with many non-metals, notably oxygen, sulphur and the halogens.

Pure iron is seldom used in industry. Steel, which is iron containing between 0.1 and 1.5% of carbon, or alloys of iron with other metals, are used instead. The production and metallurgy of steel and alloys of iron will be discussed only briefly here.

Crude iron is extracted from its principal ore, iron(III) oxide (haematite), Fe_2O_3, by reduction in a blast furnace, which is essentially a vertical furnace with a temperature gradient ranging from approximately 1500 °C (1 773 K) at the base to 400 °C (673 K) at the top. A mixture of the ore, coke and limestone, $CaCO_3$, is fed into the top of the furnace and air, preheated to 600 °C (873 K), is forced in at the bottom. Many reactions, some highly complex, occur in the blast furnace. The more important, together with the approximate temperatures at which they take place, are outlined below.

Reaction	Temperature/°C (K)
(a) $Fe_2O_3 + 3CO \rightleftharpoons 2Fe + 3CO_2$	400–700 (673–973)
(b) $C + CO_2 = 2CO$	700–900 (973–1173)
(c) $CaCO_3 = CaO + CO_2$	
(d) $Fe_2O_3 + CO = 2FeO + CO_2$	
(e) $CaO + SiO_2 = CaSiO_3$ (slag formation)	
(f) $FeO + C = Fe + CO$	1100 (1373)
(g) $Fe_2O_3 + 3C = 2Fe + 3CO$	
(h) $2C + O_2 = 2CO$	1500 (1773)

The combustion of coke at the bottom of the furnace, equation (h), provides both the energy and the principal reducing agent, namely carbon monoxide, for the production of iron. Reaction (a) is reversible and exothermic in the forward direction, and is therefore favoured by relatively low temperatures and a high ratio of carbon monoxide to carbon dioxide (§8.3.3). In the lower, hotter regions of the furnace other iron-forming reactions occur (equations (f) and (g)). Iron is produced in the semi-molten state and sinks to the bottom of the furnace from where it is drawn off periodically. Calcium oxide, formed by the decomposition of limestone, equation (c), combines with the earthy and sandy impurities present in the ore to give a molten slag, represented in simplified form by equation (e). The slag, also, sinks to the base of the furnace but floats on top of the molten iron and is periodically tapped from a separate hole.

Crude iron from the blast furnace contains up to 5% of impurities, chiefly carbon, phosphorus and sulphur, which tend to make the metal hard and rather brittle. While it is satisfactory for use as cast iron, for most applications the crude iron is refined to produce various steels. The steel making process entails the removal of most of the impurities by oxidation at high temperatures. Other metals, e.g. manganese, chromium, nickel, tungsten or vanadium, may be added to the steel to produce alloys with special properties, such as hardness or resistance to corrosion.

18.7.2 THE DIVALENT STATE (+2 oxidation state) $3d^6$

Many compounds of iron(II) are known, ranging from simple binary compounds to those containing complexes of iron(II). Most iron(II) compounds are reducing agents, since they are easily oxidised to the corresponding iron(III) compounds.

The binary compounds of iron(II)

Iron(II) oxide is a black powder which is prepared by heating iron(II) ethanedioate in the absence of air:

$$(COO)_2Fe = FeO + CO + CO_2$$

The compound is *pyrophoric*, i.e. it inflames in air to form iron(III) oxide.

Iron(II) sulphide, FeS, is prepared by heating a mixture of iron filings and sulphur, or by introducing sulphide ions into a neutral iron(II) solution.

The anhydrous halides of iron(II) are obtained by heating iron in a stream of the appropriate hydrogen halide:

$$Fe + 2HX = FeX_2 + H_2 \quad (X = F, Cl, Br \text{ or } I)$$

Iron(II) iodide can also be made by heating iron with iodine, because iron(III) iodide does not exist. The hydrated halides are prepared by adding water to the anhydrous compounds, or by dissolving iron in the appropriate hydrohalic acid. Iron(II) halides are essentially ionic.

The complexes of iron(II)

Iron(II) forms a large number of complexes, the commonest being the pale green hexaaquairon(II) ion, $[Fe(H_2O)_6]^{2+}$. Several crystalline salts containing this ion are known, e.g. $FeSO_4{\cdot}7H_2O$ and $FeSO_4{\cdot}(NH_4)_2SO_4{\cdot}6H_2O$. Other examples of six coordinate iron(II) complexes include the ion $[Fe(CN)_6]^{4-}$ and the complex $[FeCl_2(H_2O)_4]$ which is present in $FeCl_2{\cdot}6H_2O$. Haemoglobin is also an iron(II) complex.

The oxidation of iron(II)

All iron(II) compounds are capable of being oxidised to iron(III) species. The ease with which oxidation occurs depends on various factors, such as the pH of the solution and the ligands that are coordinated to the iron(II) ion. Most crystalline iron(II) salts are susceptible, at the very least, to superficial aerial oxidation, and some, e.g. $FeSO_4{\cdot}7H_2O$, are also efflorescent. However, the crystalline double salt, ammonium iron(II) sulphate hexahydrate, $FeSO_4{\cdot}(NH_4)_2SO_4{\cdot}6H_2O$, is sufficiently stable to be used as a primary standard in volumetric analysis.

In acidic solution the hexaaquairon(II) ion is oxidised only slowly by the atmosphere, despite the favourable $E^{\ominus}$ values for the half-reactions:

$$[Fe(H_2O)_6]^{3+} + e^- = [Fe(H_2O)_6]^{2+} \qquad E^{\ominus} = +0.77 \text{ V}$$

$$\tfrac{1}{2}O_2 + 2H^+(aq) + 2e^- = H_2O \qquad E^{\ominus} = +1.23 \text{ V}$$

However, more powerful oxidants, such as permanganate or dichromate ions, or hydrogen peroxide, rapidly oxidise iron(II) to iron(III) in acidic solution.

In alkaline conditions, where the insoluble hydrated hydroxide, $Fe(OH)_2(aq)$, exists, iron(II) is oxidised more readily than in acidic solution. This is indicated by the less positive $E^{\ominus}$ value for the half-reaction:

$$\tfrac{1}{2}Fe_2O_3(aq)(s) + \tfrac{3}{2}H_2O + e^- = Fe(OH)_2(aq)(s) + HO^- \quad E^{\ominus} = -0.56 \text{ V}$$

When alkali is added to a solution containing $[Fe(H_2O)_6]^{2+}$ ions, a *white* precipitate of iron(II) hydroxide is obtained in the complete absence of oxygen. On exposure to air iron(II) hydroxide is rapidly oxidised to a dark green iron(III) compound, formulated as FeO(OH), and finally to a brown sludge of hydrated iron(III) oxide, $Fe_2O_3(aq)$. Interestingly, iron(II) hydroxide is weakly amphoteric, since it dissolves in acids to form iron(II) salts and in *concentrated* alkalis to yield the ferrate(II) ion, $[Fe(OH)_6]^{4-}$. Iron(II) oxide, by contrast, appears to be exclusively basic.

The iron(II)–iron(III) system provides some excellent examples of the effect of ligands on the relative stability of different valency states. The water ligands of the hexaaquairon(II) ion can easily be substituted by six cyanide ions to form the yellow hexacyanoferrate(II) ion, $[Fe(CN)_6]^{4-}$. The $E^{\ominus}$ values show that the iron(II) cyanide complex is oxidised to the corresponding iron(III) complex, i.e. $[Fe(CN)_6]^{3-}$, more readily than the hydrated iron(II) ion is oxidised to the hexaaquairon(III) ion:

$$[Fe(CN)_6]^{3-} + e^- = [Fe(CN)_6]^{4-} \qquad E^{\ominus} = +0.36 \text{ V}$$

Thus, by changing water ligands for cyanide ligands, the trivalent state of iron is stabilised relative to the divalent state, i.e. the $[Fe(CN)_6]^{4-}$ ion is a better reducing agent than $[Fe(H_2O)_6]^{2+}$. Despite this, the hexacyanoferrate(II) ion, in contrast to the aqua-complex, is not oxidised by the atmosphere. This suggests that a kinetic effect may possibly be operating, i.e. the cyanide complex is *able* to be oxidised by oxygen but the *rate* of reaction is extremely low. The high stability of the cyanide complex may be one reason for this. The stability of the $[Fe(CN)_6]^{4-}$ ion is illustrated by the fact that, unlike the hexaaqua ion, it gives no precipitate when treated with either sulphide or hydroxide ions.

Another example is provided by the complex $[Fe(phen)_3]^{2+}$, where *phen* represents the bidentate ligand 1 : 10-phenanthroline. This complex is more stable to oxidation than both hexaaquairon(II) and hexacyanoferrate(II) ions, as shown by the potential:

$$[Fe(phen)_3]^{3+} + e^- = [Fe(phen)_3]^{2+} \qquad E^{\ominus} = +1.12 \text{ V}$$

18.7.3 THE TRIVALENT STATE (+3 oxidation state) $3d^5$

Energetically, the most stable oxidation state of iron is +2. However, because iron(II) is oxidised relatively easily to iron(III) by many oxidants, including atmospheric oxygen, it often appears that the most stable state is +3. The iron(III) ion is isoelectronic with the manganese(II) ion, but there are few similarities between them.

The binary compounds of iron(III)

Trivalent iron does not appear to form a hydroxide. When alkali is added to an aqueous iron(III) salt a brown precipitate of hydrated iron(III) oxide, $Fe_2O_3(aq)$, is obtained:

$$2Fe^{3+} + 6HO^- = Fe_2O_3(aq) + 3H_2O$$

Anhydrous iron(III) oxide is ionic and can be prepared as a red-brown powder by heating either the hydrated oxide or iron(II) sulphate:

$$2FeSO_4 \cdot 7H_2O = Fe_2O_3 + SO_2 + SO_3 + 14H_2O$$

Iron(III) oxide is amphoteric. It dissolves in acids to form iron(III) salts, and with difficulty in *concentrated* alkali solutions to form the ferrate(III) ion, $[Fe(OH)_6]^{3-}$.

$FeCl_3$

The anhydrous fluoride, chloride and bromide of iron(III) are prepared by heating iron in a stream of the appropriate halogen. The iodide does not appear to exist in the pure state. Iron(III) chloride, the most important halide of iron(III), is an almost black hygroscopic solid. It is a covalent compound, with a complex polymeric structure at room temperature. When heated, iron(III) chloride melts and vaporises with the formation of the dimer, Fe_2Cl_6, whose molecules possess a similar structure to that of the Al_2Cl_6 molecule (§13.2.6). Stronger heating produces trigonal planar molecules of the monomer, $FeCl_3$, and finally, by decomposition, iron(II) chloride and chlorine. Iron(III) chloride dissolves in water to form yellow chloro-complexes (see below).

Sulphide ions reduce hydrated iron(III) species:

$$2Fe^{3+}(aq) + S^{2-} = 2Fe^{2+}(aq) + S(s)$$

Thus, when hydrogen sulphide is passed into a neutral solution of an iron(III) salt, a black precipitate of *iron(II)* sulphide and a yellow precipitate of sulphur are obtained together. Iron(III) species also oxidise iodide ions, which accounts for the instability of iron(III) iodide:

$$2Fe^{3+}(aq) + 2I^- = 2Fe^{2+}(aq) + I_2$$

The complexes of iron(III)

Many complexes of iron(III) are known, particularly with ligands that coordinate through an oxygen atom. The pale violet hexaaquairon(III) ion, $[Fe(H_2O)_6]^{3+}$, occurs in iron alum (§13.2.7). In solution, however, hydrolysis occurs to give yellow hydroxo-species:

$$[Fe(H_2O)_6]^{3+} + H_2O \rightleftharpoons [Fe(OH)(H_2O)_5]^{2+} + H_3O^+ \quad pK_1 = 2.22$$

$$[Fe(OH)(H_2O)_5]^{2+} + H_2O \rightleftharpoons [Fe(OH)_2(H_2O)_4]^+ + H_3O^+$$

Only in strongly acidic solutions, therefore, does the hexaaquairon(III) ion exist. Hydrolysis occurs because the small, highly charged iron(III) ion distorts the coordinated water molecules. The addition of hydroxide, carbonate or hydrogencarbonate ions to a solution of hydrated iron(III) ions produces a precipitate of hydrated iron(III) oxide (cf. chromium and aluminium, §18.5.3 and §13.2.4),

$$\text{e.g.} \quad 2[Fe(OH)_2(H_2O)_4]^+ + CO_3^{2-} = Fe_2O_3(aq) + 10H_2O + CO_2$$

Iron(III) carbonate is not formed because the carbonate ion simply acts as a base towards the hydrated iron(III) ion. The hexaaquairon(II) complex is far less acidic than the corresponding iron(III) species, because the iron(II) ion distorts the coordinated water molecules to a lesser extent than the iron(III) ion. Consequently, iron(II) salts react with carbonates or hydrogencarbonates to form a white precipitate of hydrated iron(II) carbonate, $FeCO_3(aq)$.

In solutions of hexaaquairon(III) ions that contain chloride ions, the yellow chloro-complexes $[FeCl(H_2O)_5]^{2+}$ and $[FeCl_2(H_2O)_4]^+$ are formed. In the presence of a large excess of chloride ions the orange-yellow tetrahedral tetrachloroferrate(III) ion is produced:

$$Fe^{3+} + 4Cl^- \rightleftharpoons [FeCl_4]^-$$

Other important complexes of iron(III) include the orange coloured hexacyanoferrate(III) ion, $[Fe(CN)_6]^{3-}$, and the deep red thiocyanate complexes, e.g. $[Fe(SCN)(H_2O)_5]^{2+}$, which are formed when thiocyanate ions, SCN^-, are added to hydrated iron(III) ions. This reaction may be used as a sensitive test for the presence of hydrated iron(III) ions, since iron(II) does not form coloured complexes with the thiocyanate ion.

The hexacyanoferrate ions

Potassium salts containing these ions are used in the laboratory in testing for iron(II) and iron(III). The appropriate reactions are summarised below.

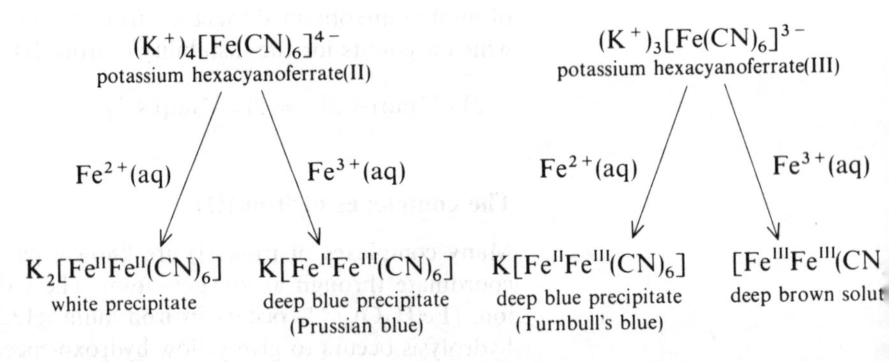

Turnbull's blue and Prussian blue appear to be similar compounds.

Compounds containing iron in two valency (oxidation) ***states***

The only important compound in this category is triiron tetraoxide, or iron(II) diiron(III) oxide, Fe_3O_4, a black solid which occurs naturally as the mineral magnetite. It can be prepared by passing steam over iron at 800 °C (1 073 K).

$$3Fe + 4H_2O \rightleftharpoons Fe_3O_4 + 4H_2$$

Triiron tetraoxide is a mixed oxide, which dissolves in acids to form iron(II) and iron(III) salts in a 1:2 molar ratio:

$$Fe_3O_4 + 8H^+(aq) = Fe^{2+} + 2Fe^{3+} + 4H_2O$$

Some reactions of $[Fe(H_2O)_6]^{2+}$

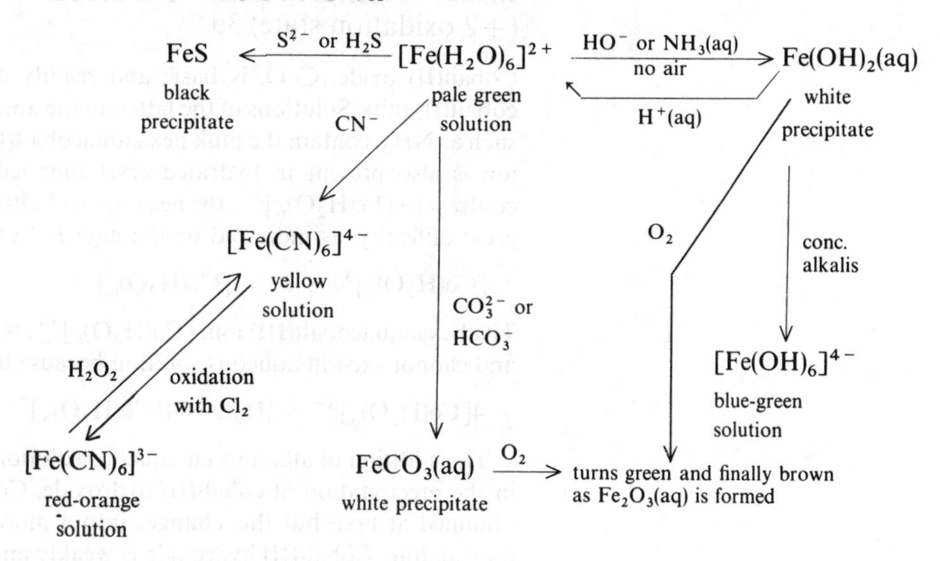

Some reactions of $[Fe(H_2O)_6]^{3+}$

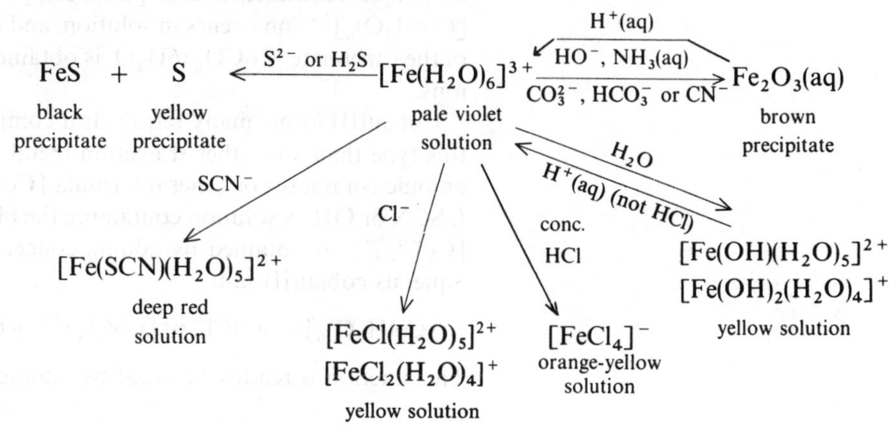

18.8 Cobalt

Outer electronic configuration: $3d^7\ 4s^2$. Principal valencies: 2, 3.

18.8.1 THE ELEMENT

Cobalt can be obtained by the electrolysis of an aqueous solution of a cobalt(II) salt. It is a hard metal and is used principally in the production of special alloys. Chemically, cobalt resembles iron but is less reactive. It dissolves only slowly in non-oxidising mineral acids, with the liberation of hydrogen and the formation of cobalt(II) salts, and is rendered passive by concentrated nitric acid.

18.8.2 THE DIVALENT STATE (+2 oxidation state) $3d^7$

Cobalt(II) oxide, CoO, is basic and readily dissolves in acids to form cobalt(II) salts. Solutions of the latter, in the absence of complexing ligands such as NH_3, contain the pink hexaaquacobalt(II) ion, $[Co(H_2O)_6]^{2+}$. The ion is also present in hydrated crystalline salts, e.g. $CoSO_4 \cdot 7H_2O$. In contrast to $[Fe(H_2O)_6]^{2+}$, the hexaaquacobalt(II) ion is oxidised only with great difficulty, as indicated by the high $E^{\ominus}$ value:

$$[Co(H_2O)_6]^{3+} + e^- = [Co(H_2O)_6]^{2+} \qquad E^{\ominus} = +1.82 \text{ V}$$

The hexaaquacobalt(III) ion, $[Co(H_2O)_6]^{3+}$, is in fact a powerful oxidant, and cannot exist in aqueous solution because it oxidises water to oxygen:

$$4[Co(H_2O)_6]^{3+} + 2H_2O = 4[Co(H_2O)_6]^{2+} + 4H^+(aq) + O_2$$

The addition of alkali to an aqueous solution of a cobalt(II) salt results in the precipitation of cobalt(II) hydroxide, $Co(OH)_2(aq)$. A blue form is obtained at first, but this changes into a more stable pink modification on standing. Cobalt(II) hydroxide is weakly amphoteric, since it dissolves in acids to give cobalt(II) salts and in *concentrated* alkalis to form the deep blue cobaltate(II) ion, $[Co(OH)_4]^{2-}$. Very little hydrolysis of the $[Co(H_2O)_6]^{2+}$ ion occurs in solution, and consequently a pink precipitate of the carbonate, $CoCO_3 \cdot 6H_2O$, is obtained on the addition of carbonate ions.

Cobalt(II) forms many tetrahedral complexes; in fact, it forms more of this type than any other transition metal ion. By far the commonest are anionic complexes of general formula $[CoX_4]^{2-}$, where X may be Cl, Br, I, SCN or OH. A solution containing the blue tetrachlorocobaltate(II) ion, $[CoCl_4]^{2-}$, is obtained by adding concentrated hydrochloric acid to an aqueous cobalt(II) salt:

$$[Co(H_2O)_6]^{2+} + 4Cl^- \rightleftharpoons [CoCl_4]^{2-} + 6H_2O$$

The reaction is readily reversed by adding water or by heating.

18.8.3 THE TRIVALENT STATE (+3 oxidation state) $3d^6$

Hydrated cobalt(III) oxide, $Co_2O_3(aq)$, is obtained by oxidising cobalt(II) hydroxide, suspended in an alkaline solution, with oxygen. Apart from this and a few other simple compounds, the chemistry of cobalt in the trivalent state is entirely that of complexes.

The relative stability of the divalent and trivalent states of cobalt depends on the coordinated ligands. With water as ligand the divalent state is the more stable, but the trivalent state becomes dominant when the ligands are ammonia or nitrite ions.

NH_3 ligands

The hexaamminecobalt(II) ion, $[Co(NH_3)_6]^{2+}$, in contrast to the hexaaquacobalt(II) ion, can be easily oxidised to a cobalt (III) ammine complex.

This can be shown by adding aqueous ammonia to a solution of hydrated cobalt(II) ions. The precipitate of cobalt(II) hydroxide, which is obtained initially, dissolves in excess ammonia to form the red-brown hexaamminecobalt(II) ion. The overall equation is:

$$[Co(H_2O)_6]^{2+} + 6NH_3 = [Co(NH_3)_6]^{2+} + 6H_2O$$

The colour slowly darkens on standing in air as oxidation to a cobalt(III) complex occurs. To obtain the orange-yellow hexaamminecobalt(III) ion, $[Co(NH_3)_6]^{3+}$, the oxidation must be performed in the presence of an activated charcoal catalyst, otherwise other complexes are obtained, such as, in the presence of chloride ions, the red pentaamminechlorocobalt(III) ion, $[CoCl(NH_3)_5]^{2+}$.

$$[Co(H_2O)_6]^{2+} \xrightarrow{\text{excess } NH_3(aq)} [Co(NH_3)_6]^{2+} \begin{cases} \xrightarrow{O_2 \text{ or } H_2O_2,\ \text{charcoal}} [Co(NH_3)_6]^{3+} \\ \xrightarrow{O_2 \text{ or } H_2O_2,\ \text{no charcoal},\ Cl^- \text{ present}} [CoCl(NH_3)_5]^{2+} \end{cases}$$

NO_2^- ligands

Oxidation to the trivalent state occurs when ethanoic acid and excess sodium nitrite are added to a cobalt(II) salt. In this case the nitrite ion acts both as an oxidising agent and a ligand:

$$[Co(H_2O)_6]^{2+} + 7NO_2^- + 2H^+(aq) = [Co(NO_2)_6]^{3-} + NO + 7H_2O$$

$Na_3[Co(NO_2)_6]$

Sodium hexanitrocobaltate(III), $Na_3[Co(NO_2)_6]$, is water soluble and is used as a qualitative test for potassium ions. When a solution of this compound is added to an aqueous potassium salt a yellow precipitate of formula $K_2Na[Co(NO_2)_6]$ is obtained.

A large number of other six coordinate complexes of cobalt(III) are known. One particularly important example is vitamin B_{12}.

Some reactions of $[Co(H_2O)_6]^{2+}$

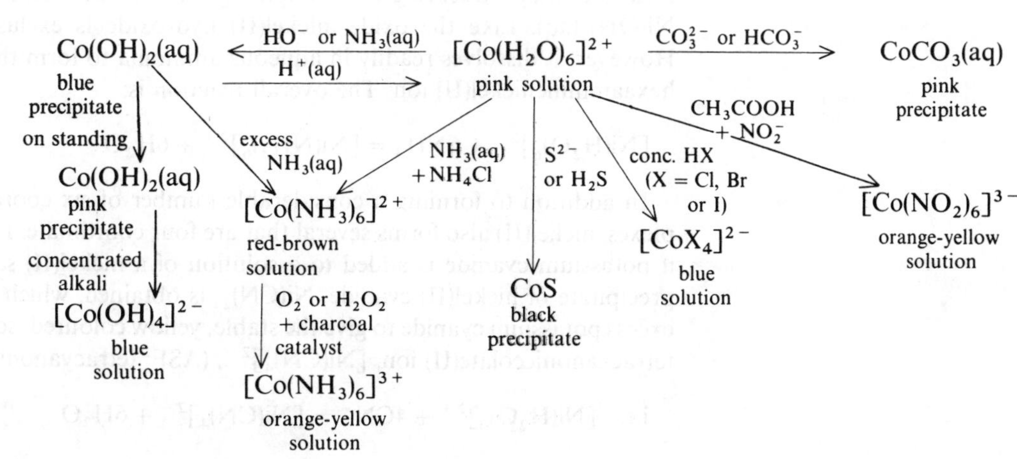

18.9
Nickel

Outer electronic configuration: $3d^8\ 4s^2$. Principal valencies: 2, 3.

18.9.1 THE ELEMENT

Nickel is isolated by the electrolysis of an aqueous solution of a nickel(II) salt. It is a silvery-white metal which is resistant to attack by water or air at room temperature, but reacts with non-oxidising mineral acids to yield hydrogen and nickel(II) salts:

$$Ni + 2H^+(aq) = Ni^{2+} + H_2$$

Contrary to popular belief, nickel is not passivated by concentrated nitric acid, but dissolves readily in the acid, particularly if heated, to produce nickel(II) nitrate and nitrogen dioxide. Alkalis have no effect on nickel.

18.9.2 THE DIVALENT STATE
(+2 oxidation state) $3d^8$

With increasing atomic number, the first series of transition elements displays a marked increase in the stability of the divalent state, coupled with a decline in the stability of higher states. This is clearly evident with nickel, for the divalent state is the only one that is stable and important.

The principal binary compounds of divalent nickel include the oxide, sulphide and anhydrous halides. Nickel(II) oxide, NiO, is a green solid which is prepared by heating the hydroxide, carbonate or ethanedioate of nickel(II). Anhydrous nickel(II) halides are prepared by direct combination of the heated elements.

The most important species of nickel(II) are its complexes. Nickel(II) oxide, which is basic, dissolves in acids to form the green hexaaquanickel(II) ion, $[Ni(H_2O)_6]^{2+}$. A number of crystalline hydrated salts are known which contain this ion, e.g. $NiSO_4 \cdot 6H_2O$ and the double salt $(NH_4)_2SO_4 \cdot NiSO_4 \cdot 6H_2O$. The addition of alkali to a solution of a nickel(II) salt produces a green precipitate of hydrated nickel(II) hydroxide, $Ni(OH)_2(aq)$. Like the oxide, nickel(II) hydroxide is exclusively basic. However, it dissolves readily in aqueous ammonia to form the deep blue hexaamminenickel(II) ion. The overall reaction is:

$$[Ni(H_2O)_6]^{2+} + 6NH_3 = [Ni(NH_3)_6]^{2+} + 6H_2O$$

In addition to forming a considerable number of six coordinate complexes, nickel(II) also forms several that are four coordinate. For example, if potassium cyanide is added to a solution of a nickel(II) salt, a yellow precipitate of nickel(II) cyanide, $Ni(CN)_2$, is obtained, which dissolves in excess potassium cyanide to give the stable, yellow coloured, square planar tetracyanoniccolate(II) ion, $[Ni(CN)_4]^{2-}$, (ASE: tetracyanonickelate(II)),

i.e. $$[Ni(H_2O)_6]^{2+} + 4CN^- = [Ni(CN)_4]^{2-} + 6H_2O$$

The red coloured precipitate formed when 2,3-butanedione dioxime (dimethylglyoxime) is added to a neutral or slightly alkaline solution of a nickel(II) salt is also a square planar complex,

i.e.

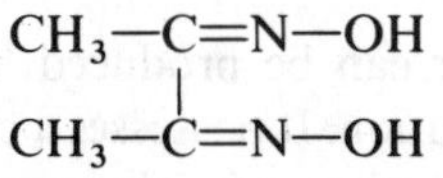

$CH_3-C=N-OH$
$CH_3-C=N-OH$

2,3-butanedione dioxime

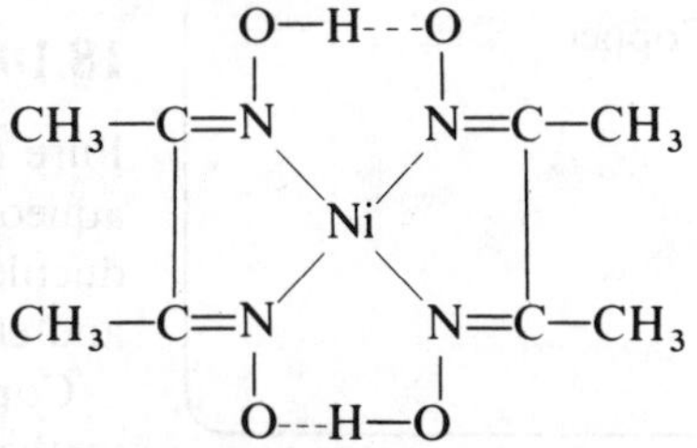

O—H- - -O
$CH_3-C=N$ $N=C-CH_3$
Ni
$CH_3-C=N$ $N=C-CH_3$
O- - -H—O

bis(2,3-butanedione dioximato)nickel(II)
--- represents hydrogen bonds

This reaction is used as a test for nickel(II) ions.

Some four coordinate complexes of nickel are tetrahedral, e.g. the blue tetrachloroniccolate(II) ion, $[NiCl_4]^{2-}$, which is formed when chloride ions and nickel(II) ions react together in ethanol. In water hydrolysis occurs:

$$[NiCl_4]^{2-} + 6H_2O = [Ni(H_2O)_6]^{2+} + 4Cl^-$$

18.9.3 THE OTHER VALENCY (OXIDATION) STATES OF NICKEL

A black insoluble nickel(III) compound, NiO(OH), forms one electrode in the nickel–iron ('nife') cell.

A number of complexes of nickel *atoms* are known, i.e. complexes which contain nickel in a zero oxidation state, e.g. the tetrahedral $[Ni(CO)_4]$.

Some reactions of $[Ni(H_2O)_6]^{2+}$

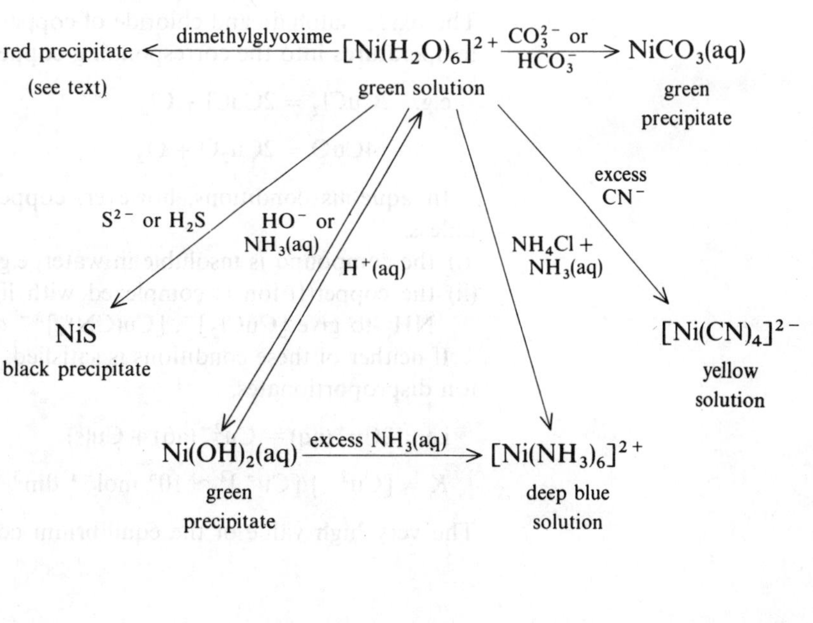

red precipitate (see text) $\xleftarrow{\text{dimethylglyoxime}}$ $[Ni(H_2O)_6]^{2+}$ green solution $\xrightarrow{CO_3^{2-} \text{ or } HCO_3^-}$ $NiCO_3(aq)$ green precipitate

$[Ni(H_2O)_6]^{2+}$ $\xrightarrow{S^{2-} \text{ or } H_2S}$ NiS black precipitate

$[Ni(H_2O)_6]^{2+}$ $\xrightarrow{HO^- \text{ or } NH_3(aq)}$ $Ni(OH)_2(aq)$ green precipitate; $Ni(OH)_2(aq)$ $\xrightarrow{H^+(aq)}$ $[Ni(H_2O)_6]^{2+}$

$[Ni(H_2O)_6]^{2+}$ $\xrightarrow{NH_4Cl + NH_3(aq)}$ $[Ni(NH_3)_6]^{2+}$ deep blue solution

$[Ni(H_2O)_6]^{2+}$ $\xrightarrow{\text{excess } CN^-}$ $[Ni(CN)_4]^{2-}$ yellow solution

$Ni(OH)_2(aq)$ $\xrightarrow{\text{excess } NH_3(aq)}$ $[Ni(NH_3)_6]^{2+}$

18.10
Copper

Outer electronic configuration: $3d^{10}\ 4s^1$. Principal valencies: 1, 2.

18.10.1 THE ELEMENT

Pure copper can be produced by the electrolysis of a copper(II) salt in aqueous solution. It possesses a characteristic pinkish colour, and is tough, ductile and rather soft. The pure metal is an excellent conductor of heat and electricity, being exceeded in this respect only by silver.

Copper is not attacked by water, steam or dilute non-oxidising acids in the absence of air. Oxidising acids, e.g. nitric acid or concentrated sulphuric acid, readily attack copper (§15.2.9 and §16.2.6). Ammonia or cyanide solutions in the presence of either air or hydrogen peroxide also dissolve the metal, with the formation of complexes,

$$\text{e.g.}\quad Cu + 2NH_3 \xrightarrow{O_2} [Cu(NH_3)_2]^+ \xrightarrow{O_2} [Cu(NH_3)_4(H_2O)_2]^{2+}$$

At red heat copper combines with oxygen to form copper(II) oxide, while at higher temperatures copper(I) oxide, Cu_2O, is obtained. Copper reacts with the halogens (except iodine) to give copper(II) halides, and with sulphur on heating to yield copper(I) sulphide, Cu_2S.

18.10.2 THE MONOVALENT STATE (+1 oxidation state) $3d^{10}$

Because the 3d sub-shell is full, copper(I) compounds are often colourless. Many copper(I) compounds, especially if they are insoluble, appear to be more stable than the corresponding copper(II) compounds. For example, the addition of iodide ions or cyanide ions to aqueous copper(II) salts produces a white precipitate of a copper(I) compound:

$$2Cu^{2+}(aq) + 4I^- = 2CuI + I_2$$

The oxide, sulphide and chloride of copper(II) all decompose at elevated temperatures into the corresponding copper(I) compounds,

$$\text{e.g.}\quad 2CuCl_2 = 2CuCl + Cl_2$$

$$4CuO = 2Cu_2O + O_2$$

In aqueous conditions, however, copper(I) compounds are unstable unless:

(i) the compound is insoluble in water, e.g. CuCl, CuCN or CuI;

(ii) the copper(I) ion is complexed with ligands, such as Cl^-, CN^- or NH_3 to give $[CuCl_2]^-$, $[Cu(CN)_4]^{3-}$ or $[Cu(NH_3)_2]^+$ respectively.

If neither of these conditions is satisfied, the simple hydrated copper(I) ion disproportionates,

$$\text{i.e.}\quad 2Cu^+(aq) \rightleftharpoons Cu^{2+}(aq) + Cu(s)$$

$$K = [Cu^{2+}]/[Cu^+]^2 \approx 10^6\ mol^{-1}\ dm^3$$

The very high value of the equilibrium constant indicates that the disproportionation of copper(I) in aqueous solution may be regarded as complete.

Disproportionation is favoured because the hydration enthalpy of the copper(II) ion ($-2\,284$ kJ mol^{-1}) is higher than that of the copper(I) ion (-482.4 kJ mol^{-1}). This arises from the relatively high surface charge density of the Cu^{2+} ion. The enthalpy of atomisation of copper is $+339$ kJ mol^{-1}, and the first and second ionisation enthalpies are $+751$ kJ mol^{-1} and $+1\,960$ kJ mol^{-1} respectively.

$\therefore$ for $Cu(s) \longrightarrow Cu^{+}(aq)$

$$\Delta H^{\ominus} = +339 + 751 - 482.4 = +607.6 \text{ kJ mol}^{-1}$$

and for $Cu(s) \longrightarrow Cu^{2+}(aq)$

$$\Delta H^{\ominus} = +339 + 751 + 1\,960 - 2\,248 = +802.0 \text{ kJ mol}^{-1}$$

Therefore, for the reaction:

$$2Cu^{+}(aq) = Cu^{2+}(aq) + Cu(s)$$

$$\Delta H^{\ominus} = -2(607.6) + 802.0 = -413.2 \text{ kJ mol}^{-1} \qquad (\S2.3.1)$$

Disproportionation of copper(I) ions is thus exothermic and likely to occur (§2.4.2). The same conclusion may be reached by considering standard electrode potentials (§2.5.3).

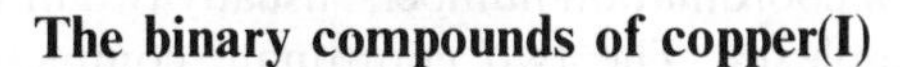

The binary compounds of copper(I)

Cu_2O

Copper(I) oxide

Copper(I) oxide, Cu_2O, is prepared as an orange-yellow precipitate by reducing an alkaline solution of a copper(II) salt with glucose, sodium sulphite or an aldehyde. The reaction, which is conveniently represented as:

$$2Cu^{2+}(aq) + 2HO^{-} + 2e^{-} = Cu_2O + H_2O$$

forms the basis of the Fehling's test for aldehydes and reducing sugars (§14.10.6).

Copper(I) oxide dissolves in concentrated hydrochloric acid or aqueous ammonia to form copper(I) complexes, but with oxoacids, such as sulphuric acid, disproportionation occurs on warming:

$$Cu_2O + 4HCl = 2[CuCl_2]^{-} + H_2O + 2H^{+}(aq)$$

$$Cu_2O + 4NH_3 + H_2O = 2[Cu(NH_3)_2]^{+} + 2HO^{-}$$

$$Cu_2O + H_2SO_4 = CuSO_4 + Cu + H_2O$$

CuCl

Copper(I) halides

Copper(I) chloride, CuCl, is the commonest halide of copper(I). A brown solution containing the dichlorocuprate(I) ion is obtained by reducing a solution of a copper(II) salt with copper or sulphur dioxide, in the presence of chloride ions,

i.e. $$Cu^{2+} + 4Cl^{-} + Cu = 2[CuCl_2]^{-}$$

$$2Cu^{2+} + 4Cl^{-} + 2H_2O + SO_2 = 2[CuCl_2]^{-} + SO_4^{2-} + 4H^{+}(aq)$$

A white precipitate of copper(I) chloride is formed when the brown solution is added to oxygen-free water. If the water contains dissolved oxygen, the precipitate rapidly turns green as oxidation to copper(II) occurs.

In the solid state copper(I) chloride possesses a covalent polymeric structure, but the heated vapour contains a dimeric species, Cu_2Cl_2. Copper(I) chloride is insoluble in water, but dissolves in solutions of chlorides, cyanides and thiosulphates to form complexes (see below). The dissolving of copper(I) chloride in aqueous ammonia gives the diammine-copper(I) complex, $[Cu(NH_3)_2]^+$, which absorbs carbon monoxide to form an addition compound. Ammoniacal copper(I) chloride also reacts with ethyne to give a red precipitate of copper(I) acetylide, Cu_2C_2, (ASE: copper(I) dicarbide):

$$2[Cu(NH_3)_2]^+ + C_2H_2 = Cu_2C_2 + 2NH_3 + 2NH_4^+$$

Copper(I) bromide, CuBr, closely resembles the chloride in its preparation and properties. Copper(I) iodide is obtained as a white precipitate by the addition of excess iodide ions to aqueous copper(II) ions (see above). This reaction is used in the analysis of aqueous copper(II) salts, for the liberated iodine can be titrated with a standard solution of sodium thiosulphate. Copper(I) fluoride is unknown.

The complexes of copper(I)

Low coordination numbers, usually two or four, are a feature of copper(I) complexes. The two coordinate complexes, e.g. $[Cu(S_2O_3)_2]^{3-}$ and $[Cu(NH_3)_2]^+$, are linear.

When a soluble cyanide is slowly added to an aqueous copper(II) salt, a toxic gas, cyanogen, $(CN)_2$, is produced and copper(I) cyanide is precipitated:

$$2Cu^{2+} + 4CN^- = 2CuCN + (CN)_2$$

The white precipitate, however, readily dissolves in excess cyanide to produce the tetrahedral tetracyanocuprate(I) ion, $[Cu(CN)_4]^{3-}$.

Copper(I) chloride is readily soluble in solutions of chloride ions to form complexes such as $[CuCl_2]^-$, $[CuCl_3]^{2-}$ and $[CuCl_4]^{3-}$, depending on the concentration of chloride ions.

18.10.3 THE DIVALENT STATE (+2 oxidation state) $3d^9$

Because the copper(II) ion has an incomplete 3d sub-shell, the complexes are coloured.

CuO and copper(II) halides

The binary compounds of copper(II)

Black copper(II) oxide, CuO, is prepared by heating the hydroxide, carbonate or nitrate of copper(II).

Only three halides of copper(II) exist, namely the fluoride, chloride and bromide. The anhydrous compounds are made by heating copper with the appropriate halogen. The fluoride is ionic and white, while the chloride and bromide are covalent (§6.4.3) and coloured yellow-brown and black respectively. At high temperatures the chloride and bromide decompose into copper(I) halides and the free halogen,

e.g. $2CuBr_2 = 2CuBr + Br_2$

Hydrated salts, e.g. the green $CuCl_2{\cdot}2H_2O$, are obtained by adding the anhydrous halides to water, or by dissolving copper(II) oxide in the appropriate hydrohalic acid.

Copper(II) sulphide, CuS, is obtained as a black precipitate by passing hydrogen sulphide through a solution of a copper(II) salt:

$$Cu^{2+}(aq) + H_2S = CuS + 2H^+(aq)$$

The complexes of copper(II)

Copper in the divalent state behaves as a typical transition element in forming a large number of coloured complexes. The blue hexaaqua-copper(II) ion, $[Cu(H_2O)_6]^{2+}$, is perhaps the most familiar complex of copper. This ion possesses a tetragonally distorted octahedral structure, in which two of the water molecules are situated further from the copper ion than the other four. The copper ion lies, therefore, in the centre of a square plane, defined by the four 'near' water molecules, with the 'far' molecules situated one above and one below this plane,

i.e.

H_2O (above)
H_2O — OH_2
Cu
H_2O — OH_2
H_2O (below)
$[\ ]^{2+}$

Solutions of the hydrated copper(II) ion are weakly acidic, due to hydrolysis:

$$[Cu(H_2O)_6]^{2+} + H_2O \rightleftharpoons [Cu(OH)(H_2O)_5]^+ + H_3O^+$$

The addition of alkali produces a bright blue precipitate of hydrated copper(II) hydroxide, $Cu(OH)_2(aq)$, which dehydrates at 100 °C (373.K) into copper(II) oxide. Copper(II) hydroxide is weakly amphoteric, and dissolves in concentrated alkalis to form deep blue hydroxo-complexes of variable composition. Crystalline copper(II) sulphate, $CuSO_4{\cdot}5H_2O$, also contains hydrated copper(II) ions (Fig. 6.24). The pentahydrate loses four molecules of water on heating to 100 °C (373 K) to form the monohydrate, $CuSO_4{\cdot}H_2O$. The remaining water molecule is lost at approximately 350 °C (623 K).

The precipitate of copper(II) hydroxide which is formed when a copper(II) salt is treated with aqueous ammonia dissolves in excess of that reagent to form the deep blue tetraamminediaquacopper(II) ion, $[Cu(NH_3)_4(H_2O)_2]^{2+}$. Structurally, this ion is related to the hydrated copper(II) ion, in that the four ammonia ligands form a square plane around the central ion. Crystalline salts, e.g. $[Cu(NH_3)_4]SO_4{\cdot}H_2O$, can be obtained by the addition of ethanol to solutions of the copper(II) ammine complexes. The hexaamminecopper(II) ion, $[Cu(NH_3)_6]^{2+}$, is obtained only when a copper(II) salt is treated with liquid ammonia.

When excess concentrated hydrochloric acid is added to an aqueous copper(II) salt, a yellow solution containing the tetrachlorocuprate(II) ion is obtained. The reaction is reversible:

$$\underset{\text{blue}}{[Cu(H_2O)_6]^{2+}} + 4Cl^- \underset{H_2O}{\overset{HCl}{\rightleftharpoons}} \underset{\text{yellow}}{[CuCl_4]^{2-}} + 6H_2O$$

At intermediate acid concentrations the solution is green in colour owing to the presence of both copper(II) complexes (blue + yellow = green). In crystalline salts the tetrachlorocuprate(II) ion may have one of two geometries; e.g. in $(NH_4)_2[CuCl_4]$ it is square planar, but in $Cs_2[CuCl_4]$ it has the form of a flattened tetrahedron.

Because the hexaaquacopper(II) ion hydrolyses in solution, normal copper(II) carbonate is unknown. The addition of a soluble carbonate to a copper(II) salt produces a green precipitate of a basic carbonate (§14.2.3).

Some reactions of $[Cu(H_2O)_6]^{2+}$

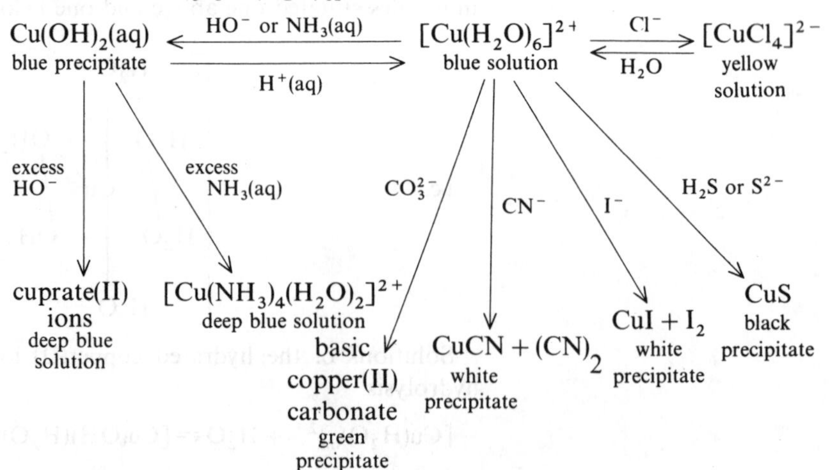

18.11 Zinc

Outer electronic configuration: $3d^{10}\ 4s^2$. Valency: 2.

18.11.1 THE ELEMENT

The principal ores of zinc are the sulphide, ZnS (zinc blende), and the carbonate, $ZnCO_3$ (calamine). In the extraction of zinc the ore is roasted in air to form the oxide, which is then reduced to the metal by heating with coke:

$$2ZnS + 3O_2 = 2ZnO + 2SO_2$$

$$ZnCO_3 = ZnO + CO_2$$

$$ZnO + C = Zn + CO$$

The crude zinc so obtained is separated from impurities, chiefly cadmium, by fractional distillation.

Zinc, a white metal, is decidedly 'non-transitional' in its properties. For instance, it possesses much lower melting and boiling temperatures than the transition elements, owing to the lack of 3d electron participation in metallic bonding. It shows only one valency state, and its compounds are colourless, unless the anion is coloured. However, it does form a range of complexes. Although its outer electronic configuration ($4s^2$) compares with that of the group 2A elements, zinc bears little resemblance to these elements apart from the stoicheiometries of its compounds.

Zinc is fairly reactive and dissolves in non-oxidising acids, with the liberation of hydrogen, to give zinc salts. It reduces nitric acid (§15.2.9). Hot, concentrated sulphuric acid is also reduced by zinc (§16.2.6). The metal dissolves readily in solutions of alkalis to produce zincate(II) ions, which are anionic hydroxo–zinc complexes, formulated as $[Zn(OH)_4]^{2-}$:

$$Zn + 2HO^- + 2H_2O = [Zn(OH)_4]^{2-} + H_2$$

18.11.2 THE COMPOUNDS OF ZINC

Zinc oxide, ZnO, a white powder, is prepared by heating the metal in air, or by heating zinc carbonate or zinc hydroxide. It is amphoteric:

$$ZnO + 2H^+(aq) = Zn^{2+}(aq) + H_2O$$

$$ZnO + 2HO^- + H_2O = [Zn(OH)_4]^{2-}$$

When zinc oxide is heated it loses a small amount of oxygen and turns yellow, but on cooling it regains lost oxygen and becomes white again.

Hydrated zinc hydroxide, $Zn(OH)_2(aq)$, is also amphoteric and is obtained as a white gelatinous precipitate on adding an alkali to an aqueous zinc salt:

$$Zn^{2+}(aq) + 2HO^- = Zn(OH)_2(aq)$$

$ZnCl_2$

Zinc chloride is the most important halide of zinc. The dihydrate, $ZnCl_2{\cdot}2H_2O$, is prepared by dissolving zinc, or its oxide or carbonate, in hydrochloric acid. Anhydrous zinc chloride cannot be obtained by heating the hydrated salt because, like hydrated magnesium chloride, it undergoes hydrolysis to give a basic chloride:

$$ZnCl_2{\cdot}2H_2O = ZnCl(OH) + HCl + H_2O$$

Anhydrous zinc chloride is prepared by heating zinc in a stream of dry hydrogen chloride or chlorine, and is extremely deliquescent.

The complexes of zinc

Zinc has less tendency than the transition elements to form complexes. Four coordinate (tetrahedral) and six coordinate (octahedral) complexes are known, the former being more common. Solutions of zinc salts, which contain the hexaaquazinc(II) ion, $[Zn(H_2O)_6]^{2+}$, and possibly the tetraaquazinc(II) ion, $[Zn(H_2O)_4]^{2+}$, are acidic because of hydrolysis:

$$[Zn(H_2O)_6]^{2+} + H_2O \rightleftharpoons [Zn(OH)(H_2O)_5]^+ + H_3O^+$$

A hydrated basic carbonate, $ZnCO_3{\cdot}2Zn(OH)_2$ or $Zn_3CO_3(OH)_4$ is obtained if a soluble carbonate is added to an aqueous zinc salt. The normal carbonate, $ZnCO_3$, is obtained if a hydrogencarbonate is used.

Common four coordinate complexes of zinc include the tetraamminezinc(II) ion, $[Zn(NH_3)_4]^{2+}$, which is produced when zinc hydroxide dissolves in aqueous ammonia, and the tetracyanozincate(II) ion, $[Zn(CN)_4]^{2-}$, formed by adding an excess of cyanide ions to a zinc salt.

Some reactions of $[Zn(H_2O)_6]^{2+}$

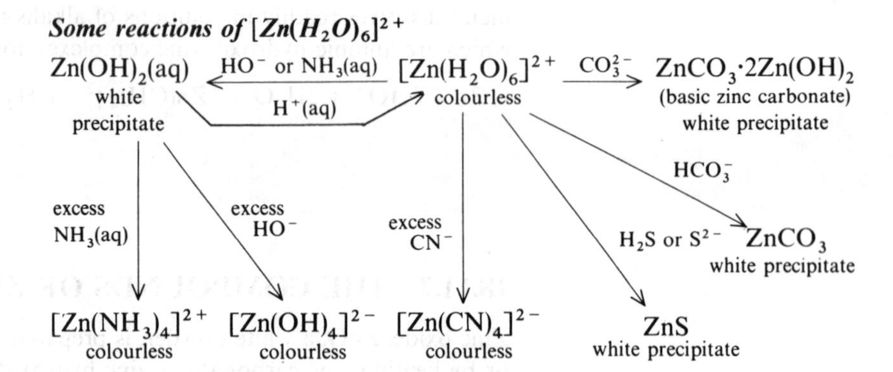

EXAMINATION QUESTIONS ON CHAPTER 18

1. (a) Listed below are the commonly occurring oxidation states in aqueous solution of the first row transition elements Sc to Zn inclusive. Complete the table by writing in the symbol(s) of the elements against the appropriate oxidation states.

Common oxidation states	Number of elements with these oxidation states	Symbol(s) of element(s)
+1, +2	1	
+2 only	2	
+2, +3	2	
+2, +6, +7	1	
+3 only	1	
+3, +4	1	
+3, +6	1	
+3, +4, +5	1	

(b) Describe any changes which occur in the stability (with respect to oxidation) of the +2 oxidation state in moving across the first row of transition elements from Ti to Zn. (JMB)

2 What do you understand by the phrase *first transition series of elements*?

Give the names and symbols of as many as possible of these elements, and discuss, with illustrative examples, *three* properties which are general among the elements and/or their compounds.

Outline a brief explanation for the properties you describe.

Describe the special features of the chemistry of any **one** of these elements. (SUJB)

3 (a) Complete the following to give the electronic configuration of:

(i) the silicon atom $1s^2 2s^2 2p^6$

(ii) the titanium atom $1s^2 2s^2 2p^6 3s^2 3p^6$

(b) Give the maximum possible oxidation state for each of these elements and, for each element, one example of a compound containing the element in this oxidation state.

(c) Why is titanium classed as a transition element but silicon is not?

(d) Give two examples to illustrate how the chemistry of silicon compounds differs from that of carbon compounds. (JMB)

4 (a) In aqueous solution, the colours of certain vanadium ions are as follows: VO_2^+ (i.e., V(V)) yellow, V^{4+} blue, V^{3+} green. When sulphur dioxide is passed through an acidic solution of ammonium vanadate (NH_4VO_3), the solution turns green. How would you find out experimentally whether the green colour was due to V^{3+}, or to a mixture of vanadium in oxidation states IV and V?

(b) The reaction between peroxodisulphate ions, $S_2O_8^{2-}$, and iodide ions is catalysed by some transition metal ions. Describe how you would compare the catalytic effect of Cr^{3+} and Fe^{3+} ions on this reaction.

(c) In fact, Fe^{3+} is found to be a catalyst, but not Cr^{3+}. Suggest a possible mechanism, in view of the following standard electrode potentials:

$Fe^{3+}(aq)\ Fe^{2+}(aq)$	+0.77 V
$S_2O_8^{2-}(aq)\ 2SO_4^{2-}(aq)$	+2.01 V
$I_2(aq)\ 2I^-(aq)$	+0.54 V

What prediction can be made about the standard electrode potential of $Cr^{3+}(aq)\ Cr^{2+}(aq)$ ions in view of the observation that Cr^{3+} ions do not catalyse the reaction? (L)

5 (a) State **four** general characteristics of the transition elements. Illustrate each characteristic chosen with **two** suitable examples.

(b) Explain the following:

(i) addition of alkali to potassium dichromate(VI) solution produces a yellow solution, and addition of barium chloride solution to potassium dichromate(VI) solution produces a yellow precipitate,

(ii) the compound $CrCl_3{\cdot}6H_2O$ can exist in three forms, all of which in aqueous solution produce a precipitate with silver nitrate solution but in different mole proportions,

(iii) when sulphur dioxide is passed through an aqueous solution of copper(II) sulphate and sodium chloride in equimolar proportions, a white precipitate is formed, the sulphur dioxide at the same time being oxidized to sulphuric acid. (AEB)

6 A pale green chloride of chromium, X, was investigated with the following results:

(a) when 1.00 g was heated in air to constant mass, water and hydrogen chloride were evolved and a residue of 0.286 g of chromium(III) oxide, Cr_2O_3, remained,

(b) when 1.00 g of X was dissolved in water, acidified with dilute nitric acid and treated with silver nitrate solution, 1.077 g of silver chloride (AgCl) was immediately precipitated. On boiling the solution until no further change occurred, a further 0.538 g of silver chloride was precipitated,

(c) on drying in a desiccator over concentrated sulphuric acid, 1.00 g of X lost 0.068 g of water. Elucidate as far as possible from this information the formula of the compound and comment on its stereochemistry.

What isomers of X would you expect to exist? State which one of these isomers would be soluble in ether and why.

(H = 1, O = 16, Cl = 35.5, Cr = 52, Ag = 108) (L)

7 When potassium chromate(VI), sodium chloride and concentrated sulphuric acid are gently heated together, a red liquid R is produced which contains only chromium, chlorine and oxygen. When the red liquid R is added to aqueous sodium hydroxide, a yellow solution S is obtained.

(i) The slow addition of silver nitrate solution to a neutralised sample of solution S produced, first, a white precipitate and then a red one.

(ii) Acidification of the yellow solution S with dilute sulphuric acid gave an orange solution T which contained chromium in oxidation state +6. T liberated iodine from aqueous potassium iodide, leaving a green solution containing chromium in oxidation state +3.

(iii) Quantitatively, 0.155 g of liquid R gave a solution S which needed 2.0 mmol of silver nitrate to produce the first trace of red colour; also 0.155 g of R gave enough T to liberate 1.5 mmol of iodine (I_2) from aqueous potassium iodide.

Deduce the formula of R and explain the reactions which have occurred. What is the structural formula of R and to what class of compound does it belong?

(Cr = 52, Cl = 35.5, O = 16) (L)

8 (a) Consider the following reaction sequence:

$$CrCl_3(s) \xrightarrow{\text{excess } H_2O} \mathbf{A} \xrightarrow{\text{aq NaOH}/H_2O_2} \mathbf{B} \xrightarrow{\mathbf{X}} Cr_2O_7^{2-} \xrightarrow{\text{aq BaCl}_2} \mathbf{C}$$

(i) What is the formula of the compound **A**?

(ii) What is the formula of the anion formed in **B**?

(iii) What is the reagent **X**?

(iv) Name the precipitate **C** and state its colour.

(b) Chromium forms three oxides. What is the formula and nature of each of these oxides? Give the oxidation number of chromium in **each** case.

(c) Iron(II) ions in the presence of excess hydrochloric acid are oxidized by potassium dichromate to iron(III) ions according to the following equations:

$$Cr_2O_7^{2-} + 14H^+ + 6e^- \longrightarrow 2Cr^{3+} + 7H_2O$$
$$Fe^{2+} \longrightarrow Fe^{3+} + 1e^-$$

(i) What volume of dichromate solution (concentration 0.1 mol dm^{-3}) would be required to oxidize 25.0 cm^3 of iron(II) solution (concentration 0.6 mol dm^{-3})?

(ii) Why would potassium manganate(VII) not be suitable in the presence of hydrochloric acid for this quantitative oxidation?

(iii) How would you distinguish qualitatively between the Fe^{2+} and Fe^{3+} ion in solution? (AEB)

9 (a) Describe how you would prepare a specimen of potassium manganate(VII) in the laboratory. Give essential experimental details and explain the chemistry involved.

(b) Chromium(III) chloride forms three different hexahydrates, **X**, **Y** and **Z** each of molecular formula $CrCl_3 \cdot 6H_2O$. 8.88 g of **X** and 8.88 g of **Y** were dissolved separately in water and the resulting solutions were diluted to 1.00 dm^3. When 25 cm^3 portions of these solutions were titrated against a standard aqueous solution of silver nitrate of concentration 0.100 mol dm^{-3}, it was found that the solution of **X** required 25.0 cm^3 of the silver nitrate solution whereas the solution of **Y** required 16.7 cm^3. Calculate the number of mols of chloride ion in each dm^3 of the solutions of **X** and **Y** and explain your results.

If the salt **Z** were subjected to exactly the same procedures what volume of the silver nitrate solution would be needed for the titration? Explain the reason for your answer. (O)

10 (a) Manganese is a typical transitional element. Illustrate **four** properties of the element and/or its compounds which support this statement giving brief explanations for each property you mention.

(b) Acidified potassium manganate(VII) (permanganate) will oxidise iron(II) ethanedioate (oxalate).

(i) Write the ionic half equation for the oxidation of the iron(II) ion.

(ii) Similarly write the ionic half equation for the oxidation of ethanedioate (oxalate) ion, assuming carbon dioxide to be the only product.

(iii) Construct the equation for the reaction between an excess of acidified potassium manganate(VII) and iron(II) ethanedioate (oxalate).

(c) How many cm^3 of 0.01 M potassium manganate(VII) solution would be required to oxidise completely 50.0 cm^3 0.01 M iron(II) ethanedioate (oxalate) in dilute sulphuric acid?

(d) How would you confirm the presence of Mn^{2+} and Ni^{2+} ions in a solution of their mixed sulphates? (SUJB)

11 Iron is a transition element with atomic number 26.

(a) Using s, p, d notation, write the detailed electronic structures of:

(i) an iron atom,

(ii) the iron(II) ion,

(iii) the iron(III) ion.

Explain why iron is referred to as a *transition* element.

(b) State **three** general characteristics of a *transition* element, illustrating them by examples from the chemistry of iron.

(c) Why has a solution of an iron(III) salt in water a pH less than 7?

(d) Describe the uses of:

(i) an iron compound in qualitative analysis,

(ii) an iron compound in titrimetric analysis. (AEB)

12 When anhydrous copper(II) sulphate is added to water, solution A is obtained. Treatment of solution A with an excess of aqueous sodium chloride gives solution B which, when treated with an excess of sulphur dioxide followed by dilution with water, gives a precipitate C. This precipitate, when filtered off and washed with distilled water, dissolves in aqueous ammonia to give a colourless solution D which rapidly becomes a blue-violet solution E upon standing in air.

(a) State the colours of solutions A and B and write formulae for the predominant copper-containing species contained in them.

(b) State the colour and formula of precipitate C.

(c) What is the formula of the predominant copper-containing species in E?

(d) Explain briefly the process occurring when solution D changes to solution E. (JMB)

13 Copper forms a *complex salt* with the formula $Cu(NH_3)_4SO_4 \cdot H_2O$, and a *double salt* $CuSO_4 \cdot (NH_4)_2SO_4 \cdot 6H_2O$.

(a) **Outline** the preparation of both of these salts starting from copper(II) sulphate pentahydrate.

(b) Give the systematic name for the *cation* in the complex salt and indicate the type of bonding present.

(c) If 10^{-2} mole of each salt is separately dissolved in 1 kg of water, how would these solutions differ if at all in their:

(i) pH value,

(ii) freezing point,

(iii) colour,

(iv) reaction to hydrogen sulphide.

Explain your answers as fully as possible. (AEB)

14 Give an account of complex ions, paying particular attention to bonding, stereochemistry, stability, stoichiometry and different ligand types. (L)

15 Describe and explain how the following properties change along the first series of transition elements: (a) atomic radius, (b) maximum oxidation state, (c) boiling point, and (d) enthalpy of fusion.

As far as possible relate your answers to the electronic structure of the elements and their bonding. (L)

16 What are the common oxidation states in aqueous solution of titanium, chromium, cobalt and nickel?

For each element, give one example of a species containing each of the oxidation states which you quote.

Suggest reasons why higher oxidation states predominate at the left hand side of the transition series whereas lower oxidation states predominate at the right hand side. Using cobalt and nickel complexes as examples, illustrate the following types of reactions of transition metal hexa-aqua ions:

(a) a ligand exchange reaction,
(b) a reaction in which there is a change in oxidation state,
(c) a reaction in which there is a change in coordination number,
(d) a reaction in which a tetrahedral nickel complex is formed,
(e) a reaction in which a square planar nickel complex is formed.

(JMB)

17 Answer the following questions about the reactions of chromium which are illustrated in the reaction sequence:

$$CrCl_3(s) \xrightarrow{\text{excess } H_2O} A \xrightarrow[H_2O_2]{NaOH/} B \xrightarrow{?} Cr_2O_7^{2-} \xrightarrow{BaCl_2} C$$

(*A*, *B* and *C* all contain chromium)
(a) What is the formula of the cation in *A*?
(b) What is the formula of the anion in *B*?
(c) What type of reaction is the conversion *A* to *B*?
(d) (i) What reagent brings about the change *B* to $Cr_2O_7^{2-}$?
(ii) Write an equation for this conversion.
(e) What is the formula of *C*?
(f) What are the oxidation states of the chromium in *A*, *B* and *C*?

(JMB)

18 (a) The atomic number of chromium is 24. Distinguishing between s, p and d electrons, give the electronic configuration of chromium in each of the oxidation states 0, +3, +6.

(b) Give, with reasons, the oxidation state of chromium in
(i) $K_4[Cr(CN)_6]$,
(ii) $Cr(CO)_6$,
(iii) $K_2Cr_2O_7$.
(c) Give one chemical test for the Cr^{3+} ion.
(d) Give the name and structure of one cationic complex of chromium and suggest how it might be formed.
(e) Explaining your reasoning, identify as far as possible each of the following compounds of chromium.
(i) Addition of concentrated sulphuric acid to an aqueous solution of **A**, followed by crystallisation, gave orange crystals. Crystallisation of a solution of these crystals in aqueous ammonia gave yellow crystals.
(ii) **B** is a deliquescent crystalline solid, 1.00 mol of which, on heating, gave a green solid and 3.75 mol of a mixture of gases. Ignition of a mixture of this residual green solid and aluminium powder caused the evolution of considerable heat.

(AEB)

19 (a) When manganese(IV) oxide is fused with potassium hydroxide in the presence of potassium nitrate and the cooled residue extracted with dilute sulphuric acid, a purple solution **A** is obtained. **A** becomes a green solution **B** when treated with concentrated potassium hydroxide solution but this green solution becomes a purple solution **C** when chlorine is bubbled through it.

(i) What are the formulae of, and oxidation states of manganese in, the species responsible for the colours in solutions **A** and **B**?

(ii) Write equations for the conversions of solution **A** into solution **B** and of solution **B** into solution **C**.

(b) When manganese(IV) oxide is heated with concentrated hydrochloric acid, a gas is evolved. Identify this gas and write an equation for the reaction. (JMB)

20 (a) Show, by stating essential reagents and conditions only, how the following conversions may be brought about:

(i) $Mn^{2+}(aq) \longrightarrow MnO_2(s)$
(ii) $MnO_2(s) \longrightarrow MnO_4^{2-}(aq)$
(iii) $MnO_4^{2-}(aq) \longrightarrow MnO_4^-(aq)$
(iv) $Mn^{2+}(aq) \longrightarrow MnO_4^-(aq)$
(v) $Cr^{3+}(aq) \longrightarrow CrO_4^{2-}(aq)$
(vi) $CrO_4^{2-}(aq) \longrightarrow Cr_2O_7^{2-}(aq)$

(b) Sketch the structures of the ions CrO_4^{2-} and $Cr_2O_7^{2-}$.

(c) Both potassium manganate(VII) and potassium dichromate(VI) may be used as oxidising agents in volumetric analysis. What are the advantages and disadvantages of each?

(d) A solution containing 0.124 g of arsenic(III) oxide (As_2O_3) required 25.0 cm^3 of acidified 0.02 M potassium manganate(VII) for complete oxidation. What was the oxidation state of the arsenic after oxidation? (JMB)

21 (a) Describe the crystalline structures of sodium chloride and caesium chloride.

Explain:

(i) the differences between the two structures,

(ii) why the melting point of sodium chloride (1081 K) is higher than that of caesium chloride (918 K).

(b) **Outline** the laboratory preparations of crystalline samples of:

(i) the double salt $CuSO_4(NH_4)_2SO_4{\cdot}6H_2O$

(ii) the complex salt $[Cu(NH_3)_4]SO_4{\cdot}H_2O$

(c) The salt $[Cu(NH_3)_4]SO_4{\cdot}H_2O$ contains a complex ion of copper.

(i) Give the formula of this complex ion.

(ii) What do you understand by the term *complex ion*?

(d) How does a complex salt differ from a double salt? (AEB)

22 Discuss the principles underlying the procedure for obtaining zinc metal from its sulphide ore, zinc blende, by conversion to and reduction of the oxide.

Zinc is included in the d-block elements. Outline similarities and differences in the typical chemistry of zinc (atomic number 30) and a characteristic member of the d-block, manganese (atomic number 25), showing how these similarities and differences are related to the respective electronic configurations. (WJEC)

23 Choose **either** zinc (atomic number 30) **or** mercury (atomic number 80) and answer each of the following questions about it.

(a) Distinguishing between s, p, d and f electrons, give the electronic configuration of the element.

(b) Give what you consider to be the principal oxidation state of the element and justify your choice. What is the electronic configuration of the element in this oxidation state? In what other oxidation states does the element exist?

(c) Describe briefly how you would obtain in the laboratory samples of (i) the oxide MO, (ii) the anhydrous chloride MCl_2.

(d) Give the name of **one** complex containing the element and describe briefly its formation in the laboratory.

(e) Give one organic reaction in which the element or one of its compounds plays a part. (AEB)

24 State three aspects of the chemistry of copper which are characteristic of transition metal chemistry.

Give brief experimental details to indicate how, starting from copper(II) sulphate pentahydrate, you would prepare samples of (a) copper(I) oxide, (b) copper(I) chloride.

Describe briefly and interpret what you would observe in the aqueous reactions between:

(i) copper(I) oxide and dilute sulphuric acid,
(ii) copper(I) chloride and concentrated hydrochloric acid,
(iii) copper(II) sulphate and potassium iodide. (JMB)

MISCELLANEOUS EXAMINATION QUESTIONS

1 (a) What is meant by the statement that ammonium chloride is hydrolysed in aqueous solution?

(b) An aqueous solution of ammonia is added to an aqueous solution containing ammonium chloride, aluminium ions and manganese(II) ions. Explain why aluminium hydroxide is precipitated, while manganese(II) hydroxide is not.

(c) Explain the chemistry of a test for any one metal ion in aqueous solution which depends on complex ion formation. (O)

2 'The particles in a substance (which may be atoms, molecules or ions) may be held together by ionic, covalent or hydrogen bonds, or by van der Waals' forces.'

(a) Illustrate this statement by reference to carbon, carbon dioxide, water, sodium chloride and a noble gas. Electron configurations should be given wherever possible.

(b) Relate the physical properties of the substances named in (a) to the type of bonding present.

(c) Suggest a reason why 2-nitrophenol, [structure: benzene ring with NO_2 and OH on adjacent carbons], is more volatile than its isomer, 4-nitrophenol, [structure: benzene ring with NO_2 and OH on opposite carbons]. (L)

3 Given a molar solution of silver nitrate in water and any equipment necessary, describe how *you* would determine the standard electrode potential of silver.

From Table 44 of your data book, write down the standard electrode potentials of the elements silver, copper, hydrogen, tin and zinc in this order. Give **two** comments on the list so obtained.

Potassium manganate(VII) ($KMnO_4$) and hydrogen peroxide react in the presence of acid with the evolution of oxygen to give a clear solution. Show how to use the information in Table 44 to determine the stoicheiometric (balanced) ionic equation of the reaction.

What volume of potassium manganate(VII) of concentration 0.200 mol dm^{-3} would be required to react with 100 cm^3 of hydrogen peroxide of concentration 0.0100 mol dm^{-3} (with excess acid present), and what volume of oxygen (at 20 °C and 1 atm pressure) would be evolved? (OC)

4 (a) State Graham's law of diffusion and explain its theoretical basis.

Under the same conditions equal volumes of the gases **A**, **B** and oxygen each separately diffuse through a small pinhole in 42.4 s, 42.4 s and 119.9 s respectively. A mixture of **A** and oxygen reacts explosively when a spark is passed but there is no reaction when a spark is passed through a mixture of **B** and oxygen. Use these observations to identify gas **A** and gas **B**.

(b) Why do all the molecules in a gas not have the same velocity at a given temperature?

Discuss the way the magnitude of the velocities is distributed and explain the following observations.

(i) A liquid can evaporate below its boiling point.

(ii) A small change in temperature can cause a large change in the rate of a reaction. (JMB)

5 (a) Give one example in each case of the formation of hydrogen from (i) a named acid, (ii) a named alkali, (iii) water.

(b) Describe the structures of (i) sodium hydroxide, (ii) water, (iii) the oxonium ion (H_3O^+).

(c) Explain why:

(i) water has a higher boiling point than hydrogen sulphide,

(ii) hydrogen iodide is a stronger reducing agent than hydrogen chloride,

(iii) the acidity of the hydrides of the elements from sodium to chlorine in the periodic table increases from left to right.

(d) Deuterium, D, is an isotope of hydrogen. Starting with deuterium oxide, describe briefly how you would prepare (i) ND_3, (ii) C_2D_2.
(AEB)

6 The s- and p-block elements of the periodic table gradually become less electronegative in each group with increase in atomic number. How do you account for this?

Illustrate this change in character by considering **each** of the following, giving the chemistry on which the illustration is based and explaining any anomalies:

(a) the hydrides of the group nitrogen to bismuth,
(b) the ionisation of water and hydrogen sulphide,
(c) the oxidising power of the halogens,
(d) the reactivity of magnesium and calcium with water. (WJEC)

7 (a) Explain the reactions which occur in (i) to (iv) below, write ionic equations and say what observations could be made and how any gases could be detected.

(i) To an aqueous solution of copper(II) sulphate, a solution of potassium iodide is added, followed by a solution of sodium thiosulphate ($Na_2S_2O_3$).

(ii) Aluminium powder is warmed with a slight excess of dilute potassium hydroxide. When reaction is complete, the mixture is cooled and filtered. An excess of dilute sulphuric acid is then added and finally the mixture is gently evaporated and cooled.

(iii) An aqueous solution of ammonium nitrate is warmed with an excess of sodium hydroxide solution. When reaction is complete, the mixture is cooled and then warmed with zinc powder.

(iv) A solution of sodium sulphate(IV) (sulphite) is warmed with a few drops of dilute nitric acid. A solution of barium nitrate is then added.

(b) By deducing reacting quantities, show how reaction (i) can be used to determine volumetrically the concentration of the copper solution. (Cu = 63.5) (SUJB)

8 For **three** of the following pairs of compounds, compare the physical and chemical properties of one member of the pair with those of the other, in each case giving reasons for any differences in behaviour which you quote:

(a) carbon dioxide and silicon(IV) oxide (silicon dioxide),
(b) trioxygen (ozone) and sulphur dioxide,
(c) beryllium chloride and aluminium chloride,
(d) calcium chloride and chromium(II) chloride. (OC)

9 Starting from an aliphatic carboxylic acid, and giving equations and necessary conditions, show how you could form (a) an acid chloride, (b) an acid anhydride, (c) an ester.

Under what conditions does **each** react with water, and what are the products?

Explain why a carboxylic acid, although apparently containing a carbonyl group, will not generally react with nucleophiles unless they are very strong.

Starting from the 1-haloalkane RCH_2X, (where R is an alkyl group, and X a halogen atom), suggest how you could synthesise the two carboxylic acids, RCOOH and RCH_2COOH. (WJEC)

10 Describe the different types of isomerism which occur in carbon compounds. Illustrate your answer by reference to isomers of each of the following: C_4H_{10}, C_3H_7Cl, C_4H_8 and $CH_3CH(OH)COOH$. State what differences (if any) in chemical behaviour and in physical properties you would expect the various isomers of each formula to show.

A hydrocarbon C_5H_{10} reacts with sulphuric acid and the product is hydrolysed to give an alcohol $C_5H_{12}O$. This alcohol can be separated into two enantiomeric forms (optical isomers). Suggest **one** possible structure for the hydrocarbon. (JMB)

11 A pure compound **A** was treated with hydrogen bromide. The product was a racemate from which optically active compound **B** was obtained. **B** on treatment with excess ammonia gave a further optically active compound **C** which contained 65.7% C, 15.1% H and 19.2% N, and had a relative molecular mass of 73. Treatment of **C** with nitrous acid gave compound **D** which was oxidised to **E** with acidified dichromate. **E** reacted with 2,4-dinitrophenylhydrazine to form a yellow crystalline derivative. **E** also gave the iodoform reaction but did not give a silver mirror with silver diammine nitrate solution (ammoniacal silver nitrate).

Explain the reactions occurring above giving **two** possible alternatives for the process **A** → **B**.

Give the structures of the compounds **A**–**E**.

Suggest how you would decide between your alternative suggestions for compound **A**. (WJEC)

12 Comment critically on **three** of the following statements. Elaborate upon them where you agree with them and offer improvements where you do not.

(a) Halogenoalkanes (alkyl halides) are susceptible to nucleophilic substitution because of the polarity of the carbon–halogen bond—the more polar the bond, the more readily will halogeno-alkanes undergo this type of reaction.

(b) The polarity of the carbonyl group induces a dipole in the adjacent (α-) CH bonds of carbonyl compounds. [$\overset{\gamma}{C}H_3\overset{\beta}{C}H_2\overset{\alpha}{C}H_2CO\sim$]. The α-hydrogens of these compounds are therefore more acidic than are the hydrogens of alkanes.

(c) Ethanoyl (acetyl) chloride is more reactive towards water than is chloroethane (ethyl chloride) because the carbonyl group in the former is more polar than the CH_2 group in the latter.

(d) Alkenes react readily with bromine water, whereas alkanes do not, because the carbon–carbon double bond is weaker than the carbon–carbon single bond.

(e) Water and petrol do not mix because like will only dissolve like. (OC)

13 Comment on, or explain, the physico-chemical principles underlying **three** of the following.

(a) Both red and white (yellow) phosphorus yield white phosphorus on distillation in an inert atmosphere.

(b) A constant-boiling mixture of hydrogen chloride in water (hydrochloric acid) may be used as a primary standard in volumetric analysis.

(c) Copper(II) sulphate solution is blue yet it turns litmus paper red and the surface of a piece of zinc immersed in the solution becomes pink in colour.

(d) Although the solubility of calcium carbonate in water at room temperature is only approximately *half* that of calcium fluoride, its solubility product is approximately *one hundred times* greater:

	Solubility	Solubility product
$CaCO_3$	6.9×10^{-3} g dm^{-3}	4.8×10^{-9} mol^2 dm^{-6}
CaF_2	1.7×10^{-2} g dm^{-3}	4.2×10^{-11} mol^3 dm^{-9}

C = 12, O = 16, F = 19, Ca = 40 (SUJB)

14 (a) Outline the Born-Haber cycle for the formation of sodium chloride from its elements, giving, for each step, an equation and the name and sign of the energy change. Show how the lattice energy may be calculated from the cycle.

(b) Describe and discuss the crystal structures of sodium chloride and of caesium chloride. Indicate briefly how these structures were determined. (JMB)

15 The following statements may be incomplete or only partly correct. Criticise **three** of them.

(a) Lattice energies enable the stability of solid compounds to be predicted.

(b) Standard electrode potentials enable the positions of equilibrium of all ionic reactions to be predicted.

(c) Hydrogen is always a univalent element with a strong tendency to form covalent bonds with other elements.

(d) The pH of a solution can be determined by titration with alkali. (OC)

16 List the names and formulae of **three** organic compounds which contain the $-NH_2$ group, selecting your examples from different types of compounds.

Give a general account of the methods available for preparing such compounds.

Certain of these compounds can be identified by making and purifying benzoyl derivatives and finding their melting points. Describe, as to a student carrying it out, how you would do this experiment. (L)

17 (a) Survey, giving examples, three general methods available for the extraction of metals from their purified ores. Give reasons for the choice of method in the examples you quote.

(b) Explain the following facts:

(i) when tin reacts with concentrated nitric acid, hydrated tin(IV) oxide is obtained,

(ii) cobalt dissolves in concentrated hydrochloric acid to give a blue solution but it dissolves in dilute sulphuric acid to give a pink solution,
(iii) lead dissolves in dilute nitric acid but the solution gives a white precipitate when dilute hydrochloric acid is added. (JMB)

18 (a) Discuss the ways in which hydrogen is bonded in inorganic compounds giving **one** example in each case. Indicate how the type of bonding present affects the properties of the compound.
(b) State and account for the relative acidic strengths of dilute aqueous solutions of hydrogen chloride and hydrogen fluoride. (AEB)

19 Give an account of the chlorides of the elements from sodium to chlorine in the periodic table, paying particular attention to:
(a) methods of preparation, including brief experimental details,
(b) the nature of the bonding,
(c) reactions with water.
0.800 g of a chloride of sulphur was hydrolysed with water and the solution diluted to 100 cm^3. 25.00 cm^3 of this solution was titrated with 0.100 M silver nitrate solution of which 29.60 cm^3 were required. Determine the empirical formula of the chloride.
(S = 32.0, Cl = 35.5) (L)

20 Describe, with essential experimental details, a laboratory method for preparing benzenecarbonyl (*benzoyl*) chloride from benzenecarboxylic (*benzoic*) acid.
Show, by means of equations, how the acid chloride can be used to prepare (a) benzenecarboxamide (*benzamide*), and (b) phenyl benzoate (*benzenecarboxylate*).
Describe what you would observe when benzenecarbonyl chloride is warmed with an excess of aqueous sodium hydroxide, and write a mechanism for the reaction which takes place.
Explain why Terylene and nylons are degraded by sodium hydroxide whereas poly(ethene) (*polythene*) is inert under the same conditions. (JMB)

21 (a) Explain, and illustrate with one example in each case, the meaning of any **four** of the following:
(i) ozonolysis,
(ii) catalytic cracking,
(iii) an addition–elimination reaction,
(iv) heterolytic cleavage,
(v) esterification.
(b) How may ozonolysis be used to help in the elucidation of the structure of certain compounds? (O)

22 Starting with iodomethane as the only organic starting material, describe a synthesis of ethyl ethanoate. For each stage of the synthesis, give reagents and conditions of reaction and write a balanced equation.
Give the conditions of reaction of ethyl ethanoate with ammonia and write an equation.

Suggest a structure for the product formed in the reaction between ethyl ethanoate and hydroxylamine.

Using the following bond energy values in kJ mol^{-1}, calculate the enthalpy change for the formation of gaseous ethyl ethanoate from gaseous atoms:

C—H 414 C—O 360 C—C 347 C=O 736 (AEB)

23 Complete the following calculations and give for each a statement of the rule or law on which it is based.

(a) A current of 1 ampere flowing for 10 minutes liberated 0.006 22 mol of silver from a silver nitrate solution. What mass in grams, of aluminium, would be liberated from a suitable aluminium compound, using the same quantity of electricity?

(b) When 10 cm^3 of a hydrocarbon was sparked with 50 cm^3 of oxygen (an excess) and the residual gases cooled to room temperature, a contraction of 20 cm^3 occurred. A further contraction of 20 cm^3 took place when the residual gases were subjected to aqueous sodium hydroxide solution. All the volumes were measured at 20 °C. Deduce the formula of the hydrocarbon.

(c) Water from a certain spring gave on boiling a gaseous mixture containing 21.0% oxygen and 43.6% nitrogen, the remaining volume being carbon dioxide. Calculate the percentage by volume of the mixture of gases in which the spring water had been in contact under conditions of s.t.p. (The absorption coefficients for oxygen, nitrogen and carbon dioxide at s.t.p. are 0.04, 0.02, 1.79.) (AEB)

24 (a) Distinguish clearly between the following terms: *ionisation energy*; *electron affinity*; *electronegativity*.

(b) Consider the elements of the second period of the periodic table (lithium to neon):

(i) Explain the variations in the first ionisation energies of these elements including particular reference to a comparison of the first ionisation energies of the following pairs of atoms: lithium and beryllium; beryllium and boron; nitrogen and oxygen.

(ii) Account for the trend in the electronegativities of these elements.

(iii) Explain why fluorine is the most powerful oxidising agent.

(iv) Excluding beryllium, describe the bonding in the hydrides of these elements. (WJEC)

25 (a) 60 cm^3 of oxygen was added to 10 cm^3 of a gaseous unsaturated hydrocarbon. After explosion and cooling to room temperature the gases occupied 45 cm^3 and after absorption by aqueous potassium hydroxide solution 15 cm^3 of oxygen remained. Calculate the molecular formula of the hydrocarbon (all measurements were taken under the same conditions of temperature and pressure).

(b) By means of equations indicate how the above hydrocarbon reacts with:

(i) bromine in diffuse light,

(ii) cold dilute alkaline aqueous potassium manganate(VII) (permanganate) solution,

(iii) hydrogen chloride in concentrated aqueous solution,
(iv) ozone followed by warming with dilute acid,
(v) bromine water.
(c) **Outline two** methods by which this hydrocarbon could be prepared in the laboratory. (AEB)

26 A solid organic compound $C_7H_7NO_2$ is thought to be either

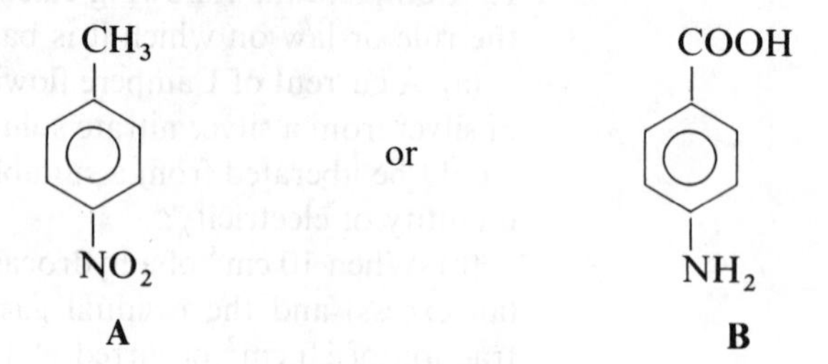

(a) Predict which of the two compounds has the higher melting point and briefly explain your prediction.
(b) Suggest one simple chemical test-tube reaction which would distinguish between **A** and **B**, giving the appropriate equation.
(c) Show, by giving the appropriate reagents, how **A** can be converted into **B** via compound **X**, $C_7H_5NO_4$.
What is the structure of **X**?
(d) The pK_a value of benzenecarboxylic (benzoic) acid is 4.17 and that of **X** is 3.43. Explain qualitatively the reason for the difference in the pK_a values. (JMB)

27 Select any **three** of the following mixtures of compounds and answer parts (a) and (b) below. (Details of experimental procedures are not required but you must state essential reagents and conditions to obtain full marks.)
(i) sodium sulphate(VI) and sodium thiosulphate(VI),
(ii) ammonium chloride and ammonium iodide,
(iii) iron(II) ethanoate (acetate) and copper(II) ethanoate,
(iv) potassium bromide and potassium nitrate.
(a) For each of the mixtures which you have selected, give qualitative tests which would confirm the presence of the ions present in the mixture.
(b) For each of the mixtures which you have selected, describe in outline a method of determining the relative proportions of the compounds known to be present in the mixture. (OC)

28 (a) What type of reaction is represented by the following equilibrium?
$[M(H_2O)_6]^{n+} + mH_2O \rightleftharpoons [M(OH)_m(H_2O)_{6-m}]^{(n-m)+} + mH_3O^+$
Discuss, with examples, how the position of the equilibrium will be influenced by the charge and size of the ion M^{n+} and show how you could promote the forward reaction.
(b) Discuss ligand substitution reactions of hydrated transition metal ions by showing, with one example in each case, how to bring about the following:
(i) ligand substitution with no change in coordination number,

(ii) ligand substitution with a change in coordination number,
(iii) ligand substitution which favours a change in oxidation state of the transition metal. (JMB)

29 An organic compound **A**, with a relative molecular mass of 178, contains 74.2%C, 7.9% H, and 17.9% O. Boiling **A** with aqueous alkali gave a volatile compound **B** which did not give a positive haloform test. Acidification of the alkaline solution gave **C** which was soluble in aqueous sodium carbonate solution. A 0.100 g sample of **C** neutralised 7.35 cm^3 of aqueous sodium hydroxide solution (0.100 mol dm^{-3}). Reduction of either **A** or **C** with lithium tetrahydridoaluminate ($LiAlH_4$) yielded **D** which reacted with ethanoic (acetic) anhydride to form **E**, $C_{10}H_{12}O_2$. Oxidation of **D** yields benzene-1,4-dicarboxylic acid.
Identify compounds **A** to **E**, giving your reasoning. (OC)

30 The neutral organic compound **A** undergoes a sequence of reactions shown in the flow diagram below.

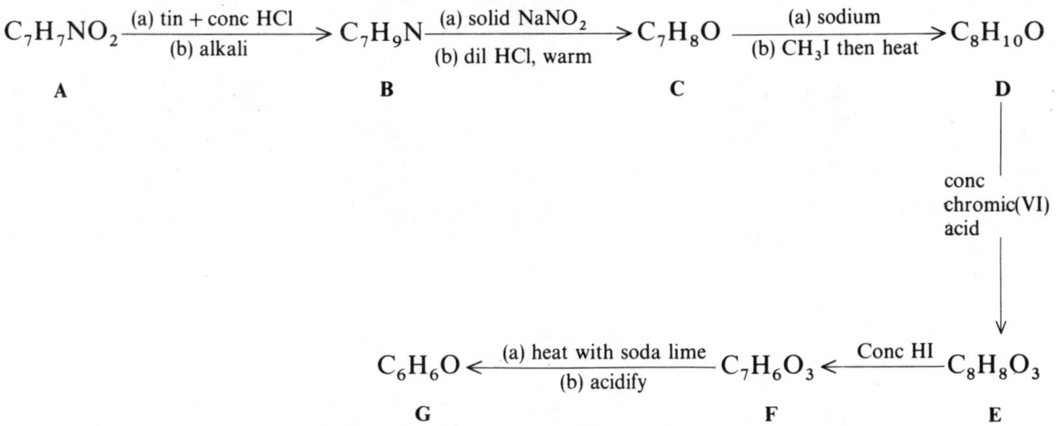

$$\underset{\mathbf{A}}{C_7H_7NO_2} \xrightarrow[\text{(b) alkali}]{\text{(a) tin + conc HCl}} \underset{\mathbf{B}}{C_7H_9N} \xrightarrow[\text{(b) dil HCl, warm}]{\text{(a) solid NaNO}_2} \underset{\mathbf{C}}{C_7H_8O} \xrightarrow[\text{(b) CH}_3\text{I then heat}]{\text{(a) sodium}} \underset{\mathbf{D}}{C_8H_{10}O}$$

$$\mathbf{D} \xrightarrow{\text{conc chromic(VI) acid}} \mathbf{E}$$

$$\underset{\mathbf{G}}{C_6H_6O} \xleftarrow[\text{(b) acidify}]{\text{(a) heat with soda lime}} \underset{\mathbf{F}}{C_7H_6O_3} \xleftarrow{\text{Conc HI}} \underset{\mathbf{E}}{C_8H_8O_3}$$

Both **C** and **G** give a violet colour with iron(III) chloride solution. **B** is soluble in acid whereas **E**, **F** and **G** are soluble in alkali.
(a) Write structural formulae for substances **A** to **G**.
(b) Write equations for the reactions taking place in the conversion of (i) **A** to **B**, (ii) **B** to **C**, (iii) **C** to **D**, (iv) **E** to **F**. (SUJB)

Index